Climate System Modeling provides a thorough grounding in the tools necessary for an appreciation of climate change and its implications. It discusses not only the primary concepts involved but also the mathematical, physical, chemical and biological basis for the component models and the sources of uncertainty, the assumptions made and the approximations introduced.

It is a comprehensive text which will appeal to students and researchers concerned with any aspect of climatology and the study of related topics in the earth and environmental sciences.

During the last few centuries, human society has entered into a new and momentous relationship with the global environment. For the first time in history, we have become a geological agent comparable to some of the other forces influencing our planet. We have altered the face of the Earth by clearing forests, building cities and converting wild lands into agriculture. During the past century, industrial production has gone up by more than 100 and the use of energy has risen by a factor of about 80. We have changed the composition of the Earth's hydrosphere and atmosphere through the use of fossil fuels, the expansion of agriculture and the production and release of industrial compounds. Almost without recognizing it, we have embarked on an enormous, unplanned, planetary experiment that poses unprecedented challenges to our wisdom, our foresight and our scientific capability.

We are just beginning to understand this marvelously complex planet that we live on and the ways in which we are changing it, and we need to devote our best efforts to the task. There is no time to waste.

Dr. D. Allan Bromley
Director of the Office of Science and Technology Policy,
Executive Office of the President,
in a presentation at the National Press Club,
Washington, D.C.,
18 September 1989

Climate System Modeling

Climate System Modeling

Edited by

Kevin E. Trenberth

CAMBRIDGE
UNIVERSITY PRESS

Published by the Press Syndicate of the University of Cambridge
The Pitt Building, Trumpington Street, Cambridge CB2 1RP
40 West 20th Street, New York, NY 10011-4211, USA
10 Stamford Road, Oakleigh, Victoria 3166, Australia

First published 1992

Printed in Great Britain by Butler & Tanner, Frome and London

A catalogue record for this book is available from the British Library

Library of Congress cataloguing in publication data

Climate system modeling / edited by Kevin E. Trenberth.
     p. cm.
  Includes bibliographical references and index.
  ISBN 0 521 43231 6
    1. Climatology – Mathematical models. 2. Atmospheric physics –
Mathematical models. 3. Ocean–atmosphere interaction – Mathematical
models. I. Trenberth, Kevin E.
QC981.C65 1993
551.6′01′5118–dc20 92-22275 CIP

ISBN 0 521 43231 6 hardback

Tag

Contents

PART 1
INTRODUCTION

Contents

PART 2
THE SCIENCE: SUBSYSTEMS AND PROCESSES

3 The atmosphere
Murry L. Salby

Contents

8 Marine biogeochemistry   241
Raymond G. Najjar

PART 3
MODELING AND PARAMETERIZATION

9 Climate system simulation: basic numerical & computational concepts   283
James J. Hack

Contents

Contents

Preface

Global warming. Those words encapsulate for the public all the complex issues involved in anticipated future climate change associated with observed greenhouse gas increases. Carbon dioxide is the best known of the greenhouse gases, and it is known to be increasing because of the burning of fossil fuels and other human activities. But it is only one of several radiatively active important gases found to be increasing in concentration in the atmosphere. What will be the change in climate caused by these anthropogenic influences?

The only quantitative tools we have for predicting future climate are climate models which include as their central components atmospheric and oceanic General Circulation Models (GCMs). However, these models involve many assumptions and approximations that are not always appreciated when interpreting their results. Moreover, the GCMs are so complex that it is not easy to understand their results, and simpler models are often needed. Consequently, a hierarchy of models of increasing complexity is required. Each model may be very useful in its own right as long as it is applied appropriately with proper appreciation of its assumptions and limitations.

Climate models that have been developed thus far for application to the greenhouse gas problem have focused on the physical climate system. Typically, the concentrations of constituents of the atmosphere, including radiatively important species such as ozone and carbon dioxide, are either fixed or specified as varying functions of time. In such scenarios, the concentrations of the gases do not depend on any of the climate changes going on in the model at all, whereas it is known that changes in rainfall or temperatures may profoundly affect the sources and/or sinks of some of the greenhouse gases. Similarly, land surface processes have been greatly oversimplified in the GCMs and no biological, ecological or chemical processes have been included.

In climate *system* models, the subject of this book, the intent is to go beyond the climate models to include all aspects of the climate system: the atmosphere, the ocean, the cryosphere (including snow, sea ice, glaciers and ice caps), the biosphere and terrestrial ecosystems, other land surface processes, and additional parts of the hydrosphere including rivers, and all the complex interactions between these components. The biogeochemical cycles in both the atmosphere and the ocean must be dealt with in some detail, potentially allowing the carbon cycle, for instance, to be treated with some veracity. Instead of projecting and specifying what future atmospheric concentrations of carbon dioxide and methane might be, the goal of these models is to deal comprehensively with the carbon cycle and to predict the future evolution of greenhouse gas concentrations, as well as the impact of those changes on the physical climate. At present, we do not know how to do this, but we must learn. All of these issues, and many more, are at the core of this book.

It is now widely recognized that human activities are transforming the global environment. The issues in global change, such as the burgeoning world population, industrialization, agricultural practices, and so on, involve complex scientific, social, economic, and political factors; most of which extend well beyond the scope of this book. However, one increasingly prominent component of the climate system, that involving the actions of humans, is not predictable in any deterministic sense, so future projections using any climate models must necessarily contain a so-called "what if" scenario. If the resulting projection proves to indicate sufficiently adverse effects on climate, then it is generally presumed that mankind will act to prevent such a scenario from being realized. This is not a foregone conclusion, if only because there are many views on what constitutes "sufficiently adverse" but it is clear that a central role of climate system models will be to help determine possible impacts and help guide possible future policies.

Entire books have been written on each component of the climate system but the major challenge of pulling all the essential ingredients from the many different disciplines together in order to comprehend and provide a basis to advance the modeling of the entire climate system has not been faced until now. This book is intended to serve at least two different audiences. The first consists of those students and scientists interested in the entire climate system or components of the climate system that are fully interactive and whose changes may feed back and affect the climate. The second audience consists of those whose interest is in environmental and societal impacts of climate change and who therefore need to understand the future climate projections and what they may mean, as well as better appreciating the limitations of the climate models. For both audiences, it is necessary to understand the primary concepts involved, the mathematical, physical, chemical and biological basis for the component models and the sources of uncertainty, the assumptions made, and the approximations introduced. Advances come from reducing those sources of uncertainty.

The text is especially intended for graduate students. It is expected that professors who are experts in one disciplinary area will be able to use this text to teach a course in climate system modeling that introduces students to the other disciplines and the issues involved in coupling among the components of the climate system. In this way, it is hoped that students will recognize the intellectual, analytical and research challenges that exist in climate system modeling and, because of the practical and societal importance of the endeavor, it is further hoped that students will be attracted into pursuing climate system modeling as a career goal.

The book is divided into six parts with 23 chapters. Part 1 (Chapters 1 and 2) introduces the climate system and some of the important concepts and interactions involved. The climate system includes one major unpredictable part, the actions of mankind, which are also discussed in Part 1.

Part 2 (Chapters 3–8) describes what has to be modeled, the components of the climate system as observed, the important dynamical, physical, chemical and biological processes involved and the nature of the interactions among the components. It introduces the reader to the sources and sinks, storage, and movement of mass, heat, energy, moisture, radiatively active trace gases, and momentum. It also introduces the fundamental physical principles expressed in the form of mathematical equations. Separate chapters describe the atmosphere including the stratosphere, the ocean and sea ice, the land surface, terrestrial ecosystems, atmospheric chemistry of both the troposphere and stratosphere, and marine biogeochemistry. In each case it has been necessary to focus on those elements of each subsystem as they relate to climate, especially where there are two-way interactions. Unfortunately, the whole issue of one-way impacts of changes in climate on the subsystems has mostly been left out.

Models of the primary individual components of the climate system are described in Part 3 (Chapters 9–16) in the form of modules that might be interconnected. While it is intended to describe the state of the art and the outstanding issues in each case, a major emphasis has been placed on conceptual and/or simple models that provide a basis for interpreting the results. The first chapter briefly examines practical issues that are central to success in carrying out climate system modeling. These include numerical methods and computational approaches. As these areas are rapidly evolving, the approach is to discuss the issues and possible future directions without being comprehensive about the latest developments. The other chapters in this part introduce atmospheric models, ocean models, sea ice models, land ice models including those of the glacial ice caps of Antarctica and Greenland, land surface processes including surface biology and terrestrial ecosystem depictions, atmospheric chemical tracer models, and models of the biogeochemistry of the oceans. Models of other components exist, and some are described in Part 2, but they are not yet sufficiently developed to be coupled into the full climate model.

Part 4 (Chapters 17 and 18) deals with the physical climate system coupling and, as dealt with in climate models in the past, mainly focuses on the coupling of the atmosphere and the ocean, which can give rise to phenomena such as the El Niño–Southern Oscillation (ENSO) and is likely to be a key to understanding major climate changes. One chapter is devoted to explaining how ENSO variations arise as a natural mode of interannual variability in climate from coupled interactions between the atmosphere and ocean. Another describes coupling strategies: how an atmospheric and oceanic GCM which appear to work well in isolation go astray when coupled, artificial "fixes" that have been temporarily used to address this, and the "thermohaline catastrophe" in which two different climates with very different oceanic circulations appear possible for the same external conditions.

Applications of and results from the models to date are introduced in Part 5 (Chapters 19–22). Expected changes in climate due to increases in greenhouse gases are compared with changes arising from natural variations in climate. Also of note are simulations of the climates in the distant past, thousands or millions of years ago. Paleoclimate studies provide evidence concerning past climates when conditions, such as surface geography, atmospheric constituents, and astronomical factors were very different from today. This information can therefore provide a test for the validity of the model, which, in turn, may provide useful clues of things to look for in the fossil record and other proxy data that indicate past climate conditions. Future prospects are discussed in Part 6 (Chapter 23).

Kevin E. Trenberth
National Center for Atmospheric Research
Boulder, Colorado
January, 1992

Acknowledgments

This book grew out of the concern of many individuals over the need to attract large numbers of competent students, from multiple disciplines, to study and advance the science of modeling the entire climate system. This concept was encapsulated into part of the Climate System Modeling Program of the University Corporation for Atmospheric Research (UCAR) and was realized by a grant from the Electric Power Research Institution (EPRI) to UCAR, under research contract RP2333-12, to produce this volume. I want especially to thank and acknowledge Rick Anthes and Wayne Shiver of UCAR and Chuck Hakkarinen of EPRI for their continuing interest and support. Michael Ghil and Guy Brasseur were helpful in deciding on an outline for the book. I want also to thank the National Center for Atmospheric Research (NCAR), which is sponsored by the National Science Foundation, for allowing me, and a number of other authors on the NCAR staff to participate in producing this volume. Special thanks go to Dorene Howard who prepared the typescript for publication, and to Justin Kitsutaka and the NCAR Graphics section who drafted all the figures.

While the individual authors listed were each primarily responsible for their chapter, I, along with other authors and many reviewers, contributed to each chapter. The chapters are not self contained; for instance, a single consolidated reference list appears at the back of the book. Redundancy has been minimized and there is no duplication of figures. Notation has been standardized to the maximum extent possible. While the form of each chapter has also been standardized, there are differences in writing style that have been retained in each chapter and that add to the color of the book. I thank the authors for their indulgence in trying to make the book a complete entity rather than simply a collection of individual contributions, perhaps at the expense of the completeness of each chapter.

A draft of this book was used as a text at a workshop on Coupled Climate System Modeling, held at the University of Wisconsin 14–27 July 1991. I am especially grateful to the lecturers, faculty, and students who participated for their critique of the fledgling work.

The following chapter reviewers provided insightful comments and

Acknowledgements

thoughtful suggestions for which we thank them very much: E. Barron, J. Birks, F. Bryan, D. Hartmann, R. Jahnke, W. Large, R. Madden, G. Maykut, R. Najjar, R. Pielke, D. Pollard, C. Raymond, D. Schimel, E. Sarachik, R. Stouffer, P. Swarztrauber.

The cover design was produced by NCAR graphics. The top photograph is courtesy Stephen Eric Levine, remaining photographs courtesy NCAR/UCAR.

Kevin E. Trenberth
National Center for Atmospheric Research
Boulder, Colorado
January, 1992

The authors

The following is a listing of the chapter authors, their affiliations and any special acknowledgments for individual chapters.

Chapter 1
Stephen H. Schneider

National Center for Atmospheric Research

Boulder, Colorado 80307–3000

Chapter 2
Michael H. Glantz

National Center for Atmospheric Research

Boulder, Colorado 80307–3000

Jerrold H. Krenz

University of Colorado

Boulder, Colorado 80309

Chapter 3
Murry L. Salby

Lady Davis Visiting Professor

Hebrew University of Jerusalem, and

Center for Atmospheric Theory and Analysis

University of Colorado

Boulder, Colorado 80309

Acknowledgments: I am grateful to R. Dickinson, D. Hartmann, and K. Trenberth for providing useful comments on earlier versions of the manuscript and to H. Hendon for helping with the construction of figures. Figures copyrighted by Academic Press, Inc. were derived from the *Fundamentals of Atmospheric Physics* by Murry L. Salby as published by Academic Press, Inc.

Chapter 4
Pearn P. Niiler

Scripps Institution of Oceanography

La Jolla, California 92093

Chapter 5
Robert E. Dickinson

Institute of Atmospheric Physics

University of Arizona

Tucson, Arizona 85721

Chapter 6
John D. Aber

Institute for Study of Earth, Oceans and Space

University of New Hampshire

Durham, New Hamspire 03824-3525

Chapter 7
Richard P. Turco

Department of Atmospheric Sciences

University of California at Los Angeles

Los Angeles, California 90024-1565

Chapter 8
Raymond G. Najjar

National Center for Atmospheric Research

Boulder, Colorado 80307-3000

Acknowledgments: I appreciate the comments of Scott Doney, Kevin Trenberth, Rick Jahnke, Dave Erickson, Greg. Crawford, Paty Matrai, and Ralph Keeling.

Chapter 9
James J. Hack

National Center for Atmospheric Research

Boulder, Colorado 80307-3000

The authors

Chapter 10
Jeffrey T. Kiehl
National Center for Atmospheric Research
Boulder, Colorado 80307–3000

Chapter 11
Dale B. Haidvogel
Institute of Marine and Coastal Sciences
Rutgers University
New Brunswick, New Jersey 08903-0231
Frank O. Bryan
National Center for Atmospheric Research
Boulder, Colorado 80307–3000

Chapter 12
William D. Hibler, III
Gregory M. Flato
Thayer School of Engineering
Dartmouth College
Hanover, New Hampshire 03755

Chapter 13
Cornelis J. van der Veen
Byrd Polar Research Center
Ohio State University
Columbus, Ohio 43210
Acknowledgments: This is Byrd Polar Research Center contribution No. 753.

Chapter 14
Piers J. Sellers
National Aeronautics and Space Administration
Goddard Space Flight Center
Greenbelt, Maryland 20771
Acknowledgments: This paper summarizes work and discussions that I have had with many researchers over the last few years, including Nobuo Sato, David Randall, Yogesh Sud, Yale Mintz, Bob Dickinson, J. Shukla, Ed Schneider, Jeff Dorman, Forrest Hall and many others at COLA, the University of Maryland, and NASA/GSFC. Margaret Rae is warmly thanked for typing and editing the first version of the paper.

Chapter 15
Guy P. Brasseur
Sasha Madronich
National Center for Atmospheric Research
Boulder, Colorado 80307–3000
Acknowledgments: The authors are very grateful to Piotr Smolarkiewicz, Stacy Walters, Kevin Trenberth, Phil Rasch, Rolf Müller, and Paul Ginoux for useful comments on the manuscript.

Chapter 16
Jorge L. Sarmiento
Atmospheric and Oceanic Sciences Program
Princeton University
Princeton, New Jersey 08544

Chapter 17
Gerald A. Meehl
National Center for Atmospheric Research
Boulder, Colorado 80307–3000

Chapter 18
Mark A. Cane
Lamont-Doherty Geological Observatory
of Columbia University
Palisades, New York 10964
Acknowledgments: I am especially grateful for the critical comments of Kevin Trenberth, Ed Sarachik, Peter Gent, and Nick Graham, which markedly improved this chapter. Thanks to Virginia DiBlasi-Morris for help in preparing the manuscript. GrantNA 87-AA-D-AC081 from the U.S. TOGA Project Office of NOAA partially supported this work.

Chapter 19
Ngar-Cheung Lau
Geophysical Fluid Dynamics Laboratory/NOAA
Princeton, New Jersey 08542
Acknowledgments: I wish to thank A. J. Broccoli, T. L. Delworth, R.A . Madden, S. Manabe, R. J. Stouffer and K. E. Trenberth for offering insightful comments on a preliminary version of this chapter. I am particularly indebted to S. Manabe for several helpful discussions on the subject matter of this chapter, and for allowing me to cite in Sec. 19.4 some of the results from the 200-year coupled GCM experiment prior to their formal publication.

Chapter 20
Warren M. Washington
National Center for Atmospheric Research
Boulder, Colorado 80307–3000

Chapter 21
John E. Kutzbach
Center for Climatic Research
University of Wisconsin-Madison
Madison, Wisconsin 53706

Chapter 22
Robert E. Dickinson
Institute of Atmospheric Physics
University of Arizona
Tucson, Arizona 85721

Chapter 23
Lennart O. Bengtsson
Max Planck Institute for Meteorology
2000 Hamburg 13
Germany

Acknowledgments: The author is grateful to Kevin Trenberth for useful comments on the manuscript

Acronyms

ADI	Alternating-Direction Implicit	ENSO	El Niño–Southern Oscillation
AGCM	Atmospheric General Circulation Model	EOF	Empirical Orthogonal Function
ALE	Arbitrary-Lagrangian-Eulerian	EPRI	Electric Power Research Institute
Alk	Alkalinity	ERB	Earth Radiation Budget
ALPEX	Alpine Experiment	ERBE	Earth Radiation Budget Experiment
ALU	Arithmetic and Logical Unit	ESS	Earth System Science
AOU	Apparent Oxygen Utilization	FAO	Food and Agriculture Organization
APE	Available Potential Energy	FCL	Fixed Cloud Level
ASHRAE	American Society of Heating, Refrigeration, and Air-Conditioning Engineers	FFT	Fast Fourier Transform
		FRH	Fixed Relative Humidity
AVHRR	Advanced Very High Resolution Radiometer	FTCS	Forward in Time and Centered in Space
BATS	Biosphere Atmosphere Transfer Scheme	GARP	Global Atmospheric Research Program
B.P.	Before Present	GATE	GARP Atlantic Tropical Experiment
BVE	Barotropic Vorticity Equation	GCM	General Circulation Model
CAS	Canopy Air Space	GCOS	Global Climate Observing System
CAC	Climate Analysis Center	GEWEX	Global Energy and Water Cycle Experiment
CCM	Community Climate Model	GFDL	Geophysical Fluid Dynamics Laboratory
CCN	Cloud Condensation Nuclei	GFLOP	Giga Floating Point Operations Per Second
CCS	Community Composition and Structure	GHG	Greenhouse Gas
CE	Cold Event	GISS	Goddard Institute for Space Studies
CEES	Committee on Earth and Environmental Sciences	GLAS	Goddard Laboratory for Atmospheric Science
		GOES	Geostationary Orbiting Environmental Satellite
CFCs	Chlorofluorocarbons	GSFC	Goddard Space Flight Center
CFL	Courant–Friedrichs–Lewy	GWE	Global Weather Experiment
CGCM	Coupled General Circulation Model	HCFCs	Hydrogenated Chlorofluorocarbons
CLIMAP	Climate Mapping and Prediction	HPE	Hydrostatic Primitive Equations
CLW	Cloud Liquid Water	I/O	Input/Output
CME	Community Modeling Effort	IPCC	Intergovernmental Panel on Climate Change
COHMAP	Cooperative Holocene Mapping Project	IR	Infrared Radiation
COLA	Center for Ocean–Land–Atmosphere Interactions	ISCCP	International Satellite Cloud Climatology Project
CPU	Central Processing Unit	ITCZ	Intertropical Convergence Zone
CZCS	Coastal Zone Color Scanner	JJA	June–July–August
DIC	Dissolved Inorganic Carbon	ka	thousand years ago
DIM	Dissolved Inorganic Matter	KE	Kinetic Energy
DJF	December–January–February	LAI	Leaf Area Index
DMS	Dimethyl Sulfide	LCL	Lifting Condensation Level
DSMP	Dimethyl Sulfonium Propionate	LFC	Level of Free Convection
DOM	Dissolved Organic Matter	LIMS	Limb Infrared Monitor of the Stratosphere
DON	Dissolved Organic Nitrogen	LSP	Land Surface Parameterization
DU	Dobson units	LW	Longwave
EBM	Energy Balance Model	LWC	Liquid Water Content
ECMWF	European Centre for Medium Range Weather Forecasts	MAA	Moist Adiabatic Adjustment
		MIMD	Multiple Instruction Multiple Data
ENIAC	Electronic Numerical Integrator and Computer	MONEX	Monsoon Experiment

MOS	Model Output Statistics		RH	Relative Humidity
MSA	Methane Sulphonic Acid		rms	root mean square
MYL2	Mellor–Yamada Level-2 Model		SATOBS	Satellite Observations
my	million years		SH	Southern Hemisphere
NASA	National Aeronautics and Space Administration		SiB	Simple Biosphere
			SISD	Single Instruction Single Data
NCAR	National Center for Atmospheric Research		SIMD	Single Instruction Multiple Data
NDVI	Normalized Difference Vegetation Index		SLP	Sea Level Pressure
NESDIS	National Environmental Satellite, Data, and Information Service		SMS	Sources Minus Sinks
			SO	Southern Oscillation
NH	Northern Hemisphere		SOI	Southern Oscillation Index
NNMI	Nonlinear Normal Mode Initialization		SPCZ	South Pacific Convergence Zone
NMC	National Meteorological Center		SSM/I	Special Sensor Microwave Imager
NMHC	Nonmethane Hydrocarbon		SST	Sea Surface Temperature
NOAA	National Oceanic and Atmospheric Administration		SW	Shortwave
			TEM	Transformed Eulerian Mean
NPP	Net Primary Production		TFLOP	Tera Floating Point Operations Per Second
NSF	National Science Foundation		TIROS	Television Infrared Operational Satellite
NWP	Numerical Weather Prediction		TKE	Turbulent Kinetic Energy
ODP	Ozone Depletion Potential		TOGA	Tropical Oceans Global Atmosphere
OGCM	Oceanic General Circulation Model		TOMS	Total Ozone Measurement Spectrophotometer
OLR	Outgoing Longwave Radiation		TOVS	TIROS Operational Vertical Sounder
OSU	Oregon State University		UCAR	University Corporation for Atmospheric Research
PAR	Photosynthetically Active Radiation			
PBL	Planetary Boundary Layer		UKMO	United Kingdom Meteorological Office
PBO	Paleobioclimatic Operator		UNEP	United Nations Environment Programme
PE	Primitive Equation		UNESCO	United Nations Educational, Scientific, and Cultural Organization
PGE	Planetary Geostrophic Equations			
PIM	Particulate Inorganic Matter		USGS	U.S. Geological Survey
POM	Particulate Organic Matter		UTC	Universal Time Coordinate
ppbv	parts per billion by volume		UV	Ultraviolet
ppmv	parts per million by volume		VLSI	Very Large Scale Integration
pptv	parts per trillion by volume		WCRP	World Climate Research Programme
PSC	Polar Stratospheric Cloud		WE	Warm Event
PSI	Pollutant Standard Index		WMO	World Meteorological Organization
QG	Quasi-geostrophic		WOCE	World Ocean Circulation Experiment
RCM	Radiative-Convective Model		ZC	Zebiak–Cane

Basic notation

There are many different uses for symbols in the climate subdisciplines. The following is a partial list of what we consider as the basic notation. The many departures from this list should be clearly distinguishable in the text. All vectors or tensors are in boldface.

Lower Case

a	mean earth radius
c	specific heat, or wave velocity
e	(water) vapor pressure
f	Coriolis parameter
g	gravity
h	Planck's constant
$\mathbf{i}, \mathbf{j}, \mathbf{k}$	unit vectors to east, north, and upwards
k	gas transfer coefficient
k, ℓ, m	wavenumber in x, y, z
m	mass
n	molecular abundance
p	pressure
q	specific humidity
r	mixing ratio, or Redfield ratio
s	dry static energy
t	time, or transmissivity
u, v, w	eastward, northward, vertical (up) velocity
x, y, z	eastward, northward, vertical (up) direction

Upper Case

A	area, or amplitude
B	blackbody radiation
C	condensation, or, with subscript, transfer coefficient
D	drag, diffusion or dissipation
E	evaporation rate
F	longwave radiation flux
G	heat flux into ground
H	scale height of atmosphere, or depth of ocean
I	intensity of radiation
J	Jacobian, or blackbody source function
K	eddy diffusion coefficient
L	latent heat (of vaporization) or loss of species
M	molecular weight
N	static stability (Brunt–Väisälä frequency)
O	order
P	precipitation rate, or production of species
Q	heating rate
R	gas constant, or runoff
S	solar (shortwave) radiation, or salinity, or sources
T	temperature
U	internal energy
V	volume, or wind speed
W	water column amount

Greek Alphabet

α	albedo
β	gradient of Coriolis parameter df/dy
γ	c_p/c_v, or rate constant
δ	divergence, or increment
ϵ	$0.622 = M_v/M_d$, or emissivity
ζ	relative vorticity
η	absolute vorticity, or imaginary part of index of refraction
θ	potential temperature
κ	R/c_p, or diffusion (eddy or molecular) or thermal conductivity coefficient
λ	longitude, or wavelength
μ	$\sin\phi$
ν	$h\nu$ photon of light, or vertical diffusivity
ρ	density
σ	Stefan–Boltzmann constant or vertical coordinate
τ	stress, optical depth, or time constant
ϕ	latitude
χ	solar zenith angle, or velocity potential
ψ	stream function
ω	vertical p-velocity
Γ	lapse rate
Δ	increment
Φ	geopotential
Ω	Earth angular velocity

Physical constants

Universal gas constant (R)	8.3143×10^3 J K^{-1} kmol^{-1}
Planck's constant (h)	6.6262×10^{-34} J s
Stefan–Boltzmann constant (σ)	5.67×10^{-8} W m^{-2} K^{-4}
Velocity of light (c)	2.998×10^8 m s^{-1}

Sun–Earth

Average Sun-Earth distance	1.50×10^{11} m
Solar constant (irradiance) (S_o)	1370 W m^{-2}
Earth radius: mean (a)	6.371×10^6 m
equatorial	6378.4×10^6 m
polar	6356.9×10^6 m
Earth rotation (angular velocity) (Ω)	7.292×10^{-5} s^{-1}
Acceleration due to gravity (g)	9.80616 m s^{-2}
(standardized to $45°$)	
Mean Earth topographic height	237.3 m
Mass of atmosphere (mean)	5.1352×10^{18} kg
Mean surface atmospheric pressure	984.4 mb
Mean ocean depth	3795 m
Ocean surface	361.1×10^6 km^2 (70.8%)
Land surface	148.9×10^6 km^2 (29.2%)

Dry air

Average composition: Nitrogen 78.09%, Oxygen 20.95%, Argon 0.93% Carbon dioxide 0.03%	
Apparent molecular weight	28.97
Gas constant (R_d)	287 J kg^{-1} K^{-1}
Density at $0°$C 1000 mb (ρ_a)	1.275 kg m^{-3}
Specific heat at constant pressure (c_p)	1004 J kg^{-1} K^{-1}
Specific heat at constant volume (c_v)	717 J kg^{-1} K^{-1}
Thermal conductivity at $0°$C	2.40×10^{-2} J m^{-1} K^{-1} s^{-1}
Mass of dry air	5.122×10^{18} kg
Surface pressure due to dry air	981.9 mb

Water substance

Molecular weight	18.016
Gas constant water vapor (R_v)	461 J kg^{-1} K^{-1}
Density for liquid water at $0°$C	1.000×10^3 kg m^{-3}
Density for ice at $0°$C	0.917×10^3 kg m^{-3}
Specific heat of water vapor at constant pressure	1952 J kg^{-1} K^{-1}
Specific heat of water vapor at constant volume	1463 J kg^{-1} K^{-1}
Specific heat of liquid water $0°$C	4218 J kg^{-1} K^{-1}
Specific heat of ice $0°$C	2106 J kg^{-1} K^{-1}
Latent heat of vaporization at $100°$C	2.25×10^6 J kg^{-1}
Latent heat of fusion at $0°$C	3.34×10^5 J kg^{-1}
Mass of water vapor in atmosphere	1.3×10^{15} kg
Mean surface pressure due to water vapor	2.5 mb

PART 1

INTRODUCTION

1

Introduction to climate modeling

Stephen H. Schneider

1.1 Context of global climate change

The Earth's climate changes. It is vastly different now from what it was 100 million years ago, when dinosaurs dominated the planet and tropical plants thrived at high latitudes; it is different from what it was even 18,000 years ago, when ice sheets covered much more of the Northern Hemisphere. In the future it will surely continue to evolve. In part the evolution will be driven by natural causes, such as fluctuations in the Earth's orbit. But future climatic change, unlike that of the past, will probably have another source as well: human activities. We may already be feeling the climatic effects of having polluted the atmosphere with gases such as carbon dioxide. Many other activities associated with human economic development driven by growing populations using technology and organization to improve their standard of living have altered the physical and chemical environment in ways that modify the natural stocks and flows of energy and materials in the environment. When these modifications become large enough it is natural to expect significant global changes. Indeed, human activities that release carbon dioxide, chlorofluoromethanes, nitrous oxide, methane, atmospheric particles (aerosols), and heat, or that use land for urbanization, agriculture or deforestation, are examples of such modifications to the stocks and flows of processes or materials related to the maintenance of environmental services. (Chapter 2 discusses human impacts on the climate system in more detail.) Table 1.1 (modified from Schneider and Londer, 1984) summarizes briefly a range of such activities, their potential climatic effects and an estimate of the scale and importance of the potential effects. The list is comprehensive but not exhaustive, and considerable uncertainty surrounds the predicted effects in specific cases. Nevertheless as Table 1.1 suggests, growing human numbers using technology and organization to increase per capita levels of consumption could have a substantial impact on climatic or other environmental systems. The fact that such global changes could be of considerable importance to human and natural systems is what motivates the need for quantitative evaluation of the potential impact of human activities in creating global change, and that evaluation of such

Table 1.1. Summary of principal human activities that can influence climate (modified after Schneider and Londer, 1984).

Activity	Climatic effect	Scale and importance of the effect
Release of carbon dioxide by burning fossil fuels.	Increases the atmospheric absorption and emission of terrestrial infrared radiation (greenhouse effect), resulting in warming of lower atmosphere and cooling of the stratosphere.	Global: potentially a major influence on climate and biological activity.
Release of methane chlorofluoromethanes, nitrous oxide, carbon tetrachloride, carbon disulfide etc.	Similar climatic effect as that of carbon dioxide since these, too, are infrared-absorbing and fairly chemically stable trace gases.	Global: potentially significant influence on climate.
Release of particles (aerosols) from industrial and agricultural practices. Sulfur dioxide is of primary concern since it photochemically converts to sulfuric acid particles.	These sunlight scattering and absorbing particles (especially soot) could decrease albedo over land, causing a warming and could (especially sulfate) increase albedo over water causing a cooling; they also change stability of lower atmosphere; net climatic effects still speculative, although net cooling effect seems more likely.	Largely regional, since aerosols have an average lifetime of only a few days, but similar regional effects in different parts of the world could have nonnegligible net global effects; stability increase may suppress convective rainfall, but particles could affect cloud properties with more far-reaching effects.
Release of aerosols that act as condensation and freezing nuclei. Again, released soot or sulfur dioxide by industrial activities is of primary concern.	Influences growth of cloud droplets and ice crystals; may effect amount of precipation or albedo of clouds in either direction.	Local (at most) regional influences on quantity and quality of precipitation, but unknown and potentially important change to Earth's heat balance if cloud albedo is altered. Some calculations suggest SO_2 released between 1950 and 1980 opposed much of the Northern Hemispheric warming trend that otherwise would have been experienced from rapid buildup of greenhouse gases during those decades.
Release of heat (thermal pollution).	Warms the surface layers directly.	Locally important now; could become significant regionally; could modify circulation.

Table 1.1 (Continued)

Activity	Climatic effect	Scale and importance of the effect
Upward transport of chlorofluoromethanes and nitrous oxide into the stratosphere.	Photochemical reaction of their dissociation products probably reduces stratospheric ozone.	Global but uncertain influence of ozone depletion on climate; less total stratospheric ozone allows more solar radiation to reach the surface but compensates by reducing greenhouse effect as well; however, if ozone concentration decreases at high altitudes, but increases comparably at lower altitudes, this would lead to potentially large surface warming; could cause significant biological effects from increased exposure to ultraviolet radiation if total column amount of ozone decreases.
Release of trace gases (e.g., nitrogen oxides, carbon monoxide, or methane) that increase tropospheric ozone by photochemical reactions.	Large atmospheric heating occurs from tropospheric ozone, which enhances both solar and greenhouse heating of lower atmosphere.	Local to regional at present, but could become a significant global climatic warming if large-scale fossil fuel use leads to combustion products that significantly increase tropospheric ozone. Contact with ozone also harms some plants and people.
Patterns of land use, e.g., urbanization, agriculture, overgrazing, deforestation. etc.	Changes surface albedo, evapotranspiration and runoff and causes aerosols.	Largely regional: net global climatic importance still speculative.
Release of radioactive Krypton-85 from nuclear reactors and fuel reprocessing plants.	Increases conductivity of lower atmosphere, with possible implications for Earth's electric field and precipitation from convective clouds.	Global: importance of influence is highly speculative.
Large-scale nuclear war.	Could lead to very large injections of soot and dust causing transient surface cooling lasting from weeks to months, depending on the nature of the exchange and on how many fires were started.	Could be global, but initially in midlatitudes of Northern Hemisphere. Darkness from dust and smoke could disrupt photosynthesis for weeks with severe effects on both natural and agricultural ecosystems of both combatant and noncombatant nations. Transient freezing outbreaks could eliminate some warm season crops in midlatitudes or weakening of monsoon rainfall could be devastating to any vegetation in tropics or subtropics. Details still speculative.

change is central to any potential policy responses to mitigate those potential changes (e.g., Schneider, 1990).

How can human societies prepare for so uncertain a climatic future? Clearly it would help to predict that future in some detail, but therein lies a problem: the processes that make up a planet's climate are too large and too complex to be reproduced physically in a laboratory. Fortunately they can be simulated mathematically with the help of a computer. In other words, instead of building a physical analogue of the land–ocean–atmosphere system, one can devise mathematical expressions for the physical principles that govern the system – laws of thermodynamics and Newton's laws of motion – and then allow the computer to calculate how the climate will evolve in accordance with these laws. For a variety of reasons to be detailed in this chapter and volume, mathematical climate models cannot simulate the full complexity of reality. They can, however, reveal the logical consequences of plausible assumptions about how the climate system operates. The critical scientific task is to formulate, build and then validate the models.

1.2   Mechanisms of climatic change

1.2.1   The climate system

Climate is typically the average state of the atmosphere observed as the weather over a finite time period (e.g., a season) for a number of different years. Thus, we can speak of the climate of a day–night cycle, month, season, year, decade, or even longer period. Climate is usually defined by the mean state together with measures of variability or fluctuations such as the standard deviation or autocorrelation statistics for the period (e.g., Mearns et al., 1990).

Although the same physical laws usually are applied to the most comprehensive tools for both climate and weather prediction, the climatic prediction is complicated by considering complex interactions between, as well as changes within, all the components of the climate system – the atmosphere, oceans, land, ice and snow, and terrestrial and marine biota. Plate 1 (Earth System Science, 1986. See also Plate 2 for a simpler version) is an attempt to represent schematically the interacting physical, chemical and biological processes which, on time scales up to centuries, control global changes. Although a weather forecaster need not consider, for example, the small day-to-day changes in ice, temperature, or circulation of the sea, such changes affecting the lower atmosphere must be considered by the predictor of atmospheric changes from one season or one decade to another. On the other hand, change of the Earth's orbit occurs during thousands of years and is negligible when considering climate changes during less than a few millennia (Shackleton and Imbrie, 1990).

As mentioned above, the following components interact to make the observed climate state the result: atmosphere, oceans, cryosphere, and land/biosphere. The atmosphere and the oceans are two fluid components of the system, each containing organized circulation, chaotic motions, and random turbulence. They react to perturbations on very different time scales. Interactions between and within them occur on many scales and tend to be concentrated close to their boundary as well as internally where gradients of physical properties, such as temperature or density, can be large. These interactions will be introduced in this chapter briefly, but discussed in depth in subsequent chapters.

The chemical composition of the atmosphere also affects climate. Aerosols, water vapor, carbon dioxide, and ozone directly affect the atmosphere's absorption and transmission of solar radiation, which provides almost all the energy for the entire system. Further, aerosols (e.g., dust or sulfate particles) may cause clouds to form and precipitation to fall. For example, Twomey et al. (1984) have argued that increased sulfuric acid aerosols (e.g., from SO_2 injections from coal or oil burning or possibly photoplankton emissions) could increase cloud brightness in unpolluted areas (see Chapter 7 for more details). Wigley (1989) updated the old suggestion (e.g., SMIC, 1971) that this may have offset some fraction of anticipated CO_2-induced Northern Hemispheric warming since 1950, see also Charlson et al. (1991). However, any such effect would be highly regional, would have diminished as SO_2 controls were applied in the 1970s to combat acid rain, and would saturate as background pollution or other aerosol levels increase.

Other complex processes in the climate system include the salinity of oceans (see Chapter 4 for more details), which affects water density, and thus circulation of the oceans. The exchange between air and surface of such absorbers of radiation as water vapor, carbon dioxide, methane and nitrogen and sulfur oxides is another example, and is determined by such physical processes as winds, rainfall, or runoff and biological processes in forest or from photoplankton productivity.

The third component of the climate system is the cryosphere (see Chapters 12 and 13), which includes the extensive ice fields of Antarctica and Greenland, other continental snow and ice, and sea ice. Continental snow and sea ice vary seasonally and interannually, causing large annual variations in continental heating and upper ocean mixing and in energy exchange between the surface and atmosphere. Although the large continental ice sheets do not change rapidly enough to cause seasonal or annual climatic anomalies, they play a major role in climatic changes during hundreds to thousands of years such as the glacial and interglacial cycles that have occurred repeatedly for at least the past one million years (Shackleton and Imbrie, 1990).

The land and its biomass constitute a fourth component of the climate system, as depicted schematically on Plate 1. This component includes the slowly changing extent, position, and orography of the continents and the more rapidly varying characteristics of lakes, rivers, soil moisture, and vegetation. Thus, the land and its biomass are variable parts of the climate system on all time scales. Proper inclusion of the biophysics of energy and materials exchange between the atmosphere and land biosphere is important to simulation of the effects of deforestation (e.g., Henderson-Sellers et al., 1988)

The entire climate system involves the interaction of the biota, air, sea, ice and land, with solar radiation providing the energy that drives it. Variations of gaseous and particulate constituents of the atmosphere, along with changes in the Earth's position relative to the Sun, vary the amount and distribution of sunlight received. The temperature of the oceans has a marked influence on the heating and moisture content of the atmosphere. The unreflected radiant energy drives the atmospheric circulation, and by wind stress and heat transfer it drives the circulation of the oceans. The atmosphere and oceans are both influenced by the extent and thickness of the ice covering the land and sea as well as by the shape and composition of the land surface itself. Since each of these components has a different range of response times, the whole system evolves continuously, with some parts lagging or leading other parts.

The system also contains feedback loops between the interacting components, as illustrated in Plate 1. These amplify (positive feedback) or damp (negative feedback) perturbations. For example, any increase in the area of polar ice or snow from a forced cooling reflects more of the incoming solar radiation, leaving less to be absorbed by the surface. If snowfall is adequate, this further lowers surface temperature, increasing ice and snow cover in a positive feedback loop. One might expect, however, that increasing snow cover and associated coldness of a continental interior could gradually limit the overlying atmosphere's ability to import moisture into the region. This eventually decreases snowfall and limits growth of the snow cover in a negative feedback loop.

1.2.2 Radiation balance and the greenhouse effect

The Sun radiates energy corresponding roughly to an ideal (black) radiator with a temperature of about 5,800 K. This implies that 90% of the radiant energy lies in the interval with wavelengths from 0.4 to 4 μm, with a maximum intensity in the green portion of the visible spectrum at 0.48 μm.

About 30% of the incoming solar energy is reflected back to space and is unavailable to warm the Earth. This reflected fraction is called the planetary albedo. Reflection occurs from the clouds, the Earth's surface, and from molecules and particles present in the atmosphere. The clouds contribute the largest part of the albedo, reflecting about 25% of the incoming radiation when averaged over a long period of time, but due to the natural variability of cloudiness over the globe, the Earth's albedo can change substantially from day to day and also from season to season.

The cloudless part of the Earth comprises the remaining 5% of the global albedo. The albedo of the cloudless part of the Earth is determined by the surface albedo and by reflection from atmospheric molecules and suspended particles. The latter, though contributing at most a small percentage to the total albedo, can be of practical importance, since such particles are a factor that can be biased by human activities. Of the incoming radiation, about 25% is absorbed by gases, clouds, and particles in the atmosphere, 30% is reflected to space as discussed above, and the remainder is absorbed at the Earth's surface. This identifies another factor that can be affected by human activities. Since humanity has significantly altered the character of the Earth's surface, it has indirectly affected the climate (at least in limited regions) by disturbing the heat and water budgets through changes in the character and albedo of the surfaces.

To maintain equilibrium, the incoming solar energy that is absorbed by the Earth–atmosphere system must be balanced by an equal amount of outgoing radiant energy. Otherwise, the temperature of the Earth would undergo a continuous change until the "energy balance" is restored. The Earth emits radiant energy, as do all physical things, in proportion to its absolute temperature. But, since the wavelength of maximum radiant energy is inversely proportional to the temperature of the radiator, the Earth emits radiation primarily in the longwave or infrared region, with most of the energy residing in the wavelengths from 4 to 100μm. Figure 1.1 (Goody, 1964) shows radiant energy spectra for solar and terrestrial "blackbody" radiators, along with a representation of the absorption of radiation by gases in the atmosphere.

If we calculate the total solar energy absorbed in the Earth–atmosphere system and equate this to the escaping infrared radiation, then we can determine an "effective radiation temperature" of the planet from the Stefan–Boltzmann law relating the flux of radiant energy to temperature. This has been observed from space to be about $-18°$C (255 K) for the effective radiative temperature of the Earth, whereas we know the average surface temperature to be about $+15°$C.

The $33°$C difference in the two temperature values is, of course, due to the presence of our atmosphere. The optically active gases, principally water vapor, carbon dioxide, methane and ozone, absorb and re-emit infrared radiation in selective "bands" of the infrared spectrum. Clouds and particles also affect the infrared radiation, with the clouds (except thin cirrus clouds) absorbing nearly all the infrared radiation they receive throughout the infrared spectrum, and the particles absorbing or scattering *relatively* little infrared radiation, depending upon the character of the particulate material.

The average surface temperature is higher than the effective radiative temperature primarily because the atmosphere is semitransparent to solar

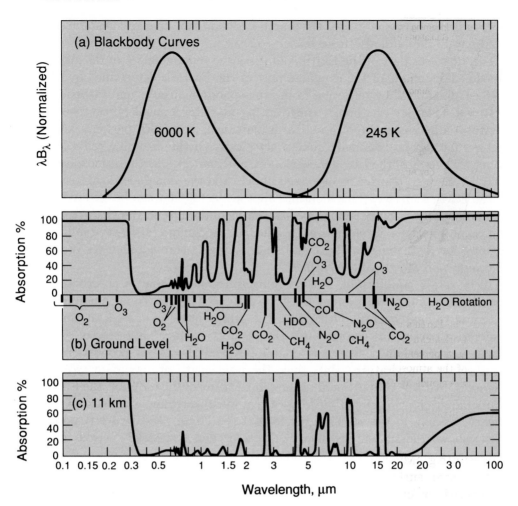

Fig. 1.1 (a) Spectral distribution of longwave emission from blackbodies at 6,000 K and 245 K, corresponding approximately to the mean emitting temperatures of the Sun and Earth, respectively, and (b) atmospheric absorption spectrum for a beam of radiation reaching the ground; (c) the same for a beam reaching the tropopause in temperate latitudes. Notice the comparatively weak absorption of the solar spectrum and the region of weak absorption from 8 to 12 μm in the longwave spectrum.

radiation but nearly opaque to infrared radiation as a result of absorbing gases and clouds. Thus, the surface, which absorbs nearly half (e.g., see Fig. 1.2; Schneider, 1990) the solar radiation, becomes a heat source for the lower atmosphere, which on the average cools steadily with increasing altitude to about 10 km. This part of the atmosphere is called the troposphere. The tropospheric vertical temperature *lapse rate*, $-\partial T/\partial z \approx 6.5$ K km^{-1}, is affected by both radiative heating and vertical convective

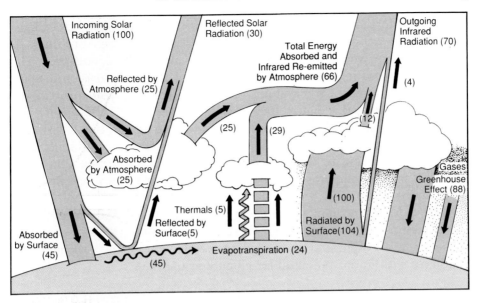

Fig. 1.2 The Earth's radiation energy balance, which controls the way the greenhouse effect works, can be seen graphically here. The numbers in parentheses represent energy as a percentage of the average solar constant – about 340 W m^{-2} – at the top of the atmosphere. Note that nearly half the incoming solar radiation penetrates the clouds and greenhouse gases to the Earth's surface. These gases and clouds re-radiate most (i.e., 88 units) of the absorbed energy back down toward the surface. This is the mechanism of the greenhouse effect.

processes (e.g., Manabe and Wetherald, 1967).

The warm surface layer emits infrared radiation, most of which is intercepted by optically active atmospheric gases, clouds, and particles. These constituents re-emit radiation both up to space and back down to the surface, the latter reducing dramatically the net loss of heat from the surface. Since the atmospheric emitters are colder than the surface, they emit proportionally less radiant energy. Because of this, the total outgoing infrared radiation from the Earth–atmosphere system is less than the radiant energy emitted by the surface alone, and the effective radiation temperature of the Earth is influenced more by the temperature of the cooler atmospheric gases and cloud tops (which emit radiation roughly like a blackbody with the temperature of the atmosphere at the cloud tops) than the warmer surface below. This phenomenon has often been called the *greenhouse effect*.

Should the amount of an infrared-absorbing gas in the atmosphere be increased, it would then intercept a larger fraction of the infrared energy coming upward from the warm layers near the surface. Thus, the outgoing infrared flux to space would be reduced by adding infrared-absorbing gases to the atmosphere. Furthermore, as seen on the right-hand arrows on Fig. 1.2, this would also increase the downward infrared flux in the lower

11

atmosphere, further warming the surface. The net result of the greenhouse effect is that an increase in the concentration of infrared absorbers in the atmosphere would lead to a rise in the surface temperature. This would be required in order to maintain constant infrared emission to space by the Earth–atmosphere system, assuming that the planetary albedo remained unchanged. Since 1900, enough extra CO_2, CH_4, N_2O and chloroflurocarbons (CFCs) have been added to trap an additional 3 W m^{-2} or so at the Earth's surface. This is not controversial (e.g., Ramanathan et al., 1985), but how to translate 3 W m^{-2} of extra surface heating into x degrees of surface temperature rise is complicated, since it involves modeling many feedback processes within the climatic system. For example, Fig. 1.2 shows that 5% of incoming solar energy is transferred at the surface to the atmosphere as sensible heat (i.e., thermals) and some 24% as latent heat in the form of evaporated or transpired water. The latter also influences cloudiness with major potential for climate feedback.

1.2.3 *Climate models of the radiation balance*

Consider the simplest possible climate model, a radiation balance for the Earth.

$$S = F \tag{1.1}$$

is an energy balance which requires that the amount of solar energy input to the system S is, in equilibrium, exactly balanced by the amount of terrestrial radiation energy F that leaves the system.

Translating this radiation balance into a simple climate model involves the energy balance equation (1.1) the Stefan–Boltzmann law, $F = \sigma T_p^4$, where σ is the Stefan–Boltzmann constant, and T_p is the effective blackbody radiative temperature of the Earth–atmosphere system in K. Rewriting (1.1) into the simplest possible climate model takes the form

$$Q(1 - \alpha) = \sigma T_p^4 . \tag{1.2}$$

In the left-hand term, $Q = S_o/4$ represents the incoming solar radiation of approximately 343 W m^{-2} averaged over the Earth's surface (i.e., this is S_o the "solar constant" divided by 4 since the area of the Earth's surface is 4 times the area of the Earth's disk which intercepts sunlight); α is the albedo of the planet. We can define a sensitivity parameter (see Chapter 10, Sec. 10.2.1 for details)

$$\lambda = \frac{1}{\frac{dF}{dT} - \frac{dS}{dT}} \tag{1.3}$$

which represents how much temperature change the planet would undergo if either the solar output S or the longwave radiation back to space F were to change. If one differentiates both sides of (1.2) with respect to T and rearranges terms, one can show that for the simple climate model

$$\lambda_p = \frac{T_p}{4S}. \tag{1.4}$$

Observations by infrared sensors on satellites suggest that the effective planetary temperature for Earth infrared radiation to space is equal to that of a blackbody with absolute temperature of approximately 255 K. Therefore, λ_p from (1.4) is approximately equal to 0.27 K W^{-1} m^2. In simpler terms this means that if the Sun were to somehow increase its energy output by 1%, the Earth's radiative temperature would eventually (i.e., in equilibrium) warm up by 0.64 K.

1.2.4 *Climatic feedback mechanisms*

Let us consider next what would happen if there were feedback in the system. That is, $\alpha_p = \alpha(T_p)$. Substituting α_p into (1.2) and differentiating both sides with respect to T one can obtain λ_p^*, a modified expression for λ_p in the case of feedback

$$\lambda_p^* = \frac{\lambda_p}{1 + \lambda_p \left(\frac{1}{1-\alpha_p}\right) \frac{\partial \alpha_p}{\partial T_p} S} \tag{1.5}$$

where λ_p is the sensitivity parameter obtained when albedo is constant (i.e., Eq. 1.4). It is an instructive exercise to provide some intuitive feeling for what this equation means. Supposing for example $\alpha_p = a - bT_p$ where a is constant and b is 0.01. What that implies is that if the Earth's temperature were to increase by 1°C then albedo would decrease by 0.01. Thus, for $\alpha_p = 0.3$ at $T_p = 255$ K and $b = 0.01$, this implies $\lambda_p^*/\lambda_p \approx 10$, indicating that a seemingly "small" feedback ($b = 0.01$) can have dramatic impact on the sensitivity of the system!

Earth satellites have suggested that for temperatures observed on Earth a linear relationship between outgoing infrared radiation, F_{ir} and surface temperature, T_s, is not a bad first approximation (e.g., Warren and Schneider, 1979). Let us rewrite our simple climate model then with this linearized form assuming α is a constant,

$$Q(1 - \alpha) = A + BT_s. \tag{1.6}$$

Consequently

$$\lambda_s = \frac{1}{B}. \tag{1.7}$$

Empirically B has been found from satellite observations to have a value of 1.83 W m^{-2} K^{-1} (Warren and Schneider, 1979). For $Q = 340$ W m^{-2} and $\alpha = 0.3$ then $\lambda_s = 0.55$ K W^{-1} m^2. This is twice λ_p, which suggests that empirically derived values of outgoing infrared radiation as a function of surface temperature lead to amplifying feedbacks.

The most often postulated feedback that could describe this amplification is the *water vapor–greenhouse effect* feedback. It is well known that increasing surface temperature increases evaporation, because evaporation increases nonlinearly with surface temperature through the Clausius–Clapeyron relationship between vapor pressure of water and temperature (e.g., see Chapter 3). In the midlatitudes, for example, although relative humidity is fairly constant from one season to the next, absolute humidity can increase by a large factor from winter to summer owing to this water vapor pressure–temperature relationship. Some recent empirical information from satellites strongly suggests that the water vapor–greenhouse feedback is indeed positive (e.g., Raval and Ramanathan, 1989), and may very well account for the substantially enhanced sensitivity of the linearized semi-empirical model (1.6) relative to the original blackbody model (1.2).

However, water vapor–greenhouse effect positive feedback is not the only potentially important feedback in the system. For example, consider

$$\alpha(T_s) = \delta + \gamma T_s \,, \tag{1.8}$$

where δ and γ are empirical constants. Plugging (1.8) into (1.6) and using (1.3) gives

$$\lambda_s = \frac{1}{B + \gamma Q}. \tag{1.9}$$

What does this mean? If γ is a positive number it means that when surface temperature increases albedo increases. This means that whatever causes surface temperature to increase would cause an increase in the reflectivity of the planet which would limit that original surface temperature rise, and is a negative feedback. If γ is negative then (1.9) implies positive feedback, since B is a positive number and the denominator of (1.9) would be reduced. If, in addition, the absolute value of γQ is greater than B, this system would become unstable. The greatest challenge in climate modeling (demonstrated here in linearized, zero-dimensional formalism) is to determine what the sum $B + \gamma Q$ is, based upon the many processes that interact in the climate system. Moreover, these processes which affect the feedback parameter γ do not occur uniformly across the globe; rather the global average values represented by the simple expressions derived so far are the manifestations of many local or regional changes that could be larger, smaller or even of opposite sign to the net global effect. For example, Raval and Ramanathan (1989) have shown empirically strong water vapor–greenhouse effect feedback, whereas Ramanathan and Collins (1991) have shown empirically strong negative cirrus cloud feedback in the part of the world with tropical cumulus clouds and ocean surface temperatures greater than 303 K. Nevertheless, this exercise given by (1.9) does show how sensitive the Earth's climate can be to seemingly small changes in these feedback processes.

1.2.5 Transient response

But what if we are interested in the transient response of the system to some global change forcing where the balance in (1.1) is perturbed? In this case our energy equation is rewritten to include an extra term for energy storage rate. That is, the rate of energy storage equals solar energy input to the Earth minus infrared radiant energy out. In symbolic terms

$$R\frac{\partial T_s}{\partial t} = Q(1-\alpha) - F_{ir} \tag{1.10}$$

where R is the "thermal inertia" of the system, i.e., the effective heat capacity of the atmosphere, oceans, land, etc. This, in turn, is proportional to the mass of each of the components of the climate system times their respective specific heats. If α is constant, or (to first order) its effects are lumped into the linear term B of (1.6), then (1.10) can be rewritten

$$R\frac{\partial T_s}{\partial t} = Q(1-\alpha) - (A + BT_s) \tag{1.11}$$

$$= \hat{Q}(t) - BT_s.$$

The solution to such an ordinary differential equation is well known from classical theory, and takes the form $T_s = T_{\text{inhomo}} + T_{\text{homo}}$. The homogeneous solution, which we could call T_{pert}, would be of the form

$$T_{\text{pert}} \quad \propto \quad \exp(-t/\tau). \tag{1.12}$$

where

$$\tau = R/B. \tag{1.13}$$

τ is the response time of the system to a step function forcing in Q, and is simply the heat capacity of the system divided by the radiative damping. B, the radiative damping coefficient, is also intimately involved in the sensitivity of the system. Recall, from (1.7) and (1.9), that the larger B is the less sensitive the system is to external forcing Q. What does this large damping mean physically? If B is a large number then the outgoing infrared radiation to space increases by a substantial amount if surface temperature increases. That is, the input of additional heat to the system would be damped out very effectively to space if B is a large number. That would limit the sensitivity of the system to an input of energy of any kind. For example, if a few W m^{-2} of additional energy were input by an increase in CO_2 or sunlight, then a large B would mean that only a small temperature change would be necessary to damp that extra heat back out to space. On the other hand a small B would imply very high sensitivity to small amounts of energy input, since the temperature would have to go up a great deal to damp that extra few W m^{-2} of heating back to space.

For the response time, τ, of the system, a large B implies a short response, since rapid damping means the system would approach its reduced equilibrium response more quickly than if there were little damping associated with a small B, as (1.13) shows. Thus, the feedback factors, which in linear form aggregate into B, not only affect inversely the overall sensitivity of the planet to forcing, but also its response time.

It is also required by (1.13) that the response time would be larger if the heat capacity were larger. This is intuitively obvious since a more thermally massive planet (i.e., one with a large portion of oceans or with oceans mixed deeply) would take longer to respond to global forcing than would one with relatively little heat capacity. The time-dependent model derived so far (i.e., (1.10)) is for an Earth with a fixed-heat capacity R.

The real world, of course, does not have a fixed-heat capacity, but consists of a mix of multiple-heat capacities, capacities which in fact change with the climatic state (e.g., Thompson and Schneider, 1979). For example, the heat capacity of the climate system over land is relatively small, consisting largely of the atmosphere itself and a few centimeters of soil. The response time is thus on the order of a month or so. The effective heat capacity of the middle of tropical oceans is largely governed by the depth of the oceanic mixed layer which is in contact with the atmosphere. Although that mixed layer is slowly ventilated from below, the dominant term in the heat capacity is that 50–70 m deep mixed layer. Nevertheless, the thermal inertia of the center of tropical oceans is at least an order of magnitude or so larger than that of the center of continents. However, polar seas, such as the Norwegian or Weddell Seas, in which oceanic convection causes the mixed layer to penetrate thousands of meters in depth, have effective thermal inertias another order of magnitude or so larger than that in the tropical oceans. Thus, to predict the transient response of the climate to global change forcings (e.g., CO_2 or other greenhouse gas buildups) will require not simply a global average model extended to include time dependence, but a model that has enough spatial and temporal resolution to capture the important nonglobal nature of the regionally heterogeneous effects of physical processes and heat capacities (e.g., Schneider and Thompson, 1981); this is explored in more detail in Chapter 17.

1.2.6 Hierarchy of models

The complexity of the climate system means that a hierarchy of models is necessary for studying the full response of the climate system to external forcing. The zero-order models we have described here are at the simpler end of the hierarchy, and three-dimensional, coupled atmospheric–oceanic–soil–vegetation–ice and chemistry models are at the more complex, more comprehensive end of that hierarchy. Many models in between the simple and complex ends can and have been constructed (see Chapter 10), with the virtue of the simpler models being their capacity to help us to understand the relative importance of interacting processes. The strength of the more comprehensive models is (hopefully) fidelity of simulation skill as well as their capacity to make regional, time-evolving projections of the response of the climate system to changing global forcing (simple models usually cannot resolve regional changes).

Many scientists believe that the ultimate goal of climate modeling should be fully comprehensive, three-dimensional models of all elements of the climate system including very high resolution and as much detail as possible. While such a goal is clearly appropriate for the distant future, practical considerations require compromises, as discussed in the next section.

1.3   Climate predictions

1.3.1   *Empirical statistical versus "first principles"*

Climate prediction, like most other forecasts of complex systems, can involve extrapolation. We attempt to determine the future behavior of the climate system from knowledge of its past behavior and present state, basically taking two approaches. One, the "empirical statistical", uses empirical statistical methods, such as regression equations with past and present observations, to obtain the most probable extrapolation. The other uses "first principles": equations believed to represent the physical, chemical, and biological processes governing the climate system for the scales of interest. The latter approach is usually called "climate modeling." Since the statistical approach depends on historical data, it is obviously limited to predicting climates that have been observed or are not caused by new processes. The statistical method cannot easily answer "what if?" questions, such as the effects of rapidly increased atmospheric carbon dioxide. Thus, the more promising approach to climate prediction for conditions or forcings different from the present or from historic precedent is climate modeling. Then, the validation of the predictions of such models becomes a chief concern.

Climate models vary in their spatial resolution, that is, the number of dimensions they simulate and the spatial detail they include. A simple model calculates only the average temperature of the Earth, independent of time, as an energy balance among the Earth's average reflectivity and the average "greenhouse" properties of the atmosphere. Such a model is zero-dimensional: it reduces the real temperature distribution on the Earth to a single point, a global average. In contrast, three-dimensional climate models reproduce the variation of temperature with latitude, longitude, and altitude. The most complex models, the General Circulation Models (GCMs), predict the time evolution of temperature plus humidity, wind, soil moisture, sea ice, and other variables through three dimensions in space (e.g., Washington and Parkinson, 1986).

Although GCMs are usually more complex than simpler models in their physical, chemical or biological detail, they are also more expensive to design, run, and interpret. The optimal level of complexity for a model depends on the problem and the resources available. More is not necessarily better. Often it makes sense to attack a problem first with a simple model and then employ the results to guide research at higher resolution. In other words, deciding how complicated a model to use for a task and whether to trade completeness and accuracy for tractability and economy is more an intuitive judgment than a scientific choice subject to explicit, logical criteria (e.g., Land and Schneider, 1987).

1.3.2 Grids and parameterization

Even the most complex GCM is limited in the spatial detail it can resolve. No computer is fast enough to calculate climatic variables everywhere on the Earth and in the atmosphere in a reasonable time. Instead, calculations are executed at the widely spaced points of a three-dimensional grid at and above the surface. For a typical example, divide the surface of the Earth into a grid of 1,920 squares, each 4.5° latitude by 7.5° longitude. At 40° latitude each square is 500 by 640 km. Then divide the atmosphere above each square into nine strata. The calculation of a year of simulated "weather" in these 17,280 grid boxes by 30 minute increments takes some 10 hours on a Cray X-MP computer.

Wide spacing creates a problem. Many climatic phenomena occur over smaller scales than an individual "box" of the grid. For example, clouds reflect much incident sunlight back to space, and they also block the escape of infrared radiation from below, thus influencing the greenhouse effect. Therefore, they help to determine the temperature on the Earth. Predicting changes in cloudiness is, therefore, an essential part of climate simulation. No GCM now available or likely to be available in the next few decades, however, has a grid fine enough to resolve individual clouds, which tend to shade a few kilometers rather than a few hundred kilometers.

Subgrid-scale phenomena like clouds are represented collectively by parameterization (short for parametric representation) rather than individually. A parameterization could, for example, be based on climatological data to derive statistical relations between variables that are resolved by the grid and ones that are not. For instance, the average temperature and humidity over, say, the large area beneath one box, can be related to the average cloudiness over the same area; to make the equation work one introduces a parameter or proportionality factor derived empirically from the cloudiness, temperature, and humidity data. Since a model can calculate the temperature and humidity over a box from physical principles, the semi-empirical parameterization predicts the average cloudiness in the box even though it cannot predict individual clouds. Modelers, of course, strive to keep their parameterizations as physical and nonempirical and scale-independent as practical. Thus, the validity of parameterization and overall model performance as well, depends ultimately on empirical tests, not only on the inclusiveness of the first principles. In other words, even our most sophisticated "first principles" models contain "empirical statistical" elements within the model structure.

1.3.3   Climate sensitivity and scenarios

Uncertainty about parameterizations of feedback mechanisms like clouds or sea ice is one reason the goal of climate modeling – reliable, verified forecasting of key variables such as temperature and rainfall on a regional, time-evolving basis – is not attainable yet. Another source of uncertainty external to the models is human behavior. Forecasting, for example, the impact of carbon dioxide on climate requires knowing how much carbon dioxide is going to be emitted (e.g., Nordhaus and Yohe, 1983) and how that emission will be distributed or removed by the physical, chemical, and biological processes of the carbon cycle as already noted on Plate 1.

What the climate models can do well is analyze the sensitivity of the climate to uncertain or even unpredictable variables. In the case of carbon dioxide, one could construct plausible scenarios of economic, technological and population growth to project growth of CO_2 emission and model the climatic consequences (e.g., as done by IPCC, 1990a). Such uncertain climatic factors as cloud-feedback parameters could be varied over a plausible range or alternative parameterization formulations can be used. The calculations would indicate which uncertain parameter or formula are most important in making the climate more or less sensitive to carbon dioxide increase. One could then concentrate research on narrowing the uncertainty surrounding those factors. The results of such sensitivity tests would also suggest the range of climatic futures that ecosystems and societies may be forced to adapt to and at what rates. How to respond to such information, of course, is a value-laden issue, which is examined in other chapters and has been addressed by the author elsewhere (Schneider, 1990).

1.3.4   Theoretical issues

Although the atmosphere–earth–ice–ocean system is complex, we can describe the known physical laws mathematically, at least in principle. In practice, however, solving these equations in full, explicit detail is impossible. First, the possible scales of motion in the atmospheric and oceanic components range from the submolecular to the global. Second are the interactions of energy transfers among the different scales of motion. Finally, many scales of disturbance are inherently unstable; small disturbances, for example, grow rapidly in size if conditions are favorable. Thus, seemingly small differences between two similar atmospheric or oceanic states cause later divergence. Meteorologists have found from theoretical considerations and from experience that useful detailed weather prediction of beyond about 10 days is impossible using current observations (e.g., Somerville, 1987). Is the longer-term climate prediction thus a hopeless task?

Several reasons, however, make prediction of climate, in contrast to weather, feasible for comparatively long periods. For one thing, although day-to-day weather is not predictable far in advance, some success can be obtained in predicting average conditions for an extended period. The situation is approximately analogous to the statistical-mechanical theory of gases: although we cannot predict the behavior of individual molecules, we can accurately predict the expected mean state and variance of an ensemble of molecules for some sets of conditions.

Another reason that climate predictions for longer periods may be possible is that the climate system is subject to forcing processes that may be of overriding importance for some time or space scales. An obvious example is the annual variation in the global distribution of solar radiation. The strength of this forcing causes the seasons to follow each other predictably, although differences (anomalies) in seasons are, of course, important from year to year (e.g., Namias, 1972; Trenberth et al., 1988).

Some forcing mechanisms are predictable whereas others, such as volcanic activity, are largely unpredictable today. Also, the atmosphere is forced by such mechanisms as oceanic surface temperatures that themselves respond gradually to atmospheric forcing, a complicating feedback. These feedbacks are commonly referred to as "internal forcings", in contrast to the straightforward "external forcings" from outside such as solar radiation. What is internal on long-time scales may be external on shorter ones, depending on what processes are included in the climatic system defined for an investigation.

In any event, the presence of forcing implies that some aspects of climate may be predictable on those time scales where the forcing and its response are important. In fact, one would expect the degree of climate response, and hence predictability, to be related to both the amplitude and the period of forcing, or, for aperiodic cases, the time scale of the forcing. Although this is true in general for external forcing, nonlinear feedbacks within the system can produce unexpected results such as periodic or aperiodic oscillations that come or go with small changes in the values of internal model parameters (e.g., see Ghil, 1981; and Harvey and Schneider, 1984). For instance, internal interactions can cause internal stochastic oscillations (e.g., Hasselmann, 1976) on different time scales or even chaotic behavior (e.g., Lorenz, 1968).

For many longer time scales, it is widely believed the global-scale climate system responds deterministically to current "boundary conditions" and has little memory of its history. In other words, the climate system will move to a unique equilibrium after a transient adjustment for these longer times, as Imbrie and Imbrie (1979) assumed when modeling glacial–interglacial cycles – just as others typically assume in CO_2 doubling experiments. Thus, deterministic models provide an equilibrium "snapshot" appropriate for the time scale of the external forcing. The validity of such equilibrium experiments, however, depends on the existence of a unique equilibrium for the given boundary conditions, which is still a debated assumption for the Earth's climate system (e.g., Lorenz, 1968; Schneider and Gal-Chen, 1973; North, 1975).

1.3.5 Scale transition

Finally, within the foreseeable future even the highest resolution three-dimensional GCMs will not have a grid much less than 100 km. They will not, therefore, be able to resolve individual thunderstorms, or the important local or mesoscale effects of hills, coastlines, lakes, vegetation boundaries, or heterogeneous soils. For regions that have relatively uniform land surface characteristics, such as a thousand kilometer scale savannah or a tropical forest with little elevation change, GCM grid-scale parameterizations of surface albedo, soil type, and evapotranspiration could adequately be used to estimate local changes. Alterations in climate predicted within a box would probably apply fairly uniformly across such nicely behaved, homogeneous areas. On the other hand, steep topography over watersheds smaller than GCM grids can mediate regional climate. Therefore, even if GCM predictions were accurate at grid scale, they would not necessarily be appropriate to local conditions.

Large-scale observed climatic anomalies are translated to local variations in Fig. 1.3 (Gates, 1985). This analysis of the local climatic variability for the state of Oregon was based on several years of data using a technique known as *empirical orthogonal functions*[1]. The north–south Cascade Mountains translate a simple change in the frequency or intensity of westerly winds into a characteristic climatic signature of typically wetter on the west slope and drier on the east or vice versa. In other words, a GCM producing altered westerlies in response to say, CO_2 doubling, could be applied to the map on Fig. 1.3 to determine the impact in a local watershed. Such a map, constructed from variations of climate observed over several years, seems an ideal way to translate information at a GCM grid scale to the local or mesoscale. Because empirical data have been used, however, such a relation would only be valid if the causes of recent climatic variations or oscillations which gave rise to Fig. 1.3 carried forward and included the effects of climatic changes forced by trace gases. It is not obvious that the signature of climatic change from increases in trace gases will be the same as

---

[1] Empirical orthogonal functions (EOFs) are widely used to analyze spatial fields into their principal patterns of temporal variability, see Morrison, 1976, Chapter 8.

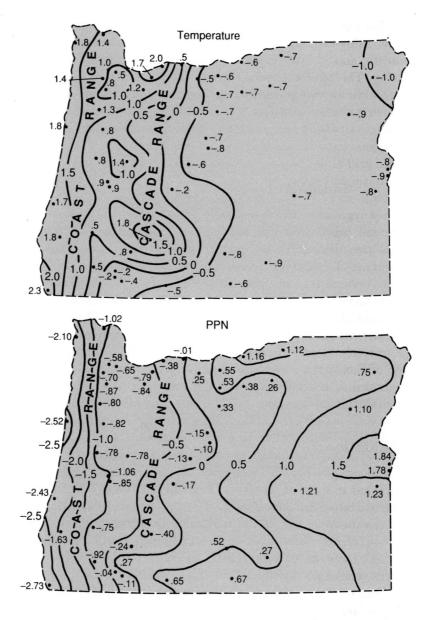

Fig. 1.3 The distribution of the relationship between large-scale (area-averaged) and local variations of the monthly mean surface air temperature (above) and precipitation (below), as given by the first empirical orthogonal function determined from thirty years of observational monthly means at 49 stations in Oregon in comparison with the state-wide average.

past vacillations, many of which could have been internal oscillations within the climate system, not created by external forcing such as changes in trace gases. Thus other translations of scale need to be considered.

22

One might embed a high resolution "mesoscale model" within a few boxes of a GCM, using as boundary conditions for the mesoscale model the wind, temperature, and so forth predicted by the GCM at the mesoscale model's boundaries. The mesoscale model could then account for local topography, soil type, lakes and vegetation cover, and translate GCM forecasts to local scale. Such embedding techniques have shown considerable early promise (e.g., Giorgi, 1990). But for such a method to have any reasonable hope of success, however, the GCM must produce accurate climatic statistics for the special grid area. To return to the Oregon case in Fig. 1.3, for example, if the climatic average of the GCM's winds in the unperturbed case (i.e., the "control" case) has the wrong westerly component, the local climate change in a region where topography amplifies any such error in the wind direction will probably be misrepresented. A prerequisite to performing scale transition through embedding of local or regional models, therefore, is assurance that the control climate of the GCM for the important variables is accurate enough that it makes sense to take the next step of imposing a scenario of trace gas increase on the GCM to estimate how the local-scale climate might change.

Practically, while testing scale transitions in steep topography and other rapidly varying local features proceeds, modelers should examine the behavior of their models using grid boxes that are much less pathological. That is, examine boxes where local features are relatively homogeneous and where translation of local-to-grid scales should prove a less serious obstacle.

1.4 Validation

The most perplexing question about climate models is whether they can be trusted to provide grounds for altering social policies, such as those governing carbon dioxide emissions. How can models so fraught with uncertainties be verified? There are actually several methods. Although none is sufficient alone, they can together provide significant, albeit circumstantial, evidence of a model's credibility.

The first verification method is checking the model's ability to simulate today's climate. The seasonal cycle is one good test because the temperature changes are larger, on a hemispheric average, than the change from an ice age to an interglacial period (i.e., 15°C seasonal range versus 5°C glacial–interglacial cycle, respectively, in the NH). GCMs map the seasonal cycle well, which suggests that their surface temperature sensitivity to large-scale radiative forcing is not way off. The seasonal test, however, does not indicate how well a model simulates such slow processes as changes in deep ocean circulation or ice cover, which may have an important effect on the decade to century time scales over which CO_2 is expected to double.

A second verification technique is isolating individual physical components of the model, such as its parameterizations, and testing them against reality. For example, one can check whether the model's parameterized cloudiness statistics matches the observed cloudiness statistics of a particular box. But this technique cannot guarantee that the complex interactions of individual model components are properly treated. The model may be good at predicting average cloudiness but bad at representing cloud feedback. In that case, simulation of overall climatic response to, say, increased carbon dioxide is likely to be inaccurate. A model should reproduce to better than, say, 10% accuracy the flow of thermal energy between the atmosphere, surface, and space (see Fig. 1.2). Together, these energy flows comprise the well-established *greenhouse* effect on Earth and constitute a formidable and necessary test for all models. A model's performance in simulating these energy flows is an example of physical verification of model components. As another example, Raval and Ramanathan (1989) compared the enhanced infrared heat trapping with increasing surface temperature by using satellite observations (e.g., see Chapter 10). They compared observed water vapor–greenhouse effect feedback calculations in GCMs against satellite observations.

A third method for determining overall, long-term simulation skill is the model's ability to reproduce the diverse climates of the ancient Earth (e.g., see Chapter 21) or even of other planets (e.g., Kasting et al., 1988). Paleoclimatic simulations of the Mesozoic Era, glacial–interglacial cycles, or other extreme past climates help in understanding the coevolution of the Earth's climate with living things. As verifications of climate models, however, they are also crucial to estimating both climatic and biological future (Schneider, 1987).

Overall validation of climatic models thus depends on constant appraisal and reappraisal of performance in the above categories. Also important are a model's response to such century-long forcings as the 25% increase in carbon dioxide and other trace greenhouse gases since the Industrial Revolution. Indeed, most climatic models are sensitive enough to predict that warming of 1°C should have occurred during the past century. The precise "forecast" of the past 100 years also depends upon how the model accounts for such factors as changes in the solar constant or sulfur dioxide emissions or volcanic dust (e.g., Schneider and Mass, 1975, or Hansen et al., 1981). Indeed as Fig. 1.4 (Wigley and Raper, 1990) shows using a highly simplified one-dimensional climate model of atmosphere and oceans, the typical prediction of an 0.5 to 1°C warming over the twentieth century is broadly consistent with, but somewhat larger than, observed. Possible explanations for the discrepancy include (see Gilliland and Schneider, 1984): (a) the sensitivity of state-of-the-art models to trace gas greenhouse increases have been overestimated some twofold; (b) modelers have not properly accounted for such competitive external forcings as volcanic dust or changes in solar energy output; (c) modelers have not accounted for other external forcings such as regional tropospheric aerosols from biological, agricultural

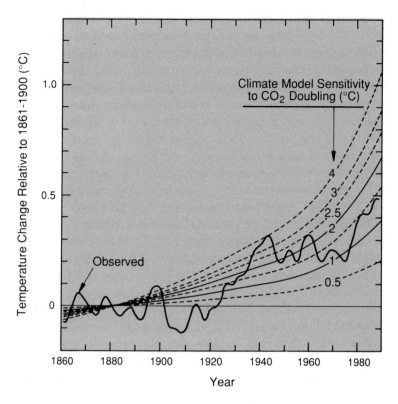

Fig. 1.4 Observed global-mean temperature changes (1861–1989) compared with predicted
values. The observed changes are given in Sec. 7 of the IPCC (1990a). The data
have been smoothed to show the decadal and longer time scale trends more clearly.
Predictions are based on observed concentration changes and concentration–forcing
relationships as given in Sec. 2 of the IPCC (1990a) and have been calculated using
the highly simplified upwelling-diffusion climate model of Wigley and Raper (1990).

and industrial activity; (d) modelers have not properly accounted for the
large heat capacity of the oceans taking up some of the heating of the
greenhouse effect and delaying warming of the atmosphere; (e) both present
models and observed climatic trends could be correct, but models are
typically run for equivalent doubling of carbon dioxide whereas the world
has only experienced a quarter of this increase and nonlinear processes have
been properly modeled and produced a sensitivity appropriate for doubling
but not for 25% increase; (f) the incomplete and inhomogeneous network of
thermometers has underestimated warming; and (g) there may have been a
natural cooling trend of up to 0.3–0.5°C during the twentieth century.

Despite this litany of excuses why observed global temperature trends in
the past century and those anticipated by most GCMs (i.e., +2 to 5°C for a
CO_2 doubling) disagree somewhat, the approximately twofold discrepancy

is not fundamental. Most climatologists do not yet proclaim the observed temperature records to have been caused beyond doubt by the greenhouse effect. Depending upon what assumptions one makes [e.g., (a)–(g) above], the twentieth century surface temperature trend could be consistent with an equivalent CO_2 doubling, equilibrium temperature response of 0.5–5.0°C (Wigley and Raper, 1990)! Thus, a greenhouse effect signal cannot yet be said to be unambiguously detected in the record. It is still possible that the observed trend and the predicted warming could be chance occurrences. One cannot easily rule out that other factors, such as natural fluctuation or solar constant variations or anthropogenic dust, simply have not been adequately accounted for over the past century – except during the past decade when adequate instruments have been measuring the last two. Nevertheless, this empirical test of model predictions against a century of observations certainly is consistent with a rough factor of 3. This test is reinforced by the good simulation by most climatic models of the seasonal cycle, diverse ancient paleoclimates, hot conditions on Venus, cold conditions on Mars (both well simulated), and the present distribution of climates on Earth. When taken together, these verifications provide a strong circumstantial case that the modeling of sensitivity of temperature to greenhouse gases is probably valid within threefold (as also was suggested by IPCC, 1990a).

Another decade or two of observations of trends in Earth's climate, of course, should produce signal-to-noise ratios sufficiently obvious that almost all scientists will know whether present estimates of climatic sensitivity to increasing trace gases have been predicted well or not. Unfortunately, the global change "experiments" now underway are not merely academic exercises in the microchips of supercomputers, but are being performed (as noted nearly four decades ago by Revelle and Suess, 1957) on the "laboratory" that we and all other living things share – Earth.

2

Human components of the climate system

Michael H. Glantz and Jerrold H. Krenz

2.1 Introduction

The purpose of this chapter is to introduce readers to the interrelationships between human activity and physical processes related to climate change. Brief overviews are presented of the environmental processes of primary concern, followed by discussion of the anthropogenic causes behind them. It is important to understand those socioeconomic forces that are contributing (often driving) the rates of climate-related environmental change. Only then can societies hope to modify if not control those rates by providing guidance to decision-makers seeking to influence the interactions of climate and society.

There are several global change and climate change schematic diagrams attempting to depict in a simplified way the complex linkages among subsystems. Plate 2 is one example schematically depicting the complexity of the interactions of the components of the atmosphere with other elements of the biosphere; a more complex version is given in Plate 1. It was chosen specifically because it is one of the few that makes explicit the importance of the human subsystem in the climate system. Aspects of many of the interactions in Plate 2 are either presently unknown or poorly understood, underscoring the need for a broad-based scientific research program to reduce uncertainties in our knowledge of those interactions. However, human activities, which in many instances form an integral part of atmospheric–biospheric interactions, are often treated in a cursory manner and sometimes incorrectly displayed in these diagrams and in their accompanying texts as well. Thus, this major forcing function has not as yet received the attention in global change research that it deserves, notwithstanding some recent attempts to correct for this situation (ISSC, 1990; U.S. Congress, 1990). Human activities that affect the atmosphere (e.g., industrialization, deforestation, desertification, land transformation, and so forth) will most likely become a central part of a global change/climate change research program.

To understand and model the causes and the impacts of physical changes at global to local scales, as well as the feedback mechanisms that, in turn, affect those physical processes (that is, enhance or weaken them), there must be an improved understanding of the human activities that directly or indirectly initiate or abet such changes. Supporting this contention, a recent Committee on Earth and Environmental Sciences (CEES) report (1991) noted that

> Without an understanding of human interactions in global environmental change that is based both on empirical observations of human behavior and on a better understanding of the consequences of human actions, the models of physical and biological processes of change will be incomplete (p. 52).

This chapter, therefore, focuses on underlying processes as well as rates of change on the human side of the climate change equation. The processes we discuss are: fossil fuel and chemical use, tropical deforestation and desertification. The final section provides some examples where human intervention, such as decisions by policymakers and individuals, has actually had an effect on altering the rates of production of greenhouse gases and tropical deforestation. These examples are cited to reinforce the authors' belief that policies *can* make a difference in whether, how, and to what extent the enhancement of the greenhouse effect continues or abates.

2.2 Greenhouse gases emissions

The idea that certain gases in the atmosphere behave like the panes of glass in a greenhouse was first proposed in 1827 by French scientist Jean Baptiste-Joseph Fourier. Many scientists continue to compare the atmosphere to a greenhouse in order to convey in the form of shorthand complex processes to nonscientific audiences. While the greenhouse analogy is useful for educating policymakers and the general public about complex processes, it is not useful for generating an improved understanding of the processes involved or for generating new scientific hypotheses. In reality the analogy does not hold because the greenhouse does not permit convective heat exchange, as opposed to the greenhouse gases blocking the radiative loss of heat.

The consequences of the greenhouse effect, which originally focused on the effects of increases in atmospheric carbon dioxide (CO_2), were regarded until the 1960s to be either beneficial to humanity or neutral. Both Arrhenius (1908), Chamberlin (1899), and, later, Callendar (1938) viewed them as beneficial to humanity because a warming of the atmosphere would serve to forestall the onset of the next (and imminent) Ice Age. Revelle and Suess (1957) in the mid-1950s observed that "human beings are now carrying out a large-scale geophysical experiment of a kind that could not be reproduced in the future." In the late 1960s and in the 1970s attention again focused on the possible anthropogenic enhancement of the greenhouse effect. Unlike

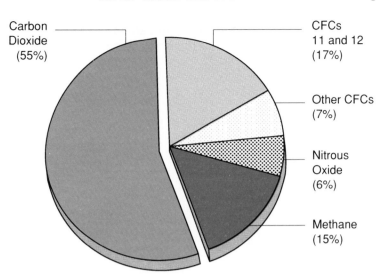

Fig. 2.1 The contribution from each of the human-made greenhouse gases to the change in radiative forcing from 1980 to 1990. The contribution from ozone may also be significant, but cannot be quantified at present. Source: IPCC, 1990a.

in the past, this time the potential effects on ecosystems and society were considered to be harmful.

The natural and anthropogenic sources and sinks of other greenhouse gases (GHGs) are also being identified, including nitrous oxide (N_2O), chlorofluorocarbons (CFCs), methane (CH_4) and tropospheric ozone. In addition to being important radiatively active trace gases, CH_4, N_2O and CFCs have adverse impacts on stratospheric ozone. Increased emissions of GHGs have resulted from industrial processes, recently referred to as "industrial metabolism" (NRC, 1990), as well as from changes in land use. Although briefly highlighted in the following section, the chemistry of the major greenhouse gases is discussed in greater detail in Chapter 7.

Figure 2.1 depicts the contribution to the increase in radiative forcing during the 1980s by the major greenhouse gases (IPCC, 1990a).

2.2.1 Carbon dioxide (CO_2)

Carbon dioxide is considered the major greenhouse gas of concern with respect to climate change. CO_2 levels have been changing throughout the Earth's history, see Fig. 21.3, and these changes have been associated with warm epochs as well as ice ages and with major changes in sea levels. Concern with CO_2 today centers on its role in global warming. The atmospheric content of carbon dioxide has been measured in detail continuously since 1957 at Mauna Loa, Hawaii, (Fig. 2.2), and earlier estimates can be made from bubbles of air trapped in ice cores. Note the

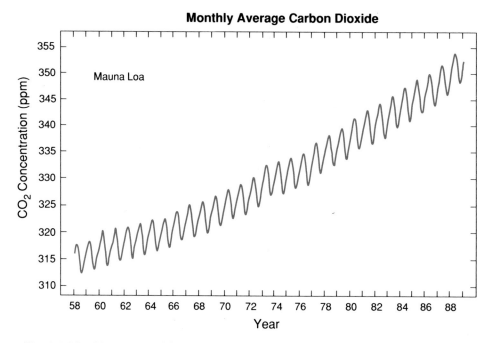

Monthly Average Carbon Dioxide

Fig. 2.2 Monthly average CO_2 record from Mauna Loa, Hawaii in parts per million by volume (after Keeling et al., 1989).

strong annual cycle superposed on a trend; the former also arises from the seasonal cycle in photosynthesis over the Northern Hemisphere (NH). As a result of human activities, carbon dioxide has increased by about 25% since the onset of the Industrial Revolution at the end of the eighteenth century (Fig. 2.3a). Societies are dependent on various forms of fossil fuels in order to intensify their industrialization processes and to achieve their economic development objectives. Deforestation and the burning of biomass are also responsible for a relatively large fraction of the anthropogenically produced CO_2.

CO_2 is a by-product of industrial processes dependent on fossil fuels. Fossil fuels are used in transportation, electric power generation, heating and cooling processes and in manufacturing. The burning of coal yields the highest amount of carbon dioxide per unit of energy produced; next comes oil; natural gas is the cleanest of the three in terms of CO_2 emissions. Woodburning also produces CO_2.

Developing countries' share of the global production of CO_2 has been relatively lower in the past than that of the industrialized countries. Current trends, however, suggest that early in the next century the contribution of greenhouse gases from the developing countries will equal or surpass that of the developed countries, mainly as a result of the energy and food demands of increasing populations and from tropical deforestation, and

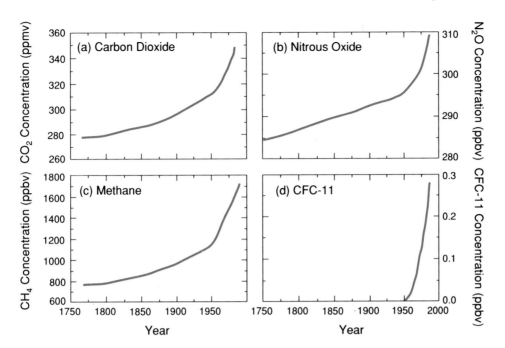

Fig. 2.3 Concentrations of carbon dioxide and methane after remaining relatively constant up to the eighteenth century, have risen sharply since then due to human activities. Concentrations of nitrous oxide have increased since the mid-eighteenth century, especially in the last few decades. CFCs were not present in the atmosphere before the 1930s. Source: IPCC, 1990a.

not necessarily from increases in per capita energy consumption in these countries.

Many developing countries view the national level of CO_2 emissions as an indicator of a country's level of industrialization, which is in turn viewed as an indicator of the level of economic development. Developing countries claim that they are not yet full-fledged parties to industrialization processes. In many instances these countries expect to rely on the only energy sources available to them (e.g., fossil fuels and wood) in order to develop their economies and to raise their standards of living. They believe that concerns about global warming and proposed policy actions expressed by some industrialized countries to limit carbon dioxide emissions will inhibit any chance for development that they might have. They also believe they are being asked to carry the burden of an environmental solution to an environmental problem that they did not create. This perspective has recently been referred to as "environmental colonialism" (Agarwal and Narain, 1991).

Thus, it is easy to show that estimates of future levels of atmospheric CO_2 are very dependent on human activities and are, therefore, subject to change. Reliable projections of future atmospheric CO_2 levels will, for example, depend on reliable estimates of the rates of population growth, per capita energy consumption, and of the types of energy used. Estimates of future levels of economic development of societies around the globe are also important. As societies become more affluent, they tend to consume more energy per capita. Also relevant to these projections are present attempts to develop more efficient ways to use energy and to change the national mix of energy sources.

2.2.2 Nitrous oxide (N_2O)

Nitrous oxide is produced in the natural environment and as a result of human activities. Information on the sources and sinks of nitrous oxide is still considered unreliable (IPCC, 1990a). The primary concern today with N_2O relates to emissions resulting from the widespread and increasing application of chemical fertilizers, tropical deforestation in general, and more specifically the process of the conversion of forest to pasture or farmland.

According to a recent report by the U.S. Department of Energy (DOE, 1990), "Ice core data indicate that the concentration of N_2O was stable for approximately 3,000 years ... [but] began to increase beginning about 150 years ago." The annual increase in N_2O is estimated between 0.2 to 0.3% (Fig. 2.3b). While it is difficult to quantitatively account for the source of the current increase in atmospheric abundance of nitrous oxides, it is thought to be man-made (IPCC, 1990a). Recent estimates of the annual contribution of N_2O from combustion and biomass burning suggest that it is much less than originally believed. Agricultural practices, other than fertilizer use, may stimulate emissions of N_2O from soils and, therefore, may prove to be important.

The use of N_2O in fertilizers is associated with food production and food security issues in developing and industrialized countries. With much of the world's arable land already in production, attempts to increase crop yields may provide one of the few ways (without attempting to cultivate lands known to be marginal to agricultural production) to increase food production in order to meet the nutritional needs of expanding populations during the next several decades.

2.2.3 Methane (CH_4)

The atmospheric content of methane has been increasing at a rate of 0.8 to 1.0% per year (Fig. 2.3c). Increases in methane track well with population growth rates. Human activities such as wet rice cultivation, domestic ruminant breeding, landfills, biomass burning, coal mining and natural gas venting and even leaks from natural gas pipelines have contributed to the atmospheric concentration of methane (IPCC, 1990a). In addition, there are natural sources of methane emissions such as termites and wetlands, both of which can be influenced by human activities. The quantitative contribution of each of these factors to the observed total methane increase is not well known at present and the contribution of each source to the total CH_4 burden carries with it a wide band of uncertainty. Nevertheless, there are research activities presently planned or under way to reduce methane emissions from rice paddies (e.g., wet rice production), despite the fact that the precise role of wet rice production in methane emissions is yet to be determined.

2.2.4 Chlorofluorocarbons (CFCs)

Chlorofluorocarbons (CFCs), discovered in the 1920s, are used as cleaning solvents, foam blowing agents, refrigerants, fire retardants, and as propellants in aerosol sprays. Before 1950, CFCs were not widely used commercially. Once commercial use started, however, it increased rapidly (Fig. 2.3d).

CFCs are chemically inert to lower atmospheric cleansing processes, including chemical reactions, enabling them to reach the stratosphere. CFCs have a long residence time in the atmosphere, estimated at 60 to 200 years. In the stratosphere, they are subjected to photodissociation, after which the free chlorine combines and recombines with ozone to form other compounds. This process makes CFCs a very efficient "ozone eater." In addition to their ozone destruction powers, CFCs are also efficient at trapping longwave radiation, thereby enhancing the greenhouse effect.

A recent scientific assessment summarized the important characteristics of these key greenhouse gases, as shown in Table 2.1.

2.2.5 Estimating future greenhouse gas concentrations

There are many factors that affect the activities that directly and indirectly produce greenhouse gases. Combinations and permutations of these factors can alter scenarios about future greenhouse gas emissions. These variable factors include but are not limited to population growth, gross national product, energy supply and demand, technological development and transfer, deforestation and land clearing, CFC production, changes in agricultural production and practices, and so on. Also, the environmental consequences of human conflicts, such as the oil well fires in Kuwait resulting from the Iraqi retreat from that country in early 1991, can add to atmospheric concentrations of greenhouse gases. Different rates of change for these factors (based on optimistic or pessimistic perceptions about them) can alter greenhouse gas projections in a major way.

Clearly, making projections about future rates of increases in greenhouse gases is fraught with difficulties, because human behavior can change or be changed as the result of effective education about the environmental

Table 2.1. Summary of key greenhouse gases influenced by human activities[1]. Source: IPCC, 1990a.

Parameter	CO_2	CH_4	CFC-11	CFC-12	N_2O
Pre-industrial atmospheric concentration (1750–1800)	280 ppmv[2]	0.8 ppmv	0	0	288 ppbv[2]
Current atmospheric concentration (1990)[3]	353 ppmv	1.72 ppmv	280 pptv[2]	484 pptv	310 ppbv
Current rate of annual atmospheric accumulation	1.8 ppmv (0.5%)	0.015 ppmv (0.9%)	9.5 pptv (4%)	17 pptv (4%)	0.8 ppbv (0.25%)
Atmospheric lifetime[4] (years)	(50–200)	10	65	130	150

[1] Ozone has not been included in the table because of lack of precise data.

[2] ppmv = parts per million by volume; ppbv = parts per billion by volume; pptv = parts per trillion by volume.

[3] The current (1990) concentrations have been estimated based upon an extrapolation of measurements reported for earlier years, assuming that the recent trends remained approximately constant.

[4] For each gas in the table, except CO_2, the "lifetime" is defined here as the ratio of the atmospheric content to the total rate of removal. This time scale also characterizes the rate of adjustment of the atmospheric concentrations if the emission rates are changed abruptly. CO_2 is a special case since it has no real sinks, but is merely circulated between various reservoirs (atmosphere, ocean, biota). The "lifetime" of CO_2 given in the table is a rough indication of the time it would take for the CO_2 concentration to adjust to changes in the emissions.

implications of human activities, technological innovations, or as the result of effective policymaking (or lack thereof). These difficulties notwithstanding, there has been a plethora of projections, predictions, and scenarios of future greenhouse gas emissions. Such scenarios must be used with great care, making explicit their strengths and shortcomings.

Along with industrial metabolism, deforestation has been identified as a major source of greenhouse gases. Desertification is another result of inappropriate land use that can affect atmospheric processes but not necessarily CO_2. The societal processes underlying deforestation of the tropical rainforests and desertification are discussed in the following section.

2.3 Deforestation

In the last decade attention has increasingly focused on the fate of tropical rainforests for a variety of reasons, including the plight of the indigenous indian tribes, the loss of biological diversity, loss of soil fertility following land conversion, international debt pressures and, not the least, because of the contribution of deforestation to reductions in regional rainfall and to the enhancement of the greenhouse effect.

The spatial extent of the remaining tropical rainforests is estimated to be as follows (Anon., 1990):

> Latin America contains the greatest part of the world's tropical forests (59%), the Amazon basin constituting by far the largest forest of its kind (about half the world's total tropical forest surface). Next comes Asia (22%), where the forest is mainly concentrated in Indonesia (about 10% of the world tropical forest), Malaysia, the Philippines, the Indochinese peninsula and the extreme south of the Chinese coast. Finally, about 19% ... is located in Africa, essentially in Zaire and the Congo (over 10% of the world total).

Of those, depletion of the Amazon rainforest has by far received a major share of that attention.

As of today, estimates of the rates of land clearing are not very reliable. In addition, the driving forces behind the clearing vary from one location to the next. Improving our understanding of the rates and purposes of land clearing in different locales throughout the rainforests can provide valuable insights to policymakers who can alter processes and rates of environmental changes, through their policy-making powers.

On a planetary scale, these forests are important to the atmosphere, because during their growth they sequester carbon from the atmosphere and store it. When the forests are cut down and burned, the stored carbon will be released as carbon dioxide into the atmosphere. The contribution of the destruction of tropical rainforests to the atmospheric content of CO_2 has been estimated by different authors as being between 5% and 20% (Table 2.2). Debate over the rate and extent of deforestation in the Brazilian Amazonia (e.g., Fearnside, 1990) and other locations (e.g., Agarwal and Narain, 1991) continues unabated.

Table 2.2. Share of individual countries in CO_2 emissions caused by the clearing of tropical rainforests by fire. Source: German Bundestag, 1989. (No measures of uncertainty were provided for this table in the report.)

Country	Share (%)
Brazil	20
Indonesia	12
Columbia	7
Ivory Coast	6
Thailand	6
Laos	5
Nigeria	4
Philippines	3
Burma	3
Peru	3
Other tropical countries	31

In recent years there has been an increased research effort to understand the interactions between the rainforests and atmospheric chemistry. Convincing linkages between rainforests and regional climate have been identified (see Chapter 22).

There are several societal factors leading to rainforest destruction. In the Brazilian situation the development of plantations, large dam construction, mining operations, logging and colonization schemes have been identified as major underlying causes of rainforest destruction (Lutzenberger, 1987). The relative importance of each of these factors varies, however, from one country to the next. More generally, observers blame poverty, high population growth rates, government development strategies, and inappropriate international development assistance from such renowned organizations as the World Bank. For example, many urban poor throughout Brazil, seeking their own land in order to feed their families, migrate to the rainforest in the hopes of establishing a livelihood by farming and raising cattle. These activities involve land clearing with the hope of growing subsistence and cash crops, often with little long-term success.

Government policies, too, have been blamed for the wanton destruction of the Amazon. In past decades such policies have actively encouraged urban populations to abandon the city for a new life as settlers along the Brazilian frontier. In addition, government tax incentives have encouraged land speculation in the Amazon. As Fearnside (1986b) notes, "Small farmers, both within and outside of planned settlement schemes, are being replaced by other types of family and corporate enterprises. Other areas are settled directly by large enterprises engaged in cattle ranching, plantation agriculture, and forest exploitation." By clearing land and putting it into grazing areas for cattle under the guise of land improvement, land speculators could sharply increase profits on their investments. As a result, cattle ranching has been the major land use in the Brazilian Amazon.

Another pressure on the Brazilian government to exploit the Amazon is its dire need for foreign exchange and debt relief. A source of hard currency derives from the demand for a variety of tropical hardwoods in the international marketplace. Trade in hardwoods generates sorely needed foreign exchange that can be used to further development prospects within Brazil. Japan is one of the major importers of tropical hardwoods not only from Brazil but from the rainforest countries of Southeast Asia.

Migration to the Amazon has also been fostered by policies of international development agencies as well as by government. For example, much of the deforestation in Rondonia, a part of the Amazon that has served as a "showcase" for rainforest destruction in the 1980s, has been attributed to the construction of Brazilian Highway BR-364 (Fearnside, 1986a) (Plate 3). World Bank funds were used to construct this highway in order to transport hardwoods and other products out of the region. What the Bank's experts apparently failed to take into account was that the road would not only be used to export commodities but would also attract settlers into the region. As a result, in the early 1980s the population of Rondonia increased exponentially. There, rates of land clearing were extremely high, as were the percentage of farming failures, due primarily to the rapid drop in soil fertility, among other factors.

In the Peruvian Amazon, the rainforest is also seen as both a barrier to and a frontier for food production. Its forests are subjected to similar pressures, but mostly from a need to increase food production to meet the needs of its growing population. Yet, the failure rate of new farms in the region is also high. According to a recent estimate (Ruiz Murrieta and Saavedra Andaluz, 1990), only about a fifth of the forest-converted-to-farmland remains in production, "the rest having been rendered a wasteland."

A proxy indicator of land clearing in forested areas is fire. Large areas are set on fire to make removal of vegetation, especially woody species, an easier task. With regard to the Brazilian Amazon, Shoumatoff (1990) noted that "In 1987 ... there were 5,000, 6,000, 7,000, or 8,000 fires in a single day. A total of 175,000 fires, each one bigger than one km^2, were set over the entire burning season ... ."

The Amazon is not the only region experiencing the use of fire to convert tropical rainforests into agricultural lands. During the 1982–83 ENSO event and its drought-related impacts in Southeast Asia, it was estimated that more than 3–4 million hectares of tropical rainforest were destroyed by fire in East Kalimantan (Borneo). At first the occurrence and, later, the geographic scope of the fire were denied by the Indonesian government. However, the extent of the Kalimantan fires was confirmed by satellite imagery. The cause of the fire was initially blamed on severe drought, but others have also blamed population pressure and the demand for the conversion of the forested area into agricultural lands. According to Malingreau (1984) "Fire is a common occurrence in swidden agricultural systems of the tropics, a natural ally of the farmer. In a sense, the drought was seen as a unique opportunity to engage in a drastic land clearing campaign which was to facilitate future agricultural activities."

Although the spatial extent of fires can be monitored from space, rates of change with regard to deforestation are still very difficult to determine with any degree of reliability. Although the quantitative assessments themselves are not very reliable, they do suggest an order of magnitude of destruction and, therefore, of the seriousness of this environmental problem. As Lovejoy (1990) recently noted, "While one could argue about the precision of numbers, the [remote sensing] images made it impossible to ignore the problem ... ."

2.4 Desertification

The concept of desertification, the creation of desert-like conditions where none had existed in the recent past, was first mentioned in a study of the dry forests in sub-Saharan Africa in regions with annual mean rainfall between 700 mm and 1,500 mm (Aubreville, 1949). Today, the concept of desertification, a concept originally applied to arid and semiarid areas like the Brazilian Nordeste, has been applied to parts of the deforested Amazon basin. As Fearnside (1986b) noted, "The issue of 'desertification' is an emotional one, especially in Brazil with reference to the Amazon. A tendency toward decrease in rainfall in the region, even if not crossing the threshold of annual precipitation that defines a desert in climatological terms, is a possible consequence of deforestation that cannot be dismissed."

During the prolonged, severe drought in the early 1970s in the West African Sahel, a debate developed as to whether desertification resulted primarily from human activities or from natural processes. In fact, it proved to be a mix of both, with the proportions varying over time by region and by land-use activities. Drought often exposes seemingly benign land-use practices that are harmful to the vegetative cover and to soil fertility but whose adverse effects are often hidden during wet periods. Desertification processes, however, can occur in the absence of drought, and droughts are not always accompanied by accelerated rates of desertification.

Desertification in arid, semiarid and subhumid environments influences atmospheric processes at local, regional and, perhaps, hemispheric levels, through changes in the albedo at the land's surface, changes in soil moisture and evaporation rates, and reduction in the source and amount of biogenic nuclei, and an increase in the amount of aerosols as a result of wind action.

There has been considerable scientific speculation about how changes in albedo in the West African Sahel could reduce rainfall (e.g., Charney et al., 1975). This hypothesis generated various modeling efforts of albedo changes and whether such changes could bring about perennial drought-like conditions, e.g., increasingly arid conditions (Anthes, 1984; Coakley and Cess, 1985). Research findings suggest that such changes in albedo could reduce rainfall locally but not necessarily on larger spatial scales. Similar albedo change studies have been carried out for higher rainfall regions (Chapter 22).

Interest in desertification was sharply increased in the early 1970s as a result of five successive years of devastating drought in the arid/semiarid West African Sahel. In this region along the southern edge of the Sahara Desert, rainfall is highly variable [in arid areas rainfall is skewed to dryness (Glantz and Katz, 1987)]. The main causes of desertification resulting from human activities in this region are known: inappropriate dryland farming practices (dryland farming has involved clearing the land of its natural vegetation in preparation for the planting of row crops), conversion of rangelands to agriculture, overgrazing of rangelands, firewood gathering, and inappropriate irrigation practices. The underlying causes for each of these processes is well documented (UNCOD, 1977).

Deterioration of the rangelands results from too many animals grazing the sparse and fragile vegetative cover of the rangelands. Many of the sources of and remedies for rangeland destruction in developed as well as developing countries have been identified. In many cases, however, solutions to protect the rangelands from degradation have not been applied, usually because of political, social, and financial reasons (Glantz, 1980). One major problem is herd sizes too large for the land's carrying capacity. Yet, in many developing countries there is an important reason for keeping large herds of cattle, goats or camels; to many they serve as a form of insurance for herders against recurrent drought conditions. In case of severe, prolonged drought, cattle can be traded as a last resort for necessities such as food. While some researchers have called this "perverse" economics (keeping the animals while they are healthy and valuable and selling them when they are in poor condition), it makes good sense from the perspective of destitute herders who must rely on their own devices to survive recurrent drought conditions.

The conversion of rangelands to farmland is a process that has been under way for decades. During rainy periods (like the 1950s and 1960s), governments as well as individuals tend to misunderstand regional rainfall conditions, often basing their perception on a few wet years and assuming that rainfall favorable for agriculture will continue. Yet, interannual rainfall in arid areas is highly variable and the possibility of crop failures in any given year is highly likely. For example, in the 1970s rangelands that had been converted to farmland during the wet decades of the 1950s and 1960s had to be abandoned after repeated crop failures in the dry decades of the 1970s and 1980s. Devoid of vegetative cover, the cleared and abandoned land was left open to wind and water erosion, and eventual desertification.

Trees and woody species have disappeared around many African villages and towns. The scarcity of firewood in sub-Saharan Africa has been called "the other energy crisis" (Eckholm, 1977). As a result, women and children must spend several hours a day in search of firewood with which to cook the evening meal or to keep warm. Wood is also used in construction of fences and homes. With the removal of woody species in arid and semiarid areas, the soils are robbed of plant cover, of shade and of nutrients and become exposed to wind and water erosion.

Beginning in the early 1970s various rates of desertification, especially in the West African Sahel, were presented at scientific meetings and to the media. The estimated rates at which the Sahara Desert was expanding southward into the Sahel varied from a few kilometers to more than 100 km per year. This last estimate of desertification rates has since been revised downward, with recent statements suggesting that the rate is between 5 and 30 km per year. Part of the problem has been that there is no generally accepted definition of desertification (e.g., see Glantz and Orlovsky, 1983), and there is no widely accepted set of desertification indicators (e.g., Reining, 1978). In addition, there has been no systematic monitoring of desertification processes; this would require continual satellite imagery, aerial photography and ground-truthing. Comparing land surface conditions over short periods of time, such as a year or two, yields inaccurate assessments because of the confounding effect of the large interannual variability of rainfall in areas at highest risk to desertification (Grainger, 1990).

Grainger correctly suggests that "The shortage of reliable data on the extent and rate of spread of desertification is a major reason why there are so many differing views on the phenomenon and why there has been little effective action to bring it under control .... The subjective judgments of a few experts are insufficient evidence for such a major component of global environmental change."

2.5   Changes in human activities that led to changes in GHG emissions

In this section we present some examples of decisions that have altered the rates and processes of the production of greenhouse gases. Three issues are briefly addressed: energy consumption, CFC production, and tropical deforestation. The purpose of these examples is to illustrate that decisions can be taken, if the will to do so exists, that can either increase or decrease the anthropogenically induced emissions of greenhouse gases.

2.5.1   *The energy crisis of the 1970s*

At present, the per capita energy consumption rate of the U.S. and Canada is considerably greater than that of other highly industrialized nations. Part of this difference may be attributed to the fact that the U.S. had a relative abundance of fossil fuels thus minimizing the need for imported fuels. The changing patterns of energy usage in the U.S. and other industrialized nations, and in the Third World, will play a key role in assessing future CO_2 emissions (Fig. 2.4).

Prior to the Arab oil embargo of 1973, world energy usage as well as that of the United States had been steadily increasing. During the 1940 to 1970 interval, world consumption increased at an average yearly rate of 5% while that of the U.S. increased 3.5% per year. Following the 1973 embargo, however, U.S. energy usage remained relatively constant, with consumption in 1986 being essentially that of 1973 (Fig. 2.5).

World energy consumption has increased at a much slower rate following the 1973 oil embargo – per capita consumption has remained more or less constant (Fig. 2.6).

In 1979, the world experienced another energy shock. The cartel of oil producers, OPEC, succeeded in enforcing a second oil price increase (partially as a result of the cutoff of Iranian supplies). As a result, there

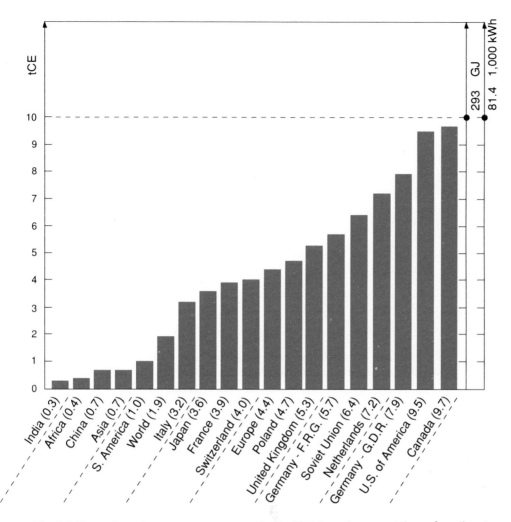

Fig. 2.4 Per capita primary energy consumption in 1986 in various countries and continents, as well as worldwide. Data are given in metric tonnes of coal equivalents (tCE) (left-hand scale and figures at top of bars), as well as in gigajoules (GJ) and 1,000 kilowatt-hours (1,000 kWh) (right-hand scale). Source: German Bundestag, 1989.

was a tenfold price increase for crude oil over the 1973–1980 period. Prior to 1973, energy prices had been stable and low – so low that decisions on energy usage (including conserving energy) were seldom based on energy costs. Other costs, such as the initial cost of an automobile or home, were considerably more important (Krenz, 1980; Lovins et al., 1981).

In light of the pre-1973 yearly increases in U.S. energy consumption, the post-1973 trend has been unprecedented. During the 1970s, none of the then-recognized energy experts came even close to anticipating, let alone predicting, such an occurrence. Projections in the early 1970s by the National Petroleum Council, The Ford Foundation, the Federal Energy

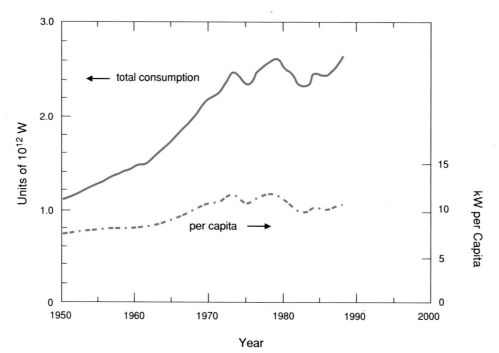

Fig. 2.5 U.S. overall and per capita energy consumption rates, 1950–1988. Source: Dupre et al., 1976, and Energy Information Administration, 1989.

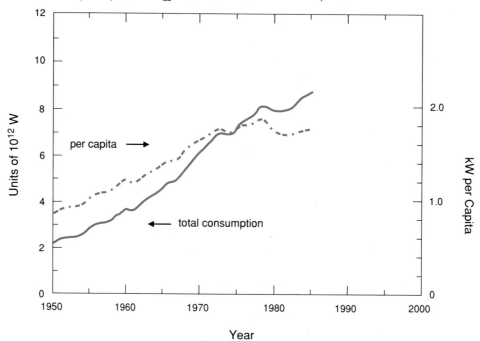

Fig. 2.6 World overall and per capita energy consumption rates, 1950–1986. Data from U.N. statistics that include only commercial energy. Source: U.N. Department of Economic and Social Affairs, 1976.

Administration and others widely missed the actual 1985 level of energy usage. These projections were 25% to 75% too high. No one, however, erred on the low side!

It is obvious that societal energy usage was not well understood and that we may still be far from an improved understanding of this interaction. Since approximately 90% of commercial energy is from fossil fuels, which on combustion emit CO_2, energy usage patterns play a crucial role in projections of atmospheric concentrations of CO_2.

The unprecedented change in U.S. energy usage after 1973 may be attributed to three factors: (1) projections of resource depletion, (2) growing levels of energy-related air pollution, and (3) OPEC-induced price increases. Detailed studies by M. King Hubbert (first published in the mid-1950s) predicted that crude oil and natural gas production in the continental U.S. (excluding Hawaii and Alaska) would reach a peak in the early 1970s and then decline (Hubbert, 1969, 1972). The "conventional" wisdom in the 1950s was that production could continue to increase at least throughout the twentieth century – a vestige of this thinking remains today. Not only was a peak reached, but Hubbert's predicted and the actual production rates showed remarkably good agreement. While the subsequent decline has not been as rapid as Hubbert predicted, it is nevertheless taking place. Hubbert's predictions based on detailed exploratory drilling data are about the only energy-related predictions that have come close to matching actual events.

With the passage of the 1970 Clean Air Act, the U.S. Environmental Protection Agency was established. In a sense, this constituted a societal recognition that energy producing and consuming activities, as well as other industrial activities, result in emissions that are injurious to human health. Emission limits were enacted for vehicles and an ambient pollutant standard index (PSI) was developed. A higher level of consciousness about the polluting effects of energy usage emerged; through the morning radio and TV announcements of expected pollutant levels, people are advised to curtail driving. While this awareness is directly associated with the harmful health effects of pollutants, it also serves to heighten the awareness of the side effects of energy usage. Furthermore, even though this awareness is not primarily directed toward the effect of greenhouse gases, concern about pollutants has been preparing the way for an emerging concern about the global warming consequences of fossil fuel usage.

Finally, price increases, particularly of petroleum products, have had a direct impact on energy usage. For most economic goods, demand tends to decline with increased prices. Societies, however, cannot rapidly adjust to increased energy costs. While one may, for example, initially curtail some automobile usage as a result of increased gasoline costs, it is primarily through the purchase of more efficient automobiles or the development of alternative modes of transportation that overall demand is eventually moderated. This is the situation for energy usage throughout society.

Better insulated homes, more efficient appliances, more efficient industrial processes and restructured commercial activities reduce energy usage. Such changes tend to depend on capital investments and improved consumer durable goods that require several years before a significant societal impact occurs. Hence, the moderation of U.S. energy usage in the 1980s (moderation in the sense that it did not continue to increase) can be attributed to the price increases of the 1970s. The most recent increases in energy usage (that is, from 1987 onward) may be attributable to the declining energy prices of the 1980s and governmental policies which no longer encouraged more efficient energy usage.

Following the landmark U.S. Energy Policy and Conservation Act of 1975, several other legislative actions occurred which have had an impact on U.S. energy consumption. In particular, tax credits were provided for solar energy systems and insulation, and minimum fuel efficiency standards were introduced for automobiles. Another impact on energy usage for residences and commercial buildings was a standard (Standard 90-75) propagated by the American Society of Heating, Refrigeration, and Air-conditioning Engineers (ASHRAE) in 1975. While the standard itself established what could be described as "good engineering practice" for its members, its impact was considerably greater. Building standards are generally established by local codes, which set minimum requirements. As a result of local governments adopting the ASHRAE standard, new residential and commercial structures are now considerably more energy efficient than prior to the 1973 embargo. Rather than the efficiency standards utilized in the U.S., other industrialized nations have adopted an aggressive tax policy to moderate fossil fuel consumption (the option of energy taxes during the 1980s has been all but excluded from an "acceptable" U.S. policy).

Nuclear power produced by the fission of uranium is frequently suggested as an alternative for fossil fuels in generating electricity. Presently nearly 20% of the electricity generated in the U.S. comes from the atom; other industrialized nations have higher percentages (but smaller absolute quantities). In France, for example, nuclear power accounts for approximately 70% of the electrical demand and in Japan the portion is 35%. While nuclear power does not directly emit any greenhouse gases, economic, safety, and waste considerations have resulted in an uncertain future for nuclear power. Waste products of the nuclear fuel cycle consist of a combination of highly radioactive fission fragments and transuranic elements. From a global environmental perspective, nuclear power replaces a partially known effect, that is, the effect of greenhouse gases, with the mostly unknown effects of radiation contamination. Although it is recognized that nuclear waste products need to be isolated from the biological environment for essentially perpetuity, progress in carrying out this task, especially given the present inventory of waste products, has been minimal.

It would be highly speculative at this point to suggest that energy usage by the industrialized nations may decrease, rather than increase as has been the historical trend. However, the technology to achieve this is either at hand or could readily be developed (AIP, 1975; 1985). For example, the energy required per dollar of economic activity in the United States has been declining – there was economic growth during the 1980s without a commensurate growth in energy consumption.

While the population of the world has been increasing, so too has the overall rate of energy consumption. The per capita energy consumption rate, however, has remained fairly uniform since 1972. The 1986 per capita consumption rate of 1.75 kW is equivalent to a yearly coal consumption of approximately 2,000 kg. While this quantity of energy, if used in a highly efficient manner, could result in a reasonable level of well-being, a more equitable distribution of the availability of resources must first be achieved. Energy usage by the world's inhabitants is far from equitable (Fig. 2.4) – nearly 80% of the world's population subsists at energy consumption rates considerably below the average. The largest groups on the chart [North America (primarily the U.S.), and the USSR] account for more than 50% of the world's energy usage even though they have less than 12% of the world's population. If the energy usage of these nations were to be reduced by 50% (a reduction that could be achieved based upon readily available technologies with no decrease in overall societal well-being), an average per capita energy consumption rate of 1.6 kW could be realized for the others without an overall increase in energy supplies. Even a more modest reduction of 25% could have a significant impact.

2.5.2 The spray can ban in the mid-1970s

In the mid-1970s concern was raised by atmospheric chemists about the possible adverse impacts on stratospheric ozone of CFCs (especially CFC-11 and CFC-12). Molina and Rowland (1974) and Stolarski and Cicerone (1974) suggested that CFCs were a serious threat to stratospheric ozone, which protects living organisms at the Earth's surface from harmful ultraviolet radiation. The linkage of ozone depletion to serious human health effects (e.g., skin cancer) prompted intergovernmental support for an assessment of the environmental and health effects of CFCs. It was pointed out later that CFCs were also an important greenhouse gas.

Despite uncertainties inherent in atmospheric models, scientists and policymakers believed that they could undertake "no-regrets" strategies to reduce CFC emissions. Nonessential uses of CFCs, e.g., as propellants in aerosol spray cans, became the target of early restrictions by some governments. A large portion of CFCs were used in spray cans for deodorants, hairsprays, and other personal care items, each considered a nonessential use. At first those restrictions were voluntary, calling

on manufacturers and consumers to protect the ozone layer by avoiding
the purchase of spray cans. Later, as the scientific findings remained
unchallenged by new research, and clear observational evidence was
produced confirming the role of CFCs in depleting ozone and leading to
the "ozone hole" over Antarctica, regulations on the manufacture, use, and
export of CFCs were enacted by some countries.

Canada, the U.S., and Sweden, among others, took the CFC/ozone
depletion issue seriously from the outset. As a result, restrictions on CFCs
were enacted early (Stoel et al., 1980). Other countries, however, such as
the United Kingdom, Japan, and the USSR, called for additional study but
did not entertain a call for regulation for a variety of reasons, some of which
were scientific and others, political and economic (Stoel et al., 1980).

Attempts in the late 1970s to carry regulations on CFCs even further,
attacking essential uses such as refrigerants, were blocked by industrial
lobbies, noting that there were still scientific uncertainties surrounding
health effects and (the real reason) the lack of substitutes for CFCs in some
important industrial processes. Nevertheless, those countries that did curtail
their production and use of CFCs had a noticeable impact on global CFC
sales figures (Fig. 2.7). Since the early 1980s, however, use has increased

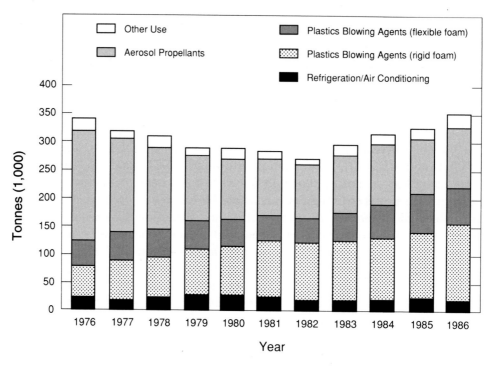

Fig. 2.7 Tonnages of CFC-11 sold worldwide between 1976 and 1986, by application. Source:
German Bundestag, 1989.

as the production activities in other countries increased to make up the reductions resulting from earlier CFC restrictions.

2.5.3   The Vienna Convention and the Montreal Protocol

With a ban by a few countries on some of the nonessential uses of CFCs, concern about CFCs and stratospheric ozone depletion did not abate.   The United Nations Environment Programme (UNEP) set up a Coordinating Committee on the Ozone Layer in 1977 to monitor the manufacture of CFCs globally. Scientific evidence of harmful CFC impacts on the environment (on ozone destruction and as a nontrivial contributor to global warming) continued to mount, as did CFC production figures.

In 1981, UNEP established a working group to draft a "Global Framework Convention for the Protection of the Ozone Layer." This activity produced the Vienna Convention for the Protection of the Ozone Layer, adopted in early 1985.   This convention was in essence an umbrella treaty to which specific protocols could be added. At that time there was an unsuccessful attempt to produce a protocol focused on CFCs.

Increased production and use of CFCs in the mid-1980s, and the lingering scientific research findings about their deleterious environmental and health effects, once again prompted serious consideration of a protocol for reducing CFCs to protect stratospheric ozone. At the end of 1986 and in early 1987, serious discussion on levels of reduction and a timetable was in progress. These negotations ended with the signing of the Montreal Protocol in 1987 (Benedick, 1991). A followup meeting in London in 1990 called for stepping up the timetable to phase out CFCs, given the strength of existing scientific evidence and the recent implication of CFCs in the deepening of the Antarctic ozone hole.

Clearly, the Montreal Protocol will ultimately have a major impact on the global production and use of CFCs.   CFCs are, as noted earlier, an effective greenhouse gas, in addition to being an effective "ozone-eater." With mounting scientific evidence about global warming and its potential environmental and societal (especially health) consequences, CFC emissions will surely be reduced as policymakers worldwide respond to scientific information and public pressure to protect the global environment.

2.5.4   Deforestation of the Brazilian Amazon

The systematic destruction of the Brazilian Amazon rainforest was prompted by a change in government development policy following the military takeover in 1964.   The policy objective was one of rapid economic development at any cost (with little or no consideration given to the environment).   These policies to develop the Amazon unleashed uncontrolled, chaotic impacts on Brazil's natural environment.

The destruction of Brazil's tropical rainforest has received considerable attention as a result of concern expressed by international environmental groups. As the largest remaining tropical rainforest in the world and because of its role in the global warming issue, and species extinction, and because of the rapid rates of deforestation and forest fires that can be seen from space, Brazil has had to endure the wrath of many governments and nongovernmental organizations around the world. In 1988, under increasing pressure from the international community, the government initiated a new program designed to end tax incentives for land speculators.

Attention was further focused on these rainforests by the murder in December 1988 of environmental activist Chico Mendes. He opposed land owners who destroyed the forest in order to establish grazing areas for cattle or small farms for influxes of poor settlers seeking a better way of life than they could find in urban areas. Mendes had called for the establishment of extractive reserves, so that rubber tappers could continue to selectively exploit the rainforest without destroying it. Ranchers and other land speculators opposed his views and organized his assassination.

Shortly after his murder, Brazil's Sarney government, sympathetic to those who sought to "develop" the Amazon by clearing the land (to settle it or to speculate), was voted out of office and the new government of Collor de Mello, which took office in mid-1990, has stated its intention to preserve the Amazon Basin and to use it only in sustainable ways. It is too soon to judge whether the rates of deforestation in the Amazon can be sharply curtailed. As mentioned previously, many of the problems that underlie deforestation have their roots in poverty. This tends to tie the international debt problem directly to rates of deforestation in the Amazon.

Elsewhere in the world, such as in Thailand and in Malaysia (especially in Sarawak), people are fighting against the wanton destruction of tropical forests by entrepreneurs seeking to make big profits in a short period of time, with little concern about the ensuing environmental degradation. They have been joined by an international network of individuals and organizations to slow down, if not arrest, the process of rapid, indiscriminate destruction of the tropical rainforests.

2.6 Conclusions

There are many examples of how decisions by individuals and by governments have changed the way human activities can affect the atmosphere. Today, the biggest concern relates to greenhouse gases and, more specifically, to carbon dioxide and the possible relationship of this gas with a global warming of the atmosphere. Although many observers suggest that it will be very difficult for the international community to negotiate a meaningful climate convention and control measures on the production of CO_2, there are encouraging signs that the level of awareness has sharply increased, as witnessed by the high level of government involvement in the IPCC process (IPCC, 1990a, 1990b, 1991). Much of this heightened awareness can be linked to a general concern among the public and policymakers about the deterioration of the Earth's environment and the central role of human activities in that deterioration.

While prevention of global warming may be an impossible task under present political, social and economic conditions, actions can be taken to slow down the current projected rate of global warming (estimated at about $0.3°C$ per decade [IPCC, 1990a]). Any lowering of this rate can be beneficial by slowing the rate of climate change and allowing scientists to improve the reliability of the model-based regional impacts scenarios. In addition, it can "buy" time for improving the ways that societies adjust to environmental changes.

A frontal attack on the CFCs is in progress. It began in the mid-1970s and reached a milestone with the development of the Montreal Protocol in 1988. Progress on curtailing and then eliminating CFCs made another leap forward at a London meeting in mid-1990 at which it was decided that the phasing out of CFCs should be accelerated. But it is easy to see that eliminating CFCs is a different problem, a relatively easier task, than eliminating carbon dioxide production. In a way, everyone in the world contributes to the CO_2 problem. Nevertheless, progress from an environmental standpoint can be made in the selection of the sources of energy that societies come to rely on, as well as on the per capita amounts of energy that are eventually consumed.

Tropical deforestation is under closer scrutiny by the international community than ever before. And the underlying causes that encourage it will be increasingly addressed by the international community. There are also efforts to undertake worldwide afforestation programs to increase this CO_2 sink. Such activities, appropriately undertaken, will be environmentally beneficial, regardless of their overall impact on atmospheric CO_2 levels.

Researchers are also focusing on reducing some of the sources of emissions of other greenhouse gases such as nitrous oxide and methane. These will involve changing the way people use agricultural land and the way they grow food. To make such changes, however, may prove to be easier in theory than in practice. How, for example, does one convince a rice grower in Southeast Asia that wet rice production practices carried out successfully for the past millennia must be changed in order to save the global atmosphere?

Clearly, the human aspects of the climate system are important to understanding climate change issues. Societies need to identify those human activities that have to be modified in order to reduce the emissions of the key greenhouse gases. Not to integrate the human component into the climate system would force policymakers to make a dangerous leap of inference from good physical and biological science to potentially inappropriate policy responses (that is, to provide the wrong solution).

PART 2

THE SCIENCE:
SUBSYSTEMS AND PROCESSES

3

The atmosphere

Murry L. Salby

3.1 Introduction

The atmosphere lies at the heart of many of the important issues facing climate research. Radiative processes involving water vapor, CO_2, and clouds, figure centrally in the Earth's energy budget, which is critical to the Earth's climate. Chemical processes are also a key factor in climate because they determine the atmosphere's composition and provide an important link to human activities. Dynamical processes are the third in a triad of physical mechanisms influencing climate. The global circulation plays a key role in determining distributions of radiatively and chemically active species, cloud formation, and exchanges of heat and moisture between the atmosphere and oceans. The physical processes mentioned above are woven together in the atmosphere into a complex fabric of radiation, chemistry, and dynamics, which shapes the climate of the Earth. This chapter develops the fundamental principles underlying those processes.

The atmosphere is a fluid system and therefore capable of supporting a wide range of motions. The mobility of fluid systems makes their description complex. Like any fluid system, the atmosphere is governed by the laws of continuum mechanics. These are derived from laws governing a discrete system by generalizing those laws to a continuum of such systems. In the atmosphere, the discrete system to which these laws apply is an infinitesimal fluid element or "air parcel", defined by a fixed collection of matter.

Two frameworks are commonly used to describe atmospheric behavior. The "Eulerian" description represents atmospheric behavior in terms of field properties, which are governed by partial differential equations and are convenient for numerical purposes. The "Lagrangian" description represents atmospheric behavior in terms of individual air parcels. Since it focuses on transformations of properties within an air parcel and exchanges between that system and its environment, the Lagrangian description is attractive conceptually and is used below to develop the laws governing atmospheric behavior.

3.1.1 Composition and structure

The atmosphere consists of a mixture of gases, mostly molecular nitrogen (78% by volume) and oxygen (21% by volume); see Table 3.1. Water vapor, carbon dioxide, ozone, and other minor constituents represent the remaining 1% of the atmosphere. Because they appear in very small abundances and are highly variable, trace species like water vapor and ozone are treated separately from the primary atmospheric constituents, which are referred to simply as "dry air."

The starting point for describing atmospheric behavior is the ideal gas law

$$pV = nR^*T = mRT \qquad (3.1a)$$

$$p = \rho RT = \frac{RT}{v}, \qquad (3.1b)$$

which constitutes the equation of state for a mixture of gases. Here, $p, T, \rho, v, R = R^*/M$, and M denote the pressure, temperature, density, specific volume, specific gas constant, and mean molar weight of the mixture, and V, m, and n refer to the volume, mass, and molar abundance of a fixed collection of matter, e.g., an air parcel.[1]

Because of their involvement in radiative and chemical processes, individual constituents of air must be quantified. The absolute concentration of the ith species is measured by its density ρ_i. The partial pressure p_i, that pressure the ith component would exert in isolation at the same temperature and volume as the mixture, is also used to quantify the concentration of a species. The relative concentration of the ith species is measured by its "mass mixing ratio"

$$r_i = \frac{m_i}{m_d}, \qquad (3.2)$$

where the subscript d refers to dry air. The mass mixing ratio is a dimensionless quantity (e.g., expressed in g kg^{-1} for tropospheric water vapor and in parts per million (ppm) for stratospheric ozone). The "volume mixing ratio" provides similar information and is distinguished by dimensions like ppmv for stratospheric ozone.

Unlike measures of absolute concentration, which change through expansion and compression, mixing ratio is linked directly to individual fluid elements. A property that is invariant for an individual air parcel is said to be "conserved." Such a property serves as a *material tracer* because particular values tag individual fluid elements and therefore trace out their motion. If the ith species does not undergo a phase transition or a chemical reaction, its mixing ratio is conserved since the masses in (3.2) are fixed for a given body of air. Under these circumstances, following particular values of

---

[1] Values of M and R are collected in the table on physical constants along with other properties of dry air.

Table 3.1. Composition of the atmosphere. Constituents are listed with an indication of whether they are radiatively active, with a mixing ratio representative of the troposphere (trop.) or stratosphere (strat.), how they are distributed vertically, and controlling processes.

Constituent	Tropospheric mixing ratio	Vertical distribution (mixing ratio)	Controlling processes
N_2	0.7808	Homogeneous	Vertical mixing
O_2	0.2095	Homogeneous	Vertical mixing
* H_2O	$\leq$0.030	Decreases sharply in trop. Increases in strat. Highly variable	Evaporation, condensation, transport Production by CH_4 oxidation
Ar	0.0093	Homogeneous	Vertical mixing
*CO_2	345 ppmv	Homogeneous	Vertical mixing Production by surface and anthropogenic processes
* O_3	10 ppmv[$]	Increases sharply in strat. Highly variable	Photochemical production in stratosphere Destruction in troposphere Transport
* CH_4	1.6 ppmv	Homogeneous in trop. Decreases in middle atmos.	Production by surface processes Oxidation produces H_2O
* N_2O	350 ppbv	Homogeneous in trop. Decreases in middle atmos.	Production by surface and anthropogenic processes Dissociation in middle atmos. Produces NO Transport
* CO	70 ppbv	Decreases in trop. Increases in strat.	Production anthropogenically and by oxidation of CH_4 Transport
NO	0.1 ppbv[$]	Increases vertically	Production by dissociation of N_2O Catalytic destruction of O_3
* CFC-11 * CFC-12	0.1 ppbv	Homogeneous in trop. Decreases in strat.	Industrial production Mixing in troposphere Photodissociation in stratosphere
ClO	0.1 ppbv[$]	Increases vertically	Production by photodissociation of CFCs Catalytic destruction of O_3

* Radiatively active   [$] Stratospheric value

r_i provides a Lagrangian description of how air and other conserved species are rearranged by the circulation.

a. Stratification of mass

Of the factors influencing atmospheric behavior, gravity is the single most important. The strong body force exerted by gravity confines atmospheric mass to a shallow layer above the Earth's surface. If vertical accelerations are neglected, Newton's second law of motion applied to the column of air between some level at altitude z at pressure p and a level

Hydrostatic Balance

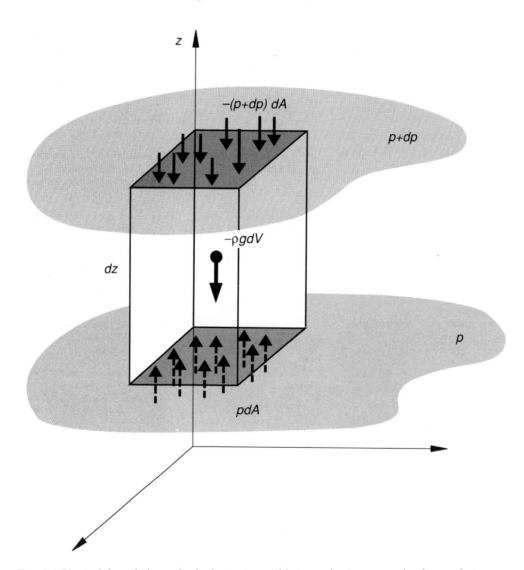

Fig. 3.1 Vertical force balance for hydrostatic equilibrium. An incremental column of air bounded by two isobaric surfaces at pressures p and $p+dp$, has weight $\rho g dV$, which must be balanced by the net pressure force acting on the column, $dp dA$. Because vertical displacements of air are small, this simple form of mechanical equilibrium is a good approximation even when the atmosphere is in motion. Reprinted from Salby (1992) by permission of Academic Press.

incrementally higher at pressure $p + dp$ reduces to a balance between the weight of that column and the net pressure force acting on it. As illustrated in Fig. 3.1, analysis along these lines leads to the balance

$$dp = -\rho g dz. \tag{3.3}$$

Known as *hydrostatic balance*, this simple form of mechanical equilibrium is a good approximation even if the atmosphere is in motion because vertical displacements of air are small. Applying the same analysis between the pressure p and the top of the atmosphere, where pressure vanishes, demonstrates that p at any level must equal the weight of the atmospheric column of unit cross-sectional area above that level. The compressibility of air makes the density in (3.3) dependent upon the pressure through the gas law. Substituting (3.1) into (3.3) and integrating between the surface pressure p_s and the pressure p yields

$$\frac{p}{p_s} = e^{-\int_0^z \frac{dz'}{H(z')}}, \tag{3.4a}$$

where

$$H(z) = \frac{RT(z)}{g} \tag{3.4b}$$

is the pressure "scale height."

As shown in Fig. 3.2, the global-mean pressure and density decrease exponentially with altitude. Pressure decreases from about 1000 mb or 10^5 Pa at the surface to only 10% of this value at an altitude of 10 km. From hydrostatic balance, it follows that 90% of the atmosphere's mass lies beneath this level, confined to a depth less than one tenth of one percent of the Earth's radius. The mean density also decreases from its surface value of about 1.2 kg m^{-3} at about the same rate. The sharp decrease with altitude of pressure implies that isobaric surfaces ($p = $ const) are quasi-horizontal. Vertical deflections of these surfaces introduce the relatively small horizontal variations of pressure that drive atmospheric motions.

b. Thermal and dynamical structure

The atmosphere is categorized according to its global-mean thermal structure, also shown in Fig. 3.2, which determines dynamical properties of individual layers. A more complete picture of the thermal structure is provided by the zonal-mean temperature (i.e., averaged longitudinally), shown in Fig. 3.3 as a function of latitude and altitude. In the lowest 10–15 km, temperature decreases with altitude at a nearly constant "lapse rate" (defined as the rate of *decrease* of temperature with altitude) of about 6 K km^{-1}. This layer immediately above the Earth's surface is known as the "troposphere", which means *turning sphere* and symbolizes the convective overturning of air characterizing this region. In the troposphere, temperature maximizes near the equator and decreases towards each pole.

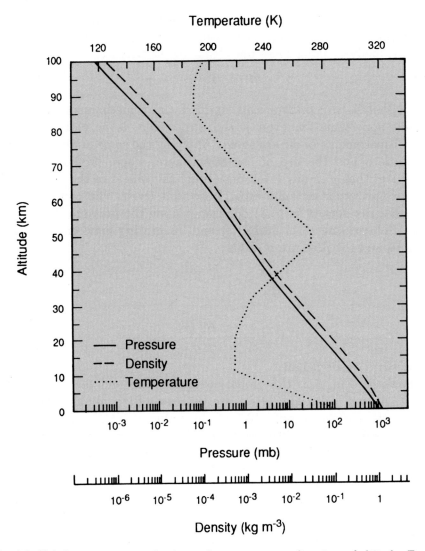

Fig. 3.2 Global-mean pressure, density, and temperature as functions of altitude. From the *U.S. Standard Atmosphere* (1976). Reprinted from Salby (1992) by permission of Academic Press.

The troposphere contains most of what we know as weather and is driven ultimately by surface heating. The upper boundary of the troposphere, or *tropopause*, is marked by a sharp change in lapse rate. The tropopause is highest in the tropics (∼16 km) and is lowest in polar regions (∼8 km). A break in the tropopause is observed at midlatitudes, where a sharp change in lapse rate is not evident.

The region from the tropopause to about 85 km has come to be known as the *middle atmosphere*. From the tropopause to about 50 km, temperature increases with altitude in the "stratosphere", which means *layered sphere*

Temperature

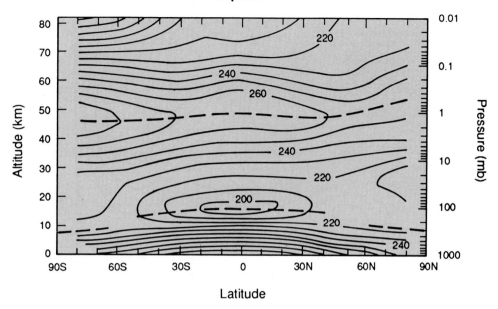

Fig. 3.3 Zonal-mean temperature during January, as a function of latitude and altitude. After Fleming et al. (1988). Reprinted from Salby (1992) by permission of Academic Press.

and is symbolic of properties at these altitudes. Unlike the troposphere, the stratosphere involves only weak vertical motions and is dominated by radiative processes. The increase of temperature with altitude reflects ozone heating, which arises from the absorption of solar ultraviolet (UV) radiation. Temperatures are highest over the summer pole and decrease steadily to a minimum over the winter pole. From the *stratopause* up to about 85 km, temperature decreases with altitude again in the "mesosphere", where ozone heating diminishes. Convective motions and radiative processes are both important in the mesosphere. Temperatures are coldest over the summer pole and increase steadily to a maximum over the winter pole. Higher altitudes are characterized by steadily increasing temperature in the "thermosphere", but virtually all of the processes relevant to climate occur below the *mesopause*.

The zonal-mean circulation, shown in Fig. 3.4, follows in large part from the thermal structure in Fig. 3.3. The troposphere is characterized by subtropical jets which are westerly[2] and intensify with altitude up to the tropopause. Above the tropopause, the zonal flow intensifies again, but is no longer westerly in both hemispheres. In the winter hemisphere, westerlies

---

[2] In meteorological parlance, "westerly" refers to motion *from the west* and "easterly" to motion *from the east*.

59

Zonal Velocity

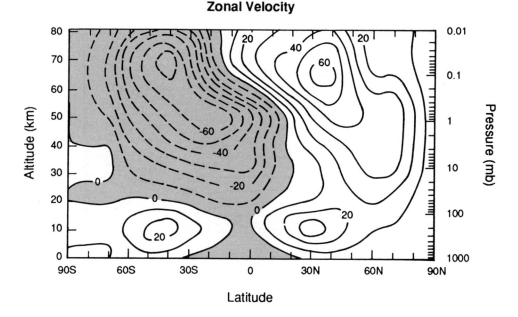

Fig. 3.4 Zonal-mean wind during January in m s^{-1}, as a function of latitude and altitude. After Fleming et al. (1988). Reprinted from Salby (1992) by permission of Academic Press.

increase above the subtropical jet. Winds in the *polar-night jet* attain speeds of 60 m s^{-1} near the stratopause. In the summer hemisphere, the zonal flow becomes easterly and also intensifies up to the stratopause. Easterly flow in the summer stratosphere merges with weak easterlies in the tropical troposphere, where zonal-mean wind speeds of 5 m s^{-1} are typical.

c.   Trace constituents

Although they exist in very small abundances, certain trace species like water vapor and ozone play key roles in radiative and chemical processes in the atmosphere. These constituents are highly variable because they are not simply redistributed by atmospheric motions, which would eventually homogenize them. Instead, they are continually produced in some regions and destroyed in others. The atmospheric circulation transports these transitive species from their source regions to their sink regions and makes their distributions dynamic.

Water vapor is perhaps the single most important trace species because it is responsible for much of the atmosphere's opacity in the infrared (IR) and because its frequent changes of phase lead to cloud formation and transfers of latent heat from the oceans. The zonal-mean distribution of water vapor is shown in Fig. 3.5 as a function of latitude and pressure. Water vapor is confined almost exclusively to the troposphere. Its mean mixing ratio decreases steadily with altitude from a maximum of about 20 g kg^{-1} at the

Water Vapor (g · kg⁻¹)

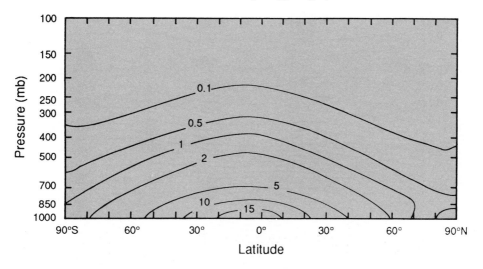

Fig. 3.5 Zonal-mean specific humidity (approximately equal to water vapor mixing ratio) for January, as a function of latitude and pressure. From European Centre for Medium Range Weather Forecasting (ECMWF) analyses: 1986–1989; courtesy K. Trenberth.

surface near the equator to just a few ppm at the tropopause. The mixing ratio also decreases with latitude, to under 5 g kg^{-1} poleward of 60°. These characteristics of water vapor reflect its production at the surface of warm tropical oceans and redistribution by the atmospheric circulation. From (3.2), the density of the ith constituent is given

$$\rho_i = r_i \rho_d. \qquad (3.5)$$

Since ρ_d decreases exponentially with altitude, water vapor is concentrated near the Earth's surface even more sharply than suggested by Fig. 3.5. Most of the global-mean water vapor is actually confined to the lowest 2 km of the atmosphere. While the mean distribution in Fig. 3.5 is fairly smooth, the distribution of water vapor on individual days is highly variable. Atmospheric motions arrange water vapor into a complex pattern of convective cells and large-scale features that evolve on time scales of hours to a day.

Like water vapor, ozone is a radiatively active trace gas. Beyond its radiative importance, ozone plays a vital biological role by intercepting harmful UV radiation. The mean distribution of ozone mixing ratio is shown in Fig. 3.6 as a function of latitude and pressure. Whereas water vapor is confined chiefly to the troposphere, ozone is concentrated in the stratosphere. Ozone mixing ratio increases sharply above the tropopause, reaching a maximum of about 10 ppmv near 30 km (∼10 mb), before

Ozone (ppmv)

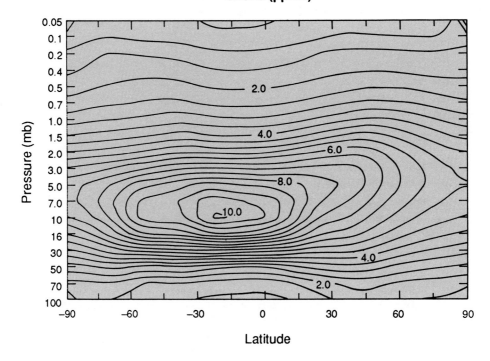

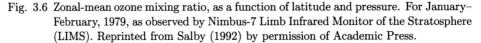

Fig. 3.6 Zonal-mean ozone mixing ratio, as a function of latitude and pressure. For January–February, 1979, as observed by Nimbus-7 Limb Infrared Monitor of the Stratosphere (LIMS). Reprinted from Salby (1992) by permission of Academic Press.

decreasing at higher altitudes. Ozone mixing ratio maximizes in the tropics, where O_3 is produced photochemically. As is true of water vapor, the absolute concentration of ozone depends on the air density through (3.5). For this reason, ρ_{O_3} decreases sharply above the tropopause, where r_{O_3} first increases, so most of the Earth's ozone is actually concentrated at altitudes of 10–20 km.

As is true for water vapor, the atmospheric circulation plays a key role in determining the mean distribution of O_3 and makes the instantaneous distribution dynamic and more complex than suggested by Fig. 3.6. The horizontal distribution of ozone is described by the "column abundance"

$$\Sigma_{O_3} = \int_0^\infty \rho_{O_3} dz. \tag{3.6}$$

Also known as *total ozone*, Σ_{O_3} is expressed in Dobson units (DU), which represent in thousandths of a centimeter the depth the ozone column would have at standard temperature and pressure. The distribution of Σ_{O_3} over the Northern Hemisphere (NH) on an individual day is shown in Plate 4(a). Values of 300 DU are typical, so the entire ozone column measures less than one half of one centimeter at standard temperature and pressure. However,

Σ_{O_3} varies considerably across the hemisphere, from 225 DU in the tropics to in excess of 600 DU over the polar cap. Even though most stratospheric O_3 is produced near the equator, the largest column abundances are found at high latitudes. Anomalies introduced by the circulation cause total ozone to vary locally by as much as 100% and on a time scale of order a day. The distribution over the Southern Hemisphere (SH) (Plate 4b) contains similar features, but is distinguished by the Antarctic ozone hole, wherein Σ_{O_3} is severely depleted each year during Austral spring, see Chapter 7.

Several other trace gases are also involved in chemical and radiative processes in the atmosphere. These include carbon dioxide (CO_2), methane (CH_4), nitrous oxide (N_2O), and industrial halocarbons such as the chlorofluorocarbons (CFCs): $CFCl_3$ (CFC-11) and CF_2Cl_2 (CFC-12). All of these species are produced at the Earth's surface and are long-lived in the troposphere. For this reason, they are well mixed, e.g., their mixing ratios are nearly homogeneous in the troposphere. Reflecting anthropogenic sources, their concentrations are increasing gradually with time. Carbon dioxide is produced naturally. However, its steady increase since the dawn of the industrial age has been attributed to human activities and has prompted concerns over global warming due to the role CO_2 plays in the global energy budget. Methane, nitrous oxide, and CFCs are also radiatively active. The latter have only anthropogenic sources and pose a significant hazard to stratospheric ozone. Their chemical stability and water insolubility make CFCs immune to normal scavenging mechanisms in the troposphere. Hence, they survive long enough to be transported by the circulation into the stratosphere, where they are photodissociated by UV radiation. Free radicals of chlorine produced in this manner are now recognized as being responsible for the formation of the Antarctic ozone hole evident in Plate 4(b).

Aerosols are another atmospheric component relevant to climate. Suspensions of small liquid and solid particulates are vital to atmospheric behavior because they promote cloud formation. Aerosol particles serve as condensation nuclei for water droplets and ice crystals that do not form readily in their absence. These particles arise naturally from surface processes (e.g., as dust, sea salt, and volcanic debris) and anthropogenically through combustion and industrial processes. Because they are radiatively active, aerosols figure in the energy balance of the Earth, wherein their influence is thought to be as great as that of CO_2. Aerosols also play a key role in several chemical processes. One of the more notable is heterogeneous chlorine chemistry, which is thought to be responsible for the formation of the ozone hole shown in Plate 4(b) and is discussed further in Chapter 7.

d. Clouds

For a number of reasons, clouds are a critical ingredient of climate. About half of the Earth is cloud-covered at any instant. The cloud field is highly dynamic because of its relationship to the atmospheric circulation. The radiative properties of clouds make them a key component of the Earth's energy balance, wherein their role is an order of magnitude greater than that of CO_2. Shallow layered clouds, such as marine stratus, are especially important because they cover a large fraction of the Earth, as is apparent in the visible image in Plate 5 that was taken from the geostationary satellite GOES-E; see also Plate 6 for the mean cloudiness.

Beyond its importance to radiative processes, convection plays a key role in the dynamics of the atmosphere and in its interaction with the oceans. Deep convection in the tropics liberates large quantities of latent heat that are released to the surrounding air when water vapor condenses and precipitates back to the surface. Derived from heat exchange with the oceans, latent heating represents a major source of energy for the atmosphere. For this reason, deep cumulus clouds are used as a proxy for atmospheric heating. Plate 7(a) shows a nearly instantaneous image of the Earth's cloud field. The grey scale emphasizes the highest objects emitting strongly in the IR. Bright features correspond to cold high clouds, whereas dark features correspond to warm surfaces in cloud-free regions. Consequently, shallow stratus that are prominent in the visible image in Plate 5 are not as apparent in the IR image. The highest (brightest) clouds in Plate 7(a) are found near the equator, especially over the tropical landmasses: South America, Africa, and the *maritime continent* of Indonesia. This band of organized deep convection, known as the "Inter Tropical Convergence Zone" (ITCZ), figures importantly in the dynamics of the global circulation and in exchanges of heat and moisture between the atmosphere and oceans.

The ITCZ emerges prominently in the time-mean cloud field, shown in Plate 7(b). Over maritime regions, the time-mean ITCZ appears as a narrow band parallel to the equator, which reflects the convergence of surface air from the two hemispheres. Over tropical landmasses, time-mean cloud cover expands due to the additional influence of surface heating, which triggers convection diurnally. Convective activity associated with the ITCZ migrates north and south annually with the sun and includes the monsoons over Southeast Asia and northern Australia during the solstices. Accompanying the ITCZ in Plate 7(b) are regions of time-mean high cloudiness at midlatitudes. These regions mark convective activity organized by cyclonic weather systems, like those evident in Plate 7(a).

Clouds are also important in chemical processes. Condensation and precipitation function as a removal mechanism for many chemical species. Gaseous pollutants that are water soluble are absorbed in cloud droplets and eliminated when those droplets precipitate to the surface. *Rain out* also scavenges aerosol pollutants, which serve as condensation nuclei for cloud droplets. Heterogeneous chlorine chemistry takes place inside very high polar stratospheric clouds (PSCs), which form over the Antarctic and which play a key role in the springtime depletion of ozone evident in Plate 4(b).

3.1.2 Radiative equilibrium of the planet

The driving force for the atmosphere is the absorption of solar energy at the Earth's surface. Over time scales long compared to those controlling the redistribution of energy, the Earth–atmosphere system is in thermal equilibrium. Consequently, the absorption of solar radiation, which is concentrated at visible wavelengths and termed "shortwave (SW) radiation", must be balanced by the emission to space of infrared or "longwave (LW) radiation" by the planet's surface and atmosphere.

A simple balance of SW and LW radiation (Sec. 1.2.3) predicts an equivalent blackbody temperature for Earth of $T_e = 255$ K. This value is some 30 K colder than the global-mean surface temperature $T_s \approx 288$ K. The difference between T_s and T_e follows from the "greenhouse effect", which results from the different ways the atmosphere processes SW and LW radiation. Although transparent to SW radiation (wavelengths $\lambda \sim 1$ μm), the atmosphere is almost opaque to LW radiation ($\lambda \sim 10$ μm) re-emitted by the planet's surface; see Fig. 1.1. Hence, SW radiation passes freely to the Earth's surface, where it can be absorbed, but LW radiation re-emitted by the planet's surface is captured by the atmosphere. By trapping radiant energy, that must eventually be rejected to space, the atmosphere's opacity elevates the surface temperature over what it would be in the absence of an atmosphere.

3.1.3 The global energy budget

Since it follows from a simple radiation balance, the equivalent blackbody temperature provides some insight into where energy is ultimately emitted to space. The value $T_e = 255$ K corresponds to the middle troposphere, where clouds are abundant. Most of the energy received by the atmosphere is supplied from the Earth's surface in connection with the absorption of SW radiation. Infrared cooling of air in the middle troposphere and warming of air through transfers from the surface constitute a heat sink and a heat source for the atmosphere, which behaves as a global heat engine in the energy budget of the Earth.

As illustrated in Fig. 1.2, energy absorbed at the Earth's surface must be transferred to the middle troposphere, where it can be rejected to space. From the time it is absorbed at the surface until it is eventually emitted to space, energy assumes a variety of forms. In addition to LW radiation, the transfer of energy from the Earth's surface involves thermal conduction, referred to as the transfer of "sensible heat", and the transfer of "latent heat", when water vapor absorbed by the atmosphere condenses and precipitates back to the surface.

The global-mean energy balance, illustrated in Fig. 1.2, is normalized by the average incoming flux at the top of the atmosphere

Net Radiaton

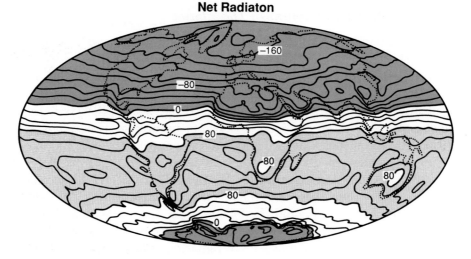

Fig. 3.7 Net radiation (SW absorbed–LW emitted to space) for December–February, as measured by Nimbus-7 Earth Radiation Budget Experiment (ERBE) narrow-field-of-view instrument. Adapted from Hartmann et al. (1986). Reprinted from Salby (1992) by permission of Academic Press.

$Q = S_0/4 = 343$ W m^{-2}. Of 100 units of incident SW radiation, 70 are absorbed by the Earth–atmosphere system and must be re-emitted to space to preserve thermal equilibrium. The surface receives 45 units through direct absorption of SW radiation and another 88 through absorption of LW radiation emitted by the atmosphere. At a mean temperature of 288 K, the surface emits 104 units. Together, these radiative contributions result in a *net heating of the surface* of 29 units, which must be compensated by transfers of sensible and latent heat to the atmosphere.

The energy budget of the atmosphere must also balance to zero. The atmosphere absorbs 25 units of the incident SW radiation, while it absorbs 100 of the 104 units emitted by the surface. A total of 154 units is radiated by the atmosphere, 66 to space and 88 down to the surface. Together, these radiative contributions result in a *net cooling of the atmosphere* of 29 units, which is balanced by transfers of sensible and latent heat from the surface.

The global-mean budget discussed above involves only vertical transfers of energy. Nonuniform optical properties controlling the absorption of SW radiation and the emission of LW radiation lead to nonuniform heating of the planet. As shown in Fig. 3.7, the net radiation absorbed is positive at low latitudes (heating) and negative at middle and high latitudes (cooling). To preserve thermal equilibrium, energy must be transferred from low latitudes to high latitudes. About 60% of that energy transfer is accomplished by the general circulation of the atmosphere.

3.1.4   *The general circulation*

The general circulation is maintained against frictional dissipation by a conversion of potential energy, associated with the distribution of atmospheric mass, to kinetic energy, associated with the motion of air. Radiative heating expands the atmospheric column at low latitudes and raises the center of mass, whereas radiative cooling compresses the atmospheric column at high latitudes and lowers the center of mass. The uneven distribution of mass that results introduces an imbalance of pressure forces which drives a meridional overturning of air, with rising motion at low latitudes and sinking motion at middle and high latitudes.

The simple meridional circulation described above is modified importantly by the Earth's rotation. Air motions in the large-scale circulation are almost parallel to isobars, as evidenced by the instantaneous circulation at the 500 mb level shown in Fig. 3.8(a). The radiative heating in Fig. 3.7 leads to a time-mean thermal structure wherein isotherms and isobars are approximately parallel to latitudes circles. Consequently, the time-mean circulation, shown in Fig. 3.8(b), is nearly circumpolar at middle and high latitudes, with only a small meridional component to transfer heat between the equator and poles. For this reason asymmetries in the instantaneous circulation in Fig. 3.8(a), that deflect air meridionally, play a key role in transferring energy between low latitudes and high latitudes. Most of the heat transfer at middle and high latitudes is accomplished by transient "synoptic weather systems." Marked by centers of low pressure in Fig. 3.8(a) and frontal cloud systems in Plate 7(a), these disturbances transfer heat meridionally by exchanging air between tropical and polar regions.

At low latitudes, the Earth's rotation has a smaller influence on air motions. There, kinetic energy is associated with "thermally direct circulations" that are forced by the geographical distribution of atmospheric heating. Latent heat release inside the ITCZ drives a meridional overturning, known as the "Hadley circulation", wherein air rises near the equator and sinks at subtropical latitudes. Subsiding air in the descending branch of the Hadley circulation inhibits cloud formation and is responsible for maintaining deserts that are prevalent in the subtropics. The nonuniform distribution of land and sea introduces zonal asymmetries in heating, which drive an east–west overturning, known as the "Walker circulation", wherein air rises at longitudes of the heating and sinks at other longitudes. The concentration of latent heating over Indonesia reinforces the Pacific Walker circulation, illustrated in Fig. 3.9, which maintains easterly trade winds across the surface of the Pacific.

Latent heat release is but one of several mechanisms that introduce asymmetry in the general circulation. Orographic features at the Earth's surface also disturb the zonal flow. Mountain ranges, such as the Himalayas, the Alps, and the Rockies, displace atmospheric mass vertically. Work

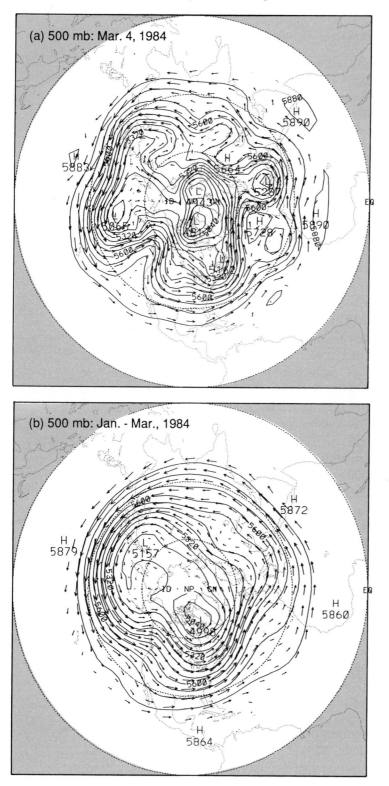

Fig. 3.8 (a) Geopotential height (contours) of and circulation (vectors) on the 500 mb isobaric surface for 4 March 1984. (b) January–March mean altitude of and circulation on the 500 mb isobaric surface. From National Meteorological Center (NMC) analyses. Reprinted from Salby (1992) by permission of Academic Press.

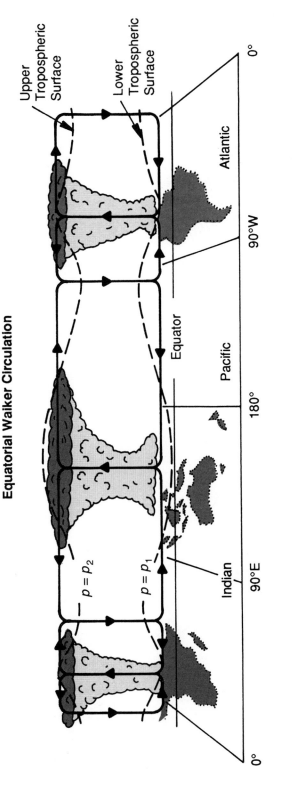

Equatorial Walker Circulation

Fig. 3.9 Pacific Walker circulation, as function of longitude and altitude over the equator. Adapted from Webster (1983). Reprinted from Salby (1992) by permission of Academic Press.

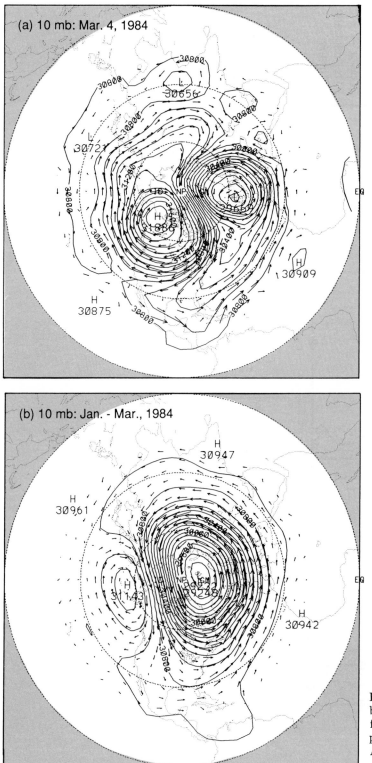

(a) 10 mb: Mar. 4, 1984

(b) 10 mb: Jan. - Mar., 1984

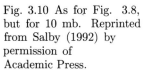

Fig. 3.10 As for Fig. 3.8, but for 10 mb. Reprinted from Salby (1992) by permission of Academic Press.

performed on the atmosphere is radiated away in the form of "planetary waves." Of global dimension, these disturbances are evidenced by gradual undulations of the jet stream in Fig. 3.8(a). Since they are generated by steady forcing mechanisms, planetary waves contain a large quasi-stationary component that remains even in the time-mean circulation in Fig. 3.8(b). Planetary waves are also generated by unsteady latent heating, e.g., in the ITCZ. Large-scale disturbances excited through either of these mechanisms propagate the influence of their forcing horizontally and vertically.

The synoptic weather systems that disturb the zonal circulation in the troposphere are not evident in the stratosphere. Above the tropopause, planetary-scale features prevail. The circulation at 10 mb, shown in Fig. 3.10(a) for the same day as in Fig. 3.8(a), is normally characterized by strong circumpolar (westerly) motion associated with the polar-night jet, as typifies the time-mean circulation (Fig. 3.10b). On the particular day shown, the circulation is disturbed by a wavenumber 1 planetary wave that has propagated upward out of the troposphere and displaced the westerly circulation well off the pole. By redistributing air over the hemisphere, disturbances like this force the stratosphere mechanically and exert an important influence on constituents like ozone.

3.2 Atmospheric thermodynamics

The link between the atmospheric circulation and the transfers of energy described above is thermodynamics. We develop the principles of atmospheric thermodynamics for a discrete system, namely an air parcel moving about in the circulation. Later, in Sec. 3.6, we generalize these principles to a continuum, valid for the atmosphere as a whole.

The thermodynamic state of a system is determined by the values of all of its properties. For a *pure substance* like air the thermodynamic state is fixed by any two properties, which are referred to as "state variables." All other properties follow from these through constitutive relations like the ideal gas law (3.1). The "state space" of a system consists of all possible values of any two state variables, e.g., as represented graphically in the $p - v$ plane. A transformation of the system from one state to another, known as a "thermodynamic process", then describes a path in the state space. Infinitely many processes connect two thermodynamic states. However, changes in state variables are *path independent*, i.e., they depend only on the initial and final states of the system.

3.2.1 The first law

The first law of thermodynamics for our system is given by

$$du = \delta Q - pdv \qquad (3.7a)$$

where u is the internal energy of the system, Q is the heat transfer into the system, and pdv is the incremental expansion work performed by the system on its environment. Lower case symbols will be used to denote specific properties, which refer to a unit mass, Q being exceptional. The incremental change of a quantity that is path independent (e.g., a state variable), is denoted by d whereas the incremental change of a quantity that is path dependent (e.g., Q) is denoted by δ. An equivalent form of the first law involving the enthalpy $h = u + pv$ is

$$dh - vdp = \delta Q. \qquad (3.7b)$$

For a pure substance, u and h are functions of T only. Their incremental changes are given by

$$du = c_v dT \qquad (3.8a)$$

$$dh = c_p dT, \qquad (3.8b)$$

where c_v and c_p are the specific heats at constant volume and pressure respectively. It follows that

$$c_p - c_v = R, \qquad (3.9)$$

which leads to the dimensionless constants for dry air

$$\gamma = \frac{c_p}{c_v} = 1.4$$

$$\kappa = \frac{R}{c_p} \approx 0.286.$$

With (3.8), the first law may be written

$$c_v dT + pdv = \delta Q \qquad (3.10a)$$

$$c_p dT - vdp = \delta Q. \qquad (3.10b)$$

3.2.2 Adiabatic processes

An "adiabatic" process is one in which the heat transfer vanishes. Processes involving heat transfer are referred to as "diabatic." For an adiabatic system, the first law (3.10) may be integrated along with the differential form of the ideal gas law

$$\frac{dp}{p} = \frac{d\rho}{\rho} + \frac{dT}{T} \qquad (3.11)$$

to obtain the identity

$$Tp^{-\kappa} = \text{const.} \tag{3.12}$$

Known as *Poisson's equation*, (3.12) describes adiabatic paths in the state space of an ideal gas.

Consider an air parcel moving vertically between two levels. If vertical motion occurs on a time scale short compared to that for heat transfer, the process may be regarded as adiabatic. Under these circumstances, the change in temperature is related directly to the change in pressure through (3.12). Adiabatic motion turns out to be a good approximation for many purposes because the time scale for radiative transfer is of order a week and therefore long by comparison with the characteristic time scale of air displacements.

The "potential temperature" θ is defined as the temperature assumed by an air parcel if compressed adiabatically to the pressure $p_o = 1000$ mb. Substituting this into (3.12) gives

$$\frac{\theta}{T} = \left(\frac{p_o}{p}\right)^\kappa. \tag{3.13}$$

A function of state variables, θ is a state variable. Because p varies sharply with altitude, surfaces of constant θ tend to be quasi-horizontal, like isobaric surfaces. From (3.12), it follows that θ *is invariant for an adiabatic process*. Hence, the potential temperature of an air parcel is conserved for adiabatic motion. Under these circumstances, an air parcel initially on a surface of constant θ remains on that surface.

Under diabatic conditions, an air parcel moves vertically across surfaces of constant θ. Combining (3.10) and (3.13) leads to

$$\delta Q = c_p T d \ln\theta, \tag{3.14}$$

so the change of θ is related directly to the heat exchange of the parcel.

3.2.3 The second law

A "reversible process" is one that leaves the system and its environment unchanged. A process that is executed slowly, with the system only infinitesimally out of thermodynamic equilibrium (e.g., the process involves only infinitesimally small pressure and temperature differences between the system and its environment), is reversible.

The second law of thermodynamics is inspired by the empirical observation that the quantity $\frac{\delta Q}{T}$ is path independent for a reversible process. Thus, the entropy s, defined by

$$ds = \left(\frac{\delta Q}{T}\right)_{rev}, \tag{3.15}$$

is a state variable. The second law of thermodynamics can then be expressed

$$ds = \left(\frac{\delta Q}{T}\right)_{rev} \geq \left(\frac{\delta Q}{T}\right) \tag{3.16}$$

For an adiabatic process, (3.16) reduces to

$$ds_{ad} \geq 0, \tag{3.17}$$

with equality applying if the process is also reversible. Thus, a reversible adiabatic process is an isentropic process (s = const).

For a reversible process, the equality in (3.16) may be substituted for the heat term in the two forms of the first law (3.10), to yield

$$du + pdv = Tds \tag{3.18a}$$

$$dh - vdp = Tds. \tag{3.18b}$$

Known as the *fundamental relations*, equations (3.18) involve only state variables. Consequently, these identities must hold regardless of path.

Entropy is related closely to potential temperature. Both are conserved under reversible adiabatic conditions. Under diabatic conditions, the fundamental relation (3.18a) may be combined with the ideal gas relation (3.11) to obtain

$$d\ln\theta = \frac{1}{c_p}ds \tag{3.19}$$

Thus, surfaces of constant θ are also isentropic surfaces. According to (3.14), the displacement across these quasi-horizontal surfaces is a direct measure of the heat exchange of an air parcel.

3.2.4 Heterogeneous systems

The discussion above focuses on a homogeneous system, where only a single phase is present. For such a system, thermodynamic equilibrium requires *mechanical equilibrium*: no pressure difference between the system and its environment, and *thermal equilibrium*: no temperature difference between the system and its environment. For a heterogeneous system, wherein multiple phases are present, thermodynamic equilibrium also requires *chemical equilibrium*: no diffusion of mass from one phase to another.

We consider first a single-component system involving only water, but in any of three phases that are in equilibrium with one another. The state of this system can be represented as a surface $p = p(v, T)$, shown in Fig. 3.11. The state surface is discontinuous in slope at the boundaries between regions where the system is homogeneous (only a single phase is present), and where it is heterogeneous (multiple phases are present). In a homogeneous region, two state variables are required to specify the third and hence the state of the system. For instance, both temperature and specific volume are needed

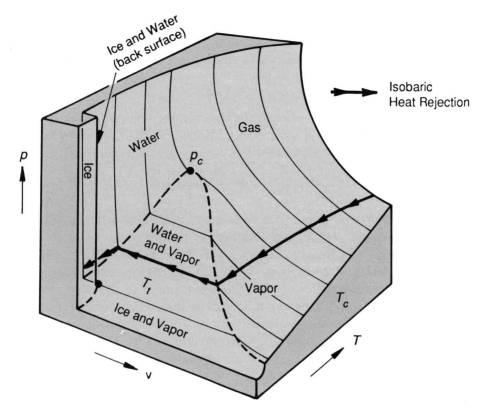

Fig. 3.11 State surface for a (single-component) mixture of water involving multiple phases in equilibrium with one another. Thermodynamic process accompanying isobaric heat rejection also indicated. Adapted from Irabarne and Godson (1981) by permission of Kluwer Academic Publishers. Reprinted from Salby (1992) by permission of Academic Press.

to determine the pressure in the vapor region.

In a heterogeneous region (e.g., vapor and water), the surface has a simpler form. There, isobars are coincident with isotherms, so

$$p = p(T). \tag{3.20}$$

Thus, specifying only the temperature of a mixture of vapor and water, at equilibrium with one another, determines the pressure of the mixture. Consider the process of isobaric heat rejection, indicated in Fig. 3.11. When the vapor–water region is encountered, the slope of the path changes discontinuously because the system's temperature no longer decreases. Instead, the temperature of the mixture remains constant and only the volume decreases as vapor condenses. This behavior continues until all of the vapor has condensed and the system is again homogeneous. At that point, the temperature of water decreases again in response to heat rejection.

For a heterogeneous state involving vapor and a condensed phase, the system pressure in (3.20) defines the "equilibrium vapor pressure", denoted *e*. Requiring chemical equilibrium of the mixture leads to the following relationship between the equilibrium vapor pressure and temperature

$$\frac{d\ln e}{dT} = \frac{L}{R_v T^2},$$ (3.21)

where L is the latent heat of conversion between the phases present and R_v is the specific gas constant of water vapor. Known as the *Clausius–Clapeyron equation*, (3.21) may be integrated to obtain

$$\log_{10} e_w \approx 9.4 - \frac{2.35 \times 10^3}{T},$$ (3.22)

where *w* refers to equilibrium with respect to water and *e* is in mb. A similar expression holds for equilibrium with respect to ice.

If the pressure of a mixture of vapor and water is below the equilibrium value (3.22), water will evaporate until the system's pressure reaches e_w, at which point the transfer of mass from one phase to the other vanishes and the system is in thermodynamic equilibrium. The equilibrium vapor pressure in (3.22) depends exponentially on temperature. Therefore, the equilibrium vapor concentration is much greater at high temperatures than at low temperatures.

3.2.5 Moist air

We now consider a two-component system of air and water, with the latter in vapor and possibly a condensed phase. The equation of state for the mixture must account for variable moisture content. The latter is usually absorbed in the *virtual temperature* T_v, defined by $R_d T_v = RT$, so the fixed gas constant for dry air R_d can be used in analyses for moist air (see Irabarne and Godson, 1981). However, differences between T and T_v are small enough to be ignored for many purposes because water vapor appears in only trace abundances. The vapor satisfies the equation of state

$$e v_v = R_v T,$$ (3.23a)

where *e* and v_v denote the partial pressure and specific volume of the vapor, respectively,

$$R_v = \frac{R_d}{\epsilon}$$ (3.23b)

and

$$\epsilon = \left(\frac{M_v}{M_d} \right) \approx 0.622.$$ (3.23c)

Several quantities are used to quantify moisture content. The "specific humidity"

$$q = \frac{m_v}{m}$$ (3.24)

is related closely to the water vapor mixing ratio

$$r = \frac{m_v}{m_d}$$
$$= \frac{q}{1-q} \approx q, \tag{3.25}$$

since q is a few percent at most.[3] In the previous section, we saw that a single-component mixture of water, containing vapor and a condensed phase in equilibrium with one another, has only one vapor pressure for a given temperature. The same holds for a two-component mixture of water and dry air because the water component behaves as though it were in isolation. Then the equilibrium vapor pressure corresponds to the partial pressure of vapor at which diffusion of mass between phases of water vanishes. In that state, no more vapor can be absorbed in solution and the air is said to be *saturated*. The equilibrium vapor pressure in (3.22) defines the "saturation vapor pressure" $e_w(T)$, which is a function of temperature alone. At vapor pressures e below e_w, air is unsaturated and can absorb moisture through evaporation. The saturation moisture content can also be expressed in terms of mixing ratio. Incorporating (3.23) leads to the relationship

$$r_w = \frac{1}{\epsilon}\left(\frac{e_w}{p}\right). \tag{3.26}$$

Two other measures of vapor content are referenced to saturation values. The "relative humidity"

$$RH = \frac{r}{r_w} \tag{3.27}$$

reflects the proximity of moist air to saturation. The same is true of the "dew point" T_d, which is the temperature to which air must be cooled *isobarically* to reach saturation. A low dew point indicates low moisture content, whereas a high dew point indicates only a small amount of cooling is needed to achieve saturation.

Because they are functions of temperature and pressure, the saturation properties (e.g., r_w) are state variables. In contrast, the actual vapor content (e.g., r) is fixed, so long as condensation and evaporation do not occur. Because of the exponential dependence in the Clausius–Clapeyron equation (3.22), the saturation mixing ratio is a strong function of temperature. The ability of air to hold vapor in solution diminishes sharply with decreasing temperature. A consequence of moist thermodynamics, this property of air is responsible for cloud formation and for confining water near the Earth's surface, as will be seen below.

---

[3] The symbol r without a subscript is understood to refer to water vapor.

3.2.6   Vertical displacements

We are now in a position to apply the thermodynamic principles developed above to an air parcel moving vertically. For the moment, consider a dry air parcel. Under adiabatic conditions, the potential temperature of the parcel is fixed. The parcel's temperature varies with pressure according to (3.10) with $\delta Q = 0$, wherein expansion work is compensated by a change of internal energy. As the parcel moves from one level to another, it adjusts immediately to the pressure of its surroundings to maintain mechanical equilibrium. Hence, the parcel's pressure equals the ambient pressure, which is in hydrostatic balance. Substituting (3.3) into (3.10b) leads to the linear variation with altitude of parcel temperature

$$\frac{dT}{dz} = -\Gamma_d, \qquad (3.28a)$$

where

$$\Gamma_d = \frac{g}{c_p} \qquad (3.28b)$$

is the "dry adiabatic lapse rate", which has a value of about 10 K km^{-1}. Adiabatic changes of state follow paths in state space known as *dry adiabats,* that are characterized by $\theta = $ const. If the parcel is moist, the linear lapse with altitude of parcel temperature still holds because water vapor modifies the thermal properties of air only slightly. Likewise, the parcel's mixing ratio r and potential temperature θ are fixed, *so long as the parcel remains unsaturated.*

Once condensation occurs, this is no longer true. Condensation of vapor results in a release of latent heat to the surrounding air molecules, which warms the parcel as though it contained an internal heat source. Consider an unsaturated parcel with initial mixing ratio $r = r_o < r_w$. An upward displacement of that parcel leads to a decrease of temperature according to (3.28). Then (3.22) indicates the saturation mixing ratio r_w decreases with altitude approximately exponentially. However, the mixing ratio r remains fixed at r_o, as long as the parcel is unsaturated. As illustrated in Fig. 3.12, a sufficiently large displacement will result in r_w decreasing to the actual moisture content r_o, at which point the parcel is saturated. The altitude where this occurs is known as the "lifting condensation level" (LCL) and marks the base of convective clouds. Ascent above the LCL results in r_w decreasing below r_o and condensation of the excess vapor ($\Delta r = r_o - r_w$) to preserve chemical equilibrium. An incremental displacement above the LCL is attended by a release of latent heat

$$\delta Q = L_v dr_w, \qquad (3.29)$$

where L_v is the latent heat of vaporization. Hence, both r and θ vary for the parcel above the LCL. The mixing ratio is equal to the saturation

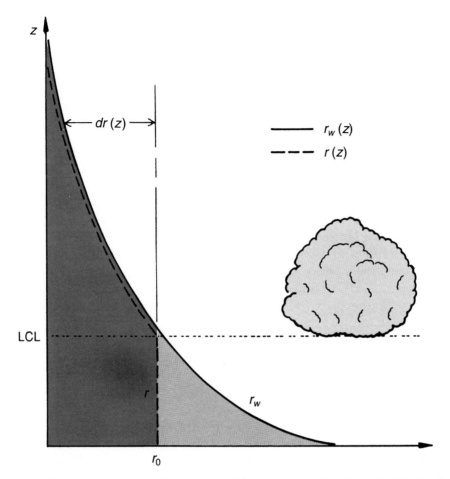

Fig. 3.12 Saturation mixing ratio r_w and mixing ratio r, as functions of altitude, for an air parcel displaced vertically. Lifting condensation level (LCL) also indicated. Reprinted from Salby (1992) by permission of Academic Press.

mixing ratio because just enough vapor condenses to maintain chemical equilibrium. The potential temperature increases due to the release of latent heat (3.29). Substituting into (3.14) and integrating from a level where the parcel's potential temperature is θ to the top of the atmosphere, where all of the latent heat must be released, obtains

$$\left(\frac{\theta_e}{\theta}\right) = e^{\frac{L_v r_w}{c_p T}} . \tag{3.30}$$

Known as the "equivalent potential temperature", the quantity θ_e represents the value of θ the parcel would assume if all of its vapor condensed. The equivalent potential temperature of a parcel is conserved both above and below the LCL. Changes of state above the LCL follow paths in state space known as *pseudo-adiabats* (also referred to as *saturated adiabats* and *moist*

79

adiabats), that are characterized by $\theta_e = \text{const.}$[4] Although temperature does not decrease linearly with altitude, it is still convenient to define a "saturated lapse rate" Γ_s, which is less than the dry adiabatic lapse rate

$$\Gamma_s < \Gamma_d$$

due to the release of latent heat. Although dependent upon the moisture content, Γ_s has a value of ~ 6.5 K km^{-1} for values representative of the troposphere.

The variations of state described above are best illustrated with an example using the *pseudo-adiabatic chart*, shown in Fig. 3.13, wherein the parcel's state space is represented as a function of altitude. Dry adiabats are indicated by heavy solid lines and labeled according to the constant values of θ characterizing them. Isopleths of saturation mixing ratio are indicated by dashed lines and labeled in units of g kg^{-1}. Pseudo-adiabats are indicated by light solid lines and are labeled according to the constant values of θ_e characterizing these paths. Consider an air parcel initially at 900 mb with a temperature of 15°C and a mixing ratio of 6 g kg^{-1}. Locating the parcel's initial state on the chart (labeled 0) determines its relative humidity

$$RH = \frac{r}{r_w} = \frac{6}{12} = 50\% \ .$$

Its dew point is determined by cooling the parcel isobarically until $r_w = r$ (state 1), wherein the parcel is saturated:

$$T_d = 4°\text{C} \ .$$

Suppose the parcel is displaced upwards in a moist thermal. As it rises, the parcel evolves along the dry adiabat passing through the initial state ($\theta = 297$ K), until the LCL is reached. At that point $r_w = r$ (state 2) and

$$p_{LCL} = 770 \text{ mb} \ .$$

Above the LCL, the parcel evolves along the pseudo-adiabat passing through the LCL, which gives the equivalent potential temperature $\theta_e = 315$ K. At 650 mb (state 3), the parcel's potential temperature has increased to

$$\theta_{650} = 303 \text{ K}$$

and its mixing ratio has decreased to

$$r = 4 \text{ g kg}^{-1}.$$

3.3 Hydrostatic equilibrium

The changes of state described above come about through the vertical distribution of mass, which is determined ultimately by gravity. Because it exerts such a strong body force, gravity must be treated with some care. The

---

[4] A pseudo-adiabatic process presumes all of the condensate is removed by precipitation. Further, the definition of θ_e in (3.30) applies only above the LCL, where $r = r_w$.

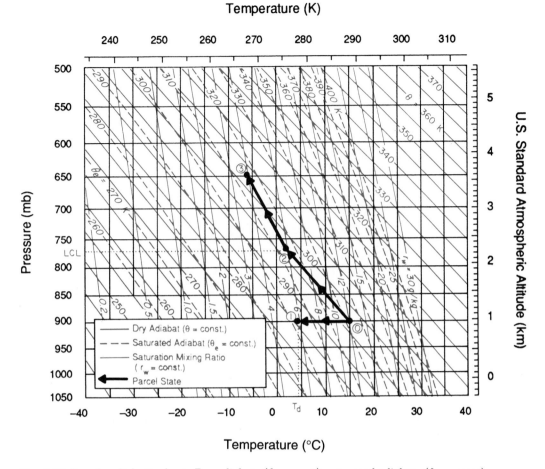

Fig. 3.13 Pseudo-adiabatic chart. Dry adiabats (θ = const), saturated adiabats (θ_e = const), and isopleths of saturation mixing ratio (r_w = const), as functions of temperature and pressure (altitude). Initial state of an air parcel is indicated by 0. Also shown is the change to state 1: isobaric cooling to the dew point T_d, to state 2: adiabatic expansion to the LCL, and to state 3: pseudo-adiabatic expansion to the 650 mb level. Reprinted from Salby (1992) by permission of Academic Press.

"effective gravity" $\mathbf{g}_{eff}$ actually includes contributions from several sources:

(i) radial gravitation

(ii) centrifugal acceleration due to rotation of the Earth

(iii) departures of the planet from sphericity;

see Fig. 3.14. Contributions (ii) and (iii) introduce horizontal components which must be balanced by other forces that unnecessarily complicate the equations governing horizontal motion. For this reason, it is useful to introduce a modified vertical coordinate that explicitly accounts for departures of $\mathbf{g}_{eff}$ from the vertical.

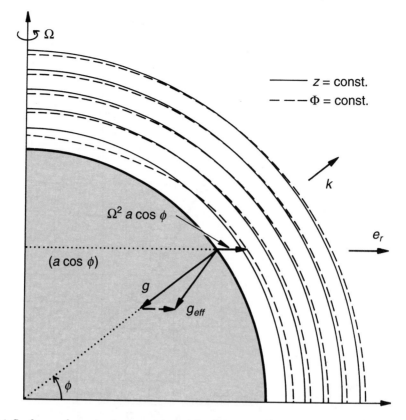

Fig. 3.14 Surfaces of constant geometric height (z = const) and geopotential height (Φ/g = const). Geopotential surfaces, which define the effective gravity, account for centrifugal acceleration and departures of the Earth from sphericity. Reprinted from Salby (1992) by permission of Academic Press.

The "geopotential" Φ is defined as the work performed to displace a unit mass in the Earth's gravitational field

$$d\Phi = -\mathbf{g} \cdot d\mathbf{r} = gdz, \qquad (3.31)$$

where $d\mathbf{r}$ is the displacement and z is measured along the line of effective gravity, which is denoted hereafter simply $\mathbf{g}$. Closely related is the "geopotential height" or

$$
\begin{aligned}
Z &= \frac{1}{g_0} \int_0^z gdz \\
&= \frac{1}{g_0}\Phi(z),
\end{aligned}
\qquad (3.32)
$$

where g_0 is a reference value usually taken as 9.8 m s^{-2}. Geopotential height accounts for variations in gravity, described above, so the constant value g_0 can be used in place of g in atmospheric force balances. For most applications, Z is close enough to geometric height to be used interchangeably with z, as is done hereafter. Defining z in this manner replaces the coordinate surfaces of constant geometric altitude with surfaces of constant geopotential height, as illustrated in Fig. 3.14.

3.3.1 Hydrostatic balance

With (3.31), hydrostatic balance (3.3) becomes

$$d\Phi = -\frac{dp}{\rho} \tag{3.33a}$$

or with the gas law (3.1)

$$d\Phi = -RT\,d\ln p. \tag{3.33b}$$

Integrating (3.33b) between isobaric surfaces $p = p_1$ and $p = p_2$ and incorporating (3.31) results in

$$\Delta z = z_2 - z_1 = \overline{H}\ln\left(\frac{p_1}{p_2}\right) \tag{3.34a}$$

where $\overline{H}$ is given by (3.4) with

$$\overline{T} = \frac{\int_{p_1}^{p_2} T\,d\ln p}{\ln(p_2/p_1)}. \tag{3.34b}$$

Known as the "hypsometric relation", (3.34) states that the *thickness* of an atmospheric layer bounded by two isobaric surfaces is proportional to the mean temperature of that layer and the pressure change across it. Applying (3.34) to a series of such layers indicates that the atmospheric column is thicker in warm regions than in cold regions. For this reason, the altitude of the 500 mb isobaric surface in Fig. 3.8 slopes downward towards the pole.

Equation (3.34) implies a one-to-one relationship between altitude and pressure. Therefore, the geopotential of an isobaric surface (e.g., that shown in Fig. 3.8) may be treated similarly to the pressure on a constant altitude surface. For instance, low values of Φ correspond to low values of p and vice versa. The horizontal pressure gradient force acting on an air parcel may be expressed either in terms of pressure as $-\frac{1}{\rho}\nabla p$ or in terms of geopotential as $-\nabla\Phi = -g\nabla z$.

3.3.2 Vertical displacements

From the discussion in Sec. 3.2.6, the temperature T' of a displaced air parcel evolves according to the dry adiabatic lapse rate Γ_d under unsaturated conditions and according to the saturated adiabatic lapse rate Γ_s under saturated conditions. On the other hand, the ambient temperature T varies according to the *environmental lapse rate*

$$\Gamma = -\frac{dT}{dz}, \tag{3.35}$$

which is independent of the parcel's temperature under adiabatic and pseudo-adiabatic conditions.

The vertical force balance for an air parcel subjected to an incremental vertical displacement dz' may be expressed

$$\frac{d^2 z'}{dt^2} = g \left(\frac{T' - T}{T} \right) \tag{3.36}$$

(e.g., Irabarne and Godson, 1981). In terms of lapse rate, this becomes

$$\frac{d^2 z'}{dt^2} = \frac{g}{T}(\Gamma - \Gamma')dz' . \tag{3.37}$$

The right-hand side, which represents the buoyancy force acting on the parcel, is proportional to the difference between the parcel's lapse rate and that of the environment.

3.3.3 Hydrostatic stability

Under unsaturated conditions, the displaced parcel evolves according to the lapse rate $\Gamma' = \Gamma_d$. As illustrated in Fig. 3.15, three possibilities exist:

(i) $\Gamma < \Gamma'$: The environmental temperature decreases with altitude more slowly than that of the parcel. Then

$$\frac{d^2 z'/dt^2}{dz'} < 0$$

and the parcel experiences a buoyancy force $\mathbf{f}_B$ *opposite to* the displacement dz'. This buoyancy reaction constitutes a positive restoring force, one which tends to restore the parcel to its original position. Under these conditions, the atmospheric layer is said to be "hydrostatically stable" and the environmental lapse rate is of "positive stability."

(ii) $\Gamma = \Gamma'$: The environmental temperature decreases with altitude at the same rate as that of the parcel. Under these circumstances, a displaced parcel experiences *no restoring force* through buoyancy. The atmospheric layer is said to be "hydrostatically neutral" and the environmental lapse rate is "neutral" or of "zero stability."

(iii) $\Gamma > \Gamma'$: The environmental temperature decreases with altitude more rapidly than that of the parcel. In this case,

$$\frac{d^2 z'/dt^2}{dz'} > 0 .$$

The parcel experiences a buoyancy reaction *in the same sense as* the displacement dz', which drives the parcel further from its original position and therefore constitutes a negative restoring force. Under these conditions, the atmospheric layer is said to be "hydrostatically unstable" and the environmental lapse rate is of "negative stability."

The degree of stability or instability is measured by the departure of the

Stability Categories

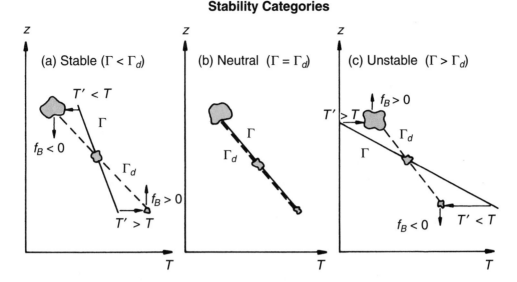

Fig. 3.15 Categories of hydrostatic stability for an unsaturated air parcel, according to the local environmental lapse rate Γ (solid) and referenced to the dry adiabatic lapse rate Γ_d (dashed). (a) *Stable* ($\Gamma < \Gamma_d$): environmental temperature decreases with altitude more slowly than that of a displaced air parcel, which therefore experiences a buoyancy force $\mathbf{f}_B$ that opposes the displacement and tends to restore the parcel to its original position. (b) *Neutral* ($\Gamma = \Gamma_d$): environmental temperature decreases with altitude at the same rate as that of a displaced air parcel, which therefore experiences no buoyancy force. (c) *Unstable* ($\Gamma > \Gamma_d$): environmental temperature decreases with altitude faster than that of a displaced air parcel, which therefore experiences a buoyancy force $\mathbf{f}_B$ that reinforces the displacement and drives the parcel further from its original position. Similar criteria hold for a saturated parcel, but with Γ_s in place of Γ_d. Reprinted from Salby (1992) by permission of Academic Press.

environmental lapse rate from the parcel lapse rate, e.g., by $(\Gamma_d - \Gamma)$.

A similar analysis may be applied to a saturated parcel (e.g., above the LCL). In that case, the conditions for stability are the same as those above, but with the saturated lapse rate Γ_s in place of Γ_d. Since $\Gamma_s < \Gamma_d$, hydrostatic stability is violated more easily under saturated conditions than under unsaturated conditions. This property of saturated ascent follows from the liberation (absorption) of latent heat, which reinforces the displacement of an air parcel by adding (removing) buoyancy.

The two reference lapse rates Γ_d and Γ_s define the range of environmental stability relative to a moist air parcel. An environmental lapse rate greater than Γ_d is "absolutely unstable", whereas an environmental lapse rate less than Γ_s is "absolutely stable." For intermediate lapse rates

$$\Gamma_s < \Gamma < \Gamma_d,$$

the environmental lapse rate is "conditionally stable." Under conditional stability, a sufficiently large displacement will lead to a parcel becoming

Conditional Instability

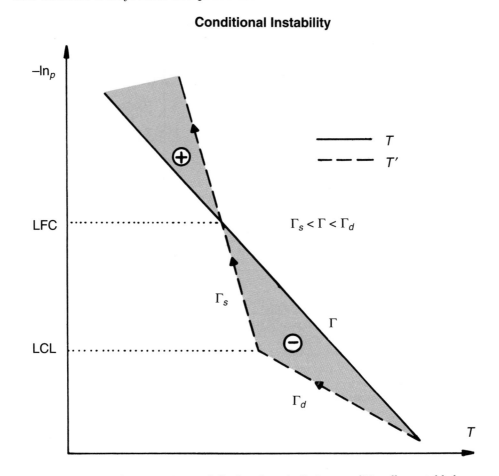

Fig. 3.16 Evolution of a moist air parcel displaced vertically in a conditionally unstable layer. Below its LCL, the parcel's temperature changes at the dry adiabatic lapse rate Γ_d and therefore more rapidly than the environment. Above its LCL, the parcel's temperature changes at the saturated adiabatic lapse rate Γ_s and therefore less rapidly than the environment. A sufficiently large displacement will carry the parcel above its level of free convection (LFC), where it is warmer and lighter than its environment and therefore accelerates under the action of buoyancy. Reprinted from Salby (1992) by permission of Academic Press.

warmer than its environment and hence positively buoyant according to (3.36); see Fig. 3.16. The level at which this occurs is known as the "level of free convection" (LFC). Most of the tropical troposphere is conditionally unstable, a property that favors deep convection.

The vertical force balance for a displaced air parcel assumes a simpler form when cast in terms of θ. Then the stability criteria translate into:

(i) $\frac{d\theta}{dz} > 0$ (stable)

(ii) $\frac{d\theta}{dz} = 0$ (neutral)

(iii) $\frac{d\theta}{dz} < 0$ (unstable)

Similar criteria hold under saturated conditions with θ_e in place of θ. These categories of hydrostatic stability reflect the vertical distribution of mass (e.g., whether heavier air lies over or underneath lighter air), taking into account changes in temperature that accompany expansion and compression.

Together with the dependence of θ on heating, these categories reveal how the stability of an atmospheric layer is established. According to (3.14), heating the base and cooling the top of a layer decrease $\frac{d\theta}{dz}$ and thus destabilize the layer. These are precisely the roles of absorption of SW radiation at the Earth's surface and emission of LW radiation from the middle troposphere, which continually destabilize the troposphere. On the other hand, cooling the base and heating the top of a layer increase $\frac{d\theta}{dz}$ and thus stabilize the layer. This is the effect of LW emission near the tropopause and the absorption of solar UV by ozone near the stratopause, which stabilize the stratosphere.

3.3.4 Buoyancy reactions

The vertical force balance may be written

$$\frac{d^2 z'}{dt^2} + N^2 z' = 0, \qquad (3.38a)$$

where

$$N^2 = \frac{g}{\theta}\left(\frac{d\theta}{dz}\right) = \frac{g}{T}(\Gamma_d - \Gamma) \qquad (3.38b)$$

defines "Brunt–Väisälä" or "buoyancy frequency" N, which is another measure of stability. Under conditions of positive stability, $N^2 > 0$ and (3.38) has solutions of the form

$$z'(t) = ae^{iNt} + be^{-iNt}. \qquad (3.39a)$$

Equation (3.39a) describes a stable oscillation that is supported by the positive restoring force of buoyancy. Under unstable conditions, $N^2 < 0$ and (3.38) has solutions of the form

$$z'(t) = ae^{Nt} + be^{-Nt} . \qquad (3.39b)$$

The first term corresponds to unstable ascent or descent. A disturbed air parcel will accelerate under the negative restoring force of buoyancy, evolving with other parcels into convective motion with finite displacements and a net rearrangement of air.

3.3.5 Vertical motions

By influencing vertical motions, hydrostatic stability controls the environment's ability to support convection and turbulent dispersion. Convective motions are inhibited in regions of positive stability, where the restoring force of buoyancy suppresses vertical displacements. By suppressing vertical motions, positive stability also suppresses complementary horizontal motions in turbulent eddies that are responsible for dispersing atmospheric constituents. In contrast, convective overturning and turbulent mixing are favored in regions of negative stability. Under these conditions, small displacements are reinforced by buoyancy until finite amplitude convection develops, which results in efficient mixing. Since θ is conserved for individual air parcels, vertical mixing tends to homogenize θ, driving the thermal structure towards neutral stability: $\frac{d\theta}{dz} = 0$ and $\frac{dT}{dz} = -\Gamma_d$. Under saturated conditions, convective overturning homogenizes θ_e, driving the thermal structure towards moist neutral stability: $\frac{d\theta_e}{dz} = 0$ and $\frac{dT}{dz} = -\Gamma_s$.

Having a lapse rate of about 6 K km^{-1}, the global-mean troposphere is weakly stable for unsaturated conditions. Under these circumstances, vertical displacements of unsaturated air (e.g., stimulated by surface heating) are met with relatively little opposition. For saturated conditions, the global-mean lapse rate is almost neutral. This property of the troposphere is no accident. Adjacent to a reservoir of moisture, convective overturning maintains the global-mean troposphere near neutral moist stability. Deep convection, reinforced by latent heat release, is responsible for a steady lapse of temperature through most of the tropical troposphere and for the high and very cold tropical tropopause. Relatively efficient vertical exchange (e.g., in cellular and sloping convection like that apparent in Plate 7) leads to a characteristic time scale for an air parcel to travel between the surface and the tropopause of order days.

By contrast, increasing temperature with altitude makes the stratosphere strongly stable. Vertical motions are suppressed, as is evidenced by the tops of deep cumulonimbus towers, whose growth is abruptly halted upon reaching the tropopause. The relative absence of vertical mixing favors a layered structure, as is typical of properties in the stratosphere. In sharp contrast to the troposphere, the characteristic time for air to be exchanged vertically in the stratosphere is of order months to a couple of years. This property is evidenced occasionally by volcanic debris which is injected into the stratosphere and which is observed to remain intact for long durations.

3.4   Radiative transfer

The thermal structure and stratification of mass discussed above are brought about in large part by radiative transfer; see Fig. 1.2. Figure 1.1 shows blackbody emission spectra for temperatures representative of the Sun and the Earth, along with the atmospheric absorptivity as a function of wavelength. The large difference in temperature leads to a wide separation of SW and LW radiation, with little overlap. At visible wavelengths (0.3– 0.7 μm, where SW radiation is concentrated, the atmosphere is virtually transparent. Only the tails of the solar spectrum are attenuated by absorption lines of oxygen and water vapor in the near IR and by ozone in the UV. At wavelengths of 5–100 μm, absorption of IR by water vapor, CO_2, and ozone makes the atmosphere nearly opaque to LW radiation. A variety of other trace gases, such as methane, nitrous oxide, and CFCs are also radiatively active at these wavelengths.

These properties of the atmosphere suggest the following simple model of radiative transfer. A "grey atmosphere" is one which is transparent to SW radiation but opaque to LW radiation with a specific absorption coefficient k (dimensions of area/mass) in the IR that is independent of wavelength, temperature, and pressure. We consider below the propagation of LW radiation, integrated over wavelength, in a grey atmosphere.

The three-dimensional propagation of radiant energy is described by the "radiance" or "intensity" $\mathbf{I}$, which represents the rate of energy flow per unit area in an increment of solid angle $d\Omega$ (dimensions of power/area·steradian), as illustrated in Fig. 3.17. The "irradiance" or "flux" of radiant energy crossing a surface with unit normal $\mathbf{n}$ in the $+\mathbf{n}$ direction (dimensions of power/area) is obtained by integrating the intensity over solid angle in the half-space towards which $\mathbf{n}$ points

$$F_n^+ = \int_{2\pi} \mathbf{I} \cdot \mathbf{n} d\Omega^+ . \tag{3.40a}$$

A similar definition holds for the flux in the $-\mathbf{n}$ direction F_n^-. The *net flux* crossing the surface is then

$$F_n = F_n^+ - F_n^- . \tag{3.40b}$$

Applying (3.40) to three coordinate planes gives the flux vector $\mathbf{F} = (F_x, F_y, F_z)$. For isotropic radiation, wherein $\mathbf{I}$ is independent of direction, the flux in any direction follows from (3.40) as simply

$$F = \pi I . \tag{3.41}$$

Consider a *plane parallel atmosphere*, wherein properties vary only in the vertical. The upward propagation of radiant energy is governed by the *radiative transfer equation*

$$\cos\theta \frac{dI}{d\tau} = I - J, \tag{3.42a}$$

where I denotes the intensity at an angle θ from the upward normal $\mathbf{k}$, the dimensionless quantity

$$\tau = \int_z^\infty \rho k dz \tag{3.42b}$$

is the "optical depth", which is zero at the top of the atmosphere and increases downward, and

$$J = \frac{1}{\pi} B(T) \tag{3.42c}$$

Propagation of Radiant Energy

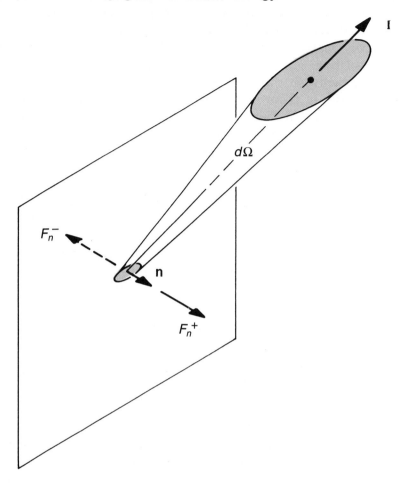

Fig. 3.17 Radiant intensity **I** (power/area·steradian) flowing through an increment of solid angle $d\Omega$. Radiant fluxes (power/area) F_n^+ and F_n^- crossing the surface with unit normal **n**, in the forward and backward directions respectively, follow by integrating **I** over the half-space of solid angle in the each direction. Reprinted from Salby (1992) by permission of Academic Press.

is the blackbody source function with[5]

$$B(T) = \sigma T^4. \tag{3.42d}$$

Equation (3.42a) states that the increase of flux across an incremental layer of optical depth $-d\tau$ equals the emission from that layer, $J d\tau$, minus the absorption in that layer, $-I d\tau$.

---

[5] The emitted radiance (3.42d) is the quantity measured from space to produce the IR imagery in Plate 7. Owing to its direct link to temperature, which varies with altitude approximately linearly in the troposphere, measurements of IR radiance can be used to infer the altitudes of the highest emitters.

The three-dimensional description of radiation (3.42) can be reduced to a one-dimensional description of vertical energy transfer by applying a *two-stream approximation* based on horizontally isotropic radiation (see Liou, 1980). Integrating over solid angle in the upper and lower half-spaces reduces (3.42) to

$$\frac{dF^\uparrow}{d\tau} = F^\uparrow - B(T) \tag{3.43a}$$

$$-\frac{dF^\downarrow}{d\tau} = F^\downarrow - B(T) \tag{3.43b}$$

where

$$F^{\uparrow\downarrow} = \int_{2\pi} \mathbf{I} \cdot \mathbf{k} d\Omega^{\uparrow\downarrow} \tag{3.44}$$

denote the upward and downward fluxes and τ is given by (3.42b) times a factor of 1.5.

3.4.1 Radiative equilibrium

Consider an incremental layer of depth dz bounded above and below by isobaric surfaces. The first law of thermodynamics (3.10b) with $dp = 0$ gives per unit volume

$$\rho c_p \frac{dT}{dt} = -\nabla \cdot \mathbf{F}$$
$$= -\frac{d}{dz}(F^\uparrow - F^\downarrow). \tag{3.45}$$

Equation (3.45) applies to a series of isobaric layers, which corresponds to using p as a vertical coordinate, as is discussed in Sec. 3.6. Thermal equilibrium requires $\frac{dT}{dt} = 0$, so the net vertical flux

$$F = F^\uparrow - F^\downarrow = \text{const.} \tag{3.46}$$

It may be shown (e.g., Goody and Yung, 1989) that under these conditions (3.43) has solution

$$B(\tau) = \frac{F_0}{2}(\tau + 1), \tag{3.47}$$

where $F_0 = (1 - \alpha)S_0/4$ is the effective SW flux incident on the top of the atmosphere, as shown in Fig. 3.18 along with $F^\downarrow$ and $F^\uparrow$.

The emission profile (3.47) determines the radiative equilibrium temperature T_R through (3.42d). If the atmosphere is hydrostatic, density decreases with altitude approximately exponentially and hence so does T_R through (3.47) and (3.42b). Further, most of the opacity in (3.42b) is contributed by water vapor and clouds, which are concentrated in the lowest levels. Therefore, the linear variation of B with optical depth, predicted by radiative equilibrium, translates into a steep decrease of temperature with

Radiative Equilibrium

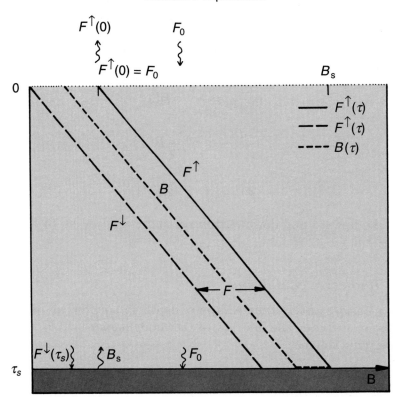

Fig. 3.18 Radiative equilibrium fluxes, as functions of optical depth τ. The upward and downward LW fluxes $F^{\uparrow}$ and $F^{\downarrow}$ vary linearly with optical depth. Since these preserve a fixed difference, the net vertical flux $F = F^{\uparrow} - F^{\downarrow}$ is constant and therefore divergenceless. At the top of the atmosphere, the upward LW flux is balanced by the incident SW flux F_0. For the surface, the upward emission B_s is balanced by the absorption of SW flux F_0 and of downward LW flux $F^{\downarrow}$ emitted by the atmosphere. Reprinted from Salby (1992) by permission of Academic Press.

altitude. For representative values, temperature decreases with altitude in some neighborhood of the surface more rapidly than the saturated adiabatic lapse rate. Hence, the radiative equilibrium atmosphere is conditionally unstable in this surface layer.

In a framework that accounts for atmospheric motion, convection develops spontaneously in this layer to neutralize the instability introduced by radiative transfer. The competing influences of convection and radiative transfer lead to two regions of distinctly different character. In the surface layer, which constitutes the troposphere of this simple atmosphere, the thermal structure is controlled by convective overturning, which is driven by radiative heating and its continual destabilization of the mass distribution. Since convection operates on a time scale much shorter than that of radiative transfer, the thermal structure of this region is maintained close to the saturated adiabatic lapse rate and a state of moist neutral stability. The region above the surface layer constitutes the atmosphere's stratosphere. There, the lapse rate of the radiative equilibrium temperature is less than Γ_s, so radiative heating stabilizes the mass distribution and maintains the thermal structure close to radiative equilibrium. Although a grey atmosphere captures the salient characteristics of energy transfer in the atmosphere, actual radiative transfer is modified importantly by scattering and by variations of absorptivity with wavelength. As discussed in Sec. 3.5, these processes play an especially important role in the interaction of radiation with clouds.

3.4.2 *Thermal dissipation*

Departures from the radiative equilibrium temperature T_R, such as those accompanying atmospheric motion, are damped by radiative transfer. Deep atmospheric circulations (e.g., those having vertical dimensions much larger than a scale height) are dissipated radiatively primarily by IR cooling to space, which behaves approximately as Newtonian cooling (Dickinson, 1973). The time scale for thermal dissipation can be estimated by considering a small perturbation from the radiative equilibrium temperature $\Delta T(t) = T(t) - T_R$ within a finite layer. Analysis of the energy budget of that layer leads to a characteristic time scale for radiative relaxation of the temperature perturbation

$$\frac{1}{\tau_R} = \frac{1}{\Delta T}\frac{d\Delta T}{dt} = \frac{4k\sigma T_R^3}{c_p}. \tag{3.48}$$

Values of $T_R = T_e = 255$ K, $(\rho k)^{-1} = 5$ km, and $\rho = 1.0$ kg m^{-3} yield a characteristic radiative damping time of $\tau_R \sim 16$ days. In the stratosphere, where water vapor and clouds are absent, τ_R becomes as short as 3–5 days.

3.4.3 *The greenhouse effect*

The thermal structure implied by radiative equilibrium also illustrates how the atmosphere influences the energy budget of the Earth. It may be shown that the surface emission is given by

$$B_s = \frac{F_0}{2}(\tau_s + 2), \tag{3.49}$$

where τ_s is the optical depth of the atmosphere. Since B_s is related to the surface temperature T_s through (3.42*d*), (3.49) is a statement of the greenhouse effect. In particular, the surface temperature is greater than it would be in the absence of an atmosphere ($\tau_s = 0$) and increases in proportion to the optical depth.

The optical depth of the atmosphere depends on the concentrations of radiatively active gases like water vapor, CO_2, and others mentioned in Sec. 3.1. Most of these gases are introduced through surface processes, e.g., through evaporation of the oceans and oxidation of organic matter. The dependence of those processes on surface temperature introduces the possibility of feedback between the two sides of (3.49). For instance, the concentration of atmospheric water vapor depends sensitively upon T_s. According to the Clausius–Clapeyron equation (3.22), the saturation vapor pressure increases sharply with temperature. A positive perturbation of T_s will increase the water vapor content of the atmosphere, which increases τ_s, which in turn increases T_s further through (3.49).

Were this positive feedback the only mechanism operating, small perturbations in surface temperature (e.g., those stimulated by changes in atmospheric composition associated with human activities and occasional volcanic eruptions) would result in large changes in surface temperature. However, other forms of feedback, such as those involving clouds, can mediate the water vapor–temperature feedback. For instance, increased moisture will tend to increase cloudiness, which acts to lower T_s by reflecting SW radiation. The role clouds play in the Earth's energy budget is more complex than simply increasing the planetary albedo. As discussed in the next section, clouds also act to warm the Earth by increasing the atmosphere's opacity to LW radiation and thus τ_s in (3.49).

3.5 Clouds

Clouds play a key role in the radiative energy budget of the Earth and in transfers of energy between the surface and the atmosphere. They form when air becomes saturated. This results most commonly from adiabatic cooling when air ascends above its LCL.

Clouds are generally classified in three major categories:

(i) *Stratus* or layered clouds form through large-scale lifting of an atmospheric layer. Vertical motions associated with stratus cloud formation are of order cm s^{-1} and the characteristic lifetime of these clouds is about a day.

(ii) *Cumulus* clouds form from localized air parcels or plumes that ascend buoyantly. Having updrafts of order m s^{-1} and characteristic time scales of minutes to hours, cumulus clouds are dynamic. They grow through positive buoyancy (e.g., supplied by surface heating and by the release of latent heat) and are dissipated through mixing with surrounding air.

(iii) *Cirrus* clouds are whispy in appearance and are found at high altitudes, where they are composed of ice particles. Cirrus clouds form through the same mechanisms responsible for stratus and cumulus clouds. Although thin, these high clouds have radiative properties

Mountain Wave Complex

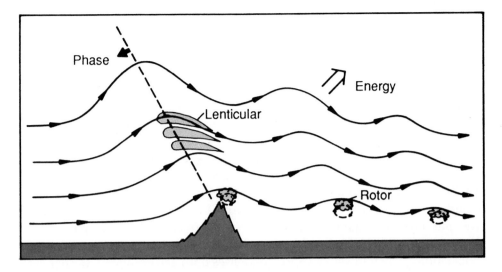

Fig. 3.19 Mountain wave complex generated by flow over elevated terrain. Westward tilt with altitude indicates upward propagation of wave activity. Air displaced above its LCL in the stationary wave pattern leads to the formation of lenticular clouds immediately downstream of the ridge. Similar behavior is found further downstream in multiple cycles that can produce wavy cloud patterns far from the source region. Since they are fixed with respect to the surface terrain leading to their formation, orographic clouds like these do not indicate the motion of air, which flows through them. Reprinted from Salby (1992) by permission of Academic Press.

that make them important in the Earth's energy budget.

One mechanism by which air is made to ascend is forced lifting over elevated terrain. Clouds produced in this manner are termed *orographic* and do not necessarily correspond to a fixed collection of air. The "lenticular" or "wave cloud", which is common in the lee of mountain ranges, is fixed relative to surface features leading to its formation, while air (displaced above its LCL) flows through the cloud, as illustrated in Fig. 3.19. Similar features are produced by propagating buoyancy waves discussed in Sec. 3.8.1. Differences in motion between these wavy clouds and the air passing through them is one reason why clouds are not a reliable indicator of velocity.

Not all clouds are formed through lifting. Low-level stratus form when moist air comes into contact with a cold surface and cools below its dew point. Such clouds are common over the ocean over much of the globe; see Plate 5.

3.5.1 Cloud formation

Clouds form under *supersaturated* conditions, i.e., when the relative humidity exceeds 100%. Cloud droplets (diameters less than 200 μm) can form in two ways. "Homogeneous nucleation" involves spontaneous condensation when water vapor molecules collide to form a small embryonic droplet that is large enough to remain intact. The survival of such a droplet is determined by a balance between condensation and evaporation, the latter being promoted by latent heat released during condensation. Under subsaturated conditions, evaporation is more efficient than condensation, so droplets are not favored. Under supersaturated conditions, the same is true for small droplets. However, condensation becomes more efficient than evaporation beyond a critical radius, so individual droplets survive long enough to grow.

Laboratory experiments indicate that droplets with radii smaller than 0.1 μm require supersaturation greater than 1% (e.g., Wallace and Hobbs, 1977). However, such levels of supersaturation are rarely observed, whereas cloud droplets with radii smaller than 0.1 μm are common. Hence, the mechanism actually responsible for producing cloud droplets is not homogeneous nucleation, which involves only water, but "heterogeneous nucleation", wherein water vapor condenses onto aerosol particles that serve as "condensation nuclei." Aerosol pollutants serving in this capacity are scavenged from the atmosphere when they precipitate out with the droplets that form around them. The same is true of gaseous pollutants that dissolve into the droplets.

Once they form, cloud droplets grow and decay through condensation and evaporation. Cloud droplets also grow through "coalescence", when droplets collide. Droplets with radii less than 20 μm have very small fall velocities. Hence, they move in unison with the air, making collisions infrequent and condensation the primary growth mechanism. Larger droplets move relative to the air, so they sweep out a volume of small droplets in a given time. As shown in Fig. 3.20, the efficiency of coalescence increases sharply with droplet radius and becomes the dominant growth mechanism for radii much greater than 20 μm.

3.5.2 *Radiative processes in clouds*

Cloud particles interact strongly with SW and LW radiation. The scattering efficiency of cloud particles makes them highly reflective to visible radiation. Figure 3.21 shows the reflectance of 3 km thick stratus for a range of microphysical properties. Most droplet size distributions and number densities lead to reflectances in the visible well over 0.5. In fact, a simple treatment of SW radiative transfer (e.g., Houghton, 1986) predicts a cloud albedo of 0.9 for optical depths typical of stratiform clouds. Thus, even shallow stratus have a high albedo (cf. marine stratus and deep cumulus in Plate 5). Because of this property and their extensive coverage of the Earth, marine stratus are a major component of the planetary albedo.

Clouds are also efficient at absorbing IR radiation, so they modify the

Droplet Growth

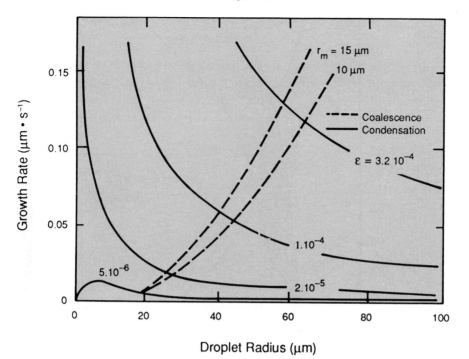

Fig. 3.20 Radial growth rate $\frac{dr}{dt}$ of droplets in a cloud with liquid water content 1.0 g m^{-3} due to condensation (solid) and to coalescence (dashed), as functions of droplet radius r. Growth rate due to condensation shown for levels of supersaturation $\epsilon = \frac{\rho_w}{\rho_{H_2O}}(RH - 1)$, where ρ_w is the saturation vapor density and ρ_{H_2O} is the density of liquid water. Growth rate due to coalescence shown for droplet radii r_m, which corresponds to the population maximum in the cloud droplet distribution. Adapted from Matveev (1967). Reprinted from Salby (1992) by permission of Academic Press.

atmospheric opacity and hence the LW energy budget. As shown in Fig. 3.21, the absorptance of 3 km thick stratus exceeds 0.80 for wavelengths longer than 2 μm over a wide range of microphysical properties. Since the optical depth τ_s is proportional to the cloud depth, deep cumulus are effective at warming the Earth according to (3.49). By comparison, shallow stratiform clouds (e.g., half a kilometer thick) do not increase the atmospheric opacity substantially above that already provided by water vapor. Yet, their high reflectance to SW radiation makes such clouds effective at cooling the Earth.

Satellite measurements (Ramanathan et al., 1989) reveal that the net cloud forcing of the radiation budget (viz. warming due to absorption of LW radiation – cooling due to reflection of SW radiation) is almost zero in the tropics. (While the effects of clouds on the SW and LW radiation

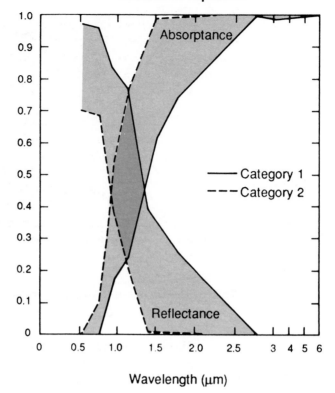

Fig. 3.21 Reflectance and absorptance of 3 km thick stratus, as functions of wavelength for incident radiation at zero zenith angle and for a range of microphysical properties that is bounded by Categories 1 and 2. Category 1 is based on a droplet distribution having a characteristic radius of 6 μm, a number density of 10^2 cm^{-3}, and liquid water content of 0.297 g m^{-3}. Category 2 involves rain with a droplet distribution having a characteristic radius of 600 μm, a number density of 0.001 cm^{-3}, and liquid water content of 2.1 g m^{-3}. Source: Cox (1981). Reprinted from Salby (1992) by permission of Academic Press.

budgets nearly cancel in the ITCZ, organized convection responsible for those clouds transfers large quantities of latent heat from the oceans to the atmosphere.) Outside the tropics, cloud radiative forcing is negative. Therefore, the overall radiative influence of clouds is to cool the Earth. How clouds function in the current climate and how they will respond to perturbations is one of the major outstanding issues of climate research.

3.6 Atmospheric dynamics

The principles developed above for a discrete system (i.e., an air parcel) must now be extended to the atmosphere as a whole, which may be regarded as a continuum of such systems. The field or "Eulerian" description of fluid motion involves specifying a property ψ at each point in the domain for all instants of time. The "Lagrangian" description of fluid motion involves specifying the properties of individual fluid elements and their positions for all instants of time. Since the field property at a given position and time must equal the property of the fluid element occupying that position, these two descriptions are closely related.

The link between the Eulerian and Lagrangian descriptions of fluid motion is the "material derivative"

$$\frac{d\psi}{dt} = \frac{\partial\psi}{\partial t} + u\frac{\partial\psi}{\partial x} + v\frac{\partial\psi}{\partial y} + w\frac{\partial\psi}{\partial z} = \frac{\partial\psi}{\partial t} + \mathbf{v}\cdot\nabla\psi, \qquad (3.50)$$

which represents the time rate of change of the property ψ *following a fluid element*. Here $\mathbf{v} = (u, v, w)$ is the three-dimensional velocity. Also known as the "Lagrangian derivative", (3.50) includes the local time rate of change $\frac{\partial\psi}{\partial t}$, which accounts for transience of the field ψ, plus an advective contribution $\mathbf{v}\cdot\nabla\psi$, which accounts for the parcel moving along the gradient of the field property ψ.

3.6.1 Equations of motion in an inertial reference frame

Applying these concepts to particular properties of a material volume of finite dimension (e.g., Dutton, 1976) transforms the equations governing a discrete system into those governing a continuum.

a. Mass

By definition, the mass of the material volume must be conserved. This requirement leads to the partial differential equation

$$\frac{d\rho}{dt} + \rho\nabla\cdot\mathbf{v} = 0. \qquad (3.51)$$

Referred to as the "continuity equation", (3.51) represents a constraint on the mass field.[6]

b. Momentum

Newton's second law of motion for the material volume is transformed into "Euler's equations of motion"

$$\rho\frac{d\mathbf{v}}{dt} = \rho\mathbf{f} + \nabla\cdot\tau, \qquad (3.52)$$

where $\mathbf{f}$ is the net body force per unit mass and τ is the stress tensor acting on the surface of the material volume. The body force of relevance in the

---

[6] Strictly, conservation of mass should also account for changes in composition, e.g., those associated with changes in water vapor and other trace species. However, these changes are small enough to be ignored in the overall mass budget and are accounted for in the conservation equations for individual species.

atmosphere is gravity $\mathbf{g}$. The stress tensor can be decomposed into a normal stress, exerted by pressure, and frictional shear stresses, which are denoted $\mathbf{D}$ for *drag*. Then the equations of motion become

$$\rho \frac{d\mathbf{v}}{dt} = \rho \mathbf{g} - \nabla p + \rho \mathbf{D}, \qquad (3.53)$$

which are known as the "momentum equations."

c. Internal energy

The first law of thermodynamics (3.10a) for the material volume is transformed into the "thermodynamic equation"

$$\rho c_v \frac{dT}{dt} + p\nabla \cdot \mathbf{v} = -\nabla \cdot \mathbf{F} + \nabla \cdot (\kappa \nabla T) + \rho Q, \qquad (3.54)$$

where $p\nabla \cdot \mathbf{v}$ represents the rate of expansion work performed by the material volume, the heat flux has been separated into conductive and radiative components, with κ denoting the thermal conductivity and $\mathbf{F}$ the flux of radiative energy, and Q has been redefined to denote any internal heating.

3.6.2 Equations of motion in a rotating reference frame

The five partial differential equations (3.51), (3.53), and (3.54), referred to collectively as the "equations of motion", govern the flow of a compressible stratified fluid in an inertial reference frame. However, the Earth's surface is rotating and therefore noninertial. Artificial forces in that reference frame modify these equations importantly. The first modification is a centrifugal acceleration, which is absorbed in the effective gravity (Sec. 3.3). Rotation also introduces a Coriolis acceleration, $2\mathbf{\Omega} \times \mathbf{v}$, that acts orthogonal to the local motion in the rotating reference frame.

For large-scale motions, the governing equations are simplified by introducing (i) the *hydrostatic approximation*, in which only the pressure gradient and gravitational forces are retained in the vertical momentum balance, and (ii) the *shallow atmosphere approximation*, in which the radial distance r in spherical geometry is replaced by $(a + z)$, where the altitude z is much smaller than the radius of the Earth a. For consistency, these approximations retain only the components of the Coriolis acceleration that are proportional to the vertical component of planetary rotation, i.e., the components $f\mathbf{k} \times \mathbf{v}$, where

$$f = 2\Omega \sin \phi \qquad (3.55)$$

is the "Coriolis parameter" and represents the magnitude of the vertical component of planetary vorticity.

In addition to these simplifications, it is convenient to express the equations of motion in *pressure coordinates*, wherein coordinate surfaces of constant geopotential altitude are replaced by isobaric surfaces (e.g., Dutton, 1976). In that framework, altitude is measured by p, the geopotential height of isobaric surfaces replaces pressure as a dependent variable, and all other dependent variables are evaluated on isobaric surfaces. The resulting scalar equations are then

$$\left(\frac{du}{dt}\right)_p - \left(f + u\frac{\tan\phi}{a}\right)v = -\frac{1}{a\cos\phi}\frac{\partial\Phi}{\partial\lambda} + D_\lambda \tag{3.56a}$$

$$\left(\frac{dv}{dt}\right)_p + \left(f + u\frac{\tan\phi}{a}\right)u = -\frac{1}{a}\frac{\partial\Phi}{\partial\phi} + D_\phi \tag{3.56b}$$

$$\frac{\partial\Phi}{\partial p} = -\frac{RT}{p} \tag{3.56c}$$

$$\frac{1}{a\cos\phi}\frac{\partial u}{\partial\lambda} + \frac{1}{a\cos\phi}\frac{\partial}{\partial\phi}(v\cos\phi) + \frac{\partial\omega}{\partial p} = 0 \tag{3.56d}$$

$$c_p\left(\frac{dT}{dt}\right)_p - \frac{RT}{p}\omega = Q, \tag{3.56e}$$

where λ and ϕ represent longitude and latitude respectively,

$$\left(\frac{d}{dt}\right)_p = \frac{\partial}{\partial t} + \frac{u}{a\cos\phi}\left(\frac{\partial}{\partial\lambda}\right)_p + \frac{v}{a}\left(\frac{\partial}{\partial\phi}\right)_p + \omega\frac{\partial}{\partial p} \tag{3.56f}$$

denotes the material derivative in pressure coordinates,

$$\omega = \frac{dp}{dt} \tag{3.56g}$$

is a measure of vertical velocity in this coordinate system, and Q represents the net diabatic heating. Equations (3.56) are referred to as the "primitive equations" because they constitute the starting point for most dynamical studies. To describe constituents like water vapor and ozone, they are complemented by conservation equations of the form

$$\frac{dr_i}{dt} = P_i - D_i, \tag{3.57}$$

where P_i and D_i denote production and destruction of the mixing ratio of the ith species r_i.

In practice, it is more convenient to treat certain derivatives of the equations of motion, in place of the horizontal momentum equations. Applying the two-dimensional curl and divergence to the horizontal velocity $\mathbf{v}_h = (u, v)$ in Eqs. (3.56a) and (3.56b) leads to

$$\frac{\partial\zeta}{\partial t} + \mathbf{v}_h \cdot \nabla_p(\zeta + f) + \omega\frac{\partial\zeta}{\partial p} = -(\zeta + f)\nabla_p \cdot \mathbf{v}_h + \mathbf{k} \cdot \left(\frac{\partial\mathbf{v}_h}{\partial p} \times \nabla_p\omega\right) + \mathbf{k} \cdot \nabla_p \times \mathbf{D} \tag{3.58a}$$

$$\frac{\partial\delta}{\partial t} + \nabla_p \cdot (\mathbf{v}_h \cdot \nabla_p\mathbf{v}_h) + \nabla_p\omega \cdot \frac{\partial\mathbf{v}_h}{\partial p} + \omega\frac{\partial\delta}{\partial p} = -\nabla_p^2\Phi - \nabla_p \cdot (f\mathbf{k} \times \mathbf{v}_h) + \nabla_p \cdot \mathbf{D}, \tag{3.58b}$$

which represent the budgets of vorticity $\zeta = \mathbf{k} \cdot (\nabla_p \times \mathbf{v}_h)$ and divergence $\delta = \nabla_p \cdot \mathbf{v}_h$, respectively. Governing horizontal motion, these are the equations most commonly solved by general circulation models.

3.7 Classes of motion

3.7.1 Scales of motion

The equations of motion describe a wide range of dynamical phenomena. Even those relevant to the atmosphere represent a broad spectrum of motions. Figure 3.22 shows the spectrum of kinetic energy in the zonal velocity for the troposphere and near the surface. In the free atmosphere, kinetic energy is concentrated at periods of days to weeks, which corresponds to the synoptic and planetary-scale disturbances apparent in Fig. 3.8. Power is also concentrated in the seasonal cycle. Some of the large-scale kinetic energy cascades to smaller scales in the form of mechanical turbulence, which is evidenced by power at periods shorter than one day and at periods of minutes. However, small-scale turbulence represents a relatively minor component of the kinetic energy of the free atmosphere.

Near the surface, turbulent motions account for a much greater portion of the kinetic energy. Nearly half of the power is concentrated at periods of minutes, which corresponds to mechanical turbulence and thermal convection. These turbulent motions are efficient at transferring momentum, heat, and constituents like water vapor between the surface and the free atmosphere. Because they destroy large-scale gradients, such motions represent a major source of dissipation for the general circulation.

The remainder of this section focuses on large-scale motions, in which most of the kinetic energy of the atmosphere is concentrated. For such motions, the governing equations may be simplified. In many applications, some quadratic terms in (3.56a) and (3.56b) are small, so the horizontal momentum equations may be written

$$\frac{d\mathbf{v}_h}{dt} + f\mathbf{k} \times \mathbf{v}_h = -\nabla_p \Phi + \mathbf{D}. \tag{3.59}$$

The horizontal momentum equation may be simplified further for certain classes of large-scale motion. Although idealized, these simple classes of motion provide insight into more complex circulations and capture a large part of observed behavior.

3.7.2 Geostrophic motion

A scale analysis for large-scale motions reveals that the dominant terms in (3.59) are the pressure gradient force and the Coriolis force. The simplest treatment of such motions is on an "f-plane", wherein spherical geometry is approximated by a Cartesian plane tangent to the Earth's

Kinetic Energy

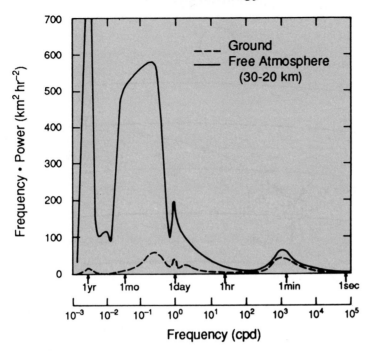

Fig. 3.22 Mean kinetic energy of the zonal wind component in the troposphere (solid) and near the surface (dashed). After Vinnichenko (1970). Reprinted from Salby (1992) by permission of Academic Press.

surface with f constant. In the limit of large horizontal scale, slowly evolving flow, and away from the equator, (3.59) reduces to the "geostrophic balance"

$$f\mathbf{k} \times \mathbf{v}_g = -\nabla_p \Phi, \qquad (3.60a)$$

where

$$\mathbf{v}_g = \frac{1}{f}\mathbf{k} \times \nabla_p \Phi \qquad (3.60b)$$

is the "geostrophic velocity."

According to (3.60), the geostrophic velocity is parallel to contours of geopotential of an isobaric surface (which reflects the distribution of pressure on a constant height surface; see Sec. 3.3), and its speed is proportional to the gradient of Φ. Thus, $\mathbf{v}_g$ is orthogonal to the pressure gradient force $-\nabla_p \Phi$, exactly perpendicular to the direction of motion normally observed in an inertial reference frame. This peculiarity of large-scale atmospheric motion is best illustrated with a simple example.

Consider a hypothetical air parcel in the geopotential field shown in Fig. 3.23, which is characterized by nearly straight contours of Φ. If the air parcel is initially motionless, it experiences a pressure gradient force $-\nabla_p \Phi$

103

Geostrophic Equilibrium

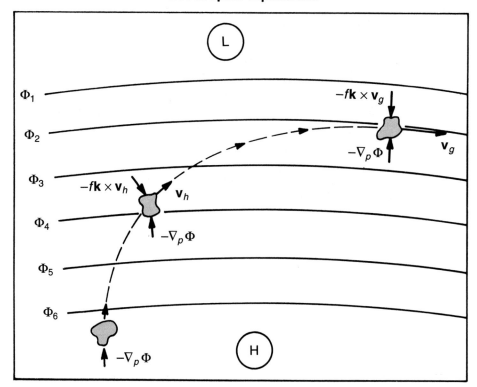

Fig. 3.23 Evolution of a hypothetical air parcel, initially motionless, in a pressure field that is characterized by nearly straight contours of geopotential. At $t = 0$, the parcel accelerates towards low pressure under the action of the pressure gradient force $-\nabla_p\Phi$. Once horizontal velocity $\mathbf{v}_h$ develops, the parcel's motion is deflected to the right (in the NH) by the Coriolis force. When the parcel's velocity has veered parallel to contours of Φ, the Coriolis and pressure gradient forces are in direct opposition. Geostrophic equilibrium results if the velocity equals $\mathbf{v}_g$ given by (3.60b). Reprinted from Salby (1992) by permission of Academic Press.

that accelerates it towards low geopotential (i.e., low pressure). As soon as a velocity $\mathbf{v}$ develops, the parcel experiences a Coriolis force $-f\mathbf{k}\times\mathbf{v}$ that acts perpendicular and to the right (left) of the parcel's motion in the Northern (Southern) Hemisphere. The effect of the Coriolis force is to deflect the parcel's motion towards lines of constant Φ. As the parcel accelerates, it veers to the right (left) and so does the Coriolis force acting on it, until the parcel's velocity is parallel to contours of Φ. At that point, the Coriolis force acts in direct opposition to the pressure gradient force, both perpendicular to the velocity. An equilibrium may be reached if the parcel's velocity equals $\mathbf{v}_g$ given by (3.60b). The parcel then moves parallel to contours of Φ, with low geopotential (i.e., low pressure) on the left in the NH. Once geostrophic equilibrium is established, the parcel's speed adjusts to gradual variations in $\nabla_p \Phi$ in inverse proportion to the spacing of contours.

Geostrophic equilibrium predicts circular motion about centers of high and low pressure. In the NH, the motion about low pressure is counterclockwise, which is in the same sense as the vertical component of planetary vorticity $f\mathbf{k}$. In the SH, the motion about low pressure is clockwise, but still in the same sense as $f\mathbf{k}$. In both cases, the motion is referred to as "cyclonic", which implies atmospheric vorticity in the same sense as the planetary vorticity. Motion about high pressure is clockwise (counterclockwise) in the Northern (Southern) Hemisphere. Such motion is in a sense opposite to the vertical component of planetary vorticity in both hemispheres and is termed "anticyclonic."

Under the conditions described above, the geostrophic velocity (3.60b) has zero horizontal divergence and is termed simply *nondivergent* in meteorological parlance. Hence, contours of Φ are streamlines of $\mathbf{v}_g$. Then the continuity equation (3.56d) indicates that geostrophic flow involves no vertical motion. Vertical motion is produced by mechanisms that drive the horizontal velocity ageostrophic and introduce horizontal divergence. The mechanism most important in this regard is friction.

3.7.3 *Frictional geostrophic motion*

In the lowest kilometer of the atmosphere, the drag $\mathbf{D}$ in (3.59) cannot be ignored. Turbulence in the "planetary boundary layer" transports momentum between the surface and the free atmosphere by exerting a shear stress on adjacent layers. In the presence of drag, horizontal forces no longer balance with $\mathbf{v}_h$ parallel to contours of Φ. Drag reduces $\mathbf{v}_h$, which reduces the Coriolis force, which allows the pressure gradient force to drive $\mathbf{v}_h$ across isobars towards low geopotential (i.e., towards low pressure).

A cross-isobaric component of motion is observed near the surface, where turbulent stresses are large. The attending divergence introduces vertical motion in centers low and high pressure. As pictured in Fig. 3.24, frictional convergence into surface low pressure is compensated by rising motion, which favours cloud formation and precipitation in cyclonic weather systems. By contrast, frictional divergence out of surface high pressure is compensated by subsidence, which inhibits cloud formation. Because it inhibits convective overturning, subsidence generated in this manner also tends to trap pollutants near their sources at the surface.

3.7.4 *Vertical shear*

As was indicated in Sec. 3.1, the global circulation is related closely to the thermal structure of the atmosphere. This relationship comes about through vertical shear of the geostrophic wind. Differentiating (3.60) with

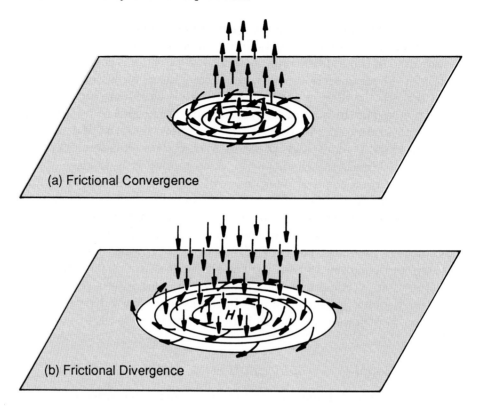

Fig. 3.24 Vertical motion introduced by (a) frictional convergence into surface low pressure, (b) frictional divergence out of surface high pressure. Reprinted from Salby (1992) by permission of Academic Press.

respect to p and incorporating hydrostatic balance (3.33) leads to

$$-\frac{\partial \mathbf{v}_g}{\partial \ln p} = \frac{R}{f}\mathbf{k} \times \nabla_p T \qquad (3.61)$$

Known as "thermal wind balance", this relationship implies that the change of horizontal velocity across an incremental layer of depth $-\partial \ln p$ is proportional to the horizontal gradient of temperature in that layer.

The zonal-mean fields in Figs. 3.3 and 3.4 verify this fundamental relationship. Throughout the troposphere, the temperature gradient is poleward[7] in both hemispheres. Hence, $-\frac{\partial \mathbf{v}_g}{\partial \ln p}$ is positive and westerlies intensify with altitude to form the subtropical jets located near the tropopause in each hemisphere. In the stratosphere, the temperature gradient is still poleward in the winter hemisphere. Consequently, the westerly flow continues to strengthen with altitude up to the stratopause, where it forms the polar-night jet. By contrast, the temperature gradient

---

[7] The convention adopted in meteorology and in this book is to refer to the temperature gradient as poleward if low latitudes are warm and high latitudes are cold.

in the summer hemisphere is equatorward. In accord with (3.61), the zonal flow becomes easterly and intensifies up to the stratopause.

3.8 Atmospheric waves

The atmosphere supports a variety of wave motions, all of which are represented in the equations of motion. Two important types are "buoyancy waves" and "Rossby waves."[8] Buoyancy waves, also referred to as "gravity waves", are made possible by the positive restoring force supplied by buoyancy under stable stratification. These waves are typically small scale, meters to a few tens of kilometers, have characteristic time scales of minutes to hours, and are responsible for the wavy cloud patterns common in the lee of mountains and in stratiform clouds; see Fig. 3.19. Rossby waves occur on large spatial scales, 1,000 km and greater, and have characteristic time scales of days and longer. On the largest dimensions, they are referred to as "planetary waves" and are manifested by gradual undulations in the global circulation; see Fig. 3.8(b). The smaller-scale synoptic weather systems in Fig. 3.8(a) are also Rossby waves, but they originate in instability of the zonal circulation. Both gravity waves and planetary waves are excited by flow over elevated terrain and by latent heat release in organized convection.

3.8.1 Gravity waves

A simple model of buoyancy waves involves a motionless isothermal atmosphere that is perturbed by disturbances whose time scales are short compared to rotation, friction, and radiative transfer. The equations of motion may be linearized in the small perturbation quantities $(u', v', w', \frac{p'}{\bar{p}}, \frac{\rho'}{\bar{\rho}})$, where overbar denotes unperturbed quantities. We consider disturbances in the $x - z$ plane of the form

$$\psi' = \text{Re}\left\{\Psi e^{(z/2H) + i(kx + mz - \sigma t)}\right\}, \tag{3.62}$$

where ψ corresponds to any of the field properties mentioned above, H is given by (3.4), and the exponential factor accounts for conservation of energy in an atmosphere where mass is stratified hydrostatically. If we consider further the limit $k \to \infty$ with σ fixed, substituting (3.62) reduces the perturbation equations to the dispersion relationship

$$\frac{m^2}{k^2} = \frac{N^2}{\sigma^2} - 1. \tag{3.63}$$

According to (3.63), gravity waves propagate vertically only for frequencies below N. Gravity waves are *dispersive*; their phase velocities differ from their group or energy velocities. In fact, the simple gravity waves discussed

---

[8] Named after C. G. Rossby, whose pioneering work revealed their importance in weather prediction.

above have a group velocity that is orthogonal to their phase velocity, with upward energy propagation corresponding to downward phase propagation (e.g., Gill, 1982).

An important generation mechanism for gravity waves is flow over elevated terrain. Mountain complexes generate stationary wave patterns, like that pictured in Fig. 3.19, which propagate horizontally and vertically. The absorption of these waves and the momentum they carry aloft is thought to be responsible for driving the circulation out of radiative equilibrium in the mesosphere, where temperature increases from a minimum over the summer pole (in perpetual daylight) to a maximum over the winter pole (in perpetual darkness).

3.8.2 *Planetary waves*

At frequencies below 2Ω, another form of wave motion is possible. Planetary waves are large-scale Rossby waves that exist only in a rotating fluid. These waves owe their existence to the variation with latitude of the Coriolis parameter (3.55), which provides a torque on air parcels displaced meridionally similar to the positive restoring force supplied by buoyancy. Like gravity waves, planetary waves propagate horizontally and vertically and can communicate the influence of their forcing far from regions where they are excited.

The simplest model of planetary waves involves two-dimensional, horizontally nondivergent motion of an inviscid fluid with constant density. Under these circumstances, the equations of motion (3.56) reduce to

$$\frac{d\mathbf{v}_h}{dt} + f\mathbf{k} \times \mathbf{v}_h = -\nabla\Phi \tag{3.64a}$$

$$\nabla \cdot \mathbf{v}_h = 0. \tag{3.64b}$$

Applying the curl to (3.64) leads to the conservation principle

$$\frac{d(\zeta + f)}{dt} = 0, \tag{3.65}$$

where $\zeta = \mathbf{k} \cdot (\nabla \times \mathbf{v}_h)$ is the fluid vorticity. The quantity in parentheses in (3.65) represents the "absolute vorticity" of a fluid element, which equals the sum of the *relative vorticity* ζ plus the *planetary vorticity* f, both referring to vertical components. According to (3.65), the absolute vorticity of a fluid element (that seen by an observer in an inertial reference frame) is a conserved property. Air deflected equatorward will experience reduced planetary vorticity and spin up cyclonically ($\Delta\zeta > 0$) to conserve absolute vorticity. Air deflected poleward will react in the opposite sense. Torques exerted in this manner make it possible for an air parcel to cycle back and forth across an equilibrium latitude circle and form the basis for planetary waves.

If it is horizontally nondivergent, the flow may be described in terms of a streamfunction

$$\mathbf{v}_h = \mathbf{k} \times \nabla\psi, \tag{3.66}$$

where $\psi = \frac{1}{f}\Phi$. Linearizing the vorticity equation (3.65) about a constant zonal flow $\overline{u}$ on a midlatitude "beta-plane", wherein $x = \cos\phi_0\lambda$, $y = \frac{\phi}{a}$, and f is linearized about its value at latitude ϕ_0

$$f = f_0 + \beta y \tag{3.67}$$

with $f_0 = f|_{\phi_0}$ and $\beta = \frac{df}{dy}|_{\phi_0}$, leads to the *Rossby wave equation*

$$\frac{D}{Dt}\nabla^2\psi + \beta\frac{\partial\psi'}{\partial x} = 0, \tag{3.68}$$

where $\frac{D}{Dt} = \frac{\partial}{\partial t} + \overline{u}\frac{\partial}{\partial x}$. Then considering solutions of the form

$$\psi' = \mathrm{Re}\left\{\Psi e^{i(kx+ly-\sigma t)}\right\} \tag{3.69}$$

reduces (3.68) to the *Rossby wave dispersion relationship*

$$c - \overline{u} = -\frac{\beta}{k^2 + l^2} \tag{3.70}$$

where $c = \sigma/k$ is the wave phase speed. According to (3.70), Rossby waves propagate westward relative to the background flow, with phase speeds that are inversely proportional to their wavelengths. Energy from stationary waves (e.g., those forced by orographic features at the Earth's surface) propagates only if the background flow is *westerly*.

For motion that is horizontally divergent, absolute vorticity is not conserved. However, a more involved analysis reveals a conservation principle analogous to (3.65)

$$\frac{d}{dt}\left(\frac{\zeta + f}{h}\right) = 0, \tag{3.71}$$

where h measures the thickness between two material surfaces (e.g., between two isentropic surfaces under adiabatic conditions). The quantity $\left(\frac{\zeta+f}{h}\right)$ is the "potential vorticity", which is conserved for individual air parcels under adiabatic conditions. The conservation principle (3.71) accounts for horizontal divergence through the height of material tubes h. As implied by the continuity equation (3.56d), horizontal convergence is accompanied by a vertical stretching of material tubes, which leads to a spin-up of the absolute vorticity to conserve angular momentum. Conversely, horizontal divergence is accompanied by a vertical compression of material tubes and a spin-down of absolute vorticity to conserve angular momentum.

In this framework, planetary waves are described by the linearized counterpart of (3.71). Analysis similar to that above leads to the three-dimensional dispersion relationship

$$c - \overline{u} = -\frac{\beta}{k^2 + l^2 + \left(\frac{f_0^2}{N^2}\right) m^2}, \qquad (3.72)$$

where m is the vertical wavenumber. For stationary waves ($c = 0$), like those excited by flow over elevated terrain, vertical propagation ($m^2 > 0$) is possible only for a restricted range of zonal wind

$$0 < \overline{u} < \frac{\beta}{k_h^2}, \qquad (3.73)$$

where $k_h^2 = k^2 + l^2$. Thus, stationary Rossby waves propagate vertically only for weak westerly flow and even then only for the largest horizontal scales (k_h^2 small). For easterly flow, vertical propagation is excluded for all horizontal scales. Charney and Drazin (1961) used this result to explain the conspicuous absence of synoptic weather systems from the day-to-day circulation of the stratosphere (cf. Figs. 3.8a, 3.10a).

3.9 The general circulation

The differential heating of the Earth discussed in Sec. 3.1.3 must be compensated by a transfer of energy from low latitudes to middle and high latitudes. Net radiation in Fig. 3.7 tends to establish a time-mean thermal structure with isotherms and contours of geopotential parallel to latitude circles. Owing to the Earth's rotation, the steady circulation in equilibrium with this thermal structure is nearly parallel to latitude circles and therefore accomplishes very little heat transfer between the equator and poles. To preserve thermal equilibrium, zonally asymmetric disturbances known as "baroclinic waves", arise through instability of the zonal circulation. These unsteady disturbances, which underlie cyclonic and anticyclonic weather systems, transport heat meridionally by exchanging tropical and polar air. They are maintained by a conversion of zonal-mean potential energy to eddy kinetic energy.

The large-scale circulation of the atmosphere involves three basic forms of energy: (i) internal energy, associated with molecular motions, (ii) potential energy, associated with the distribution of mass, and (iii) kinetic energy, associated with air motions. The first two serve as reservoirs of energy which is imparted to the atmosphere through radiative, conductive, and latent heat transfers. As indicated in Fig.1.2, most of this energy is communicated through radiative transfer from the surface (the greenhouse effect). However, transfers of sensible and latent heat (e.g., through evaporative cooling of the oceans) also contribute significantly.

It may be shown that the internal and potential energies of an air column preserve a fixed ratio (see Irabarne and Godson, 1981). Their sum, equal

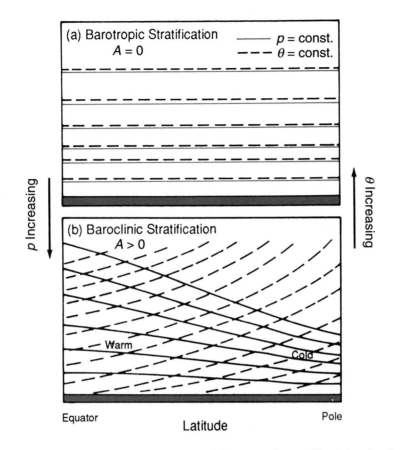

Fig. 3.25 Thermal structure corresponding to (a) barotropic stratification, wherein isobaric surfaces coincide with isentropic surfaces and the available potential energy A is zero and (b) baroclinic stratification, wherein isobaric surfaces do not coincide with isentropic surfaces and A is positive. Reprinted from Salby (1992) by permission of Academic Press.

to the column enthalpy, is referred to as the "total potential energy." Changes in kinetic energy are derived from changes in total potential energy. However, only a small fraction of the total potential energy is actually available to drive atmospheric motions. Known as the "available potential energy", that fraction is determined by the atmospheric stratification.

Consider adiabatic motions in a hydrostatic atmosphere. If isentropic surfaces coincide with isobaric surfaces, as illustrated in Fig. 3.25(a), the atmosphere is said to be *barotropically stratified*. Since motions take place adiabatically, $d\theta = 0$ for individual air parcels. Hence, an air parcel initially on an isentropic surface must remain on that surface. Under barotropic stratification, $p = p(\theta)$, so no pressure gradient exists along the isentropic surface to drive atmospheric motion. Consequently, no potential energy is available for conversion to kinetic energy.

If isentropic surfaces do not coincide with isobaric surfaces, as illustrated in Fig. 3.25(b), the atmosphere is said to be *baroclinically stratified*. Adiabatic displacements of an air parcel must still occur along an isentropic surface, but a pressure gradient now exists along that surface to drive such motion. Consequently, baroclinic stratification permits a conversion of total potential energy to kinetic energy. This form of stratification also provides a source of vorticity, e.g., as is inherent to cyclonic weather systems (see Holton, 1979). Since individual air parcels conserve θ, the adiabatic redistribution of mass that results tends to mix potential temperature horizontally and drive isentropic surfaces into coincidence with isobaric surfaces. When this state has been achieved, the atmosphere is barotropically stratified and no further potential energy is available to drive air motions. The difference between the total potential energy when the atmosphere is baroclinically stratified and that when it has been driven into barotropic stratification through an adiabatic redistribution of mass defines the available potential energy (APE).

Available potential energy is created by nonuniform heating of the atmosphere. Radiative heating at low latitudes (Fig. 3.7) increases θ and tends to lower isentropic surfaces (since θ must increase with altitude for hydrostatic stability). Radiative cooling at middle and high latitudes decreases θ and tends to raise isentropic surfaces. Hence, nonuniform heating tilts isentropic surfaces downward towards the equator, which maintains baroclinic stratification and continually produces APE. According to thermal wind balance (3.61), the poleward temperature gradient that results is accompanied by a westerly shear with altitude of the zonal flow. When the temperature contrast becomes sufficiently great, the zonal flow is "baroclinically unstable."

The synoptic weather systems evident in Fig. 3.8(a) originate through baroclinic instability. Baroclinically unstable waves develop and intensify along the jet stream, which is marked by the sharp gradient of geopotential in Fig. 3.8. According to the hypsometric relation (3.34), this sharp change of Φ coincides with a strong gradient of tropospheric temperature, the latter being associated with APE of the circulation. Synoptic disturbances evident in Fig. 3.8(a) and Plate 7(a) extract APE from the zonal circulation through sloping convection, which transports heat meridionally by exchanging cold polar air with warm tropical air. By mixing air horizontally, these eddies tend to make the conserved quantity θ horizontally uniform, driving the circulation towards a state of barotropic stratification and zero APE.

3.10 The upper atmosphere

The synoptic weather systems that are responsible for daily fluctuations in the troposphere deflect the tropopause and overlying air in the lower stratosphere. As is evident in Plate 4, accompanying changes in pressure along isentropic surfaces can result in considerable changes in ozone column abundance through (3.5) and (3.6). Beyond these influences, synoptic weather systems play a relatively minor role in the upper atmosphere. According to (3.73), synoptic-scale disturbances are excluded from the stratosphere and mesosphere by strong westerlies in the winter hemisphere and by easterlies in the summer hemisphere (Fig. 3.4), which limit the vertical propagation of Rossby waves. As a result, the wintertime circulation at 10 mb (Fig. 3.10a) is much smoother than the corresponding circulation at 500 mb (Fig. 3.8a), while the summertime circulation tends to be almost zonally symmetric. The absence of horizontal mixing by baroclinic eddies and vertical mixing by convection favors a thermal structure that is closer to radiative equilibrium than that of the troposphere. However, the wintertime circulation is still disturbed from radiative equilibrium by planetary waves that propagate upwards out of the troposphere and force the middle atmosphere mechanically. Global-scale disturbances like that in Fig. 3.10(a) play a key role in establishing the time-mean circulation and distributions of constituents like ozone.

The radiative energy budget of the middle atmosphere is determined in large part by absorption of solar UV by ozone and IR emission to space by CO_2. Thus, ozone heating is responsible for much of the thermal structure of the middle atmosphere and the attending circulation. The simplest treatment of ozone photochemistry, due to Chapman (1930), considers a pure oxygen atmosphere. As discussed in Chapter 7, Chapman chemistry is reasonably successful at predicting the observed vertical distribution of ozone (e.g., Craig, 1965). However, pure oxygen chemistry overestimates observed ozone mixing ratios, which are influenced by other species; see Sec. 7.3.3. More importantly, photochemistry alone predicts the greatest ozone abundance at low latitudes, where the UV flux is largest and where most stratospheric ozone is produced. In reality, column abundances are largest at middle and high latitudes (see Plate 4). This fundamental discrepancy can be explained only by incorporating dynamical influences.

The historical explanation for the observed distribution of Σ_{O_3} is a meridional circulation that transports air poleward and downward from the chemical source region in the tropical middle stratosphere. This meridional overturning of air is referred to as the *Brewer–Dobson circulation;*[9] see Fig. 15.3. Air is transferred between the equator and poles by this circulation on a time scale of months, which is indicative of the strong control by the Coriolis force that deflects the air stream zonally and inhibits meridional motions. In reality, the meridional circulation comes about through air motions that are more complex (e.g., Salby and Garcia, 1990).

It can be shown that net transport is accomplished only if the circulation is driven out of radiative equilibrium (WMO, 1986). Figure 3.26 displays

---

[9] Proposed by Dobson (1930) and Brewer (1949) to explain tracer measurements.

Radiative Equilibrium Temperature

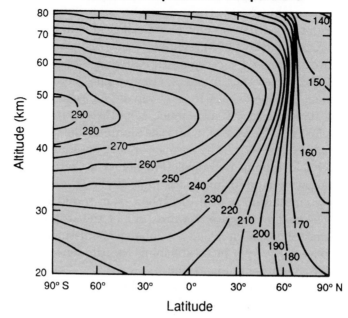

Fig. 3.26 Radiative equilibrium temperature of the middle atmosphere during solstice, as calculated in a radiative–convective–photochemical model. After Fels (1985). Reprinted by permission of Academic Press.

the radiative equilibrium temperature for the middle atmosphere. The winter hemisphere is marked by very cold temperatures in the polar night which result from IR cooling to space and which are associated with wind speeds in excess of 200 m s^{-1}. Observed temperatures in Fig. 3.3 are some 50 K warmer and zonal-mean winds (Fig. 3.4) greater than 100 m s^{-1} are rare. On the other hand, radiative equilibrium temperatures in the summer hemisphere are closer to observed values, as are the corresponding winds.

The ingredients responsible for maintaining the polar-night vortex warmer and weaker than that predicted by radiative equilibrium are planetary waves. Generated near the Earth's surface, these global-scale disturbances satisfy the Charney–Drazin criterion for vertical propagation (3.73) and radiate freely into the winter stratosphere, where they disrupt the zonal circulation; cf. Figs. 3.10(a),(b). As they propagate vertically in the winter stratosphere, planetary waves amplify due to the exponential decrease with altitude of air density. By 10 mb, meridional excursions of air are large enough to displace the airstream significantly. Air parcels in the disturbed circulation cycle back and forth between one radiative environment and another, so they spend less time in the polar night and have less opportunity to cool radiatively. Because planetary waves are excluded by easterlies, the summer circulation is much less disturbed and closer to radiative equilibrium.

During winter, planetary waves amplify sporadically and introduce dramatic changes in the circulation and the distribution of ozone. Referred to as *sudden stratospheric warmings*, these episodes are marked by an abrupt increase of temperature over the polar cap, e.g., more than 50 K in just a few of days. During a sudden warming, the polar-night vortex is displaced and distorted. At the same time, the zonal-mean temperature gradient reverses direction, i.e., from poleward to equatorward. According to thermal wind balance (3.61), the latter is attended by a reversal of the zonal-mean circulation. The dramatic sequence of events described above actually involves motions which are zonally asymmetric and which are accompanied by a large exchange of air between low latitudes and high latitudes, as is implied in Fig. 3.10(a).

These horizontal air motions can be visualized through the distribution of potential vorticity, which is conserved on time scales shorter than a week. Plate 8 shows the potential vorticity distribution near 10 mb during a stratospheric warming on the same day as shown in Fig. 3.10(a). Amplification of wavenumber 1 has displaced the polar-night vortex (air of high potential vorticity) out of zonal symmetry and drawn midlatitude air poleward in just a day or two. These air motions are associated with a horizontal redistribution of ozone, which, in combination with vertical displacements along isentropic surfaces, can result in a marked increase of Σ_{O_3} over the polar cap (Salby and Garcia, 1990); cf. Plate 4(a). Motions evident in Fig. 3.10(a) and Plate 8 are also associated with irreversible mixing of midlatitude and polar air (McIntyre and Palmer, 1983) and lead to net radiative heating that drives the slow meridional circulation described above. As noted earlier, ozone figures centrally in the radiation budget of the middle atmosphere, which determines the thermal structure and ultimately the circulation. Consequently, radiation, dynamics, and chemistry are all coupled in the recovery of the stratosphere to undisturbed conditions. An improved understanding of how these physical processes interact to establish the time-mean state is a major focus of middle atmosphere research.

4

The ocean circulation

Pearn P. Niiler

4.1 Introduction

The principal role of the oceans in maintaining the present climate system of the Earth is the creation of large reservoirs of heat in tropical latitudes and the transport of this thermal energy to the polar latitudes. Sea ice forms an insulating upper barrier to the water below and a reflecting bottom surface to the atmosphere above. This chapter introduces the fundamental observations and several theories of how the oceans circulate on basin-wide spatial scales and decadal time scales, and discusses the processes by which heat is stored in the oceans and how it is transported by these large-scale ocean circulation patterns. It begins with the wind-driven circulation in the surface layers and ends with thermally driven deep water mass movements. Several properties of ocean water masses which control the growth and decay of sea ice are introduced.

The chapter is meant to serve as an introduction to one specific area of physical oceanography in terms of physical concepts: the role of the oceans in transporting thermal energy. It is to be read together with basic textbooks in physical oceanography and fluid mechanics, such as Pond and Pickard (1978) and Gill (1982).

4.2 Basic concepts

The oceans cover 71% of the Earth's surface (see table on Physical constants for general statistics), and their water masses exert a profound influence on the habitability of the globe. The continents divide the oceans into the Atlantic, Pacific, Indian and Southern basins, with further segregation imposed by mid-ocean ridges which rise a few thousand meters from the ocean floor. The patterns of surface winds and heat and moisture exchanges with the atmosphere cause the water to circulate in this interconnected, rotating domain and determine its physical structure.

The oceans are the Earth's principal time-varying reservoirs of thermal energy and moisture. The atmosphere is a much smaller reservoir because there is a thousandfold difference in the density between tropospheric atmosphere and sea water. The month-to-month changeable heat carrying capacity of only 3.2 m of the ocean is equivalent to that of the whole atmosphere. Significant amounts of thermal energy are not stored in land because its conductivity is a thousandfold smaller than the effective turbulent conductivity of the ocean arising from mixing and convection. Thus, only shallow layers of land are important in storing heat while the ocean extends to an average depth of ∼3,800 m. If the temperature of the lower layers of the atmosphere over the oceans depart from the surface temperature of the oceans, there is a vigorous exchange of heat and moisture between the two fluids, and the temperature of the atmosphere, because of its low heat capacity, adjusts to the ocean temperature. Thus, the local changes of air temperature and moisture, or what humans consider as "climate changes" over a large part of the Earth's surface, are strongly influenced by the changing ocean temperature.

The rest of the climate system, and the atmosphere in particular, is influenced by the ocean through changes in sea surface temperature (SST). Several processes whereby the land surface is influenced by SSTs are discussed in Chapters 17 and 18. Mean SSTs (Fig. 4.1) seldom exceed 30°C and the warmest water is typically found in the tropical western Pacific. At high latitudes, where sea water freezes at −1.8°C, sea ice is found but varies greatly in extent with the annual cycle. It is therefore very important for the climate system to not only model the evolution of SST, but also to be able to understand the physical processes involved. It is readily apparent that SSTs are greatly influenced by the temperature and circulation of the entire water column of the oceans.

That the temperature distribution in the oceans is critically dependent on its circulation patterns is most vividly demonstrated by the centuries old observation that the light-filled warm surface layers of the subpolar oceans are separated from the dark, cold deeper layers by a main thermocline which lies between 100 m to 1,000 m depth. Such a vertical distribution of temperature cannot be maintained in a conducting water column over many centuries unless a volume flux of cold water is supplied to the lower layers and a net heating, or a volume flux of warm water, is supplied to the surface layers. Furthermore, conservation of heat in stratified basins which have practically an insulating bottom implies that, since there is a heat gain from the atmosphere near the surface in the tropical and subtropical areas, there must be an approximately equivalent transfer of heat back to the atmosphere in subpolar and polar latitudes. The influence of the oceans on climate, must therefore consider the ocean circulation which transports thermal energy from places where it is absorbed at the ocean surface to places where it is given back to the atmosphere. Month to year time-scale changes of the ocean temperature can occur over the entire water depth when net surface heat flux distributions are disturbed or when the ocean

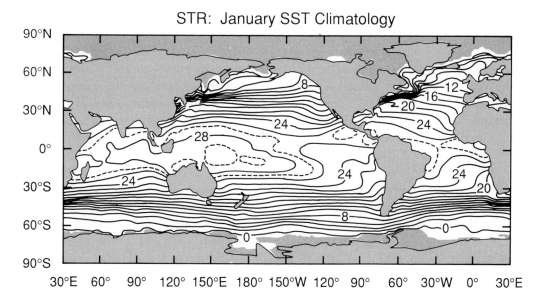

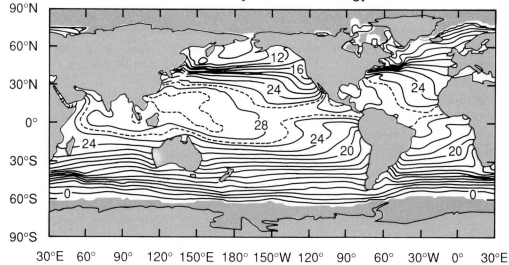

Fig. 4.1 Monthly mean SSTs for January and July. Shaded areas indicate sea ice. The contour interval is 2°C except for dashed contours of 27 and 29°C. From Shea et al. (1990).

circulation changes. These are the topics addressed by climate modeling of the oceans.

As a prelude to hydrodynamic models of the ocean circulation and temperature patterns, it is necessary to gain a perspective on the observations and dynamical causes of those features which participate significantly in the storage rate and transport of heat in the oceans. Scales of motion of a few centimeters transfer heat vertically through the stratified water column and are just as important as the scales of the entire basins, by which heat is transported horizontally by powerful ocean current systems. There are very few direct measurements of the horizontal circulation and therefore approximate dynamical models are used for constructing hypothetical circulation patterns. The currents are of the magnitude of 0.01–3 m s^{-1}. These dynamically computed patterns might well be verified by more extensive measurements in the future. The large-scale vertical component of circulation in the oceans is two to three orders of magnitude smaller than horizontal flow, and is so small that it cannot be measured directly. Accordingly, models are always used for its evaluation. An evaluation of both horizontal and vertical components of flow is needed for understanding and modeling the role of the oceans in the climate system.

The conclusion drawn from the combination of measurements and dynamical principles is that the oceans transport significant amounts of the thermal energy from the equatorial areas toward the poles. The mode of that transport in the Northern Hemisphere (NH) is warm, near surface currents flowing to the north, and deep, cold currents returning the water molecules back south. The water which is cooled in subpolar and polar latitudes sinks by thermal convection. In areas where convective motions occur, moisture exchange with the atmosphere and the salinity of the water plays an important role. In the tropical and subtropical latitudes water rises principally where wind forces a surface divergence.

Hydrodynamic modeling of this "meridional cell" in the oceans is at a very primitive stage. Realistic calculations of the heat transporting horizontal and vertical current patterns for large interconnected areas of the globe are needed. Ocean current strength and its heat transporting capability depends upon the small-scale turbulence, convectively activated plumes and flow through and over narrow passages. These physical processes are not very realistically represented in the ocean models of global scale at the present stage of computing power or conceptual parameterization. Because circulation depends on small-scale turbulence, its equilibrium time is a few centuries, and perhaps the oceans never come to equilibrium. This precludes modeling several aspects of equilibrium states of the deep waters of the oceans. Observational verification of the basin-wide heat transporting mode of the oceans has not been done because large-spatial-scale direct current measurements of the oceans have not been attempted.

4.3   Upper ocean circulation observations

The distribution of surface circulation of the oceans has been obtained from the drift of vessels near the prevailing sailing and trade routes of the world. Ship drift is computed from the difference in the position of a ship where dead reckoning and leeway would have located it and the actual position determined from a navigational aid. Traditionally, ship logs document the ship drift to within 1/2 knot (about 0.25 m s^{-1}) and assign a direction to this drift, often to 15 degrees of a compass circle. Various admiralties have collected these data through the centuries, and ship drift charts are prepared from them. A recent complete compilation by Richardson (1989) shows surface drift patterns that, at casual inspection, follow the sense of the circulation of the lower atmosphere (Fig. 4.2).

For example, in the North Pacific there is a drift current to the west in the area of the northeast trades and an eastward drift current in the latitudes of the westerlies. Two notable exceptions to this pattern occur, and these observations lie at the heart of every dynamically correct model of ocean circulation. First, the ocean current patterns are strongly intensified to the western side of every basin, regardless of the hemisphere; and secondly, in the northern tropical oceans, surface countercurrents flow against the winds in the vicinity of the atmospheric intertropical convergence zones.

The strongest persistent ship drifts are 3–4 knots (1.5–2.0 m s^{-1}) off the east coast of South Africa in the Agulhas Current. Over most of the ocean surface area the circulation is not well known because ship drift estimates are not accurate enough to discern the very slow mean flows (0.02–0.10 m s^{-1}). Also, the drafts of the ships vary and their leeway is not well known, so the drift is a not well determined vertical average of the near-surface ocean currents and the winds in the rigging. It should also be noted that very few ship drift observations exist in the circumpolar oceans where merchant vessels rarely sail. Because of the resolution which is achieved by estimating drift every day, these ship drift charts combine the space and time variations of currents rendering western boundary currents weaker and broader than they are synoptically. Geographers have given names to the major current systems, which oceanographers still use today and these are presented in Fig. 4.3.

Current meters and calibrated radio-tracked drifters are the most accurate techniques by which measurements of the near surface flow distributions are obtained. The analysis of the time variations of these measurements reveals that there is a locally wind-related flow and its daily, vector-averaged direction a few meters below the surface is very nearly to the right of the time variable surface wind in the NH. Second, these wind-related flows are confined to a layer of water which has vertically uniform temperature, the "mixed layer" (Fig. 4.4). Third, near the surface there are flows which are not directly related to the local winds and are associated with measurable sea level slopes. The major western boundary currents, which appear in the ship drift charts (Fig. 4.2), are flows in which sea level changes by 1.0–1.5 m over horizontal distances of less than 100 km. Thus, a satellite borne altimeter can also be used to measure sea level slopes caused by strong ocean currents

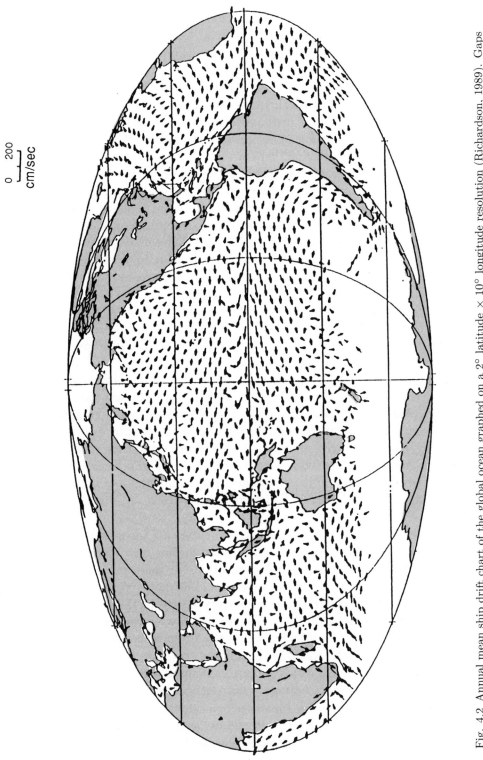

Fig. 4.2 Annual mean ship drift chart of the global ocean graphed on a 2° latitude × 10° longitude resolution (Richardson, 1989). Gaps mean there is no data.

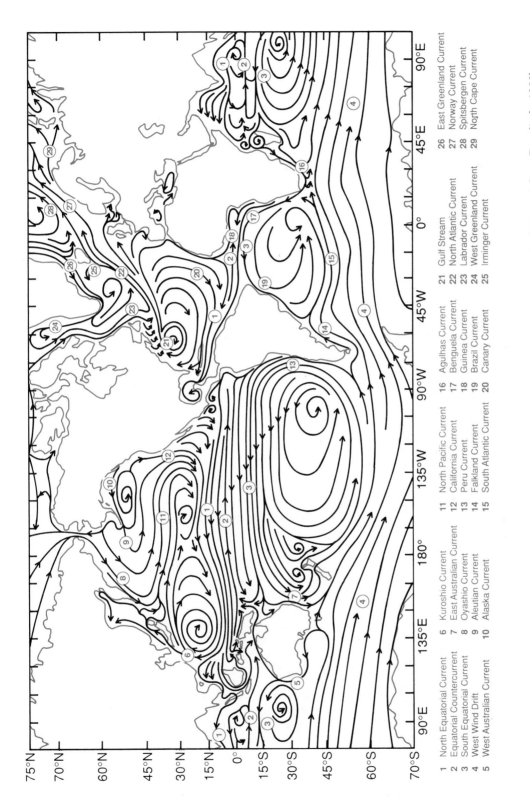

1 North Equatorial Current | 6 Kuroshio Current | 11 North Pacific Current | 16 East Greenland Current | 21 Gulf Stream | 26 East Greenland Current
2 Equatorial Countercurrent | 7 East Australian Current | 12 California Current | 17 North Atlantic Current | 22 Benguela Current | 27 Norway Current
3 South Equatorial Current | 8 South Equatorial Current | 13 Peru Current | 18 Guinea Current | 23 Labrador Current | 28 Spitsbergen Current
4 West Wind Drift | 9 Oyashio Current | 14 Falkland Current | 19 Brazil Current | 24 West Greenland Current | 29 North Cape Current
5 West Australian Current | 10 Alaska Current | 15 South Atlantic Current | 20 Canary Current | 25 Irminger Current

Fig. 4.3 Major traditionally named ocean current systems of the world (Northern Hemisphere winter). (Adapted from Bowditch, 1966).

123

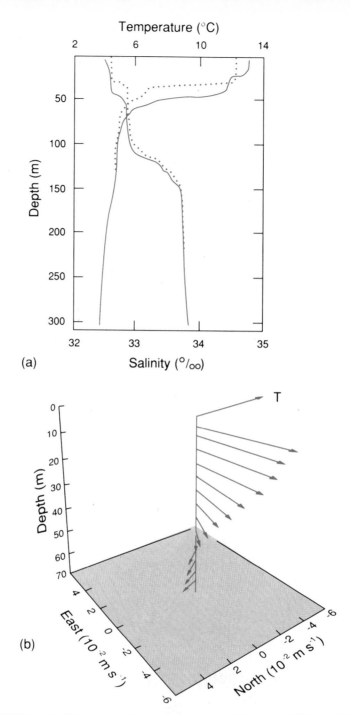

Fig. 4.4 (a) Vertical distribution of temperature and salinity at 50°N., 145°W. in early September, 1977. The solid lines are before a storm and the dotted lines are after a storm, which depict the vertical mixing above the seasonal thermocline. The main thermocline, or pycnocline in this area is between 110 m and 160 m depth. (b) Time-averaged velocity for a 25 day summer period at an open ocean site southwest of Bermuda. Current meter measured velocity is referenced to 70 m. The topmost dashed vector is the time-averaged wind stress (Price et al., 1986).

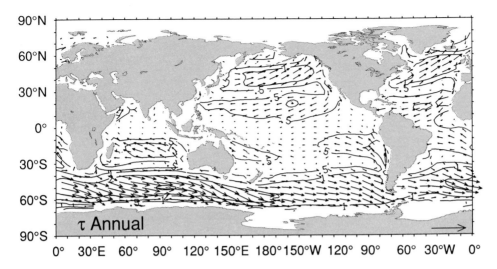

Fig. 4.5 Annual mean surface wind stress over the oceans depicted as vectors. The arrow at lower right corresponds to 5 dyn cm^{-2} and contours of magnitude 0.5, 1, 2, and 3 dyn cm^{-2} are plotted. From Trenberth et al. (1990).

and these data can be used to study their spatial and temporal variabilities (Tai, 1990). In fact, most of the deep circulation patterns have been mapped by measuring the pressure variations or density variations which are caused by these currents, which is commonly known as the "geostrophic method" (see Sec. 3.7.2).

It is evident from ship drift data and marine meteorological observations that air moves across the ocean faster than the strongest oceanic surface flows, and there is a momentum transfer from the atmosphere to the ocean. This imposes a surface wind stress on the ocean (Fig. 4.5). The annual mean wind stress is strongest over the southern oceans. The wind stress is strongest in winter and undergoes a distinctive annual cycle (see Trenberth et al., 1990). It is apparent that *winds drive ocean currents*. But from the surface drift charts and the surface wind charts, it is also obvious that no simple, single relationship can be found between atmospheric momentum input and the ship drift in the oceans.

4.4 Forces which maintain the surface currents

The density of ocean water is influenced mainly by temperature and salinity. However, density variations within the ocean are small and, for many purposes, a constant value ρ_o can be assigned. This approximation is termed the "Boussinesq approximation", and it recognizes that it is possible for practical purposes to neglect density variation effects on mass conservation but that the influences on weight and buoyancy should be retained, see Chapter 11 [and Eqs. (4.3) and (4.7)] for more detail.

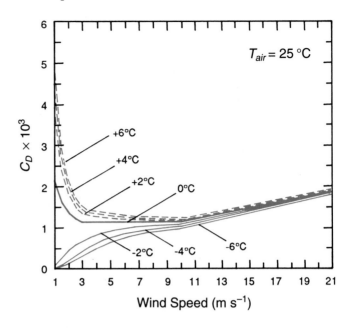

Fig. 4.6 The wind stress is computed from the wind speed, V, according to the formula $\tau = \rho_a C_D V \mathbf{v}$ and it is in the direction of the wind. Shown is the drag coefficient $(\times 10^3)$, C_D , as a function of wind speed and atmospheric stability, as measured by the air–sea temperature differences, based on Large and Pond (1981), as given by Trenberth et al. (1989). Values are for an air temperature of 25°C and dashed lines indicate air less than sea temperatures (unstable).

The momentum flux, or horizontal stress, from the wind on the ocean (Fig. 4.5) is caused by small horizontal vertical and temporal scales of turbulent motions in the atmosphere. These three-dimensional motions can be measured directly and the vertical flux of horizontal momentum into the ocean from the atmosphere can be computed directly. It is found that this momentum flux into, or stress on, the water is roughly proportional to the square of the wind speed and is in the direction of the wind (Fig. 4.6). It is reduced if the vertical stratification in the air column near the ocean surface increases. This flux is transported vertically into the water until the strong stratification below the mixed layer inhibits its further penetration. Thus, the momentum from the wind is stored in the mixed layer (Davis et al., 1981) and a net turbulent force is produced on the mixed layer column, proportional to the rate at which it is stored.

When the turbulent vertical convergence of horizontal momentum is balanced by the Coriolis force arising from the currents, the layer is said to be in an "Ekman layer" balance (Ekman, 1905). The equations for the turbulent Ekman layer are,

$$f\mathbf{k} \times \mathbf{v} = \frac{\partial}{\partial z}(\tau/\rho_o) \qquad (4.1)$$

where f is the Coriolis parameter (see Eq. 3.55), $\mathbf{v}$ is the horizontal velocity, and $\mathbf{k}$ is unit normal in z direction. Here τ is the turbulent vertical transport rate of horizontal momentum, or stress, in water, whose value on the surface, τ_s, is the wind stress. Additional equations are developed which relate the turbulent stress in the water to the wind, the water motion and the stratification variabilities (Kantha et al., 1989).

If a vertical integral of (4.1) is taken from a level where the stress is much reduced by stratification to the ocean surface, where the stress is the wind stress, a wind-driven transport per unit meter of water is seen to occur to the right of the surface winds in the NH. It is proportional to the wind stress and inversely proportional to the Coriolis parameter:

$$< \mathbf{k} \times \mathbf{v} >= \frac{1}{f\rho_o}\tau_s \qquad (4.2a)$$

or, alternatively

$$< \mathbf{v} >= \frac{1}{f\rho_o}\tau_s \times \mathbf{k} \qquad (4.2b)$$

where $< \; >$ is the vertical integral over the turbulent Ekman layer depth. This propensity is independent of how the stress is related to the details of the turbulent motions in the Ekman layer. In the theory of Ekman layer flows, it is found that the Coriolis force, as well as vertical stratification, restricts the vertical momentum transfer to a finite thickness layer. In a model of the upper ocean with constant density and constant vertical diffusivity of momentum, ν, the depth of the Ekman layer is proportional to $(\nu/f)^{1/2}$.

The observational fact that wind-driven surface transports are forced to the right of the wind in the NH, implies that there must be a poleward ocean flow component under the trades and an equatorward component under the westerlies. Dynamically, the atmospheric stresses on the ocean (Fig. 4.5) force the surface layer of water together in the subtropics, and pull it apart in the subpolar areas and on the equator (due to the changing in sign of the Coriolis parameter). The kinematical effect of this mixed layer motion is that there must be a downward vertical motion at the base of the mixed layer, w_E, in the subtropics and an upward vertical motion in the subpolar areas and on the equator (Fig. 4.7). This equatorial upwelling causes the low SSTs in the eastern Atlantic and Pacific along the equator (Fig. 4.1).

The equation for mass conservation in the ocean is

$$\rho_o\left(\nabla \cdot \mathbf{v} + \frac{\partial w}{\partial z}\right) = 0. \qquad (4.3)$$

where ∇ is the two-dimensional Laplacian operator. w_E at the base of the Ekman layer can be computed from the vertical integral of (4.3)

$$w_E = \nabla \cdot < \mathbf{v} > \qquad (4.4)$$

and can be related to the surface wind stress by inserting the horizontal

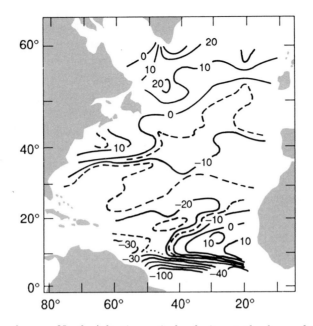

Fig. 4.7 Annual mean North Atlantic vertical velocity at the base of the Ekman layer. Positive is upward and the units are in 10^{-8} cm s^{-1}. Note that off South America the contours between -60 and -100 are left out because of crowding (Leetmaa and Bunker, 1978).

transports from (4.2)

$$w_E = \mathbf{k} \cdot \nabla \times (\boldsymbol{\tau}_s/\rho_o f) \,. \tag{4.5}$$

Consider now the dynamical consequences of this vertical circulation on the deeper water column.

4.5 Thermocline circulation and western boundary currents

The most direct effect of the atmosphere on the deep oceans is to force vertical effluxes of mass from the mixed layers on the space scales of the atmospheric wind patterns. Where the turbulent atmospheric boundary layer is divergent, the oceanic turbulent boundary layer is forced to be convergent. The most direct dynamical effect of the large-spatial-scale oceanic downwelling can best be understood in terms of the equations which govern steady, large-scale balance of momentum and mass in the main thermocline. Here, because of the persistent strong stratifications, vertical turbulent transfers of momentum are expected to be weak. The motions are essentially geostrophic

$$f\mathbf{k} \times \mathbf{v} = -\frac{1}{\rho_o}\nabla p \tag{4.6}$$

128

and hydrostatic

$$\frac{1}{\rho_o}\frac{\partial p}{\partial z} = -\frac{\rho}{\rho_o}\,g\,. \tag{4.7}$$

Eliminating pressure by taking the vertical component of the curl of (4.6) and using (4.3), results in the equation for the vertical component of vorticity in the water column below the Ekman layers,

$$\beta v - f\frac{\partial w}{\partial z} = 0 \tag{4.8}$$

where $\left(\beta = \frac{1}{a}\frac{\partial f}{\partial \phi}\right)$ is the variation of Coriolis parameter with latitude. Now integrate (4.8) vertically from the base of the mixed layer, where $w = w_E$, to below the main thermocline to a depth $z = -H$ where $w = w_{-H}$. Recall that there is a flux of water being forced out of the mixed layer by the wind. The result is,

$$\beta[v] = f\mathbf{k}\cdot\nabla\times(\tau_s/\rho_o f) - fw_{-H} \tag{4.9}$$

where [ ] is the vertical integral from the base of the Ekman layer to $z = -H$. Because w_{-H} is much smaller than w_E it can generally be neglected when considering transports in the thermocline.

If there is a surface sinking of water, as occurs under the trade winds, there must be a geostrophically balanced flow of water to the south in the NH. Adding the meridional geostrophic transport from (4.9) and the Ekman transport from the meridional component of (4.2) gives the "Sverdrup equation"

$$\beta\{v\} = \mathbf{k}\cdot\nabla\times(\tau_s/\rho_o) \tag{4.10}$$

where { } is the integral from the ocean surface to $z = -H$. If this north–south flow is summed from the east coast of a basin, the wind driven "Sverdrup transport" results. In Chapter 11, Sec. 11.4, these concepts are applied to a simple wind-driven model of the ocean.

In the North Atlantic, from the coast of West Africa to the coast of Florida, about 30×10^6 m^3 s^{-1} flows south, and this is about ten times larger than the directly wind-driven Ekman transport in the mixed layer to the north. The immediate consequence of such a basin-wide subsurface flow to the south via (4.5), is that there must be a deep horizontal pressure gradient below the main thermocline.

If, furthermore, this southward flow decreases with depth, the main thermocline is forced to become deeper on the western side of the ocean. By taking the vertical derivative of (4.6) and using (4.7) an expression for the east–west slope of the ocean density can be given

$$-\frac{g}{a\cos\phi}\frac{\partial(\rho/\rho_o)}{\partial\lambda} = f\frac{\partial v}{\partial z}\,. \tag{4.11}$$

A deepening of the main thermocline from east to west is observed. Its

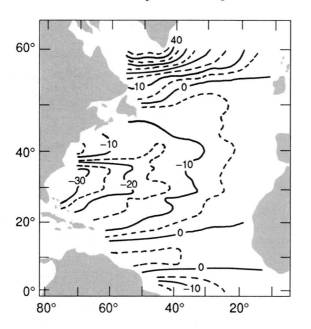

Fig. 4.8 Contours of the annual mean North Atlantic Sverdrup volume transport. Units are 10^6 m s^{-2} (Leetmaa and Bunker, 1978).

east–west slope is very close to that predicted by (4.11) if v goes to zero in a linear fashion at 700 m and the total transport is given by (4.10) (Stommel et al., 1978).

This massive wind-driven flow to the south above the main thermocline must recirculate to the north approximately above the depth of the thermocline on the western side of the ocean, giving birth to the swift western boundary current (Stommel, 1948). The volume transport of the Florida Current at the latitude of 27°N has been measured with transport dropsondes to be 30×10^6 m^3 s^{-1} to the north, which is very close to that predicted by the Sverdrup transport calculation (Figs. 4.8 and 4.9; Stommel et al., 1978; see Trenberth et al. (1990) for recent global estimates). As nearly half of the 24°N Sverdrup transport forms an Antilles Current almost half of the Florida transport comes from the South Atlantic (Schmitz et al., 1992). The northeastward transport of the Gulf Stream increases rapidly along the Carolinas and it reaches a maximum value of 180×10^6 m^3 s^{-1} at 60°W. The increase of transport north of the Straits of Florida is thought to be caused by the rectification of mesoscale eddies which form in the western North Atlantic and drive an intensive mass recirculation in several subbasin-scale gyres there (Hogg, 1991).

The evaluation of (4.9) using the observed wind stress (Fig. 4.5) in the area of the intertropical convergence zones of the atmosphere requires a northward transport, imbedded between two southward moving areas to the north and south. The only way mass can be provided for this northward moving flow, because w_{-H} is thought to be small, is from the west, giving rise to the tropical countercurrents (Sverdrup, 1947).

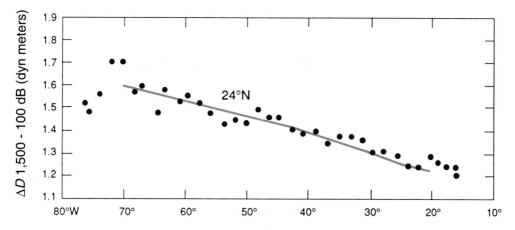

Fig. 4.9 Dynamic height (p/ρ_o) (where $H = 1,500$ m) and the base of the Ekman layer is at 100 m, plotted for individual hydrographic stations vs. longitude along 24°N in the Atlantic. The light line displays the dynamic height computed from the curl of the wind stress forcing of the geostrophic transport per unit distance (Eq. 4.9) (after Leetmaa et al., 1977).

The existence of the western boundary currents and the equatorial countercurrents is the prima facie evidence that vorticity dynamics of ocean circulation patterns, as forced by the winds, principally governs the mass transport above the main thermocline and, thus, also the heat transport patterns there. Climate changes which produce significant changes in the wind stresses over the oceans quickly change the heat transport patterns of the upper oceans (Anderson and Gill, 1975).

The first reality of climate modeling of the oceans is that realistic wind models have to be used to drive the oceans. The second reality is to seek an understanding of how deep circulation patterns are maintained, and that leads to consideration of the heating and cooling of the oceans.

4.6 How oceans exchange heat with the atmosphere

Visible solar radiation that penetrates through the cloud layers reaches the ocean surface during the daylight hours and is the principal source of thermal energy for the oceans. Over 90% of this flux, which can reach peak values of up to 1,300 W m^{-2} at summertime high noon, is absorbed by the sea water. About 50% is absorbed in the upper 5 m of the sea, and the remaining part penetrates to a depth dependent on the clarity of the water, with 99% absorption by 183 m depth in pure sea water. The biota and the turbidity of the sea water determine the absorption characteristics (Kraus, 1977). By mid-afternoon on calm and sunny tropical days the temperature of the 3–5 m layers near the surface are warmed by this direct absorption process by 2–3°C (Taft and McPhaden, 1988; Cornillon and Stramma, 1985; Price et al., 1986). On windy days the absorbed solar flux is transported by

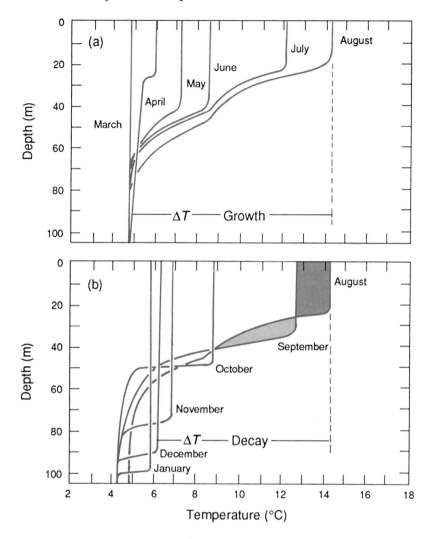

Fig. 4.10 The development of the seasonal thermocline in the eastern North Pacific in the vicinity of 50°N, 145°W. The warming of SST from March to August is due to progressively warmer layers being formed during net heating of the ocean. The cooling from August to September is due to vertical mixing, because equal volumes of cold and warm water are exchanged vertically. Cooling of this area sets in from October onward (Tully and Giovando, 1963).

turbulent motions vertically throughout the depth of the mixed layer and the entire mixed layer is warmed. Through the late spring and summer, as progressively warmer layers are laid down each day, the stratification below the mixed layer increases and wind mixing of the heat from the sun becomes progressively shallower. In this manner, a seasonal thermocline is built up in the upper 30–50 m of the ocean each summer (Fig. 4.10).

The ocean surface is radiating thermal energy much like a blackbody in the infrared band at a rate proportional to the fourth power of SST. Some of this infrared radiation is absorbed by the atmosphere and is re-radiated downward to the ocean at a rate which depends upon the concentrations of moisture and greenhouse gas constituents in the air (see Chapter 3). Through the moist tropical atmosphere only about 10% of the ocean's blackbody radiation escapes to space, while in the dry, cold winter polar nights, the atmosphere does not form such an effective barrier (Fung et al., 1984). In general, longwave infrared radiation cools the ocean.

The air a few meters above the ocean usually has a lower vapor pressure and is at a lower temperature than the air in direct contact with the water molecules at the sea surface. In the absence of air or water motion, there would be a slow molecular diffusion of water vapor and thermal energy from the ocean to the atmosphere, but, in the presence of the vigorous turbulent motions near the wavy sea surface, the evaporative and conductive, or sensible, transport processes are greatly enhanced. With evaporation, there is cooling of the sea surface, the salt concentration of the water increases and its buoyancy decreases. Like the momentum transport, the turbulent transport rates of moisture and heat increase with wind speed and decrease with air boundary layer stability. The larger the air temperature and SST differences are, the larger these turbulent heat transport processes become (Kraus, 1977). Sea fog forms if the air dew point is lower than that of the water in contact with the ocean surface.

In summary, there are four components of heat exchange between the oceans and the atmosphere: shortwave radiative flux which heats the ocean, longwave infrared flux which cools the ocean, and turbulent fluxes of moisture and heat which generally cool the ocean. Practical formulae have been developed for estimating these fluxes from marine weather observations (Liu et al., 1979; Simpson and Paulson, 1979) and the net value can be computed to an accuracy of ± 40 W m^{-2} (Talley, 1984). This error is due to the uncertainty of the empirically determined flux component formulae, the accuracy of the marine weather observations and the data that are available from which to form seasonal means. If this error in net flux is distributed over a mixed layer depth of 50 m, the accuracy of predicting SST change over a month is $\pm 0.6°$C. From these formulae it is found that in autumn and winter, the turbulent and longwave fluxes extract most of the heat that was stored in the seasonal thermocline in summer (Tully and Giovando, 1963).

The seasonal range of net heating is comparable in magnitude to the climatological mean value in western boundary current areas and selected spots of the polar oceans, which are cooled, and the tropical and coastal upwelling areas, which are heated (Fig. 4.11).

In polar oceans, sea ice forms in winter and a greater part melts in the summer (see Fig. 4.12). On the Siberian side of the Arctic, sea ice is a permanent feature. Sea ice generally reduces the heat flux because it

Net Annual Heat Gain by the Ocean W m^{-2}

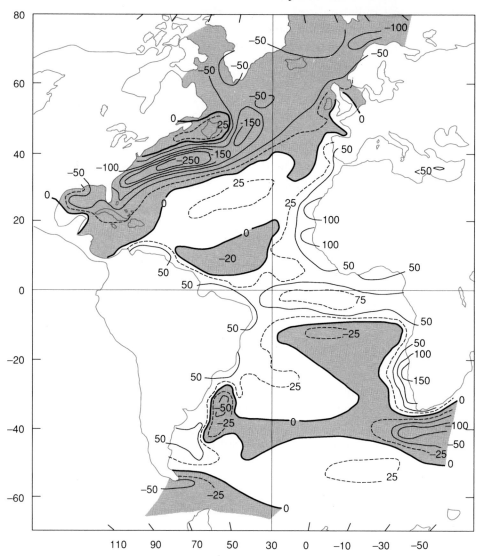

Fig. 4.11 Net annual heat gain by the Atlantic. The areas of maximum heating and cooling penetrate the water masses only few hundred meters deep. Subthermocline water is produced by weak cooling north of the 50°N. (Bunker, 1980).

forms an insulating barrier to the water surface and a reflecting surface to the atmospheric radiation above. Sea ice grows to a thickness where its bottom temperature is at the freezing temperature of the sea water, no matter how cold the air or its surface temperature gets. Its formation process increases the salt content of the water and makes the water more convectively unstable. When melting occurs, stable, fresh layers can prevent deep convective activity (Marsden et al., 1991).

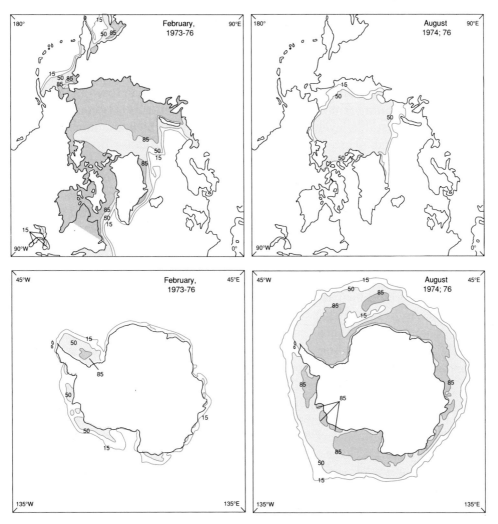

Fig. 4.12 Sea ice distributions in February and August in both hemispheres. Contours displayed are 15, 50 and 85% coverage. Adapted from Zwally et al. (1983) and Parkinson et al. (1987).

First year sea ice is typically 2 m thick. Under the forces of strong winds and underlying ocean currents, underwater ridges form which may be 10 to 20 m in depth. The ridges, although they cover a small fraction of the surface area, increase the drag between the water and ice significantly (see also Chapter 12). Leads, or breaks in the ice, occur and in these areas the turbulent and radiative heat transport out of the ocean to the polar air is several orders of magnitude greater than the heat transfer through the ice. Wind can push the ice to separate at more rapid rates than it can form, exposing "polynyas", or large expanses of open water. One such large polynya was observed for several years in the Antarctic waters in August

1974 to 1976 near the Greenwich meridian (Fig. 4.12). In polynyas deep ocean convection is very intense.

Wintertime cooling (caused by evaporation and sensible transfers and infrared radiation) and sea ice formation and winds are more important processes in determining the long-term temperature structure of the oceans, than are the summertime air–sea interactions. First, this is because warming forms shallow stable layers near the ocean surface which are quickly eroded, but cooling and brine formation during icing can cause deep convection which, once removed from the influence of the atmosphere, has a long persistence. Convection is the only known mechanism by which cold water sinks to the bottom of the world's oceans, and anomalous convection can leave behind anomalous water masses in the deep which are known to persist for a long time (Dickson et al., 1988). Second, the wind-stress curl in winter is much more pronounced due to the increased wind speed and to the nonlinearity of the stress dependence on wind speed (Fig. 4.6). A small fraction of the storms in winter are responsible for removing most of the thermal energy which is painstakingly laid down over the entire summer period (Large et al., 1986). The long-term average ocean wind-stress forcing and cooling patterns thus resemble the winter conditions.

The correct modeling of long-term changes of intensive wintertime conditions is of greater importance for producing credible models of ocean circulation climate, because it is winter when the ocean receives fluxes of heat and momentum that penetrate to the greatest depth. Also this is when the Ekman convergences are the strongest and water is forced to sink to or rise from the main thermocline. In contrast, in summer, air–sea interaction is comparatively weaker and only shallow layers are affected.

The accuracy of the net heat exchange estimate from ship or buoy observations permits a basin-wide diagnosis of local changes which might be associated with 1–2°C interannual changes of SST. These anomalies are usually established over a few months of intensive air–sea interaction (Cayan, 1990). The centers of action of the significant long-term heat flux patterns in today's climate also appear to be well determined, since they are highly localized. Where these centers of action span most of the basin, like the Atlantic, the accuracies of the net, basin-wide estimates of air–sea interaction are significant and these can then be compared with the transports of heat carried by the ocean circulation. The Pacific, however is so large compared to the centers of action that bias errors of ± 10 W m^{-2} obviate an accurate assessment of the net heating of the Pacific. To compare the ocean transports of heat with what is given up or absorbed from the atmosphere requires a diagnostic analysis of the deep circulation patterns.

4.7 Deep circulation patterns

There is both observational evidence and theoretical confirmation that the time-mean, direct wind-driven circulation over most of the ocean is restricted to the mixed layer and the upper thermocline of the oceans. However, since all density surfaces of the oceans come in contact with the atmosphere somewhere on the globe, the range of density of the ocean water and also the stratification is determined by air–sea interaction processes. The circulation in the lower three-quarters of the water column is forced by either the time-dependent eddy mixing processes that receive their energy from the time-mean wind-driven circulation, or by the effects of the deep convections that are forced by the combined effects of intensive buoyancy loss and wind mixing in the surface layers when the thermocline is eroded. At depth the ocean bottom relief also exerts a strong controlling influence. How the deep ocean circulates is suggested by the basin-wide distributions of temperature and chemical constituents of sea water (see Chapter 16). Some direct measurements of deep currents have been made with current meters and neutrally buoyant floats.

The global observations of the deep circulation indicate that the strongest deep currents are on western sides of the basins and that sills and straits between the deep basins are the control points at which volume fluxes of the bottom layer flow are regulated. The deep circulation patterns which affect the heat transports only of the Atlantic are now examined in four vertical layers of water below the thermocline. We have already discussed two wind-driven layers: the Ekman transport and the Sverdrup transport. The deep layers have contact with the atmosphere in quite distinct parts of the globe. To consider each of the other basins in such detail would require a more copious work. Also the Atlantic is the best studied and best measured ocean.

The densest water in the Atlantic can be traced to several formation locations on the continental shelf of the Antarctic Continent. When brine forms during the growth of sea ice and offshore winds force cracks, polynyas and leads in the ice, the shallow water columns on the continental shelves so rapidly lose buoyancy that they plummet to the bottom of the Ross and the Weddell Seas. Weddell Sea Bottom Water, as it sinks, flows clockwise as a gravity current on the continental rise. In its northward movement, some of it becomes entrained in the circulation of the Antarctic Circumpolar Current and some of its cold signature can be traced northward into the Argentine Basin. The portion which mixes with the circumpolar circulation moves farther north, most clearly on the western side of the Atlantic all the way to the bottom of the Sargasso Sea southeast of Bermuda (Fig. 4.13). Several choke points control its northward migration. Direct current measurements in the Vema Passage suggest hydraulic control to be important in the transport and the reservoir size of bottom water in the Argentine Basin (Schmitz and Hogg, 1983).

The next most dense layer has its origins in the northernmost parts of the Atlantic. During intensive winter stormy periods over open water in

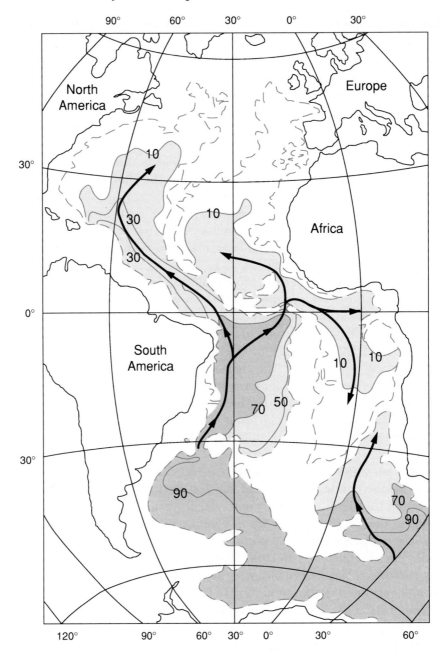

Fig. 4.13 Spreading of Antarctic Bottom Water, represented by the percentage of the Antarctic component (adapted from Wust, 1935).

the Greenland, Norwegian and Labrador Seas deep convection is formed which forces oxygen-rich and salty water to sink near the bottom of these basins. The depth to which this convection penetrates is very sensitive to the salinity of the surface water, and here in the open ocean, sea ice, because it forms before the strong storms occur, plays an important but different role than it does in the Antarctic. During winter, sea ice insulates the air–sea interaction process and its melt-water can form a stable surface lens which can further reduce the following winter's deep water production process. Thus, the increase of sea ice in the Greenland Sea is thought to reduce the rate of deep water formation. From the Norwegian Sea, deep water flows southward through the Shetland Islands Gap and over the Iceland–Faroes Ridge, and from the Greenland Sea it flows southward through the Denmark Strait, again under hydraulic control. Added to this is the Labrador Sea contribution and this North Atlantic Deep Water can now be traced southward along the western boundary of the continental rise around the Grand Banks by its oxygen and salinity concentrations (Fig. 4.14) (Worthington and Wright, 1970). Its westward intensified southward flow signature is also expressed in increased levels of tritium and chlorofluoromethanes along the western continental rise which it acquired from the atmosphere upon sinking in the polar latitudes (Jenkins and Rhines, 1980; Weiss et al., 1985). Direct measurements have been made of its flow from the Denmark Strait to east of the Bahamas. Its oxygen signature can be traced eastward half way across the Atlantic and southward to the circumpolar current which flows around Antarctica. Its salinity signature is also found in the deep waters of the Pacific (Reid, 1986).

The third layer of water from the bottom originates from the Mediterranean Sea, over which evaporation exceeds precipitation on an annual basis so that there must be a flux of fresh water through the Strait of Gibraltar. When winter Mistrals blow from the Alps onto the Gulf of Genoa, warm and saline water is forced to the bottom (Leaman and Schott, 1991). Again, through hydraulic control at the Strait of Gibraltar, this water is introduced as a gravity current into the eastern North Atlantic (Armi and Farmer, 1988). It flows northward to the Labrador and Norwegian Seas and enhances the deep convection there by preconditioning the surface water with high salinity (Fig. 4.15).

The fourth layer from the bottom inhabits the volume directly below the main thermocline and is best characterized by a high concentration of phosphates and nitrates which it retains from the Antarctic subpolar front. The processes which force it to sink to these intermediate depths are not clearly understood, as there are many mixing and air–sea interaction processes proposed for water modification in any upper ocean frontal region (Fig. 4.16).

In summary, in the subtropical North Atlantic at about 25°N the bottom water appears to be flowing northward as Antarctic Bottom Water. In the next layer the North Atlantic Deep Water flows southward; the third and fourth layers have Mediterranean and Antarctic intermediate sources and these flow northward. The topmost layers are wind-driven where the thermocline layer is forced southward as geostrophic flows proportional to

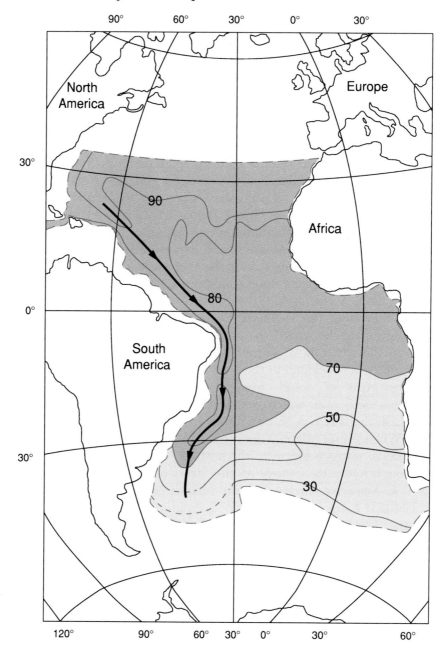

Fig. 4.14 Spreading of the middle North Atlantic Deep Water in the core layer (intermediate oxygen maximum), represented by the percentage of the North Atlantic component (adapted from Wust, 1935).

the curl of the wind stress, and the Ekman layer flows northward due to direct wind driving (see the summary schematic Fig. 4.17). Each of these six layers have different temperatures and the evaluation of the net

140

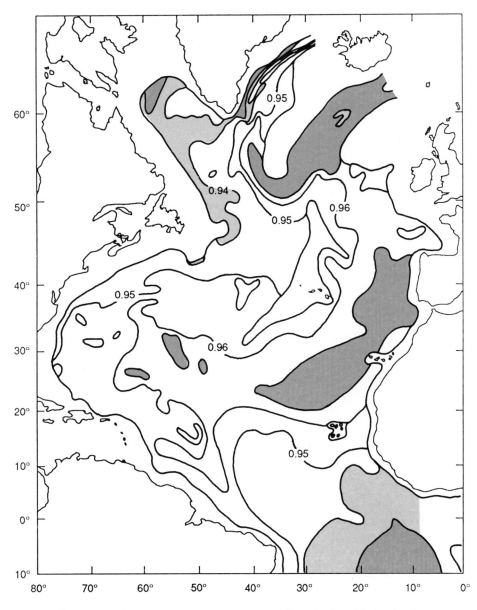

Fig. 4.15 Spreading of the Mediterranean, Norwegian, Greenland and Labrador Seas' waters, as indicated by salinity contours on the 2.8°C potential temperature surface of the North Atlantic (adapted from Worthington and Wright, 1970).

heat transport carried by this rather complicated circulation requires some conceptual manipulation of the thermodynamical equations and a thoughtful use of the very few direct measurements of currents. It is, however, apparent that the supplies of the deep waters of the world below the thermocline are regulated by rare, intense and small-scale air–sea interaction events and the

141

Phosphate

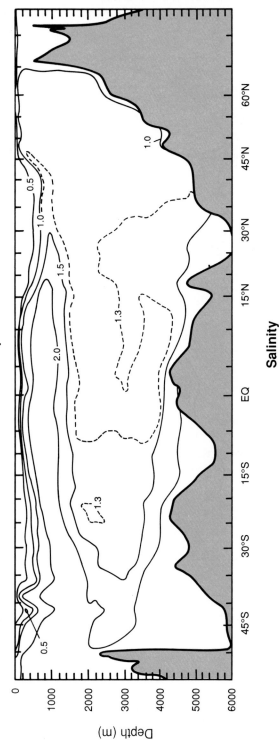

Salinity

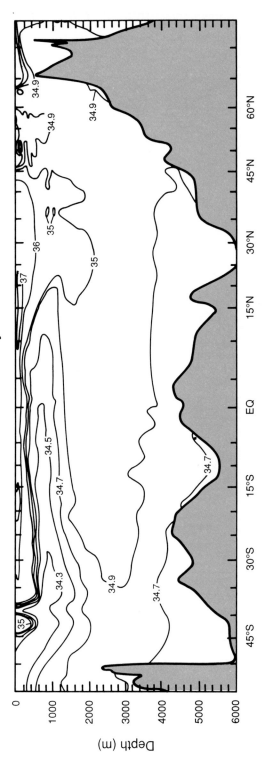

Fig. 4.16 Atlantic phosphate (upper) and salinity (lower) distributions as a function of depth and latitude on a section through the western basins (adapted from GEOSECS Atlas, Vol. 2, 1981).

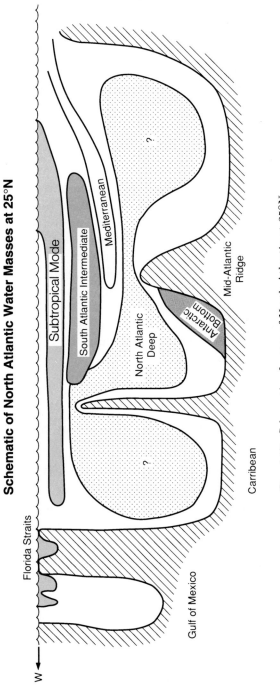

Fig. 4.17   Schematic of water of North Atlantic at 25°N.

143

convection processes that are very sensitive to formation and movement of sea ice. The spreading rates of abyssal waters from their sources are as much under nonlinear hydraulic controls as they are governed by general wind-driven and eddy-driven circulation and mixing. These processes are represented by only the most rudimentary hydrodynamics in climate circulation models of the oceans.

Deep convective flows are gravity currents near their formation or hydraulic control points, but quickly they become part of the general circulation of the deep oceans. The most robust conceptual model of the dynamics of the general deep circulation which might aid the spread of subthermocline water from their sources is due to Stommel and Arons (1960) (but see also Sec. 11.4). They showed that in an ocean the size of the Atlantic which has longitudinal boundaries, the horizontal mass transport is *twice* that which upwells into the thermocline. In this model, convectively formed fluid is transported southward along a western boundary. It leaves the boundary under frictional control and upwells into the main thermocline. But this model also predicts that the transport of the western boundary current is twice that of the source at the sinking region and the interior basin recirculates an additional volume flux. Source–sink flows due to convection, which are produced on a rotating globe, can have large horizontal transports associated with them, far in excess of the source and sink strength, so deep ocean horizontal circulation amplitude cannot be directly predicted from the strength of the formation rate of the deep and bottom water without a carefully constructed hydrodynamical model.

4.8 Heat transport by ocean circulation patterns

Almost a century ago, United States Coast Guard Captain Elliott Pillsbury (1912) recognized that: "The currents of the ocean are the great transporters of sun's heat and moisture from the torrid zone to temper the climate of more polar regions." But it was not until a decade ago that oceanographers made calculations of the rate at which this transport was taking place. Pillsbury's problem, as it is still a problem today, was that there is a plethora of data on the water velocity of the oceans. The most accurate method of calculating the northward heat flux through a section of the ocean which is bounded by continents to the east and west is provided by Hall and Bryden (1982). This technique is based on computing a heat flux normal to a section whose vertically averaged temperature at each location along the path of the section across the ocean is a constant.

The heat flux Q normal to a vertical ocean section is denoted by

$$Q = \{\overline{\rho_o c_p \Theta v_n}\} \tag{4.12}$$

where Θ is the potential temperature, v_n is the velocity normal to the section, the vertical and horizontal integral from the bottom is { } and the integral across the basin is denoted by an overbar. Divide the potential temperature and normal velocity of the water column into a vertical mean and a deviation from that mean, designated by a prime:

$$v_n = \{v_n(\phi, \lambda)\} + v'_n(\phi, \lambda; z)$$

$$\Theta = \{\Theta(\phi, \lambda)\} + \Theta'(\phi, \lambda; z). \tag{4.13}$$

Upon substituting from (4.12) into (4.13) and noting that $\{v'_n\} = 0$, and $\{\Theta'\} = 0$, the expression for the heat flux normal to a section which has a constant vertical average temperature, Θ_o, is

$$Q = \rho_o c_p \Theta_o \{\overline{v}_n\} + \{\overline{\rho_o c_p \Theta' v'_n}\}. \tag{4.14}$$

The term under the first integral on the right-hand side of (4.14) is the total volume flux through the section which, in general, is not known in the oceans because there are so few direct measurements of currents. But, Hall and Bryden noted that along a 25°N hydrographic section in the North Atlantic basin, the vertically averaged temperature was nearly constant at 5.4°C from the African continental rise to the Bahamas Banks. They used expression (4.14) for a greater portion of the section across the Atlantic. Second, direct measurements of the velocity and temperature had been made by Niiler and Richardson (1973) for the portion of the section across the Florida Straits. Thirdly, the measurement of currents on the west African continental rise revealed that the transport there could be neglected compared with that of the other segments of the section. Thus, it was evident that the total volume transport southward at 27°N in the section where the vertically averaged temperature was constant was nearly equal to the northward transport of the Florida Current. The first term could be evaluated quite accurately by substituting the Florida Current transport for $\{\overline{v}_n\}$, and that part of the flux going northward in the Florida Current could be evaluated from direct measurements of velocity and temperature.

The deviation velocity from the vertical average in the second term on the right-hand side of (4.14) can be evaluated from the relative geostrophic flow computed from hydrographic data along the section below the mixed layer and the Ekman transport in the mixed layer. The deviation velocity, v'_n, is computed by the geostrophic method relative to an arbitrary velocity at some depth (see Eq. 4.11) and integrated with depth. To this is added the Ekman transport and, when the sum is set equal to zero, the required reference velocity component normal to the section is determined. The heat transport due to this is referred to as the "baroclinic" heat transport. The summary of these calculations is reproduced in Table 4.1 from Hall and Bryden (1982). Their estimate in the Atlantic basin at 25°N is that 1.2×10^{15} W of heat is being transported northward by ocean circulation.

In locations where no measurements of western boundary currents are available, various techniques of a level of no motion or minimal motion assumptions (Wunsch and Grant, 1982) have been used to estimate the

Table 4.1. Northward heat transport by currents across 25°N in the Atlantic Ocean. Positive values are northward, units are 10^{15} W. From Hall and Bryden (1982).

Component	Barotropic	Baroclinic	Ekman	Total
Florida Straits	1.88	0.50	–	2.38
Mid Ocean: direct		−0.93	0.53	
Mid Ocean: calculated as compensation	−0.65		−0.11	−1.16
Total	1.23	−0.43	0.42	1.22

mass and heat transports across hydrographic lines of constant latitude. Estimates of the Atlantic transports of heat from oceanographic data and wind-stress data reveal that at every latitude the Atlantic circulation is transporting heat to the north (Fig. 4.18). At about 25°N warm wind-driven tropical water above 700 m flows northward in the Florida Current and cold water which originates in the deep convective zones of the polar Atlantic flows southward (Fig. 4.19).

Figure 4.18 also displays the net amount of thermal energy that is exchanged between the atmosphere and the ocean from the northernmost extent of the Atlantic at 75°N to any particular southern latitude from that location. For a basin-wide thermodynamic balance, this net surface exchange would have to equal the rate at which heat is transported north by the ocean circulation at that latitude. From this figure it appears that a general agreement between these two estimates is achieved. This agreement is perhaps as much a function of the size of the Atlantic, so biases of the flux estimates do not overwhelm the estimate, as it is a general verification that the air–sea heat flux estimates from the formulae are accurate to 25%, or better [Bunker's flux estimates (1976, 1980) have been subsequently shown to introduce biases due to several unrealistic assumptions in the practical formulae he used; Isemer and Hasse, 1991]. But when a similar estimate is repeated for the Pacific, it quickly becomes apparent that uncertainties in the effect of stability on the turbulent transfer coefficients, effects of clouds on the radiation balances and the biases in the marine meteorological observations, produce uncertainties in the net fluxes. When integrated over the Pacific basin, these uncertainties accumulate rapidly and overwhelm the signal from the centers of air–sea interaction (Talley, 1984). Also direct current measurements are not available through the western boundary currents in the Pacific. Computations of the ocean heat transport in the Pacific are based on models of the circulation with the "smallest" deep kinetic energy or "smoothest" fields of flow which reduce horizontal mixing of heat or potential vorticity: the inverse model solutions whose veracity is not known.

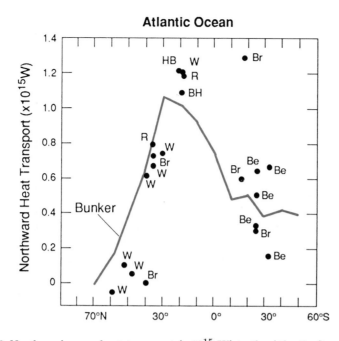

Fig. 4.18 Northward ocean heat transport ($\times 10^{15}$ W) in the Atlantic Ocean as a function of latitude, derived from aggregating Bunker's (1980) net annual ocean heat gain (see Fig. 4.11). Single points are values obtained by estimating oceanic heat transport carried by various models of oceanic circulation: Be, Bennett; Br, Bryan; BH, Bryden and Hall; HB, Hall and Bryden; R, Roemmich; W, Wunsch.

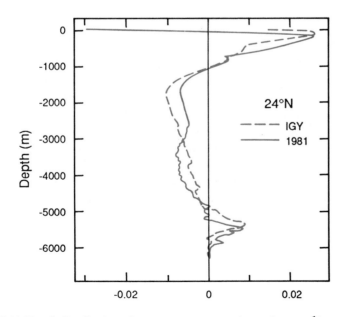

Fig. 4.19 Depth distribution of east–west average velocity (in m s^{-1}) in the North Atlantic derived from data sets 22 years apart (Roemmich and Wunsch, 1985).

147

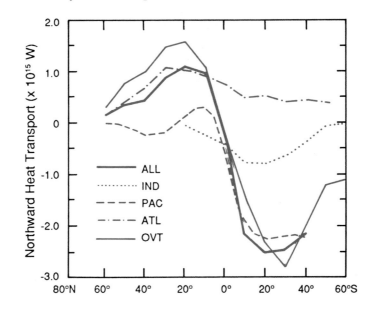

Fig. 4.20 Annual mean meridional heat transported by all oceans based on surface heat fluxes calculated using Bunker's (1980) method. For comparison, Oort and Vonder Haar's (1976) Northern Hemisphere and Trenberth's (1979) Southern Hemisphere computation of the oceanic transport as the residual between satellite measurements of the global radiation budget at the top of the atmosphere and what atmospheric motions can carry poleward is shown labelled OVT. For a detailed discussion of the uncertainties of these methodologies, see Talley, 1984.

The conclusion today is that within the broadest interpretation of oceanic circulation data, large-scale flows in the Pacific and Indian Oceans cannot transport as much heat southward south of the equator as the atmospheric budgets and surface flux analyses demand. However, several estimates of the net air–sea fluxes appear to be consistent with the atmospheric budget requirements (Fig. 4.20). Oceanographers are now examining the role of small-scale but intense flows through island passages in the western equatorial Pacific and eddy transports of heat around the tip of South Africa in providing additional southward transports in the Southern Hemisphere. The atmospheric scientists should also re-examine their data sets on the accuracy of the atmospheric transports over the southern oceans.

5

Land surface

Robert E. Dickinson

5.1 Role of the land surface in the climate system

Land has a dual role in the climate system. First, it acts as a lower boundary for approximately 30% of the atmosphere, exchanging moisture, momentum, and heat. Second, from a practical viewpoint of human requirements, land is crucial. As a lower boundary condition, land may in some ways be less important than the oceans. With regard to thermal energy, it provides much less storage and negligible horizontal transport.

On the other hand, land is more variable and changeable than the oceans for many of the important coupling processes. Most notable is moisture exchange. When wet, it can exchange water with the atmosphere more rapidly than the oceans because of greater surface roughness; but when dry, it provides no water at all to the atmosphere. Because of the relatively low heat capacity of the land surface, local thermal conditions are much more responsive to net radiation from or through the atmosphere than are the oceans. Presence or absence of clouds has a substantial effect. Temperatures can range by 10–20°C or more over the diurnal cycle. The fraction of solar radiation reflected (the albedo) varies with type of surface cover, and vegetation and soils have large spectral variations of albedo from generally low values at visible wavelengths to much higher values in the near-infrared. Another characteristic of the land system is its wide-ranging heterogeneity, with the distribution of soil properties and vegetation cover almost fractal in nature.

Aspects of land climate of concern beyond climate modeling include those that determine the environment for growth of natural and cultivated vegetation and the provision of water resources. These issues are also becoming of interest to climate modelers as we realize a strong commonality between the basic physical processes needed to describe land's climatic role and the framework needed to determine the impacts of climate change on vegetation and surface hydrology.

Much of the flux of water from land to atmosphere moves through vegetation. The rates of these fluxes are modulated by the ability of roots to extract water from the soil and by the resistances imposed by the stomatal

149

controls of leaves (see Chapter 14). Plants transpire water as a byproduct of their requirement to extract carbon from the atmosphere. On an annual basis, land vegetation removes from the atmosphere about 80 gigatonnes of carbon, compared to the six released by the burning of fossil fuels. However, about the same flux of carbon returns to the atmosphere from the respiration of live plants and decay of dead ones.

Water is supplied from the atmosphere to the land surface primarily by precipitation. Soils act as reservoirs for this supply but, in the long run, precipitation is balanced by evapotranspiration and runoff. Climatologists have often viewed runoff as simply the residual after evapotranspiration requirements have been satisfied; conversely, hydrologists have often perceived runoff as a direct response to precipitation, with evapotranspiration a residual. In reality, they both affect each other. Hydrologists with their studies of runoff on local and basin scales have demonstrated the importance of the rates of water infiltration into soils and the distribution of hillslopes in determining the rate of water movement from soils into streambeds and subsoil storage. Micrometeorologists and plant scientists have established the important controls of surface boundary layers and stomatal resistance (see Chapter 14). Climatologists have shown that, under well-watered conditions, water fluxes are controlled primarily by net radiation or a temperature surrogate.

In high latitudes and in temperate latitudes during winter, additional processes become important for land's role in the climate system. Seasonally or permanently frozen soil impedes the flow of water into the soil. Snow accumulates during the winter season (Fig. 5.1) and acts as a good insulator, allowing little heat to be exchanged between soil and atmosphere. It enhances the surface albedo and, by this reflection of solar radiation, delays spring warming. This effect can be considerably weakened by overlying vegetation. Vertical profiles of temperature measured deep into Arctic permafrost on the Alaska North Slope suggest recent climate changes (Lachenbruch et al., 1988).

5.2 Observational elements of land climate

Because of its complexity, land climate can be viewed observationally in terms of a wide variety of statistics and of variables. From the viewpoint of modeling, the rates of water and energy exchange are fundamental. Some key elements, in particular precipitation and incident solar radiation, are largely controlled by the overlying atmosphere (though significant feedbacks on these fields by land properties are known). Unrealistic fields of precipitation and incident solar energy will seriously degrade efforts to model land as a component of the climate system, and negate improvements that would otherwise be realized by better treatments of the processes specific to land.

15 Year January Snow Cover Frequency Map

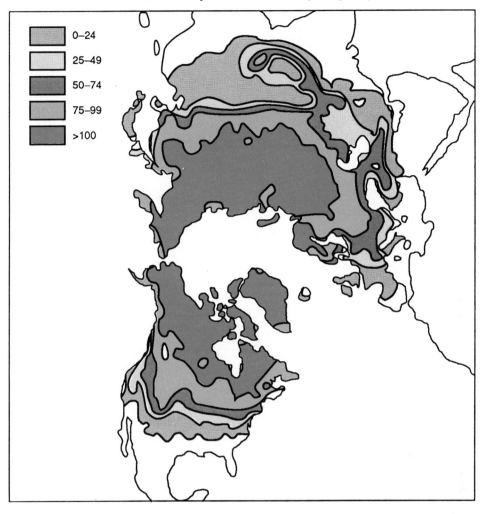

Fig. 5.1 January area covered by snow over the Northern Hemisphere (adapted from Matson et al., 1986). Frequency of occurrence is indicated by hatch pattern according to legend at top.

5.2.1 Precipitation

Climate modelers usually first compare their predicted fields of precipitation with global-scale smoothed and time-averaged patterns (Fig. 5.2). Such figures show the broad seasonal migration of the tropical rainfall belt (e.g., defined as where average precipitation exceeds 5 mm day^{-1}), including the summer monsoon in Southeast Asia and the maximum core of rainfall across tropical South America. They show regions of minimum precipitation (e.g., less than 1 mm day^{-1}) over the arid and

(a) DJF Precipitation: Observed

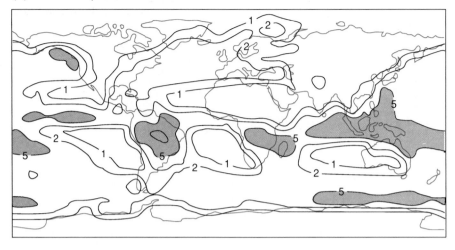

(b) JJA Precipitation: Observed

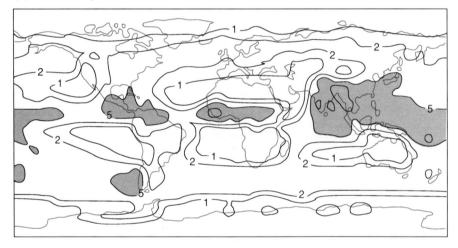

Fig. 5.2 January and July "observed" precipitation from IPCC (1990a) in mm day^{-1}. (a) Average of December through February, (b) average of June through August.

semiarid regions in the subtropics for part or all of the year, over much of the central continental regions during winter, and over all but the extreme maritime fringes of Antarctica all of the time.

Displays of the global distribution of precipitation, besides being extremely smoothed, may show magnitudes only within a factor of two. Over oceans, the mean precipitation at any point may be uncertain by that amount, but because much of the land surface is heavily instrumented with rain gauges, mean amounts over many land areas are known within a few tens of percent. Changes in precipitation by as little as ten percent would be quite significant; this is the magnitude of increase expected globally with a several degree warming from increasing greenhouse gases. The measured precipitation amounts over land have considerable spatial variability, especially in areas of large topographical variations (Fig. 5.3), that increase the uncertainty of the measurements. Some biases occur because of relative lack of coverage at higher altitudes and because of errors in instrumental measurements of precipitation.

5.2.2 *Solar radiation*

Long-term records of surface solar radiation have been kept at only a few sites worldwide, and spatial patterns have been estimated from these data for only a few countries (e.g., Fig. 5.4). Solar radiation is conceptually simple and so its attenuation through the atmosphere can be modeled with a fair degree of accuracy. The greatest uncertainty in estimating surface solar radiation is the effect of overlying clouds. Satellite observations of reflected solar radiation help to remove this uncertainty, and with the aid of a radiation model or correlative relationships, are used to estimate the global distribution of surface radiation.

The solar radiation arriving at the top of the atmosphere is the product of the solar constant S_0 and an astronomical factor $f(d)$ of annual average 1.0, proportional to d^{-2}, where d is the distance of the Earth from the Sun. Passing through the atmosphere, the solar beam undergoes wavelength- and direction-dependent absorption and scattering by atmospheric gases, aerosols, and cloud droplets. The scattered radiation reaching the surface is called "diffuse." On clear days, the diffuse component from the Rayleigh and aerosol scattering (see Chapters 7 and 10) is about 10–30% of the total incident radiation, whereas when the solar beam passes through a cloud essentially all the surface radiation is diffuse. Diffuse radiation consists of solar photons arriving from all directions of the sky, with intensities dependent on the incoming direction. For modeling land processes, it is usually adequate to assume that the diffuse radiation is isotropic (same intensity in all directions from the sky). If S is the intensity of radiation arriving at the surface from a given direction, then the amount incident per unit surface area is $S\cos Z$, where Z is the angle between the normal to the surface and the direction of the beam. All GCMs up to now have assumed land to be a horizontal surface. However, in hilly or mountainous terrains, the distribution of slopes has major effects on surface climate.

Since surface radiation can vary widely according to the frequency and optical thickness of clouds, modeling these cloud properties successfully (see Chapter 10) is important for treatment of surface energy balance and consequently evapotranspiration. Validation against observed cloudiness (e.g., Fig. 5.5, see also Plate 5) is part of this task.

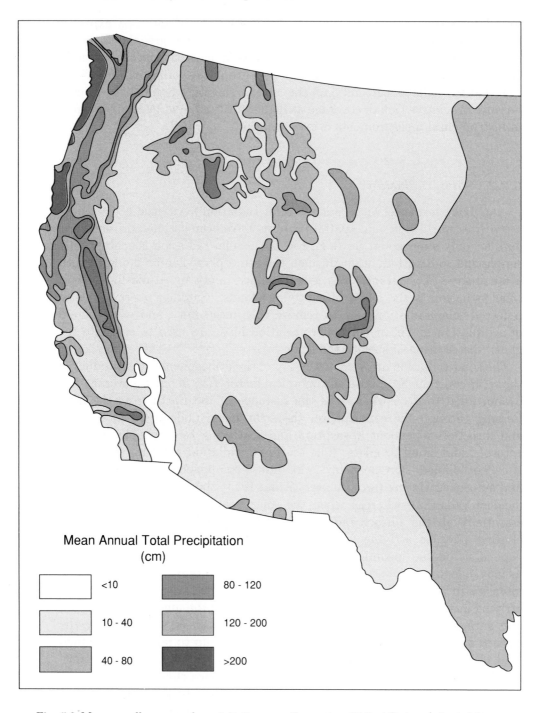

Fig. 5.3 Mean anually averaged precipitation over the western United States, (adapted from Court, 1974). Note the large spatial variation.

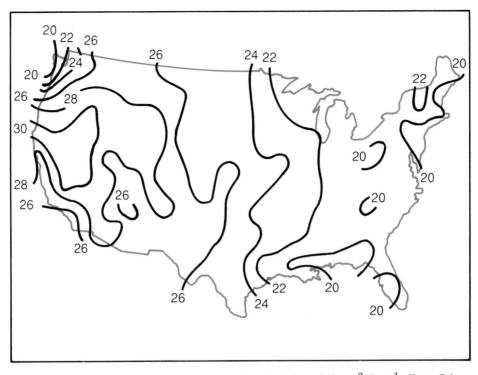

Fig. 5.4 Total solar radiation incident at surface for July in MJ m^{-2} day^{-1}. From Solar Energy Research Institute (1981).

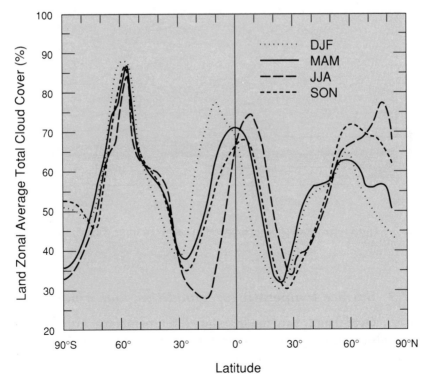

Fig. 5.5 Cloud cover from Warren et al. (1986) zonally averaged over land and shown for the four seasons.

155

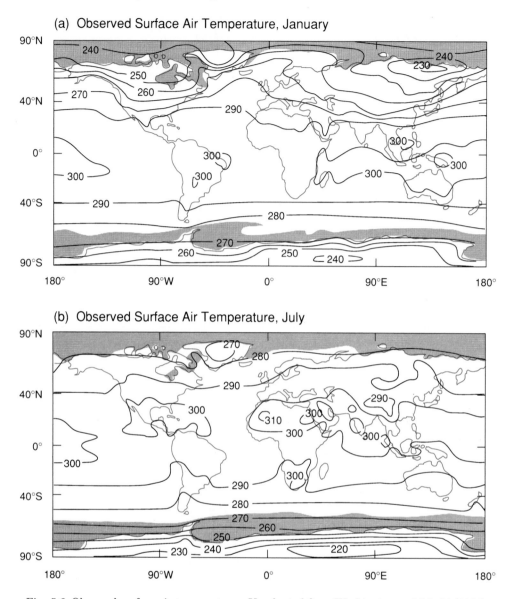

Fig. 5.6 Observed surface air temperatures, K, adapted from Washington and Meehl (1984).

5.2.3 Surface temperatures, humidities, and winds

Over land, two model-related temperatures are readily available from observation, a surface air temperature and a "skin" temperature. Surface air temperatures are generally standardized by measuring them in a well-ventilated and shaded box about a meter above a flat, grassy surface. Temperatures can be recorded either hourly or as maximum and minimum

156

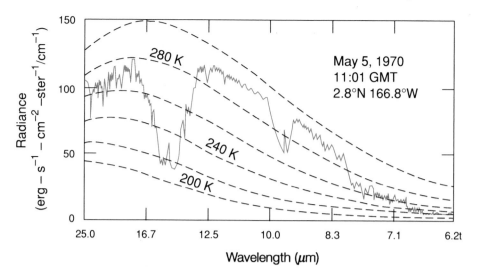

Fig. 5.7 Terrestrial thermal radiation from Hanel et al. (1972). The solid line gives the intensity of the radiation, the dashed lines, the radiation that would be emitted from a blackbody at various temperatures. The effective radiating temperature of a given wavelength can be judged on the basis of the point of intersection of the solid with the dashed line.

values, and daily averages constructed either by averaging over the 24 hours or simply with the maximum and minimum values (Fig. 5.6). Atmospheric models with their lowest mesh point below 100 m sometimes archive their lowest layer temperature as surface air temperature. This should work if errors of a few degrees or so are acceptable. Otherwise, it may be necessary to extrapolate the temperature of the lowest model layer downward to the actual altitude and underlying surface of observations to adequately compare model output with surface data. Temperature measured 1 m above a grassy surface will generally not correspond exactly to the temperature of air a meter over the dominant vegetation cover of a model grid square (as may be needed for flux calculations) because of differences in surface roughness lengths. Possible corrections for altitude and measurement surface must be addressed in relating model humidities and winds to those observed. In addition, statistics on peak wind gusts in the model will not be comparable to those observed unless subgrid-scale convective winds are parameterized and included as part of model wind statistics.

"Skin" temperature refers to the effective radiating temperature of the soil plus canopy surface. It can be inferred from satellites using channels in the 8–12 μm window region (Fig. 5.7). In climate models, it will be whatever temperature is used to determine upward thermal emission. In models without vegetation, this would simply be the temperature of the top soil surface. In more realistic parameterizations, it would be an average of canopy and soil temperature, weighted according to vegetation cover. This temperature will usually have an even larger diurnal variation than that of surface air, which should be recognized in comparisons between model and observation. Space-based observations are, furthermore, likely to have a bias toward clear skies that may degrade the comparisons.

5.3 Surface albedo

In astronomy, the term albedo has long been used to designate the fraction of solar energy reflected, on the average, by a planet. In meteorology, the same term is applied to the fractional reflection at a given point and given time (i.e., given solar geometry) and sometimes at a specific wavelength (in that case, referred to as spectral albedo). In radiation studies, the term "hemispheric reflectance" is equivalently but more precisely used, recognizing that albedo represents the radiation reflected in all directions into the upward hemisphere (assuming radiation directions are represented in spherical coordinates). An instrument with a narrow field of view, such as most satellite sensors, on the other hand, sees radiation traveling in a single direction.

Albedos of vegetated surfaces can vary from around 0.1 to 0.3 in the absence of snow. These differences appear to depend more on the light-trapping properties of the canopy architecture and leaf orientations than on the optical properties of individual leaves. A simple model for canopy albedos is presented in Dickinson (1983) and in Chapter 14. Soils can vary over an even wider range, depending largely on the mineralogical composition of the soil and on whether the soil is wet or dry. Addition of snow to surfaces generally increases albedos, since new-fallen snow can have an albedo as large as 0.8. However, shading of the snow by other elements (rocks, dead grass, trees, etc.) often implies that the surface albedo is some kind of mixture of what it would be without snow and what it would be with only snow.

The wavelength dependence of surface albedos is significant. The chlorophyll in vegetation absorbs strongly at visible wavelengths; consequently, leaves may reflect only about 0.05 of the solar radiation short of 700 nm but approximately 0.85 of longer wavelengths. Soils also increase their albedos with increasing wavelength, although more gradually, e.g., Fig. 5.8 from Tucker and Miller (1977). Snow crystals, on the other hand, absorb much more solar radiation in the near infrared because of the water vapor absorption bands over that spectral region (Fig. 5.9). The relative intensities at the surface of visible versus near-infrared radiation depend on their relative attenuations in passing through the atmosphere, which in turn depend on atmospheric water vapor and cloudiness. Thus, some inclusion of spectral variations of surface albedo may improve the accuracy of calculation of reflected solar energy in the context of atmospheric feedbacks. Surface

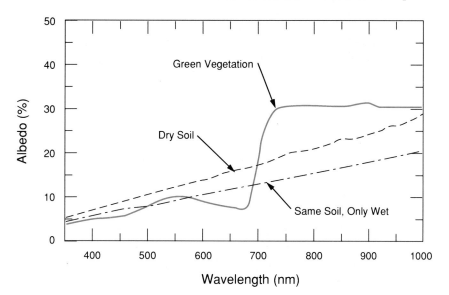

Fig. 5.8 Spectral reflectances for dry soil, wet soil, and a plot of blue grama grass (Tucker and Miller, 1977).

albedos also vary significantly with the angle of incidence of the incoming solar radiation (Fig. 5.10).

As this discussion indicates, surface albedos depend on a wide variety of variable parameters. Furthermore, observational characterizations have been sparse relative to what is needed. Thus the few observations in hand must be supplemented by theoretical modeling of surface albedos, allowing for the dependences mentioned above. Such models should be compatible not only with the needs of the climate models but also of use for the interpretation of surface albedos in terms of satellite remote sensing. The forthcoming NASA Earth Observing System satellite program contains several instruments that will look at the land surface and provide detailed angular and spectral information on surface reflectances, in part to improve the characterization of albedos in climate models.

One class of models for radiative transfer in canopies (presented in Chapter 14) is based on the assumption that the system can be viewed as a cloud of leaves, with a random distribution of leaf locations and orientations (Fig. 5.11 from Dickinson, 1983). Most simply, all the leaf positions are independent of each other, but canopy organization can be mimicked in part by assumptions about clustering of leaves. The simplest assumption about the distribution of orientations is that all are equally probable, but other distributions allowing for more vertical or more horizontal orientations are also used. Leaves are generally partially transparent to radiation, so they are characterized with optical properties as scattering objects the same as cloud particles. Little is yet known about the detailed angular distribution

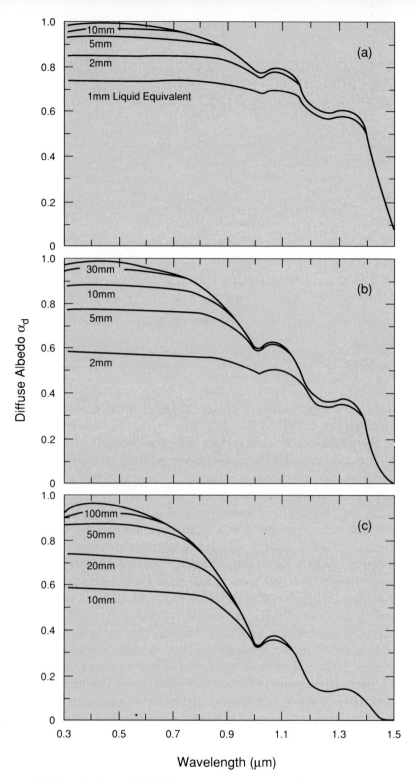

Fig. 5.9 Calculated values of albedo for snow, assuming diffuse radiation, a variety of snow
depths, and three values of grain radius: (a) 50, (b) 200, and (c) 1,000 μm, from
Wiscombe and Warren (1980).

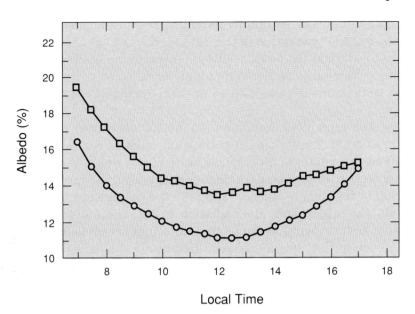

Fig. 5.10 Diurnal variation of average albedo of a tropical forest (○) and clearing (□). (Pinker et al., 1980).

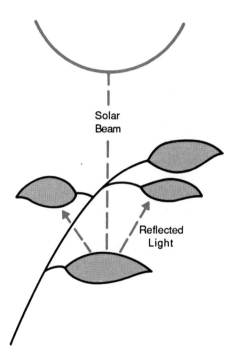

Fig. 5.11 Canopy as a cloud of leaves traps solar radiation, which, incident on leaves, is partially transmitted, partially reflected, and partially absorbed (Dickinson, 1983).

161

of scattering from a leaf, so simpler parameters are used. That is, according to the position and orientation of the leaf, intercepted light is assumed to be either absorbed, reflected or transmitted. A scattering coefficient is used to represent the fraction of the beam that is either reflected or transmitted, and radiative transfer models are used to provide tracing of multiply scattered light rays.

Because the individual leaves are so highly absorbing in the visible, reflected radiation consists almost entirely of radiation scattered only a few times. Even the assumption of "single scattering", which is using only the upward reflection of previously unintercepted radiation, can give a usable description of the canopy albedo, especially the dependence on solar zenith angle. On the other hand, the radiation in the near-infrared is so little absorbed at any one leaf that numerous scatterings are needed to resolve the canopy albedo in that spectral region. Because of the difficulty of deriving measured values of canopy albedos from first principles, some parameter adjustment is generally needed to realistically specify canopy albedos over the range of ecosystem types that might be specified in a GCM.

Another class of models most appropriate to arid and semiarid regions is based on the assumption that vegetation can be idealized as a randomly distributed array of vertical sticks (Otterman, 1985). These sticks or poles are opaque to the incident radiation so that only singly scattered radiation can be reflected to the atmosphere. In this and the leaf cloud model, allowance for interacting reflections between the soil and vegetation elements introduces an additional complexity. The albedo of soils might be modeled the same as for that of a mineral aerosol in the atmosphere. However, this approach approximates reality only for a sand, since smaller particles generally cluster together, depending on soil structure, in ways that are not easily modeled. A water film around a soil particle significantly increases its absorption of radiation. The mean albedos of soils must be established by measurement because of the strong dependence of absorption properties on various trace minerals.

At the wavelengths of terrestrial radiation, land surfaces have generally been assumed to be perfectly absorbing, (i.e., blackbodies). This is not strictly true, inasmuch as measured surface emissivities of canopies are generally less than 1.0, i.e., in the range 0.96–0.98, or, equivalently, reflectivities of 0.02–0.04. The most notable departures of land surfaces from blackbody properties is seen over quartz sands, where emissivities can be as low as 0.7. The resulting error in climate models is the difference between the reduced emission and reflected thermal radiation, which differ over vegetation by a few W m^{-2} and over quartz sand deserts by up to several tens of W m^{-2}.

5.4 Idealized canopies

The biological world is extremely complex in form and function. Inclusion of any aspects of biology in climate models (as described in the Chapter 14) must be extremely simplified compared to reality, emphasizing only those aspects that may be significant on the spatial scales treated by the models. Exchange of trace gases is considered in the following two chapters. Thus, here we emphasize the processes that determine surface energy balance and act as drag elements on atmospheric flow. Water and radiative exchanges are dominant elements of surface energy balance in moist regions. In more arid regions, conductive heat transfer into soils and convective transport to the atmosphere (called "sensible heat transport") can also be major terms.

The overall structure of vegetation canopies is characterized by their typical height and area of leaf surface (one-sided, as projected on a flat surface) per area of land surface. This quantity is referred to as the leaf area index, LAI. Over the large areas of a climate model, we may also need some description of heterogeneities in terms of relative fractions of different surfaces and their spatial scales. One such simple distinction is the fraction of a given large-scale surface that is vegetated versus that which is not.

Over moist regions, most of the water transferred from soil to atmosphere moves through vegetation. Water moves into the roots, through the plant parts to leaves, and passes into the atmosphere through little openings in the leaves called stomates. Studies of this plumbing of many dominant land plant species show that the rates at which water can move into the atmosphere are largely controlled by the extraction of water from the soil by roots and the movement through the stomates. Thus, we would like to distill the essence of the root and stomatal controls in terms of reasonable average properties and atmospheric and soil variables treated by the climate model, e.g., fields of surface radiation, atmospheric temperature, humidity, winds, and soil moisture and temperature.

The flux of water or heat between two points in a surface complex is often controlled by processes of steady-state molecular diffusion; if so, its rate is generally proportional to the difference of the concentrations of water vapor or temperature at the two points. This proportionality also frequently applies to situations where the transfer is by turbulence. The proportionality factor is called the "conductance", or its inverse the "resistance" for the given transfer process (Fig. 5.12). If several of these processes are linked, then, if in parallel, the conductances are added to get total conductance; if in series, the resistances are added to get the total resistance. In particular, the movement of water from the inside of a leaf to the outside is controlled by the leaf's stomatal conductance (or resistance). Similar leaves would act in parallel so that their conductances add. Thus, if r_s, the resistance per unit leaf area, were assumed the same for all leaves, then the resistance of all

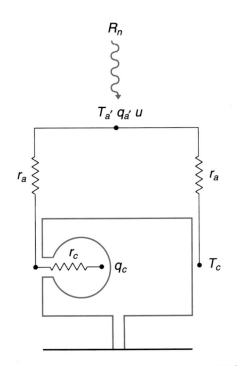

Fig. 5.12 Schematic resistances for a simple generic canopy model (Dickinson et al., 1991). The box represents either a canopy or individual leaves. Sensible heat fluxes are driven by the temperature gradient $T_c - T_s$ against the aerodynamic resistance r_a. Latent fluxes driven by the difference between water vapor inside the leaf q_c and atmospheric water vapor q_a is resisted by both r_a and the integrated stomatal resistance (canopy resistance) r_c.

the leaves in a canopy characterized by a leaf area index of LAI and acting in parallel would be $r_s/$LAI.

Both the individual stomatal resistance of leaves and the effective total resistance of the canopy can be measured observationally. There is considerable variability in the measured resistances of individual leaves, some of which depend on the age of the leaf and position in the canopy, as well as the availability of water in the soil. Plant physiologists have shown that leaves with an adequate water supply have dependences of their transpiration on temperature and light levels similar to the dependences of their carbon assimilation (Ball et al., 1987). In particular, most leaves exhibit largest stomatal opening (conductance) in the presence of full sunlight and close in the absence of sunlight. When the stomates are closed, a residual transpiration may occur through the epidermis of the leaf. The dependence of stomatal resistance on soil moisture derives from the limiting rates at which water can diffuse from the soil to the roots and be extracted by the roots. This diffusion is driven by the difference between the potential energy of water in the roots and that in the soil, the latter being a function of soil water. The rate of water extraction by roots is also proportional to the surface area of roots, a parameter difficult to determine locally, let alone over a large area.

A useful parameter that has been measured for many plants is the wilting point of a given plant in terms of soil water potential. The soil water potential is one of the soil properties in a given soil water parameterization related to soil moisture and soil texture. In terms of soil water, the wilting point varies little between different plant species and can reasonably be specified as a model constant.

For given root development, a canopy will have a maximum rate at which it can transpire before roots cannot keep up with the demand and stomates must close to compensate. This rate will be maximum under well-watered conditions and may vary approximately with the difference between soil water potential and wilting point potential. Thus the wilting point soil water potential is the same as the potential that leaves maintain to drive transpiration. The above statements can be used for a simple parameterization of the effects of root resistance (Dickinson, 1984), assuming the potential in the roots is fixed at that of the leaves. As this is a rather crude approximation, some models (e.g., the simple biosphere, or SiB, model) attempt more precise relationships between these two potentials (see Sellers et al., 1986, and Chapter 14).

Because of the variability of stomatal resistance, especially that from the closure of stomates with low light levels and the strong attenuation of light within a canopy, it is probably necessary to synthesize canopy resistance more carefully than by simply dividing the stomatal resistance by LAI. If $d\text{LAI}_i$ is an element of leaf area of the canopy with adequately constant stomatal resistance r_i, then the actual canopy resistance can be obtained by summing conductances over all these elements, $1/r_c = \sum (d\text{LAI}_i/r_i)$. Because all but the uppermost leaves are shaded enough to significantly reduce their stomatal resistance, the canopy resistance is more than twice what it would be if all its leaves could be exposed to full sun. Further complications may arise in attempting to allow for additional contributions to canopy resistance (see Chapter 14). The aerodynamic resistance across the boundary layer in the vicinity of the leaf is relatively well understood, though usually an order of magnitude smaller than the stomatal resistance and possibly within the noise level of specification of the latter. Additional resistances are sometimes included that account for turbulent transport within the canopy to some reference level in contact with the surface boundary layer overlying the canopy.

A few years ago, it was widely questioned in the communities concerned with canopy transfer processes whether observed canopy resistances would be significantly different than those simply constructed from stomatal resistances. However, because the more thorough procedures recently developed to do this synthesis (see Chapter 14) provide theoretical canopy resistances in agreement with observation, the approach just summarized is probably adequate for use with climate models and even further simplification may be warranted.

The other key resistance, in addition to canopy resistance, is the aerodynamic resistance controlling movement of heat, moisture, and momentum from some reference level in the canopy to some other reference level in the overlying air, e.g., the lowest mesh point of the atmospheric GCM. The theory generally adopted for this component is known as the Monin–Obukhov similarity theory. Aerodynamic resistance constructed according to this theory (a version of which is described further in Chapter 14) depends on the roughness length z_0 of the surface, the surface air winds, and the difference between surface and air temperature. The latter is referred to as "stability", i.e., with the surface warmer than overlying air, the boundary layer is "unstable" or vice versa. The theory is simplest under conditions of neutral stability, where the aerodynamic resistance r_a is simply $r_a = (\ln z/z_0)^2/(k^2 u)$, where u is the overlying atmospheric wind and z the level at which it is measured, and $k \sim 0.4$ is the Prandtl constant.

Otherwise, the theory may require an iterative computation to determine resistances. In some GCMs, this iterative calculation is avoided by use of a curve-fit representation of r_a in terms of the neutral value and a function of the "bulk Richardson number."

In past GCMs with simple surface parameterizations (as reviewed further in Chapter 14), only aerodynamic resistances have been included, i.e., canopy resistances were ignored. Furthermore, the aerodynamic resistances were assumed to correspond to those of a short grass surface. In this case, the aerodynamic resistances are typically as large as canopy resistances. The need to include canopy resistances is most obvious in the case of tall vegetation (such as forests), where the aerodynamic resistances are an order of magnitude smaller than the canopy resistances. Calculations of surface temperature for a forest, neglecting canopy resistances, can give very rapid evaporation and surface temperatures significantly colder than the overlying air. These conditions are close to reality only for the interception losses during and immediately after precipitation, when water freely evaporates from the canopy.

5.5 Role of soil and its hydrology

The role of soil, as needed to be represented in a climate model, varies depending on the vegetation cover. In well-vegetated regions, it serves primarily as a reservoir for the extraction of water by the roots. On the other hand, in the near absence of vegetation, it takes the brunt of daytime solar heating and consequently acts to store a significant fraction of this heat.

Incident precipitation infiltrates into the soil, but with sufficiently heavy rates of precipitation, some runs off the surface. Water within the uppermost soil layers can also run down hydraulic potentials imposed by slopes, as well as infiltrate to deeper soil layers. Vertical movement of water in soils can be addressed using well-established physical principles; their primary limitations are the presence of subgrid-scale variability in the relevant properties and lack of information on soil hydraulic properties below the surface layers. Also it is difficult to represent the horizontal runoff in a climate model.

Water in soil moves vertically down the gradient of hydraulic potential, determined by gravity and various forces attracting water to the soil particles. These forces become stronger in drier soils and give more negative potentials. In homogeneous saturated soils, only the gravitational potential drives water movement. The rate at which water moves downward through a saturated soil is proportional to the saturated hydraulic conductivity. More generally, the hydraulic conductivity is defined as the rate of water movement per unit gradient of hydraulic potential per unit area. The hydraulic conductivities and potentials of a given soil are functions of the water content of the soil. These parameters vary with different soils, largely according to soil texture and structure. Texture describes the typical particle sizes in a soil, and ranges from clay soils to sandy soils. Saturated hydraulic conductivity varies by over an order of magnitude in spanning a typical range of textures. Conductivities will usually decrease with increasing soil depth; this decrease will influence soil water storage versus channeling into hillslope runoff. Another question is the depth of the water table. A high water table can provide water back to the rooting zone and an amplification of runoff.

Evaporation from a soil surface generates a large negative potential at the surface, forcing water flow upward to supply this demand. At high rates of demand, or low soil moisture, the evaporation becomes limited by the rate at which water can diffuse to the surface. At lower rates of evaporation, supply provides whatever is the surface demand. Surface demand is determined immediately by aerodynamic resistance and by the gradient of water vapor between the overlying air and the surface, but ultimately largely by net radiative heating.

It is impractical to attempt to capture all the complexities of local surface hydrology in a GCM climate model. Furthermore, because models are necessarily much simpler than reality, different authors have emphasized different details. However, some basic concepts are necessarily incorporated in any model. One such important concept is the amount of water that can be stored in the soil. Agronomists usually estimate this quantity as the difference between the water content at "field capacity" and at the wilting point, integrated over the depth of the roots (the rooting zone). Field capacity is estimated as the water content a few days following a saturating rain when the water drainage rate has become less than some small value. This term is somewhat fuzzy, but, whatever the precise definition, it will generally be much less than the saturated water content of the soil.

Below the wilting point, evapotranspiration will not necessarily vanish, but it generally will be quite small. Also, the depth of the rooting zone is not precisely defined, since in some systems roots of certain species can extend to great depths, whereas in others they remain near the surface. Furthermore, it must include the layers below roots that can readily provide water to the roots.

Agronomists have estimated the soil water capacity (as a function of soil texture per meter of soil) to range from about 200 mm for a loam soil to around 100 mm for light and heavy textures (sand and clay). This range of values is implicit in a detailed soil-water model, whose water capacity then depends on the presumed depth of the rooting zone. This depth is typically referenced as ranging from 0.5 m for shallow soils, such as in taiga forests, to as much as 2 m for deep agricultural soils. Some systems such as tropical rainforests are especially problematical in estimation of rooting depth because they may place almost all their roots close to the surface for more efficient recycling of nutrients but extend some roots to considerable depth to avoid water stress.

The simplest bucket treatments of surface hydrology have assumed a universal 150 mm of soil water capacity. This value is probably reasonable within a factor of two over any given continental scale region in comparison with reality. It may be difficult at present to obtain much more accuracy in constructing this quantity over a region from local soil information. Yet a factor of two variation in the water holding capacity over a large region could be of considerable climatic significance. Thus the development of some large-scale observational constraints on this parameter would be highly desirable. In continental-scale regions, where soils can be nearly depleted over a drying season, it should be possible to use atmospheric water budget, precipitation, and runoff measurements to estimate the net soil depletion between the end of the wet season and the end of the drying season (e.g., Rasmusson, 1968).

5.6 Geographic data needed for land processes

In order to include vegetation and soils in a climate model, values must be assigned to each of the parameters needed for the land process subcomponents. These values vary widely from one plant to another and from one plot of soil to another (Fig. 5.13). Over larger areas, they will vary with geography and predominant vegetation cover. However, they have been measured, if at all, only for local canopies and soils. Thus, we must develop procedures to map these local measurements to the global distribution of continents. Currently, this is done by using global data sets of vegetation cover and soils to specify either the dominant cover or the distribution of cover over each GCM land grid square. Local measurements of the parameters needed in the surface parameterizations are then associated with these data sets. Descriptions of vegetation cover

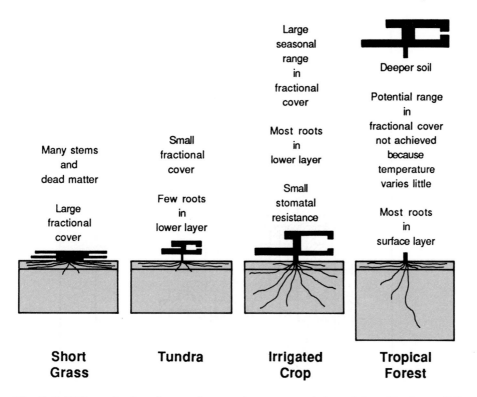

Large
seasonal
range
in
fractional
cover

Deeper soil

Potential range
in
fractional cover
not achieved
because
temperature
varies little

Many stems
and
dead matter

Small
fractional
cover

Most roots
in
lower layer

Large
fractional
cover

Few roots
in
lower layer

Small
stomatal
resistance

Most roots
in
surface layer

**Short
Grass**

Tundra

**Irrigated
Crop**

**Tropical
Forest**

Fig. 5.13 Different land surfaces and vegetation structure (adapted from Henderson-Sellers and McGuffie, 1987).

can go into considerable detail. However, these details give no information to a modeler unless different model parameter values are distinguishable according to the description of vegetation. Our current confidence in surface parameterizations and in ability to relate model parameters to descriptions of vegetation data is limited and suggests use of at most 10–20 different cover classes.

Global vegetation cover and soils have been summarized in various systems and to varying degrees of accuracy in many national atlases. Since vegetation cover continues to be changed by human activities, climate change, and other causes, it is desirable to link such descriptions to a reference date, but such is usually not available. UNESCO (United Nations Educational, Scientific, and Cultural Organization) has attempted to provide a general classification for vegetation and FAO (Food and Agriculture Organization of the United Nations) has done the same for soils. Several attempts have been made to summarize the available global information on either 0.5- or 1.0-degree meshes (summarized in Henderson-Sellers et al., 1986). These data sets, condensed to a limited number of cover descriptions, are now beginning to be used by climate modelers. Figure 5.14 suggests the minimum level of geographic detail on vegetation and soils that

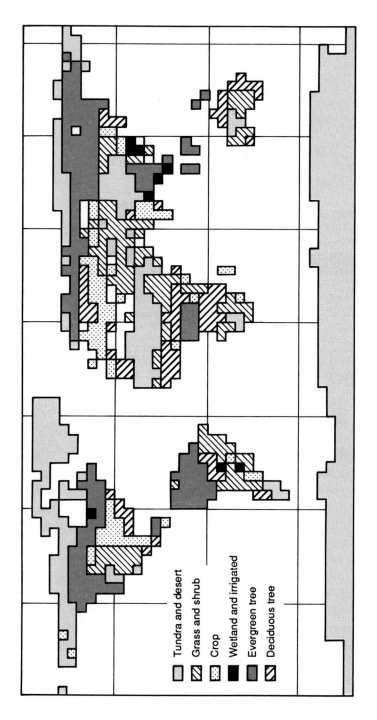

Fig. 5.14 Distribution of general ecotypes on a 5° × 5° grid (Olson et al., 1983).

Tundra and desert
Grass and shrub
Crop
Wetland and irrigated
Evergreen tree
Deciduous tree

might presently be incorporated usefully into climate models.

Part of the issue when including geographic distributions of input parameters into climate models is the question of importance or sensitivity of these specifications. Currently, we know that presence or absence of vegetation is significant, as is the distinction between tall and short vegetation (Chapters 14 and 22). These distinctions are probably adequately captured in current treatments. To the extent that further details are important, it is easy to question whether they are included properly because of inherent inaccuracies, both in geographical information and in procedures for relating model parameters to this information.

Several studies have suggested that spatial heterogeneities in soil moisture will give a significantly different description of evapotranspiration than that based on assuming uniform properties across a model grid cell. Such heterogeneities result from subgrid-scale distributions of precipitation, as well as from the vegetation and soil cover. It remains to be seen whether data at the 0.5- to 1.0-degree resolution will be adequate for the parameterizations treating these heterogeneities or alternatively whether finer-scale geographical information will be needed. Treatments of runoff based on known dominant processes are likely to require much finer resolution, at least of the topography.

6

Terrestrial ecosystems

John D. Aber

6.1 Introduction

Studies of terrestrial ecosystems examine the movement of energy and mass (especially water and nutrient or pollutant elements) through units of the landscape called ecosystems. Of central importance are the effects of temperature, radiation, and the availability of water and nutrients on the production and decomposition of organic matter by plants and microbes, and the cycling of critical elements. The convergence of interests on atmosphere–biosphere exchanges, for example, of water and CO_2, has led to increased interaction between the climate and ecosystem research communities.

However, integrating models of climate change and terrestrial ecosystem dynamics poses one of the most difficult challenges in the field of Earth system science. In no other pairing of disciplines are the temporal and spatial scales at which research is conducted so tremendously mismatched. Current General Circulation Models (GCMs) operate at time steps of minutes to hours while dividing the Earth's surface into blocks which are hundreds of kilometers on a side. In contrast, most terrestrial ecosystem models operate at monthly to annual time steps on a spatial scale measured in tens of meters. The fundamental reason for this difference is the fixed structural nature of terrestrial ecosystems. This causes significant changes in function to occur over short distances, while also providing stability in the factors controlling key processes, causing them to change relatively slowly over time at any one spot. In essence, the atmosphere is a well-mixed entity, and the degree and dynamics of that mixing are its critical components, while terrestrial ecosystems are not "well-mixed" and so require study over longer time scales and shorter spatial scales.

Still, understanding the coupled climate–land system is critical to unraveling the question of global change. There are two interactive "halves" to this problem (Fig. 6.1) of relatively equal importance. The purpose of this chapter is to discuss how atmospheric inputs (energy, water and chemical deposition) affect the structure and function of terrestrial ecosystems, and in turn, how these changes alter outputs of energy, water and, particularly,

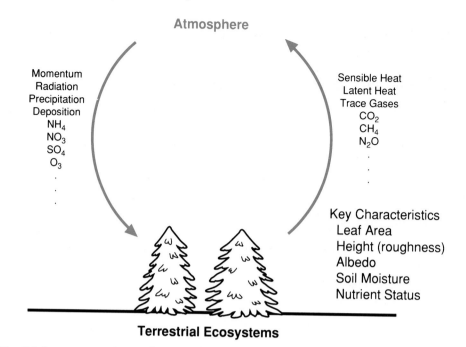

Fig. 6.1 Important exchanges between the atmosphere and terrestrial ecosystems.

radiatively active trace gases. As Chapter 5 has concentrated on the short-term energy and water balances, and the role that ecosystem structure plays in those balances, this chapter will focus on changes in ecosystem structure which can result from climatic change, and also on factors controlling trace gas fluxes. For further reading on ecosystems in the global context see texts by Aber and Melillo (1991), Schlesinger (1991) and Whittaker (1975).

Human use of the land surface also plays a major role in determining the effects of terrestrial ecosystem function on global climate. Use can be both direct, as in the conversion of tropical forests to pasture, or indirect, as in the deposition of industrial pollutants in distant rural areas. Either form of use can initiate a series of very long-term (decades to centuries) changes in both structure and function. While human use of the landscape is addressed more fully in Chapter 2, it is necessary to include some discussion of this critical factor in this chapter as well.

6.2 Climate- and human-driven changes in ecosystem structure

6.2.1 *Importance of the vegetated surface*

Terrestrial vegetation provides an important set of boundary conditions for climate models. Parameters of interest in the short term include structure (roughness) of the canopy, total leaf area and the seasonality of leaf display, albedo, and the partitioning of incoming radiation between latent and sensible heat. This partitioning depends on canopy structure and soil moisture holding capacity, as well as the time-integrated site history of precipitation and evapotranspiration (summarized in water stress effects on vegetation). Terrestrial ecosystems differ greatly in structural characteristics and physiological factors controlling transpiration. Climate-driven changes in ecosystem structure could have important feedbacks to the climate system.

6.2.2 *Potential global changes in the vegetated surface*

Both climate and human use determine the distribution and structure of ecosystem types. Climatic control over ecosystem (or "biome") distribution is relatively well understood. Human use is as diverse as the cultures and technologies from which those uses spring, and so is more difficult to describe. Still, some generalized results of the effects of human use can be identified.

a. *Shifts in distribution of native biome types in response to climatic change*

(i) *Climate–vegetation models*
One of the most straightforward questions that can be asked is: "Will climate change cause a shift in the distribution of vegetation types on the land?" If so, this would alter the characteristics of the Earth's surface which affect energy and water balances.

Among the earliest concepts in the field of Ecology was the relationship between climate and the distribution of major vegetation types, or biomes. One of the simplest presentations of this relationship describes potential vegetation as a function of mean annual temperature and precipitation (Fig. 6.2). Perhaps the most widely used system is the Holdridge (1947) classification based on "biotemperature" (annual sum of degree days between 0 and 30°C) and "humidity provinces" calculated as the ratio of potential evapotranspiration to precipitation. Holdridge's life zones (Fig. 6.3) can be combined with global-scale mean climatic data to produce large-scale maps of potential vegetation distribution (Fig. 6.4a), and other

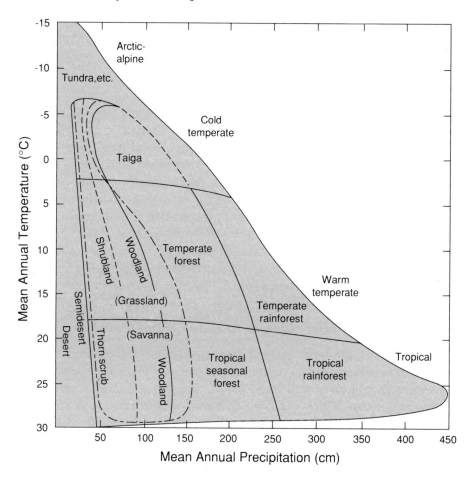

Fig. 6.2 An example of a simple classification of vegetation types of the world based on annual precipitation and mean annual temperature (Whittaker, 1975).

global ecosystem characteristics, such as the total storage of nitrogen in soils (Post et al., 1985). More complex models include information on the seasonality of temperature and precipitation, and so can distinguish between temperate grasslands which receive most of their precipitation in summer and temperate shrublands which are dry in summer and receive precipitation in winter.

Predictions of the effects of climate change on biome distribution are easily made with these systems. A first order approximation would be to use current GCMs to predict temperature and precipitation fields under various

176

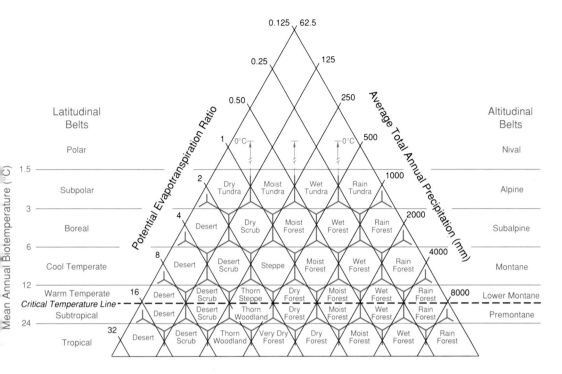

Fig. 6.3 The Holdridge system for classification of terrestrial life zones on the basis of annual "biotemperature", precipitation, and potential evapotranspiration (Holdridge 1947).

climate change scenarios, and use these to compute the new distribution of vegetation types (e.g. Fig. 6.4b). Changes in structure predicted by this process could be re-entered as new terrestrial conditions for GCM runs, and an analysis done of the importance of these changes for future predicted climate.

There are, however, several important factors which complicate this simple analysis.

First, the movement of species from one site to another is not instantaneous, and may not be rapid enough to keep pace with shifts in climate. For example, in the central portion of North America, a 3°C increase in mean temperature could lead to a northward shift of 300–. 375 km in the biome–climate equilibrium boundaries within 100 years (IPCC, 1990a). Studies on the rates of species re-invasion following the last de-glaciation suggest that species move only at the rate of 25 to 40 km per century, with a maximum of 200 km (Davis, 1981). Assuming very limited species mobility, Davis (1988) has used current and two different

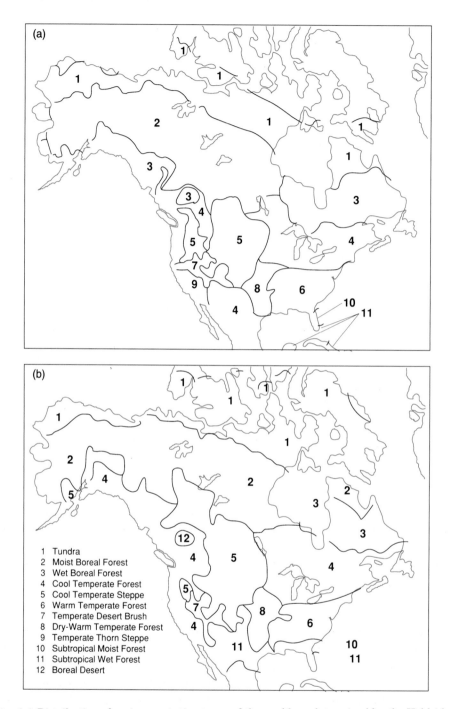

Fig. 6.4 Distribution of major vegetation types of the world, as determined by the Holdridge system: (a) under current climatic conditions, and (b) under predicted climatic conditions with a doubling of atmospheric CO_2 concentration (Emanuel et al., 1985).

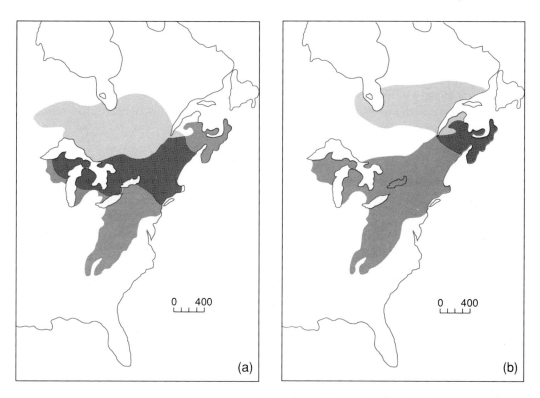

Fig. 6.5 Present distribution of eastern Hemlock (*Tsuga canadensis*), (gray) and potential future distribution (light gray), with area of overlap given by the heavy shading, under two scenarios of climate change with a doubling of atmospheric CO_2 concentration (Davis, 1988).

projections of future climatic patterns in North America to predict the possible contraction of the distribution of one species as a result of climate change (Fig. 6.5).

Second, there are also close associations between the distribution of major vegetation types and the distribution of major soil types which may hinder the free movement of biome boundaries. For example, boreal forests tend to form thick, acidic organic, surface soil horizons, with a bleached, acidic, nutrient poor upper mineral soil horizon (the classic podzolic soil, or Spodosol). In contrast, grasslands tend to form soils which are less acidic, have a relatively even and deep distribution of organic matter, and are rich in clays (the classic prairie soil, or Mollisol). At mid-continent in North America, there is a boundary between grassland and boreal forests (see Fig. 6.4a). If climate change pushed the boreal forest north, could grasslands invade the podzolic soils produced under boreal forest? Would grassland species grow as well on podzolic soils as on grassland soils?

Third, there is the relatively remote possibility that the *mix* of climatic signals (day length, temperature, length of growing season, etc.) that trigger important stages in plant reproduction and growth might change such that either distribution or migration of species would be severely reduced.

If any of these complexities become important, then climate change would *not* result in a smooth transition from one vegetation type to another, but rather to episodes of vegetation die-back followed only slowly by replacement with in-migrating species.

(ii) *Climate–physiology–vegetation models*

Answers to these complicating questions can be developed by adding an additional level of resolution to the climate–vegetation models. This step adds the physiological responses of individual species of plants to climatic variables, and then models the process of plant competition and succession to predict the steady-state vegetation expected under given climatic conditions. A series of models have been developed which do this for forest ecosystems (Botkin et al., 1972; Shugart and West, 1977; Aber et al., 1982; Pastor and Post, 1986; Solomon, 1986).

For example, the LINKAGES model of Pastor and Post (1986) describes individual species in the eastern U.S. in terms of their responses to growing season heat sum, cumulative growing season water stress, and the availability of light and nitrogen. This model accurately describes the current distribution of species in the region, and has been used to predict the future steady-state distribution under a $2\mathrm{x}CO_2$ (doubling of atmospheric CO_2) scenario. These predictions are similar to those produced from the simpler climate–vegetation models and lend support to those models.

However, LINKAGES can also predict the dynamic response to climate change; the time-course of adjustment to altered climate. The results of these individual site simulations suggest very different responses in different parts of the eastern North American deciduous forest region. Forests dominated by species at the dry or warm edge of their range are forced into decline as climate warms. In contrast, forests with dominant species well within their climatic boundaries often increase in productivity as temperatures increase and the length of the growing season increases.

While physiological models can address the complexities of species dynamics and growth within forest biomes, there are no species-level models available which can deal with or predict the type of biome-to-biome (e.g., boreal forest to grassland) switch described above. At least one model of this type is currently under development.

6.2.3 Land use and alteration of the vegetated surface

Models of vegetation dynamics under "natural" conditions apply to a rapidly decreasing fraction of the Earth's land surface. Over most of Europe and much of temperate North America, the domestication of the landscape is nearly complete, and only relatively small packets (on a global scale) remain in pre-settlement conditions. The current wave of clearing and conversion in tropical forests show that models of natural systems will have decreasing relevance in those areas as well.

The ecological research community has, for a host of reasons, been slow to undertake the study of human-manipulated ecosystems. As a result, less is known about these systems than their spatial importance requires. Still, it is possible to present some generalizations regarding the effects of human use on ecosystem structure.

In general, human use includes removal of biomass (live plant material) from the system. Removals can be small, as in the grazing of native grasslands, or can be extreme, as in the conversion of forests to agriculture. The most fundamental change with respect to water and energy balances would come in the conversion from forest to crops or pasture. In tall forest canopies, increased "roughness" increases turbulent mixing through the canopy such that stomatal resistance becomes more important than aerodynamic resistance in determining rates of transpiration and the partitioning of incoming radiant energy between latent and sensible heat. In grasslands, pastures and crops, with low, relatively dense and even canopies, mixing is reduced and aerodynamic resistance becomes more important than stomatal resistance (see Chapters 5 and 14).

In addition, any form of crop removal reduces transpiring leaf area, increases solar radiation striking the soil surface, and probably reduces rooting depth and density, all of which will tend to reduce transpiration. On the other hand, in industrial farming, the use of fertilization and irrigation can increase leaf area and prolong the growing season. These differences have been shown to be important in meso-scale models of atmosphere–biosphere interactions.

6.3 Climate- and human-driven changes in carbon balances of terrestrial ecosystems

6.3.1 *Importance of terrestrial ecosystem carbon balances*

Carbon dioxide is currently the most important greenhouse gas. With a 25% increase in concentration since the beginning of the industrial revolution (280 to 350 ppm), CO_2 accounts for about half of the increased potential of the atmosphere to trap out-going longwave radiation (see Fig. 2.1). While the distribution and extent of climate change are meteorological phenomena, human alteration of the cycles of CO_2 and other trace gases are, along with structural changes in the surface of the Earth, the forcing function driving that change.

Terrestrial ecosystems are a critical part of the global carbon cycle, with a gross annual flux roughly equal to that of the world's oceans (Fig. 6.6).

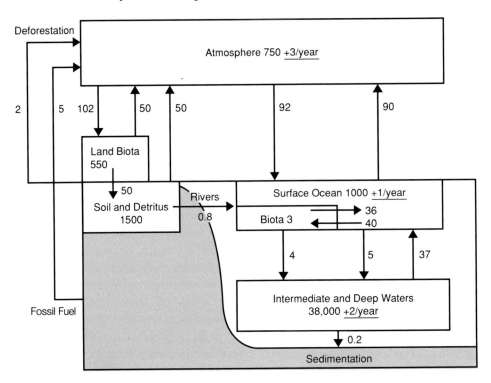

Fig. 6.6 Current view of the global cycle of CO_2 (pool sizes in 10^{15} grams and fluxes in 10^{15} grams per year, IPCC, 1990a).

Deforestation and other conversions and uses of the land surface are thought to contribute a net flux of 1–2×10^{15}g of carbon per year to the atmosphere, 20–40% of the contribution due to combustion of fossil fuels.

6.3.2 *The carbon and nutrient cycles of terrestrial ecosystems*

Carbon dioxide is consumed by terrestrial ecosystems through photosynthesis, and is produced by respiration of both plants and decomposers (primarily soil fungi and bacteria). However, this simple balance is complicated by the production, within plants, of a wide variety of organic materials. Some, such as foliage and fine roots, are generally short-lived and become available to decomposers within one to a few years of production. Others, such as woody stems in forests and root crowns of grassland species, can live for decades to centuries.

In addition, each type of tissue produced contains a mixture of relatively easily decayed carbon compounds (such as starch and sugars) and more resistant compounds (such as cellulose, a long-chain carbohydrate polymer), and lignin, a large, amorphous polymer of phenolic ring structures and complex side-chains. These different compounds decay at different rates, and to different stages of completion. Only the smallest carbohydrate molecules can be assimilated directly by microbes. Larger starch and cellulose molecules are degraded by the microbial production of extra-cellular enzymes which are exuded into the soil matrix. This is a less efficient process, as the enzymes produced can be deactivated by condensation with polyphenols or lignin. Lignin is especially resistant to decay because of its amorphous structure and the high energetic cost of initiating decay of the phenolic ring structures. Indeed, the decay of lignin is generally incomplete, and modified ligno-protein complexes are thought to compose most of the very old (hundreds to thousands of years) organic matter in soils known as humus.

It is possible to structure a model of carbon dynamics in terrestrial ecosystems which deals with all of these pools (Fig. 6.7). In such a model there are several decision steps which determine the allocation of carbon to different tissues, and within tissues, to different types of carbon compounds.

The allocation of carbon both to tissues (e.g. roots versus leaves) and to types of carbon compounds within tissues (e.g. lignin versus protein) is controlled by the genetics of the plant (i.e. inherent growth form such as herb versus tree) modulated by the relative availability of nutrients, light and water. For example, a given plant growing under suboptimal nitrogen availability will have more mobile carbohydrate relative to mobile nitrogen within the plant (see Fig. 6.7), in comparison with a plant growing with excess nitrogen availability. As a result, less of the mobile carbohydrate will be combined into amino acids (and on into proteins, enzymes, etc.), and more will be transformed into complex carbon compounds such as lignin and cellulose (Waring et al., 1985).

As these tissues undergo senescence and are shed as litter, the effect of higher cellulose and lignin concentrations will be to reduce litter decay rates (Melillo et al., 1982; Aber et al., 1990) and nitrogen release from decaying litter. This completes a positive feedback loop where low nitrogen availability leads to low nitrogen content and higher lignin content in litter, which in turn slows decomposition and reduces nitrogen availability. The feedback also works in the opposite direction where high N availability leads to litter high in nitrogen and low in lignin, resulting in faster decay rates and further increases in N availability.

These feedbacks become negative when carbon (as CO_2) is the resource being altered. Increased CO_2 concentration in the atmosphere will increase net rates of photosynthesis and water use efficiency (the quantity of CO_2 fixed in photosynthesis per unit water transpired). With more available carbon, synthesis of dry matter can increase. However, without a coincident increase in nitrogen availability, more of the carbon will go into lignin and cellulose, diluting plant tissue nitrogen to lower concentrations. Beyond certain limits, further dilution of N may not be possible, and further growth increases will not occur. When these tissues produced at elevated CO_2

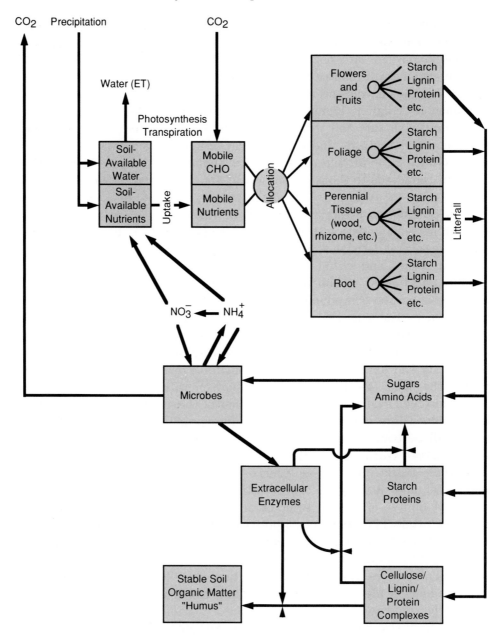

Fig. 6.7 A generalized model of carbon and nitrogen movement through terrestrial ecosystems (see text for discussion).

concentrations are shed, they will decay more slowly and reduce future N availability, perhaps reducing, rather than increasing, plant growth.

These interactions also affect the processes and substrates controlling methane and nitrous oxide fluxes of terrestrial systems. Nitrogen is mineralized from organic matter as ammonium (NH_4^+). Ammonium can

184

be taken up directly by plants, or can be oxidized by microbes to nitrate (NO_3^-). In strongly nitrogen-limited systems, plants (with fungal root symbionts together called mycorrhizae), compete effectively with free living nitrifiers such that nitrate production (nitrification) is very low and nitrate pools are low to undetectable. As discussed in the next section, soil concentrations of ammonium, along with nitrification rate and soil nitrate concentrations, are important factors determining net fluxes of methane and nitrous oxide over soils (for a more complete discussion of gross and net N mineralization and nitrification see Robertson, 1982; Schimel et al., 1989; Davidson et al., 1990).

Two field experiments will help to demonstrate carbon and nitrogen interactions and feedbacks in terrestrial ecosystems.

The first is an experimental CO_2 enrichment study in the arctic tundra. The potential for increased atmospheric CO_2 concentrations to increase net primary production (NPP, the total mass of plant material produced by plants per year) and net carbon storage on land has received a great deal of attention. The short-term effects of elevated CO_2 concentration on plants include increased rates of net photosynthesis, reduced stomatal conductance and increased water use efficiency (g CO_2 fixed per g water transpired, Mooney et al., 1991; Bazzaz, 1990). These changes lead to increased growth and carbon storage in laboratory studies where temperature and nutrient conditions are optimal. However, in the strongly nitrogen-limited tundra system, plant growth rate did not increase in response to an initial, short-term increase in photosynthesis, most likely because the nutrients required to build additional biomass were not available. Instead, there was an accumulation of starch in foliage and other tissues, a dilution of foliar nitrogen with carbon (higher C:N ratio) and acclimation of the photosynthetic system, returning CO_2 fixation rates to pre-treatment levels (Tissue and Oechel, 1987, results summarized in Mooney et al., 1991).

In the second example, consider the case of initially nitrogen-limited spruce-fir forests at high elevations in the mountains of New England. These systems have very low rates of nitrogen cycling, and the dominant tree species are evergreens with very high lignin concentrations and low N concentrations in foliage. These forests are currently subjected to a range of annual N deposition rates which varies from east to west due to proximity to sources of N pollution. Apparently in response to this gradient of N inputs, foliar lignin concentrations and lignin:nitrogen ratios have declined along with soil carbon:nitrogen ratios, while nitrogen mineralization (or net release) from soils has increased (McNulty et al., 1991). The feedback between increasing N availability, higher foliar tissue "quality" and increased decomposition of organic matter can be seen clearly in this example.

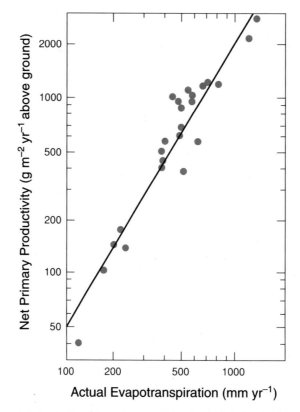

Fig. 6.8 An early, global relationship between estimated annual actual evapotranspiration and annual net primary productivity above ground (Rosenzweig, 1968).

6.3.3 Models of carbon–nutrient–climate interactions

The earliest models related to terrestrial carbon balances at the global scale derived simple relationships between summary climatic variables and net primary production (the net production of dry organic matter by plants, Fig. 6.8). Similar relationships also exist for the decay of freshly deposited litter (senescent plant material). Estimated annual actual evapotranspiration (by a simple monthly water balance calculation based on temperature and precipitation) is the variable which has been generally used for these simple relationships.

Such simple relationships are only incomplete approximations of the carbon dynamics of terrestrial ecosystems. NPP in these models generally includes above ground biomass only, excluding root production which can contribute 30% to 70% or more of total NPP depending on the ecosystem type. Decomposition is not controlled by climate alone, but also by the chemical composition of the material (Meentemeyer, 1978; Melillo et al., 1982; Aber et al., 1990), which is under both genetic and environmental control (Waring et al., 1985; Flanagan and Van Cleve, 1983).

As with the discussion of biome movement, a more complete analysis of the integrated effects of change in both climate and atmospheric chemistry on terrestrial carbon balances can be achieved through the use of more physiologically based models of ecosystem function. Several models are currently being applied to the question of carbon–nutrient–climate interactions. Four will be reviewed briefly here.

The LINKAGES model described before has also been applied to the question of nitrogen feedbacks controlling plant production and total carbon storage on land. This model accurately predicts the interaction between soil moisture holding capacity, species distribution and resulting rates of N cycling for a series of temperate forests (Pastor and Post, 1986). It has also been used to predict changes in N cycling in boreal forests as driven by cycles of disturbance and recovery, and patterns of litter quality over this sequence (Pastor et al., 1987).

A very similar model, also built on the plot, individual stem and species level, is the FORENA model of Solomon (1986). Using both the current climatic regimes, and those predicted under a $2xCO_2$ climatic scenario for 21 locations within the eastern deciduous forest region of North America, the model was run to predict changes in carbon storage in above ground biomass only (Fig. 6.9).

A similar analysis has been done for soil carbon storage in North American grasslands using the CENTURY model (Parton et al., 1988). CENTURY has a much simpler structure. Plant biomass is divided according to tissue type (e.g., foliage, stems, roots) rather than dealing with individual species or plants. Soil carbon is divided into five compartments, through which organic matter passes during the decay process (Fig. 6.10). The rate of turnover for each compartment is fixed, and the partitioning of carbon between compartments is a function of litter chemical content and soil texture.

Extensive data on net primary production in relation to monthly time-step climatic data have been used in conjunction with soil type and current and $2xCO_2$ climate scenarios to predict net changes in total carbon storage for this region (Burke et al., 1991). It is interesting to point out that in both of the last two cases (e.g., Fig. 6.9) the predicted net changes in total carbon storage show significant spatial patterning. This emphasizes the importance of both the nonlinear interactions between climate change and ecosystem processes, and the spatial pattern of edaphic (site) factors which partially control those processes (Burke et al., 1990; Schimel et al., 1990).

To date, no models have been developed which deal with the full complement of potential global change drivers of native ecosystem function. These include direct CO_2 effect on plants and feedbacks to the decomposition system, effects of temperature and precipitation on both plant growth and decomposition, the additional effects of climatic change

Net Change = -11 t ha^{-1}

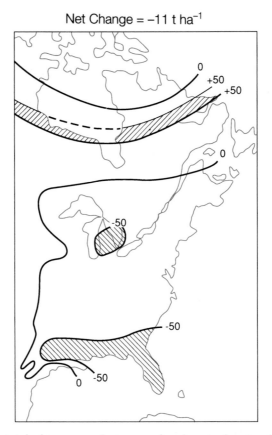

Fig. 6.9 Changes in total above ground storage of carbon in forests of the eastern U.S. in response to climatic changes predicted with a doubling of atmospheric CO_2 concentration as predicted by a species- and plot-level model (Solomon, 1986, all values in 10^6 grams per hectare).

on water balances, water stress, and effects of these on plant growth and decomposition, and the interaction of all of these with human alterations in the deposition of nutrients and pollutants to ecosystems. Dealing with changes in direct human use of the landscape will add a further dimension of complexity to the problem.

6.3.4 Role of land use in production and nutrient cycling

Human use of the landscape generally reduces carbon storage. The harvesting of forests generates CO_2 both through the decay or combustion of wood products and by increasing soil and litter decomposition and depressing primary production while the re-establishment of the vegetation is occurring. Conversion of either forests or grasslands to agriculture releases CO_2 both through the loss of vegetation and litter inputs (removed as harvest) and through plowing which increases humus decomposition

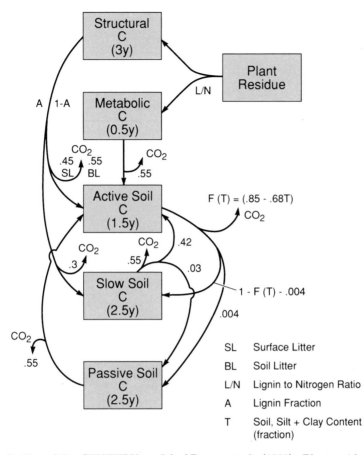

Fig. 6.10 Outline of the CENTURY model of Parton et al. (1988). Plant residue (litter) is divided into structural (cell wall) and metabolic (cell content) pools which differ both in decay rate and the degree to which each contributes to more resistant decay pools. Decay rates are affected by type of organic matter, climatic variables and soil texture.

significantly. However, re-growing forests or abandoned agricultural land reverting to native vegetation become a net sink for CO_2 as both tree biomass and soil organic matter are re-accumulated.

Entire regions will show net changes in carbon storage only if the relative distribution of land uses is changing. As a simple example (Fig. 6.11), assume a region consisting of three large blocks of forest which are harvested at 90 year intervals, but the harvests are staggered by 30 years so that different blocks are cut at different times. Each third of the area will continually pass through the full spectrum of carbon content, but as one area is losing carbon through harvest and decay, another is gaining carbon through regrowth. The total carbon storage for the region, the sum of the three curves for the subregions, varies relatively little.

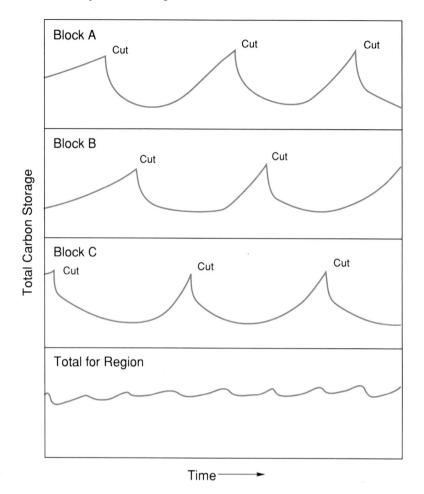

Fig. 6.11 Effects of repeated cycles of harvest and regrowth on the cumulative carbon balance of a region. This demonstrates that although individual units of the landscape may exhibit large changes in total carbon storage, the net effect at the regional level will be minimal unless the overall intensity or frequency of biomass removal over the entire region is altered (Aber and Melillo, 1991).

The same concept applies to regions which contain a mix of agricultural and native landscapes. As long as the percentage of the region in each land use category does not change, the total carbon storage for the region will not show large changes. Conversely, large changes in carbon storage will occur in regions where dramatic changes in the distribution of land use is occurring.

It is not yet possible to model the carbon balances of the complete array of forest management and agricultural practices which exist throughout the world. Rather, the most data-rich of existing "models" of global net carbon

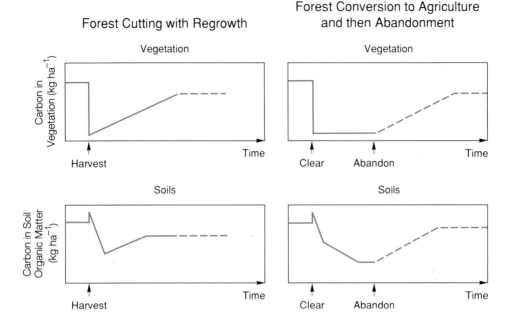

Fig. 6.12 Changes in total carbon storage in vegetation and soils for two types of land use sequences in tropical forests as described for a "population cohort-type" model of global carbon balances of terrestrial ecosystems (Houghton et al., 1983).

flux due to direct human use of terrestrial ecosystems (Houghton et al., 1983) uses essentially a population dynamic approach to units of land which pass through empirically derived patterns of total carbon storage as a function of age and land use practice (e.g. Fig. 6.12).

Historical trends in land settlement and the spread of agriculture are clearly revealed in the model projections of carbon balances at the continental scale (Fig. 6.13a). It is interesting that this model predicts that only within the last 40 years has the input of carbon to the atmosphere from fossil fuels been greater than the input from terrestrial ecosystems (Fig. 6.13b). This suggests that much of the 25% increase in the CO_2 content of the atmosphere is the result of human use of the landscape, rather than fossil fuel consumption.

The current best estimate derived from this model and others is that terrestrial ecosystems are a source of 1 to 2×10^{15} g of carbon per year to the atmosphere, or 20% to 40% of the fossil fuel input. There is general agreement that harvesting and conversion of tropical forests is by far the largest component of the carbon flux from terrestrial ecosystems (Houghton et al., 1983).

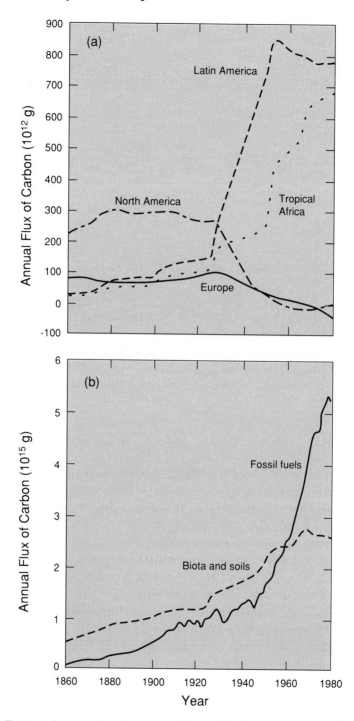

Fig. 6.13 Estimated temporal patterns in: (a) net CO_2 flux over terrestrial ecosystems at the continental scale as a result of human land use, and (b) comparison of total flux from terrestrial ecosystems with emissions due to combustion of fossil fuels (Houghton et al., 1983).

6.4 Interactions between climate and fluxes of methane and nitrous oxide

6.4.1 *Importance of net terrestrial fluxes of trace gases*

Trace gases other than CO_2 contribute roughly half of the total increase in longwave radiation absorbance in the atmosphere (Fig. 2.1). Two of the most important of these gases are methane (CH_4) and nitrous oxide (N_2O). For both of these gases, terrestrial ecosystems, including wetland systems, represent important fractions of the global cycle.

Unlike CO_2, neither CH_4 nor N_2O are important components of the annual cycles of the C and N in terrestrial ecosystems. This means that the rates of production and consumption of these two are not controlled by coarse scale interactions between climatic conditions, nutrient availability and net primary production. Instead, controls are more subtle and complex, which would, by itself, make regional-to-global prediction and modeling of the dynamics of these gases difficult.

In addition, however, fluxes of both CH_4 and N_2O are also very sensitive to the oxidation/reduction potential in the soil/sediment. Redox state is, in turn, subject to short-term fluctuations in time, due primarily to changes in water balances, and in space due to topographic and hydrographic position (i.e., rapid change in depth-to-water table over short distances). This leads to the existence of "hot spots", or a very uneven distribution of fluxes in both time and space. The sensitivity of these fluxes to potential changes in climate is thus also complex and difficult to predict. Will predicted changes in climate result in increased net fluxes to the atmosphere from terrestrial ecosystems, creating a positive feedback and further warming, or will net fluxes diminish, creating a negative feedback? This is one of the key questions being pursued at this time in the field of biosphere–atmosphere interactions.

6.4.2 *Factors controlling fluxes of methane*

a. *Reactions and controls*

There are two major pathways for the production of methane, both requiring anaerobic conditions. They are the reduction of CO_2 with hydrogen, fatty acids or alcohols as hydrogen donors, or transmethylation of acetic acid or methyl alcohol by CH_4 producing bacteria (methanogens, IPCC, 1990a, Cicerone and Oremland, 1988). Methanogenesis requires highly reduced conditions, such that production in terrestrial ecosystems is limited to wetland soils and sediments, and the digestive systems of animals. Methanogenic organisms can tolerate wide extremes of salinity and pH, but generally show increased production with increasing temperature (Cicerone and Oremland, 1988). Global methane production is thus dominated by fluxes from natural wetland and bog areas, where production rates may be related to water depth (Sebacher et al., 1986), and from agricultural wetlands such as rice paddies (Table 6.1). Estimates of both the aerial extent and flux rates of natural and agricultural wetlands are required to predict total global production rates (Aselman and Crutzen, 1989; Mathews and Fung, 1987).

Prediction of net rates of methane flux from wetland areas are complicated by the common occurrence of methane consumption in aerobic soil/sediment layers overlaying anaerobic layers. For this reason, total methane flux from natural wetlands (Bartlett et al., 1991) or rice paddies may be increased by the presence of emergent vegetation. Emissions can occur by diffusion through the stems of plants, by-passing the less anaerobic near-surface water and sediment layers (Schutz et al., 1989; Sass et al., 1990). Movement of methane from sediments to water surface in the form of bubbles may also provide a mechanism for increased methane release in the absence of emergent vegetation (Bartlett et al., 1991; Martens et al., 1991).

Aerobic soils in upland terrestrial ecosystems also consume methane. At least two distinct groups of bacteria are involved: methanotrophs and nitrifiers (Jones and Morita, 1983; Ferenci et al., 1975). The enzyme systems developed by these sets of organisms do not effectively discriminate between CH_4 and NH_4 as substrates. This means that there is a competitive inhibition of methane consumption with increasing concentrations of

Table 6.1. Estimated sources and sinks of methane (from IPCC, 1990a).

	Annual release (T_g CH_4)	Range (T_g CH_4)
Source		
Natural wetlands (bogs, swamps, tundra, etc.)	115	100 – 200
Rice paddies	110	25 – 170
Enteric fermentation (animals)	80	65 – 100
Gas drilling, venting, transmission	45	25 – 50
Biomass burning	40	20 – 80
Termites	40	10 – 100
Landfills	40	20 – 70
Coal mining	35	19 – 50
Oceans	10	5 – 20
Freshwaters	5	1 – 25
CH_4 hydrate destabilization	5	0 – 100
Sink		
Removal by soils	30	15 – 45
Reaction with OH in the atmosphere	500	400 – 600
Atmospheric increase	44	40 – 48

ammonium in soils. Two studies have demonstrated significant reductions in methane consumption as a result of experimental additions of nitrogen fertilizers to agricultural and temperate forest soils (Steudler et al., 1989; Mosier et al., 1991). The implication of this is that methane consumption in forest and agricultural soils may be declining in response to pollution-induced increases in nitrogen deposition in northern temperate and boreal forests, and fertilizer applications in agricultural regions. It is not yet known whether this represents an important contribution to the global increase in atmospheric methane concentration.

b. *Rates and locations of sources and sinks*

The current view of the global methane cycle (Table 6.1) indicates that anthropogenic sources of methane contribute 345×10^{15} grams to the atmosphere annually, compared with natural sources of 180×10^{15} grams. Natural sources include emissions from wetlands and bogs as well as from the digestion processes of termites and a rate estimated for digestion by pre-industrial levels of wild ruminant populations. Rice paddies are the most important anthropogenic source of methane, followed by biomass burning (especially in the tropics), gas and coal drilling and transmission, and anaerobic decay in landfills. The major sink for methane is oxidation in the atmosphere, with a lesser amount consumed by aerobic soils. Note that uncertainties in the sources and sinks for methane in Table 6.1 result in an imbalance that does not account for the observed atmospheric increases. Estimates of many of these fluxes are being revised continually.

c. *Interactions with climate change*

Methane concentrations in the atmosphere are increasing at a rate of nearly 1% per year. The potential contribution of this to global warming is on the order of 15% of the total due to all anthropogenic trace gas increases (Fig. 2.1). How might climate change affect methane production and consumption in terrestrial ecosystems?

Major attention is currently focused on the large wetlands and bogs of the boreal and tundra zones in the Northern Hemisphere. Much of this area is underlain by a layer of permanently frozen soil called permafrost. In many areas, deep layers of peat (partially decayed, carbon-rich organic matter) are locked in this frozen layer. One line of reasoning suggests that increased temperatures in this region will push the depth to permafrost farther down in the soil, increasing the volume of anaerobic soil in which methane production can occur. Higher soil temperatures and a longer frost-free season could also increase rates of methane production. In addition, there is a very large quantity of methane currently frozen in this large permafrost zone, as well as frozen near-shore sediments. Global warming could release part of this pool, although current estimates of the importance of this particular feedback are very low (Kvenvolden, 1988).

An opposing line of reasoning suggests that increasing temperatures in the boreal zone will alter the hydrologic balance such that water tables are lowered, resulting in a larger zone of aerobic peat over the anoxic layers below. The oxidation of methane by methanotrophs and nitrifying bacteria in this layer could reduce emissions to rates below those currently found. A regional drying might also reduce the extent of boreal wetlands, reducing their global contribution.

This remains a very active area of research.

6.4.3 *Factors controlling fluxes of nitrous oxide*

a. Reactions and controls

Nitrous oxide is produced by two processes in terrestrial ecosystems: nitrification and denitrification (Bremner and Blackmer, 1978; Firestone and Davidson, 1989). Both of these processes show important interactions with the relative degree of nitrogen limitation on biological activity.

A simplified view of nitrogen cycling in terrestrial ecosystems (Fig. 6.7) shows that the mineralization of soil organic matter leads first to the production of ammonium (NH_4^+), which can then be oxidized by soil microorganisms to nitrate (NO_3^-). However, in very nitrogen limited systems, plants can compete very effectively with nitrifying microbes, primarily through the development of mycorrhizal roots (a symbiosis between plant roots and soil fungi), and take up ammonium directly (Fig. 6.7). This eliminates one process by which N_2O is produced (nitrification) and also eliminates the substrate for the second process (denitrification). What nitrate is produced in nitrogen limited systems, can also be rapidly immobilized (taken up) by microbes (Schimel et al., 1989; Davidson et al., 1990)

The amount of N_2O produced during nitrification is a very small fraction of the total ammonium oxidized (Bremner and Blackmer, 1981). This release should be relatively constant (constrained mostly by temperature and soil water stress) within a given ecosystem over time. However, the fraction of ammonium in soils oxidized to nitrate can vary from near 0% to 100% depending mainly on the degree to which nitrogen is limiting within that system (i.e., the strength of plant and microbial demand versus N supply) and also to some extent on soil pH and soil structure and rooting patterns. Nitrification occurs only under aerobic conditions.

Denitrification requires the presence of nitrate and readily metabolizable (or labile) carbon under anaerobic conditions (Knowles, 1982). Thus, denitrification tends to occur in systems with rapid fluctuations in redox potential and where nitrogen limitations are reduced. N_2O is an intermediary product in the denitrification process, with N_2 the final product. However, the conversion of N_2O to N_2 is inhibited at the low soil pH and only weakly anaerobic conditions found in most upland terrestrial ecosystems.

b. *Rates and locations of sources and sinks*

Forest soils are thought to be the major source of N_2O (Table 6.2). Tropical systems produce more than temperate systems because of higher temperatures and longer growing seasons, but also because, as a regional generalization, tropical forests are limited more by low phosphorous availability rather than nitrogen. This results in higher rates of nitrification.

c. *Potential effects of climate change*

Currently, nitrous oxide is thought to contribute a relatively small fraction to total greenhouse gas warming (approximately 6%, Fig. 2.1). Sources of N_2O are almost entirely from existing forest ecosystems, particularly in the tropics. As most GCM climate change scenarios suggest minimal changes in tropical regions, it would seem that climate change would be unlikely to alter these rates sufficiently to produce important feedbacks to the global climate system. However, conversion of tropical forests in Amazonia to pasture may increase rates of N_2O emissions by as much as threefold (Matson and Vitousek, 1990).

In temperate zones, warmer and wetter climatic patterns would tend to increase both the rates of nitrate production between rainfall events and perhaps the length of time during which soils, or soil microsites, would be anaerobic during and after rain events. However, this would be intimately linked to effects of climate change on plant growth as well, as this would affect nitrogen demand, and the availability and production of nitrate.

Table 6.2. Estimated sources and sinks of nitrous oxide (from IPCC, 1990a).

	Range (Tg N per year)
Source	
Oceans	1.4 – 2.6
Soils (tropical forests)	2.2 – 3.7
(temperate forests)	0.7 – 1.5
Combustion	0.1 – 0.3
Biomass burning	0.02 – 0.2
Fertilizer (including ground-water)	0.01 – 2.2
TOTAL:	4.4 – 10.5
Sink	
Removal by soils	?
Photolysis in the stratosphere	7 – 13
Atmospheric increase	3 – 4.5

Another important consideration for temperate systems is the increasing occurrence of "nitrogen saturated" forests due to chronic N deposition, a component of acid deposition (Schulze, 1989; Aber et al., 1989). Regional increases in N deposition over much of the temperate and boreal forest zones will continue to alleviate nitrogen limitations in these forest types. As N limitations are removed, increased production of nitrate and perhaps N_2O will occur (e.g., McNulty et al., 1990).

6.4.4 *Models and regional estimates of fluxes*

In contrast to processes of species distribution, primary production, and other aspects of biogeochemical cycling, process-level models of nitrous oxide and methane fluxes from terrestrial ecosystems are poorly developed, although research is advancing rapidly in this area. Until such models are available, global fluxes must be estimated by multiplying the mean of acceptable measured rates for each biome or landscape unit by land surface area (e.g., Mathews and Fung, 1987; Cicerone and Oremland, 1988; Aselmann and Crutzen, 1989).

A more sophisticated method for estimating trace gas flux rates is the use of stratified or spatially explicit descriptions, within regions, of habitats, landforms or important environmental gradients related to those fluxes. For example, Burke et al. (1990) entered climatic data and soil type as data planes in a geographic information system, and used relationships between these and N_2O emissions to produce geographically explicit predictions of variation in N_2O fluxes for the Great Plains region of the United States. In contrast, Matson and Vitousek (1990) divided tropical forest zones along gradients of soil fertility, elevation and human disturbance to derive regional estimates of N_2O flux.

The current rapid accumulation of detailed measurements of flux rates over a range of conditions within certain vegetation types, and under experimentally imposed conditions (e.g., Steudler et al., 1989) will allow for the development of models based on soil physical, chemical, and biological characteristics in the near future. These models will enable finer-resolution predictions of feedbacks between climate (physical and chemical) change and trace gas fluxes of terrestrial ecosystems.

6.5 Methods of direct measurement at the regional scale

Most of the discussion to this point has focused on plot-level research and methods for extrapolating this to regional and global scales. Advances in both remote sensing and chemical analytical techniques offer entirely new methodologies for near-direct measurement of regional trace gas balances. A few examples of this rapidly developing field are offered here.

The most widely used variable available through remote sensing is the "normalized difference vegetation index" (NDVI), which uses a complex ratio of reflectance in the red and near-infrared portions of the spectrum. Reflectance in the red region decreases with increasing chlorophyll content of the plant canopy, while reflectance in the infrared increases with increasing wet plant biomass. This technique has been used most successfully with data from the Advanced Very High Resolution Radiometer (AVHRR) to chart the seasonal development and senescence of vegetation in the Sahel region of Africa (Tucker et al., 1985) as well as other parts of the world. This technique has been developed to the extent that it is used operationally to predict the degree of drought and potential famine in the Sahel region. Seasonal changes in NDVI also have been successfully related to seasonal changes in regional atmospheric CO_2 content (Tucker et al., 1986; Fung et al., 1987), and to seasonal evapotranspiration (Running and Coughlan, 1988).

On the decomposition side, both laboratory (Wessman et al., 1988a; McLellan et al., 1991) and initial field trials (Wessman et al., 1988b) suggest that emerging high-spectral-resolution sensors will allow remote determination of the biochemical quality of litter material produced by plants (e.g., lignin, cellulose, and nitrogen content). In one study (Wessman et al., 1988b), lignin concentration in forest canopies was accurately predicted by remote sensing, and lignin concentration in turn was highly correlated with measured annual nitrogen mineralization (release) from soils, such that the spatial distribution of rates of N cycling could be mapped from remote sensing data.

A very different approach is the direct measurement of trace gas fluxes over large areas through the use of airborne sensors which sample the concentration distribution of gases in the troposphere, and combine these with estimates of convective air movement away from terrestrial surfaces to calculate flux rates directly. This approach has been used very successfully over tropical and boreal systems (e.g., Harriss et al., 1990), and provides measurements of current flux rates at a level of resolution most compatible with the needs of global budgets. It also provides an invaluable check on the consistency and completeness of ground-based measurements.

6.6 Summary

Interactions between nitrogen and carbon cycles within ecosystems and the effects of direct and indirect human use of the landscape interact with climatic drivers to constrain the structure and trace gas balances of terrestrial ecosystems. Predictions of changes in structure and CO_2 balance can be made using existing models of species and biome distribution and plant productivity and decomposition. The availability of nitrogen and other limiting nutrients will also constrain the transient behavior of ecosystems during a period of adjustment to new climatic regimes. It appears likely that human use, both direct and indirect, is likely to be as important as, if not more so than, climate change in determining the future distribution and dynamics of terrestrial ecosystems.

7

Atmospheric chemistry

Richard P. Turco

7.1 Introduction

This chapter discusses atmospheric chemistry in the context of global climate change and climate modeling. Atmospheric chemistry is itself a complex and broad subject about which numerous technical volumes have been written (e.g., Brasseur and Solomon, 1986; Levine et al., 1985; Wayne, 1985; Finlayson-Pitts and Pitts, 1986; Seinfeld, 1986). Here the reader is introduced to fundamental processes and concepts in atmospheric chemistry, to provide background so that more specialized treatments may be digested, and to place in a clear perspective the role of chemistry in the overall climate system. In Sec. 7.2, some basic concepts in atmospheric chemistry are reviewed. Descriptions of the principal atmospheric gases that contribute to the climatic state of the atmosphere are provided in Sec. 7.3. Section 7.4 surveys the chemical and physical properties of atmospheric aerosols (airborne particles) and their influences on climate processes. A closing section addresses future developments in atmospheric tracer chemistry modeling that are relevant to global climate change.

7.2 Coupling of chemistry to climate

Atmospheric chemistry, in the context of climate modeling, incorporates information from a number of disciplines, including: chemical kinetics, thermodynamics, spectroscopy, radiative transfer, photochemistry, physical chemistry, and analytical chemistry. The most obvious connection between chemistry and climate is the molecular constituency of the atmosphere and the spectroscopy and thermodynamics of those constituents. Chemical processes control the composition of the atmosphere to a significant degree, although some important climate-related compounds – for example, carbon dioxide (CO_2) – are strongly influenced by biological processes with atmospheric chemistry having a negligible role in their global cycles, while others – for example, water vapor (H_2O) – are dominated by thermodynamics (in the troposphere) with no significant influence of atmospheric chemistry. On the other hand, a number of trace compounds that have been linked to climatic change also have dominant atmospheric chemical sources or sinks; these species include ozone (O_3), methane (CH_4), nitrous oxide (N_2O), and the chlorofluorocarbons (CFCs). Secondary chemical processes can also affect the behavior of chemically inert compounds in the atmosphere. For example, chemically generated soluble materials carried on airborne particles control the condensation of water vapor to form cloud droplets and hence the hydrological cycle. Radiative and thermodynamic processes couple chemically inactive compounds such as carbon dioxide to the chemistry of ozone by determining the temperature of the atmosphere.

The chemistry of climate-active gases and particles can involve a broad range of processes. Here, chemical mechanisms are broken down into five broad categories. *Photochemistry* is driven by solar radiation – mainly in the ultraviolet spectrum – and represents the principal force altering atmospheric composition. The fundamentals of photochemical processes are discussed in the next section. *Heterochemistry* here refers to chemical transformations that take place in aqueous solution or on the surfaces of particles (solid dry particles or ice crystals, for example). In atmospheric chemistry, aqueous processes occur primarily in cloud droplets, but can also occur in precipitation, fogs and other liquid aerosol particles. The term *anthropochemistry* is adopted to indicate chemistry that is directly influenced by human activities. Included in this category are processes such as engine combustion or the emission to the atmosphere of exotic compounds such as the CFCs. *Biochemistry* involves the respiration, assimilation and emission of gaseous species by living organisms, which can directly affect atmospheric composition. Natural biomass combustion may be included in this category. Finally, *geochemistry* refers to chemical processes occurring at the interface between the atmosphere and the ocean and land. Geochemical processes include dry deposition and decomposition of gases on mineral surfaces, and solution into and dissolution from oceans.

Atmospheric chemists study the photochemistry of atmospheric gases, including related thermochemical processes, and often include the hydrochemistry of clouds. A major goal of atmospheric chemistry is to understand the environmental impacts of anthropochemical processes (e.g., emissions of nitrogen oxides by aircraft engines, or of sulfur dioxide from power plants). However, in order to make accurate long term forecasts, related biochemial and geochemical processes – primarily representing sources and sinks for atmospheric constituents – are often treated in chemistry models [indeed, atmospheric chemists are often involved in basic geochemical and biogeochemical research (e.g., see Crutzen, 1987; Cicerone and Oremland, 1988) because of the important impacts these processes have on the atmosphere].

Chemical processes are influenced by, and influence, the physical processes that govern the dynamical evolution of the atmosphere. Figure 7.1

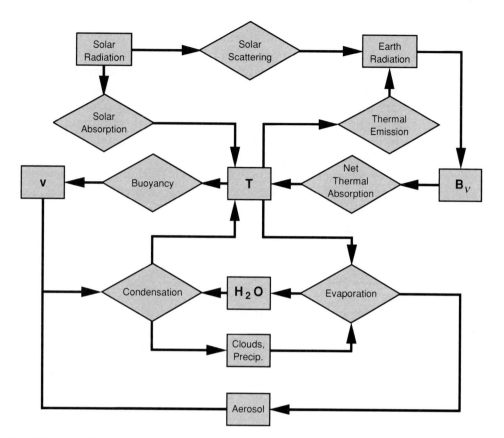

Fig. 7.1 Interactions of the key physical elements of climate models that are coupled with
chemical tracers, or modify their photochemistry or physical chemistry. The
climate parameters are indicated as boxed elements, and include solar radiation,
Earth radiation (i.e., reflected solar energy – the albedo – plus thermal emissions),
temperature (T), water vapor (H_2O), dynamics (**v**), thermal radiation (B_ν), clouds,
and aerosols. A basic set of physical processes, shown in diamonds, interconnect
these climate parameters. Chemically active species participate in most of the basic
physical processes shown. Not all possible physical processes and pathways are
given.

illustrates the key physical processes – and interrelationships – that are
treated in climate models. Likewise, these physical parameters play an
important role in the chemistry of the atmosphere. Proper allowance for
the interactions between physical and chemical processes is essential for an
accurate description of the overall climate system.

7.2.1 *Photochemical kinetics and tracer modeling*

The chemical activity of the atmosphere is driven almost exclusively
by solar ultraviolet radiation at wavelengths below about 350 nanometers.

If the atmosphere were merely in thermodynamic equilibrium (that is, solely under the influence of chemical transformations driven by heat), there would be essentially no ozone present, nor any other reactive species. The atmosphere would consist exclusively of chemically inert compounds such as N_2, O_2, CO_2 and H_2O. Ultraviolet radiation can break down these inert compounds into highly reactive fragments with a complex photochemistry.

When a molecule absorbs a photon of ultraviolet radiation, it may disassociate into two or more product species. These fragments carry away the energy of dissociation as potential chemical energy (and any excess energy as heat or internal excitation). Subsequently, the potential chemical energy drives reactions that further alter the composition of the atmosphere. Practically all chemical reactions of atmospheric interest are exothermic reactions, which expend potential chemical energy when they occur (this expended energy is converted into heat). Reactions that require additional thermal energy from the environment before they can occur are referred to as endothermic reactions, and are much less important in the atmosphere.

Molecules (or atoms) that collide may simply rebound elastically, or may react to form other species. The reaction proceeds through the formation of an activated complex, or collision complex, in which the chemical bonds that hold the individual atoms of the colliding species together can rearrange before the complex falls apart into rebounding fragments. The reaction is described by a reaction rate coefficient, k, which is similar to a collision kernel or cross section for collision. For binary (two-body) reactions, k has units of cm^3 s^{-1} molecule^{-1}; for ternary (three-body) reactions, the units are cm^6 s^{-1} molecule^{-2}. The initially interacting species are called the reactants, and the species that evolve from the activated complex, the products. A binary reaction can be written symbolically as

$$X + Y \rightarrow V + W \tag{7.1}$$

where X and Y are the reactants and V and W are the products. In a ternary reaction of the type $X + Y + M \rightarrow V + M$, the "third-body", M, is needed to balance energy and momentum between reactants and products.

Often, the reactants must overcome an energy barrier before the products can form; this barrier is referred to as the 'activation' energy for the reaction. The presence of a substantial activation energy implies a large sensitivity of the reaction to temperature; the warmer the reacting gases, the greater the probability they will have enough thermal energy to cross the activation barrier.

The chemical rate of a reaction, R, is obtained by multiplying the rate coefficient by the concentrations, n (molecules cm^{-3}), of the reactant gases. The chemical rate may be interpreted as the number of reaction events occurring per unit volume of air per unit time (i.e., molecules cm^{-3} s^{-1}),

$$R_{lij} = k_l n_i n_j. \tag{7.2}$$

Note that, in Eq. (7.2), species i could be X in (7.1), and j, Y. The effect of chemistry on a given species is determined by the sum of the rates for all the chemical processes that create or consume that species, including any radiative processes (i.e., photodissociation processes, see below).

The absorption of solar radiation (principally ultraviolet radiation) by a molecule can lead to its photodissociation, in which the molecule is split into two or more fragments of simpler composition. This process is indicated symbolically by an equation of the form

$$X + h\nu \rightarrow V + W \tag{7.3}$$

where X is the molecule undergoing photodissociation, V and W are the product species, and $h\nu$ indicates a photon of radiation [h is Planck's constant, and ν is the frequency of the radiation, related to the wavelength, λ, by $\nu\lambda = c$ (the speed of light, 3×10^8 m s^{-1})]. The photochemical rate in this case is the product of the photodissociation coefficient and the concentration of the absorbing species, X:

$$R_{li} = J_{li}n_i. \tag{7.4}$$

The photodissociation coefficient, J (which has units of s^{-1}), for such a process is calculated as follows:

$$J_{li}(\bar{x}) = \int_0^\infty I(\lambda, \bar{x})\sigma_i(\lambda)\phi_{li}(\lambda)\,d\lambda \tag{7.5}$$

where λ is the wavelength (in units of nanometers, nm, or micrometers, μm), I is the spectral flux intensity, or radiance of sunlight (having units of photons cm^{-2} s^{-1} μm^{-1}) at a given location, $\bar{x}$, in the atmosphere. I is calculated as

$$I(\lambda, \bar{x}) = I_0(\lambda)e^{-\tau(\lambda, \bar{x})} \tag{7.6}$$

where $I_0(\lambda)$ is the solar radiance at the top of the atmosphere. τ – the absorption optical depth, or dimensionless opacity, of the atmosphere – at wavelength λ, is determined by integrating the molecular absorption over the path from the point of interest to the top of the atmosphere in the direction of the sun (denoted by the differential path length, ds),

$$\tau(\lambda, \bar{x}) = \sum_{\{j\}_{abs}} \int_{\bar{x}}^\infty \sigma_j(\lambda)n_j(x')ds = \sum_{\{j\}_{abs}} \sigma_j(\lambda) \int_z^\infty n_j(z')dz'/\cos\chi. \tag{7.7}$$

In the final form of (7.7), the slant path integral has been replaced with a vertical integral assuming the atmosphere is spherically homogeneous, where χ is the solar zenith angle. The sum in (7.7) extends over all the important absorbing constituents in the atmosphere, which in practice reduces to the sum of O_2 and O_3 absorption. In (7.5) (and 7.7), the species absorption coefficient, σ_i(cm^2), is the *total* cross section that a molecule of type i presents to impinging photons, while ϕ_{li}, the quantum yield, is the *fraction* of the cross section associated with dissociation into a specific set of fragments, identified by 'l'. In these equations, the species concentrations can vary

205

in space and time, with the instantaneous distributions being used at any particular time.

The chemistry of the atmosphere is complex. To study the stratospheric ozone layer, for example, a complete model would treat at least 50 species and 200 chemical processes. The variation in the concentration of each trace species is determined by its continuity equation,

$$\frac{\partial n_\alpha}{\partial t} = P_\alpha - L_\alpha - \nabla \cdot \Phi_\alpha, \tag{7.8}$$

in which n_α, again, is the concentration (molecules cm^{-3}) of species 'α', P and L are the total photochemical production and loss rates (molecules cm^{-3} s^{-1}), and Φ is the species flux (molecules cm^{-2} s^{-1}) due to advective and diffusive processes,

$$\Phi = n\mathbf{v}_a - \mathbf{D} \cdot \nabla n \tag{7.9}$$

where $\mathbf{v}_a$ is the three-dimensional wind vector, and $\mathbf{D}$ is a diffusion tensor (cm^2 s^{-1}) corresponding mainly to model-dependent subgrid-scale 'turbulent diffusion' or eddy-scale transport (molecular diffusion is usually negligible).

The P and L terms can include expressions for chemical reactions, photodissociation processes, and species emissions, among other contributions. In general, a number of photochemical terms are required, as is indicated by the general form of the chemical kinetics equation:

$$P_\alpha - L_\alpha = \sum_{l(i+j\rightarrow\alpha+...)} R_{lij} - \sum_{l(i+\alpha\rightarrow...)} R_{li\alpha} + \sum_{l'(i\rightarrow\alpha+...)} J_{l'i}n_i - \sum_{l'(\alpha\rightarrow...)} J_{l'\alpha}n_\alpha. \tag{7.10}$$

It is obvious that, even with a modest number of chemical species, the resulting set of coupled tracer continuity equations (7.8) is highly nonlinear. These partial differential equations must therefore be formulated as finite difference equations for numerical solution.

The advective and diffusive transport terms in the tracer continuity equation can be formulated using observed winds or wind fields generated by an atmospheric General Circulation Model (GCM) (see Chapter 15 for a thorough description of tracer transport models). GCMs provide the most detailed (numerical) description of atmospheric wind systems available today. However, turbulent (eddy) diffusion coefficients can only be roughly estimated from these models. A GCM may be coupled to a tracer chemistry model and the tracers allowed to influence the atmospheric radiation field, heating rates and other parameters that force the dynamics – thereby providing feedback between the tracers and the dynamics. In this case, the simulation is said to be "interactive." In passive, or uncoupled, simulations, the atmospheric motions evolve either independently of the tracers, or under specific assumptions about the tracer distributions and their variations with time.

7.2.2 *Radiative processes*

The principal connection between atmospheric chemistry and climate occurs through the influence of atmospheric composition on the radiative balance of the Earth at all wavelengths. Chemical processes alter the concentrations of certain radiatively active species – both gases and aerosols. The principal gases that are active in the longwave, or thermal infrared, spectral region, H_2O and CO_2, are quite insensitive to chemical processes. Water vapor is controlled by hydrological processes and carbon dioxide by biogeochemical processes. However, both gases influence the amounts and distributions of other radiatively active gases such as ozone and methane. Moreover, because all of the elements of the climate system are coupled to varying degrees, subtle influences of chemistry even on the major gases can be imagined – for example, depletions of stratospheric ozone can lead to ultraviolet radiation increases that affect biological productivity, which in turn modifies the carbon cycle, altering carbon dioxide concentrations, and so on. It is not suggested that these are first-order effects. Care must be taken, nonetheless, in constructing models of a system as complex as the Earth's climate and in interpreting the coupling between elements of the system.

a. *Atmospheric heating and cooling rates*

The atmospheric energy balance, aside from the latent heat component associated with the hydrological cycle, results from the direct absorption of impinging solar radiation, and the transfer of thermal radiation and heat between the surface and the atmosphere (Fig. 7.1). Within the solar spectrum, ultraviolet wavelengths are absorbed by ozone in the stratosphere, and near-infrared wavelengths (in bands) by tropospheric water (vapor and clouds) and carbon dioxide. The residual visible and near-infrared wavelengths are absorbed by the surface or reflected to space (also refer to Fig. 1.2). Globally, cooling occurs by thermal emission to space primarily from water vapor, carbon dioxide and ozone molecules, and from clouds and surfaces (see Sec. 3.4 and Fig. 20.3). A number of other trace gases, noted earlier, also affect the heating and cooling rates.

Atmospheric radiation is the engine driving the photochemistry of gases such as ozone, methane, nitrous oxide and CFCs. Some of these gases, in turn, affect the propagation of solar and thermal radiation. Ozone, in particular, has a profound effect on solar radiation in the ultraviolet spectrum, and an important effect on longwave radiation at 9.6 μm (in the atmospheric window). The coupling between radiation and chemistry is twofold. First, radiation at shorter wavelengths directly activates photochemical processes. Second, radiation absorption and emission at longer wavelengths affects the thermal structure of the atmosphere, which influences the rates of many chemical reactions, as well as the dynamical structure, which affects the transport, distribution and removal of many species. The transport-related aspects of the coupling may be crucial to any accurate predictions of the response of the climate system to altered forcing.

b.   *Photochemical control of radiatively active species*

The microscopic nature of photochemical processing has been described in Sec. 7.2.1. Specific examples involving the principal climate-active gases are discussed in Sec. 7.3. The gas most often thought of as being controlled by photochemical processes is ozone. Indeed, ozone can be considered 'self-interactive', because ozone absorption limits the penetration of ultraviolet radiation that drives its own photochemistry. The atmospheric lifetimes of chemically stable greenhouse gases such as N_2O and the CFCs are also controlled by photochemistry. These compounds do not react with common atmospheric radicals such as OH, and must be lifted into the stratosphere above most of the ozone layer to be broken down by ultraviolet radiation. Accordingly, large-scale transport is critical in carrying these stable gases to altitudes where they can be decomposed. Some man-made compounds, e.g., CF_4, are so extraordinarily stable that they must be carried as high as the mesosphere before photons of sufficient energy are encountered to dissociate them (in general, the higher the altitude the greater the average energy of a photon).

c.   *Aerosol scattering and absorption*

An often overlooked contribution to the global atmospheric radiation balance is associated with small particles, or aerosols – that is airborne particles other than cloud water droplets or crystals (Charlson, 1988). Aerosols are ubiquitous in the atmosphere, and are particularly plentiful in the troposphere. However, they are rarely included, even empirically, in climate models. This neglect is due in part to the generally small (second-order) effects of aerosols on radiation and chemistry. However, it is also related to the extreme complexity of specifying their composition, physical and optical properties, and geographical distributions for climate modeling (see Sec. 7.4). To describe the optical properties of an aerosol population completely, one must know the size distributions, morphology, and complex indices of refraction as a function of wavelength for each component (e.g. Twomey, 1977).

Many aerosol species are produced mechanically or injected directly into the atmosphere, while others are generated in situ by chemical processes. Sea salt aerosol, dust, and smoke fall into the former category. Some aerosols are generated from vapors by chemical processes – for example, sulfate aerosol and organic particulates. All aerosols may be altered (directly or indirectly) by chemical processes during their residence in the atmosphere. Accordingly, the radiative properties of aerosols are expected to vary with time. Aerosols can also become more active as cloud condensation nuclei as they 'age' chemically in the atmosphere (see below).

Aerosols influence direct radiative forcing in two distinct ways. The aerosols scatter impinging radiation, some of which is reflected to space, reducing solar insolation and increasing the planetary albedo. Particles can also absorb solar radiation, thus transferring to the atmosphere some of the solar energy that might otherwise have been deposited at the surface. This redistribution in the heating pattern can modify the dynamics of the atmosphere which, in turn, controls the dispersion rates and residence times of the aerosols. In addition, aerosols can have a significant effect in converting direct solar radiation into diffuse radiation. To the extent that the distribution and photochemical activity of the diffuse radiation component is different from that of the direct component, detailed radiative transfer models are needed to account for this effect. Aerosols in general have a small infrared cross section, owing to their small physical sizes (typically less than one micrometer in diameter). Nevertheless, under some circumstances the infrared forcing can be significant: for example, on a regional scale in a dust storm, or on a hemispheric scale in a dense volcanic eruption cloud.

d. Cloud microphysics and radiative properties

The microphysical (and compositional) properties of a cloud determine its radiative effects. The size distribution and spatial distribution of cloud droplets are the key variables (see Sec. 3.5). In clouds containing ice particles, the crystal morphology is also important. If the quantity of liquid water in a cloud is fixed, then the reflectivity of the cloud at solar wavelengths depends on the size of the cloud droplets. The smaller the droplets, the more reflective the cloud, other factors remaining equal. When material that is highly absorbing (e.g., soot) is entrained into a cloud, the absorptivity of the cloud can increase substantially. The exact response of cloud radiative properties to aerosol chemistry depends on the number, sizes and composition of the aerosols absorbed into the cloud in a complex way.

The microphysical properties of clouds are controlled by the dynamical and thermodynamical processes leading to water condensation (Pruppacher and Klett, 1978) and by the chemical nature of the aerosols that act as cloud condensation nuclei (CCN). The more abundant are CCN, the greater the concentration of cloud particles, the smaller their average size, and the more reflective the resulting clouds for a given condensed mass of water (Twomey, 1977), notwithstanding the possible offsetting contribution of the CCN to absorption by the cloud particles. The coupling between aerosols, clouds, and radiation is a potentially powerful feedback mechanism in the climate system (e.g., Charlson et al., 1987), although the modulating effects of CCN on cloud microphysics remain largely unquantified.

An indirect connection can also be made between the chemical processes that form CCN and the albedo of clouds. It has been hypothesized, for example, that industrialization occurring over the last century has increased the concentrations of pollutant aerosols in the lower atmosphere and created a more reflective cloud cover for much of the Northern Hemisphere (NH) (Twomey, 1977). Of course, such a response would affect the interpretation of temperature records over that same period by introducing a potential cooling factor.

7.2.3 *Effects on the hydrological cycle*

The potential alterations of cloud radiative and microphysical properties due to chemical processes involving aerosols (Sec. 7.2.2, and Fig. 7.1) imply changes in the hydrological cycle, including precipitation rates. The basic hydrological mechanisms involving clouds are straightforward; water vapor, under conditions that lead to supersaturation, condenses to form cloud particles and precipitation that maintains the air within a few percent of the saturation point. However, the detailed microphysical processes involved are not well understood and cannot presently be simulated adequately in climate models (see Sec. 10.3).

If the abundance of CCN increases in a region – for example, as a consequence of increased sulfur emissions from biological or industrial sources – clouds formed in that region will tend to have a higher concentration of water droplets that are smaller in size on average (as noted in the previous section). The resulting cloud is less likely to generate precipitation (Twomey, 1977). The formation of precipitation in typical convective systems involves the autoconversion of small cloud droplets into precipitation-sized elements. The autoconversion rate is very sensitive to cloud droplet size; the smaller the droplets, the longer the autoconversion time. It is observed, for instance, that clouds formed in marine air, with low CCN counts, precipitate more frequently than clouds formed in continental air, with high CCN counts (although the difference may be enhanced by the higher moisture content of maritime air).

The precipitation–CCN coupling can be an important, although quite complex, climate feedback mechanism. Precipitation efficiency affects the humidity of the middle and upper troposphere, which can influence cloudiness over a wide area removed from convective zones.

7.2.4 *Biological feedback processes*

There are important links between biological processes, atmospheric composition, and climate (see Chapter 6 for more detailed discussion). Prominent among these is the role of terrestrial photosynthetic organisms in absorbing and sequestering carbon dioxide. These organisms may, for example, provide a buffer against CO_2 increases by responding with accelerated growth and accumulation of biomass, a process referred to as (carbon dioxide) fertilization. The carbon flux into the deep oceans also has a biological component that is sensitive to the availability of nutrients such as nitrates and phosphates, some of which are transported to the oceans via the atmosphere (Chapter 8). There are numerous subtle feedbacks in the climate system that involve the composition and chemistry of the atmosphere as they affect biological processes. Ozone depleting (N_2O) or enhancing (CH_4) chemical agents may link surface ultraviolet radiation intensities and biological productivity to the thickness of the ozone layer. The acidity of rainwater, which is strongly modified by sulfur and nitrogen oxide emissions, has important biological impacts on regional scales, although the nature of possible feedback to the atmosphere – if it exists – is uncertain.

7.3   Atmospheric chemistry of climatologically active gases

The general interactions between physical climate parameters are summarized in Fig. 7.1. Water vapor plays a dominant role in the radiative balance and hydrological cycle. Accordingly, water vapor has a special place among atmospheric trace species in climate modeling (see Chapters 3 and 10). However, other trace species can have a substantial impact on the radiation balance and cloud properties. Some aspects of the climate problem in relation to chemistry and chemical processes are explored in this section. The physical and chemical properties of the principal greenhouse gases are summarized in Table 7.1.

7.3.1   Water vapor (H_2O)

Water vapor is the most plentiful trace component of the atmosphere following the inert noble gases argon and neon. Water vapor is a principal element in the thermodynamics of the atmosphere, as explained elsewhere in this book (Chapters 3 and 10). However, H_2O also contributes absorption (and emission) in a number of bands in the infrared spectrum, including the 2.7 μm (stretching+rocking) fundamental and several overtone and combination bands in the near-infrared (solar) spectrum, and the 6.3 μm fundamental (bending) and pure rotational band (>20 μm) in the thermal infrared. In addition, water readily condenses into clouds that reflect and absorb solar radiation, thus directly affecting the radiant energy balance. The contributions of water vapor to basic physical and chemical processes in the atmosphere are summarized in Fig. 7.2 (also see Table 7.1).

a.   Climate/chemistry interactions
Greenhouse effect: Water vapor is the principal greenhouse gas in the Earth's atmosphere, accounting for about 30°C warming due to longwave radiative trapping (see Sec. 1.2). This warming effect is an order of magnitude greater than that of the second most important greenhouse

Table 7.1. Properties of radiatively active atmospheric constituents. Values in the troposphere are indicated by t and in the stratosphere by s.

Species	H_2O	CO_2	O_3	CH_4	N_2O	CFCs[d]
Present atmospheric abundance (ppmv)[a]	$\sim 1\%$[b] $\sim 3\text{–}6\ s$	360	0.01 t $\sim 10.\ s$	1.75	0.31	0.0003– 0.0005
Pre-industrial atmospheric abundance (ppmv)	$\sim 1\%$[b] $\sim 3\text{–}4\ s$	280	<0.01 t $\sim 10.\ s$	0.8	0.29	0
Rate of increase (%/year)	— ?	0.5	$\sim 1\ t$ $-0.2\ s$	0.9	0.2	~ 4
Source strength (Tg/year)[c]	$\sim 1 \times 10^{14}$	$\sim 6{,}000$[c]	—[c]	400–600	5–15	~ 1
Atmospheric lifetime (years)	~ 0.01	>100	$\sim 0.1\ t$ $\sim 1\ s$	~ 10	~ 150	65–130
Solar bands (μm)	0.9, 1.1, 1.4, 1.9, 2.7	4.3	0.25 0.65			
Longwave bands (μm)	6.3 >20	15.	9.6	7.7	7.8	8–14[d]
Chemical activity	HO_x Thermo., Aqueos Chem.	Bio- Geochem.	O (^{1}D) t O, O (^{1}D) s	HO_x, O_x, ClO_x Biochem.	NO_x Biochem.	ClO_x

[a] ppmv = parts per million by volume; 1 ppmv = 1,000 ppbv (parts per billion) = 1,000,000 pptv (parts per trillion).
[b] The tropospheric water vapor fractional abundance is given in percent rather than ppmv; tropospheric abundances vary from several % to less than 0.01%.
[c] Source strengths represent all identified sources; for water vapor, the principal source is evaporation from oceans.
[d] CFCs (chlorofluorocarbons) are a class of compounds that include CFC-11 and CFC-12, presently the principal CFC greenhouse contributors. The individual compounds in this category have infrared bands in the spectral interval from 8–14 μm but concentrated in the 8–10 μm region.

gas, CO_2. An important positive feedback in the climate system involves the exponential increase in the equilibrium vapor pressure of water with increasing temperature. Thus, as the surface temperature rises, more water is evaporated from ocean reservoirs, which intensifies the greenhouse effect and further warms the surface (Fig. 7.2, and Chapters 1 and 20). The

Radiative Components

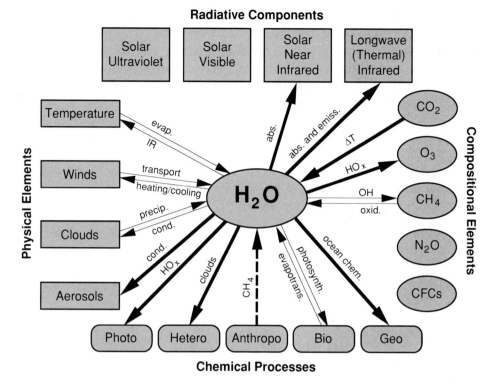

Fig. 7.2 Schematic diagram of the physical and chemical interactions involving water vapor that might be included in a comprehensive global climate model. The key physical climate parameters are shown in boxes, as in Fig. 7.1. Atmospheric chemical species are enclosed in circles. Chemical mechanisms that may involve water vapor are divided into classes, identified at the bottom of the figure, and the pathways and modes of interaction are indicated by arrows. *Photo*chemistry encompasses those chemical processes driven by solar radiation or involving the reaction of species produced when ambient air is exposed to solar ultraviolet radiation; *hetero*chemistry refers to chemical processes occurring in aqueous solutions, principally in cloud droplets, or on particle surfaces, particularly solid aerosols and ice crystals; *anthropo*chemistry includes emissions and processes associated mainly with human activities; *bio*chemistry is related to the assimilation and respiration of atmospheric constituents by living organisms; and *geo*chemistry refers to chemical processes occurring on surfaces or in media at the interface between the atmosphere and the oceans and land.

greenhouse warming caused by other trace gases can be enhanced by this positive water vapor feedback effect.

Clouds: Water vapor in condensed forms – water droplets and ice crystals suspended as clouds – has a major impact on both the solar and longwave radiation balances (see Sec. 3.5). The frequency of cloud occurrence, their altitudes and water contents, and the microphysical properties of the condensed water all influence cloud radiative properties. The subject is so complex that clouds remain one of the principal unresolved aspects of climate modeling. It is still debated whether clouds provide a weak positive or strong negative feedback in the climate system response to radiative forcing (see Sec. 3.5 and Chapter 20). The chemical composition of the atmosphere can affect cloud properties in several ways. First, the trace composition of cloud droplets determines the amount of solar visible radiation absorbed by the cloud and thus its net effect on heating or cooling (Sec. 7.2.2d). Second, chemistry occurring in the background atmosphere leads to the formation of aerosols that act as CCN. The concentration and chemical composition of CCN influence the microphysics of clouds, and hence their overall radiative properties (Sec. 7.2.2d).

Photochemistry: Water vapor is involved in fundamental photochemical processes that ultimately determine the lifetimes of a number of greenhouse-active gases (Fig. 7.2). For example, the atmospheric lifetime of methane is controlled by reaction with hydroxyl radicals derived largely from water vapor in the troposphere:

$$O_3 + h\nu \rightarrow O(^1D) + O_2$$

$$O(^1D) + H_2O \rightarrow OH + OH \tag{7.11}$$

$$OH + CH_4 \rightarrow H_2O + CH_3.$$

(The notation $O(^1D)$ refers to oxygen atoms in their first electronically excited state, denoted ^1D). According to (7.11), an increase in water vapor can lead to a net decrease in methane, another greenhouse gas. Furthermore, OH generated from water vapor is the principal scavenger of an entire class of chlorocarbons that contribute to climate change, including methyl chloroform and CFC-22; these compounds have carbon–hydrogen (C–H) bonds that are subject to chemical attack by OH. Increasing concentrations of tropospheric water vapor due to greenhouse warming would imply shorter lifetimes and lower abundances of these greenhouse gases.

Water-derived hydrogen radicals also catalyze ozone destruction in the stratosphere:

$$OH + O_3 \rightarrow HO_2 + O_2$$

$$\underline{HO_2 + O_3 \rightarrow OH + 2O_2} \tag{7.12}$$

$$O_3 + O_3 \rightarrow 3O_2 \quad \text{(net)}$$

This overall reaction cycle is "catalytic" in that the hydrogen species are recycled – there is no net loss of OH – whereas the ozone is chemically destroyed – there is a net conversion of O_3 into O_2. Accordingly, an increase in water vapor in general implies a decrease in stratospheric ozone. Ozone contributes to the greenhouse effect through its absorption of longwave radiation in the upper troposphere and lower stratosphere (see Sec. 7.3.3a). Hence, the chemical coupling of water vapor to ozone and other greenhouse gases provides an important pathway influencing global climate.

b.   Measurements and trends

Water vapor concentrations in the lower atmosphere can vary by orders of magnitude from place to place, with values as large as several percent of the total air density. This large variability is a fundamental problem in climate modeling. In the stratosphere, water vapor concentrations are much smaller (a few parts per million) and less variable. The quantity of water in the atmosphere is regulated, to first order, by the temperatures of the oceans and the thermodynamic properties of water. Chemical processes may influence water vapor concentrations through higher-order effects – such as through the composition of CCN. (Above the mesosphere, however, water vapor is very much under chemical control.)

7.3.2   Carbon dioxide (CO_2)

Carbon dioxide is the most widely discussed climate-related atmospheric gas owing to its increasing concentration, apparently connected to human activities, and its potential for future greenhouse warming (see Chapter 2). The overall interactions of CO_2 in atmospheric physical and chemical processes is summarized in Fig. 7.3 (also see Table 7.1). Carbon dioxide, in addition to inducing the well-known radiative greenhouse effect, influences the photochemistry of the atmosphere and can affect a wide range of atmospheric species. Carbon dioxide also establishes the baseline acidity for rainwater; pure water in equilibrium with atmospheric CO_2 is acidic with a pH of 5.6, which is considerably lower than the neutral pH of pure water, 7.0 (pH is the scale of aqueous acidity calculated as $-\log_{10}$ of the H^+ ion concentration in solution – each unit of pH is thus a factor of 10 in the H^+ concentration, or acidity).

a.   Climate/chemistry interactions

Even though carbon dioxide does not have any important photochemical reactions in the troposphere or stratosphere, it affects the photochemistry of other species through its control of atmospheric temperatures. In particular, an increase in CO_2 will warm the surface and cool the stratosphere (see Chapter 20). Since chemical reaction rates vary with temperature, some strongly, the composition of air that is heated or cooled will change. Moreover, the result of surface warming is to increase water vapor concentrations, which leads to further changes in photochemistry.

Temperature forcing: The greenhouse effect of carbon dioxide is well known (see Chapters 1 and 20). If carbon dioxide were to double in concentration, surface temperatures might increase by about 2–4°C on average over the globe. Such a warming of ocean surface temperatures

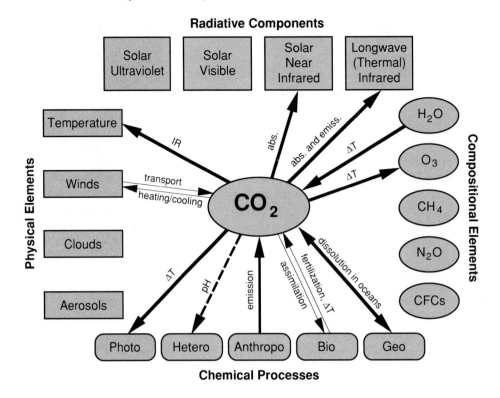

Fig. 7.3 Schematic diagram of the physical and chemical interactions involving carbon dioxide that might be included in a comprehensive global climate model. The key physical climate parameters are shown in boxes, as in Fig. 7.1. Atmospheric chemical species affected by carbon dioxide are enclosed in circles. Chemical mechanisms that may involve, or affect, carbon dioxide are divided into classes – shown at the bottom of the figure – and the pathways and modes of interaction are indicated by arrows. An explanation of the scheme for classifying chemical processes is found in Sec. 7.2, and in the caption for Fig. 7.2.

would be accompanied by an increase in average water vapor concentrations by 10–20%. However, the net sensitivity of surface temperatures and water vapor to CO_2 is uncertain. Increases in surface temperatures and H_2O concentrations might, for example, alter the efficiency of injection of water vapor at high altitudes in convective systems.

Carbon dioxide in the stratosphere cools that region by radiating heat to space through the thin overlying atmosphere. As CO_2 increases, the stratosphere cools, which affects the photochemical kinetics of the region. The concentrations of several climate-active gases are modified accordingly.

Effect on stratospheric ozone: The effect of CO_2 on ozone in the stratosphere follows from the temperature sensitivity of the chemical destruction of ozone by NO_x and other ozone-catalytic agents (also refer to Sec. 7.3.3). The following reaction sequence applies:

$$NO + O_3 \rightarrow NO_2 + O_2$$

$$\underline{NO_2 + O \rightarrow NO + O_2} \qquad\qquad (7.13)$$

$$O + O_3 \rightarrow 2O_2 \quad \text{(net)}.$$

The reaction of NO with O_3, in particular, has a large activation energy (see Sec. 7.2.1). Hence, as CO_2 increases, the stratosphere cools, the rate of depletion of ozone decreases, and the ozone concentration increases. If it is assumed that no perturbations in surface temperature and water vapor concentrations occur, the cooling effect of a CO_2 doubling alone leads to an increase of several percent in stratospheric ozone. That ozone increase can induce further greenhouse warming, in a second-order positive feedback. However, when atmospheric water vapor concentrations are allowed to increase in response to warmer surface temperatures, the enhancement in OH (see Sec. 7.3.1) is sufficient to cause a net ozone *decrease* in the lower stratosphere (7.12). Also note that reduced stratospheric temperatures can induce the formation of ice clouds causing local depletions of ozone (see Sec. 7.4.2). The coupled responses to increases in carbon dioxide must be carefully considered.

Geochemical and biochemical coupling: The natural geochemical and biogeochemical cycles of CO_2 are complex, involving land and marine biota (see Chapters 6 and 8). The principal chemistry of carbon dioxide involves its dissolution in clouds (where the pH, or acidity, of the cloud water is affected) and in the oceans, where the geochemical carbonate cycle is initiated. The atmosphere/ocean exchange of carbon involves the processes:

$$CO_2 \text{ (gaseous)} \underset{}{\overset{H_2O}{\Longleftrightarrow}} CO_2 \cdot H_2O \text{ (aqueous)}$$

$$CO_2 \cdot H_2O \Longleftrightarrow H^+ + HCO_3^- \qquad\qquad (7.14)$$

$$HCO_3^- \Longleftrightarrow H^+ + CO_3^{2-}$$

where a double arrow indicates a reversible process. The dissolution of CO_2 in oceans and clouds depends on the temperature of the water; the colder the water, the more soluble is CO_2. The dissolved carbon species, $CO_2 \cdot H_2O, HCO_3^-$ and CO_3^{2-}, can be assimilated by marine phytoplankton through photosynthesis. The bicarbonate and carbonate – i.e., HCO_3^- and CO_3^{2-}, respectively – can also be incorporated into shells or tests, which lead to the formation of carbonate deposits. The bicarbonate and carbonate stored in the deep oceans represents the major short term reservoir of CO_2. Accordingly, as increasing atmospheric abundances of CO_2 begin to modify the geochemistry and biochemistry of the oceans, compensating feedback mechanisms may be activated.

Changes in surface temperature can affect the rate and extent of biological activity at the Earth's surface. The principal biochemical process that couples atmospheric carbon dioxide to the land and oceans is green-plant photosynthesis:

$$CO_2 + H_2O \xrightarrow{\quad photosynthesis \quad} {'CH_2O'} + O_2 \qquad (7.15)$$

where $'CH_2O'$ represents a unit of organic matter. The organic matter is oxidized back into carbon dioxide when burned, or consumed by organisms as a source of energy:

$$'CH_2O' + O_2 \xrightarrow[combustion]{respiration} CO_2 + H_2O. \qquad (7.16)$$

Increased concentrations of atmospheric carbon dioxide affect the rate of photosynthesis in many plants. It follows that biochemical responses to atmospheric CO_2 variations may be significant. The seasonal oscillation in CO_2 concentrations associated with the seasonal modulation of global biological activity is obvious in measurements of CO_2 (see Fig. 2.2). It also seems reasonable that an acceleration of forest cutting and burning can exacerbate carbon dioxide accumulation in the atmosphere.

b. Measurements and trends

Atmospheric carbon dioxide concentrations over the last 2–3 centuries show an increasing trend which accelerates in the last 50–100 years (Fig. 2.3). While the accumulation of atmospheric carbon dioxide is clear, the explanation for the rate of increase is less certain. The principal source of excess CO_2 appears to be the combustion of fossil fuels. A secondary industrial source is associated with the reduction of calcium carbonate to CaO during the production of cement. The clear-cutting of forests to obtain raw material for paper pulp and for agricultural purposes (slash and burn agriculture and the creation of grazing land) is a human-linked CO_2 source of growing concern. By some estimates, deforestation contributes up to 20% of the total anthropogenic source of carbon dioxide.

Nevertheless, the role of geochemical and biogeochemical cycling and feedbacks on the upward trend in CO_2, which are discussed in more detail in Chapter 8, are not yet fully resolved. Two factors apply here. First, CO_2 is rapidly cycled through the biosphere and surface ocean reservoirs, with time scales ranging from 10 to 100 years. Second, the land and ocean reservoirs of carbon are larger than the atmospheric reservoir and, hence, have the capacity to ameliorate, or accelerate, the CO_2 change. To resolve this important issue, a more complete understanding of the chemical, physical and biological interactions between the atmosphere, biosphere and oceans must be obtained.

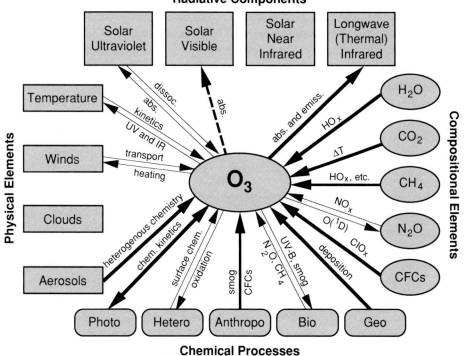

Fig. 7.4 Schematic diagram of the physical and chemical interactions involving ozone that might be included in a comprehensive climate model. The key physical climate parameters are shown in boxes, as in Fig. 7.1. Atmospheric chemical species are enclosed in circles. Chemical mechanisms that may involve, or affect, ozone are divided into classes, shown at the bottom of the figure, and the pathways and modes of interaction are indicated by arrows. An explanation of the scheme for classifying chemical processes is found in Sec. 7.2, and in the caption for Fig. 7.2.

7.3.3 Ozone (O_3)

Of all the greenhouse gases under discussion here, ozone is by far the most photochemically active. Ozone is also important because of its strong absorption of solar ultraviolet radiation. Indeed, without ozone in the atmosphere, all of the questions about greenhouse warming and climate change would be moot, since humans would not be present on the Earth to be concerned. Most of the Earth's ozone is safely stored in the stratosphere. About 10% is found in the troposphere, where in some very smoggy locations, ozone concentrations can approach those found in the lower stratosphere. Ozone is not only a radiatively active gas, but has many other important physical and chemical interactions in the atmosphere (Fig. 7.4 and Table 7.1).

219

a.   Climate/chemistry interactions

Greenhouse effect: Ozone absorbs thermal radiation in the 9.6 μm band (actually two overlapping bands), which falls within the "atmospheric window" for longwave transmission (Fig. 1.1). The effect of ozone on the energy balance is a strong function of its altitude distribution (Fig. 3.6). In particular, ozone in the middle and upper troposphere and lower stratosphere (where the coldest temperatures are found) is most effective as a greenhouse agent. Ozone in these regions is under anthropogenic influence from several sources; e.g., commercial aircraft traffic is now thought to be responsible for increasing ozone concentrations in the upper troposphere (see Sec. 7.3.3c).

Radiative heating/middle atmosphere dynamics: Ozone absorption of solar ultraviolet radiation is responsible for the heating that establishes the global-scale stratospheric temperature inversion. Hence, the stability of the lower stratosphere and its role as an upper boundary for tropospheric circulations is affected by the vertical distribution of ozone. Likewise, the dynamics of the middle atmosphere (stratosphere and mesosphere) are controlled by ozone heating balanced by carbon dioxide radiative cooling. Changes in both O_3 and CO_2 can influence the burdens and distributions of other important greenhouse gases such as N_2O and CFCs, for which stratospheric photodecomposition is the principal sink, and whose lifetimes are thus mainly limited by transport into the middle stratosphere.

Photochemistry: Ozone is formed through the photodissociation of molecular oxygen by solar ultraviolet radiation. The concentration of ozone is limited by reactions that regenerate oxygen from ozone. The simplest complete photochemical cycle for ozone is the pure-oxygen 'Chapman' scheme (Chapman, 1930),

$$O_2 + h\nu \rightarrow O + O$$
$$O + O_2 + M \rightarrow O_3 + M$$
$$O_3 + h\nu \rightarrow O + O_2 \tag{7.17}$$
$$O + O_3 \rightarrow O_2 + O_2.$$

In (7.17), the last reaction represents the ozone loss (oxygen regeneration) step. The O and O_3 are equivalent 'odd-oxygen' species in this scheme (see below). Using tabulated chemical rate coefficients and photodissociation coefficients, the equilibrium, or steady-state, concentration of ozone is readily calculated from mechanism (7.17). The predicted ozone abundances are too high, however. Additional ozone loss (recombination) processes catalyzed by trace compounds including hydrogen, nitrogen and chlorine species must be considered. Reaction cycles (7.12) and (7.13) represent two of the most important of these ozone loss processes.

Through a series of photochemical reactions, ozone contributes directly to the decomposition of some greenhouse gases, and indirectly to the decomposition of others. Accordingly, ozone plays an important role in determining their atmospheric lifetimes. The reaction sequence (7.11), for example, indirectly controls the atmospheric abundance of methane. Nitrous oxide can be directly attacked by ozone through the reaction sequence (occurring in the middle stratosphere),

$$O_3 + h\nu \rightarrow O\,(^1D) + O_2$$
$$O\,(^1D) + N_2O \rightarrow NO + NO\,. \tag{7.18}$$

The overall atmospheric lifetime of N_2O is dominated by stratospheric photolysis:

$$N_2O + h\nu \rightarrow N_2 + O(^1D). \tag{7.19}$$

Similar patterns of photochemical decomposition hold for the CFCs. The products of N_2O (and CFC) decomposition by $O(^1D)$ are key reactive atmospheric constituents (Secs. 7.3.5 and 7.3.6).

b. Stratospheric chemistry

To model the stratospheric ozone layer in complete chemical detail requires a large number of constituents ($\sim$50) and photochemical processes ($\sim$200). Heterogeneous chemistry on stratospheric aerosols of various types (sulfate aerosols, polar stratospheric clouds) should also be included (Sec. 7.4.2). Accordingly, treatments of ozone in typical climate models are greatly simplified. Because the stratosphere is critical as a sink for important climatically active compounds, it may eventually be necessary to integrate full ozone photochemistry into climate forecast models.

Many of the reactive chemical species of interest may be divided into 'families' of related compounds. The principal families of interest are the 'odd-oxygen' or ozone (O_x) family $\{O, O(^1D)$ and $O_3\}$, the odd-hydrogen (HO_x) family derived from H_2O $\{H, OH, HO_2$ and $H_2O_2\}$, the odd-nitrogen (NO_y, see Sec. 7.3.5a) family derived from N_2O $\{NO, NO_2, NO_3,$ N_2O_5, HNO_2, HNO_3 and $HO_2NO_2\}$, the methane-derived family $\{CH_3,$ $CH_3O_2, CH_3O, CH_3OOH, CH_2O$ and $CHO\}$; the odd-chlorine (Cl_x) family derived from CFCs $\{HCl, Cl, ClO, OClO, ClONO_2, HOCl, Cl_2O_2, Cl_2$ and $ClNO_2\}$, the bromine family (Br_x) derived from a number of bromine-bearing compounds $\{HBr, Br, BrO$ and $BrONO_2\}$, and the sulfur family derived from COS $\{S, SO, SO_2, HSO_3, SO_3$ and $H_2SO_4\}$. In addition, a number of other gases are important in the stratosphere (e.g., $CO, CH_3Cl,$ C_2H_6, etc.). The chemical properties of, and interactions among, these species is complex and beyond the scope of this book (e.g., see Turco, 1985).

It is important to note that most stratospheric chemical processes involve compounds derived from the greenhouse gases of primary focus in this book: H_2O, O_3, CH_4, N_2O and CFCs. Thus, it is evident that these photochemical 'source' gases can have both climatic and chemical impacts.

c. Tropospheric chemistry

The tropospheric chemistry of ozone and related species retains the features of stratospheric chemistry, but with several important complications. The chemistry of the troposphere is dominated by the Earth's surface, particularly gas exchange between the surface and the atmosphere, by clouds and aerosols, and by the nonuniform distribution of gas emissions. Hence, tropospheric chemistry is highly variable on regional scales (Isaksen, 1989). Many additional species must be considered, particularly the more complex organic compounds associated with human activities and biological activity. Cloud convection and precipitation, and heterogeneous chemical processes occurring within clouds, have a profound effect on tropospheric composition and chemistry, and introduce additional variability. The regional deposition of sulfuric, nitric, and certain organic acids in precipitation is often concentrated enough to affect biological activity.

Anthropogenic sources of hydrocarbons, nitrogen oxides and carbon monoxide enhance the formation of tropospheric ozone. In its extreme form, the chemistry is manifested as photochemical smog. The signature of increasing influences of anthropogenic pollution is seen in the substantial modification of tropospheric composition observed in the NH as compared to the still relatively pristine Southern Hemisphere (SH). The production of ozone from the simplest common hydrocarbon, methane (which has important anthropogenic sources), in the presence of high concentrations of NO (nitric oxide, which is dominated by anthropogenic emissions in many regions) can be summarized by the overall reaction:

$$\mathrm{CH_4 + OH + 9O_2 \xrightarrow{NO} CO_2 + \frac{1}{2}H_2 + 2H_2O + 5O_3.} \qquad (7.20)$$

In the upper troposphere, the injection of NO from aircraft engines can produce excess ozone through a process similar to (7.20). Such ozone increases may be particularly important to the greenhouse warming problem.

The chemistry of important tropospheric gases such as carbon monoxide and the nitrogen oxides is elaborated on below (see Sec. 7.3.4a and 7.3.7a).

d. Measurements and trends

An in-depth analysis of all existing ozone observations (Watson et al., 1988) concludes that the global stratospheric ozone burden is presently decreasing at a rate of approximately 0.2%/yr (Table 7.1). The decrease has significant geographical and seasonal dependence, being largest at high latitudes in winter and spring. The profile of ozone, likewise, is changing, with most of the ozone reduction occurring in the middle and upper stratosphere.

A deep 'ozone hole' has developed over Antarctica since the mid-1970s (Farman et al., 1985). Measured depletions in total ozone amounts (that is, the vertically integrated column of ozone) are 50% or more during austral spring. The depletion is concentrated between 15 and 25 km altitude. This ozone hole is caused by heterogeneous chemical interactions between CFC-derived chlorine species and polar stratospheric ice clouds (Sec. 7.4.2). The ozone hole deepened during the 1980s but has apparently bottomed out (that is, further reductions in total ozone will be limited roughly to the amount of ozone between 15 and 25 km). The dimensions of the hole appear to be restricted principally by the vertical and horizontal extent of the region in which stratospheric clouds form. Recent measurements in the wintertime NH indicate (relatively) small stratospheric ozone depletions at high latitudes compared to the SH. The warmer temperatures of the northern high latitudes in winter do not favor the continuous and extensive formation of polar stratospheric clouds nor the dynamical stability of the polar vortex, both of which contribute to the large ozone reductions seen in the SH.

The occurrence of the ozone hole points to the sensitive and highly non-linear relationships that arise in coupled atmospheric dynamical/chemical systems. The ozone hole requires a certain level of stability in the wintertime polar vortex circulation to create conditions favorable to sustained ice condensation. The presence of ice clouds and deep reductions in ozone, in turn, affect the radiative energy balance of the vortex and thus its stability. The coupled system is so complex that, to date, no coupled three-dimensional simulations of the ozone hole have been carried out.

Tropospheric ozone in the NH has been increasing at a rate of about 1% per year. The increase is associated with photochemical production associated with human emissions of nitrogen oxides, methane, nonmethane hydrocarbons and carbon monoxide. The SH shows only a weak ozone trend, presumably as a result of lower levels of anthropogenic pollution.

Potential changes in global ozone concentrations have been calculated for simultaneous increases in the chemically active greenhouse gases CH_4, N_2O and CFCs (e.g., Isaksen and Stordal, 1986). The assumed atmospheric increases in greenhouse gases are projections for the next 50–100 years; however, CFC emissions will presumably be controlled under the Montreal Protocol and London Amendments (see Sec. 2.5.3). Substantial ozone increases are predicted in the NH troposphere, and decreases in the middle and upper stratosphere. The tropospheric ozone increase is associated with increases in methane and enhancements in ultraviolet radiation due to ozone depletion in the stratosphere. The stratospheric ozone decrease is caused by the accumulation of nitrous oxide and CFCs. Significant climate feedbacks are implied by the ozone perturbations projected in such scenarios.

7.3.4 Methane (CH_4)

Methane is an abundant natural gas that also has significant human sources. Methane absorbs thermal radiation at about 8 μm in

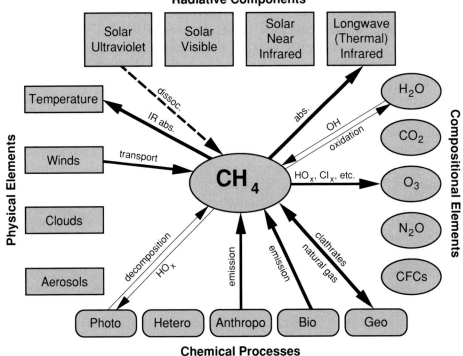

Fig. 7.5 Schematic diagram of the physical and chemical interactions involving methane that might be included in a comprehensive climate model. The key physical climate parameters are shown in boxes, as in Fig. 7.1. Atmospheric chemical species are enclosed in circles. Chemical mechanisms that may involve, or affect, methane are divided into classes, shown at the bottom of the figure, and the pathways and modes of interaction are indicated by arrows. An explanation of the scheme for classifying chemical processes is found in Sec. 7.2, and in the caption for Fig. 7.2.

the atmospheric window. The photochemistry of methane is relatively simple, and is closely connected to that of ozone. The atmospheric chemical interactions and effects of methane are indicated in Fig. 7.5. Methane concentrations are determined by a complex set of biochemical, geochemical and anthropochemical processes (Cicerone and Oremland, 1988). Some of the established properties of methane are summarized in Table 7.1.

a. Climate/chemistry interactions

Greenhouse effect: Methane directly absorbs thermal infrared radiation and modifies the energy balance. An unresolved issue pertains to the response of atmospheric methane to global temperature change. It is observed that prehistoric methane concentrations have been correlated with temperature (Fig. 21.3). Postulated connecting mechanisms include the condensation and evaporation of methane in the form of clathrates in permafrost, and the sensitivity of biological methanogenesis to environmental temperature. Methane appears to provide a positive feedback to global warming; i.e., methane concentrations increase in response to warming, causing an additional incremental greenhouse effect. On the other hand, drier climatic regimes might deplete wetlands and the anoxic environments that generate methane.

The atmospheric concentration of methane is controlled by its reaction with OH in the troposphere [see eq. (7.11)]. For a fixed source strengh, therefore, the atmospheric abundance of methane is sensitive to the same factors that affect OH, including tropospheric water vapor and ozone concentrations. Water vapor concentrations are sensitive, in particular, to surface temperatures. Global warming increases H_2O, and thus OH, and causes a decrease in CH_4. The decrease in methane, in turn, leads to a reduced methane-induced greenhouse warming (i.e., a negative feedback, estimated to be as large as 10%).

Methane decomposition leads to water vapor production (7.20). While this production of water vapor is insignificant in the troposphere, it is quite important in the middle and upper stratosphere and mesosphere. After methane is transported from the troposphere into the stratosphere, it is oxidized according to the overall process:

$$CH_4 + OH + O_2 \rightarrow 2H_2O + \frac{1}{2}H_2 + CO \qquad (7.21)$$

[here, the reacting OH molecule represents one-half of a water molecule decomposed by reacting with $O(^1D)$]. The water vapor generated in the upper atmosphere from increasing concentrations of methane may be responsible for an increased occurrence of polar stratospheric ice clouds and noctilucent clouds [the latter are composed of water ice crystals residing at the cold summer mesopause, where temperatures may fall as low as 130 K (cf. Fig. 3.3)]. The stratospheric clouds are coupled to climate through direct radiative effects and indirect effects related, for example, to the vertical distribution of ozone.

Influence on ozone: Methane has a number of photochemical effects related to ozone. In the troposphere, as discussed above, methane decomposition can lead to ozone generation (7.20). A similar smog-like process leading to ozone production can occur in the upper troposphere and lower stratosphere, particularly when NO_x is injected in substantial quantities, as by aircraft operations.

In the stratosphere, methane interferes with chemical cycles that catalyze ozone destruction. The role of CH_4 in the chlorine cycle is noteworthy. Methane moderates the attack of chlorine on ozone (see Sec. 7.3.6) through the reaction:

$$CH_4 + Cl \rightarrow HCl + CH_3 \qquad (7.22)$$

where HCl acts as an inert "reservoir" for chlorine (that is, holding it in a form that does not react directly with ozone). Reaction (7.22) may indeed lead to ozone production through the following series of reactions,

$$CH_3 + O_2 + M \rightarrow CH_3O_2 + M$$
$$CH_3O_2 + NO \rightarrow CH_3O + NO_2$$
$$NO_2 + h\nu \rightarrow NO + O \tag{7.23}$$
$$O + O_2 + M \rightarrow O_3 + M,$$

which are part of a typical "smog" reaction sequence.

b.   Sources, measurements and trends

Methane has a variety of natural and anthropogenic sources ranging from termites to rice paddies (Chapter 2). During historic times, methane concentrations have been increasing. The growth in CH_4 concentrations appears to be correlated with the expansion of the human population. Methane has roughly doubled since the mid-1800s, consistent with the doubling of the total methane source attributable to increasing anthropogenic activities. The present rate of increase in atmospheric methane is roughly 1% per year on a global scale. Methane has numerous chemical (and physical) connections to other climatically important gases such as ozone and water vapor. Accordingly, its continuing increase, and related potential climatic feedback effects, should be properly represented in climate models.

7.3.5   Nitrous oxide (N₂O)

Nitrous oxide is produced by natural and anthropogenic sources. The contribution from human activities is uncertain, but may be responsible for the observed slow rate of increase in N_2O concentrations. Nitrous oxide absorbs radiation in the thermal infrared spectrum near 8 μm wavelength. Nitrous oxide is also photochemically active, although only in the stratosphere, where it affects ozone concentrations. Some properties of N_2O are summarized in Table 7.1. The principal interactions between nitrous oxide and other physical and chemical elements of the climate system are depicted in Fig. 7.6.

a.   Climate/chemistry interactions

Greenhouse effect: N_2O contributes directly to the greenhouse effect through thermal infrared absorption in the troposphere. The tropospheric lifetime of N_2O is quite long, and its interaction with other gases, with clouds and aerosols are minimal.

Stratospheric ozone impact: N_2O is decomposed in the stratosphere,

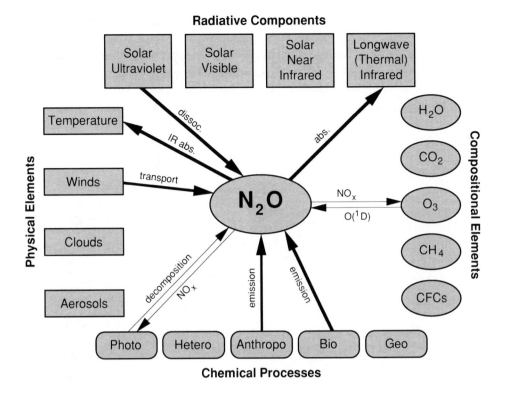

Fig. 7.6 Schematic diagram of the physical and chemical interactions involving nitrous oxide that might be included in a comprehensive climate model. The key physical climate parameters are shown in boxes, as in Fig. 7.1. Atmospheric chemical species are enclosed in circles. Chemical mechanisms that may involve, or affect, nitrous oxide are divided into classes, shown at the bottom of the figure, and the pathways and modes of interaction are indicated by arrows. An explanation of the scheme for classifying chemical processes is found in Sec. 7.2, and in the caption for Fig. 7.2.

mainly by solar photodissociation (7.19):

$$N_2O + h\nu \rightarrow N_2 + O(^1D).$$

N_2O that is decomposed in the presence of ozone also forms nitrogen oxides or NO_x as follows from (7.18)

$$O_3 + N_2O + h\nu \rightarrow NO + NO + O_2 \qquad (7.24)$$

[note: NO_x is often defined as $NO+NO_2$, but sometimes includes NO_3 and other nitrogen oxide compounds; NO_y includes NO_x plus HNO_3 and all other nitrogren-bearing acids and nitrates (Sec. 7.3.3b)]. The NO_x then reacts with and destroys ozone in a catalytic cycle that is summarized in (7.13). Present atmospheric concentrations of N_2O are responsible for about 30–40% of the total ozone destruction in the middle and lower stratosphere. Accordingly, any significant change in N_2O would cause a strong feedback on the ozone abundance and distribution. The ozone variation in turn would affect the overall energy balance and climate (see Sec. 7.3.3).

Nitrous oxide is produced by the process of microbial denitrification in anaerobic environments. Since N_2O affects ozone, and ozone affects the intensity of ultraviolet radiation at the Earth's surface, which in turn can affect biological activity, a photochemical/biochemical feedback mechanism may exist that acts to modulate the magnitude of this source of nitrous oxide. Nevertheless, the first-order responses of the climate system to changes in N_2O concentrations are related to direct radiative forcing and stratospheric ozone depletion.

b. Sources, measurements and trends

The principal atmospheric source of nitrous oxide is denitrification in natural and agricultural biomes, although the agricultural contribution is highly uncertain and may be relatively small (see Chapter 2). The oceans are another significant source of N_2O. Combustion, including biomass burning, is a secondary source. There has been an apparent increase in N_2O concentrations over the last 100 years. The present rate of increase is small, $\sim 0.2\%$ per year. However, if the trend continues through the next century, or accelerates, the enhancement in nitrous oxide levels would be very significant to ozone and climate.

7.3.6 Chlorofluorocarbons (CFCs)

The chlorofluorocarbons, or CFCs, comprise a family of compounds containing chlorine and fluorine in various combinations. These compounds are manufactured, and have no natural sources. All of the common CFCs are active greenhouse gases; that is, they have strong absorption bands in the thermal infrared spectrum, falling within the atmospheric window. Most of these compounds are chemically inert, and have long atmospheric lifetimes. The climate-related properties of the CFCs as a group are summarized in Table 7.1. The most prevalent CFCs are CFC-11 ($CFCl_3$) and CFC-12 (CF_2Cl_2). CFCs and related chlorocarbons (e.g., carbon tetrachloride, CCl_4, and methyl chloroform, CH_3CCl_3) are being closely monitored in the atmosphere (Table 7.2).

The general interactions of CFCs with other elements of the climate system are indicated in Fig. 7.7. Because the CFCs are chemically inert, their interactions with the biosphere and geosphere, and with clouds and aerosols, are negligible. The first-order effects of CFCs on the climate system involve direct radiative forcing in the longwave spectrum, and stratospheric ozone depletion through the release of chlorine at high altitudes. The interaction chart for the CFCs is very similar to that for N_2O, except for the lack of strong biological coupling in the case of CFCs.

Table 7.2. Halocarbon concentrations and trends, from IPCC (1990a).

Halocarbon	Common designation	Mixing ratio (pptv)	Rate of increase (%/year)	Atmospheric lifetime (years)
$CFCl_3$	CFC-11	280	4	65
CF_2Cl_2	CFC-12	484	4	130
CF_3Cl	CFC-13	5		400
$C_2F_3Cl_3$	CFC-113	60	10	90
$C_2F_4Cl_2$	CFC-114	15		200
C_2F_5Cl	CFC-115	5		400
CCl_4	Carbon tetrachloride	146	1.5	50
CHF_2Cl	HCFC-22	122	7	15
CH_3Cl	methyl chloride	600		1.5
CH_3CCl_3	methyl chloroform	158	4	7
CF_2ClBr	Halon-1211	1.7	12	25
CF_3Br	Halon-1301	2.0	15	110
CH_3Br	methyl bromide	~ 10		1.5

a.   Climate/chemistry interactions

Greenhouse effect: CFCs have two properties that make them, gram for gram, the most potent greenhouse gases. First, these compounds have very strong absorption bands in the spectral region from 8 to 14 μm. Accordingly, CFC mixing fractions of only one part per billion can have significant climatic consequences (compared to hundreds of parts per million of carbon dioxide, whose infrared absorption bands are highly saturated and incrementally have little impact). Second, because the CFCs are very stable in the atmosphere – the only known significant sink being photodissociation in the stratosphere – they can accumulate over long periods of time.

Stratospheric ozone depletion: The parallels between nitrous oxide and the CFCs extend to their fate and effects on ozone. Both N_2O and CFCs are decomposed mainly by solar ultraviolet photodissociation in the stratosphere. In the case of the CFCs, chlorine is released in a form (Cl_x) that can erode the ozone layer; for example,

$$CFCl_3 + h\nu \rightarrow CFCl_2 + Cl. \qquad (7.25)$$

The principal reaction cycle for the catalysis of ozone by chlorine is:

$$Cl + O_3 \rightarrow ClO + O_2$$
$$\underline{ClO + O \rightarrow Cl + O_2} \qquad (7.26)$$
$$O + O_3 \rightarrow 2O_2 \text{ (net)}.$$

229

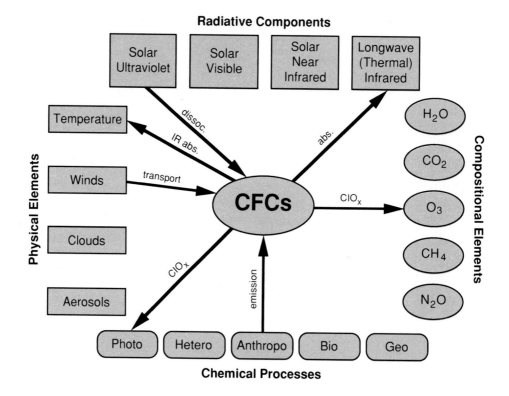

Fig. 7.7 Schematic diagram of the physical and chemical interactions involving chlorofluoro-carbons that might be included in a comprehensive climate model. The key physical climate parameters are shown in boxes, as in Fig. 7.1. Atmospheric chemical species affected by CFCs are enclosed in circles. Chemical mechanisms that may involve chlorofluorocarbons are divided into classes, shown at the bottom of the figure, and the pathways and modes of interaction are indicated by arrows. An explanation of the scheme for classifying chemical processes is found in Sec. 7.2, and in the caption for Fig. 7.2.

The chlorine released from CFCs currently dominates the chlorine cycle of the stratosphere, although natural sources of stratospheric chlorine exist (e.g., methyl chloride, or CH_3Cl, volcanoes, and meteors).

b. Measurements, lifetimes and trends

Table 7.2 summarizes the atmospheric properties of the most important chlorofluorocarbons, chlorocarbons and halons (bromine-containing compounds). The lifetimes of these compounds against photochemical decomposition ranges from about 1–2 years for the natural organic halogen gases CH_3Cl and CH_3Br to hundreds of years for the extensively fluorinated CFCs (e.g., CFC-115). The lifetime tends to increase with the degree of fluorination. The hydrogen-containing CFCs (or HCFCs) have relatively short atmospheric lifetimes (for example, ~15 years for HCFC-22).

Measurements of a large number of CFCs and related compounds show a rapid increase in their atmospheric concentrations (Table 7.2). CFC-113 (and, most likely, CFC-114 and CFC-115) is increasing at the greatest rate and has a high potential for climate change (or greenhouse warming potential, GWP), given its long atmospheric lifetime. The 1988 Montreal Protocol and 1990 London Amendments have set a timetable for the phase-out of many CFCs (see Sec. 2.5.3) while placing limits on the use of a number of other related compounds. Nevertheless, the beneficial effects of the treaty will not be apparent for a decade, and will be fully realized only if all nations comply with its provisions. Otherwise, the unrelenting increase in atmospheric CFC concentrations may become a major factor in greenhouse warming.

Alternative CFCs: A new family of fluorinated compounds is being designed that contain hydrogen–carbon bonds – the HCFCs and HFCs. These compounds have many of the desirable thermodynamic properties of the classic chlorofluorocarbons, but create fewer undesirable photochemical side effects. HCFCs in general have short atmospheric lifetimes because OH can effectively attack their C–H bonds (Table 7.2):

$$CHF_2Cl + OH \rightarrow CF_2Cl + H_2O \qquad (7.27)$$

where the fluorocarbon fragment is chemically unstable. HFCs can be produced that are free of chlorine atoms, thus eliminating the cause of ozone depletion by CFCs and HCFCs. Such specially engineered organohalogen compounds would have much smaller global impacts on radiation and chemistry than comparable emissions of CFCs. The ozone depletion potentials (ODPs) and GWPs for HCFCs are considerably smaller than those for CFCs. Nonetheless, if worldwide usage of these compounds continues to expand, particularly in the developing countries, future concentrations sufficient to cause significant chemical and climatic effects might still be reached.

7.3.7 Other greenhouse gases

Many common atmospheric gases have infrared bands that can affect the thermal energy balance. Among these species are carbon monoxide (CO), sulfur dioxide (SO_2), carbonyl sulfide (COS), and a host of organic compounds (e.g., C_2H_2, C_2H_4, C_2H_6). Generally, the abundances of these gases, and hence their direct contributions to global warming, are much smaller than those of the principal greenhouse gases discussed above. On the other hand, these trace compounds may contribute indirectly to climate change. For example, CO affects the lifetime of methane, and COS affects the abundance of reflective stratospheric aerosols (see Sec. 7.4.1*b*). Accordingly, the chemistry of these additional species is briefly reviewed. It should be noted that, by including these gases, the coupled radiation/composition relationships in climate modeling become even more complex.

a.   Climate/chemical interactions

Carbon Monoxide: CO is coupled to the chemistry of methane and ozone by a number of reactions. For example,

$$CH_4 + OH \rightarrow \cdots CO + \text{Products}$$
$$CO + OH \rightarrow CO_2 + H$$
$$H + O_2 + M \rightarrow HO_2 + M$$
$$HO_2 + NO \rightarrow OH + NO_2 \qquad (7.28)$$
$$NO_2 + h\nu \rightarrow NO + O$$
$$O + O_2 + M \rightarrow O_3 + M.$$

CO reacts with OH and can thus reduce OH concentrations, which, in turn, affects the lifetimes of a number of other gases such as sulfur dioxide and methane. Anthropogenic emissions of CO and NO also act to increase tropospheric ozone concentrations; note that the last step in sequence (7.28) results in the production of an ozone molecule.

Nitrogen Oxides: Unlike the stratosphere, where the photodecomposition of N_2O provides the major source of NO_x (and NO_y) (Sec. 7.3.5a), most tropospheric NO_x is directly emitted into the atmosphere (from soils and internal combustion engines) or is generated by lightning and other high-temperature processes. NO_y compounds are not significant contributors to the direct forcing of the radiative greenhouse effect. However, they play an important role in the atmospheric photochemistry of other greenhouse gases. The effects of nitrogen oxides on stratospheric ozone are discussed in Secs. 7.3.2a and 7.3.5a (also see 7.3.3b) and on tropospheric ozone, in Secs. 7.3.3c and 7.3.4a (and in this section). In some geographical regions, nitrogen oxides are also a major contributor to acidic rain. The removal of these compounds from the troposphere occurs through the following sequence of reactions:

$$NO + O_3 \rightarrow NO_2 + O_2$$
$$NO_2 + OH + M \rightarrow HNO_3 + M$$
$$NO_2 + O_3 \rightarrow NO_3 + O_2 \qquad (7.29)$$
$$NO_2 + NO_3 + M \rightarrow N_2O_5 + M$$
$$N_2O_5 + H_2O \xrightarrow{\text{Cloud}} 2\,HNO_3.$$

Note that, in this sequence, NO emissions are converted photochemically into nitric acid, which is readily scavenged by precipitation. The rate of removal of NO_x is controlled by OH and O_3 concentrations; in turn, any change in the NO_x residence time couples back into the chemistry of ozone.

Other gases: The photodecomposition of natural and anthropogenic nonmethane hydrocarbons (NMHCs) generates both hydrogen radicals and ozone, as does the photooxidation of methane (Eqs. 7.20 and 7.23). The yields of various secondary compounds that have a role in climate change, including ozone, are very sensitive to the abundance of NO_x. Sulfur-bearing gases, mainly SO_2, COS and $(CH_3)_2S$ (dimethyl sulfide), produce aerosols (Sec. 7.4.1*b*) with climatic implications (Sec. 7.2.2*c* and *d*).

b. Sources, measurements and trends

Carbon monoxide has a relatively short atmospheric lifetime ($\sim$2–3 months in the rural troposphere) and thus exhibits highly variable concentrations. The sources of CO include combustion (fossil fuels and biomass), which accounts for about one-half of all emissions, and the oxidation of methane (natural and anthropogenic), which accounts for the other half. Measurements show a higher concentration of CO in the NH compared to the SH, by a factor of about 2, consistent with the greater emissions in the north. CO appears to be increasing in the NH at the rate of $\sim$1% per year (Khalil and Rasmussen, 1988).

Nitrogen oxide sources are also divided between natural (soil biology, $\sim$20 Tg-N yr^{-1}; lightning, $\sim$5 Tg-N yr^{-1}) and anthropogenic (fossil fuel combustion, 20 Tg-N yr^{-1}; biomass burning, $\sim$5 Tg-N yr^{-1}) processes. The atmospheric residence time of NO_x is as short as a few days, and variability in NO_x (and NO_y) concentrations are large. The increase in anthropogenic emissions during the last century has probably led to a substantial increase in NO_x abundances in the NH, as has been the case for CO. The data base describing NO_x distributions and trends in the global atmosphere is sparse.

7.4 Atmospheric chemistry of aerosols

7.4.1 *Sources and distributions*

Atmospheric aerosols, or particulates, have a large number of sources. In general, their characterization involves the specification not only of spatial and temporal distributions, but of composition, particle size distribution, and physical properties such as morphology and indices of refraction. Tropospheric aerosols have short atmospheric residence times ($\sim$ days to several weeks) and exhibit enormous spatial variability. In addition, aerosol sources often vary sharply in space and time, leading to strong gradients (as in a dust storm, for example). Accordingly, the data describing tropospheric aerosols on a global scale are sketchy at best, and information on the distributions and properties for climate modeling are preliminary.

Stratospheric aerosols have been characterized more thoroughly than tropospheric aerosols because: (i) the particles are more homogeneous in composition, spatial distribution and temporal variation, and (ii) satellite-based instruments have been monitoring the global stratospheric aerosol layer for more than a decade (McCormick et al., 1979). There is no comparable program for tropospheric aerosols.

Atmospheric aerosols can be broadly divided into two categories – primary particles that are injected into the atmosphere, and secondary particles that are formed in the atmosphere by chemical and microphysical processes. Occasionally the distinction is fuzzy, and one must be flexible in categorizing aerosols. In the present discussion, cloud particles (droplets, ice crystals and hydrometeors) are not treated as aerosols; the topic of clouds is dealt with in Sec. 3.5. Aerosols can also be categorized by size. Large particles are those with radii exceeding about 1 μm; these quickly fall out of the atmosphere or are efficiently scavenged by clouds and precipitation. Particles having radii in the range of $\sim$0.1–1 μm are referred to as 'accumulation mode' aerosols. These particles have relatively low fallspeeds, but are still efficiently scavenged by precipitation. Some of the aerosols in this category contain soluble components such as sulfates and nitrates, which allows them to act as CCN (Sec. 7.2.2d and 7.2.3). At radii much smaller than 0.1 μm, an aerosol 'nucleation mode' is often found, consisting of new particles generated from vapors (see below). These fine particles are generally depleted by coagulation.

For radiative energy balance considerations, the most important particles are those in the accumulation mode, because of their large *specific* extinction coefficients (i.e., total scattering plus absorption cross section per unit mass) (see Sec. 7.2.2c).

a. Primary particles

Primary aerosols are those emitted directly into the atmosphere by natural and anthropogenic processes. Generally speaking, these aerosols tend to have their greatest influence on local and regional energy balances and chemistry.

The most common primary aerosols in the troposphere are sea salt, soil dust, and smoke or soot. Sea salt haze is created over the oceans by wind and wave action. The salt particles are generally large (greater than 1–10 μm radius) and typically do not travel more than $\sim$100 meters vertically into the atmosphere before falling back into the ocean. Salt haze (which should not be confused with the fog or stratus clouds that are common over oceans) is generally confined to windswept regions of the seas.

Soil dust (or dust) in the atmosphere can be raised by winds moving across arid regions. Major dust storms, in which sunlight may be blocked from the surface for days, are associated with the Earth's largest deserts (e.g., Saharan dust storms, in which dust may be carried as far as the Caribbean). Vehicles also raise large quantities of dust from arid lands. However, most of the dust mass consists of large particles (as in the case of sea salt) with sizes typically exceeding 10 μm; these particles remain suspended for a few hours, at most. A fraction of the dust mass consists of submicron particles. These smaller particles may remain in the atmosphere for days to weeks and travel thousands of kilometers (the aeolian dust). On occasion, dust from the Gobi desert is detected over Mauna Loa in Hawaii. Ocean sediment cores also reveal patterns of continental dust fallout over the western Pacific Ocean, for example.

Smoke from biomass burning, and soot from inefficient fossil fuel combustion, are primary sources of aerosols over the tropics, and over industrial regions, respectively. Massive plumes associated with the combustion of agricultural and forest debris are clearly evident in satellite photographs. Soot released in exhaust streams creates a gray pall over a number of urban regions in the NH, and darkens the Arctic haze aerosol that blankets the northern high latitudes in winter. Smoke particles typically fall within the accumulation mode and, hence, are relatively long-lived and radiatively active.

Trends in primary particle concentrations have not been discerned. Only limited historical data are available, except in isolated urban airsheds, which are not typical of the global atmosphere. Even today, very few accurate measurements of primary aerosol emission rates and properties are being collected on a regular basis.

b. Secondary (photochemical and organic) aerosols

Secondary aerosols are composed primarily of sulfate, nitrate and organic compounds. The sources of sulfur for particles include natural emissions of reduced sulfur gases, amounting to as much as 70 Tg-S yr^{-1}, and anthropogenic sulfur emissions, mainly SO_2, totaling roughly 80 Tg-S yr^{-1}. Sulfate aerosol is generated by the sequence of processes:

$$SO_2 + OH + M \rightarrow HSO_3 + M$$
$$HSO_3 + O_2 \rightarrow SO_3 + HO_2$$
$$SO_3 + H_2O \xrightarrow{M} H_2SO_4 \tag{7.30}$$
$$H_2SO_4 \xrightarrow{H_2O} \text{Sulfate aerosol.}$$

The final process represents the condensation of sulfuric acid (and related sulfate compounds) onto existing particles, or the formation of new particles by nucleation. A similar process occurs in the stratosphere, resulting in a distinct sulfate layer [or Junge layer, after its discoverer (Junge, 1963)]. However, in this case, the initial sulfur species is COS, which is photolyzed and oxidized according to the following sequence:

$$COS + h\nu \rightarrow S + CO$$
$$S + O_2 \rightarrow SO + O \tag{7.31}$$
$$SO + O_2 \rightarrow SO_2 + O.$$

Dimethyl sulfide, $(CH_3)_2S$ (or DMS), is generated by phytoplankton in the surface waters of the world oceans (see Sec. 8.4). DMS accounts for the largest emission of reduced sulfur to the atmosphere, about 40 Tg-S yr^{-1}. The photooxidation of DMS generates SO_2 and methane sulfonic acid (MSA), both of which condense to form aerosols in the marine boundary layer. These aerosols, with sea salt particles, provide the CCN upon which marine stratus clouds form.

Nitrates follow a similar chemical pathway as sulfates (7.29). Both sulfates and nitrates are often found in the form of ammonium salts in the troposphere, owing to the fact that ammonia gas, NH_3, is generated in large quantities by biological processes and reacts vigorously with sulfuric and nitric acids.

Organic aerosols may be generated when an organic precursor is oxidized into compounds of very low volatility (or vapor pressure). An example of this process is the reaction of terpenes emitted by plants (e.g., α-pinene from certain species of evergreens) with OH to form readily condensable species. The result is an organic haze resembling smog.

The chemistry and physics of secondary aerosol formation are complex. First, the photochemistry of precursor vapors leading to condensible species involve the chemistry of OH and tropospheric ozone. Second, the microphysical processes that govern the evolution of the aerosols after formation must be considered (see Sec. 7.4.3). Accordingly, empirical aerosol optical parameters are often used to explore the first-order radiative and climatic effects of aerosols. In this case, the aerosol optical depths as a function of spatial location, and perhaps time, are specified for radiative calculations in a climate model.

c. Volcanic aerosols

Major volcanic eruptions on occasion generate a global layer of sulfuric acid aerosols in the stratosphere (see Turco et al., 1982). Two recent volcanic events producing global effects were El Chichón (Mexico, 17°N, 4 April 1982) and Mount Pinatubo (Philippines, 17°N, 15–16 June 1991). The general properties of volcanic aerosols are summarized in Table 7.3. With regard to climate, stratospheric aerosols differ from tropospheric aerosols in several important ways. First, volcanic aerosols are rapidly distributed over the entire planet, affecting the global surface energy balance. Second, the aerosols from a single eruption may persist in the stratosphere for several years, much longer than they could remain in the troposphere. Additionally, volcanic aerosols can have large optical depths, substantially exceeding the average optical depth of tropospheric aerosols (except on regional or smaller scales).

Volcanic eruptions that are large enough to affect global climate are relatively rare (Lamb, 1977). In recent historical times, the largest such event was Tambora (Indonesia, 1815), which produced the notable 'year

Table 7.3. Properties of volcanic aerosols.

Composition:	Silicates; H_2SO_4/H_2O ($\sim$70%/30%); Traces of sulfates, nitrates, chlorides, etc.
Origin:	Volcanic SO_2 fumes; also, OCS, CS_2, H_2S; Photochemical oxidation to H_2SO_4 via OH
Properties:	Liquid spheres; Solid mineral particles dominant the first month; $\sim$100 ppbm; (highly variable); $\sim$10 cm^{-3} $\sim$0.3 μm radius
Distribution:	Regional (days); Zonal (weeks); Hemispheric (months); Global (year)
Mass budget:	$\sim$10–100 Tg-S; H_2O, $\times$10; CO_2, $\times$10–100
Residence time:	$\sim$1–3 years (deduced from aerosol observations)
Effects:	Shortwave radiation scattering ($\tau > 0.1$) leads to surface cooling; Longwave absorption warms the stratosphere; Injection of HCl, H_2O, etc. affects air chemistry and composition; Heterogeneous reactions on sulfate aerosol surfaces; Ozone decrease from ΔT, ΔHCl, ΔAerosol, etc.;
Influences:	Impact on ozone (possibly affected by future chlorine buildup)
Trends:	Significant eruption every $\sim$20 years on average; Major eruptions are $\sim$100 years apart

without a summer' (Stommel and Stommel, 1983). Typically, a significant eruption (i.e., one that injects at least 5–10 Tg-S into the stratosphere and produces a global scale aerosol haze with a visible extinction optical depth in the range of $\sim$0.1) occurs every one to three decades, and eruptions the magnitude of Tambora (which injected $\sim$100 Tg-S into the stratosphere), every one to three centuries (although the statistics on, and predictability of, volcanic eruptions are unreliable). In the context of global climate change, volcanic eruptions provide significant 'noise' (cooling episodes) in the temperature record that must be identified in a careful analysis of possible temperature trends. Volcanic aerosols may also catalyze certain reactions that lead to substantial ozone depletion in the stratosphere (Sec. 7.4.2): the importance of this effect might be enhanced in the future if the atmospheric concentrations of chlorine continue to increase.

7.4.2 Aerosol chemistry

The chemistry of aerosols may be divided into several distinct categories: the physical chemistry and thermodynamics of condensed (aerosol) materials; reactions of trace compounds adsorbed onto the surfaces of solid particles; and reactions of trace materials dissolved in liquid droplets. The latter category could be broadened to include reactions occurring in cloud and precipitation drops (reactions on ice crystals would fall into the second category). The physical chemistry and morphology of atmospheric aerosols is not well established. Some specific aerosol types – e.g., sulfate aerosol, soil dust, volcanic particulates – are relatively well characterized in terms of composition and physical chemistry. Other particle types, such as smoke and organic aerosols, are poorly characterized.

Surface reactions on aerosol particles have only recently been seriously considered in global chemical modeling. Interest has been boosted by the discovery of the ozone hole over Antarctica and the central role of polar stratospheric clouds (PSCs) in causing ozone depletion. PSCs, which are composed of ice and nitric acid hydrates, have been found to catalyze the following reactions:

$$ClONO_2 + HCl \xrightarrow{PSC} Cl_2 + HNO_3$$

$$ClONO_2 + H_2O \xrightarrow{PSC} HOCl + HNO_3$$

$$N_2O_5 + HCl \xrightarrow{PSC} ClNO_2 + HNO_3 \tag{7.32}$$

$$N_2O_5 + H_2O \xrightarrow{PSC} 2HNO_3.$$

These reactions generate photochemically active chlorine species (Cl_2, HOCl and $ClNO_2$) from the inert chlorine reservoir species (HCl and $ClONO_2$), while also converting active NO_x (in the form of $ClONO_2$ and N_2O_5) into the less reactive species, HNO_3. None of these reactions occur in the gas phase under stratospheric conditions. PSC ice particles provide surfaces on which the reactants can condense; the surfaces act to concentrate the reactants and reduce the thermodynamic barriers that otherwise inhibit these reactions.

Reactions occurring in aqueous solutions have been studied for many years. The application of basic aqueous reaction data to atmospheric chemistry problems is well established (Seinfeld, 1986). Among the chemical processes of interest are the oxidation of SO_2 and NO_x to sulfates and nitrates, respectively, and their subsequent deposition by precipitation. The principal oxidants in aqueous solution are hydrogen peroxide (H_2O_2), ozone, and certain metal ions.

Chemical reactions involving aerosols can influence the concentrations and distributions of climatologically important gases, such as ozone. The chemical transformations of gases into particles (e.g., SO_2 into sulfate aerosols) has both direct and indirect climate implications: direct, because the buildup of aerosols affects the scattering of radiation in the atmosphere; indirect because aerosols are involved in the formation of clouds, which affect the planetary albedo.

7.4.3 Aerosol microphysics

The evolution of aerosol particles in the atmosphere involves both chemical and microphysical processes. The microphysics of airborne particles is discussed in a number of specialized texts (e.g., Twomey, 1977; Pruppacher and Klett, 1978). The most significant microphysical processes are nucleation, condensation/evaporation, coagulation and sedimentation. In modeling aerosol physics, the size resolution of the particles and the compositional differences between aerosol types must be treated. Such details are needed to define the chemical, microphysical and radiative properties of the aerosols.

For numerical analysis, aerosols are usually divided into discrete size categories. A continuity equation is written for each discrete particle size (and type) as:

$$\frac{\partial n_{pi}}{\partial t} = S_{nuc} + S_{inj} + (P - L)_{cond} + (P - L)_{coag} - \nabla \cdot \Phi_{pi} \tag{7.33}$$

$$\Phi_{pi} = \mathbf{v}_{adv} n_{pi} - \mathbf{v}_{sed} n_{pi} \hat{z}$$

where n_{pi} is the concentration of particles in the size range i, the sources (S) of aerosols due to nucleation and direct injection are separately indicated, and the Production–Loss ($P-L$) term for each specific microphysical process is indicated (the mathematical forms of these terms are described, for example, in Twomey, 1977). The advective velocity, $\mathbf{v}_{adv}$ is obtained from observed winds, or a dynamical model, as for the gaseous tracer continuity equation (7.8); the sedimentation velocity, $\mathbf{v}_{sed}$, is a function of the size, shape and density of the particles.

Aerosols are generated by the nucleation of vapors that are supersaturated. Homogeneous nucleation involves the condensation of a vapor into a pure aerosol of that material; heterogeneous nucleation involves the deposition of a vapor onto a substrate particle (for example, dust). Condensation/evaporation refers to the deposition of a vapor onto, or evaporation from, a pre-existing aerosol containing similar material. The rate of condensation or evaporation is proportional to the difference between the vapor pressure and the partial (ambient) pressure of the condensing gas over the particle. Coagulation refers to a class of dynamical processes by which one aerosol particle dynamically encounters another and the two particles coalesce. This process is represented quantitatively by a coagulation (or collision/coalescence) kernel. Finally, the sedimentation of a macroscopic particle under the influence of gravity is described by its terminal fallspeed, for which the gravitational force is balanced by the aerodynamic drag of the atmosphere.

7.5 Future developments

The first-order climatic effects of changes in atmospheric composition have been modeled extensively with existing global climate models. The remaining issues include the importance of coupling chemically active species to climate models, and determining the potential feedbacks that may either increase or reduce the magnitude and duration of the climate response to anthropogenic forcing. Many chemical processes may turn out to have second-order effects, although several such effects, if they are reinforcing, might have a substantial cumulative impact on climate forecasts. It is also important in modeling the coupled climate system to identify new feedback mechanisms and to insure that all significant feedbacks have been accounted for. The ultimate goal of some climate modelers is to achieve a fully integrated dynamics–radiation–chemistry simulation of the climate system.

A number of groups are currently experimenting with coupled three-dimensional models of global dynamics and photochemistry (see Chapter 15). These models are still primitive, and impractical for long-term climate simulations. Little work has yet been carried out on coupled atmosphere–ocean tracer-chemistry modeling. Carbon dioxide appears to be the climatically active species most amenable to such modeling, inasmuch as its photochemistry is negligible and its geochemistry is relatively straightforward. However, biological processes that drive the carbon cycle must still be quantified. These processes are poorly understood, and adequate mathematical treatments (that might be included in a climate model) are lacking.

It will probably take a decade or longer to develop the basic scientific data upon which a fully coupled global climate model incorporating photochemistry, geochemistry and biochemistry can be built. Moreover, implicit in all discussions of future generations of chemistry–transport climate models is the assumption that much larger and faster computers will be available. It should be noted that, no matter how sophisticated our computational facilities become, there will be little science done with these computers until means are developed for processing the huge volume of output generated by the models, of evaluating model performance against observations – likely global satellite data sets – and of interpreting the results in terms of basic physical and chemical processes. There is little hope for understanding the climate system, let alone finding solutions to the daunting problems of global pollution, if researchers become intoxicated with graphics generated by numerical models and neglect the basic science.

8

Marine biogeochemistry

Raymond G. Najjar

8.1 Introduction

Marine biogeochemistry is an important component of the climate system because there are significant fluxes of several climatically relevant trace gases across the air–sea interface. The most notable of these gases are carbon dioxide (CO_2), dimethyl sulfide (DMS), and nitrous oxide (N_2O). As discussed in earlier chapters, CO_2 is a greenhouse gas of great importance. Oceanic DMS is thought to be the main source of marine non-sea-salt sulfate (NSS-SO_4^{2-}) aerosols, which scatter solar and terrestrial radiation and serve as cloud condensation nuclei (Sec. 7.4). It is thought that the net climatic effect of marine DMS emissions is to cool the Earth (Charlson et al., 1987). N_2O is both a greenhouse gas and an important player in the destruction of stratospheric ozone (Sec. 7.3.5).

Ice core records suggest that marine biogeochemistry and climate are intimately coupled. Both atmospheric CO_2 (Barnola et al., 1987; see also Fig. 21.3) and NSS-SO_4^{2-} aerosols (Legrand et al., 1991) are correlated with atmospheric temperature over the past 160,000 years. The CO_2 and NSS-SO_4^{2-} variations are significant and it appears likely that they were driven by changes in the marine carbon and sulfur cycles.

The current emissions of CO_2, sulfur, and N_2O resulting from human activities have given additional importance to the study of marine biogeochemistry. It is estimated that the oceans are currently absorbing 18 to 40% of anthropogenic CO_2 emissions (IPCC, 1990a); see also Table 16.1. Although the oceans represent a modest fraction of the total N_2O sources to the atmosphere, they cannot be ignored in current budgets since the oceanic source of N_2O to the atmosphere is approximately equal to the current rate of increase in atmospheric N_2O. Anthropogenic emissions of reduced sulfur (which are a source of aerosols) are now approximately equal to natural emissions of reduced sulfur. Given the significance of the anthropogenic perturbations of atmospheric CO_2, sulfur, and N_2O, it is important to understand the natural context in which these perturbations take place and how feedback mechanisms might be activated.

This chapter will cover the basic marine biogeochemical processes involving CO_2, DMS and N_2O. The focus will be on the marine carbon cycle for three reasons: (1) the marine carbon cycle is often thought to be the most important with regard to climate, (2) much more is known about CO_2 than N_2O and DMS in the ocean, and (3) the marine carbon cycle serves as the basis for understanding many other elemental cycles in the ocean and is of great use in understanding marine DMS and N_2O. Other climatically relevant gases that may have significant air–sea fluxes include carbon monoxide, ammonia, methyl iodide, the nonmethane hydrocarbons and carbonyl sulfide (Wolfe et al., 1991). Because relatively little is known about the marine biogeochemistry of these gases, they will not be discussed here. This chapter serves as an introduction to chapter 16.

The reader is referred to the following books and review articles for additional background material and more detailed discussions of some of the subtopics in this chapter. The review article by Wanninkhof (1992) is an excellent synthesis of air–sea gas exchange studies. The texts by Butler (1982) and Stumm and Morgan (1981) have complete discussions of the relevant aspects of inorganic carbon chemistry. Parsons et al. (1984) provide a basic text in biological oceanography. The review article by Jahnke (1990) is a recent and comprehensive status report on the observations of biological fluxes of carbon and related species. Broecker and Peng (1982) give a broad and easily read discussion of the marine carbon cycle. Andreae (1990) provides a recent review of dimethyl sulfide.

8.2 Air–sea gas exchange

8.2.1 *Air–sea interface*

One of the most important factors in air–sea gas exchange is turbulence in the near-surface ocean. The mixing of tracers by turbulent processes is much greater than by molecular diffusive processes. Because of surface tension, turbulent motions are inhibited very close to the air–sea interface. The resistance to air–sea exchange is encountered in these two layers. For the gases of interest here, the resistance of the liquid boundary layer dominates, allowing the boundary layer in the air to be ignored.
this boundary layer is quite small – on the order of tens of microns. There is also a corresponding boundary layer in the atmosphere just above the interface. The resistance to air–sea exchange is encountered in these two layers. For the gases of interest here, the resistance of the liquid boundary layer dominates, allowing the boundary layer in the air to be ignored here.

Several models have been developed to understand the processes that determine the gas flux in and near the liquid boundary layer. The simplest of these models is the *stagnant film model*, which assumes that the boundary layer is a discrete, stagnant layer in which only molecular diffusion takes place. The stagnant film sits on top of a well-mixed, purely turbulent layer

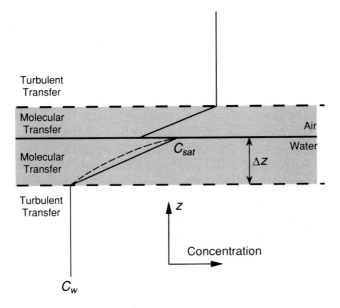

Fig. 8.1 Stagnant film model for air–sea gas exchange. The curved dashed line shows the profile of a gas that is chemically reactive (see Sec. 8.3.4d).

(Fig. 8.1). The flux of gas across the air–sea interface is equal to the flux in the stagnant film, which is given by Fick's law:

$$F = -D\frac{\partial C}{\partial z} \qquad (8.1)$$

where F is the upward flux, D is the molecular diffusivity of the gas in sea water and C is the concentration of the gas in the film. If there are no sources or sinks of the gas within the film, then the concentration profile should be linear within the film. The flux is then

$$F = \frac{D}{\Delta z}(C_w - C_{sat}) \qquad (8.2)$$

where Δz is the thickness of the film, C_w is the concentration at the base of the film and in the "bulk" fluid directly below the film, and C_{sat} is the concentration at the top of the film, which is also known as the saturation concentration. Because of the thinness of the layer, C_{sat} is not directly measurable. It can, however, be directly related to the partial pressure of the gas in the atmosphere, p_a, because the partial gas pressure is continuous across the air–sea interface. This allows the use of Henry's law:

$$\alpha p_a = C_{sat} \qquad (8.3)$$

where α is the solubility of the gas in sea water. Equation (8.2) can then be written as

$$F = \frac{D}{\Delta z}(C_w - \alpha p_a) \qquad (8.4)$$

243

or

$$F = k_w(C_w - \alpha p_a) \tag{8.5}$$

where k_w is known as the *transfer velocity* or the *piston velocity*. In this model, the piston velocity is proportional to the molecular diffusivity of the gas in sea water and inversely proportional to the thickness of the stagnant film. Δz is imagined to be primarily a function of the amount of turbulence in the fluid. As turbulence increases, the thickness of the boundary layer will decrease. Turbulence near the boundary layer is affected by the molecular viscosity of water, such that the more viscous the fluid is the lower the turbulence will be. One would expect, therefore, that the piston velocity increases as the molecular viscosity of water decreases. The competing effects of gas diffusion and fluid viscosity on the piston velocity are incorporated into the Schmidt number, Sc:

$$Sc = \frac{\nu}{D} = \frac{\mu}{\rho D} \tag{8.6}$$

where μ and ρ are the viscosity and the density, respectively, of sea water and $\nu = \mu/\rho$ is the kinematic viscosity. Gas exchange models that explicitly treat the dynamics of the boundary layer just below the air–sea interface are also cast in the form of (8.5) and predict that the piston velocity is a function of the Schmidt number. k_w is found to be proportional to $Sc^{-2/3}$ at low wind speeds (Deacon, 1977) and $Sc^{-1/2}$ at higher wind speeds (Ledwell, 1984). Experiments done in wind-wave tunnels support these theoretical assertions (Jahne et al., 1987).

Wind-wave tunnels have also been used extensively to determine other factors that affect the piston velocity. These experiments, an example of which is shown in Fig. 8.2, always reveal that the piston velocity increases with wind speed. This is not surprising since wind generates turbulence. These results are supported by experiments done on lakes (Wanninkhof et al., 1987) and in the ocean (Watson et al., 1991). Another feature commonly apparent is a nonlinear dependence of the piston velocity on wind speed – k_w has a greater sensitivity to wind at higher wind speeds and the change in sensitivity often occurs abruptly (at a wind speed of about 10 m s^{-1} in Fig. 8.2).

The nonlinearity in the wind speed dependence of the piston velocity is important when using measured or model-derived wind speeds to compute the piston velocity and the gas flux. Because winds are variable, the piston velocity at the average wind speed will be be lower than the average piston velocity. As the difference will be greater for greater temporal variability in the wind, one needs to know the wind speed statistics when attempting to compute the average piston velocity and the gas flux.

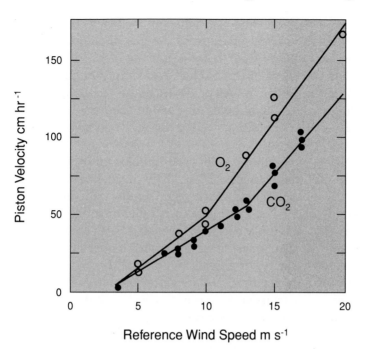

Fig. 8.2 Wind tunnel experiments of Broecker and Siems (1984) demonstrating the effect of bubbles on gas exchange. Reprinted by permission of Kluwer.

8.2.2 Bubbles

A sudden increase in the sensitivity of the piston velocity to wind speed often occurs at the onset of breaking waves. It has been speculated that air bubbles formed by the breaking waves enhance water–air gas exchange (and hence the piston velocity) by providing additional surface area for the exchange. Gas bubbles, which initially have the same composition as the overlying atmosphere, rapidly penetrate to a certain depth and then rise to the surface owing to their buoyancy. During the ascent of the bubbles, gas exchange occurs with the ambient water until the bubbles completely dissolve, evade into the atmosphere, or reach equilibrium with the surrounding water.

The total flux of a given gas due to bubbles, F_b (mol m^{-2} s^{-1}) can be expressed as

$$F_b = S_w^{-1} \sum_{i=1}^{N} f_{b_i} = S_w^{-1} \sum_{i=1}^{N} k_{b_i}^o S_{b_i}(C_w - \alpha p_{b_i}) \qquad (8.7)$$

where N is the number of bubbles per unit surface area of ocean S_w, f_b is the flux of the gas (mol s^{-1}) due to a single bubble, S_b is the surface area of the bubble, p_b is the partial pressure of the gas inside of the bubble and k_b^o is the piston velocity (m s^{-1}) for the bubble for the gas in question. k_b^o is a

245

function of the bubble velocity, the diffusion coefficient of the gas, and the bubble radius. If the partial pressure of the gas in the bubble is the same as that in the atmosphere, that is $p_b = p_a$, then (8.7) has the same form as (8.5), so that the effect of the bubbles can simply be incorporated as an enhancement of the piston velocity. To estimate the total piston velocity for the gas due to all of the bubbles, one would then need to know the total number density of the bubbles as well as the size and velocity distributions of the bubbles.

The assumption of equilibrium between bubbles and the atmosphere will not hold under all circumstances. First, if the residence time of the bubble in water is long enough, it will reach equilibrium with the ambient water. The equilibrium will be faster for gases with higher solubilities because the amount of gas that a bubble can hold decreases with increasing solubility. The equilibrium will also be faster for gases with higher diffusion coefficients because the piston velocity for a given bubble will be higher. By definition, once equilibration has occurred between the bubble and the water there is no longer any gas transfer between the water and the bubble. Thus the enhancement of the piston velocity due to bubbles will be greater for gases with lower solubilities and diffusion coefficients. Since the solubility of oxygen is much lower than that of CO_2, oxygen has a higher piston velocity than CO_2 (for the same Schmidt number) at high wind speeds (Fig. 8.2).

The second reason for disequilibrium between the atmosphere and bubbles is that bubbles have additional gas pressure due to hydrostatic effects and surface tension. For each meter of water depth, the gas pressure will increase by about 10% due to the increased pressure imparted by the overlying water. In the situation when there is no air–sea pressure gradient, if bubbles penetrate to a depth great enough to increase their pressure then there will be a gas flux from the bubble, and hence the atmosphere, to the ocean. In this case the notion of a piston velocity is unsatisfactory, since k_w would need to be infinite. This effect is more important for gases with lower solubilities because, for a given value of C_{sat}, the amount of gas in the bubble will increase with decreasing solubility. The effect can be quite significant as pointed out by Spitzer and Jenkins (1989), who found that 40% of the summertime supersaturation of argon is due to bubble injection in the North Atlantic. Because of the high solubility of CO_2, it is expected that bubbles are less efficient gas exchangers for this gas.

8.2.3 Formulations of the piston velocity

While variables other than wind speed, such as the air–sea temperature difference (an indicator of stability), the presence of surfactants, fetch and the history of the wind, may affect the piston velocity (through turbulence), most formulations to date have primarily considered the

Schmidt number and the wind speed. In general,

$$k_w \propto Sc^{-n} f(V) \tag{8.8}$$

where f is some function of the wind speed V usually normalized to a height of 10 m. For many gases the Schmidt number is known, and is a strongly decreasing function of the temperature.

The formulation by Liss and Merlivat (1986), which is based on lake, wind-wave tank and theoretical studies, has received much use in modeling the gas flux. Estimates of the piston velocity over the ocean using the purposely released tracer sulfur hexafluoride largely agree with the formulation of Liss and Merlivat, though these measurements have been made in only one area of the ocean so far. An analysis of the annual cycles of several gases in the upper ocean near Bermuda also supports the Liss and Merlivat relationship (Spitzer and Jenkins, 1989). The Liss and Merlivat relationship is significantly lower than an estimate of the piston velocity made using $^{14}CO_2$ released by the nuclear bomb tests of the 1950s and 1960s (Broecker et al., 1985b). The discrepancy perhaps indicates that it is necessary to include other parameters that affect the piston velocity besides the wind speed and the Schmidt number. Consideration of all of the estimates to date of the piston velocity of CO_2 yields a range in its average value between about 10 and 20 cm hr^{-1}.

8.2.4 *Equilibration time in the surface ocean mixed layer*

Consider a surface mixed layer of depth z_{ml} that is isolated from deeper waters by stratification. Assume that there is gas exchange across the air–sea interface and that there are no sources and sinks of the gas within the mixed layer. Then

$$\frac{\partial C_w}{\partial t} = \frac{k_w}{z_{ml}}(C_{sat} - C_w). \tag{8.9}$$

If C_w is not equal to C_{sat}, the mixed layer concentration will approach the saturation concentration asymptotically with an e-folding time of $\tau = z_{ml}/k_w$. τ is the time scale for gas phase equilibration between the mixed layer and the atmosphere. If it is assumed that the average piston velocity is 15 cm hr^{-1} (4 m day^{-1}) for a gas with $Sc = 600$, then the average time for a gas in the surface mixed layer of depth 80 m to equilibrate with the atmosphere will be $\sim$20 days. This is short enough to keep most gases close to equilibrium with the atmosphere. For chemically reactive gases, such as CO_2, the equilibration time will be longer (Sec. 8.3.4b).

8.3 The carbon cycle

During pre-industrial times the ocean contained about 65 times more carbon than the atmosphere. This is in contrast to many other atmospheric gases (e.g., oxygen, nitrogen and argon), which primarily reside in the atmosphere. The high solubility of CO_2 is partly responsible for this. CO_2 is about 30 times more soluble than most other gases of similar molecular weight. (Generally the solubility of a gas increases with molecular weight.) The main reason, however, for the relative abundance of carbon in the ocean is its chemical reactivity with water. In average surface water, for example, only 0.5% of all of the dissolved inorganic carbon (DIC) is in the form of dissolved CO_2 gas. Thus the atmosphere "sees" only a tiny fraction of the carbon present in sea water, enabling the ocean to be a relatively large reservoir of carbon. It is the large capacity of the ocean to hold carbon and the fact that it is always close to equilibrium with the atmosphere that makes the ocean an important sink of anthropogenic CO_2 and the likely source of atmospheric CO_2 variations over the past several hundred thousand years (Fig. 21.3).

The partitioning of carbon between the atmosphere and the ocean is also affected by internal oceanic processes (Fig. 8.3). Biological activity in the well-lit upper layers of the ocean (the *euphotic zone*) converts DIC into calcium carbonate and organic carbon. This material is transported to the deeper, dark layers of the water column (the *aphotic zone*) and the sediments by particle settling and the ocean circulation. In the aphotic zone and in the sediments, chemical and biological processes dissolve calcium carbonate and remineralize organic carbon to DIC. This biological cycling of carbon in the ocean is known as the *biological pump* because it maintains gradients in DIC and many other chemical species in the ocean. The rate of exchange of CO_2 gas across the air–sea interface is also important in the partitioning of carbon between the atmosphere and the ocean.

8.3.1   The inorganic chemistry of carbon dioxide

a.   The basic equations

Like any other gas, the partial pressure of carbon dioxide (pCO_2) in water depends on its solubility, α, and its concentration in water:

$$pCO_2 = \frac{[CO_2]}{\alpha}. \qquad (8.10)$$

The brackets indicate concentration in units of, for example, moles of solute per kg of solution. Equilibrium between the surface ocean and the atmosphere is established when surface ocean pCO_2 is equal to atmospheric pCO_2. The solubility of most gases, including CO_2, is higher at lower temperatures. Thus for water at equilibrium with an atmosphere of a fixed pCO_2, the concentration of CO_2 in the water will increase as the temperature (T) decreases. The solubility of CO_2 is also dependent on the salinity (S) such that CO_2 is less soluble in more saline waters. CO_2 dissolved in water hydrates to form carbonic acid, H_2CO_3, which dissociates to yield hydrated

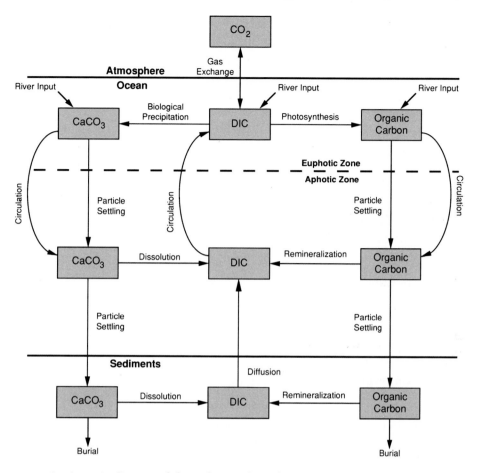

Fig. 8.3 A schematic diagram of the carbon cycle in the ocean.

protons (represented by H^+) and bicarbonate ion (HCO_3^-):

$$CO_2 + H_2O \Leftrightarrow H_2CO_3 \qquad (8.11)$$

$$H_2CO_3 \Leftrightarrow H^+ + HCO_3^- . \qquad (8.12)$$

In marine waters $[H_2CO_3]$ is very small; for all practical purposes it completely dissociates and can be ignored. Thus, reactions (8.11) and (8.12) are usually combined to form:

$$CO_2 + H_2O \Leftrightarrow H^+ + HCO_3^- . \qquad (8.13)$$

Bicarbonate ion then dissociates into hydrated protons and carbonate ion (CO_3^{2-}):

$$HCO_3^- \Leftrightarrow H^+ + CO_3^{2-} . \qquad (8.14)$$

Reactions (8.11) through (8.14), as well as the others discussed in this section, reach equilibrium extremely rapidly (in less than a few minutes).

This means that the concentrations of the species in these reactions must obey the laws of chemical equilibria. For reactions (8.13) and (8.14), respectively, the following equilibria must hold:

$$K_1 = \frac{[\text{H}^+][\text{HCO}_3^-]}{[\text{CO}_2]} \tag{8.15}$$

$$K_2 = \frac{[\text{H}^+][\text{CO}_3^{2-}]}{[\text{HCO}_3^-]} \tag{8.16}$$

where K_1 and K_2 are called the first and second dissociation constants of carbonic acid, respectively. In writing (8.15), it has been assumed that the concentration of water (H_2O) in sea water is constant, which is an excellent assumption for the weak solutions we shall be concerned with. H^+ is also involved in the ionization of water:

$$\text{H}_2\text{O} \Leftrightarrow [\text{H}^+] + [\text{OH}^-]. \tag{8.17}$$

The equilibrium of this reaction is given by

$$K_w = [\text{H}^+][\text{OH}^-], \tag{8.18}$$

where K_w is called the ionization constant of water. The final relevant relationship expresses electrical neutrality: the amount of positive charge in solution must equal the amount of negative charge. For CO_2 and its dissociation products in *pure* water this is

$$[\text{H}^+] = [\text{OH}^-] + [\text{HCO}_3^-] + 2[\text{CO}_3^{2-}]. \tag{8.19}$$

This equation states that to balance the electrical charge, the concentration of hydrogen ions must be equal to the sum of the concentrations of the hydroxide, bicarbonate and twice the carbonate ions. The factor of two is required in front of the carbonate ion because there are two negative charges for each carbonate ion. It will be convenient to use units of equivalents (eq), which is simply the number of moles of an ion multiplied by the charge of the ion.

The system of four equations (8.15, 8.16, 8.18 and 8.19) has five unknowns: CO_2, HCO_3^-, CO_3^{2-}, H^+, and OH^-. Specifying any one of these five species completely determines the rest. Often CO_2 or H^+ is measured. A more useful master variable, however, is the sum of the carbon species, which is known variously as total inorganic carbon, total CO_2 (ΣCO_2), total carbon, or dissolved inorganic carbon (DIC):

$$[\text{DIC}] = [\text{CO}_2] + [\text{HCO}_3^-] + [\text{CO}_3^{2-}]. \tag{8.20}$$

Since the equilibrium constants K_1, K_2, and K_w vary with temperature, salinity and total pressure (p), a water parcel undergoing a change of state will experience variations in CO_2, HCO_3^-, CO_3^{2-}, H^+, and OH^-. A change of state, however, will not affect DIC because carbon will simply be redistributed among the three carbon species. Therefore, changes in DIC can be used to quantitatively assess the various processes that add or remove inorganic carbon from water.

Up to now we have ignored the ions that primarily account for the salinity of sea water. The charge balance in sea water is considerably more complex than that given by (8.19), and can be expressed by

$$[H^+] = [OH^-] + [HCO_3^-] + 2[CO_3^{2-}]$$

$$+[Cl^-] + 2[SO_4^{2-}] + [Br^-] + [NO_3^-]$$

$$+[B(OH)_4^-] + [H_3SiO_4^-] + [H_2PO_4^-] + 2[HPO_4^{2-}] + 3[PO_4^{3-}] + \ldots$$

$$-[Na^+] - [K^+] - 2[Mg^{2+}] - 2[Ca^{2+}] - [NH_4^+] - \ldots . \qquad (8.21)$$

The dots emphasize that only the major ions and those that are important to this discussion have been included. This appears to have increased our number of variables significantly. Conveniently, all of the additional ions that have been added to the charge balance, except for $B(OH)_4^-$ (the borate ion), $H_3SiO_4^-$ (the silicate ion), and the ions containing phosphorus, are dissociation products of strong acids and bases. This means that the net charge of these additional ions can be determined analytically. In the ocean there is a slight excess positive charge of these ions, called the alkalinity, Alk:

$$[Alk] = [Na^+] + [K^+] + 2[Mg^{2+}] + 2[Ca^{2+}] + [NH_4^+] + \ldots$$

$$-[Cl^-] - 2[SO_4^{2-}] - [Br^-] - [NO_3^-] - \ldots . \qquad (8.22)$$

Alkalinity is usually reported in units of equivalents per kg of sea water. While the alkalinity consists of a number of species, internal source and sinks of alkalinity in the ocean are primarily due to sources and sinks of calcium ion (Ca^{2+}), nitrate ion (NO_3^-) and, to a much smaller extent, ammonium ion (NH_4^+). Alkalinity is otherwise a conservative quantity in sea water. A rigorous treatment of the charge balance, which can be found in Peng et al. (1987), should include the silicate ions and the phosphorus-containing ions in (8.21). For simplicity, and because these ions are of relatively minor importance in the charge balance, they will be ignored here. Combining the last two equations gives us an alternative expression for the alkalinity:

$$[Alk] = [HCO_3^-] + 2[CO_3^{2-}] + [OH^-] - [H^+] + [B(OH)_4^-]. \qquad (8.23)$$

These are simply the ions from weak acids and bases that must balance the excess positive charge in (8.22). In going from pure water to sea water, (8.23) has replaced (8.19) as the condition of electrical neutrality. Meanwhile, we have added two more variables: [Alk] and $[B(OH)_4^-]$. Fortunately, the borate ion concentration can be computed from $[H^+]$ and the salinity as follows. Borate is formed from the reaction of boric acid (H_3BO_3) and water:

$$[H_3BO_3] + [H_2O] \Leftrightarrow [H^+] + [B(OH)_4^-] \qquad (8.24)$$

251

whose equilibrium expression is

$$K_b = \frac{[B(OH)_4^-][H^+]}{[H_3BO_3]} \qquad (8.25)$$

where K_b is the dissociation constant of boric acid. The sum of the species containing boron is conserved in the ocean and varies only because of the freshwater flux at the ocean surface that can concentrate it (by evaporation or sea ice formation) or dilute it (by precipitation, melting of sea ice and river runoff), and so is proportional to the salinity:

$$[H_3BO_3] + [B(OH)_4^-] = cS \qquad (8.26)$$

where c is a known constant.

We have added (8.25) and (8.26) to our system but introduced only one more variable, H_3BO_3, giving us now six equations with eight unknowns. This means, in summary, that we must specify two variables (in addition to T, S and p) to completely determine the rest. In addition to DIC, the alkalinity is often chosen as a master variable because, like DIC, it is conservative with respect to changes in temperature, salinity and pressure.

b. pCO_2 as a function of DIC, Alk, T and S

The solution of equations (8.10), (8.15), (8.16), (8.18), (8.20), (8.23), (8.25) and (8.26) for pCO_2 as a function of T and S under average surface conditions ([DIC] = 2,000 μmol kg^{-1}, [Alk] = 2,300 μeq kg^{-1}, p = 1 atm) is shown in Fig. 8.4(a). This set of nonlinear algebraic equations has been solved using an iterative technique (Peng et al., 1987). The figure shows that pCO_2 increases as temperature increases and salinity increases. To examine this dependency, (8.10), (8.15) and (8.16) can be combined to give

$$pCO_2 = \frac{K_2}{\alpha K_1} \frac{[HCO_3^-]^2}{[CO_3^{2-}]}. \qquad (8.27)$$

The ratio $[HCO_3^-]^2/[CO_3^{2-}]$ is relatively insensitive to temperature. On average, about 75% of the sensitivity of pCO_2 to temperature is due to the solubility, while the remainder is due to K_2/K_1. About 50% of the sensitivity of pCO_2 to salinity is due to K_2/K_1, 30% is due to $[HCO_3^-]^2/[CO_3^{2-}]$, while the remainder is due to the solubility, on average.

Fig. 8.4(b) shows that pCO_2 increases with increasing DIC and decreasing Alk, with approximately equal sensitivity. The dependence on DIC is easy to understand, because an increase in DIC means that there is simply more carbon to be partitioned among the three carbon species, though this partitioning is itself a function of DIC (see Sec. 8.3.4b below). The effect of Alk is more subtle, and its interpretation requires making a few approximations. DIC is 89.1% HCO_3^-, 10.4% CO_3^{2-} and 0.5% CO_2 under average surface conditions (DIC, Alk, and p as above, $T = 20°C$ and $S = 35$ permil), enabling DIC to be approximated by

$$[DIC] \approx [HCO_3^-] + [CO_3^{2-}]. \qquad (8.28)$$

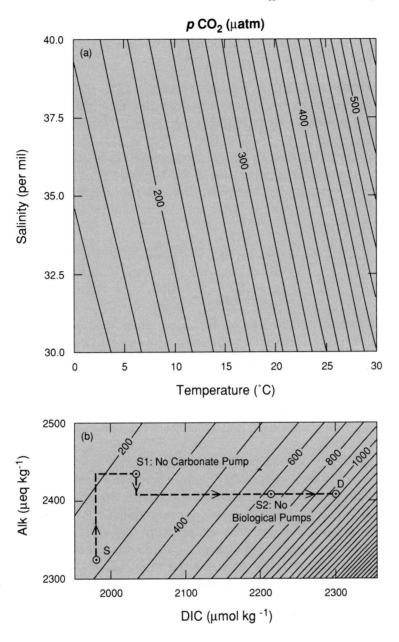

Fig. 8.4 The partial pressure of CO_2 as a function of (a) temperature and salinity at a typical surface DIC (2,000 μmol kg^{-1}) and Alk (2,300 μeq kg^{-1}) and (b) DIC and Alk at a typical surface temperature (20°C) and salinity (35 permil). The computation was done using the algorithm of Peng et al. (1987), ignoring the effect of phosphate and silicate on the charge balance. S = average surface water, D = average deep water warmed to 20°C and brought to surface pressure. Lines connecting S and D are discussed in Sec. 8.3.5.

253

Alk, as defined in (8.23), is 77.5% HCO_3^-, 18.0% CO_3^{2-}, 4.3% $B(OH)_4^-$ and 0.2% $(OH^- - H^+)$, enabling Alk to be approximated by

$$[Alk] \approx [HCO_3^-] + 2[CO_3^{2-}]. \tag{8.29}$$

For Alk to increase and DIC to remain constant, CO_3^{2-} must increase and HCO_3^- must decrease by equal amounts, in order to satisfy the last two equations. The decrease in HCO_3^- forces reaction (8.13) to the right, thereby decreasing CO_2.

DIC and Alk can decrease by dilution (the addition of fresh water) and increase by concentration (the removal of fresh water). The salinity is used to remove the effects of the freshwater flux on the distributions of DIC and Alk. Let DIC° and Alk° be DIC and Alk normalized to a salinity of 35 permil, respectively:

$$DIC° = DIC \cdot \frac{35 \text{ permil}}{S} \tag{8.30}$$

$$Alk° = Alk \cdot \frac{35 \text{ permil}}{S}. \tag{8.31}$$

Salinity variations in the ocean can be higher than 10% (see Chapters 4 and 11). The freshwater flux can therefore produce variations of at least 200 μmol kg^{-1} in DIC and 200 μeq kg^{-1} in Alk, which are comparable to variations produced by biological processes. With respect to CO_2, however, these variations cancel because of the fact that DIC and Alk have nearly equal and opposite effects on CO_2 (Fig. 8.4b). Thus DIC and Alk variations caused by evaporation and precipitation have a negligible impact on atmospheric pCO_2.

DIC° and Alk° are not uniformly distributed in the ocean. Figure 8.5 shows that there is a depletion of DIC° and Alk° in the surface ocean with respect to the deep ocean. Although these gradients are only on the order of 16% for DIC° and 4% for Alk°, they are very significant – average deep water warmed to the average surface temperature has a pCO_2 that is more than a factor of three higher than that of average surface water (Fig. 8.4b). There are three processes responsible for the variations in DIC° and Alk° seen in the ocean: the biological cycling of organic matter, the biological cycling of calcium carbonate and air–sea gas exchange of CO_2. These will be discussed in the following three sections.

8.3.2 The cycle of organic matter

Given adequate light and nutrients, phytoplankton will photosynthesize, utilizing the inorganic nutrients dissolved in sea water (mainly CO_2, nitrate and phosphate) to form their organic tissue. This is represented by the idealized equation

$$106CO_2 + 16NO_3^- + HPO_4^{2-} + 122H_2O + 18H^+ \rightarrow$$

$$(CH_2O)_{106}(NH_3)_{16}(H_3PO_4) + 138O_2 \tag{8.32}$$

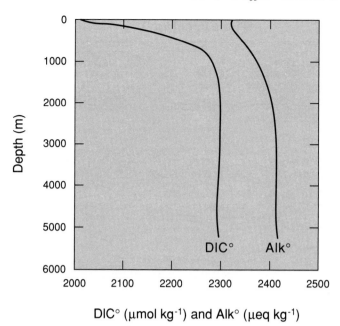

DIC° (μmol kg⁻¹) and Alk° (μeq kg⁻¹)

Fig. 8.5 Average vertical profiles of DIC and Alk, normalized to a salinity of 35 permil. Computed from data obtained during the GEOSECS and TTO cruises. During pre-industrial times, surface DIC° was about 30 μmol kg^{-1} lower.

where $(CH_2O)_{106}(NH_3)_{16}(H_3PO_4)$ represents the average composition of organic matter (Redfield et al., 1963). Respiration (or remineralization), accomplished primarily by zooplankton and bacteria, is the reverse of (8.32) and occurs throughout the water column. Since photosynthesis exceeds respiration only in the euphotic zone, there is a net sink of CO_2, phosphate and nitrate in the euphotic zone and a net source in the aphotic zone. A downward flux of organic matter and an upward flux of nutrients connects the euphotic zone sink and aphotic zone source of nutrients (see Fig. 8.3). This cycle of organic matter and nutrients has been termed the *soft tissue pump* (Volk and Hoffert, 1985), with the idea that the majority of organic matter transported out of the euphotic zone is in particulate form. Here it will be called the *organic matter pump*, recognizing the possible role of dissolved organic species in this transport.

The organic matter pump creates a surface depletion and deep enrichment of DIC. It also creates a surface enrichment and deep depletion of Alk because nitrate is removed from sea water during photosynthesis and released during respiration (8.32 and 8.22). Thus, because of its effects on both DIC and Alk, the organic matter pump decreases $p CO_2$ of the surface ocean (and hence the $p CO_2$ of the atmosphere) as it increases in strength.

a. Phosphate and nitrate as tracers of the organic matter pump

In order to trace the sources and sinks of organic matter in the ocean,

Nitrate (μmol kg⁻¹)

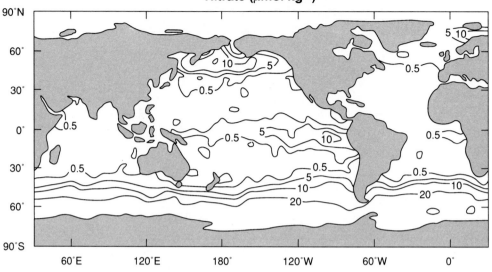

Fig. 8.6 Annual composite nitrate concentration (μmol kg^{-1}) at the ocean surface from Levitus et al. (1992).

it has long been customary to use the nutrients phosphate and nitrate. As (8.32) and its reverse reaction would suggest, carbon, nitrogen and phosphorus are cycled very similarly during photosynthesis and respiration. In their studies of chemical distributions in the water column and in plankton, Redfield et al. (1963) suggested that during organic matter cycling, carbon, nitrogen, phosphorus and oxygen are cycled in the following ratio C:N:P:$-O_2$ = 106:16:1:138. That is, for every phosphate ion taken up during photosynthesis, 16 nitrate ions and 106 molecules of CO_2 are taken up as well. In addition, 138 molecules of oxygen are produced. These ratios are known as the *Redfield ratios*. It should be noted that more recent studies (Takahashi et al., 1985) suggest somewhat different values (C:N:P:$-O_2$ = 140:16:1:172). The great utility of phosphate and nitrate is that their distributions are dominated by the cycle of organic matter, so that, given the Redfield ratio, they can be used to separate the effects (on DIC and Alk) of photosynthesis and respiration from other processes such as gas exchange and calcium carbonate precipitation and dissolution.

The distribution of surface nutrients indicates where and to what degree the organic matter pump is operating. The amount of nutrients present in the euphotic zone is a manifestation of two opposing processes: input from below the euphotic zone and biological uptake. The atmospheric input of phosphate and nitrate in precipitation is relatively minor. The input of nutrients into the euphotic zone is primarily a function of the rate of upwelling and mixing up from below the euphotic zone, as well as the nutrient content of this water. As shown in Fig. 8.6, surface nitrate concentrations are relatively high in upwelling regions along the equator and coasts, and in high-latitude convective areas. Conversely, nitrate is low where downwelling occurs, such as in the subtropical gyres, and where surface waters are stratified. (See Chapter 4 for a discussion of vertical motion in the upper ocean.) The phosphate distribution is very similar. The fact that phosphate and nitrate are not everywhere depleted in the euphotic zone tells us that the organic matter pump is not operating at peak efficiency. As discussed in Sec. 16.2, models suggest that surface nutrient concentrations, particularly in high latitudes, may be important diagnostics of the efficiency of the organic matter pump.

b. The euphotic zone

The euphotic zone depth is defined as the depth at which photosynthesis can no longer occur and, in practice, is somewhat difficult to measure. An operational definition often used is the depth at which the light intensity is 1% of its surface value. Defined in this way, the euphotic zone is about 100 m deep in the subtropical gyres and considerably shallower in the more productive coastal and upwelling regions.

The transformation of inorganic nutrient to particulate and dissolved organic matter in the euphotic zone is highly complex. It involves many types of organisms and various biological processes that respond to the physical environment of light, temperature and turbulence as well as to the chemical environment. Some of the salient features of this transformation are illustrated in Fig. 8.7. Nitrogen is used in this flow diagram as the basic element for reasons which will become apparent. Conceptually, the best way of dealing with this complex (yet still idealized) system is to break it down into a series of loops. The simplest loop begins when a parcel of sea water containing nitrate is transported into the euphotic zone. Phytoplankton exclusively consume nitrate and convert it into organic tissue by utilizing the Sun's energy. Because the source of nitrate to the euphotic zone is due only to ocean circulation, one can think of it as nitrogen that is newly available to phytoplankton. Photosynthesis due to the uptake of nitrate is therefore known as *new production*. At the end of their life cycle some of these plants become senescent and are transported out of the euphotic zone as detrital material via gravitational settling or ocean currents. If this were all that were occurring in the euphotic zone, then new production, phytoplankton mortality rate, and detrital flux at the base of the euphotic zone would all be equal, on time average. This loop closes in the aphotic zone where detritus is remineralized first to ammonium (NH_4^+) and then to nitrate. The process by which ammonium is converted to nitrate is known as *nitrification* (see Sec. 8.5). Nitrification is inhibited by light and so proceeds only very slowly in the euphotic zone, if at all.

Another loop becomes necessary within the euphotic zone when one considers that a large portion of the particle flux out of the euphotic

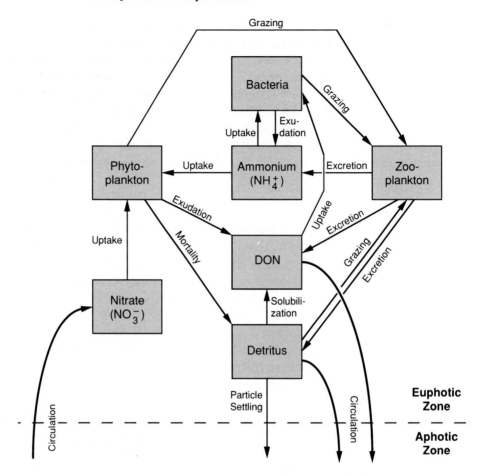

Fig. 8.7 A schematic diagram of the euphotic zone ecosystem based on the model of Fasham et al. (1990). New production is phytoplankton uptake of nitrate, and regenerated production is phytoplankton uptake of ammonium.

zone is in the form of zooplankton fecal pellets in addition to senescent phytoplankton. This places zooplankton between phytoplankton and detritus in the previous loop. Zooplankton graze upon phytoplankton and excrete dense fecal particles that have some of the highest sinking rates among organic particles in the ocean, ranging from 20 to 900 m day^{-1} (Fowler and Knauer, 1986). This distinction is important because it means that fecal pellets are relatively unaffected by ocean currents, and are transported only by gravitational settling.

A fourth component to the euphotic zone ecosystem has also been shown to be important from measurements which reveal that total phytoplankton production, or *primary production*, is usually significantly larger than new production. The difference is due primarily to the uptake of ammonium, which is produced partly by zooplankton excretion. This forms a loop within the euphotic zone whereby nitrogen is cycled from phytoplankton to zooplankton by grazing, then back to phytoplankton by ammonium excretion and uptake. Because ammonium is a product of internal processes within the euphotic zone, it is referred to as recycled or regenerated nitrogen, and its uptake by phytoplankton is known as *regenerated production*.

New and regenerated nitrogen production are both measurable quantities because it is possible to make the distinction between nitrate and ammonium experimentally. This distinction is not possible for other tracers of organic matter cycling, such as phosphorus and carbon, because the newly available and regenerated forms of these elements are identical. In this light, the value of using nitrogen as the basic element for studying organic matter cycling is clear.

New production is very important when considering the organic matter pump. Over long enough time (a year at most), new production must balance the input of nutrients into the euphotic zone as well as the export of organic matter. In the absence of new production, all of the primary production would be recycled, biological sources and sinks of CO_2 and other nutrients in the euphotic zone would exactly balance, and atmospheric CO_2 would be much higher.

Another recycling loop within the euphotic zone involves two additional components of the ecosystem: bacteria and dissolved organic nitrogen (DON). Phytoplankton exude DON, perhaps passively through their cell membranes, providing a portion of this major food source for bacteria. Like zooplankton, bacteria produce ammonium during respiration and excrete it. Thus, nitrogen passes successively through phytoplankton, DON, bacteria and ammonium, and then back to phytoplankton.

Other processes besides the few loops mentioned here are likely to be relevant to organic matter cycling in the euphotic zone. These include zooplankton excretion of DON, competition for ammonium by bacteria, grazing of bacteria by zooplankton, and the breakdown of detritus to DON (Fig. 8.7). Missing from Fig. 8.7 is the diurnal vertical migration of zooplankton and fish. These organisms feed at night in surface waters and excrete dissolved nitrogen at depth, constituting a net downward flux of nitrogen out of the euphotic zone that may be significant.

Phytoplankton, being the major biological sink of CO_2 in the ocean and lying at the foundation of the food chain in the ocean, constitute the most important component of the euphotic zone ecosystem. Their growth rate, in simplest terms, can be expressed by

$$\frac{\partial P}{\partial t} = \mu P - L \tag{8.33}$$

where P is the phytoplankton concentration in units of, for example, mol C m^{-3}, μ is the specific growth rate (s^{-1}) and L is the sum of all loss terms, including grazing by zooplankton, sinking and excretion of dissolved organic matter. Laboratory experiments show that μ increases strongly

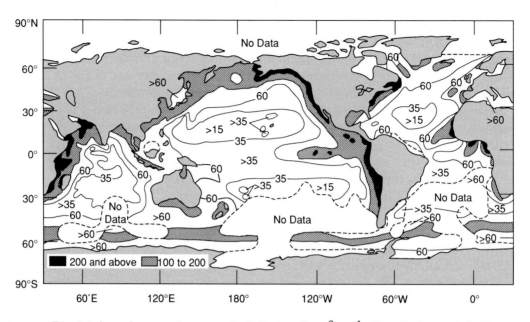

Fig. 8.8 Annual mean primary productivity in g C m^{-2} yr^{-1}. Compilation made by Berger et al. (1987).

with the light intensity and nutrient concentration. The dependency, however, differs for different phytoplankton species, making modeling phytoplankton populations a difficult task. Although phytoplankton require many nutrients, nitrogen (as nitrate and ammonium) is thought to be the primary limiting nutrient. There is generally enough phosphorus available as well as abundant amounts of DIC. Some species of phytoplankton, notably the diatoms, require silica to grow, and are limited by the availability of this nutrient in some regions. All of these nutrients are predominantly supplied to the euphotic zone from below, during upwelling and convection. Thus upwelling and convective zones are often associated with high productivity. Satellite estimates of surface temperature and chlorophyll are consistent with this (Plate 9). Cool water (associated with upwelling) is generally higher in chlorophyll. The loss terms (L) are less well known for the phytoplankton. Zooplankton grazing is probably the most important. There are many different species of zooplankton that graze particular phytoplankton species, making the system quite complex.

Phytoplankton primary production, which is essentially μP, is expected to vary with nutrient concentration and light intensity. Figure 8.8 shows the most recent global compilation of primary productivity measurements (Berger et al., 1987). The close resemblance to the nitrate distribution (Fig. 8.6) is striking. Although productivity is high in the high nutrient, open ocean environments along the equator and at high latitudes, it does not reach the high values present along some coastlines. It has been

260

speculated that iron, another essential nutrient, is not present in great enough concentration in the high latitudes and along the equator to sustain high growth rates. Recently it has been shown in laboratory incubations that phytoplankton taken from these high nutrient environments grow at much higher rates when given additional iron (Martin et al., 1990). The precise interpretation of these results is still being debated but they suggest that iron, which is predominantly supplied to the ocean from the atmosphere as windblown dust, may be an important limiting nutrient.

Primary productivity also has temporal variability. The seasonal cycle of mixed layer depth, nutrients, phytoplankton and primary productivity near Bermuda (Fig. 8.9) reveal that as the mixed layer deepens in the autumn and early winter, nutrients from below become entrained in the mixed layer. Phytoplankton and primary productivity are still quite low because, although nutrients are available, there is little light. It is not until light becomes great enough during the early spring that phytoplankton and primary productivity really take off. This *spring bloom* lasts until phytoplankton utilize all of the nutrients in the mixed layer. The cycle then repeats itself. It is widely believed that the spring bloom is initiated by the shallowing of the mixed layer, because phytoplankton then spend more time in higher light near the surface. This view is not supported by Fig. 8.9, nor by modeling studies which demonstrate that a bloom can occur without any seasonal variability in the mixed layer depth (Evans and Parslow, 1985).

The seasonality in phytoplankton growth can have a profound effect on carbon dioxide in the mixed layer. At a site farther north in the North Atlantic, the mixed layer has a similar cycle (Fig. 8.10). The spring bloom is seen in the drawdown of nitrate and DIC. The decrease in DIC causes the $p\mathrm{CO}_2$ to drop as well. The decrease in $p\mathrm{CO}_2$ cannot be due to the temperature, which is increasing and would cause the $p\mathrm{CO}_2$ to increase through the solubility and the dissociation constants.

Estimates of global marine primary productivity vary considerably. The historical compilation shown in Fig. 8.8 yields a value of 27 Gt C yr^{-1} (1 Gt $= 10^{15}$ g), about a factor of five larger than the current rate of fossil fuel combustion. It has been suggested that this estimate of primary productivity is too low by a factor of two due to metal contamination in historical measurements (Martin et al., 1987). New production is even less well known. If 20% of global primary production is new production (Eppley and Peterson, 1979), then new production may be between 5 and 10 Gt C yr^{-1} (1 to 2 mol C m^{-2} yr^{-1}). This estimate appears to be at odds with local estimates of new production of 4 mol C m^{-2} yr^{-1} based on tracer distributions in low productivity environments (e.g., Spitzer and Jenkins, 1989).

c. *The downward transport of organic matter*
    Organic matter in the ocean is often classified as large (or sinking)

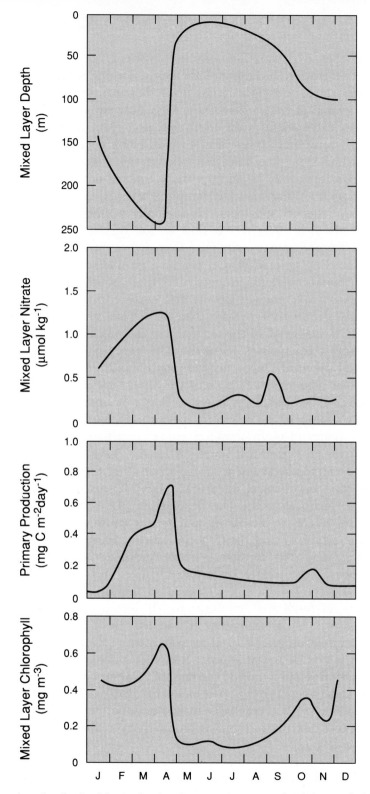

Fig. 8.9 Seasonal cycle of mixed layer depth, nitrate, primary productivity, and chlorophyll near Bermuda in 1958 (adapted from Menzel and Ryther, 1960 and Fasham et al., 1990).

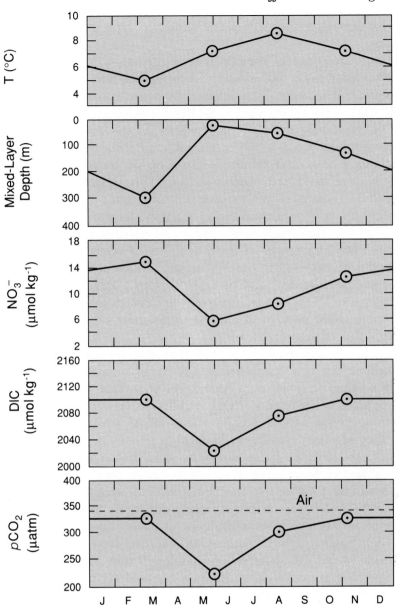

Fig. 8.10 Seasonal cycle of temperature, mixed layer depth, nitrate, DIC, and $p\mathrm{CO_2}$ at 64°N in the North Atlantic (from Peng et al., 1987).

particulate organic matter (POM), small (or suspended) POM, or dissolved organic matter (DOM). Large POM is typically greater than 50 μm in diameter, sinks rapidly in the water column ($\sim$ 100 m day^{-1}, Fowler and Knauer, 1986), and is usually sampled with sediment traps. Small POM is typically between about 1 and 50 μm in diameter, sinks very slowly (if at all), and is sampled by filtering sea water. DOM includes colloidal as well as purely dissolved material, and is operationally defined as all organic matter that will pass through a filter with a fine mesh size, typically between about 0.1 and 1 μm.

These broad categories of organic matter are important because they reflect distinct processes. For example, because the sinking rate of large POM is so great, it is generally unaffected by ocean currents, unlike small POM and DOM. This means that large POM sinking out of the euphotic zone will be degraded directly below where it was produced, thereby providing a coupling between the organisms inhabiting the euphotic zone and those living in the water column below. This distinction of large POM will also be seen in the distribution of chemical species in the ocean. That is, if large POM dominates the vertical flux of organic matter, then one would expect enhanced remineralization rates in the water column below highly productive regions, and hence, enhanced nutrient concentrations in these regions. The importance of the size classes of organic matter is also reflected in the processes that break it down. Different size zooplankton graze different size particles, and bacteria are most likely responsible for the consumption of DOM. These important differences among large POM, small POM and DOM make it of interest to determine the role of each in the downward flux of organic matter.

Until recently, it was widely thought that sinking particles were the primary medium for the downward transport of organic matter. One reason why sinking particles were deemed so important was the seeming unimportant roles of suspended POM and DOM. Vertical concentration gradients in suspended POM are far too small to provide a significant downward flux of organic matter. Vertical gradients in DOM measured with conventional techniques are also quite small in magnitude compared with those of the corresponding dissolved inorganic species. But another reason for assigning a major role to sinking particles is that sediment traps catch a respectable fraction of the primary production. In the past several years, however, new controversial techniques for measuring dissolved organic carbon (DOC) (Sugimura and Suzuki, 1988) and dissolved organic nitrogen (DON) (Suzuki et al., 1985) have been developed that yield much higher concentrations than conventional techniques, particularly in surface waters. These measurements suggest that most of the downward transport of organic matter in the ocean takes place in a dissolved phase, relegating sinking particles to a relatively minor role. Due to the controversial nature of both the DOM and particle flux measurements, the relative roles of sinking POM and DOM in the downward flux are, for the time being, sketchy at best.

d. The aphotic zone

Ultimately, the organic matter exported out of the euphotic zone is decomposed into inorganic nutrients such as phosphate, nitrate and DIC. Figure 8.11 shows some of the possible pathways for this decomposition.

Sinking POM, which has been studied more than its suspended and

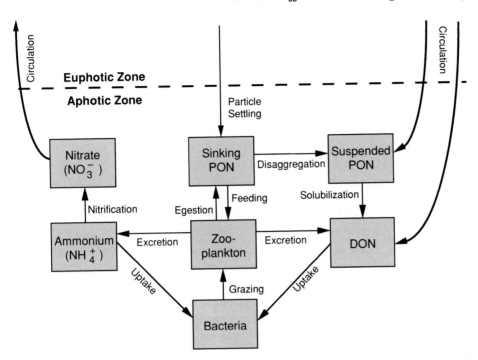

Fig. 8.11 A schematic diagram of the aphotic zone ecosystem.

dissolved counterparts, consists mainly of *marine snow*, zooplankton fecal pellets and intact organisms. The most abundant component appears to be marine snow particles, which are amorphous, heterogeneous aggregates greater than 500 μm and are composed of detrital material, living organisms and inorganic matter (Alldredge and Silver, 1988).

Sediment traps are the most commonly used devices for measuring the sinking POM flux in the aphotic zone. The particle flux data are often fitted with an expression of the form

$$F(z) = F_o \left(\frac{z}{z_o}\right)^{-a} \tag{8.34}$$

where $F(z)$ is the downward POM flux at depth z and F_o the flux at a depth z_o, typically near the base of the euphotic zone. Equation (8.34) describes a vertical flux that decreases monotonically with depth, with a more rapid decrease at shallow depths. Although the use of sediment traps to measure the particle flux is extremely suspect (Jahnke, 1990), particularly at shallow depths, it is most probable that the particle flux decreases with depth, suggesting that particle remineralization plays a role in the source of CO_2 and other nutrients in the aphotic zone.

Several processes have been proposed to account for the loss of sinking POM in the aphotic zone. It has been suggested that the feeding of zooplankton is an important process in removing sinking POM from the water column. The particle loss due to zooplankton results in an excretion of inorganic nutrients and dissolved organic matter. The latter may support the growth of some free-living microorganisms, which, in turn, excrete inorganic nutrients. Another hypothesis, not necessarily in contradiction with the above, is that particle disaggregation, due to physical or biological processes, is responsible for the decreasing POM flux with depth, and is the main source of suspended POM below the euphotic zone.

Bacterial activity has been found to be significant in the remineralization of organic matter below the euphotic zone (Cho and Azam, 1988). It appears that most bacteria below the euphotic zone are free-living (i.e., not attached to particles) suggesting that bacteria are remineralizing DOM. Although bacteria inhabit sinking particles, their numbers appear to be too low to be important in particle remineralization (Karl et al., 1988).

In summary, both zooplankton and bacteria may be important players in remineralizing the organic matter transported out of the euphotic zone. The simplest possible picture to fit the available data consists of a sinking POM flux from the euphotic zone remineralized by zooplankton, and a DOM flux from the euphotic zone (due to advection and diffusion) remineralized by bacteria. Other possible pathways that have not been quantified include zooplankton release of DOM, uptake of inorganic nutrients by bacteria and grazing of zooplankton on bacteria (Fig. 8.11).

The combined effects of aphotic zone remineralization and deep water circulation are reflected in the interbasin differences in the concentration of nutrients. Deep water flows from the Atlantic through the Indian to the Pacific Ocean as part of the ocean's so-called "conveyor belt" (see Fig. 17.12 and Plate 14). As a result of remineralization, the nitrate concentration, for example, of deep water is highest in the Pacific, lowest in the Atlantic and intermediate in the Indian Ocean (Fig. 8.12). The same is true for phosphate, DIC, Alk, and many other chemical species.

e. The sediments

A small fraction of the particulate organic matter that exits the euphotic zone survives remineralization in the water column and is deposited on the sea floor (about 4% using (8.34) with an average bottom depth of 4 km and $a = 0.858$ [Martin et al., 1987]). Most of this organic matter is remineralized in the sediments to inorganic nutrients that diffuse back into the water column – only a small fraction is buried. Recent studies suggest that a significant portion of the remineralization below 1 km occurs in the sediments (Jahnke and Jackson, 1987).

8.3.3 The cycle of calcium carbonate

Calcium carbonate ($CaCO_3$) is a mineral solid that plays an important role in the marine carbon cycle. Some plants and animals living

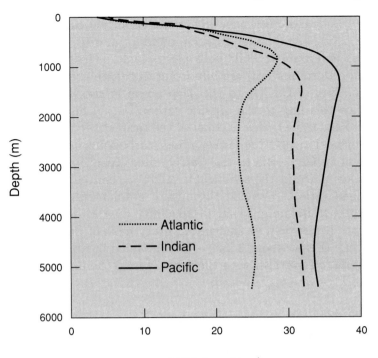

Fig. 8.12 Average vertical profiles of nitrate in the Atlantic, Indian and Pacific Oceans. Data taken from Levitus et al. (1992).

in the euphotic zone have $CaCO_3$ skeletons (known as *tests*) which they *precipitate* from dissolved calcium and carbonate ions:

$$Ca^{2+} + CO_3^{2-} \rightarrow CaCO_3 \,. \tag{8.35}$$

The $CaCO_3$ formed in this way eventually sinks and is dissolved back to calcium and carbonate ions in the deeper parts of the water column and in the sediments. Ocean circulation closes the loop by transporting these ions back to surface waters (see Fig. 8.3). This cycling of $CaCO_3$, known as the *carbonate pump* (Volk and Hoffert, 1985), creates a surface depletion and deep enrichment in DIC (8.20) and Alk (8.22). Since the variations due to the carbonate pump are twice as great on Alk as on DIC, an increase in the strength of the carbonate pump will increase atmospheric CO_2.

a. Biological precipitation

Little is known about the spatial and temporal variability of the flux of $CaCO_3$ out of the euphotic zone. Calcareous organisms are ubiquitous throughout the ocean.   They appear to dominate more often in low productivity environments, while organisms with siliceous tests tend to dominate in high productivity environments. Even more difficult to assess is the relative importance of the two mineral phases of calcium carbonate found in the ocean: calcite and aragonite. The distinction is important because calcite and aragonite have different dissolution rates in the deep ocean (Sec. 8.3.3b). Aragonite is precipitated primarily by zooplankton (pteropods and heteropods) while both phytoplankton (coccolithophorids) and zooplankton (foraminifera) are important in calcite precipitation.

Although the flux of $CaCO_3$ to the deep ocean is not well known, it is possible to scale this flux to the organic carbon (C_{org}) flux. The ratio of the C_{org} flux to the $CaCO_3$ flux is equal to the ratio of the vertical gradient (deep minus surface) in DIC due to the organic carbon flux (ΔDIC_{org}) to the vertical gradient in Ca^{2+} due to the $CaCO_3$ flux (ΔCa^{2+}), assuming that little burial takes place in the sediments. ΔDIC_{org} can be computed from the C:N Redfield ratio ($r_{C:N}$) and the nitrate gradient (ΔNO_3^-). ΔCa^{2+} can be computed from the gradient in alkalinity (ΔAlk) corrected for the nitrate gradient. It can then be shown that every mole of $CaCO_3$ that sinks to the deep ocean (below about 2 km) is accompanied by about 3 to 4 moles of organic carbon, on average, with the range due to the uncertainty in the C:N ratio.

b.   Chemical dissolution

Chemical processes play an important role in the carbonate pump. The dissolution and precipitation of $CaCO_3$ depend on its degree of saturation, Ω:

$$\Omega = \frac{[Ca^{2+}][CO_3^{2-}]}{K_{sp}} = \frac{[Ca^{2+}][CO_3^{2-}]}{([Ca^{2+}][CO_3^{2-}])_{sat}} \qquad (8.36)$$

where K_{sp} is the solubility product of $CaCO_3$. At equilibrium with pure $CaCO_3$, the product of the calcium and carbonate ion concentrations in solution must be equal to K_{sp}, which is a function of the temperature, salinity and the pressure. If $\Omega < 1$ then the solution is undersaturated and any $CaCO_3$ present will dissolve. If $\Omega > 1$ then the solution is supersaturated and $CaCO_3$ will be precipitated. Since $[Ca^{2+}]$ variations are quite small compared with $[CO_3^{2-}]$ variations, most of the variability in Ω at a given T, S and p is due to $[CO_3^{2-}]$. $[Ca^{2+}]$ variations are mainly due to the fresh water flux at the surface and therefore parallel the salinity. Because of the relative importance of CO_3^{2-}, it is convenient to define the carbonate ion saturation concentration, $[CO_3^{2-}]_{sat}$:

$$[CO_3^{2-}]_{sat} = \frac{K_{sp}}{[Ca^{2+}]} \propto \frac{K_{sp}}{S}. \qquad (8.37)$$

The degree of saturation is then given by

$$\Omega = \frac{[CO_3^{2-}]}{[CO_3^{2-}]_{sat}}. \qquad (8.38)$$

Figure 8.13 shows globally averaged vertical profiles of $[CO_3^{2-}]$ and $[CO_3^{2-}]_{sat}$. The carbonate ion concentration is computed from the average

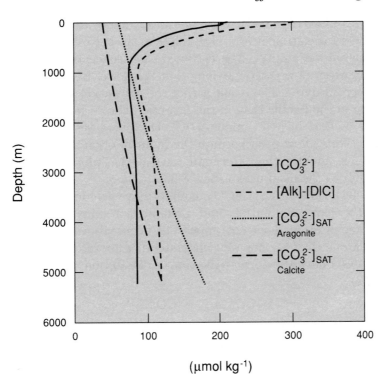

Fig. 8.13 Average vertical profiles of the carbonate ion concentration, saturation concentrations for calcite and aragonite and an approximation of the carbonate ion concentration.

vertical profiles of [DIC] and [Alk] using the equations in Sec. 8.3.1. The equilibrium constants were computed from average vertical profiles of temperature, salinity and pressure (which is simply proportional to the ocean depth). The vertical structure of $[CO_3^{2-}]$ can be understood by combining (8.28) and (8.29) to yield an approximate expression for the carbonate ion:

$$[CO_3^{2-}] \approx [Alk] - [DIC]. \qquad (8.39)$$

As Fig. 8.13 shows, this approximation parallels $[CO_3^{2-}]$, meaning that variations in CO_3^{2-} are mainly due to variations in the difference between Alk and DIC. Note that there are two curves for the carbonate ion saturation concentration corresponding to the two mineral phases of $CaCO_3$. Aragonite has a higher value of $[CO_3^{2-}]_{sat}$ than calcite, meaning that aragonite is more soluble that calcite. $[CO_3^{2-}]_{sat}$ increases with depth because of the pressure increase and, to a lesser extent, the temperature decrease.

The depth at which $\Omega = 1$ is known as the *carbonate ion saturation horizon*. By definition, waters above the saturation horizon are supersaturated in $CaCO_3$ and those below are undersaturated in $CaCO_3$. On average, the saturation horizon lies at 3.5 ± 0.5 km for calcite and $1 \pm$

0.5 km for aragonite (Fig. 8.13), with the uncertainties due chiefly to the experimental determination of K_{sp}. The depth of the saturation horizon has substantial spatial variability. The carbonate ion concentration decreases along the deep water flow path, meaning that it is lower in the deep Pacific than in the deep Atlantic. The result is that the saturation horizon is about 1 km shallower in the Pacific than in the Atlantic. The variability in $[CO_3^{2-}]$ is mainly due to that in [Alk] minus [DIC]. DIC increases more than Alk along the deep water flow path because the input of organic carbon (which increases DIC) is greater than the input of $CaCO_3$ (which increases Alk more than DIC) to the deep ocean (Sec. 8.3.3a).

The large deviations from $CaCO_3$ equilibrium seen in sea water are evidence of the slow dissolution and precipitation kinetics of $CaCO_3$. Abiological precipitation does not appear to occur above the saturation horizon for reasons that are not well understood. Laboratory studies (e.g., Acker et al., 1987) have shown that the rate of dissolution r (% day^{-1}) can be expressed as

$$r = k([CO_3^{2-}]_{sat} - [CO_3^{2-}])^n \qquad (8.40)$$

where k is the dissolution rate constant. Laboratory estimates of k vary widely and depend to a large extent on the surface area to volume ratio of the $CaCO_3$ particle. Though these experiments suggest that significant dissolution may take place in the water column, water column dissolution is supported by some sediment trap observations and refuted by others. Because of the relatively shallow saturation horizon for aragonite, there may be substantial dissolution of this phase in the water column, particularly in the Pacific where the aragonite saturation horizon is very shallow.

Regardless of how much $CaCO_3$ dissolution occurs in the water column, organic matter is remineralized at shallower depths than $CaCO_3$ is dissolved. The vertical profiles of DIC° and Alk° (Fig. 8.5) can be used to demonstrate this because DIC° is dominated by the organic carbon cycle and Alk° is dominated by the $CaCO_3$ cycle (see Fig. 8.4b). Note that the deep water enrichment of Alk° is concentrated at greater depths than that of DIC° (Fig. 8.5). As discussed in Sec. 16.2.3, the different regeneration depths of $CaCO_3$ and organic matter may be relevant with regard to atmospheric CO_2 variations.

c. The sediments

The simplest picture of sedimentary $CaCO_3$ is burial in supersaturated (shallow) sediments and dissolution in undersaturated (deep) sediments. Indeed, the $CaCO_3$ content of sediments decreases with increasing depth near the saturation horizon. The upper portion of this transition zone in the sediments in called the *lysocline*. It is now believed that significant dissolution can occur in sediments bathed in supersaturated bottom waters (Emerson and Archer, 1990). Organic carbon remineralization in the sediments (Sec. 8.3.2e) increases DIC in the pore waters, which results in a decrease of pore water CO_3^{2-} below the bottom water concentration [Equation (8.39)]. If the carbonate ion concentration drops below saturation in the pore waters then dissolution will occur. Thus it is important to know the rate of organic carbon degradation in the sediments as well as the degree of CO_3^{2-} saturation in the bottom water when considering $CaCO_3$ dissolution in the sediments.

8.3.4 Air–sea gas exchange of carbon dioxide

a. The solubility pump

A water parcel at equilibrium with air of a fixed CO_2 concentration will increase its DIC concentration as the water temperature decreases. This is because the solubility and dissociation of CO_2 are higher in colder water (Sec. 8.3.1). The deep ocean is filled with cold water that originates in high-latitude surface waters (Sec. 4.4). If these waters achieve full CO_2 gas phase equilibration with the atmosphere before sinking, then a substantial fraction (about 50%) of the vertical gradient in DIC can be accounted for by gas exchange. The process by which the ocean maintains a vertical gradient in DIC as a result of gas exchange is known as the *solubility pump* (Volk and Hoffert, 1985). In the high latitudes, equilibration is difficult to achieve because the surface ocean mixed layer is very deep. In addition, the equilibration for CO_2 is particularly long because it is chemically reactive.

b. Equilibration time in the surface ocean mixed layer

Consider the situation where the ocean is undersaturated in CO_2 with respect to the atmosphere. There will then be a flux of CO_2 from the atmosphere to the ocean. Because there is a chemical sink of CO_2 in the mixed layer, the asymptotic approach to saturation will take longer than if there were no chemical reaction. The mass balance for the mixed layer for dissolved inorganic carbon is similar to that for nonreactive gases (Sec. 8.2.4):

$$\frac{\partial[\text{DIC}]}{\partial t} = \frac{k_w}{z_{ml}}([CO_2]_{sat} - [CO_2]_w). \qquad (8.41)$$

The fraction of the CO_2 flux from the atmosphere to the mixed layer that will react to form carbonate and bicarbonate ions is characterized by the *buffer factor* or *Revelle factor*, ζ, which is the fractional change in atmospheric CO_2 divided by the fractional change in DIC after equilibrium has been established:

$$\zeta = \frac{\partial[CO_2]/[CO_2]}{\partial[\text{DIC}]/[\text{DIC}]}. \qquad (8.42)$$

ζ depends on T and [DIC] and, to much smaller extent, S and [Alk]. Under present surface ocean conditions, ζ varies between about 9 and 14 with an average of about 10. This means, for example, that if atmospheric pCO_2 changes by 10%, oceanic [DIC] will change by about 1% on average, *at equilibrium*. Combining (8.41) and (8.42) yields

$$\frac{\partial [CO_2]_w}{\partial t} = \frac{k_w}{z_{ml}} \frac{\zeta [CO_2]_w}{[DIC]} ([CO_2]_{sat} - [CO_2]_w). \qquad (8.43)$$

For relatively small atmosphere–ocean gradients, the factor $[DIC]/\zeta[CO_2]_w$ will not change very much and can be assumed constant. The equilibration time is then augmented over nonreactive gases by this factor. Assuming average surface values for [DIC], $[CO_2]_w$ and ζ of 2,000 μmol kg^{-1}, 10 μmol kg^{-1} and 10, respectively, the equilibration time for CO_2 is augmented by a factor of 20 over nonreactive gases, meaning that for every 20 molecules of CO_2 entering the ocean, 19 get converted to carbonate and bicarbonate and only one remains as CO_2, on average. The average equilibration time of CO_2 in the surface ocean mixed layer with respect to gas exchange is therefore about 400 days, assuming $k_w = 15$ cm hr^{-1} and $z_{ml} = 80$ m. The long equilibration time of CO_2 in the mixed layer allows physical and biological processes to drive CO_2 further from equilibrium than other gases. Figure 8.14 shows that deviations from equilibrium can be quite large for CO_2.

c.   The distribution of surface ocean pCO_2

The most important factors affecting the distribution of ΔpCO_2 $(pCO_{2_{oc}} - pCO_{2_{atm}})$ are the cycle of organic carbon, the surface heat flux, and the anthropogenic increase in atmospheric CO_2. The high pCO_2 of equatorial waters (Fig. 8.14) is due to upwelling of water rich in DIC, partly as a result of organic carbon remineralization. These waters are also high in DIC because they once outcropped at higher, cooler latitudes and exchanged CO_2 with the atmosphere. As equatorial surface waters travel poleward and cool, they decrease in pCO_2. Hence many middle and high latitude regions are undersaturated in pCO_2 (Fig. 8.14). Imprinted on these natural biological and physical processes is the anthropogenic invasion of CO_2 into the ocean. As shown in Sec. 16.4, the pattern of ΔpCO_2 has changed significantly as a result of the atmospheric CO_2 increase, particularly in regions of convection and upwelling.

d.   Chemical enhancement

The chemical reactivity of CO_2 in sea water may affect its piston velocity. If the ocean is undersaturated in CO_2 with respect to the atmosphere, there will be a flux of CO_2 from the atmosphere to the ocean. This additional CO_2 will react with water to form carbonic acid in the diffusive boundary layer if the time scale for diffusion of CO_2 through the boundary layer is comparable to or greater than the reaction time constant. In this case, the concentration profile within the boundary layer will not be

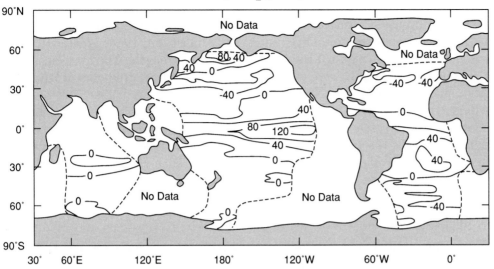

Fig. 8.14 Distribution of surface ocean pCO_2 minus atmospheric pCO_2 for January–April, 1972–1989. Adapted from Tans et al. (1990). A five-point smoother was applied before contouring.

a straight line, as we have assumed thus far. The CO_2 gradient at the air–sea interface, and hence the flux, will be enhanced by the chemical reaction (Fig. 8.1). The enhancement can also occur for CO_2 flux from the ocean to the atmosphere. The time scale for molecular diffusion in the boundary layer is $(\Delta z)^2/D$ where Δz is the boundary layer thickness. From the stagnant film model, $\Delta z = D/k_w$. Assuming average values for $k_w = 15$ cm hr^{-1} and $D = 1.6 \times 10^{-5}$ cm^2 s^{-1}, respectively, for CO_2, then $\Delta z = 40$ μm and the diffusive time scale is 0.1 s. The hydration and dehydration time constants for CO_2 are about 30 s and 10 s, respectively (Johnson, 1982). Under average conditions, therefore, chemical enhancement is probably unimportant. At low wind speeds, however, when the boundary layer thickness is greater, chemical enhancement may play a significant role.

8.3.5 Relative strengths of the carbon pumps

A way of viewing the relative strengths of the carbon pumps is shown in Fig. 8.4(b). S represents average (pre-industrial) surface water; S1 represents surface water in the absence of the carbonate pump; S2 represents surface water in the absence of the carbonate and organic carbon pumps; D represents deep water, which would be the same as surface water in the absence of all of carbon pumps. To get an idea of the relative strengths of the carbon pumps, the lines connecting the points labeled S, S1, S2, and D in Fig. 8.4(b) can be used. Moving up (down) and left (right) means an

273

increase (decrease) in CO_2 pumping from the atmosphere to the deep ocean, with the strength proportional to the distance moved. Thus the organic matter pump has a positive CO_2 pumping strength that is about 2.5 times as great as that of the solubility pump, and almost four times greater than (though opposite in sign to) the carbonate pump strength. The locations of the points S1, S2 and D were computed from the vertical gradients of $DIC°$, $Alk°$, and NO_3^- with a C:N Redfield ratio of 106/16. A higher C:N ratio will give more strength to the organic carbon pump. This highly simplified analysis can only give a rough idea of the relative strengths of the carbon pumps, because the relative importance of high latitudes has been ignored in computing the average surface values. Models, such as those discussed in Chapter 16, must be used to make a more accurate assessment.

8.3.6 *Oceanic buffers of atmospheric CO_2*

The ocean has an enormous capacity to attenuate or *buffer* perturbations in atmospheric CO_2. There are two CO_2 buffering mechanisms operating in the ocean. One involves the reaction of CO_2 with water [Eq. (8.13) and (8.14)] and therefore takes place in the water column. The other involves calcium carbonate dissolution and precipitation (Sec. 8.3.3) and therefore is important primarily in the sediments.

a. The water column buffer

The buffer factor (Sec. 8.3.4*b*) is an important quantity when considering the equilibrium response of the oceanic water column to an atmospheric CO_2 perturbation. The larger the buffer factor, the more difficult it is for the ocean to attenuate such a perturbation. Since the buffer factor increases with increasing DIC, the ocean's ability to absorb anthropogenic CO_2 will decrease with time, (i.e., as atmospheric CO_2 increases). Assuming that the oceanic carbon reservoir is 65 times larger than the atmospheric carbon reservoir (as it was in pre-industrial times) and assuming that the buffer factor has an average value of 10 (Sec. 8.3.4*b*) it is easy to show that the ocean will eventually absorb about 85% of a modest perturbation of atmospheric CO_2. By modest it is meant an initial doubling or quadrupling of atmospheric CO_2. Larger perturbations will absorb smaller fractions because of the dependency of ζ on DIC. The rate at which CO_2 is absorbed by the ocean is limited by the rate at which the deep ocean overturns, which is approximately several hundred to one thousand years.

b. The sedimentary buffer

The buffer involving calcium carbonate results from the effect of atmospheric CO_2 perturbations on the carbonate ion concentration in the ocean. An increase in atmospheric CO_2, for example, will cause oceanic DIC to increase and CO_3^{2-} to decrease [Eq. (8.39)]. This will result in a shallowing of the carbonate ion saturation horizon and a dissolution of sedimentary $CaCO_3$. The increase in calcium ion concentration will increase the alkalinity and therefore lower CO_2 (Sec. 8.3.1*b*). It is estimated that there is sufficient sedimentary $CaCO_3$ available for the complete neutralization of all fossil fuel reserves, though this would take thousands of years, due to the slow rates of deep ocean ventilation and sedimentary $CaCO_3$ dissolution (Broecker and Peng, 1982). As the anthropogenic CO_2 perturbation becomes larger with time in the deep ocean, the sedimentary $CaCO_3$ buffer will become more important.

8.4 Dimethyl sulfide

The source of dimethyl sulfide gas $[(CH_3)_2S$ or DMS] to the atmosphere is almost exclusively from the ocean. DMS is so greatly supersaturated in surface sea water that the atmospheric concentration can be ignored when computing the net flux from the ocean to the atmosphere [see Eq. (8.5)]. As will be shown, phytoplankton, zooplankton, and bacteria all play important roles in controlling the distribution of DMS in sea water.

The source of DMS in sea water was originally believed to be excretion by phytoplankton. Laboratory experiments have shown that phytoplankton are capable of producing DMS (Keller et al., 1989). The distribution of DMS in sea water is somewhat similar to that of chlorophyll, an indicator of phytoplankton biomass. Like chlorophyll, DMS often has a maximum at the surface or within the euphotic zone and rapidly decreases below the euphotic zone. Like many phytoplankton indicators, DMS often has a strong seasonal cycle with a maximum in the summer and a minimum in the winter. Despite the qualitative similarity between DMS and chlorophyll, only 30% of the variability in DMS can be accounted for by chlorophyll variability (Andreae, 1990). The weakness of the correlation is due in part to the fact that DMS production by phytoplankton is highly species-specific (Keller et al., 1989). Indeed, DMS is more highly correlated with the abundance of particular phytoplankton species than with total chlorophyll (Turner et al., 1988).

The dominant precursor of DMS is thought to be dimethyl sulfonium propionate $[(CH_3)_2$–S–CH_2–CH_2–COO or DMSP]. DMS can be formed by the enzymatic cleavage of DMSP as well as the oxidation of DMSP with OH^-, oxygen or hydrogen peroxide. Oxidation is too slow to be of significance at the pH of sea water, though enzymes, bacteria, or both may accelerate the rate of oxidation (Dacey and Blough, 1987). DMSP is present in both particulate and dissolved forms. DMS appears to be more highly correlated with dissolved DMSP than with particulate DMSP, suggesting that DMS is produced primarily from dissolved DMSP (Turner et al., 1988).

DMSP most likely originates in phytoplankton where it is believed to be an osmolyte (which serves in maintaining osmotic pressure). Consistent with this, laboratory experiments show that intracellular DMSP and the rate of DMS production of phytoplankton increase with increasing salinity (Vairavamurthy et al., 1985). Also, DMS normalized to chlorophyll generally increases with salinity in estuaries (Iverson et al., 1989). In the open ocean, however, where salinity variations are small, salinity is probably a minor variable in algal DMSP production. It is not known what biological function, if any, is served by the production of DMS from DMSP.

It has been speculated that phytoplankton prefer nitrogen-based osmolytes and will produce other osmolytes, such as DMSP, primarily when ambient nitrate concentrations are low. This is supported by preliminary laboratory experiments and field data which have shown that intracellular DMSP increases with decreasing ambient nitrate (Turner et al., 1988), and by experiments conducted in tanks with natural phytoplankton populations which have demonstrated that DMS production is low during the growth phase when nutrients are abundant and high during the senescence phase when nutrients are depleted (Nguyen et al., 1988). Also, over large spatial scales, DMS normalized to primary productivity tends to be higher in low productivity (i.e., low nitrate) environments than high productivity (i.e., high nitrate) environments (Andreae, 1990).

Zooplankton also play an important role in the production of DMS. Dacey and Wakeham (1986) demonstrated that the release of DMS in phytoplankton incubations was greater in the presence of zooplankton grazers than in their absence. These investigators estimated that DMS production in the ocean as a result of grazing may be six times greater than the passive release of DMS by ungrazed phytoplankton. The significance of grazing in DMS production is supported by observations over a seasonal cycle of a correlation between DMS and zooplankton biomass and no correlation between DMS and any phytoplankton indicators (Leck et al., 1990). The mechanism of DMS release as a result of grazing is unknown. Several possibilities exist: (1) phytoplankton cell rupture during capture and ingestion by zooplankton which may release intracellular DMS and DMSP, (2) decomposition of algal DMSP in the guts of zooplankton, and (3) decomposition of algal DMSP by bacteria in the fecal material of zooplankton.

Not all DMS production is lost to the atmosphere. Marine bacteria have been found to consume DMS at a rate that is roughly an order of magnitude greater than the loss to the atmosphere (Kiene and Bates, 1990). Other possible sinks of DMS are photochemical oxidation and adsorption onto sinking particulate matter. Little is known about the magnitude of these loss mechanisms but currently they appear to be even less important than loss to the atmosphere.

Figure 8.15 is a schematic diagram of the processes that may be important in controlling the concentration of DMS in surface sea water. At present it appears that DMS is produced mainly as the result of zooplankton grazing

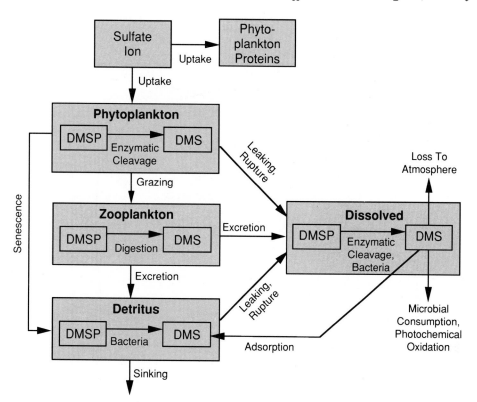

Fig. 8.15 A schematic diagram of the marine sources and sinks of DMS.

on phytoplankton and that DMS is lost mainly as a result of bacterial consumption. This model is very tentative as it is based on only a very few observations and experiments made over limited areas of the ocean. The important task of successfully modeling the distribution of DMS in the ocean and its flux to the atmosphere awaits further observations and experiments to quantify the sources and sinks of DMS in many oceanographic settings.

8.5   Nitrous oxide

The oceans are known to be a source of N_2O to the atmosphere because surface waters are nearly always and everywhere supersaturated with respect to the atmosphere. The degree of this supersaturation is, on average, about 5 to 10% but can locally reach much higher than 100%.

The subsurface distribution of N_2O has been shown to be inversely correlated with oxygen in oxygenated waters (Fig. 8.16). In very low oxygen waters N_2O has been found to decrease with decreasing oxygen (Cohen and Gordon, 1978). Generally, N_2O is near saturation at the surface, reaches a maximum at a depth of about one kilometer, and decreases slightly to the sea floor (Fig. 8.16). The correlation between N_2O and oxygen is improved

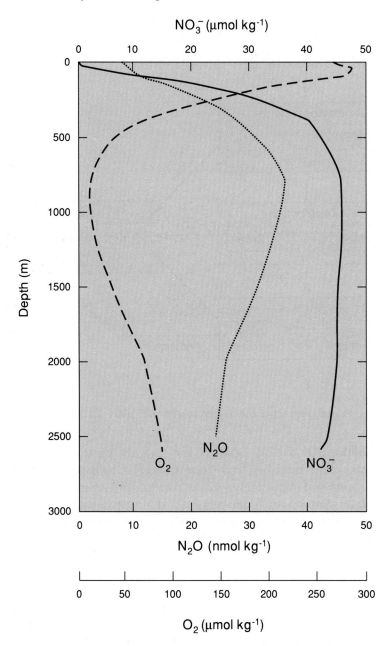

Fig. 8.16 Vertical profiles of oxygen, nitrous oxide and nitrate in the northeast Pacific Ocean. Note the correlation between nitrous oxide and nitrate, and the inverse correlation of these two species with oxygen. From Cohen and Gordon (1979).

in oxygenated waters if one compares the N_2O excess (ΔN_2O) with the apparent oxygen utilitization (AOU), where

$$[\Delta N_2O] = [N_2O] - [N_2O]_{sat} \qquad (8.44)$$

$$\text{AOU} = [O_2]_{sat} - [O_2]. \qquad (8.45)$$

$[N_2O]_{sat}$ and $[O_2]_{sat}$ are the concentrations that a water parcel at the observed temperature and salinity would have if it were at gas-phase equilibrium with the atmosphere. Since temperature and salinity are conservative in sea water below the surface, AOU and ΔN_2O are estimates of how much net oxygen consumption and net N_2O production, respectively, a water parcel has experienced since it left the surface. Plots of ΔN_2O versus AOU suggest that for every molecule of N_2O produced below the surface, about 5×10^3 to 13×10^3 molecules of oxygen are consumed (Cohen and Gordon, 1979).

The relationship between oxygen and N_2O has suggested to many investigators that nitrification, a process that usually occurs simultaneously with oxygen consumption [the reverse of (8.32)], is primarily responsible for the production of N_2O. Below the euphotic zone (see Fig. 8.11), organic nitrogen degrades to ammonium, most of which is "nitrified" by bacteria to nitrate:

$$NH_4^+ + 2O_2 \rightarrow NO_3^- + H_2O + 2H^+. \qquad (8.46)$$

Nitrification is responsible for the deep water enrichment of nitrate (Fig. 8.12). A small fraction of the ammonium, however, is thought to be converted to N_2O by bacteria according to:

$$2NH_4^+ + 2O_2 \rightarrow N_2O + 3H_2O + 2H^+. \qquad (8.47)$$

Using an O_2:N Redfield ratio and the AOU–N_2O correlation allows the estimation of the fraction of ammonium that is oxidized to N_2O. For O_2:N = 138:16, then for each molecule of N_2O produced below the euphotic zone, there are 600 to 1,500 moles of nitrate produced.

Data on the isotopic composition of N_2O do not appear to support the belief that N_2O is produced during nitrification. In bacterial cultures it has been found that nitrifying bacteria produce N_2O that is greatly depleted in the isotope [15]N (Yoshida, 1988). Marine N_2O, however, is generally enriched in [15]N. Further, laboratory experiments show that nitrifying bacteria produce about 5,000 molecules of nitrate for each molecule of N_2O (Yoshida et al., 1989), much larger than the ratio deduced from N_2O, oxygen and nitrate distributions. This suggests that nitrification is not responsible for the production of N_2O in the ocean. Some investigators have suggested that *denitrification* is the source of N_2O in the ocean. Denitrification is the process by which bacteria use nitrate instead of oxygen as an oxidant of organic matter. It takes place under low oxygen conditions and can produce nitrous oxide:

$$2NO_3^- + 2CH_2O \rightarrow N_2O + H_2O + 2CO_2 + 2OH^- . \qquad (8.48)$$

Since N_2O is found in high-oxygen environments, it has been hypothesized that denitrification occurs in microzones of low oxygen inside organic particles. It seems more likely, however, that denitrification consumes N_2O, since in regions of the eastern tropical Pacific where water-column denitrification is thought to take place, N_2O appears to be relatively depleted (Cohen and Gordon, 1978).

Estimates of the magnitude of the marine source imply that it is a significant component of the atmospheric N_2O budget. If it is assumed that 1 mole of N_2O is produced for every 1×10^4 moles of O_2 consumed, and if the average rate of oxygen consumption is 6 mol O_2 m^{-2} yr^{-1} (which would support a new production of 4 mol C m^{-2} yr^{-1}), then the marine N_2O source is $\sim 4 \times 10^{12}$g N yr^{-1} – about equal to the current atmospheric increase (IPCC, 1990a).

PART 3

MODELING AND
PARAMETERIZATION

9

Climate system simulation: basic numerical & computational concepts

James J. Hack

9.1 Introduction

The physical principles discussed in earlier chapters form the basis for our ability to model the climate system numerically. Models of the climate system can take many forms ranging from simple one-dimensional energy balance models to complex three-dimensional time-dependent circulation models of the atmosphere and ocean (e.g., see Schneider and Dickinson, 1974). The simpler mechanistic models are primarily aimed at providing insight into the relative roles of individual climate processes, while the more complex models are oriented toward the detailed simulation of observed phenomena. It is this latter class of the model hierarchy, the modern day General Circulation Model (GCM), that we will discuss in this chapter. This type of model attempts to provide the most comprehensive mathematical description of the atmosphere or ocean system, and historically has been the most numerically complicated and computationally demanding of scientific simulation tools.

We will focus on the development of the *atmospheric* GCM (or AGCM) as one of the principal model components of the Earth's climate system. The scope of this book does not permit a complete theoretical development of all the relevant numerical and computational issues associated with modern AGCMs. Therefore, we will introduce some of the more important numerical method concepts related to the construction of these models, while providing a perspective on the computational issues that have historically paced GCM development. Additional, more advanced numerical method concepts are discussed in Chapter 15. Finally, we will address the computational challenges that will face future modeling efforts as we move toward the twenty-first century, particularly the scientific and computational difficulties of coupling such complex atmospheric models with other major climate system component models.

9.1.1 Development of atmospheric simulation capabilities

Climate is generally thought of as the average behavior of the land–ocean–atmosphere–cryosphere system over some relatively long period of time; i.e., it is not associated with the detailed sequence of daily fluctuations we commonly refer to as "weather." From a statistical point of view, these detailed fluctuations might be regarded to be nothing more than high-frequency noise superimposed on the more meaningful climate signal. The dynamical processes responsible for this "statistical noise," however, are believed to be fundamentally interconnected with long-term variations in the climate signal. Thus, GCMs, which are frequently used to predict weather on a global scale, have enjoyed increasing utilization in studies of the atmospheric processes that maintain the observed climate. Because of the inherent unpredictability of atmospheric flow (e.g., Lorenz, 1969), these models are unable to predict the day-to-day sequence of weather events a century, a season, or even a few weeks in advance. Their true utility, therefore, lies in the ability to predict the statistical properties of some future climate state, which one hopes are not directly tied to the chronological order in which the intermediate atmospheric states occur.

The scientific basis for modern AGCMs, indeed the foundation for all present day atmospheric simulation capabilities, is fundamentally rooted in the laws of classical mechanics and thermodynamics that were developed during the eighteenth and nineteenth centuries (e.g., see Thompson, 1978). The first explicit statement that the prediction of atmospheric motions amounted to the solution of a properly posed initial-value problem can be found in a remarkable 1904 paper entitled *The Problem of Weather Prediction Considered from the Point of View of Mechanics and Physics* by Vilhelm Bjerknes. In this paper, Bjerknes formulated a simplified hydrodynamical system of equations consisting of Newton's equations of motion, the mass continuity equation, the ideal gas law, and the first law of thermodynamics. Even though the system of equations described only the simplest moments of atmospheric fluid motion, they could not be solved analytically. The graphical solution techniques Bjerknes proposed for their solution turned out to have little practical value. Nevertheless, he had succeeded in providing the first sound scientific approach to the problem.

Little visible progress on the topic of atmospheric prediction occurred until the 1922 publication of a book entitled *Weather Prediction by Numerical Process* written by Lewis F. Richardson. He generalized the fundamental governing system put forth by Bjerknes to include a limited set of diabatic processes, and proposed a finite difference approach to their solution. The most important attribute of his numerical approach was that it reduced to a well-defined sequence of simple arithmetic operations to integrate the equations forward in time. Its most significant drawback was the overwhelming magnitude of arithmetic required to produce even a one day global forecast. Over a period of many months, Richardson would complete a test calculation of his method by hand (a six-hour forecast for a region over central Europe). His effort produced a forecast that was totally unrealistic, predicting surface pressure tendencies two orders of magnitude larger than were observed. In retrospect, the major reasons for the failure of his forecast were the lengthy six-hour time step which violated fundamental numerical stability limits (limits later discovered by Courant, Friedrichs, and Lewey, 1928), coupled with a general inadequacy of observational data in the free atmosphere (i.e., away from the Earth's surface). Although his efforts were generally regarded as a failure, his imaginative approach to the problem of atmospheric prediction had set the stage for events that would follow nearly a quarter of a century later.

The next 25 years, which Reed (1977) refers to as the "Transitional Era" in the development of meteorological science, were marked by two important technological developments that would eventually make Richardson's approach feasible. The first was the development of the radio meteorgraph, or radiosonde, which made possible the establishment of networks of stations providing regular upper-air meteorological observations. The second was the development of numerical processors using electronic switching elements in the form of radio tubes. Another important development during this period was the introduction of the concept of the "stored-program computer," generally attributed to John von Neumann, a professor of Mathematics at the Princeton University Institute for Advanced Study.

In the late 1930s, von Neumann began a series of consulting and advisory relationships focusing on the development of a computing machine controlled by a detailed preprogrammed sequence of instructions, the Electronic Numerical Integrator and Computer (ENIAC). During his relationship with the ENIAC project, von Neumann suggested some relatively minor engineering changes, along with extensions to the instruction set, that would convert the ENIAC to a crude stored-program computer; i.e., a computer in which the memory contained both data and a set of programmed instructions that could be dynamically modified by the data as execution of the program proceeded. He singled out the problem of numerical weather prediction for special attention and succeeded in obtaining funding for the creation of a meteorology research group at the Institute for Advanced Study. The detailed events of this important period are chronicled in review articles by Platzman (1979) and Thompson (1983) which document the challenges facing the developers of the first numerical "weather" prediction model. The first forecast experiments, which involved the numerical solution of the nondivergent barotropic vorticity equation (Charney et al., 1950), were completed in early 1950, marking the dawn of numerical weather prediction.

The success of the ENIAC experiments, prompted the scientific community to develop models that went beyond the principle of absolute vorticity conservation, allowing for thermal advection and the physical process of cyclogenesis. Because of computational limitations, most of these early baroclinic models were very simple regional two-layer or two-parameter numerical models based on the quasi-geostrophic system of equations. Phillips (1956) was the first to attempt to address the atmospheric general circulation, employing a two-level quasi-geostrophic "hemispheric" model which crudely included the effects of diabatic heating and frictional dissipation. The statistical properties of the baroclinically unstable flow for a simulated period of about three weeks bore a remarkable resemblance to the observed behavior of the real atmosphere. This work, which might be characterized as a proof of concept, was a significant departure from applications of the quasi-geostrophic model to short-term forecasts. More importantly, it was instrumental in pointing the way for applying numerical prediction techniques to the investigation of climate.

Before Phillips' pioneering work could be extended to more complete experiments on the atmospheric general circulation, a number of serious shortcomings in his approach would have to be addressed. One problem was that his numerical method appeared to experience some form of nonlinear instability (a subject we will discuss later) resulting in an unplanned and rather abrupt termination of the numerical integration. The most serious deficiency, however, concerned the use of the quasi-geostrophic system itself which provides an inadequate dynamical framework for motions in the vicinity of the equator. Consequently, the general circulation modeling activities that were to follow would be based on the more complete meteorological primitive equations, in effect returning to Richardson's original approach to the problem of atmospheric prediction.

One of the more important complications of using the primitive equations is that they admit a class of high-frequency solutions known as inertia-gravity waves, one of several types of fluid motion that are filtered in the quasi-geostrophic system. Although these motions generally account for only a very small fraction of the overall energy of the atmosphere, their mere existence has profound computational significance. Among the more immediate consequences of using this system of equations was an increased model sensitivity to small imbalances in the initial conditions, an issue of greatest concern for the initial-value problem of numerical weather prediction (e.g., Charney, 1955b). The short time scale of the gravitational solutions would also prove to be a severe constraint on the size of an explicit model time step, making a large class of investigations computationally intractable. The development of more powerful computer systems coupled with innovative numerical integration techniques have since helped to minimize many of these early limitations. Nevertheless, computational capability remains a limiting factor in the development of AGCMs as we will show.

In addition to the numerical complexities and computational demands of integrating the primitive equations, numerical simulations using this system exhibited a marked sensitivity to poorly formulated treatments of nonresolvable diabatic processes, otherwise known as physical parameterizations. Many of these processes, such as radiative transfer, surface energy exchanges, and moist convection, are relatively unimportant to the short-term evolution of the large-scale flow, as demonstrated by the early success of the quasi-geostrophic forecast experiments. The proper representation of these processes is crucial, however, to the quality of extended-range integrations, and remains one of the principal scientific thrusts of atmospheric modeling initiatives today.

9.1.2 Development of computational capabilities

In the four decades since the first computer forecasts, numerical simulation capabilities have rapidly become more sophisticated, primarily due to rapid advances in computer technology. Indeed, since the inception of the stored-program computer in the late 1940s, internal computational performance has increased by more than five orders of magnitude, as shown in Fig. 9.1. In terms of applied computational performance, memory system improvements, advances in input/output technology, and the development of advanced operating system and compiler technology, have easily multiplied these gains by another order of magnitude or more. Such improvements meant that more complete formulations of the meteorological equations could be used with more complex treatments of physical processes, for larger domains, at higher resolutions, and for longer simulation periods. An example of how improvements in computational capability have paced the development of atmospheric simulation capabilities is shown in Fig. 9.2. This figure schematically depicts the improvement in forecast skill that accompanied the introduction of an increasingly complex set of numerical models at the U.S. National Meteorological Center during the period 1955–1990. Also shown is the acquisition of successive generations of computer hardware that enabled the introduction of improved operational modeling capabilities.

Early increases in processor performance were based largely on technological developments, particularly the invention of the transistor and the accompanying improvements in component packaging technology. Computer systems based on solid-state technology began to appear around 1960, and provided an order of magnitude improvement in computational performance over their vacuum tube predecessors. Improvements to the central processing unit (CPU) were complemented by another very important "architectural" change which was the introduction of input/output (I/O) channels. These data channels permitted an asynchronous concurrency between I/O and CPU operations, even though the I/O was still under the control of the CPU. At about the same time high-level computer languages

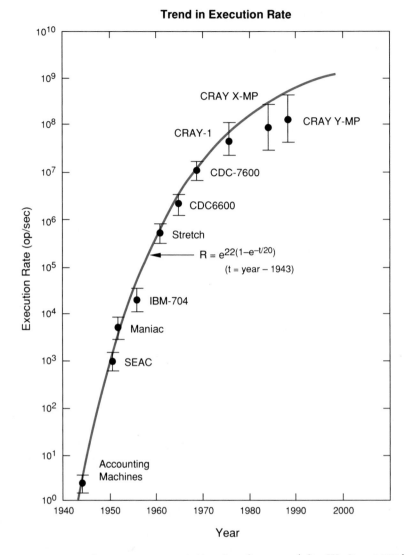

Fig. 9.1 Trend in single processor computational performance (after Worlton, 1987 [personal communication]).

(e.g., FORTRAN) began to appear along with multiprogrammed operating systems, both of which helped to reduce the complexity of implementing application programs. The later introduction of integrated circuit technology contributed to improved performance along with higher reliability, while enabling innovative improvements in machine organization, i.e., the implementation of the central processor architecture.

The late 1960s and early 1970s saw the introduction of vector extensions to the von Neumann architecture, further improving processor utilization. Two different approaches emerged during this period which

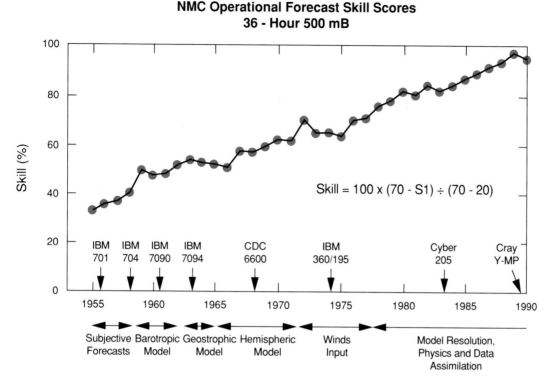

NMC Operational Forecast Skill Scores
36 - Hour 500 mB

Skill = 100 x (70 - S1) ÷ (70 - 20)

Fig. 9.2 36 hour 500 mb forecast skill over North America at the National Meteorological
Center (after Kálnay et al., 1991).

were the parallel vector processor and the pipelined vector processor.
These architectural extensions represented a significant departure from
technological and machine organization improvements since improved
performance relied directly on algorithmic design and programming skill.
Pipelined vector processors, which were the first to be associated with
the term supercomputer, have been the workhorse of large-scale scientific
computing activities throughout the decade of the 1980s. But, as Fig. 9.1
suggests, the more recent generations of these very expensive and complex
computer systems have provided only modest improvements in single-
processor performance.

Although reductions in high-performance machine cycle times continue
to be realized, they have been coming at a much slower pace in
recent years. This trend is expected to become more acute as signal
propagation constraints, which limit the physical size of the central
processing unit, become the limiting technological factor. Improvements
in both machine organization and architecture also appear to be reaching
practical technological and application limits. These limitations have
sparked renewed interest in approaches to high-performance computing that
use many processors to work cooperatively on a single problem at the same

time. The principle of these parallel architectures is that the performance of a p processor configuration (in units of individual processor performance) on an application that can be partitioned into p quasi-independent but comparable tasks should ideally be linearly proportional to the number of processors. At this point, highly parallel or massively parallel computational approaches appear to provide the only solution to the large increases in computational capability required for climate simulation.

9.2 Numerical methods used in global climate models

Most, if not all, AGCMs are based on the meteorological primitive equations [see Chapter 3, Eqs. (3.56–3.58)]. These highly nonlinear partial differential equations do not have closed-form solutions, and must be integrated numerically. All numerical methods necessarily involve some type of discrete approximation to the continuous differential equations. For atmospheric modeling these techniques generally fall into the broad categories of finite difference, spectral, and finite element methods. At this point, finite difference and spectral methods are the most dominant in atmospheric general circulation modeling. Below we introduce some of the more basic concepts associated with these numerical solution techniques, particularly those relevant to current and future computer systems. We refer the interested reader to more complete developments of numerical methods in Haltiner and Williams (1980), Mesinger and Arakawa (1976), Machenhauer (1979), and Staniforth and Côté (1991).

9.2.1 Numerical approximation concepts

a. Finite difference methods

Finite difference methods make use of a discrete set of points within the domain of interest. The dependent variables are initially defined on this grid (sometimes equivalently called a mesh or a lattice) and subsequently evaluated at these points. Most often, an Eulerian approach is used in which both space and time coordinates are chosen as independent variables where the spatial distribution of the finite difference grid remains fixed. Derivatives appearing in the equations are approximated using differences of dependent variables over finite space and time intervals, hence the term finite difference method. These derivative approximations are used to construct a system of algebraic equations which are then solved using an electronic computer.

Consider an arbitrary function, f, of a single independent variable x

$$f = f(x) \tag{9.1}$$

which we want to approximate in some region $0 \leq x \leq L$. The simplest approach to the problem is to divide the region into an integer number of intervals J, of equal length Δx, called the grid length. Thus, the function

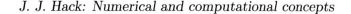

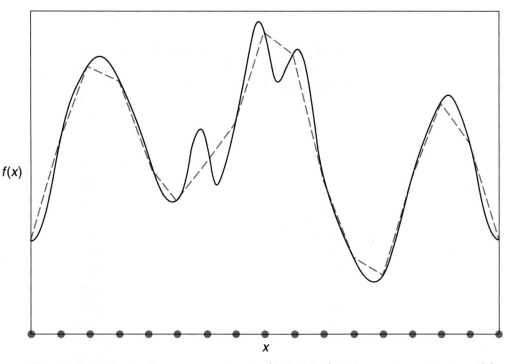

Fig. 9.3 Example of a discrete approximation (dashed line) of the continuous function $f(x)$ (solid line).

f is approximated at the discrete points $x_j = j\Delta x$ where $j = 0, 1, 2, \ldots, J$. It is important to point out that the smallest feature (i.e., the shortest wavelength) that can be represented on such a discrete grid is $2\Delta x$ in size. For example, Fig. 9.3 shows a function $f(x)$ which has been sampled on a finite difference grid. Note how the smaller-scale features (i.e., features smaller than $2\Delta x$ on this particular finite difference grid) do not exist in the discrete representation, while the larger features are more faithfully approximated. Thus, the interval $2\Delta x$ defines a lower resolution limit, although not necessarily a sufficient resolution for the proper treatment of such scales of motion, as we will demonstrate shortly.

Finite differences are discrete approximations to continuous deriviatives and are calculated over one or more of the intervals Δx (where x refers to any independent variable, including time). Formally, the first derivative of $f(x)$ at a point x_0 is defined as

$$\frac{d}{dx}f(x_0) \equiv f_x(x_0) = \lim_{\Delta x \to 0} \frac{f(x_0 + \Delta x) - f(x_0 - \Delta x)}{2\Delta x}. \qquad (9.2)$$

Finite difference approximations to the first derivative may be derived by using Taylor series expansions of f about a point x_0 such that

$$f(x_0 + \Delta x) = f(x_0) + f_x(x_0)\Delta x + f_{xx}(x_0)\frac{\Delta x^2}{2} + O(\Delta x^3), \qquad (9.3)$$

$$f(x_0 - \Delta x) = f(x_0) - f_x(x_0)\Delta x + f_{xx}(x_0)\frac{\Delta x^2}{2} + O(\Delta x^3). \qquad (9.4)$$

Uncentered difference operators may then be defined by

$$f_x(x_0) = \frac{f(x_0 + \Delta x) - f(x_0)}{\Delta x} - f_{xx}(x_0)\frac{\Delta x}{2} + O(\Delta x^2), \qquad (9.5)$$

$$f_x(x_0) = \frac{f(x_0) - f(x_0 - \Delta x)}{\Delta x} + f_{xx}(x_0)\frac{\Delta x}{2} + O(\Delta x^2). \qquad (9.6)$$

where the terms following the leading term on the right-hand side define the truncation error for the scheme, a measure of the difference quotient accuracy for small values of Δx. The order of accuracy for the scheme is defined by the lowest power of Δx appearing in the truncation error. For the uncentered operators defined above, the leading term in the truncation error is $O(\Delta x)$, which makes the scheme first-order accurate. A centered difference operator can be derived by averaging the uncentered operators such that

$$f_x(x_0) = \frac{f(x_0 + \Delta x) - f(x_0 - \Delta x)}{2\Delta x} + O(\Delta x^2). \qquad (9.7)$$

Note how the leading term in the truncation error is much smaller for this second-order accurate scheme. For the one-dimensional centered case, the second derivative can be obtained by simply applying the first derivative twice

$$f_{xx}(x_0) \approx \frac{1}{\Delta x}\left[\frac{f(x_0 + \Delta x) - f(x_0)}{\Delta x} - \frac{f(x_0) - f(x_0 - \Delta x)}{\Delta x}\right]$$
$$= \frac{f(x_0 + \Delta x) - 2f(x_0) + f(x_0 - \Delta x)}{\Delta x^2}. \qquad (9.8)$$

Another measure of the accuracy of a finite difference scheme can be illustrated by deriving what is sometimes referred to as the response function for the discrete operator. We begin by assuming that

$$f(x) = A\sin\left(\frac{2\pi x}{L}\right), \qquad (9.9)$$

for which the analytic derivative can be written as

$$f_x(x) = \frac{2\pi}{L}A\cos\left(\frac{2\pi x}{L}\right). \qquad (9.10)$$

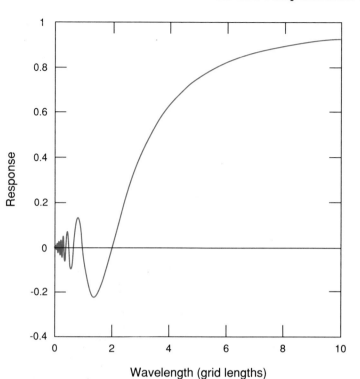

Wavelength (grid lengths)

Fig. 9.4 The response function for a centered difference approximation to the function $f(x) = A \sin \left(\frac{2\pi x}{L} \right)$.

Using (9.7), the centered difference approximation of $f_x(x)$, $\delta_x f$, can be written as

$$\delta_x f = \frac{A}{2\Delta x} \left[\sin \left(\frac{2\pi}{L}(x + \Delta x) \right) - \sin \left(\frac{2\pi}{L}(x - \Delta x) \right) \right]$$

$$= \frac{A}{\Delta x} \cos \left(\frac{2\pi x}{L} \right) \sin \left(\frac{2\pi \Delta x}{L} \right) \tag{9.11}$$

$$= \frac{L}{2\pi \Delta x} \sin \left(\frac{2\pi \Delta x}{L} \right) \left[\frac{2\pi}{L} A \cos \left(\frac{2\pi x}{L} \right) \right].$$

The response function for this centered difference approximation to $f_x(x)$, the ratio of the numerical to the analytic estimate, can then be derived as

$$R(L, \Delta x) = \frac{L}{2\pi \Delta x} \sin \left(\frac{2\pi \Delta x}{L} \right). \tag{9.12}$$

The solution, R, is graphically depicted in Fig. 9.4. This figure clearly illustrates how derivative estimates depend on the resolution of the finite difference grid, i.e., in this case, how many discrete points are contained in a single wave. Despite the smooth, periodic nature of $f(x)$, eight or more grid intervals are required before obtaining a reasonably accurate estimate of the first derivative using a centered difference approximation. Even a cursory inspection of the governing system should underscore the importance of accurate derivative estimates in predicting the time evolution of large-scale atmospheric flow.

b. Spectral methods

Spectral methods represent an entirely different approach to the numerical approximation of continuous differential equations. Let us begin by considering the Fourier expansion for our arbitrary function $f(x)$ which can be written as

$$f(x) = \frac{a_o}{2} + \sum_{k=1}^{\infty} \left[a_k \cos\left(\frac{2\pi kx}{L}\right) + b_k \sin\left(\frac{2\pi kx}{L}\right) \right] \qquad (9.13)$$

where the expansion coefficients are given by

$$a_k = \frac{2}{L} \int_0^L f(x) \cos\left(\frac{2\pi kx}{L}\right) dx$$

$$k = 0, 1, 2, \ldots . \qquad (9.14)$$

$$b_k = \frac{2}{L} \int_0^L f(x) \sin\left(\frac{2\pi kx}{L}\right) dx$$

What makes this series expansion possible is that the expansion basis, $\sin\left(\frac{2\pi kx}{L}\right)$ and $\cos\left(\frac{2\pi kx}{L}\right)$, is orthogonal in the interval $(0,L)$, i.e.,

$$\frac{2}{L} \int_0^L \cos\left(\frac{2\pi kx}{L}\right) \cos\left(\frac{2\pi \ell x}{L}\right) dx = \begin{cases} 0 & k \neq \ell \\ 1 & k = \ell \neq 0 \end{cases}$$

$$(9.15)$$

$$\frac{2}{L} \int_0^L \sin\left(\frac{2\pi kx}{L}\right) \sin\left(\frac{2\pi \ell x}{L}\right) dx = \begin{cases} 0 & k \neq \ell \\ 1 & k = \ell \neq 0. \end{cases}$$

This fundamental property of orthogonality ensures that each component of the expansion basis is unique and will not project onto any other component of the expansion basis. Thus, if our function $f(x) = \cos\left(\frac{2\pi x}{L}\right)$, only a single expansion coefficent, a_1, would have amplitude associated with it; i.e., $f(x)$ would project entirely and uniquely on the $k = 1$ cos expansion component.

The Fourier expansion in (9.13) is written as an infinite series. In practical applications, however, the spectral representation of an arbitrary function can include only a relatively small number of terms in the expansion. When a limited number of terms in an infinite series are retained (e.g., $k \leq K$), the expansion is said to be truncated, i.e., the contribution of the basis functions above the wavenumber truncation limit is ignored. Retaining more terms in the expansion (i.e., the higher the spectral truncation) generally results in a more accurate spectral representation. We illustrate this property with the following two examples. Let us first assume the top-hat function

$$f(x) = \begin{cases} 1 & 0 \leq x \leq L/2 \\ 0 & L/2 \leq x \leq L \end{cases} \tag{9.16}$$

for which the spectral expansion coefficients can be analytically determined[1] as

$$a_o = 1, \qquad a_k = 0 \qquad k = 1, 2, 3, \ldots$$

$$b_k = \frac{1}{k\pi} \left(1 - \cos(k\pi)\right) \qquad k = 1, 2, 3, \ldots. \tag{9.17}$$

Figure 9.5 shows $f(x)$ along with several approximations to $f(x)$ in which a varying number of terms have been retained in the spectral expansion. As a second example we assume a function of the form

$$f(x) = \begin{cases} \sin^2\left(\frac{2\pi x}{L}\right) & 0 \leq x \leq L/2 \\ 0 & L/2 \leq x \leq L \end{cases} \tag{9.18}$$

for which the spectral expansion coefficients can also be analytically determined as

$$a_o = 1/2, \qquad a_2 = -1/4, \qquad a_k = 0 \quad k = 1, 3, 4, 5, \ldots$$

$$b_2 = 0$$

$$b_k = \frac{1}{4\pi} \left\{ \frac{2}{k} [1 - \cos(k\pi)] + \frac{1}{(2-k)} [1 - \cos((2-k)\pi)] \right. \tag{9.19}$$

$$\left. - \frac{1}{(2+k)} [1 - \cos((2+k)\pi)] \right\} \quad k = 1, 3, 4, 5, \ldots.$$

As before, Fig. 9.6 shows the analytic $f(x)$ along with several truncated approximations. There are two noteworthy attributes of spectral methods contained in these diagrams. First, the convergence of the truncated expansion is clearly more rapid in the case of the smoother function. This is a general characteristic of spectral schemes and is linked, in part, to the use of a smooth global expansion basis. Thus, spectral methods are most effective when dealing with relatively smooth, well-behaved mathematical functions. A second attribute of spectral methods illustrated in these figures is the oscillation appearing in the vicinity of sharp gradients for the truncated expansions. This behavior is known as the Gibbs phenomenon or ringing, and will appear to various degrees in most truncated expansions. The severity of the undershooting and overshooting depends on the gradients

---

[1] In practice, expansion coefficients are most often determined via numerical integration.

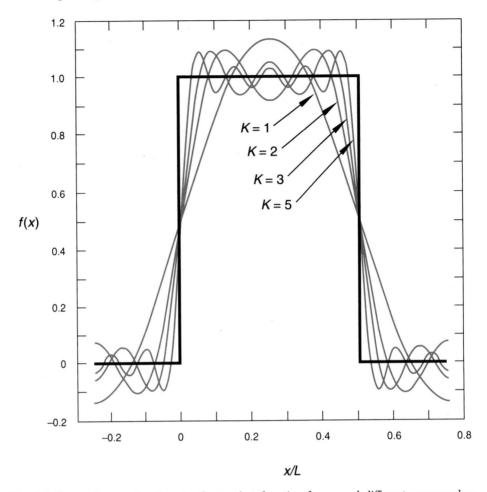

Fig. 9.5 Spectral approximations to the top-hat function for several different wavenumber truncations.

contained in the continuous function (e.g., it is most severe in the case of a discontinuity, such as in the first example) as well as on the number of terms retained in the expansion. This property can be most undesirable when representing scalar fields that exhibit sharp gradients but for physical reasons must maintain a positive definite value (e.g., water vapor). The computational problems associated with ringing have provided a strong motivation for treating selected fields with finite difference based shape preserving nonoscillatory schemes in spectral AGCMs (e.g., Williamson, 1990; Rasch and Williamson, 1991).

In contrast, with the finite difference method, derivatives are obtained by differentiating the functions comprising the expansion basis. Thus, if $f(x)$

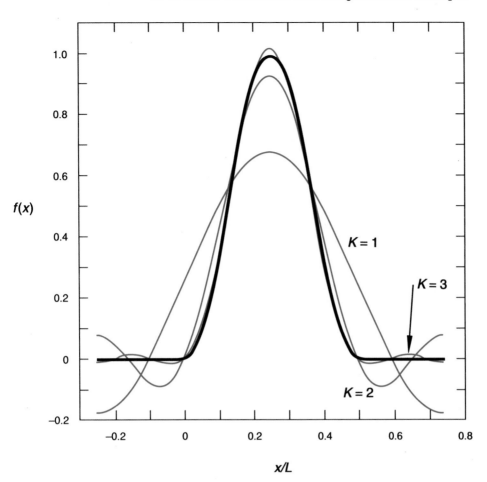

Fig. 9.6 Spectral approximations to the function given by (9.18) for several different wavenumber truncations.

is represented by the expansion

$$f(x) = \frac{a_o}{2} + \sum_{k=1}^{K} \left[a_k \cos \left(\frac{2\pi k x}{L} \right) + b_k \sin \left(\frac{2\pi k x}{L} \right) \right] \qquad (9.20)$$

f_x can be written as

$$f_x(x) = \frac{2\pi}{L} \sum_{k=1}^{K} k \left[b_k \cos \left(\frac{2\pi k x}{L} \right) - a_k \sin \left(\frac{2\pi k x}{L} \right) \right]. \qquad (9.21)$$

Similarly, f_{xx} can be written

$$f_{xx}(x) = - \left(\frac{2\pi}{L} \right)^2 \sum_{k=1}^{K} k^2 \left[a_k \cos \left(\frac{2\pi k x}{L} \right) + b_k \sin \left(\frac{2\pi k x}{L} \right) \right]. \qquad (9.22)$$

Accordingly, derivative estimates of $f(x)$ are *exact* up to the truncation limit, K.

Let us review what has been presented thus far by comparing a spectral and finite difference approach to the solution of the one-dimensional *linear* advection equation

$$\frac{\partial f}{\partial t} = -c\frac{\partial f}{\partial x}\,. \qquad (9.23)$$

Using a spectral approach $f(x, t)$ would be represented as

$$f(x, t) = \sum_{k=0}^{K} a_k(t)e^{ikx} \qquad (9.24)$$

which, using centered time differences, would require solution of K equations of the form

$$a_k(t + \Delta t) = a_k(t - \Delta t) - 2c\Delta t(ika_k(t))\,. \qquad (9.25)$$

A centered finite difference approach, however, would require the solution of J equations of the form

$$f_j(t + \Delta t) = f_j(t - \Delta t) - \frac{c\Delta t}{\Delta x}[f_{j+1}(t) - f_{j-1}(t)], \qquad (9.26)$$

where j refers to the jth mesh point and J is the number of gridpoints in the finite difference grid. The major computational advantage of the spectral approach is that if $f(x, t)$ is a well-behaved function (i.e., smooth, etc.), K is generally very much less than J for comparable solution accuracy. Additionally, the spectral method evaluates linear advection exactly (except for time truncation error), so there is no computational dispersion; i.e., the advection phase speed does not vary with wavelength as it can with finite difference solution techniques. Although the computational advantage can be clearly demonstrated for many linear problems, the conventional application of the spectral method to nonlinear problems is considerably more complicated and computationally expensive because of the nonlinear interaction of the expansion coefficients (e.g., the advection terms). The introduction of the spectral-transform procedure (Eliasen et al., 1970; Orszag, 1970), however, would make the spectral approach computationally competitive with finite difference schemes when applied to nonlinear equations, as is discussed in the next section.

To introduce two very important conceptual issues related to numerical stability, we make use of a finite difference framework, although the principles are also applicable to other numerical techniques. For simplicity, assume a sinusoidal function of the form

$$u(x, t) = U(t)\sin(kx)\,. \qquad (9.27)$$

Substitution into the nonlinear advection equation

$$\frac{\partial u}{\partial t} = -u\frac{\partial u}{\partial x} \qquad (9.28)$$

yields

$$\frac{\partial u}{\partial t} = -\frac{1}{2}U^2(t)k\sin(2kx).$$ (9.29)

As discussed earlier, the shortest wave that can be represented on a finite difference grid is $2\Delta x$ in length, which corresponds to a maximum wavenumber

$$k_{max} = \frac{\pi}{\Delta x}.$$ (9.30)

The nonlinear advective term on the right-hand side of (9.28), however, can contain a wave with twice the maximum wavenumber (or half the minimum wavelength) that can be represented for the function u. Such a wave cannot be properly resolved on the finite difference grid and will be misrepresented in terms of the permissible waves (see Fig. 9.7). This type of error is known as aliasing error. The most serious consequence of aliasing errors is a spurious accumulation of energy in high wavenumber scales of motion. Over time, the energy at these scales grows beyond physically reasonable bounds leading to nonlinear instability in the numerical integration. This is, in fact, the type of instability that was encountered by Phillips (1956) in his pioneering general circulation modeling experiment. Phillips (1959) later correctly interpreted the nature of the instability and verified his explanation by periodically eliminating the high-frequency components of the vorticity field in a long integration of the nondivergent vorticity equation. His approach suppressed the instability by what amounted to a severe numerical damping of the shortest waves. It was later recognized that the same result could be accomplished by using a dissipative numerical integration scheme, or, in a more flexible way, by including an adjustable dissipative term in the equations. More elegant approaches to the problem of nonlinear instability have since been developed by Arakawa (1966, 1972) who proposed finite difference schemes for the advective terms that are totally free of the spurious cascade of energy to smaller scales, thus eliminating the need for an artificial numerical dissipation.

We introduce the second stability concept in the context of the linear advection equation (9.23). Assuming an initial condition of the form

$$f(x,0) = Ae^{ikx}$$ (9.31)

gives the time-dependent analytic solution

$$f(x,t) = Ae^{ik(x-ct)}.$$ (9.32)

We assume the centered finite difference solution given in (9.26) which can also be written as

$$f(x, t + \Delta t) = f(x, t - \Delta t) - \frac{c\Delta t}{\Delta x}[f(x + \Delta x, t) - f(x - \Delta x, t)].$$ (9.33)

Substitution of the analytic solution (9.32) into the centered difference

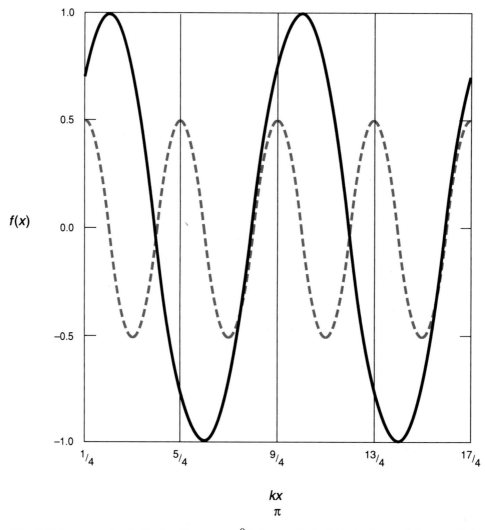

Fig. 9.7 An example of aliasing the term $u\frac{\partial u}{\partial x}$ (heavy dashed line) where $u(x) = \sin(kx)$ (heavy solid line) and k is the maximum permissible wavenumber allowed by the finite difference grid (denoted with dashed vertical lines). Note how $u\frac{\partial u}{\partial x}$ incorrectly appears as a constant amplitude function on the finite difference grid.

solution reveals that a numerically stable (i.e., nonamplifying) solution exists if and only if

$$\frac{c\Delta t}{\Delta x} \leq 1. \tag{9.34}$$

This stability bound is most commonly known as the Courant–Friedrichs–Lewy or CFL criterion, named after Courant, Friedrichs, and Lewy (1928) who discovered that this restriction was required for stable solutions to a certain class of finite difference problems. This relationship bounds the maximum time step by the ratio of the spatial resolution to the fastest signal propagating in the model (in this case a wave with the phase speed c). Thus, as the model resolution increases, the time step must necessarily decrease proportionally, further increasing the computational demand.

9.2.2 Horizontal and vertical coordinate systems

a. Horizontal representation

The numerical simulation of the Earth's atmosphere requires application of the concepts presented in the previous section to the solution of the governing equations in three space dimensions and time. The earliest global atmospheric modeling initatives modified accepted numerical methods to accommodate the solution of the meteorological equations in spherical coordinates. A large component of this effort was directed toward finite difference approaches, many of which are reviewed by Williamson (1979). As the modeling community would discover, the adaptation and development of computational methods for solving partial differential equations in spherical geometry is complicated by the unique characteristics of the coordinate system itself. Since longitude is multivalued at the pole, any nonzero vector function will have multivalued or discontinuous components, even though the same function will have smooth properties in Cartesian coordinates. This characteristic of the spherical coordinate system presents great difficulty for all numerical schemes, particularly for the simpler finite difference approaches, and is generally known as the "pole problem."

As one example, the use of a uniformly distributed latitude–longitude grid requires an unnecessarily small time step (to satisfy the local CFL stability criteria) or some form of empirical filtering of longitudinal waves near the poles because of the convergence of longitude lines. Nevertheless, the use of such grids dominate modern day AGCMs based on the finite difference method. Attempts to produce more homogeneous finite difference grids have so far enjoyed limited success. Phillips (1957a) approached the problem by using two polar stereographic grids that overlapped with a Mercator projection in the tropics. The three coordinate systems were coupled using an interpolation procedure. Phillips (1962) later showed that the approach could give good results with careful definition of the finite difference scheme and interpolation procedure, but the method failed to gain popularity. More recently, a variation of Phillips' approach, called the composite-mesh method (Starius, 1980) has been examined in the framework of the shallow water equations (Browning et al., 1989). Although the technique appears promising, it requires further study in the context of long atmospheric integrations. Other attempts at producing an isotropic grid system were made by Sadourny et al. (1968) and Williamson (1968) who used a spherical geodesic grid based on an icosahedron. Williamson (1971) developed and tested conserving finite difference schemes for the shallow water equations but found that truncation errors dominated the mass flux estimates because of the small nonuniformities in the grid. Many other approaches to the pole problem, such as equal area grids and skipped grids have also exhibited numerical deficiencies that are particularly detrimental to long-time integrations of the primitive equations.

The spectral method presents a more natural solution to problems introduced by spherical geometry. Spectral techniques were first used for the solution of meteorological equations by Silberman (1954) who solved the nondivergent barotropic vorticity equation using an interaction coefficient method to calculate the nonlinear advection terms. This procedure proved to be computationally tractable for only a small number of waves. Modest increases in spectral resolution gave rise to a rapid increase in the number of interaction coefficients, resulting in prohibitively large storage and computational requirements. Another major drawback to the method was its inability to include "local" diabatic physical processes. The invention of the Fast Fourier Transform (Cooley and Tukey, 1965), and the introduction of the spectral transform method (Eliasen et al., 1970; Orszag, 1970) would eventually make spectral techniques competitive.

The basic idea behind the spectral transform method is to locally evaluate all nonlinear terms (including diabatic physical processes) in physical space on an associated finite difference-like grid, called the transform grid. These terms are then transformed back into wavenumber space to obtain tendencies for the spectral coefficients for each of the discrete levels in the vertical. Because of the efficiency of the Fast Fourier Transform (FFT), the procedure enjoys a computational cost comparable to the most efficient of finite difference procedures for comparable accuracy. The technique also offers a number of other computational advantages for global atmospheric models, among which is the ease with which semi-implicit time differencing[2] can be incorporated (due to the simple form of the ∇^2 operator in wavenumber space when using a spherical harmonic basis), and the absence of a "pole problem" when formulating the problem in terms of vorticity and divergence (i.e., by raising the order of the system). A more complete discussion of the history and numerical characteristics of this procedure as applied in global atmospheric models can be found in Machenauer (1979).

Spectral atmospheric GCMs most frequently make use of a spherical harmonic basis for the horizontal expansion of scalar fields. The spherical harmonic functions, $Y_n^m(\lambda, \mu) \equiv P_n^m(\mu)e^{im\lambda}$, used in the spectral expansion are the eigensolutions of the barotropic wave equation in spherical coordinates and, as such, constitute a complete and orthogonal expansion basis. Although alternative orthogonal expansion bases with more desirable theoretical properties exist (e.g., Kasahara, 1977), they are far too expensive

---

[2] A time integration technique that treats the high-frequency gravitational modes with an implicit method, which is absolutely stable, and the lower-frequency Rossby modes with an explicit method, which is subject to the CFL stability limits discussed earlier.

computationally for use in a production mode. The discrete spherical harmonic transform pair can be written as

$$\xi(\lambda_l, \mu_j) = \sum_{m=-M}^{M} \sum_{n=|m|}^{N(m)} \xi_n^m P_n^m(\mu_j) e^{im\lambda_l}, \qquad (9.35)$$

and

$$\xi_n^m = \sum_{j=1}^{J} \left\{ \frac{1}{2M} \sum_{l=1}^{2M} \xi(\lambda_l, \mu_j) e^{-im\lambda_l} \right\} P_n^m(\mu_j) w_j, \qquad (9.36)$$

where λ and μ respectively denote the zonal and meridional independent variables, P_n^m are the associated Legendre functions for which m denotes zonal wavenumber, $n - m$ denotes a form of meridional wavenumber, M and N are the spectral truncation limits, and J is the order of the north–south transform grid, a function of the truncation parameters. The analysis component, (9.36), consists of a forward Fourier transform, which can be completed using an FFT procedure, followed by a forward Legendre transform which, in this case, is completed using Gaussian quadrature (where w_j are the Gaussian weights). The synthesis component, (9.35), consists of an inverse Legendre transform, followed by an inverse Fourier transform. A very nice summary of the properties of Legendre polynomials and Gaussian quadrature can be found in Appendix B of Washington and Parkinson (1986).

The resolution of a spectral model is often presented either in terms of the wavenumber truncation or in terms of the resolution of the associated spectral transform grid. Although it is conceptually easier to think of model resolution in terms of a finite difference grid, it is an inappropriate basis for subjectively comparing spectral and finite difference models. The spectral transform grid is generally not indicative of the true model resolution since a significant amount of information calculated on the grid is thrown away to avoid problems with aliasing, i.e., $2\Delta x$ structures on the transform grid are totally ignored by most spectral schemes. Additionally, the transform grid can maintain structures not included in the underlying spectral representation (e.g., structures associated with a decrease in the zonal grid distance as the pole is approached). A better basis for comparing model resolution is by the smallest feature that can be *accurately and uniformly* represented by the model, although what constitutes comparable accuracy can be ambiguous due to the nature of finite difference response functions (see Fig. 9.4).

In a spectral model, the smallest feature size that can be accurately represented is unambiguously defined by the spectral truncation limits. In two-dimensional spherical geometry, the wavenumber truncation denotes both the type and size of the truncation. Two types of truncation are most commonly used in global spectral models, rhomboidal (denoted with

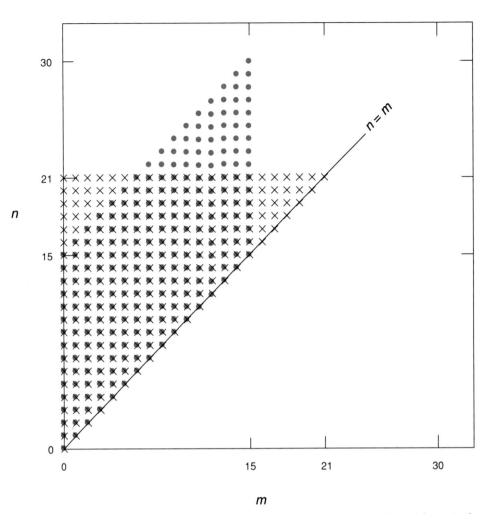

Fig. 9.8 Examples of rhomboidal and triangular wavenumber truncation with a similar
number of degrees of freedom.

an R) and triangular (denoted with a T). These two truncation approaches
are schematically illustrated in Fig. 9.8. The major advantage of triangular
truncation over rhomboidal truncation is that the solution in a triangularly
truncated system is invariant to an arbitrary coordinate rotation. Another
way to characterize this mathematical property is that information is not lost
if the flow field were to be oriented differently within the truncated spherical
coordinate system. Most of the early spectral climate models made use of
a fifteen wave rhomboidal truncation (R15), whereas more recent models
have adopted triangular truncations with a comparable or larger number of
degrees of freedom (e.g., T21, T31, T42, etc.). This is in sharp contrast
with operational forecast models, which generally make use of the highest
resolutions possible using existing dedicated computer technology. These

models currently make use of resolutions ranging from T63 to T213.

The spectral transform grid, on which nonlinear and diabatic forcing terms are evaluated, is chosen to allow exact unaliased transforms of quadratic (i.e., nonlinear) terms. Consequently, the resolution of the transform grid and the spectral truncation (or spectral resolution) are very tightly coupled. These grids are generally quite similar to the uniform latitude–longitude grids commonly used in finite difference formulations of the primitive equations, although more recent work has demonstrated that the longitudinally dense coverage near the poles is unnecessary (Hortal and Simmons, 1991). Slight irregularities in the grid most often arise due to the meridional placement of the gridpoints to allow for exact numerical integration in the north–south direction (e.g., the Legendre transform in Eq. 9.36). The specific wavenumber truncation in spectral models is most often determined by matching the aliasing requirements of the truncation with the FFT requirements of the transform grid. For example, a triangularly truncated model with maximum wavenumber M requires at least $3M + 1$ gridpoints on the transform grid in the east–west direction to avoid nonlinear aliasing. The number of points in the east–west direction must generally be factorable in powers of 2, 3, and 5 to allow the use of an FFT in the spherical harmonic transform. Thus, the most frequently used spectral truncations represent an optimum balance between the aliasing requirements of the truncation and the FFT requirements of the transform grid.

b. *Vertical coordinate*

In principle, there are many different ways of treating the vertical coordinate in AGCMs such as z, p, and θ coordinates (e.g., Kasahara, 1974). Since large-scale atmospheric motions are for the most part hydrostatic, the vertical coordinate can be expressed in terms of pressure (see Eq. 3.56). The pressure or isobaric coordinate has many analysis advantages including a much simpler form for the continuity equation, and the elimination of density from the equations. However, in a numerical model with an irregular lower boundary (i.e., mountains), these advantages are offset by the numerical difficulties of treating a time-dependent lower boundary condition in which constant p surfaces intersect the bottom topography. This is a characteristic of most vertical coordinate systems for which some degree of computational difficulty is encountered when dealing with a nonuniform lower boundary. For example, the z coordinate suffers from numerical problems that arise from the intersection of constant z surfaces with mountains. Consequently, a modification of the pressure coordinate, called the transformed pressure or σ coordinate, is most frequently used in large-scale atmospheric models (Phillips, 1957b). In its simplest form the σ coordinate is defined as

$$\sigma = \frac{p}{p_s} \tag{9.37}$$

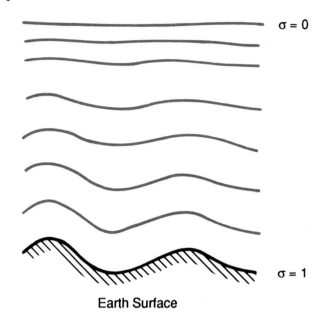

Fig. 9.9 Schematic representation of the σ coordinate system.

where p_s is the surface pressure, a function of the horizontal coordinates and time. Note that $\sigma = 1$ when $p = p_s$, i.e., $\sigma = 1$ everywhere along the lower boundary. Thus, the σ surfaces follow the lower boundary, as opposed to intersecting the lower boundary, an attractive computational property (see Fig. 9.9). Despite this computational advantage, the σ coordinate leads to a less than ideal formulation of the horizontal pressure gradient, which in the vicinity of very steep mountains is represented by the small difference of two very large terms. This relationship is a source of numerical approximation error in models and is one reason why atmospheric GCMs smooth the bottom topography to be consistent with the model's horizontal and vertical resolution.

As in the horizontal, the vertical coordinate must be approximated for purposes of numerical integration. Most AGCMs make use of finite difference and numerical integration (quadrature) techniques for approximating vertical derivatives and integrals, even if the horizontal discretization is spectral. The vertical domain is most frequently limited to the troposphere and lower stratosphere omitting regions above 20 km or so to minimize computational expense. Although this truncation of the vertical domain has historically been regarded to be adequate, recent studies (e.g., Boville and Baumhefner, 1990) have demonstrated strong systematic errors associated with the treatment and placement of the upper boundary. This work suggests the importance of explicitly treating the region between 20 and 40 km to avoid spurious systematic errors in the simulation, which clearly implies a significant additional computational burden.

Generally, the vertical differencing method and the computation of vertically integrated quantities (e.g., in the hydrostatic equation) is done in such a way that total energy is conserved for adiabatic inviscid motions (Arakawa and Lamb, 1977; Williamson, 1988). It is not simply absolute energy conservation that is important, but the way in which energy conservation is obtained (Takacs, 1988). Although a particular vertical difference scheme may accurately conserve energy, it may give relatively poor derivative estimates, particularly for fields in which the vertical gradient is large (e.g., water vapor). This problem is often aggravated by the use of unevenly distributed grids in the vertical which can reduce the formal accuracy of a differencing scheme. Conversely, schemes that are capable of providing accurate derivative estimates, may have poor energy conservation properties. Consequently, there are many different approaches to vertical finite differencing, the details of which we leave to the interested reader (e.g., Sundqvist, 1979; Arakawa and Suarez, 1983).

9.2.3 *Representation of parameterized physics*

One of the most challenging aspects of modeling the climate system is the treatment of nonresolvable physical processes, otherwise known as parameterization. Atmospheric processes operate over a very wide range of time and space scales. Because of computational cost, however, numerical integrations of the governing meteorological equations explicitly resolve only the primary energetic and phenomenological scales of motion. Nevertheless, there are significant interactions between the explicitly resolved large-scale flow and the truncated scales of motion which must be accounted for in some way. Parameterization techniques seek to express the statistical contribution of these nonresolvable processes to the time evolution of the explicitly resolved motions. At the same time, the contribution of these nonresolvable motions are diagnosed as functions of the large-scale fields.

The treatment of the water budget is the most difficult component of the parameterization problem. Convective-scale motions (i.e., motions on the order of several km) are responsible for most of the phase change and associated precipitation occurring in the atmosphere. These processes occur well below the resolvable scales of motion in a GCM, but represent a very large, and often dominant, local energy source/sink in the climate system. Their effect on the time evolution of the large-scale motion fields are represented by what is most frequently referred to as a moist convective parameterization. Phase change associated with these convective motions also plays an important role in the Earth's radiative energy budget through the formation of clouds, another process that must be represented in terms of the large-scale state variables. Examples of other physical processes that must be parameterized are the transfer of longwave and shortwave radiation in the atmosphere, surface energy exchanges, atmospheric boundary layer processes, vertical and horizontal diffusion processes, and the effects of vertically propagating gravity waves. These are discussed further in Chapter 10.

The computational complexity of physical parameterizations for the same process can vary by an order of magnitude or more. Consequently, there can be large differences in the computational load associated with different parameterization techniques which are directly related to the level of sophistication with which the physical processes are treated. This characteristic is fundamental to all physical parameterizations.

Certain parameterized process can be treated completely independently from all other nonresolvable processes. Other processes tend to be more tightly coupled with each other (e.g., the planetary boundary layer, convection, cloud formation, and radiative transfer). In most cases, however, the treatment of parameterized physics is a highly local column-oriented computational process, i.e., the collection of parameterized processes requires large-scale state information for a single vertical column, where each vertical column is independent from all other vertical columns. This is desirable in the context of a parallel computing environment, but is complicated by the local iterative nature of many solution techniques. We will discuss this characteristic further in the context of computer architecture.

9.3 Computational requirements and constraints

Because of a large natural variability in the time averaged state of the climate system, very long numerical integrations are often required to establish statistical significance in climate sensitivity studies. As a point of reference, a 10 year global atmospheric simulation using a state-of-the-art GCM can require several tens of hours on supercomputer systems capable of sustaining more than 10^9 floating point operations per second (i.e., 1 GFLOP).

The quality of these extended range simulations is crucially dependent upon the sophistication of the physical parameterizations incorporated in the model. Additionally, both forecast and climate investigations (Boville, 1991; Simmons, 1990) have begun to quantitatively demonstrate the added importance of model resolution to the quality of the numerical simulation. An example of the role of resolution is shown in Plate 10 which schematically contrasts the crude representation of topography in the majority of current climate models with that of operational forecast models. Modest increases in model resolution, increases in the sophistication of model physics, and the incorporation of new capabilities (e.g., interactive coupling with ocean circulation models, biogeochemical models, etc.) can increase the computational requirements for numerically simulating the climate system by several orders of magnitude. Accordingly, advanced numerical models of the climate system are likely to demand a sustained computational performance exceeding 10^{12} floating point operations per second (i.e., 1 TFLOP) to achieve reasonable turnaround. This kind of increase in applied computational performance is unlikely to be achieved without exploiting some form of parallel processing on a large scale. An illustration of the differences in parallel architectures using a very simple taxonomy is given below. This simple characterization of parallel architectures provides a framework in which to introduce the algorithmic challenges of using such systems in future climate system models.

9.3.1 Basic concepts in computer systems architecture

The earliest stored-program computers were based on the von Neumann architecture which, in its simplest form, consists of a control unit, an arithmetic and logical unit (or ALU), and a memory system (see Fig. 9.10). The memory system contains instructions along with the initial, intermediate, and final data associated with the instructions. The control unit decodes (i.e., interprets) and issues instructions, resulting in their execution. Finally, the ALU is where a specific set of the instructions issued by the control unit is executed. It is conceptually convenient to think of the combination of the control unit and the ALU as the central processing unit (or CPU). A simple instruction execution cycle in such an architecture would proceed as follows:

(1) The control unit requests and fetches the next instruction to be executed from the memory system.

(2) The control unit decodes the instruction, i.e., it determines the kind of instruction and its operand (data) requirements.

(3) The control unit may issue a command to transfer an operand between the memory system and the ALU's internal registers, or if a memory transfer is not required, i.e., the operands are already contained in the ALU registers, control is transferred to the ALU to allow an operation on the data to take place. An important property of the von Neumann model is that the resulting arithmetic/logical operation may determine the address of the next instruction.

(4) Return to step 1.

Note that each instruction deals with at most a single pair of operands. For this reason, this type of machine architecture is categorized as Single Instruction Single Data (SISD), after Flynn (1972) who classified machine architectures according to the characteristics of their instruction and data streams. An instruction stream is the sequence of instructions as executed by the computer system. Similarly, a data stream is the sequence of data manipulated by the instruction stream. The two other relevant members of

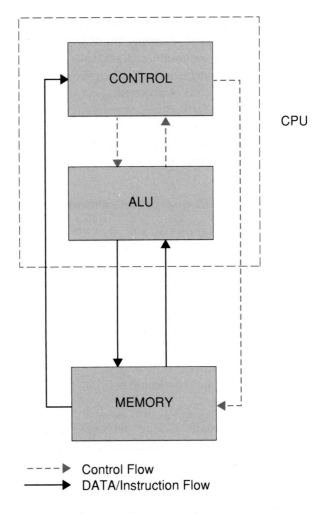

Fig. 9.10 Schematic illustration of the von Neumann architecture.

Flynn's processor classification are Single Instruction Multiple Data (SIMD), and Multiple Instruction Multiple Data (MIMD) architectures.

The first two decades of digital computer design focused on improving the machine organization, i.e., the efficiency, of the SISD architecture. Early implementations of the von Neumann architecture were for the most part sequential. For example, if an instruction decode step was underway, no other processor activity would be permitted to occur. Since each of the steps in the instruction execution cycle can require many machine cycles (i.e., ticks of the internal clock that synchronize processor operations), latency associated with even the most basic of processor activities dominated the total time required to execute a program. Improvements in performance throughout the 1950s and 1960s centered around reducing latency through technological means, or hiding it entirely through a process known as

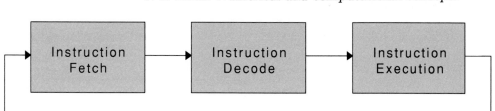

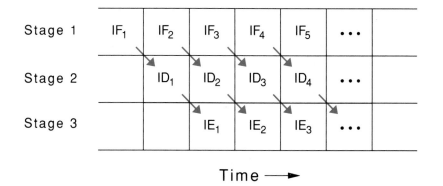

Fig. 9.11 Schematic illustration of a pipelined instruction unit (see text for further discussion).

pipelining. As one example, the functionally disjoint stages of the instruction execution cycle can be overlapped in the form of a pipeline (e.g., see Fig. 9.11). This approach ensures that the different stages of the instruction execution cycle are idle only when absolutely necessary (e.g., during an instruction branch). Through the use of sophisticated pipelining techniques at the instruction execution cycle level and below, computer designers continue to implement additional levels of concurrency within the basic von Neumann architecture (see Kogge, 1981). Average instruction execution rates in some of the more advanced machines of this class are currently approaching one instruction per machine cycle.

Hiding latency through sophisticated pipelining techniques is basically a form of parallel processing at a very low level. Beginning in the late 1960s, SIMD architectures would take this concept one step further in which a single control unit fetches and controls an instruction stream that can operate on a multitude of operands. Pipelined vector processors and parallel vector processors are the two most common architectures in this class. These machines make use of an extended set of instructions, called vector instructions, which eliminate the overhead of loop control instructions. Pipelined vector processors, typified by machines like the Cray Research CRAY–1, consist of a handful of highly sophisticated pipelined ALUs, each of which can operate independently and concurrently through another form

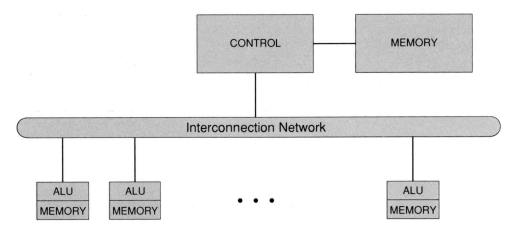

Fig. 9.12 Schematic illustration of a parallel vector processor.

of overlapping operations called chaining. The performance bottleneck of a vector processor is the speed with which data can be made available to the ALUs, i.e., they are limited by the speed and bandwidth of the main memory system. Parallel vector processors, typified by machines like the ILLIAC IV (Slotnick, 1967) and the Connection Machine (Hillis, 1985), generally consist of an array of relatively simple ALUs (or processing elements) under the control of a single instruction unit, each of which has a small local memory, and communicates with its neighbors via an interconnection network (see Fig. 9.12). The performance bottleneck using this approach is the efficiency with which interprocessor communication of operand information can occur. Otherwise, the performance of an SIMD array processor is, in principle, limited only by the size of the processing element configuration. Under optimal circumstances SIMD architectures are capable of executing the SISD equivalent of more than one instruction per clock cycle.

MIMD architectures generally consist of a homogeneous array of processing elements, each with its own program control, that can operate in an asynchronous concurrent manner (see Fig. 9.13). Each processor sees a separate instruction stream, i.e., many different instruction cycles may be active at a given instant, each independently fetching and operating on instructions and data. Main storage for these systems may be a large, directly addressable shared memory (e.g., the CRAY Y-MP), or may be distributed as local memory across the processor network (e.g., Intel Touchstone machines). In principle, the rate at which instructions are executed by machines of this class is bounded only by the size of the processor configuration, although, as in the case of SIMD architectures, the algorithmic properties of the target application can play an equally important role in limiting computational performance.

Shared-memory MIMD architectures generally employ a small number

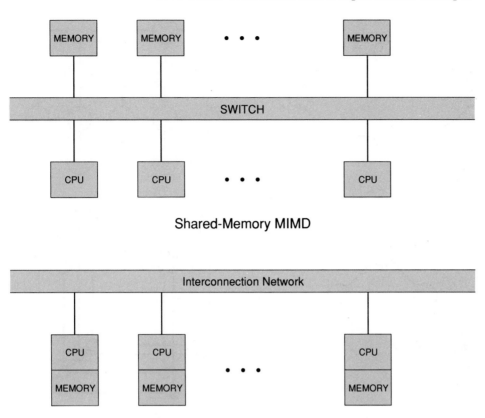

Fig. 9.13 Schematic illustration of shared-memory and distributed-memory, message passing, MIMD architectures.

of high-performance SISD or SIMD processors, that share, and to some extent communicate via, a common memory system. These processor configurations have existed commercially since the middle 1960s and were initially motivated by the need to ensure high availability of the computing system, i.e., to provide redundancy in the event of a hardware failure. Since that time, the rationale has slowly gravitated toward increasing system performance, first in terms of throughput and, more recently, in terms of turnaround. As in the case of vector processors, a major performance bottleneck is the speed with which data in the globally accessible memory can be made available to each of the processing units in the parallel configuration. A fully connected network, sometimes called a crossbar switch, is very expensive since its complexity grows with the square of the number of processors. Consequently, the configuration size of shared-memory systems is fundamentally limited by the cost and complexity of memory technology.

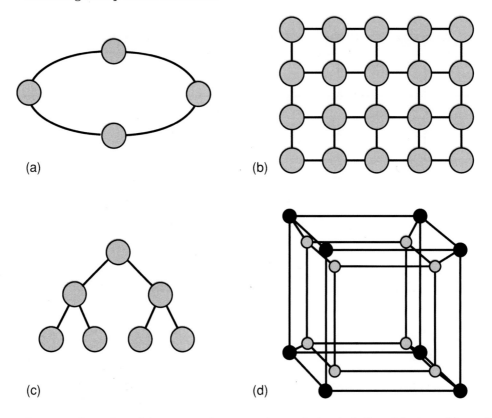

Fig. 9.14 Examples of interconnection network topologies including (a) ring, (b) mesh, (c) tree, and (d) hypercube networks.

Distributed-memory, message passing, multiprocessors are a relatively new class of MIMD architectures, a product of Very Large Scale Integration (VLSI) technology which made the wholesale reproduction of complex hardware economically feasible. At the moment, these systems employ from a few to more than one thousand homogenous SISD or SIMD processing nodes which have small private local memories, and coordinate their activities by sending messages to each other via some kind of interconnection network (see Fig. 9.14). The most familiar of the many interconnection topologies is the hypercube network (e.g., Broomwell and Heath, 1983) a design that has already enjoyed limited commercial success in both MIMD and SIMD architectures.

In general, both SIMD and MIMD architectures require a high degree of parallelism in the application program to realize even a small fraction of their performance potential, most often expressed in terms of peak floating-point performance (Hack, 1986). Many scientific applications achieve reasonably efficient utilization of present-day shared-memory supercomputer systems. The most difficult challenges arise when using distributed-memory

314

SIMD and MIMD systems for which the programming paradigm changes from managing a large global memory to managing many separate local memories. This kind of explicit memory management currently carries with it a significant programming burden associated with synchronization and communication between processors. Additionally, many existing numerical algorithms are less than ideal for efficient utilization of such architectures. At the moment, the programming and algorithmic challenges tend to increase with additional degrees of hardware parallelism.

9.3.2 Algorithmic implications

The goal of sustaining 10^{12} floating point operations per second (i.e., 1 TFLOP) will require a four order of magnitude increment in applied computational performance over existing single-processor capability. Hence, the engineering solution will require anywhere from several thousand to tens of thousands of very-high-performance processors. Accordingly, the class of scientific problems to be addressed with such a machine will minimally be required to exhibit a comparable degree of parallelism in their data structures and algorithms. Since the additional complexities of system resource management have yet to be seriously addressed, empirically developed techniques applicable to smaller processor configurations will necessarily have to scale with the size of the processor configuration.

To make the engineering task feasible, distributed-memory SIMD and MIMD architectures will provide the framework for the design of such machines. Whether these systems will be uniquely SIMD or MIMD, or whether they are homogeneous or heterogeneous collections of processors remains to be determined. The key phrase in the description of these future "massively parallel" systems, however, is that they are likely to be implemented as *distributed-memory* architectures, (even though they will ultimately have to make use of a shared-memory programming paradigm). Distributed-memory implies the need for some degree of explicit communication between processors by both the operating system and the application program. In addition to periodic synchronization requirements of MIMD architectures, communication is required to transfer data between processors as necessitated by the underlying numerical algorithm. Communication with a neighboring processor generally takes the least amount of time, and is often dominated by the startup latency associated with building and sending a message. Communicating with any but a neighboring processor requires additional time which is related to the type of interconnection network (see Fig. 9.14). As one example, the communication diameter for a hypercube network is $\log_2 p$ (where p is the number of processors in the parallel configuration). That is, to communicate information to the farthest reaches of the processor network costs $\log_2 p$ as much as the cost of a nearest neighbor communication.

The latency associated with any kind of interprocessor communication represents a computational overhead that can rapidly diminish the performance of a highly parallel processing configuration (Hack, 1989). Consequently, efficient utilization requires that the programmer find ways to hide communication latency with other useful work, and/or minimize communication requirements by using an alternative numerical algorithm. The objective of minimizing interprocessor communication is one reason why there is renewed interest in alternatives to spectral methods (which generally exhibit global communication characteristics) for the horizontal representation of the time-dependent fields in atmospheric models.

The essence of the communication problem can be seen by examining the discrete spherical harmonic transform pair introduced in (9.35) and (9.36). Whatever the parallel partitioning strategy, both the synthesis and analysis components require information about the spherical harmonic functions (the global expansion basis), which for higher-order truncations may be too large to replicate in each processor's local memory. The only alternative is to transfer the information as needed across the processor network. The inconsistency between the physical space data structures and the wavenumber space data structures can require periodic redistribution of spectral and gridpoint information within the processor network at various stages in the transform procedure. Finally, the need for reproducibility implies need for data movement so that the synthesis component can avoid the sequential bottleneck associated with an ordered sum (since $X_1 + X_2 + X_3$ does not necessarily equal $X_1 + X_3 + X_2$ due to roundoff).

The global communication problems illustrated above are not necessarily unique to the spectral method. Gridpoint based schemes, which are generally considered to be local algorithms, can exhibit similar communication requirements arising, for example, from the solution of the elliptic problem introduced by a semi-implicit time differencing algorithm. Thus, the more important characteristic of an algorithm appropriate to massively parallel architectures is its communication properties, not whether it belongs to particular class of solution methods.

In MIMD architectures, the control, synchronization, and scheduling of processors is a particularly difficult problem for both the operating system and the application program. This is not the case with SIMD architectures for which basic control and synchronization is automatically achieved through the use of a single instruction stream. Although this is a desirable programming attribute, it reduces the flexibility available to the application program. As discussed in Sec. 9.2.3, many of the components comprising complex geophysical models exhibit irregular computational characteristics. For example, the local iterative convergence properties of many physics components cannot be treated as efficiently in an SIMD architecture as they might be in an MIMD architecture. Under such circumstances, the performance of an SIMD architecture will be limited by the slowest local convergence rate, i.e., the entire processor network can be required to wait if even a single point in the domain requires additional time to achieve convergence. MIMD architectures are not as severely constrained when dealing with a situation like this since each processor is capable of executing its own instruction stream. This flexibility, however, requires additional work by the programmer to ensure that processors do not remain idle, which may not always be possible due to unavoidable imbalances in the partitioned computational load.

Numerical modeling frequently involves making compromises when selecting the most appropriate algorithms for a particular scientific problem. Historically, the computational characteristics of the target computing environment have played a relatively minor role in this selection process. The efficient utilization of massively parallel systems, however, will necessarily play a growing role in how numerical algorithms are selected. Given the additional algorithmic constraints associated with these architectures, it is important to ensure that the fundamental conservation and accuracy properties of a numerical algorithm continue to play an important role in the selection process (e.g., see Williamson et al., 1992).

9.4 Modularization and coupling

Historically, scientific progress in atmospheric modeling has been slowed by the technical difficulties of incorporating and testing physical parameterizations in different large-scale numerical models. Such problems are fundamentally linked to the fact that most codes are not modular in their design and often make use of very different data structures. Recognizing this problem, a number of scientists from major atmospheric modeling institutions have adopted a set of coding rules to make physics packages more "plug compatible" in order to facilitate their easy exchange (Kálnay, et al. 1989). Although this set of rules is specifically intended for physics packages that do not have a large number of their own prognostic variables, the conceptual approach is appropriate, and will ultimately prove necessary, for coupling complex climate system components (ocean circulation models, chemical models, biosphere models, etc.). Some of the more important concepts contained in this coding standard are:

(1) Each component should refer only to its own subprograms and a limited set of standard intrinsic functions.
(2) Each component should provide for its own initialization of static information and initial data through a single initialization entry point.
(3) All communication between packages shall take place through the argument list associated with a single unique entry point into each package.

This approach to coupling major model components is quite reasonable.

Consider, for example, how one might couple an atmospheric and ocean GCM, such as in the work of Washington and Meehl (1989). Each of these component systems is relatively self-contained, for which the physical mechanisms governing their interaction are well defined and relatively small in size. The ocean component requires information on the surface wind stress, the net surface energy balance (i.e., the sum of the sensible and latent heat fluxes with the difference between the incoming and outgoing radiation fluxes), and the net influx of fresh water (generally, precipitation minus evaporation). The atmospheric component also requires surface energy exchange information as well as the sea surface temperature. In principle, this is a small amount of information that can be computed in an interface procedure and/or passed between the two component models through the calling list. Complications arise, however, due to differences in the data structures employed in each of the models (e.g., different spatial resolution) and differences in the time scales that are appropriate to each component system (e.g., manifested in the form of different time steps). Techniques for synchronizing such component models, and resolving differences in spatial representation (e.g., see Meehl 1990a and Chapter 17) are much easier to address, however, once the physical means by which the two component systems interact have been identified and isolated in the implementation. Thus, the conceptual guidelines outlined above would seem to represent a minimum requirement for facilitating the routine coupling of complex climate system component models.

10

Atmospheric general circulation modeling

Jeffrey T. Kiehl

10.1 Introduction

10.1.1 Historical review

Attempts to mathematically model the Earth's atmosphere and its response to internal or external perturbations are at least a century old. One of the earliest mathematical "models" of climate change was that of Arrhenius (1896), in which a surface energy balance model was used to deduce that a 2°C warming would result from a doubling of atmospheric carbon dioxide. The first description of the necessary physics for modeling the atmosphere was presented by V. Bjerknes (1904). Basically two approaches to modeling the atmosphere have developed in parallel over the past century. One approach is based on energy balance requirements for the climate system, while the other approach depends on dynamical considerations of the general circulation (e.g., role of eddy transport). An early energy balance study by Emden (1913), indicated the importance of vertical and horizontal air motions in determining the thermal structure of the atmosphere. A quantitative picture of the various energy components of the global climate system appeared in the work of Dines (1917, 1929). Richardson (1922) presented a thorough analysis of the necessary equations and components to numerically simulate the air motions over short time periods (few days). Simpson (1928) adopted an energy budget approach to analyze the climate system, and was interested in the question of the stability of the climate system to perturbations. Indeed, Simpson's analysis of climate stability foreshadowed the work in the 1970s.

The most comprehensive mathematical models of the atmosphere appeared with the advent of electronic computers (see Chapter 9). In the 1950s, numerical solutions of a simplified version of the atmospheric equations of motion were carried out at Princeton by von Neumann and colleagues. The culmination of these efforts appeared in the work of Smagorinsky et al. (1965), which was the first global climate simulation employing the complete form of the equations of motion and model physics.

Modeling and parameterization

At the same time in the early 1960s, one-dimensional radiative-convective models were developed for greenhouse climate studies. The 1970s and 1980s also witnessed wide use of one-dimensional energy balance models for feedback analysis. With ever increasing computational speed, atmospheric modeling of the circulation has grown in complexity.

10.1.2 Hierarchy of atmospheric models

Understanding how the atmosphere works has been aided by a hierarchy of climate models. The conceptual formulation for a number of these models is based on averaging the equations describing the atmosphere (the so-called primitive equations, Chapter 3, Sec. 3.6.2) over either vertical and/or horizontal directions. To aid in constructing the model hierarchy certain averaging operations are employed. Performing a density weighted average of any variable, $\xi(z, \lambda, \phi, t)$, over the entire depth of the atmosphere is denoted by

$$\hat{\xi} = \int_0^\infty \xi(z, \lambda, \phi, t)\rho dz = \frac{1}{g} \int_0^{p_s} \xi(p, \lambda, \phi, t)\, dp \tag{10.1}$$

where p_s is the surface pressure. Averaging along longitude, λ, results in the *zonal* mean,

$$[\xi] = \frac{1}{2\pi} \int_0^{2\pi} \xi(z, \lambda, \phi, t)\, d\lambda \tag{10.2}$$

while area averaging over both latitude, ϕ, and longitude, λ, yields

$$\langle \xi \rangle = \frac{1}{2\pi} \int_0^{\pi/2} \int_0^{2\pi} \xi(z, \lambda, \phi, t) \cos\phi\, d\lambda d\phi \tag{10.3}$$

where the $\cos\phi$ factor is an area weight. Finally, averaging over a time interval, τ, is defined by

$$\bar{\xi} = \frac{1}{\tau} \int_0^\tau \xi(z, \lambda, \phi, t)\, dt. \tag{10.4}$$

Consider the flux form of the thermodynamic equation for the atmosphere (Chapter 3, Eq. 3.56e),

$$c_p \frac{\partial T}{\partial t} = -c_p \nabla \cdot (\mathbf{v}T) - c_p \frac{\partial(\omega T)}{\partial p} + c_p \frac{\kappa \omega T}{p} + \tilde{Q}_{rad} + \tilde{Q}_{con} \tag{10.5}$$

where T is the temperature, $\mathbf{v}$ is the horizontal wind vector, ω the vertical p-velocity (i.e. $\frac{dp}{dt}$), $\tilde{Q}_{rad}$ net radiative heating, and $\tilde{Q}_{con}$ the heating due to condensational processes. The third term on the right-hand side of (10.5) represents the conversion of kinetic to potential energy. Averaging this equation over horizontal and vertical dimensions yields

$$c_p \frac{\partial \langle \hat{T} \rangle}{\partial t} = \langle S \rangle - \langle F \rangle \tag{10.6}$$

where S is the net solar radiation absorbed by the Earth–atmosphere system

320

and F is the longwave radiation emitted to space. For a long time mean and for a stable climate to prevail

$$c_p \frac{\partial \langle \hat{T} \rangle}{\partial t} = 0 \,, \tag{10.7}$$

stating that a balance must exist between the absorbed solar and emitted longwave radiation,

$$\langle S \rangle - \langle F \rangle = 0 \,. \tag{10.8}$$

This is a zero-dimensional climate model, and has served as an important tool for understanding the climate system (see Chapter 1).

Returning to (10.5), if averaging is performed only over horizontal directions, then

$$c_p \frac{\partial \langle T \rangle}{\partial t} = \langle \tilde{Q}_{rad} \rangle + \langle \tilde{Q}_{con} \rangle \tag{10.9}$$

where all quantities are a function of p. Given expressions for $\tilde{Q}_{rad}$ and $\tilde{Q}_{con}$, this model yields a globally averaged vertical profile of T. This type of model is known as a one-dimensional radiative-convective model (RCM). Once again, if (10.5) is averaged in the vertical and in the longitudinal direction then

$$c_p \frac{\partial [\hat{T}]}{\partial t} = -\frac{c_p}{a \cos \phi} \frac{\partial ([\widehat{vT}] \cos \phi)}{\partial \phi} + [S] - [F] \tag{10.10}$$

where v is the meridional (northward) wind velocity, and a the radius of the Earth. This equation yields an expression for T as a function of latitude, and requires that the first term on the right-hand side, the poleward heat transport, be related to T. This type of model is a one-dimensional energy balance model (EBM), and has been used to study climate feedback and stability problems.

Two-dimensional models can be obtained by averaging (10.5) in longitude only

$$c_p \frac{\partial [T]}{\partial t} = -c_p \frac{1}{a \cos \phi} \frac{\partial ([vT] \cos \phi)}{\partial \phi} - c_p \frac{\partial ([\omega T])}{\partial p} + c_p \kappa \left[\frac{\omega T}{p} \right] + \left[\tilde{Q}_{rad} \right] + \left[\tilde{Q}_{con} \right] \tag{10.11}$$

which is a model for $T(\phi, p)$. This type of model has been very popular for studying chemical and dynamical issues of the middle atmosphere. It requires a relationship for the poleward and vertical heat transports in terms of the zonal mean temperature. Averaging (10.5) over pressure, but not over the horizontal domain, results in

$$c_p \frac{\partial \hat{T}}{\partial t} = -c_p \nabla \cdot (\widehat{\mathbf{v}T}) + S - F \,. \tag{10.12}$$

This is a two-dimensional energy balance model, and has been used to study paleoclimate problems. Solution of this model requires a relation between the horizontal heat transport term, $c_p \nabla \cdot (\widehat{\mathbf{v}T})$, and $\hat{T}$.

Finally, there are models that do not average (10.5) over any spatial dimensions, yielding a model for T as a function of altitude, latitude and longitude. This equation along with the momentum equations, continuity equation for dry air, the equation of state, and a continuity equation for moist air [see Chapter 3, Eq. (3.56)] result in the atmospheric *primitive equations* (PEs),

$$\frac{\partial \mathbf{v}}{\partial t} = -\mathbf{v} \cdot \nabla \mathbf{v} - \omega \frac{\partial \mathbf{v}}{\partial p} + f \mathbf{k} \times \mathbf{v} - \nabla \Phi + \mathbf{D}_M \tag{10.13}$$

$$\frac{\partial T}{\partial t} = -\mathbf{v} \cdot \nabla T + \omega \left(\frac{\kappa T}{p} - \frac{\partial T}{\partial p} \right) + \frac{\tilde{Q}_{rad}}{c_p} + \frac{\tilde{Q}_{con}}{c_p} + D_H \tag{10.14}$$

$$\frac{\partial q}{\partial t} = -\mathbf{v} \cdot \nabla q - \omega \frac{\partial q}{\partial p} + E - C + D_q \tag{10.15}$$

$$\frac{\partial \omega}{\partial p} = -\nabla \cdot \mathbf{v} \tag{10.16}$$

$$\frac{\partial \Phi}{\partial p} = -\frac{RT}{p}, \tag{10.17}$$

where $\mathbf{D}_M = (D_\lambda, D_\phi)$ are dissipation terms for momentum and D_H and D_q are diffusion terms for heat and moisture, respectively, q is the specific humidity, E and C are the rates of evaporation and condensation due to cloud processes, and Φ is the geopotential. It is common terminology in the general circulation modeling community to refer to all nondynamical processes as the *model physics*. Thus, $\tilde{Q}_{rad}$, $\tilde{Q}_{con}$, E and C are the model physics of the GCM.

In this chapter, various applications of the hierarchy of climate models are explored and, more importantly, the assumptions required to apply these models to climate problems are discussed. Focus will be on the most comprehensive model, the three-dimensional *atmospheric General Circulation Model* (AGCM). But much insight is gained from the lower dimensional climate models, and they serve as a practical introduction to atmospheric climate modeling. There are a number of very useful reviews of atmospheric models. North et al. (1981) provide an excellent review of one-dimensional energy balance models, Ramanathan and Coakley (1978) review RCMs, see Saltzman (1978) for two-dimensional tropospheric models and Brasseur and Solomon (1986) for chemistry applications. For AGCMs, see the introductory text by Washington and Parkinson (1986) and an up-to-date review of many aspects in Schlesinger (1988a).

10.2 Simple models of the atmosphere

10.2.1 Zero-dimensional models

The simplest model of the climate system, as introduced in Chapter 1, is based on the balance of incoming and outgoing radiant energy, (10.8). Consider this condition for the present climate,

$$\langle N \rangle = \langle S \rangle - \langle F \rangle \,, \tag{10.18}$$

where $\langle N \rangle$ is the net radiative flux. The globally averaged absorbed solar energy is

$$\langle S \rangle = Q(1 - \alpha_p) \tag{10.19}$$

where $Q = S_0/4$ and S_0 is the solar constant, and α_p is the top of atmosphere planetary albedo, which represents the ratio of incoming solar flux to reflected outgoing solar flux. The factor of 4 accounts for the fact that while a surface area of $4\pi a^2$ radiates longwave radiation to space (a being the radius of the Earth), an area of only πa^2 intercepts solar radiation.

For the system to be in equilibrium,

$$\langle N \rangle = 0 \,. \tag{10.20}$$

Now assume that the system has been perturbed in some manner (e.g., by an increase in atmospheric carbon dioxide), then the initial balance will be perturbed,

$$\Delta\langle N \rangle = \Delta\langle S \rangle - \Delta\langle F \rangle \tag{10.21}$$

where Δ represents the perturbation to the system. For a balance to be re-established, it is assumed that the surface temperature changes by $\Delta\langle T_s \rangle$, thus,

$$\Delta\langle T_s \rangle \left[\frac{\Delta\langle F \rangle}{\Delta\langle T_s \rangle} - \frac{\Delta\langle S \rangle}{\Delta\langle T_s \rangle} \right] = -\Delta\langle N \rangle = G \tag{10.22}$$

where G is called the *direct radiative forcing* to the climate system (e.g. for a doubling of CO_2, $G = 4$ W m^{-2}). Equation (10.22) yields a relation between the change in globally averaged surface temperature and the initial radiative perturbation to the climate system,

$$\Delta\langle T_s \rangle = \lambda G \tag{10.23}$$

where λ is the *climate sensitivity factor*, and is defined as

$$\lambda = \frac{1}{\dfrac{\Delta\langle F \rangle}{\Delta\langle T_s \rangle} - \dfrac{\Delta\langle S \rangle}{\Delta\langle T_s \rangle}} \,. \tag{10.24}$$

Note that λ determines the magnitude of the response of the climate system. Furthermore, the dependence of λ on other climate processes is evident upon expanding the longwave and shortwave factors in terms of a Taylor series,

$$\frac{\Delta\langle F \rangle}{\Delta\langle T_s \rangle} = \frac{\partial\langle F \rangle}{\partial\langle T_s \rangle} + \frac{\partial\langle F \rangle}{\partial\langle W \rangle}\frac{\Delta\langle W \rangle}{\Delta\langle T_s \rangle} + \frac{\partial\langle F \rangle}{\partial\langle A_c \rangle}\frac{\Delta\langle A_c \rangle}{\Delta\langle T_s \rangle} + \cdots \tag{10.25}$$

and,

$$-\frac{\Delta\langle S\rangle}{\Delta\langle T_s\rangle} = Q\left[\frac{\partial\langle\alpha_p\rangle}{\partial\langle T_s\rangle} + \frac{\partial\langle\alpha_p\rangle}{\partial\langle A_c\rangle}\frac{\Delta\langle A_c\rangle}{\Delta\langle T_s\rangle} + \frac{\partial\langle\alpha_p\rangle}{\partial\langle A_v\rangle}\frac{\Delta\langle A_v\rangle}{\Delta\langle T_s\rangle} + \cdots\right] \quad (10.26)$$

where $\langle W\rangle$, $\langle A_c\rangle$, and $\langle A_v\rangle$ are the globally averaged column water vapor amount, cloud amount, and vegetation coverage, respectively. Note that other terms can also affect λ.

Consider the minimal number of feedbacks. As a body's temperature increases it emits more longwave radiation, which is a statement of the Stefan–Boltzmann law,

$$\langle F\rangle = \epsilon\sigma T_s^4\,, \quad (10.27)$$

where ϵ is the emissivity of the atmosphere and is assumed to be constant. σ is the Stefan–Boltzmann constant. Neglecting all other feedbacks in (10.25) and (10.26) except for the first term in (10.25) leads to the *no-feedback* climate sensitivity factor (see also Chapter 1),

$$\lambda_{NF} = \frac{1}{\frac{\partial\langle F\rangle}{\partial\langle T_s\rangle}} = \frac{\langle T_s\rangle}{4\langle F\rangle} = 0.3 \text{ K W}^{-1}\,\text{m}^2\,. \quad (10.28)$$

A more sophisticated zero-dimensional model can be constructed by employing two features of the observed climate system. First, observational data of F and T_s can be used (North, 1975) to deduce the following empirical relation

$$F = 1.55\, T_s - 212\,. \quad (10.29)$$

Note that the climate sensitivity from this formulation of $F(T_s)$ is

$$\lambda = 0.6 \text{ K W}^{-1}\,\text{m}^2 \quad (10.30)$$

which is twice as large as the no-feedback λ. The observed F must include the dependence on factors such as amount of H_2O and clouds. Since these terms increase the climate sensitivity factor they are referred to as positive feedbacks to the climate system.

The second observed feature of the climate system concerns the planetary albedo, α_p. Due to the presence of ice on the planet, α_p should depend on T_s. In a globally averaged model, for sufficiently low temperatures, the planet should possess a relatively high albedo. For sufficiently warm temperatures the planet should possess no ice, and have a fairly low albedo. The simplest parameterization of this physical process is given by the following,

$$\alpha_p = \begin{cases} 0.3 & T_s > 270 \text{ K} \\ 0.3 - 0.4\frac{T_s-270}{40} & 230 \le T_s \le 270 \text{ K}\,, \\ 0.7 & T_s < 230 \text{ K} \end{cases} \quad (10.31)$$

where the coefficients are empirically chosen. The easiest way to solve the system (10.8), (10.29) and (10.31) is by graphical means. Figure 10.1 shows a plot of both (10.29) and (10.19) on the same figure, where

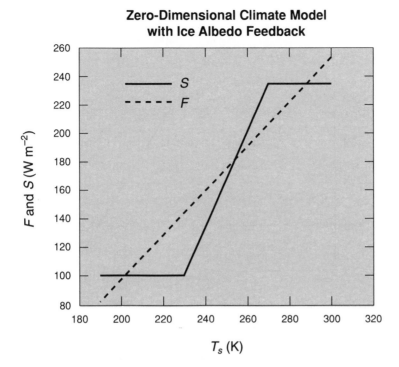

Zero-Dimensional Climate Model with Ice Albedo Feedback

Fig. 10.1 Graphical solution to the zero-dimensional climate model with ice-albedo feedback. The outgoing longwave flux, F, and absorbed solar flux, S, in W m^{-2} are plotted versus the globally averaged surface temperature, T_s (K).

$S_0 = 1{,}340$ W m^{-2}.[1] The intersection of the two lines are equilibria and it is apparent that three equilibria exist for this simple zero-dimensional climate model. The warmest equilibrium temperature is 288 K and can be considered the present climate state, the coldest equilibrium is at 202 K and represents a completely ice covered planet, while the mid-point equilibrium lies at 254 K. This simple model exhibits a phenomenon known as "multiple equilibria," i.e., the existence of a number of climate states in the system, and suggests there is the potential for the climate system to reside in one of these states.

Much insight into the climate system has been gained from this very simple zero-dimensional climate model. Perhaps the most important "parameter" to arise from this analysis has been λ, the climate sensitivity factor for it determines the magnitude of the temperature change due to an initial perturbation, G. It also indicates whether the climate equilibrium point is stable, and yields information on the adjustment time scale of the climate system (see Chapter 1).

---

[1] The best recent estimates place the solar constant higher than 1,340 W m^{-2}, but this was the value used by North et al, 1981, whose results we will cite.

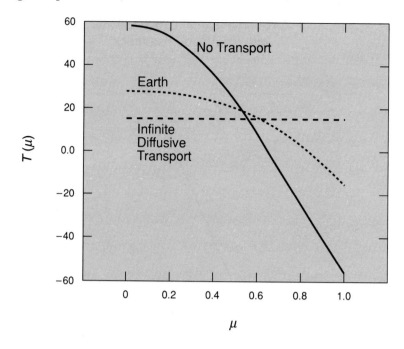

Fig. 10.2 Zonally averaged surface temperature (K) as a function of the sine of latitude, μ, observed and for cases of no horizontal heat transport, and infinite horizontal heat transport. From North et al. (1981).

10.2.2 One-dimensional models

a. The energy balance model

Extension of the zero-dimensional energy balance model to one dimension is achieved through (10.10). In order to use this as a climate model, the horizontal heat transport term must be "parameterized" in terms of T_s. Excluding the heat transport term, the energy balance relation can be used to determine T_s given the latitudinal distribution of solar insolation, $S(\phi)$. Figure 10.2 shows this radiative equilibrium temperature distribution along with the observed zonally averaged temperature. The radiatively determined temperature is much too warm in the equatorial regions and far too cold in the polar regions. Horizontal transport of heat acts to reduce this radiatively determined pole to equator temperature gradient.

The earliest conceptual model for the heat transport term assumed that the dynamics acts in a diffusive manner to remove gradients in temperature. It is now recognized that dynamical transport is fundamentally nondiffusive in nature. However, as a simple approximation for the one-dimensional energy balance climate model, the assumption is made that the correlation of v and T can be replaced by

$$[\widehat{vT}] = D' \frac{\partial T_s}{\partial \phi} \qquad (10.32)$$

326

where D' is an "eddy" diffusion coefficient and accounts for all atmospheric processes that transport heat poleward (e.g., mean meridional motions, mid-latitude storms, ...). For equilibrium conditions (i.e., $\frac{\partial T}{\partial t} = 0$), substitution of (10.32) into (10.10), with $D = c_p D'/a$ and $\mu = \sin\phi$, and replacing T_s with F through (10.29) yields

$$-\frac{\partial}{\partial\mu} a\left(1-\mu^2\right) D\frac{\partial F}{\partial\mu} + F = QS(\mu)\,a(\mu,\mu_s)\,. \tag{10.33}$$

For simplicity it is assumed the atmosphere is symmetric about the equator. The observed mean annual latitudinal distribution of insolation can be fit very well by the following function,

$$S(\mu) = 1 + s_2 P_2(\mu)\,, \tag{10.34}$$

with $s_2 = -0.482$ and

$$\int_0^1 S(\mu)\,d\mu = 1\,. \tag{10.35}$$

$a(\mu,\mu_s)$ is the co-albedo $1 - \alpha_p(\mu,\mu_s)$, and μ_s is the sine of the latitude of the ice extent in the model. The functional form of the co-albedo is taken to be

$$a(\mu,\mu_s) = \begin{cases} b_0 & \mu > \mu_s \\ a_0 + a_2 P_2(\mu) & \mu < \mu_s \end{cases} \tag{10.36}$$

where $b_0 = 0.38, a_0 = 0.697$ and $a_2 = -0.0779$, and $P_2(\mu)$ in (10.34) and (10.36) is the second-order Legendre polynomial,

$$P_2(\mu) = \frac{3\mu^2 - 1}{2}\,. \tag{10.37}$$

The final piece of information required to define this model climate is a definition for the ice line. It is assumed to occur at $T_s = 263$ K or, in terms of the outgoing longwave flux, $F_s = 195.7$ W m^{-2}. The solution of (10.33), like any differential equation, requires boundary conditions, and for these it is assumed that the poleward heat flux vanishes at the pole and the equator, i.e., the first term on the left-hand side of (10.33) is zero at $\mu = 0,1$. Solution of this model could now proceed by applying finite differencing techniques to (10.33), but there is a more elegant method of solution. The Legendre polynomials are *eigenfunctions* of the following operator,

$$\frac{d}{d\mu}\left(1-\mu^2\right)\frac{dP_n(\mu)}{d\mu} = -n(n+1)\,P_n(\mu)\,. \tag{10.38}$$

This differential operator is equivalent to the dynamical heat transport operator in (10.33). If the solution $F(\mu)$ is expanded in a series of Legendre polynomials, then the differential operator for the heat transport can simply be replaced with a series of these polynomials. This method of solution is known as a *spectral technique* and is used to solve more complex climate models, such as GCMs, see Chapter 9. The fundamental procedure is to consider the differential operator appearing in the equations describing the

327

climate, then identifying whether there are certain functions, e.g., Legendre polynomials, that are eigenfunctions of the differential operator, if so then the solution variables are expanded in a series of these functions.

To proceed, the solution variable F is represented as

$$F(\mu) = \sum_n F_n P_n(\mu) \tag{10.39}$$

where the sum is only taken over even integers for a symmetric annual mean model. Substitution of this series into (10.33), along with (10.38), multiplying both sides by $P_m(\mu)$ and then integrating μ from 0 to 1 yields

$$[n(n+1)D + 1] F_n = QH_n(\mu_s) \tag{10.40}$$

where

$$H_n(\mu_s) = (2n+1) \int_0^1 S(\mu) a(\mu, \mu_s) P_n(\mu) d\mu. \tag{10.41}$$

To obtain (10.41) the orthogonality relation of the Legendre polynomials has been employed, namely,

$$\int_0^1 P_n(\mu) P_m(\mu) d\mu = \begin{cases} \frac{1}{2n+1} & n = m \\ 0 & n \neq m \end{cases}. \tag{10.42}$$

Equation (10.40) is the solution of the one-dimensional climate model as it yields F_n for $n = 0, 2, 4, 6, \ldots, N$, and substitution of these coefficients into (10.39) gives $F(\mu)$.

The lowest order model that includes some latitudinal structure is for $N = 2$. First, consider the solution of (10.33) for $n = 0$,

$$F_0 = Q \int_0^1 S(\mu) a(\mu, \mu_s) d\mu, \tag{10.43}$$

which is a statement of global energy balance. The $n = 2$ solution is

$$(6D + 1) F_2 = 5Q \int_0^1 S(\mu) a(\mu, \mu_s) P_2(\mu) d\mu \tag{10.44}$$

where F_2 is related to the equator to pole temperature gradient. For the present climate system the ice line is at approximately $72°$, or $\mu_s = 0.95$. As $Q = 335$ W m^{-2} and using the above expressions for $S(\mu)$ and $a(\mu, \mu_s)$ in (10.43),

$$F_0 = 233 \text{ W m}^{-2} \tag{10.45}$$

where $H_0(\mu_s) = 0.698$ has been used, which implies a globally averaged temperature of $T_s = 288$ K. Now the definition of $F(\mu)$ at the ice line is

$$F(\mu_s) = F_0 + F_2 P_2(\mu_s) \tag{10.46}$$

and, using $F(\mu_s) = 195.7$ W m^{-2}, implies that $F_2 = -44.02$ W m^{-2}, or $T_2 = -28.4$. Finally, the value of D can be determined from (10.44), giving $D = 0.382$.

Figure 10.3 shows a comparison of the two mode solution with observed temperatures, as well as a four mode solution. The two mode model agrees

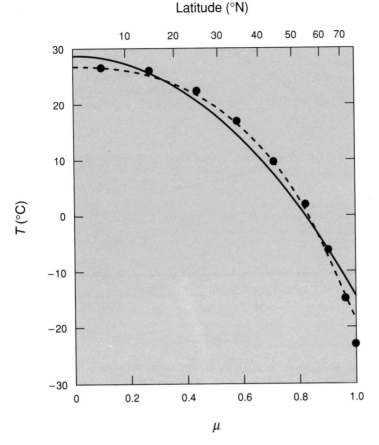

Fig. 10.3 Solution to the one-dimensional energy balance model. Surface temperature versus sine of latitude, μ. Circles denote observations. The solid line is the two mode solution, while dashed line denotes the four mode solution. From North (1975).

fairly well with the observations, but fails to capture the "flat" structure in T_s near the equator. Of course this agreement has been obtained by a judicious choice of the co-albedo coefficients. These results show that the observed latitudinal dependence of T_s can be "fit" with a series of Legendre polynomials. What is more useful is to employ (10.33) for *predictive* studies of the sensitivity of T_s to changes in factors such as Q, or α_p. Solution of (10.33) for Q as a function of μ_s is shown in Fig. 10.4. This figure is similar to the zero-dimension model results, in that for the present climate $(Q/Q_0 = 1)$ there are three equilibria. A stability analysis of these states shows that the $\mu_s = 0.95$ and 0 solutions are stable, while the intermediate state at $\mu_s = 0.4$ is unstable.

The one-dimensional EBM has been very popular in climate modeling, and has served as a tool to investigate dependency on poleward heat transport mechanisms, ice age climates and present coupled ocean–atmosphere problems (North et al., 1981, Harvey, 1988).

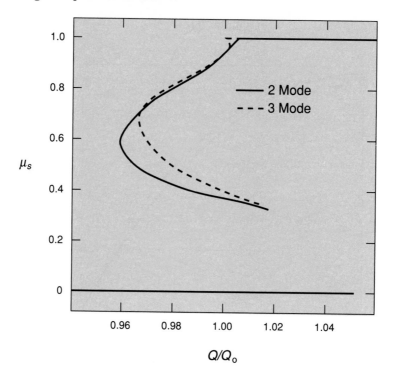

Fig. 10.4 Position of the ice sheet, denoted by the sine of the latitude of the ice sheet μ_s, as a function of the fractional change in the incident solar flux (Q/Q_0). From North (1975).

b.   The radiative-convective model

Averaging the thermodynamic equation over all horizontal dimensions leads to (10.9). Since the equilibrium temperature profile depends solely on a balance between net radiative heating and convective heating, the model is called a one-dimensional *radiative-convective model* (RCM). To determine the equilibrium vertical profile, $T(z)$, computationally affordable methods for evaluating the radiative and convective heating as a function of altitude must be employed. Parameterizations of $\tilde{Q}_{rad}$ and $\tilde{Q}_{con}$ developed for RCMs can also be used in more complex models, such as GCMs. Methods to determine the net radiative heating rate are discussed in Sec. 10.3.1. Comprehensive models of convective heating have been developed requiring information that is available only from three-dimensional models. However, the simplest technique for evaluating $\tilde{Q}_{con}$, namely a "convective adjustment" technique, is used in most RCMs. This method along with more complex convection schemes is discussed in Sec. 10.3.2.

The surface temperature is obtained by assuming net energy balance (see Eq. 17.1). Adopting a radiation and convection scheme, (10.9) is solved iteratively to obtain $T(z)$. Since these models generally employ sophisticated

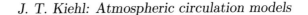

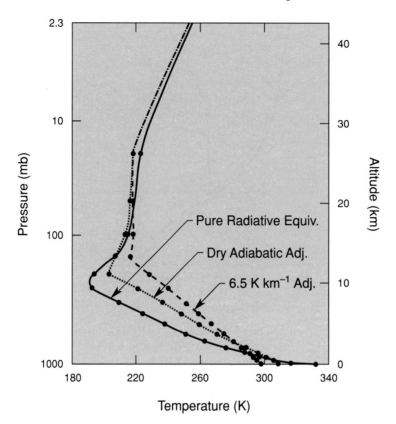

Fig. 10.5 Vertical temperature profile obtained from the one-dimensional radiative-convective model of Manabe and Strickler (1964). Shown are temperature profiles for the pure radiative equilibrium case, convectively adjusting to a dry adiabat and a lapse rate of 6.5 K km^{-1}.

radiation schemes, they have been used extensively in studies of greenhouse gases and climate change. The RCM is also useful in considering the effects of clouds and aerosols on the climate system. A typical radiative-convective profile is shown in Fig. 10.5, along with a pure radiative equilibrium profile. In general, the radiative-convective temperature profile is in good agreement with the globally averaged thermal structure, with a tropopause near 15 km. Above 15 km the temperature steadily increases. This stabilization is due to solar heating of ozone, the importance of which is shown in Fig. 10.6, where ozone heating has been removed from the RCM.

Sensitivity of the surface temperature to low, mid-level and high cloud amount is shown in Fig. 10.7. Note that increases in low and mid-level clouds decrease the surface temperature, while increases in high cloud actually increase the surface temperature. For the low and middle clouds, the increase in albedo due to increased cloud amount dominates over the longwave warming effect of the cloud. For high cloud cover, the longwave warming effect of the clouds dominates the albedo effect.

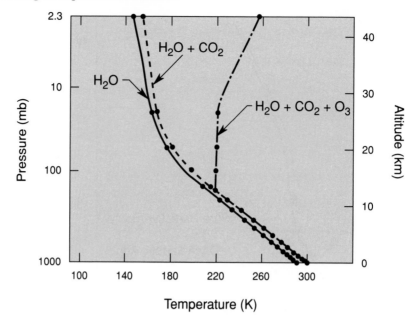

Fig. 10.6 Vertical profile of temperature from Manabe and Strickler (1964) for the case of a pure H_2O, $H_2O + CO_2$, and $H_2O + CO_2 + O_3$ atmosphere.

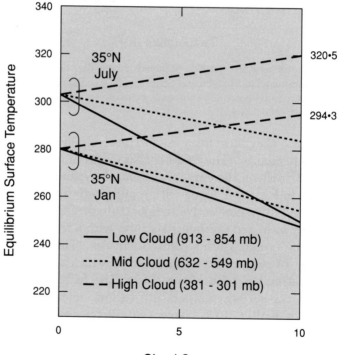

Fig. 10.7 Surface temperature obtained from a one-dimensional radiative-convective model versus fractional cloud cover. Results are shown for low, mid-level and high cloud cover. From Stephens and Webster (1981).

10.2.3 *Two-dimensional climate models*

a. Zonally averaged models

The averaging over longitude is usually performed in two steps. First, all prognostic variables, u, v, ω, T and Φ, in the primitive equations are expressed in terms of a zonal mean component and an "eddy" term, which is just the deviation of any variable from the zonal mean,

$$\xi = [\xi] + \xi^* . \tag{10.47}$$

Second, the equations are zonally averaged, resulting in five equations for $[u], [v], [\omega], [T]$ and $[\Phi]$. For example, the prognostic equation for the zonal mean wind is

$$\frac{\partial [u]}{\partial t} + [v] \frac{1}{a \cos \phi} \frac{\partial ([u] \cos \phi)}{\partial \phi} - f[v] + [\omega] \frac{\partial [u]}{\partial p} = [D_M] + E_u \tag{10.48}$$

where,

$$E_u = -\frac{1}{a \cos^2 \phi} \frac{\partial ([v^* u^*] \cos^2 \phi)}{\partial \phi} - \frac{\partial ([\omega^* u^*])}{\partial p} . \tag{10.49}$$

The E_u term represents the effects of "eddies" on the mean zonal wind. Assuming $E_u = 0$ results in a system of equations for an axially symmetric atmospheric climate. The axially symmetric model has been used for studying the Hadley circulation (Lindzen and Hou, 1988; Farrell, 1991). Alternatively, the eddy fluxes, i.e., $[v^* u^*]$ and $[\omega^* u^*]$, can be specified from observational data (Schneider, 1984), or can be related to the mean variables through an eddy diffusion assumption thus allowing for dynamical feedback. All of these approaches have limitations. Fixing the eddy flux terms does not allow the eddies to change if the mean flow (i.e., $[u]$), changes, while relating the eddy terms to the mean flow does not satisfactorily model many of the features of the general circulation. Finally, the parameterizations may constrain the response of the model to perturbations. These deficiencies can be addressed only by using a more comprehensive climate model.

The zonally averaged models have also been extensively used for middle atmosphere studies (e.g., Garcia and Solomon, 1983; Ko et al., 1985). Because their computational cost is small compared to three-dimensional models, they can be coupled to chemistry and radiation models to study stratospheric chemical–radiative–dynamical interactions. Early middle atmosphere models fixed the eddy terms, while latter models used an eddy diffusion approach. Recently, more sophisticated dynamical parameterizations have been developed to relate the eddy terms to the mean flow (Hitchman and Brasseur, 1988).

b. The energy balance model

Averaging the thermodynamic equation over the depth of the atmosphere results in a two-dimensional, (λ, ϕ), energy balance model.

North et al. (1983) used this model to study the climate of the ice ages by replacing the two-dimensional dynamical heat transport term with a two-dimensional diffusion operator,

$$\nabla \cdot (\mathbf{v}T) = \kappa \nabla^2 \hat{T}. \tag{10.50}$$

For the two-dimensional EBM, the temperature is expanded in terms of *spherical harmonics*

$$\hat{T}(\lambda, \phi, t) = \sum_{l=0}^{L} \sum_{m=-l}^{+l} \hat{T}_{lm}(t) Y_l^m(\lambda, \phi). \tag{10.51}$$

The spherical harmonics are the eigenfunctions of the ∇^2 operator on the sphere. They are a product of the associated Legendre polynomials $P_l^m(\phi)$ and $e^{im\lambda}$. L defines the horizontal resolution of the climate model, larger L leading to higher spatial resolution. The model is solved using the spectral technique as discussed for the one-dimensional EBM and in more detail in Chapter 9.

10.3 Atmospheric general circulation models (AGCMs)

Zero-, one- and two-dimensional climate models present a qualitative picture of how the atmospheric climate system works. However, these models either neglect various processes that are known to be important in the atmosphere, or they use simple mathematical representations for these atmospheric processes. This parametric representation of these processes is called *parameterization*. For example, in the energy balance models the approximation of horizontal heat transport by a diffusive process is a parameterization of large-scale eddies in the atmosphere. To accurately account for the general motions of the atmosphere requires the solution of the complete set of equations (10.13) through (10.17). The solution of these equations on the sphere, given realistic boundary conditions, defines the AGCM.

Implementation of an AGCM requires: a numerical solution technique, algorithms for the various physical parameterizations, and boundary data sets for predetermined vertical and horizontal resolutions. The solution of the system of primitive equations and parameterizations proceeds as outlined in Fig. 10.8. Assuming initial data are available for the prognostic variables, the model calculates initial fluxes for use in the planetary boundary layer (PBL) and surface components of the model. These, along with the thermodynamic and moisture profiles at each gridpoint, are used to test whether the atmospheric column is stable or unstable (A). If unstable, a convection parameterization is used to determine the convective heating and moistening terms. Otherwise, if saturated, the stable condensation process is invoked. Based on the type of condensation process, cloud fractions are assigned to model layers. Condensational heating and cloud

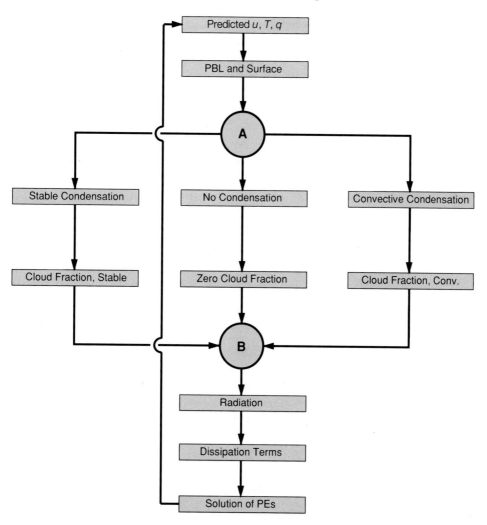

Fig. 10.8 Diagram of the procedures employed in an AGCM.

amounts are stored for further use (B). Radiative fluxes and heating rates are then calculated based on the thermal, moisture and cloud profiles in the atmosphere. Mechanical dissipation terms are then determined. At this point all forcing and dissipative terms for the PEs are available, and a numerical solution technique (see Chapter 9) is applied to obtain new values for the prognostic variables. Note that the order displayed in Fig. 10.8 is not unique. The model physics could be calculated in a different sequence.

The time length of the integration, or number of iterations of the above procedure, is usually determined by the problem under consideration, which in turn is governed by time scales inherent to the various physical processes. For example, it is known that the tropical tropopause region has a long time scale ($\sim$80–100 days) due to radiative processes. Thus, any problem that requires detailed knowledge of these temperatures will require model simulations in excess of 100 days. Problems involving soil moisture changes possess long time scales, and require multiyear integrations. Another requirement for model integrations is the need for sufficiently long time series to address climate variability and signal-to-noise questions (see Chapter 1). Each climate modeler must develop her or his own criteria for length of integration and spatial resolution.

Ideally, the only fixed boundary conditions that should be specified in a climate system model are the distribution and height of the continental terrain. All other factors such as sea surface temperature, surface albedo (which is determined by the surface vegetation), distribution of sea ice, ozone mixing ratio, etc., would be predicted by other components of the climate system model and would be available to the atmospheric model. At present few of these components have been coupled to an AGCM. This is the challenge of climate system modeling. Many atmospheric models must prescribe these other factors. The data prescription is typically based on climatological observations. A summary of the prescribed boundary data used by many AGCMs is presented in Table 10.1.

Although AGCMs do not require the parameterization of the dynamical motions of the atmosphere, they cannot resolve all the physical processes that occur on spatial scales that are smaller than the resolution employed to solve the model equations. In AGCMs a finite resolved horizontal scale must be specified (e.g., $\Delta\lambda = 2°$ and $\Delta\phi = 2°$), as well as a vertical scale (typically $\Delta z \sim 2\,\text{km}$). Any physical process, such as radiation or cumulus convection, that occurs on scales smaller than this resolution, must be represented via parameterization. Virtually all physical processes operating in the atmosphere require parameterization. The only model variables available for the parameterizations are the large-scale fields predicted by the model. Relating the subgrid processes to these large-scale values usually depends on knowledge of the fundamental physics involved in the process. The following sections describe a number of the physical parameterizations employed by AGCMs. The continued success of atmospheric modeling, to a large degree, depends on improving these various parameterizations.

Table 10.1. Boundary data sets required for an atmospheric general circulation model.

Boundary data	Parameterization	Dimensions
Sea surface temperatures	Radiation and PBL	latitude, longitude, time
Surface type (land, ocean...)	Surface temperature	latitude, longitude
Surface roughness length	Surface	latitude, longitude
Land hydrology	Surface	latitude, longitude
Surface albedos	Radiation	latitude, longitude
Ozone mixing ratio	Radiation	latitude, longitude
Orography	Dynamics	latitude, longitude
Subgrid variance of orography	Gravity wave drag	latitude, longitude

10.3.1   *Parameterization of atmospheric radiation*

General circulation models require the following radiative quantities: the net (upward and downward) radiative fluxes at the top of the atmosphere and surface, and internal atmospheric heating rates. The surface fluxes are components of the surface energy balance and contribute to the determination of the surface temperature (see 17.1). The top of atmosphere fluxes are required for determining the overall energy budget of the surface–atmosphere system. The net radiative heating profile is needed for the thermodynamic tendency calculation (10.14) and is the sum of radiative heating by visible and near-infrared wavelength radiation and cooling by longwave radiation

$$\tilde{Q}_{rad} = \tilde{Q}_s + \tilde{Q}_l. \tag{10.52}$$

The heating terms are proportional to the divergence of the net radiative flux (Chapter 3, Sec. 3.4.1) and in terms of K s^{-1},

$$Q_s = \frac{\tilde{Q}_s}{c_p} = \frac{g}{c_p}\frac{dF_s^{net}}{dp} \quad \text{and} \quad Q_l = \frac{\tilde{Q}_l}{c_p} = \frac{g}{c_p}\frac{dF_l^{net}}{dp} \tag{10.53}$$

where

$$F_s^{net} = F_s^{\downarrow} - F_s^{\uparrow} \quad \text{and} \quad F_l^{net} = F_l^{\uparrow} - F_l^{\downarrow}. \tag{10.54}$$

$F^{\uparrow}$ and $F^{\downarrow}$ are the up and down radiative fluxes.

The object of a radiative transfer parameterization is to calculate F_s^{net} and F_l^{net} for clear and/or cloudy conditions at each gridpoint of the model domain. The transfer of radiant energy through the atmosphere is described by the *radiative transfer equation* (Chapter 3, Sec. 3.4). The transfer equation is a conservation equation for radiant energy (or number of photons). The upward and downward radiative fluxes are hemispheric averages of the radiant energy in these two directions. Photons can be either absorbed or scattered as they propagate through the atmosphere. The processes that need to be parameterized are essentially of molecular scale for gaseous absorption and micron scale for particulate scattering. Since the source of radiation is quite different for solar (i.e., the Sun) and longwave radiation (i.e., thermal emission from the Earth), these two processes are considered separately.

a.   *Solar radiation*

Molecular absorption occurs when a photon excites an energy mode of the molecule. The modes can be an electronic transition, molecular rotation, molecular vibration or a combination of these. The necessary energy for an excitation is discrete. Hence, the absorption spectrum for a molecule is composed of a series of discrete absorption lines. Owing to molecular collisions and translations these discrete absorption lines are

337

smeared out. The discrete absorption for a given wavenumber ν is defined in terms of an absorption coefficient, k_ν and amount of absorber, w, as

$$A_\nu(w) = 1 - e^{-k_\nu w} . \tag{10.55}$$

The absorption coefficient is a function of temperature and pressure. The contribution of each narrow absorption line must be accounted for to model the transfer of radiation. For either the longwave or solar spectral region, there are tens of thousands of such lines arising from all the absorbing gases in the atmosphere. Thus, to include all lines in a parameterization of absorption would require an explicit summing over all lines at each model level and horizontal location. These types of calculations can be performed on present day supercomputers and are called line-by-line models. However, this approach is far too cumbersome for climate modeling purposes.

To account for molecular absorption for climate models an implicit integration over the absorption lines must be carried out,

$$a(w) = \frac{\int_{\nu_1}^{\nu_2} S_\nu^0 A_\nu(w) d\nu}{\int_{\nu_1}^{\nu_2} S_\nu^0 d\nu} \tag{10.56}$$

where S_ν^0 is the solar flux for a given wavenumber interval and $a(w)$ is the fractional absorption of solar flux. This process leads to functional relations for absorption for fairly broad spectral intervals. An example of this functional relation for ozone absorption is shown in Fig. 10.9 for the visible and ultraviolet spectral regions. This function assumes a homogeneous (constant temperature and pressure) atmosphere and fixed absorber amount. The real atmosphere is neither homogeneous in temperature or pressure. Temperatures vary from near 300 K at the surface to 180 K in the mesosphere, while pressure decreases by six orders of magnitude from the surface to the mesopause. Temperature and pressure effects of the fractional absorption arise through k_ν defined as

$$k_\nu = s_\nu(T) \, f_\nu(p, T) , \tag{10.57}$$

where $s_\nu(T)$ is the line strength, which is determined from quantum mechanics, and $f_\nu(p, T)$ is the line shape factor.

Methods to account for the inhomogeneous structure of the absorption path have been based on factoring all temperature and pressure dependence into an effective gas amount and replacing w in (10.56) with the following,

$$\tilde{w} = \int_{z_1}^{z_2} \left(\frac{p}{p_0}\right)^n \left(\frac{T_0}{T}\right)^m \rho_a dz , \tag{10.58}$$

where n and m are empirical constants. This approach has been used, in particular, for absorption by water vapor and carbon dioxide. Temperature and pressure effects for ozone absorption are much smaller than for the near-infrared absorbers.

Molecular scattering occurs in the atmosphere and is called Rayleigh scattering. This type of scattering is fairly easy to account for in climate

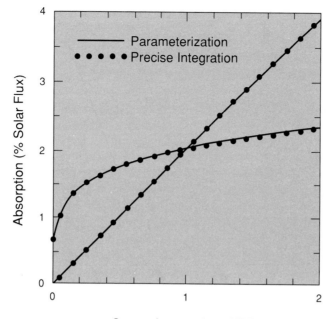

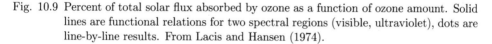

Fig. 10.9 Percent of total solar flux absorbed by ozone as a function of ozone amount. Solid
  lines are functional relations for two spectral regions (visible, ultraviolet), dots are
  line-by-line results. From Lacis and Hansen (1974).

models (e.g., Lacis and Hansen, 1974). Scattering of radiant energy by
particles (e.g., cloud droplets or aerosol particles) is also very important
to the transfer of solar energy through the atmosphere. The scattering
process is determined by a function describing the angular distribution
of scattered photons (the phase function), and the efficiency with which
photons are absorbed by a single scattering event. If there are particles of
various sizes present in the atmosphere, an integration over the distribution
of particle sizes must be carried out. As with molecular absorption,
detailed calculations that account for each of these processes can be
performed. However, for climate modeling applications, these techniques
are too computationally expensive.

One approach for including scattering in a climate model is called the *ray
tracing* method. It is assumed that solar radiation can be represented by
beams propagating through the atmosphere. An example of this method is
illustrated in Fig. 10.10. Consider the case where there is a single cloud layer
overlying the Earth's surface. Assume that the albedo of the cloud top and
base is α_c, the transmissivity of the cloud is t_c and that the surface albedo
is α_s. Also, assume that the atmosphere above and below the cloud does
not absorb any solar radiation. Part of the initial beam of solar radiation
from the top of the atmosphere will be reflected back to space, α_c, while
the remaining fraction will pass through the cloud and be reflected from the

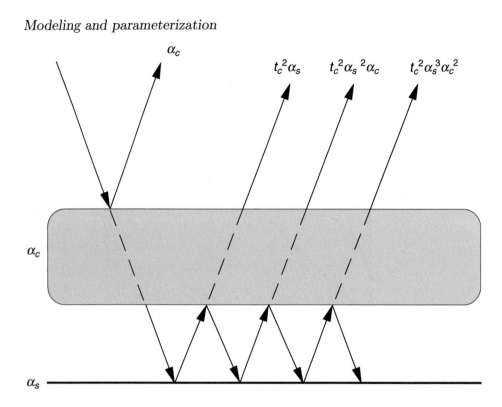

Fig. 10.10 Diagram of a solar beam reflecting from a system composed of a single cloud layer and surface.

surface, $t_c\,\alpha_s$. This beam is then partially reflected and transmitted through the cloud. The beam can be followed through a number of reflections and transmissions. Considering the first few terms in this scattering process yields a planetary albedo of,

$$\alpha_p = \alpha_c + t_c^2\alpha_s + t_c^2\alpha_s^2\alpha_c + t_c^2\alpha_s^3\alpha_c^2 + t_c^2\alpha_s^4\alpha_c^3 + \ldots \qquad (10.59)$$

which can be factored as

$$\alpha_p = \alpha_c + t_c^2\alpha_s[1 + \alpha_s\alpha_c + (\alpha_s\alpha_c)^2 + (\alpha_s\alpha_c)^3 + \ldots] \qquad (10.60)$$

or,

$$\alpha_p = \alpha_c + \frac{t_c^2\alpha_s}{1 - \alpha_s\alpha_c}. \qquad (10.61)$$

The amount of shortwave radiation absorbed by the cloud–surface system is just

$$SW_{abs} = Q(1 - \alpha_p) \qquad (10.62)$$

while the amount absorbed for clear sky conditions is,

$$SW_{abs-clear} = Q(1 - \alpha_s) \qquad (10.63)$$

where Q is the amount of solar flux incident at the top of the Earth's atmosphere. Assuming this is 340 W m^{-2}, the two absorption terms as

**Solar Radiation Absorbed by
Atmosphere-Surface System**

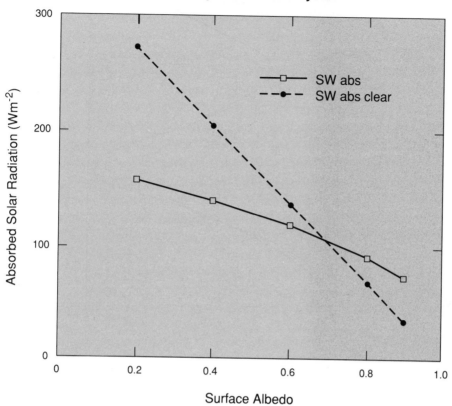

Fig. 10.11 Absorbed solar flux for the case in Fig. 10.10. Absorbed flux is shown for the case of cloud (solid) and clear sky (dashed) as a function of surface albedo.

a function of surface albedo are shown in Fig. 10.11. Note that for a highly reflecting surface the combined albedo of cloud and surface is actually lower than that of clear sky. Thus, the presence of cloud over highly reflective surfaces actually increases the amount of solar radiation absorbed by the system. This apparently occurs when stratus clouds overlay snow and ice, such as in the northern or southern polar regions. The ray tracing method is easy to construct for simple cases, such as a single cloud layer. For multiple cloud layer cases it quickly becomes burdensome to evaluate. A more sophisticated approach is to obtain closed form solutions to the original radiative transfer equation. Solutions can be obtained if it is assumed that the radiation can be divided into a constant upward and downward stream of radiant energy (Meador and Weavor, 1980). These so-called "two-stream" methods result in simple relations for the reflected and transmitted radiation by an atmospheric layer. To combine multiple layers for the entire atmosphere requires a procedure known as the "adding method".

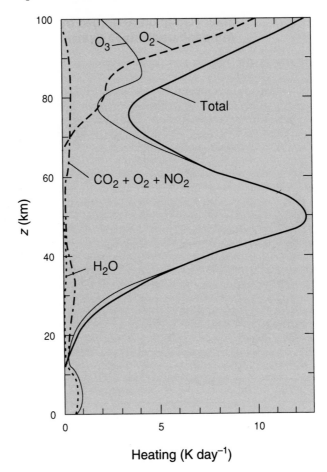

Fig. 10.12 Vertical profile of radiative solar heating K day^{-1} for various atmospheric constituents. From London (1980).

When these methods are applied to a model atmosphere, fluxes are obtained for the interface model layers. The fluxes can then be used in (10.53) to determine Q_s, as shown in Fig. 10.12. As previously mentioned, water vapor heating dominates the troposphere with maximum heating of 2.0 K day^{-1} near 700 mb. Ozone heating dominates in the stratosphere with peak heating at the stratopause of around 15 K day^{-1}. Carbon dioxide heating is only a few tenths of a degree per day throughout the atmosphere. The presence of clouds in an atmosphere can severely alter this clear sky heating distribution, as shown in Fig. 10.13 for high and low cloud cases. The low cloud heating peaks near cloud top at 4 K day^{-1} while the high cloud heating is over 6 K day^{-1}.

Heating Rate Comparison LBL and δ - Eddington

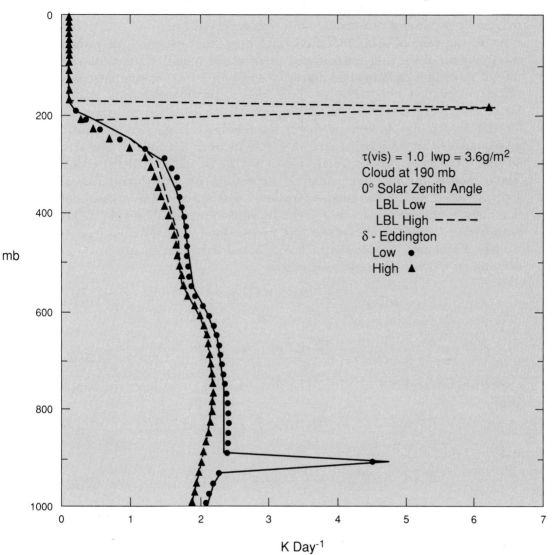

Fig. 10.13 Vertical profile of solar heating in the troposphere for the case of a single cloud layer in the lower troposphere (solid) and in the upper troposphere (dashed) [Line-by-line (LBL) results from Ramaswamy and Friedenreich, 1991]. Dots and triangles for the same respective cases but using the two-stream δ-Eddington approach. From Briegleb (1992).

b. Longwave radiation

The most important absorber of longwave radiation in the Earth's atmosphere is water vapor. The second most important gas in determining Q_l is carbon dioxide, which has dominant effects in the upper stratosphere.

The third most important absorber for longwave radiative cooling is ozone. Gases such as CH_4, N_2O and CFCs (the other *trace gases*) are also radiatively significant and should be included in AGCMs.

As in the case of solar radiation, including the absorption of radiant energy by all of the above mentioned gases would require integration over tens of thousands of individual narrow absorption lines. For computational efficiency, implicit integration of these lines is provided by *band models.* Band models are based on the assumption that a closed form expression for the absorption A_ν can be found for relatively large (10–100 cm^{-1}) wavenumber interval widths, $\Delta\nu$. For wavenumber intervals of less than 50 cm^{-1}, the most popular band models are the so-called random models. These models assume some analytic expression for the distribution of spectral lines, and a line shape function. For specific line distribution functions, exact integration over ν can be carried out to obtain $A_{\Delta\nu}$. The most frequently used models of this sort are the Goody and Malkmus random models (Kiehl and Ramanathan, 1983). Broad band models are based on defining an *absorptivity* and *emissivity* as

$$\alpha(p, p') = \frac{\int_0^\infty A_\nu(p, p') \frac{d\pi B_\nu(p')}{dT} \, d\nu}{\frac{d\pi B}{dT}} \tag{10.64}$$

and

$$\epsilon(p, p') = \frac{\int_0^\infty A_\nu(p, p') \pi B_\nu(p') \, d\nu}{\pi B}, \tag{10.65}$$

respectively, where $\pi B = \sigma T^4$. The flux equations now have the simple form

$$F^-(p) = \sigma T^4(p_0) \, \epsilon(p, p_0) + \int_{p_0}^p \alpha(p, p') \, d\sigma T^4(p') \tag{10.66}$$

and

$$F^+(p) = \sigma T^4(p_s) - \int_p^{p_s} \alpha(p, p') \, d\sigma T^4(p'). \tag{10.67}$$

The advantage of (10.66) and (10.67) is that integration over ν has been implicitly carried out. A number of methods exist to formulate accurate "closed form" functional fits of α and ϵ to benchmark line calculations.

Clear-sky cooling longwave rates for a model tropical atmosphere profile are shown in Fig. 10.14. As previously suggested, cooling by H_2O dominates the troposphere with peak cooling near the surface of –3 K day^{-1}, while CO_2 and O_3 are most important for stratospheric cooling. The large H_2O cooling near the surface is not due to the line absorption process considered so far. An additional form of molecular absorption can occur and is called *continuum absorption.* The k_ν absorption coefficient for continuum absorption exhibits a fairly smooth variation with ν, and the path length for the continuum is linearly dependent on the partial pressure of water vapor. Thus, for moist environments (e.g., the near surface tropical atmosphere) the absorption and re-emission of radiation by the H_2O continuum in the

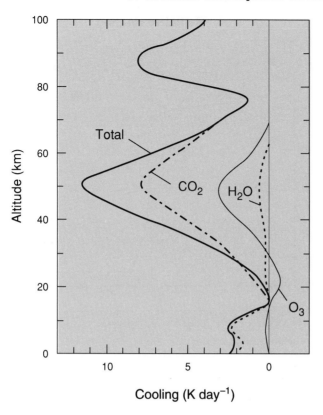

Fig. 10.14 Vertical profile of radiative longwave heating for various atmospheric constituents. From London (1980).

8–12 μm spectral region plays a significant role in atmospheric longwave radiative transfer, and is responsible for the peak near-surface cooling in Fig. 10.14.

The inclusion of clouds in (10.66) and (10.67) has traditionally been accomplished by writing the total flux in a layer as,

$$F = F_{clr}(1 - A_c) + F_o A_c \qquad (10.68)$$

where A_c is the fractional cloud amount, F_{clr} is the contribution from clear sky and F_o is the contribution from the overcast sky. F_o is evaluated from (10.66) and (10.67) by replacing the surface and top boundary temperatures, $T(p_s)$ and $T(p_0)$, with cloud top and cloud base temperatures. Radiative fluxes from clouds above and below the region of interest are computed assuming these clouds are randomly overlapped. Equation (10.68) and the random overlap assumption are very simplistic compared to real cloud–radiation interactions; nevertheless, they are extensively used in climate models. Results for a case where low and high clouds are present in the tropical atmosphere are shown in Fig. 10.15. The first feature to note is that longwave processes tend to warm cloud base, while they cool cloud

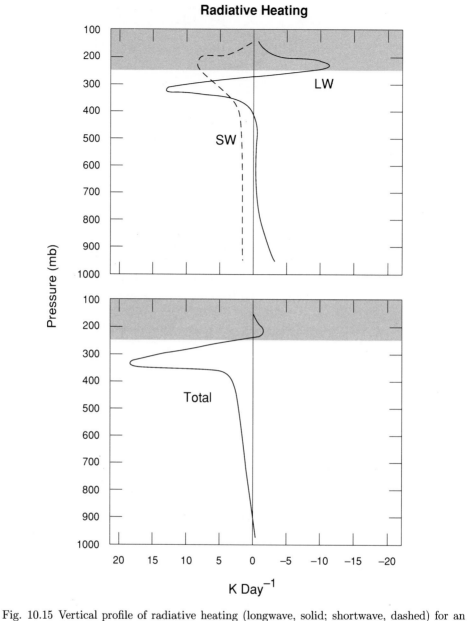

Fig. 10.15 Vertical profile of radiative heating (longwave, solid; shortwave, dashed) for an upper tropospheric cloud. From Webster and Stephens (1980).

tops. This results from the net convergence of radiation from the warm surface into the cold cloud base, while the cloud top essentially loses radiant energy to space, which cools the top. This same process has a marked dependence on cloud altitude. Higher clouds display more marked cloud top cooling and base warming (~ 10 K day^{-1}).

10.3.2 Convective processes in atmospheric models

As discussed in Chapter 3, Sec. 3.3.3, the thermal structure resulting from pure radiative equilibrium,

$$c_p \frac{\partial \langle T \rangle}{\partial t} = \tilde{Q}_{rad} \, , \tag{10.69}$$

is buoyantly unstable to small vertical displacements of air. Radiation in general destabilizes the atmospheric thermal structure, which results in the onset of convective motions. In a completely dry atmosphere the equilibrium thermal profile would follow a dry adiabat with lapse rate, Γ_d. But in most places in the Earth's atmosphere, the presence of moisture alters the dry convective process.

a. Cumulus parameterization

As moist parcels begin to rise, cloud formation processes take place that considerably complicate the convective overturning of air. The physical processes and scales of vertical motion involved in the moist convective process range from microns to a few kilometers. This range in scales is much smaller than the spatial scale resolved in any climate model. Hence, the effects of convective motions on the evolution of the thermal, moisture and even dynamical properties in the model atmosphere must be parameterized. Since the meteorological term for convective clouds is "cumulus" (Chapter 3, Sec. 3.5), the problem is called *cumulus parameterization* in the atmospheric modeling community. These parameterizations are, in general, for deep convective clouds that occur in regions of moisture convergence. Over the years, several deep convective parameterizations have been developed for AGCMs. Like any parameterization, the unresolved properties must be linked to the large-scale model variables. This "linking" process is known in the convective parameterization field as the "closure assumption."

To see how the parameterization of moist convection enters the atmospheric climate models it is best to start with the prognostic equations for temperature and moisture, (10.14) and (10.15). Assume that these equations are applied to an area that contains a number of convective clouds of varying height and area. Further assume that the atmospheric variables can be separated as

$$\xi = \bar{\xi} + \xi' \, , \tag{10.70}$$

where the overbar represents a spatial average over the large region and the prime represents cloud scales, which are typically two orders of magnitude less than the large-scale region. Note, do not confuse these averaging symbols with the previous averaging symbols used in Sec. 10.2.

Applying this "scale separation" assumption to the tendency equations for dry static energy, $s \equiv c_p T + \Phi$, and moisture, given by specific humidity

q, and neglecting small-scale horizontal transport terms yields

$$\frac{\partial \overline{s}}{\partial t} + \overline{\mathbf{v}} \cdot \nabla s + \overline{\omega} \frac{\partial \overline{s}}{\partial p} = -\frac{\partial \left(\overline{\omega' s'} \right)}{\partial p} + L(\overline{C} - \overline{E}) + \overline{Q}_{rad} \qquad (10.71)$$

and

$$\frac{\partial \overline{q}}{\partial t} + \overline{\mathbf{v}} \cdot \nabla q + \overline{\omega} \frac{\partial \overline{q}}{\partial p} = -\frac{\partial \left(\overline{\omega' q'} \right)}{\partial p} - (\overline{C} - \overline{E}) \qquad (10.72)$$

where C is the rate of condensation, E the rate of evaporation, and L the latent heat of condensation. Terms on the left-hand side of these equations represent variables on the large spatial scales that are resolved by the atmospheric climate model. Terms on the right-hand side represent processes that occur on scales smaller than the climate model resolution and hence must be parameterized. Determination of Q_{rad} has already been discussed in the previous section. The condensation and evaporation rates are usually defined in terms of the large-scale specific humidity, q, in a model layer relative to the saturation specific humidity, q_s, i.e., the relative humidity of the layer. Various approaches exist to partition the excess moisture into rain that falls to the surface or is re-evaporated back into the atmosphere. The vertical transport of heat and moisture by the subgrid-scale processes is given by the first terms on the right-hand side of (10.71) and (10.72), respectively. It is the formulation of $\overline{\omega' s'}$ and $\overline{\omega' q'}$ in terms of the large-scale variables that is the basis of cumulus parameterization.

The earliest attempt to account for cumulus effects on the large scale was the *moist adiabatic adjustment* scheme of Manabe et al. (1965), which was based on the convective adjustment idea discussed in Manabe and Strickler (1964). The idea of *convective adjustment* is that when the computed lapse rate, Γ, of the model atmosphere exceeds some critical lapse rate Γ_{crit}, the model temperatures are adjusted back to a lapse rate of Γ_{crit}. The process assures that the column integral of s and q are conserved. As discussed in Chapter 3, Sec. 3.3.3, for dry atmospheres the lapse rate cannot exceed the dry adiabat. Hence, dry adiabatic adjustment assumes $\Gamma_{crit} = \Gamma_d$. The adjustment process also insures that the column integral of internal energy, $c_p T$, does not change. Manabe and Strickler noted that the observed globally averaged lapse rate is more like 6.5 K km^{-1}, so for their one-dimensional radiative convective model they chose this for Γ_{crit}. Figure 10.5 shows the radiative equilibrium temperature profile along with a profile that has undergone adjustment to $\Gamma_{crit} = 6.5$ K km^{-1}. The adjustment procedure acts to cool the surface and warm the upper troposphere relative to the radiative equilibrium temperature profile. Thus, the adjustment can be viewed as a process that mixes warmer air near the surface up through the atmosphere. The simplicity of this approach has made it a very popular convection parameterization for one-dimensional radiative-convective models.

The dry adjustment scheme is also widely used in AGCMs as an upper bound on Γ. For most regions of the atmosphere the presence of moisture suggests that the critical lapse rate should be Γ_m, the moist adiabat. This is the assumption made for the *moist adiabatic adjustment* scheme, which has been employed in a number of AGCMs over the past 25 years. Moist adiabatic adjustment is performed on any atmospheric layer that has a lapse rate greater than Γ_m and a specific humidity exceeding the saturation value, q_s. The excess moisture in the layer is assumed to rain out of the layer, hence q in the layer is adjusted as well. Since the combined energy $c_p\Delta T + L\Delta q$ must be conserved, the following condition is applied to the column that has undergone adjustment

$$\int (c_p\Delta T + L\Delta q)\,dp = 0\,. \tag{10.73}$$

The process is iterative in that, after layers in a column are sequentially adjusted, the integral constraint (10.73) is checked and the column is readjusted until (10.73) obtains.

Although moist adiabatic adjustment is conceptually simple and easy to implement in AGCMs it possesses a number of unphysical features. First, the adjustment procedure is instantaneous. The temperature profile is altered in a given model time step, and yet cumulus clouds are known to go through a life cycle. Second, the adjustment process is local and does not explicitly depend on important processes such as the planetary boundary layer, surface effects, or horizontal transport of moisture by large-scale motions. A number of variants on the above adjustment procedure have appeared. For example, the relative humidity in the layer need not exceed 100% for the adjustment process to be invoked. "Soft" adjustment techniques have also been developed (e.g., Betts and Miller, 1986) that include a time scale and a reference profile other than Γ_m for adjustment.

An alternative approach to moist adiabatic adjustment was proposed by Kuo (1974). The Kuo scheme relates cumulus effects on the large-scale temperature and moisture structure to the column integral large-scale moisture convergence. The moisture convergence into a model column is defined as

$$M_q = -\frac{1}{g}\int_{p_{trop}}^{p_s} \nabla\cdot\mathbf{v}q\,dp + E_s\,, \tag{10.74}$$

where E_s (see 10.97) is the surface evaporation of moisture and the integral extends from the surface to the tropopause. The Kuo scheme then assumes that a fraction, b, of this moisture is available to moisten the atmosphere, while the remaining fraction, $(1-b)$, condenses and rains out. The vertical distribution of the heating and moistening is determined by assuming that these terms are proportional to the difference between in-cloud variables and

Modeling and parameterization

environment variables,

$$-\frac{\partial \left(\overline{\omega' s'}\right)}{\partial p} = \gamma_s (s_c - \bar{s}) \qquad (10.75)$$

and

$$-\frac{\partial \left(\overline{\omega' q'}\right)}{\partial p} = \gamma_q (q_c - \bar{q}), \qquad (10.76)$$

where γ_s and γ_q are defined as

$$\gamma_s = \frac{L(1-b)M_q}{\int_{p_t}^{p_b} (s_c - \bar{s}) \, dp} \quad \text{and} \quad \gamma_q = \frac{bM_q}{\int_{p_t}^{p_b} (q_c - \bar{q}) \, dp} \qquad (10.77)$$

and the integration is performed over the cloud depth. q_c is assumed to be equal to the saturation specific humidity, q_s. Thus the base and height of the cloud as well as its temperature are required. These quantities are obtained from a one-dimensional cloud model. The only remaining parameter to be specified in the Kuo scheme is b. Relating b to the large-scale variables, such as s and q, defines the closure hypothesis for the Kuo scheme. A number of ways to calculate b have been published (Anthes, 1977; Donner et al., 1982; Krishnamurti et al., 1983), but most relate b to the large-scale relative humidity, *RH*. For example, Donner et al. (1982) define b to be

$$b = 1 - \frac{1}{0.6} \int_{0.4}^{1.0} RH(p) \, d\left(\frac{p}{p_s}\right) \qquad (10.78)$$

where the integration is performed over the lowest 60% of the atmospheric column.

The Kuo scheme has also been a popular cumulus parameterization for climate models, due to its relative simplicity. It is more explicitly tied to the large-scale environment through the surface fluxes and the moisture convergence. However, the assumptions (10.75) and (10.76) must be viewed as very approximate compared to how realistic cumulus activity affects the environment. Furthermore, the uniqueness in defining the b parameter is problematic.

The mass flux approach to cumulus parameterization relates the subgrid-scale heat and moisture fluxes to basic cloud physical processes. The mass flux approach has become very popular and variants of it are currently used in a number of AGCMs. The method is based on the conceptual model shown in Fig. 10.16. The large-scale region is assumed to be populated by a range of cumulus clouds. Each cloud transports heat, s, and moisture, q. The parameterization assumes that the collective behavior of this spectrum of clouds can be represented by a bulk cumulus cloud (see Fig. 10.17).

The cloud mass flux is defined as the amount of air transported in the vertical direction by the cloud (e.g., Yanai et al., 1973). In terms of the vertical p-velocity (note that upward motion occurs when $\omega < 0$), and cloud

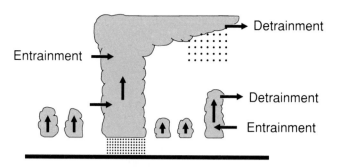

Fig. 10.16 Schematic of an ensemble of cumulus clouds. From Yanai et al. (1973).

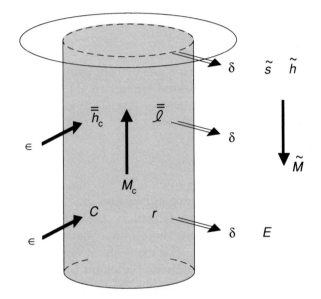

Fig. 10.17 Depiction of the cloud mass flux model (Eq. 10.84 and 10.85). ϵ denotes entrainment rate of environmental air into the cloud, δ denotes detrainment of cloud properties into the surrounding environment, h_c is the cloud moist static energy, ℓ is the cloud liquid water, $\tilde{s}$ is the environmental dry static energy, and $\tilde{h}$ is moist static energy of the environment. M_c is the cloud mass flux and $\tilde{M}$ is the mass flux of the air outside the cloud. r is rainout of cloud liquid water, E is the evaporation rate of rain and C is the condensation rate in the cloud. From Yanai et al. (1973).

fractional area, A_c, this flux is

$$M_c = -A_c \omega_c \tag{10.79}$$

while for the surrounding environment,

$$M_e = -(1 - A_c)\omega_e \,, \tag{10.80}$$

such that the total mass flux for the large scale region is

$$\overline{M} = M_c + M_e \,. \tag{10.81}$$

Modeling and parameterization

It can then be shown that the subgrid-scale flux of heat and moisture can be expressed as

$$-\overline{\omega' s'} = M_c \left(s_c - \overline{s}\right) \tag{10.82}$$

and

$$-\overline{\omega' q'} = M_c \left(q_c - \overline{q}\right) \tag{10.83}$$

where it has been assumed that the s and q of the environment can be approximated by the large-scale $\overline{s}$ and $\overline{q}$, which are predicted from the AGCM. Thus, M_c, s_c and q_c need to be determined. These quantities are determined from balance equations for mass, heat, moisture, and cloud liquid water for the bulk cloud model.

For example, at any level within the cloud, the mass balance equation is

$$\frac{\partial M_c}{\partial p} = \epsilon - \delta \tag{10.84}$$

where ϵ is the *entrainment rate* of environmental air into the cloud, while δ is the *detrainment rate* of cloud air out to the environment. Note that a steady-state assumption is made in (10.84) (i.e. $\frac{\partial M_c}{\partial t} = 0$). The balance equation for moisture is

$$\frac{\partial (M_c q_c)}{\partial p} = \epsilon \overline{q} - \delta q_c - C \tag{10.85}$$

where C is the condensation rate. The four balance relations (mass, heat, moisture, and liquid water) depend on the fundamental cloud processes of entrainment, detrainment, condensation, and conversion of water vapor to liquid water. In order to solve these four equations the fundamental cloud properties must be specified. For example, it could be assumed that ϵ is a constant, while δ is nonzero only at the cloud top.

At present, these bulk cloud properties are highly 'tunable.' A specific set of assumptions for these parameters can be found in Tiedtke (1989). The solution of the balance equations also requires a boundary condition at cloud base, $M_c(p_b)$. This boundary condition serves as the closure condition for the mass flux scheme. Tiedtke (1989) following the idea of Kuo [see (10.74)] defines the cloud base mass flux as

$$M_c(p_b) = -\frac{1}{g} \int_{p_b}^{p_s} \nabla \cdot \mathbf{v} q \, dp + E_s \, . \tag{10.86}$$

The mass flux scheme is attractive because it relates the cloud effects on the large-scale temperature and moisture to more fundamental cloud properties. Hence, it could be argued that it is a more *physically* based parameterization. The scheme is also easily generalized to include more detailed cloud dynamics. For example, it is known that cloud motions are composed of both updrafts and downdrafts. The mass flux scheme can be extended to explicitly include both of these motions (see Tiedtke, 1989). Also, once the cloud mass flux is determined it can be used to transport other atmospheric quantities, such as a chemical tracer or momentum. The weakness of the bulk mass flux scheme is the lack of knowledge concerning the fundamental cloud properties.

A more comprehensive mass flux model is that of Arakawa and Schubert (1974). This approach represents the spectrum of cumulus clouds in a more explicit manner. It is far more mathematically and computationally intensive than the bulk mass flux model. It is currently used operationally in only a few AGCMs (e.g. Randall et al., 1989). A concise description of this method is given by Tiedtke (1988).

Results from three of these schemes (moist adiabatic adjustment, Kuo, and Arakawa–Schubert) have been compared by Tiedtke (1984). The contribution of cumulus heating (K day^{-1}) in the thermodynamic equation from these schemes is shown in Fig. 10.18(a). Heating due to the mass flux scheme for a different model simulation is presented in Tiedtke (1989), and is reproduced in Fig. 10.18(b). All schemes exhibit maximum heating in the tropics, with smaller peak heating associated with extratropical storm track regions. The vertical distribution of the heating is quite different between the moist adiabatic adjustment and other two schemes, with Kuo and Arakawa–Schubert having peak heating near 300 mb. The heating peak in the mid to upper troposphere is in better agreement with observations, and is one reason these schemes are preferred over moist adjustment.

The cumulus parameterizations discussed include some aspects of observed cumulus convection. However, the present schemes still neglect a wide range of observed features in cumulus systems. It is known that cumulus systems generate mesoscale circulations (Houze, 1989), and these in turn can alter the cumulus development. These circulations also link to the larger scale, perhaps negating the simple scale separation assumed in (10.70) The presence of saturated downdrafts in cumulus systems are also known to exist, and may need to be included in the parameterizations. Finally, cumulus convection that is not explicitly linked to the boundary layer is known to occur in extratropical frontal systems. This convection is associated with wind shear and is not buoyantly driven, like the convection discussed in this section.

Most of the parameterizations described so far are for deep cumulus clouds occurring in regions of large-scale moisture convergence. Yet it is known that shallow cumulus clouds are ubiquitous. Shallow convection can occur in regions of large-scale divergent flow, or subsidence. Perhaps the most important shallow cumulus clouds are those associated with the trade wind inversion. These clouds are known to efficiently transport moisture out of the boundary layer, thus making this moisture available for horizontal transport into the tropical convergence zone, where deep cumulus clouds occur. One example of a shallow scheme is described by Tiedtke et al. (1988). The flux terms in (10.71) and (10.72) are assumed to depend on the vertical gradient of large-scale variables. Hence, a diffusion assumption is made concerning the transport by shallow clouds.

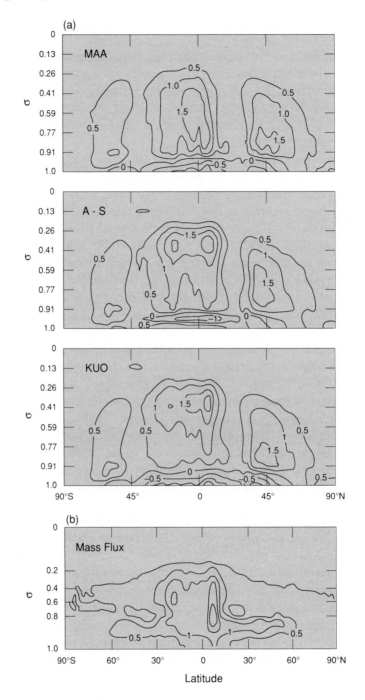

Fig. 10.18 (a) Zonally averaged distribution of heating as a function of $\sigma \equiv p/p_s$ due to the moist adiabatic adjustment scheme (MAA), the Arakawa–Schubert scheme (A-S), and the Kuo scheme (KUO) in K day^{-1}. From Tiedtke (1984). (b) as in (a) but for the mass flux scheme of Tiedtke (1989). Note (a) and (b) are based on different numerical integrations.

354

b. Stable condensation

Heating due to condensation can also occur in thermodynamically stable environments ($\partial\theta/\partial z > 0$) when the model layer becomes supersaturated ($q > q_s$). The specific humidity and temperature are both adjusted whenever stable condensation occurs. The new specific humidity is obtained from the old specific humidity through

$$q_{new} = q_{sat}(T_{old}) + \frac{dq_{sat}}{dT}\bigg|_{T_{old}}(T_{new} - T_{old}) \qquad (10.87)$$

while the new temperature is given by,

$$(T_{new} - T_{old}) = \frac{L}{c_p}(q_{old} - q_{new}). \qquad (10.88)$$

These equations are solved iteratively. The heating due to stable condensation, $\tilde{Q}_{con}$, is the temperature difference in (10.88) divided by Δt, the time difference between the new and old calculation.

10.3.3 Planetary boundary layer and surface processes

The interaction of the atmosphere with the Earth's surface involves the exchange of heat, momentum, moisture and chemical species and hence is an important component of climate system modeling. The layer of the atmosphere between the Earth's surface and approximately two kilometers is known as the *planetary boundary layer* (PBL) (or atmospheric boundary layer). Unlike the atmosphere above the PBL, which is called the *free atmosphere*, the PBL's structure is dominated by surface processes that generate turbulent motions on a range of spatial scales. The PBL is further delineated into a number of sublayers (see Fig. 10.19). The surface layer (typically the lowest 100 to 200 meters) is defined by the condition of near constant *fluxes* of heat, momentum and moisture with height. The PBL structure exhibits a strong diurnal dependence due to surface solar heating. It plays an important role in cloud processes, as was noted in the previous section. Because the PBL is the physical "link" between the surface and the free atmosphere, it must be included in climate models.

Since the typical height of the PBL is 2 km or less, a large number of vertical layers are required to properly resolve the layer structure depicted in Fig. 10.19. For most climate models, the computational cost of these additional layers is considered too large. Thus, the effects of the PBL are parameterized in large-scale models (Driedonks and Tennekes, 1981). The parameterization should account for the dependence of the PBL height on surface conditions, and be sufficiently robust to respond to different surface types (see Chapter 5). It should also distinguish conditions between

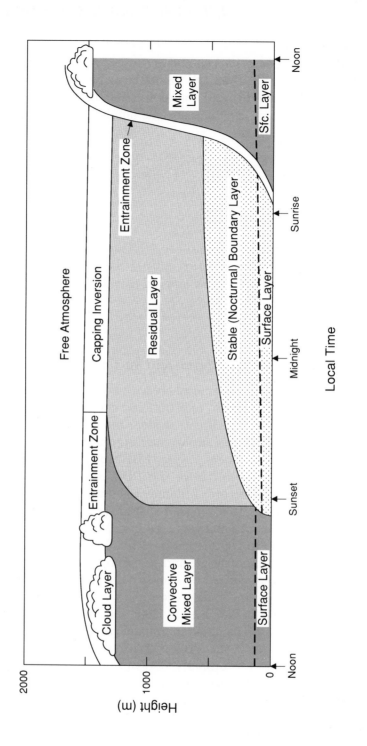

Fig. 10.19 Schematic of the atmospheric boundary layer. Note the presence of the surface layer, the mixed layer and a nocturnal stable layer. Also note the vertical scale of these atmospheric layers. From Stull (1988).

356

where the thermal structure is unstable, neutral or stable (see Chapter 3, Sec. 3.3.3); and it should couple in a consistent manner with the other parameterizations used in the free atmosphere domain of the model. To date, few AGCM PBL parameterizations meet all these requirements.

a. Planetary boundary layer

If only one or two levels are available near the surface, then a bulk model approach is usually adopted for the surface and PBL parameterization. As noted previously, the spatial scales that are important in the PBL are much smaller than those present in the rest of the atmosphere. This means that the prognostic variables such as u, v, ω, T, and q can be separated, as was done in (10.70), into mean and subgrid-scale terms. Just as in the cumulus case, large-scale terms can be collected on one side of the prognostic equation leaving all "turbulent" scale terms on the right-hand side. The correlation terms like $\overline{u'\omega'}$ are called eddy flux or *stress* terms and are denoted by τ. The mathematical development of these "Reynolds stress" terms and a discussion of the "closure" of the equations is given in Chapter 11, Sec. 11.2.2. Above the surface layer, the frictional terms in the momentum equations (10.13) $\mathbf{D}_M = (D_\lambda, D_\phi)$ are related to these stresses through

$$D_{\lambda,\phi} = -g \frac{\partial \tau_{\lambda,\phi}}{\partial p}. \tag{10.89}$$

In the thermodynamic equation (10.14)

$$D_H = -\frac{g}{c_p} \frac{\partial H}{\partial p}, \tag{10.90}$$

while in the moisture equation (10.15)

$$D_q = -\frac{g}{c_p} \frac{\partial Q}{\partial p} \tag{10.91}$$

where $H = \rho \overline{w'T'}$ and $Q = \rho \overline{w'q'}$ are the eddy fluxes of heat and moisture. The simplest approach of obtaining these eddy flux or stress terms is to assume they are described by an eddy diffusion process (Smagorinsky et al., 1965),

$$\{\tau_\lambda, \tau_\phi, H, Q\} = -g\rho K \frac{\partial}{\partial p} \{u, v, T, q\} \tag{10.92}$$

where K is an eddy diffusion coefficient. Although this is a straightforward model for the PBL, a constant K coefficient is insufficient to model many observed states of the PBL. More elaborate methods exist for specifying K as a function of local stability, and have been employed in AGCMs. Another approach to modeling the PBL is to use bulk mixed layer models that include a prediction for the depth of the PBL (Suarez et al., 1983).

b. Surface layer

As stated above, the surface layer is defined by the near vertical constancy (variations less than 10%) of the heat, momentum and moisture fluxes. For the surface layer, the momentum stresses are often based on

$$\tau_\lambda = -\rho_h C_D V u_h \tag{10.93}$$

and

$$\tau_\phi = -\rho_h C_D V v_h \,, \tag{10.94}$$

where

$$V = \sqrt{u^2 + v^2} \tag{10.95}$$

is the wind speed, the subscript h denotes the first atmospheric level above the surface, and C_D is the drag coefficient. For heat and moisture, the surface flux relations are given by the bulk formulae

$$H = \rho_h C_H V (T_s - T_h) \,, \tag{10.96}$$

and

$$E_s = D_w \rho_h C_D V (q_s(T_s) - q_h) \,. \tag{10.97}$$

C_H and C_Q are bulk transfer coefficients for heat and moisture, respectively. D_W is a wetness factor that accounts for varying evaporation efficiencies over different surface types. The bulk transfer coefficients depend on the wind shear and static stability of the atmosphere above the surface, and the roughness of the surface. In general, C_D, C_H and C_Q increase in magnitude as the atmosphere becomes more unstable.

c. Surface and subsurface temperatures

Traditionally, AGCMs have calculated land surface temperatures from either a surface energy balance condition (17.1) or a finite heat capacity surface model (14.7). Recently, however, land surface models are being employed within AGCMs to evaluate the surface and subsurface temperatures (see Chapter 14).

Over oceans, the surface temperature must be either specified or calculated from an interactive ocean model (see Chapter 11). Many GCM studies of the atmosphere have employed fixed sea surface temperatures (SSTs). The daily SSTs are prescribed by linearly interpolating between mid-month observed SSTs and thus allow for seasonal variation. Although this enables the AGCM to be run with great computational efficiency, it denies the atmosphere any realistic interaction over ocean regions. For example, surface fluxes are irrelevant to SSTs. It also implies that the total surface–atmosphere system need not be in energy balance, i.e., the globally and annually averaged top of atmosphere net radiation balance need not be zero! Typically, fixed SST models are "tuned" in some manner to guarantee this balance condition. But any perturbation to the "tuned" model, such as a change in the model physics, will cause an imbalance to occur for the new climate state. Over sea ice regions, surface temperatures need to be evaluated from a realistic sea ice model (Chapter 12).

10.3.4   Cloud prediction schemes for atmospheric models

Prediction of cloud amount is of great importance to climate modeling. As discussed in Chapter 3, clouds play a fundamental role in controlling the amount of solar and infrared radiation available to the climate system. It is important to remember that the effect of *cloud amount* enters into the climate model only via the radiation. Clouds range in size from hundreds of meters to hundreds of kilometers in radius, which means that most clouds are smaller in area than the typical grid resolution of climate models. Unlike other climate variables (e.g., momentum, temperature or moisture), there is no *fundamental* prognostic equation for cloud fraction. The cloud amount for a grid region must be related to the other predicted climate variables. However, the cloud fraction must be prescribed in association with the occurrence of condensation. Thus when stable or convective condensation occurs a cloud amount is assigned to the same grid box. Many cloud parameterizations differentiate between cloud amount for convective clouds versus stable clouds.

Historically, the most popular variable to use in cloud amount parameterizations is the large-scale relative humidity, RH (Xu and Krueger, 1991). The parameterization can be as simple as

$$A_c = \begin{cases} 1 & RH \geq RH_{crit} \\ 0 & RH < RH_{crit} \end{cases}, \tag{10.98}$$

where A_c is the cloud amount, RH is the grid box relative humidity, and RH_{crit} is a critical relative humidity (all variables expressed as fractions). RH_{crit} is usually assumed to be 0.8 or 1.0. This parameterization would totally fill the grid box with cloud whenever the the relative humidity exceeds the critical value. The value of 1 for $RH \geq RH_{crit}$ can also be viewed as a "tunable" constant, i.e., it can be set to a value of less than 1. More "complex" cloud fraction parameterizations may assume the cloud fraction is a linear function of RH (Smagorinsky, 1960)

$$A_c = a + b(RH - RH_{crit}), \tag{10.99}$$

or a quadratic functions of RH (Slingo, 1987)

$$A_c = \left(\frac{RH - a}{b}\right)^2, \tag{10.100}$$

where a, b, and RH_{crit} are viewed as "tunable" values. Convective cloud cover is typically assigned a smaller value. Similar parameterizations for cloud amount associated with cirrus and marine stratus clouds are also used in GCMs (Ramanathan et al., 1983; Slingo, 1987). All of these parameterizations must be viewed as extremely simplistic compared with the real atmospheric processes that lead to cloud cover. In particular, some of these schemes result in cloud "blinking." When condensation occurs a cloud is formed at that time step of the model. But in the next time step the relative humidity may be less than RH_{crit} and the cloud fraction will be zero. Inspection of time series of cloud amount from this approach indicates high-frequency "on–off" cloud activity. The "tunable" (a, b, RH_{crit}) constants are certainly dependent on the grid scale of the model, but to date this dependence is not well understood.

Current efforts in cloud parameterization are focused on predicting the amount of cloud liquid water (CLW) in a GCM grid box. These methods are based on prognostic equations for liquid water formulated in terms of more fundamental physical processes. However, as prognostic formulations for cloud liquid water continue to be implemented in GCMs, there is still the need to relate this CLW to a cloud amount or some related radiative quantity.

10.3.5 *Mechanical dissipation mechanisms in the free atmosphere*

The need for mechanical dissipation in the free atmosphere becomes evident upon considering the kinetic energy (KE) spectrum of the atmosphere. The KE spectrum describes the contribution to the kinetic energy for each spatial wave scale. Analysis of observed KE spectrum shows the largest contribution arising from planetary scale waves, with a steady decrease in KE for smaller and smaller scales. If dissipation in the numerical model for the smallest resolvable scales is insufficient, the KE actually increases for the smallest scales. This is termed spectral blocking because it is associated with the spectral resolution of the model and the absence of still smaller scales to which the energy would normally pass through nonlinear interactions. Thus, the $\mathbf{D}_M = (D_\lambda, D_\phi)$ in (10.13) need to be specified not only for the PBL, but also above this region. To prevent spectral blocking in the numerical models, dissipation terms are introduced to act strongest on the smallest model scales. Mathematical functions that meet this requirement are either the harmonic operator or the bi-harmonic operator. For a spectral general circulation model formulated in terms of the absolute vorticity, $\eta \equiv \zeta + f$, and the divergence, δ, the dissipation terms, $\mathbf{D}_M$, have the form

$$D = \kappa_2 \left[\nabla^2 \eta + \left(\frac{2\eta}{a^2} \right) \right] \tag{10.101}$$

and

$$D = -\kappa_4 \left[\nabla^4 \eta - \left(\frac{2\eta}{a^2} \right)^2 \right]. \tag{10.102}$$

The $2\eta/a^2$ term in (10.101) and (10.102) insures that solid body rotation of the atmosphere is not damped by the dissipation term D. The coefficient, κ_2 or κ_4, is "tuned" so that the KE spectrum does not exhibit spectral blocking for the smallest model scales. The form of these terms and the magnitude of κ are difficult to justify on physical principles. It is most often argued that in some way they represent unresolved turbulent motion.

More recently another form of mechanical dissipation for climate models has been introduced. Many climate models exhibit strengthening of zonal winds in both the troposphere and the stratosphere as the model horizontal resolution increases. Increased horizontal resolution should mean that more scales are being resolved and the dynamical flow would be more like the observed flow. The strong zonal winds suggest that some mechanical dissipative process may be missing in the climate models.

Lilly (1972) pointed out that vertically propagating gravity waves generated by air flowing over mountains could be an important source of drag (or dissipation) on the winds in the free atmosphere. The gravity waves grow in amplitude as they propagate upwards, eventually becoming unstable at some height; at or near this region, the waves deposit the momentum gained at the surface. The actual dynamics of these waves in three dimensions is quite complex. To date simple linear models of the problem (e.g., Lindzen, 1981; McFarlane, 1987) are used to parameterize the effects of gravity waves on the mean flow. The stress generated at the surface is proportional to the variance of the orography (a measure of the "hilliness" of the region). The main problem of gravity wave drag parameterization is determining how the momentum is deposited throughout the free atmosphere. Breaking gravity waves are also important in the middle atmosphere, especially the mesosphere. All existing gravity wave drag parameterizations contain constants that cannot be deduced from fundamental physics. This makes the parameterizations highly tunable at present.

10.4 Simulation and validation of AGCMs

a. Simulation

Once the atmospheric climate model simulation is complete, the more challenging problem arises of interpreting the model results. This involves choosing an appropriate set of diagnostics. Perhaps the simplest diagnostics to consider are the zonal (10.2) and time mean (10.4) averages of the model variables. Figures 10.20(a)–(b) show time mean, zonal mean averaged T and u for a January simulation from a troposphere–stratosphere version of NCAR Community Climate Model (Boville, 1991). The CCM is a spectral model, and this version uses a T63 horizontal resolution. Note that the structure in the zonal mean is very similar to the observed zonal flow (Chapter 3, Sec. 3.1.1). The midlatitude tropospheric jets are close to their observed position and strength. The tropical easterlies are also well simulated.

The physical contributions to the temperature tendency (10.14), due to radiation, condensation adjustment plus subgrid-scale effects are shown in

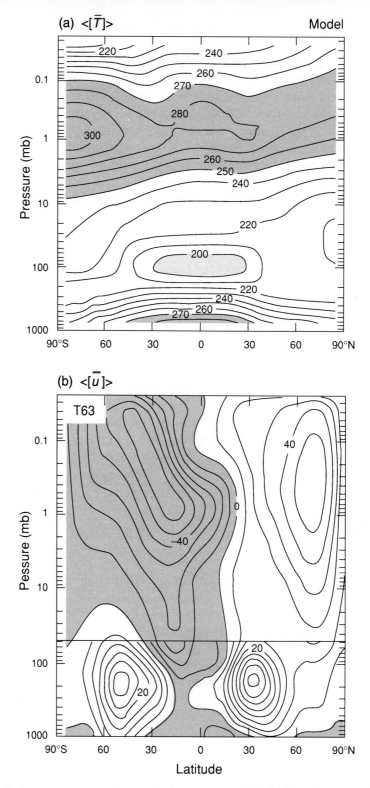

Fig. 10.20 (a) January average zonally averaged temperature (K) (b) Zonally averaged zonal wind (m s^{-1}) from Boville (1991).

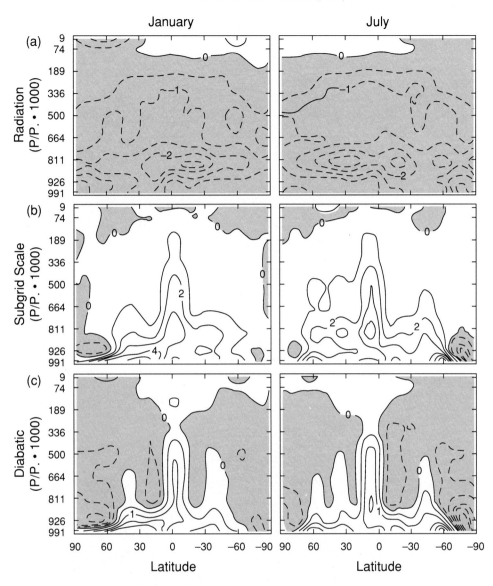

Fig. 10.21 (a) January and July mean zonally averaged net radiative heating (K day^{-1}) from the NCAR CCM1 (Boville, 1985), (b) as in (a) but net condensational heating, (c) total diabatic heating which is the sum of (a) and (b). Contour interval is 0.5 K day^{-1} for (a) and (c), and 1 K day^{-1} for (b).

Fig. 10.21(a)–(b). The net radiative effect is to cool the troposphere and warm the upper stratosphere. The convective heating looks similar to the results in Fig. 10.18. The sum of these physical processes is shown in Fig. 10.21(c). The convective heating dominates in the tropics leading to net heating. This net "diabatic" heating is balanced by upward motion which causes "adiabatic" cooling.

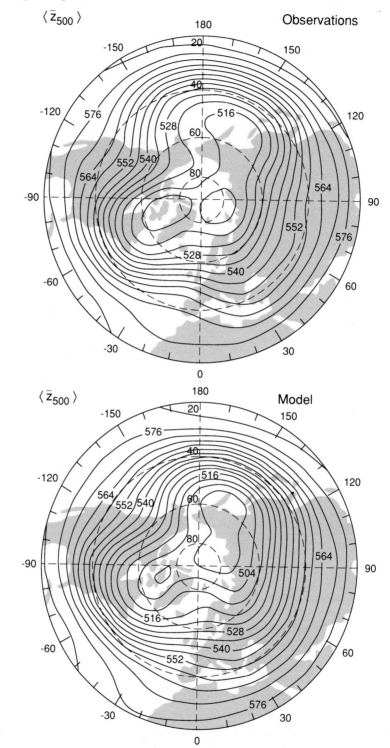

Fig. 10.22 January mean 500 mb height field (in decameters) from observations and a model simulation from the NCAR CCM1. From Boville and Randel (1986).

Sea Level Pressure

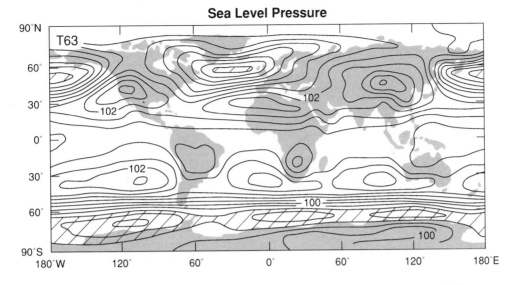

Fig. 10.23 January mean sea level pressure (in decabars) from a version of the NCAR CCM1. From Boville (1991).

Although zonal means are useful diagnostics for analyzing the general circulation model, the zonal mean can mask model biases. Therefore, consideration must also be given to the latitudinal and longitudinal morphology of model variables. Figure 10.22 shows the Northern Hemisphere 500 mb geopotential height field (see Chapter 3, Sec. 3.3) and the observed height field. Figure 10.23 shows the distribution of sea level pressure, where regions of high pressure denote sinking (or subsidence) motion. The radiative budget is determined by the outgoing longwave radiation (OLR) (Fig. 10.24a) and the planetary albedo (Fig. 10.24b). Regions of high cold cloud cover over regions of strong upward motion are apparent in the OLR figure.

Another diagnostic to consider is the deviation from both time and space means. Consider the time mean of a model variable, ξ, and deviations from this mean, denoted by ξ',

$$\xi = \bar{\xi} + \xi'. \tag{10.103}$$

Using (10.47) and (10.103), the time mean, zonal mean meridional transport of any quantity ξ is just

$$[\overline{v\xi}] = [\bar{v}][\bar{\xi}] + [\overline{v^* \bar{\xi}^*}] + [\overline{v'\xi'}]. \tag{10.104}$$

The first term on the right-hand side represents transport of ξ by the *mean meridional circulation*, the second term transport by the *stationary eddies*, and the last term by the *transients*. The model simulated total eddy (stationary plus transient) contribution to the meridional transport of zonal momentum is shown in Fig. 10.25.

365

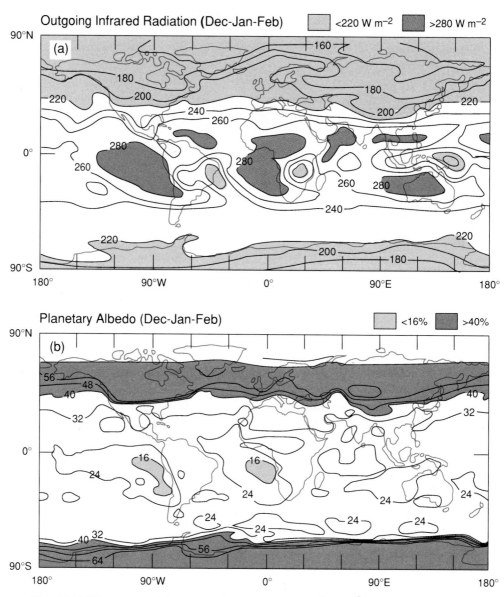

Fig. 10.24 Winter mean (a) outgoing longwave radiation (W m^{-2}) and (b) planetary albedo from CCM1. Based on Smith and Vonder Haar (1991).

b. Validation

The validation of an AGCM can be done for a range of spatial and temporal scales to test both the dynamical and the physical aspects of the model climate. AGCM validation is, however, limited by the spatial and temporal coverage of available observations. Satellite data, which give the

366

$$[\, u^* \, v^* \,] + [\, u' \, v' \,]$$

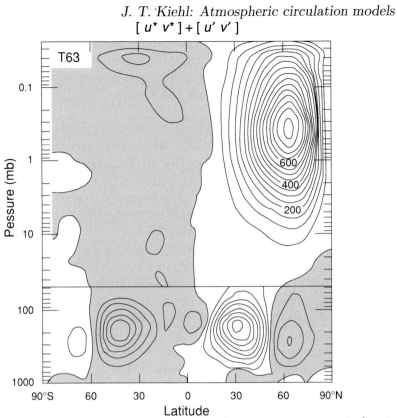

Fig. 10.25 January mean zonally averaged meridional momentum transport due to stationary and transient eddies. From Boville (1991).

greatest spatial coverage, is limited by temporal coverage (many satellite instruments having only 2–5 year lifetimes). While surface observations are, by their very nature, biased toward land, the tropics and the Southern Hemisphere, in particular, suffer from a paucity of surface-based data. Despite these limitations, a diverse observational data base is available for AGCM validation.

The standard network of rawinsondes provide twice-daily vertical profiles of temperature, moisture and winds. Although nonuniform and sparse in places, there are valuable data available in the tropical regions, for example. These data are useful for validating convective and radiative parameterizations. There is also a wide network of near-surface (so-called screen level) observations which can be used to validate surface parameterizations in GCMs.

Both the National Meteorological Center (NMC) in the U.S. and the European Centre for Medium Range Weather Forecasts (ECMWF) provide twice-daily atmospheric analyses. These analyses incorporate observational data from both the surface and from satellites into a four-dimensional data assimilation system that uses a numerical weather prediction model to carry forward information from previous analyses, giving global uniformly gridded data. The data include temperature and winds on specified pressure levels, and surface pressure. These data are valuable for validating the dynamical features of an AGCM. However, it must be kept in mind that the analyses contain a mixture of observations and model. This is especially the case for the tropical circulation.

Satellite data provide near-global coverage of top of atmosphere radiative quantities. The OLR and planetary albedo data are available from Nimbus-7 and from the Earth Radiation Budget Experiment (ERBE). ERBE also provides measurements of clear sky longwave and shortwave absorbed fluxes, which are valuable for validating radiation parameterizations in the AGCMs. The International Satellite Cloud Climatology Project (ISCCP) provides global, 3-hourly radiances for the 10 μm and visible spectral regions. ISCCP also archives various cloud statistics based on these radiances. These radiances can be easily produced in AGCMs and compared with satellite observations, thereby serving as another validation tool. The SSM/I instrument provides data over ocean regions for total integrated atmospheric water vapor (precipitable water) and integrated cloud liquid water. These data are used to validate the hydrological aspects of the model. Several other satellite data sets are also being developed.

Finally, there are data available from field programs. Although the spatial and temporal scales of data from field programs are much smaller than the AGCM resolution, field programs are invaluable for development and validation of physical parameterizations.

10.5 Future improvements for AGCMs

Improvement in simulation of the atmospheric general circulation will occur in four areas: physical parameterization, numerical methods, computational software, and hardware. The last three areas are discussed in Chapter 9, and new advances should lead to increased computational power needed for higher spatial resolution and longer model integrations. To date no a priori method exists to determine the optimal spatial and temporal resolution for climate modeling. To ensure that the effects of the smallest scales are explicitly included and evaluated, substantial increases in computational resources are required to study this modeling issue. The second area, numerical integration, is most important for the solution of the dynamical transport equations for moisture and chemical species. Overall improvement to climate modeling will benefit most from improved physical parameterizations. These are not independent of resolution. As noted in this chapter, many of the parameterizations currently used are quite simplistic. The most important physical processes that require improvement are:

a. *Clouds*

Tremendous improvements are required in this area of AGCM

parameterization. Most cloud amount prediction schemes are far too simplistic to accurately model the real spatial and temporal distribution of cloud cover. Further attempts need to be made in linking more fundamental cloud properties (e.g., liquid water content, or drop size distribution) to the large-scale variables. Cumulus parameterizations need to account for mesoscale circulation effects. Better shallow convection parameterizations are also needed for the AGCMs. Cloud radiation effects should be a part of the cloud schemes. The present linear sequential approach (see Fig. 10.8) needs to be replaced with a more integrated cloud scheme that simultaneously accounts for radiative, dynamical, and thermodynamical effects.

b. Momentum sources and sinks

Present approaches to the parameterization of momentum sinks in the atmosphere are far too crude given the importance of this process in determining the general circulation. Improved methods to incorporate the effects of gravity waves on the mean flow are desperately needed. These methods should employ the fewest possible "free" parameters. Incorporation of momentum transport by cumulus clouds should also be investigated.

c. Planetary boundary layer

Present PBL parameterizations need improving. Many climate system components exist within the PBL (e.g., plant canopy, source for chemical species, ocean surface). The coupling of these components with an AGCM will only be as good as the PBL parameterization. Additional computational resources may lead to explicit PBL models for the AGCMs.

Many of these issues have been held back by lack of observational data and/or robust computational studies. It is hoped that future field programs will lead to improved understanding of the fundamental physics governing these processes. However, another approach would be to use a hierarchy of three-dimensional models to improve the parameterizations. A coordinated effort that employs micro-, meso-, and large-scale models, in tandem, could lead to considerable improvement in physical parameterization of clouds, gravity wave drag formulations, and the PBL.

d. Interactions with the rest of the climate system

This area deals with the interactions between the AGCM and all the other components of the climate system, as given by other chapters in this book. It includes the need for improved land surface processes, hydrology and biology, snow cover, sea ice, and ocean–atmosphere interaction.

11

Ocean general circulation modeling

Dale B. Haidvogel and Frank O. Bryan

11.1 Introduction

Modeling of the oceanic general circulation has much in common with its atmospheric counterpart. In particular, many of the component processes, the relevant equations of motion, and the applied numerical techniques are similar to those discussed in Chapters 9 and 10. Nonetheless, there are important physical and technical issues unique to the oceanic environment. For example, whereas the atmosphere is forced thermally throughout its volume, the ocean is forced both thermally and mechanically primarily at its surface. In addition, the geometry of ocean basins is very complex. Together, these characteristics have several immediate practical consequences as described in Chapter 4: oceanic mean states are quite complicated, having narrow but important boundary layers on nearly all bounding surfaces as well as within the oceanic interior; boundary conditions on ocean flows are difficult to define and to parameterize; and numerical simulation of the ocean circulation is computationally very demanding.

Ocean modeling is also severely constrained by the existence in much of the world's oceans of mesoscale eddies with time and space scales, respectively, of weeks to months and tens to hundreds of kilometers. Although called "mesoscale" by many oceanographers, dynamically these nearly geostrophic turbulent eddies are the oceanographic counterpart of the atmospheric synoptic scale. Nevertheless, there are important differences. First, ocean eddies are not perturbations on an energetic mean flow. The greater part by far of oceanic kinetic energy is bound up in the eddy scale, and their scales of motion may play an important role in the poleward transport of heat. Second, they are relatively small in horizontal extent so that ocean climate models, which must have the same overall exterior dimensions as atmospheric GCMs (i.e., global), may require as much as 20 times the latitudinal and longitudinal resolution as atmospheric models if the eddies are to be explicitly resolved.

Perhaps the greatest limitation on ocean modeling, however, as described in Chapter 4, is that the ocean is more difficult to observe, and data are relatively more sparse, than in the atmosphere. It is estimated, for example, that the amount of data available to ocean scientists in the decade of the 1990s – including those data expected to be available from satellite sea surface temperature, wind and surface height measurements – will still be an order of magnitude fewer than in the atmosphere (Ghil and Malanotte-Rizzoli, 1991). Also, the data are not only sparse, but nonuniform (being most plentiful at or near the sea surface and in the Northern Hemisphere) and indirect (i.e., typically providing measures of the mass field rather than of velocity).

Indeed, modeling the composite ocean/climate problem on global space scales and climatic time scales with the resolution required to resolve explicitly and simultaneously all the energetic oceanic phenomena would be prohibitively expensive. In addition, there remain fundamental uncertainties on key ocean modeling issues, such as how to parameterize convection, diffusion, and the turbulent mixing processes in the surface layer. In the interest of efficient progress, therefore, a diverse set of more manageably sized and somewhat specialized ocean models have evolved, including basin- and global-scale GCMs, regional ocean models, and idealized, process-oriented models.

The most conceptually simple of these, the process models, use idealized geometries and (often) simplified boundary conditions, and concentrate on interactions between physical mechanisms (the effects of rotation, topography, stratification, nonlinearity) or on a single fundamental physical process in isolation. A process modeling approach has been used throughout the history of ocean modeling, e.g., to explore the nature of the oceanic thermohaline circulation (Cox, 1989), to develop eddy-resolving basin-scale models of the wind-driven ocean (Holland, 1985), and to study mid-ocean, mesoscale turbulence (Rhines, 1977).

The world's oceans have great geographical heterogeneity in geometry and dynamics. It is not surprising, therefore, that various regional models have been developed as part of the scientific evolution toward a fully global climate model. Though overlapping somewhat in intent and design with the process models, regional models are typically subbasin-scale and more realistic in their treatment of regional factors such as geometry, topography, etc. As such, they are more amenable to direct validation against available observations. The subbasin-scale extent of these models does introduce considerable uncertainty in the proper treatment of the "open boundaries" – edges of the numerical domain which coincide with ocean rather than with a continental or other solid boundary. Nonetheless, simulations of the Gulf Stream system, the northeast Pacific Ocean, and the equatorial regions of the world ocean, as well as other regions, have shown some success in reproducing observed features of the ocean circulation (e.g., see Chapter 18 for simulations of the tropical Pacific).

Any atmospheric climate model used for integrations of duration longer than a season will, of course, have to be coupled to a *global* model of at least the upper ocean. Ocean GCMs of varying completeness and complexity exist. Basin- and global-scale models have traditionally come in two varieties: ocean circulation models which include much of the detail of the ocean (e.g., realistic coastline, topography, observed surface forcing) but no mesoscale eddies, and eddy-resolving GCMs, capable of including mesoscale eddies at the expense of physical and geometrical simplicity. The global ocean models being readied for use in coupled climate models are only now being run in eddy-resolving mode (Semtner and Chervin, 1988), thus merging these two traditional varieties of GCMs.

Other recent summaries of results obtained with these three classes of ocean models, as well as further information on the many related issues in numerical methods and model design, can be found in Abarbanel and Young (1987), Anderson and Willebrand (1989), O'Brien (1986), Robinson (1983), and Warren and Wunsch (1981).

11.2 Equations of motion

Given the variety of important oceanic processes and phenomena, and the wide range of space and time scales covered by them, it is perhaps not surprising that a similarly wide variety of mathematical models of the ocean circulation have been developed. The starting point for many of these has been the primitive equations of motion, which are now briefly reviewed. A more complete derivation can be found in Veronis (1973).

11.2.1 *The hydrostatic primitive equations*

The formulation of the equations of motion for a frictionless (inviscid) and unforced ocean proceeds in a generally similar fashion to that described in Chapter 3 (Sec. 3.6) for the atmosphere. In particular, both the hydrostatic and shallowness approximations are appropriate in the oceanic case. Additional simplifications, specific to the ocean, are also possible however. For example, in the ocean the spatial and temporal variations in density are small compared to its mean value. Therefore, as noted in Chapter 4,

$$\rho_{tot} = \rho_o + \rho(x, y, z, t) \tag{11.1}$$

where ρ_o represents the mean density of the ocean. Recognizing that $\rho_o \gg |\rho|$, there is justification in replacing ρ by its mean value, ρ_o, everywhere except in the buoyancy term, $-\rho g$, and in the thermodynamic density conservation equation. This is the *Boussinesq approximation*.

With these approximations, the resulting hydrostatic primitive equations

(HPE) in Cartesian coordinates are:

$$\frac{du}{dt} = fv - \frac{\partial \Phi}{\partial x} \tag{11.2}$$

$$\frac{dv}{dt} = -fu - \frac{\partial \Phi}{\partial y} \tag{11.3}$$

$$\frac{\partial \Phi}{\partial z} = -\frac{\rho}{\rho_o} g \tag{11.4}$$

$$\frac{dT}{dt} = 0 \tag{11.5}$$

$$\frac{dS}{dt} = 0 \tag{11.6}$$

and

$$\frac{\partial u}{\partial x} + \frac{\partial v}{\partial y} + \frac{\partial w}{\partial z} = 0 \tag{11.7}$$

where

$$(x, y, z) = (\text{Cartesian coordinates}),$$

$$(u, v, w) = \left(\frac{dx}{dt}, \frac{dy}{dt}, \frac{dz}{dt} \right)$$

$$\frac{d}{dt} = \frac{\partial}{\partial t} + u \frac{\partial}{\partial x} + v \frac{\partial}{\partial y} + w \frac{\partial}{\partial z},$$

and $\Phi = p/\rho_o$ (not to be confused with geopotential, as used in Chapter 3); $f = 2\Omega \sin \phi$, the Coriolis frequency; and T and S are the temperature and salinity, respectively. To close the resulting equations, an equation of state, $\rho = \rho(T, S, p)$, must be added. In global-scale problems, the analogous equations written in spherical geometry are used (see, e.g., Bryan, 1989).

11.2.2 *Turbulent friction and the closure "problem"*

Thus far, the ocean is assumed to be inviscid. This is approximately true in the oceanic interior away from both horizontal and vertical boundaries. However, friction cannot be totally ignored for several reasons. First, the global ocean is forced at its surface by fluxes of momentum, heat and moisture, and accompanying injections of energy and vorticity. Given these inputs, the global balances of energy and *enstrophy* (the square of vorticity) require dissipative sinks if a forced-damped equilibrium is to be reached. Second, the surfacial driving forces may themselves be frictional in origin (e.g., the surface wind stress). Lastly, though weak throughout the bulk of the ocean, frictional forces must be consequential in certain narrow regions (e.g., along the western boundary) in order to insure local dynamical balance. An example of this requirement is given below when a simple model of the wind-driven ocean is considered.

It is true that molecular viscous forces are entirely insignificant everywhere in the ocean. Yet, it has just been said that friction is essential somewhere – how can this be? The point is that large-scale motions in the ocean (and atmosphere) do not exist in isolation. Nonlinearities in the equations effectively link all scales of motion and lead to systematic exchanges, or cascades, of energy and enstrophy. In particular, large-scale current systems often serve as a source of turbulent energy and/or enstrophy for smaller scales of motion. Unfortunately, no complete theory exists to describe the effective viscous forces experienced by the large-scale ocean circulation due to these nonlinear cascades. This leads to the vexing issue of how to account for, or parameterize, the effects of subgrid-scale motions in discrete numerical model solutions.

Suppose the total three-dimensional vector velocity field is divided into the large-scale components of interest and smaller-scale motions, as in (10.70),

$$\mathbf{v} = \overline{\mathbf{v}} + \mathbf{v}' , \tag{11.8}$$

where the overbar represents an average over the excluded spatial scales, defined such that

$$\overline{\mathbf{v}'} = 0 . \tag{11.9}$$

The evolution equation for the large-scale motion, $\overline{u}$ say, is obtained by substituting (11.8) into the u momentum equation and then averaging

$$\frac{\partial}{\partial t}\overline{u} + \mathbf{v} \cdot \nabla \overline{u} - f\overline{v} = -\frac{\partial}{\partial x}\overline{\Phi} - \frac{\partial}{\partial x}\overline{u'u'} - \frac{\partial}{\partial y}\overline{v'u'} - \frac{\partial}{\partial z}\overline{w'u'} , \tag{11.10}$$

where use has been made of the continuity equation (11.7). Equation (11.10) shows that the momentum fluxes associated with the fluctuating, small-scale motions look like stresses applied to the larger-scale flow. These are the *Reynolds stresses*, which result from the averaging of the equations.

Dropping the overbar, and using the conventional notation for the Reynolds stresses $\tau_{xy} = -\rho \overline{u'v'}$, and so forth, the momentum equations for the large-scale flow become:

$$\frac{du}{dt} - fv = -\frac{\partial \Phi}{\partial x} + \frac{1}{\rho}\left[\frac{\partial}{\partial x}(\tau_{xx}) + \frac{\partial}{\partial y}(\tau_{xy}) + \frac{\partial}{\partial z}(\tau_{xz})\right] \tag{11.11a}$$

and

$$\frac{dv}{dt} + fu = -\frac{\partial \Phi}{\partial y} + \frac{1}{\rho}\left[\frac{\partial}{\partial x}(\tau_{yx}) + \frac{\partial}{\partial y}(\tau_{yy}) + \frac{\partial}{\partial z}(\tau_{yz})\right] . \tag{11.11b}$$

The resulting equations (11.11) properly incorporate the effects of the small-scale motions on the larger. However, since the Reynolds stress terms are themselves unknown at this point, further information – in the form of additional constraints on the stresses – is necessary before (11.11) can be solved. This is the so-called *closure problem*. An analogous closure problem exists for the temperature and salinity equations.

A crude assumption often made is to assume that the Reynolds stress terms are proportional to the spatial gradients of the large-scale circulation, e.g., $\tau_{xz} = \rho(\nu_v \frac{\partial u}{\partial z} + \nu_h \frac{\partial w}{\partial x})$, where the *turbulent eddy viscosity coefficients*, ν_h and ν_v, are assumed constant. The resulting momentum equations are:

$$\frac{du}{dt} - fv = -\frac{\partial \Phi}{\partial x} + \nu_h \nabla_h^2 u + \nu_v \frac{\partial^2 u}{\partial z^2} \qquad (11.12)$$

and

$$\frac{dv}{dt} + fu = -\frac{\partial \Phi}{\partial y} + \nu_h \nabla_h^2 v + \nu_v \frac{\partial^2 v}{\partial z^2} , \qquad (11.13)$$

where ∇_h^2 is the horizontal Laplacian. In a similar way, the temperature and salinity equations are often written:

$$\frac{dT}{dt} = \kappa_h \nabla_h^2 T + \kappa_v \frac{\partial^2 T}{dz^2} \qquad (11.14)$$

and

$$\frac{dS}{dt} = \kappa_h \nabla_h^2 S + \kappa_v \frac{\partial^2 S}{\partial z^2} \qquad (11.15)$$

where κ_h and κ_v are constant eddy diffusivity coefficients. Although convenient, the closure assumptions leading to (11.12–11.15) are difficult to justify. Recognizing this, modelers of the oceanic general circulation have experimented with a wide variety of alternate subgrid-scale parameterizations. One alternative, often used in eddy-resolving models, are viscous terms of biharmonic form (e.g., $\nu_{4h} \nabla_h^4 u = \nu_{4h} \nabla_h^2 \nabla_h^2 u$) [See also (10.101) and (10.102)]. Greater scale-selectivity of the viscous effects can be achieved in this way.

Convective overturning occurs on very small horizontal scales that cannot be explicitly resolved in climate models and at which the hydrostatic approximation may not be valid. This process must also be parameterized in any thermally forced model. The simple vertical eddy diffusivity term $\kappa_v \frac{\partial^2 \rho}{\partial z^2}$ can be modified to accomplish this:

$$\kappa_v = \begin{cases} \kappa_v^\circ & \frac{\partial \rho}{\partial z} \leq 0 \\ \infty & \frac{\partial \rho}{\partial z} > 0 , \end{cases}$$

where κ_v° is the vertical eddy diffusivity for stably stratified conditions. In practice, convection is often parameterized by an "adjustment" procedure that restores the static stability of the water column after all other terms have been computed (cf., 10.3.2 for the atmosphere).

Many alternate formulations of these mixing terms have also been proposed (Holloway, 1989). Unfortunately, the results of large-scale ocean circulation models are often quite sensitive to both the choice of mixing formulation and to the assumed values of the eddy mixing coefficients (F. Bryan, 1987). In addition, the mixing values required for stable numerical integration of the large-scale models often exceed the observationally inferred values (Gregg, 1987).

11.2.3 Surface, lateral, and bottom boundary conditions

Equations (11.12–11.15) – with the hydrostatic and continuity equations, (11.4) and (11.7), appended – represent, in principle, a closed system from which the evolution of the variables (u, v, w, Φ, T, and S) may be determined. To do so, however, two additional kinds of information are needed. The first are initial conditions – that is, a complete specification of the values of the five variables at the initial instant, e.g., $u(x, y, z, t) = 0$, etc. Unless they are being used for the purpose of short-term prediction, ocean models are often initialized with an appropriate mean stratification from which the pressure field is also prescribed fully by integration of the hydrostatic equation. The initial velocity field is typically chosen in one of two ways: either it is assumed to vanish (i.e., $u = v = w = 0$) or it is chosen to be in geostrophic balance with the horizontal pressure gradients (see Chapter 3, Sec. 3.7.2). (The two choices are equivalent if the initial density field is level, that is, a function of z only.)

Although the particular choices made for the initial field values have important consequences – for example, they dictate in part the length of initial integration necessary before the model reaches an approximate equilibrium with the applied forcing – the specification of a suitable set of spatial boundary conditions is even more subtle, in that they express the coupling between the model ocean, on the one hand, and the atmosphere and the solid earth, on the other. For the closure assumptions leading to equations (11.12–11.15), a boundary condition on each of the four variables u, v, T and S is needed on each of the surface, lateral, and bottom boundaries.

Suitable boundary conditions on solid lateral and bottom boundaries are the least controversial, though alternate approaches do exist. It has been traditional, for example, to assume that the flux of temperature and salinity through these solid walls vanishes. For constant eddy diffusivities, this corresponds to the assumption that the normal gradient of T and S goes to zero on the lateral walls and bottom. In a similar fashion, constraints on the normal fluxes of u and v on the ocean bottom are often prescribed in terms of an assumed bottom drag law. Finally, the sidewalls have typically been treated as being either no-slip (in which case the tangential, as well as the normal, velocity component vanishes) or as free-slip (for which the normal stress is assumed to vanish). These traditional assumptions disguise the fact that the real ocean has no sidewalls; rather, the ocean bottom rises (more or less) smoothly to join the continental boundaries. Depending on the choice of vertical coordinate (see Sec. 11.3.2), numerical ocean models may approximate the true situation with more or fewer vertical sidewall segments.

Sensitivity of model solutions to this choice is not well understood.

The construction of appropriate boundary conditions for nonsolid ("open") lateral boundaries in subglobal-scale ocean models is a matter of substantial current debate and effort. A variety of approaches are being explored, including boundary treatments based upon radiation conditions, frictional "sponge" layers, and the relaxation of near-boundary field values to their prescribed climatological values (Røed and Cooper, 1986).

The surface boundary condition on momentum is generally taken to be the stress exerted by the wind:

$$\rho_o \nu_v \frac{\partial \mathbf{v}}{\partial z}\bigg|_{z=0} = \boldsymbol{\tau}_s$$

though the wind is sometimes applied as a body force within a layer near the surface. Computation of the stress from measured wind speeds and conditions at the air–sea interface is based on empirical formulae with considerable uncertainties, however (see Sec. 4.4 and Fig. 4.6). As a result, there can be significant differences between ocean simulations using different wind-stress climatologies. There also remains a choice as to what range of frequencies should be retained in the wind forcing. A number of studies suggest that high-frequency wind fluctuations may be rectified into changes in the time mean circulation.

There is much variety in the choice of surface boundary conditions for temperature in ocean modeling studies. In a coupled ocean–atmosphere model, all quantities required to compute the heat flux across the air–sea interface are predicted as part of the simulation. In an uncoupled ocean simulation, however, some assumptions must be made in specifying the flux (or another condition on temperature) to close the problem. Perhaps the simplest choice is to specify the sea surface temperature itself. This has been done in a number of coarse resolution studies. One justification for this choice is that the sea surface temperature can be measured more accurately than the heat flux and hence may provide a more reliable forcing of the interior circulation. This choice is less appropriate in high-resolution studies since the temporal evolution and spatial scales of the imposed temperature distribution and the internally generated features of the model (fronts, eddies) will generally not be well matched.

Perhaps the most common technique for specifying the boundary condition on temperature is to compute the surface heat flux based on the model-predicted sea surface temperature T_s and a prescribed atmospheric state (surface air temperature, humidity, wind speed, cloudiness, etc). Since the surface air temperature and sea surface temperature are usually quite close to one another, this is often done with a simple linear approximation to the complete surface energy balance. The heat flux across the sea surface Q is approximated as

$$Q = Q_I + \frac{\partial Q}{\partial T}\bigg|_{T=T_a} (T_a - T_s) \tag{11.16}$$

where T_a is the observed surface air temperature and Q_I is the solar radiation penetrating the sea surface. This can be simplified to

$$Q = \frac{\partial Q}{\partial T}(T^* - T_s) \tag{11.17}$$

where

$$T^* = T_a + Q_I / \frac{\partial Q}{\partial T} \tag{11.18}$$

is called an "effective atmospheric temperature" (Haney, 1971). The largest contribution of the direct solar radiation Q_I to T^* is in the tropics. Although this general form is used extensively, the choices of $\frac{\partial Q}{\partial T}$ and T^* vary widely. Typical values for $\frac{\partial Q}{\partial T}$ range from 20 to 50 W m^{-2} K^{-1} and are primarily a function of wind speed. For a 50 m thick surface layer, a value of 50 W m^{-2} K^{-1} implies an adjustment time scale of approximately 45 days.

On the basis of this line of reasoning, a frequently used ad hoc approach is to relax the model-predicted SSTs towards the observed values, T_{obs}, on a time scale γ^{-1} of one to two months by adding an additional term to the temperature equation at the surface

$$\frac{\partial T_s}{\partial t} = \ldots + \gamma(T_{obs} - T_s) \ . \tag{11.19}$$

An advantage of this formulation over specifying the SST exactly is that it allows the model to develop structures and small-scale features not present in the forcing data. A disadvantage is that, in the case of time-dependent forcing, the amplitude of the annual cycle will be diminished and the phase will lag that of the observed SSTs. Further, note that in the case that the model predicts the sea surface temperature exactly, we have the unrealistic situation of the surface heat flux vanishing everywhere. As in the case of selecting atmospheric data for wind forcing, the choice of the time scales retained in the data used in any of these formulations for the temperature boundary condition will influence the model solution and range of phenomena that can be accurately simulated.

The situation for salinity is somewhat different. Whereas there are strong feedbacks between SST and the surface heat flux, there are essentially no feedbacks between surface salinity and its forcing – evaporation E, precipitation P, runoff from continents R, and the rate of change of sea ice thickness $\frac{\partial h_I}{\partial t}$. Thus it is not possible to formally derive relations analogous to (11.17) or (11.19) for the salinity equation. The surface salinity forcing may be specified as a flux independent of the predicted surface salinity,

$$\kappa_v \frac{\partial S}{\partial z}\Big|_{z=0} = S_0 \left(E - P - R + \frac{\partial h_I}{\partial t} \right) , \tag{11.20}$$

where S_0 is a (constant) reference salinity. Nevertheless, perhaps the most frequently used condition, in practice, is a relaxation term of the form (11.19) with the time scale set to the same value used for temperature. As discussed in later sections, the choice of the type of boundary condition used on salinity can have profound effects on the simulated circulation.

11.2.4 The surface mixed layer

Simple constant eddy viscosity parameterizations may well be crudely appropriate for the bulk of the oceanic interior. However, more elaborate mixing parameterizations are required at the ocean surface, where a complex set of processes mediates the exchanges of momentum, heat, and fresh water with the atmosphere. These processes include a complex surface energy budget, turbulent mixing, convection, and advection. Together they control the spatial and temporal evolution of the *surface mixed layer*, often a region of strong vertical shear and of strong thermal gradients at its base (see Chapter 4, especially Fig. 4.4). The physics of the mixed layer is sufficiently complex that a variety of approaches has been applied to its modeling.

Apart from the introduction of a surface heating term, the equations describing the vertical structure of temperature and velocity in the surface mixed layer are, in principle, similar to those already encountered. Neglecting horizontal variations, as is often done in such models, the equations are:

$$\frac{\partial u}{\partial t} - fv = -\frac{\partial}{\partial z}\overline{u'w'}$$

$$\frac{\partial v}{\partial t} + fu = -\frac{\partial}{\partial z}\overline{v'w'}$$

$$\frac{\partial T}{\partial t} = -\frac{\partial}{\partial z}\overline{T'w'} + \frac{1}{\rho_o c_p}Q(z,t)$$

$$\frac{\partial S}{\partial t} = -\frac{\partial}{\partial z}\overline{S'w'}$$

(11.21)

and $\nabla \cdot \mathbf{v} = 0$, where $Q(z,t)$ is the solar heating absorbed at depth in the water column and c_p is the specific heat of water.

Equations (11.21) are deceptively simple. The problem once again is how to "close" the equations by providing parameterized forms for the turbulent correlation terms. Two general approaches exist: *differential* and *bulk models* (Henderson-Sellers and Davies, 1989). Differential models solve explicitly for the detailed vertical structure of the mixed layer, having first provided a closure to (11.21), and are thus able to predict the existence, as well as the depth, of the mixed layer. Bulk models assume a priori the existence of a homogeneous mixed layer, which allows the equations (11.21) to be simplified considerably by vertical integration over the depth of the mixed layer.

a. Differential models

An example of a differential surface mixed layer model is the Mellor Yamada level-2 model (MYL2). The MYL2 model (Mellor and Yamada,

1982) utilizes equations (11.21) in the form:

$$\frac{\partial u}{\partial t} - fv = \frac{\partial}{\partial z}\left[(K_M + \nu_M)\frac{\partial u}{\partial z}\right]$$

$$\frac{\partial v}{\partial t} + fu = \frac{\partial}{\partial z}\left[(K_M + \nu_M)\frac{\partial v}{\partial z}\right] \tag{11.22}$$

$$\frac{\partial T}{\partial t} = \frac{\partial}{\partial z}\left[(K_H + \nu_H)\frac{\partial T}{\partial z}\right] + \frac{1}{\rho_o c_p}Q(z,t)$$

$$\frac{\partial S}{\partial t} = \frac{\partial}{\partial z}\left[(K_H + \nu_H)\frac{\partial S}{\partial z}\right].$$

Here K_M and K_H are eddy coefficients for vertical diffusion within the mixed layer, and ν_M and ν_H are coefficients for ambient diffusion below the mixed layer.

The boundary conditions that drive (11.22) are

$$\left.(K_M + \nu_M)\frac{\partial u}{\partial z}\right|_{z=0} = \frac{\tau_{s_x}}{\rho_o}$$

$$\left.(K_M + \nu_M)\frac{\partial v}{\partial z}\right|_{z=0} = \frac{\tau_{s_y}}{\rho_o} \tag{11.23}$$

$$\left.(K_H + \nu_H)\frac{\partial T}{\partial z}\right|_{z=0} = \frac{1}{\rho_o c_p}Q_s$$

and

$$\left.(K_H + \nu_H)\frac{\partial S}{\partial z}\right|_{z=0} = (E - P)S|_{z=0} ,$$

where τ_{s_x} and τ_{s_y} are the components of the surface wind stress, and Q_s is the surface heat flux.

For the MYL2 turbulence closure scheme, mixing within the mixed layer is governed by the eddy coefficients K_M and K_H. These, in turn, are given by

$$K_M = lqS_M$$

$$K_H = lqS_H$$

where l is a turbulent length scale, $q^2/2$ is the turbulent kinetic energy (TKE), and S_M and S_H are stability functions which depend on the stratification and vertical shear of the flow. The value of q is determined from a quasi-equilibrium form of the TKE equation in which shear production, buoyancy production, and dissipation are in local balance. Estimation of the turbulent length scale l is perhaps the most problematic aspect of this approach; see Gaspar et al. (1990) for a recent discussion.

b. Bulk models

The Niiler model is a bulk or integrated model in which the mixed layer is treated as a well-defined, homogeneous layer. The equations for the momentum, temperature, and salinity of the mixed layer are then obtained by integrating (11.21) over the mixed layer, giving

$$h\frac{\partial u_m}{\partial t} + \Delta u\frac{\partial h}{\partial t} = \frac{1}{\rho_o}\tau_{s_x} - \nu_M\left.\frac{\partial u}{\partial z}\right|_h + hfv$$

$$h\frac{\partial v_m}{\partial t} + \Delta v\frac{\partial h}{\partial t} = \frac{1}{\rho_o}\tau_{s_y} - \nu_M\left.\frac{\partial v}{\partial z}\right|_h - hfu \qquad (11.24)$$

$$h\frac{\partial T_m}{\partial t} + \Delta T\frac{\partial h}{\partial t} = \frac{1}{\rho_o c_p}Q_s - \nu_H\left.\frac{\partial T}{\partial z}\right|_h + \frac{1}{\rho_o c_p}(Q - Q_h)$$

$$h\frac{\partial S_m}{\partial t} + \Delta S\frac{\partial h}{\partial t} = S_m(E - P) - \nu_H\left.\frac{\partial S}{\partial z}\right|_h \, ,$$

where h is the mixed layer depth, and the subscripts m and h denote values within and just below the mixed layer, respectively. The delta denotes the jump at the base of the mixed layer, i.e., $\Delta( ) = ( )_m - ( )_h$. The profiles of momentum, heat, and salt below the mixed layer are maintained by using (11.22) with $K_M = K_H = 0$.

The equation for the entrainment at the base of the mixed layer for the Niiler model, which is derived from the TKE equation, has the general form:

$$\left(\frac{1}{2}\frac{\Delta\rho}{\rho_o}gh\right)\frac{\partial h}{\partial t} = m_o u_*^3 + m_s A + D + m_c C \, ,$$

where $u_* = (\tau/\rho_o)^{0.5}$ is the surface friction velocity, τ is the magnitude of the stress, and A, D, and C are specified (though complicated) functions of the environmental and forcing parameters, which parameterize the energetic effects of surface fluxes, solar radiation, work from internal waves, and dissipation. The constant m_o is the part of the surface flux of TKE from wave action used for mixing (generation minus dissipation), m_s is a bulk critical Richardson number which regulates the TKE generated by shear production, and m_c is the fraction of the TKE generated by buoyancy dissipated when there is convection due to a negative surface buoyancy flux.

c. Model evaluation

The MYL2 and Niiler models, as well as several other differential and bulk models, have been intercompared on mixed layer datasets taken in both the North Atlantic and North Pacific Oceans (Martin, 1985). All models tested showed some degree of skill in predicting the depth, structure, and temporal variability of the mixed layer; however, no single model was shown to retrieve all the important observed features. One significant difference among the models was that the differential models required, on average, about ten times more computer time to apply than did the bulk (integrated) models.

11.3 Solving the equations of motion

The equations of motion discussed, even in their most simplified form, are multi-dimensional nonlinear partial differential equations for which, in general, exact solutions are unavailable. It is therefore necessary to seek approximate solutions to the dynamical equations being used. In particular, a finite approximation of some form is needed, since typically the resulting approximate equations are solved on a computer. A brief introduction to the concepts of discrete numerical approximation has been presented (Chapter 9, Sec. 9.2.1). Here, a brief elaboration on three important concerns for the configuration of numerical ocean models is also presented.

11.3.1 *Horizontal discretization*

On a horizontal finite difference grid, there exist several alternative approaches to the arrangement of dynamic variables, each of which has subtly different numerical properties (Arakawa and Lamb, 1977). On one of the grids (the "A" grid), the variables are all located at common points, whereas the variables are staggered in various ways on the remaining grids. The resulting properties of a discrete approximation will depend on which gridding scheme is chosen and on how the gridpoint values of $u, v,$ and h are averaged in the discretized equations (Bryan, 1989). In particular, grids "B" and "C" (the most prevalent choices among ocean modelers) have quite different inertial-gravity wave propagation characteristics. Whereas the "C" grid behaves in a manner closest to the continuous solutions for grid spacing fine compared to the deformation radius $\lambda = (gH)^{\frac{1}{2}}/f$, the "B" grid has, on average, better properties when the grid is coarser than λ. Comparable dependencies of the propagation properties of free Kelvin and planetary (Rossby) waves on the specifics of the horizontal grid system have also been documented.

11.3.2 *Vertical coordinates*

There are three categories of vertical coordinate that have become popular for ocean circulation modeling. Perhaps the simplest system of all uses depth z as the vertical coordinate. A disadvantage of this representation is that the bottom of the ocean is approximated as a series of steps. This may lead to spurious transfer of energy from the barotropic to the baroclinic components of motion. Another disadvantage of the z-coordinate system is in the treatment of horizontal mixing of momentum, heat and salinity, which may be responsible for excessive cross-isopycnal mixing. This problem can be avoided, in part, by rotating the mixing tensor so that mixing is primarily aligned along isopycnal surfaces, i.e. surfaces of constant density.

In most atmospheric models used in numerical weather prediction, pressure normalized by surface pressure is used as the vertical coordinate (see Chapter 9, Sec. 9.2.2). The analogue in ocean circulation models is a vertical coordinate normalized by the depth of the bottom, z/H. The advantage of this "sigma" coordinate is that it allows a better representation of processes related to sloping topography. One disadvantage, however, is that systematic errors can arise in calculating the pressure force. Also, mixing along sigma coordinate surfaces can cause a spurious cross isopycnal mixing similar to that encountered in z-coordinate models. Nonetheless, GCMs based upon the sigma coordinate transformation have proven valuable for regional ocean modeling studies, particularly in the coastal ocean (Haidvogel et al., 1991a), and may soon be extended to basin-scale and even global ocean simulations.

Equations of motion based on isopycnal coordinates are in many ways most natural for the large-scale ocean circulation. Throughout much of the oceanic interior, mixing by eddies is thought to occur primarily along isopycnal surfaces. Parameterization of such fluxes is, therefore, most easily handled directly in an isopycnal formulation. Also, an isopycnal model is already in a very convenient form for analysis and thus provides greater ease of interpretation. The disadvantage is that a semi-Lagrangian coordinate is inherently more complex to work with than fixed coordinates. This is particularly true over complex topography. However, rapid progress is being made in the development of very general isopycnal ocean circulation models and their application to basin-scale simulations (Bleck and Smith, 1990).

11.3.3 *Conservation and positivity properties*

In the absence of either forcing or damping, the continuous equations of motion have many analytic properties. Among these are that solutions which are initially single-signed (positive, say, like density) should remain so. In their continuous form, the equations also show that many quantities (mass, energy, enstrophy, etc.) are strictly conserved in the inviscid and unforced limit. Unfortunately, approximate solutions of the same equations obtained using any of the mentioned techniques are not guaranteed to retain these important properties.

The choice of space time discretization for advective equations is particularly complicated by the fact that no single algorithm currently exists which simultaneously meets the requirements that numerical dissipation and dispersion error be minimized, that positivity of fields be maintained, and that important conservation properties be preserved. For example, methods are available (e.g., the flux-corrected transport and semi-Lagrangian techniques) which preserve positivity, but at the expense of significant numerical dissipation. Although, generally speaking, it has proven to be advantageous to use higher-order methods (fourth-order finite difference, and spectral and finite element techniques) and schemes which are formally conservative of lower-order properties (mass, energy, enstrophy, etc.), even these approaches often fail to meet some of the above criteria (e.g., positivity of fields). Some of these points are reviewed in detail in Arakawa and Lamb (1977), O'Brien (1986) and Rood (1987).

11.4 Simple models of the ocean circulation

11.4.1 The wind-driven ocean

Although climate system models will undoubtedly require sophisticated ocean model components, much physical intuition about the ocean circulation may be gained from simpler models. Consider the classical example of a homogeneous ocean ($\rho = \rho_o = $ constant) of uniform depth H driven at its surface by a specified wind stress $\boldsymbol{\tau}_s = (\tau_{s_x}, \tau_{s_y})$, as shown in Fig. 11.1. In this idealized setting, and after appropriate simplification of the equations of motion, the steady response of the ocean to the applied stress can be determined directly. Among other things, this reveals the important role of surface and lateral boundary layers.

On an f-plane (i.e., with $f = f_o$), and, for the moment, ignoring temporal and lateral variations ($\frac{\partial}{\partial t} = \frac{\partial}{\partial x} = \frac{\partial}{\partial y} = 0$), an important integral property of the surface *Ekman boundary layer* may be determined as described in Chapter 4, Sec. 4.4. In particular, the continuity equation, together with the requirement that vertical velocity vanish at the surface (presumed to be a rigid lid), implies that w vanishes everywhere; hence, as given in (4.2), the momentum equations become:

$$-f_o v = \frac{1}{\rho_o} \frac{\partial \tau_{s_x}}{\partial z} \qquad (11.25a)$$

and

$$f_o u = \frac{1}{\rho_o} \frac{\partial \tau_{s_y}}{\partial z} \ . \qquad (11.25b)$$

Among the properties of the Ekman layer is that the integrated mass flux in the surface Ekman layer is proportional to the applied stress:

$$\int_{-\infty}^{0} (\rho_o v) dz = -\tau_{s_x}/f_o \qquad (11.26a)$$

$$\int_{-\infty}^{0} (\rho_o u) dz = \tau_{s_y}/f_o \ . \qquad (11.26b)$$

and that there is a net vertical velocity, or *Ekman pumping*, at the base of the layer which is related to the curl of the wind stress:

$$w_E = \mathbf{k} \cdot \nabla \times \frac{\boldsymbol{\tau}_s}{\rho_o f_o} \ , \qquad (11.27)$$

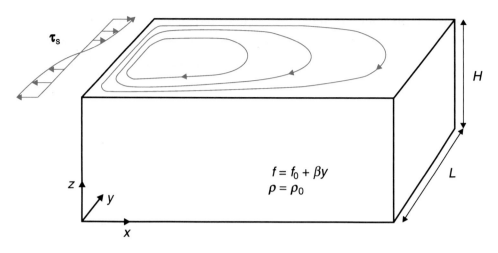

Fig. 11.1 Schematic of the barotropic model of the wind-driven circulation of the subtropical gyre.

see also Eq. (4.5). Expressions analogous to (11.27) may also be derived for the net effects of an *Ekman bottom boundary layer*. There, the Ekman pumping is proportional to the vorticity of the interior flow. See, for example, the complete treatment given by Pedlosky (1987a). The vertical mass fluxes at the ocean surface and bottom, though weak, are an essential element of the ocean's response to applied wind stresses.

In Cartesian geometry, and with a constant eddy viscosity closure, the HPE for a homogeneous ocean can be scaled to yield a simple, single equation which contains much information about the response of the ocean to an imposed wind stress (Pedlosky, 1987a). For a basin-scale flow in the interior of the ocean, and with L, H and U_o representing typical length, height, and velocity scales respectively, the resulting scaled equations are characterized by several nondimensional parameters: the Rossby number $(Ro = U_o/f_oL)$, nondimensional beta $(\hat{\beta} = \beta L/f_o)$, and the horizontal and vertical Ekman numbers $(E_h = \nu_h/f_oL^2$, $E_v = \nu_v/f_oH^2)$, where, $f_o = 2\Omega \sin \phi_o$ and $\beta = f_o \cot \phi_o/a$. If it is now assumed $Ro \approx E_v^{\frac{1}{2}} \approx \hat{\beta} < 1$ and $E_h << 1$, the lowest-order interior flow is found to be geostrophic, horizontally nondivergent, and depth independent.

At the next order in Rossby number, the evolutionary equation is:

$$\frac{d}{dt}\left[\zeta + \left(\frac{\hat{\beta}}{Ro}\right)y\right] = \left(\frac{\partial}{\partial t} + u\frac{\partial}{\partial x} + v\frac{\partial}{\partial y}\right)\left[\zeta + \left(\frac{\hat{\beta}}{Ro}\right)y\right] \quad (11.28)$$

$$= w_{z=0} - w_{z=-H} + \left(\frac{E_h}{Ro}\right)\nabla^2\zeta \,,$$

where the relative vorticity $\zeta = \frac{\partial v}{\partial x} - \frac{\partial u}{\partial y}$. Equation (11.28), a form of the well-known *barotropic vorticity equation* (BVE), shows how the evolution of the vorticity in the interior of the ocean is directly influenced by vorticity produced by vertical velocities at the surface ($w_{z=0}$) and/or bottom ($w_{z=-H}$). Using the values for w_E at the top and bottom, the BVE, in dimensional form, is then:

$$\frac{d}{dt}(\zeta + \beta y) = \mathbf{k} \cdot \nabla \times \left(\frac{\boldsymbol{\tau}_s}{\rho_o H}\right) - r\zeta + \nu_h \nabla^2 \zeta ,$$

where r^{-1} is a frictional time scale related to the vertical eddy viscosity (ν_v). Since the lowest-order flow field is horizontally nondivergent, a streamfunction $\psi(x, y)$ such that $(u, v) = (-\partial\psi/\partial y, \ \partial\psi/\partial x)$ may also be introduced. Then:

$$\left[\frac{\partial}{\partial t} + J(\psi, \ )\right](\nabla^2\psi + \beta y) = \mathbf{k} \cdot \nabla \times \frac{\boldsymbol{\tau}_s}{\rho_o H} - r\nabla^2\psi + \nu_h\nabla^4\psi \qquad (11.29)$$

where J is the Jacobian.

In the ocean interior, the dominant terms in (11.29) reflect a balance between the input of vorticity by the wind stress curl and the vorticity tendency associated with north–south movement on the β-plane:

$$\beta v = \mathbf{k} \cdot \nabla \times \frac{\boldsymbol{\tau}_s}{\rho_o H} . \qquad (11.30)$$

This is called the *Sverdrup balance* (see Chapter 4, Sec. 4.5). Suppose the wind stress takes the form $\tau_{s_x} = -\tau_o \cos(\pi y/L)$ in a square basin of length L. Then, in the interior, the north–south velocity

$$\frac{\partial\psi}{\partial x} = v = -\left(\frac{\tau_o\pi}{\rho_o HL\beta}\right)\sin\left(\frac{\pi y}{L}\right) . \qquad (11.31)$$

However, the circulation pattern is not fully determined. [To see this, try to integrate (11.31) to determine a $\psi(x, y)$ field that satisfies the constraint that ψ vanish simultaneously at $x = 0$ and L, corresponding to the requirement that there be no normal flow through any point on either the eastern or western boundaries.] In particular, since the interior flow is directed southwards at any interior latitude, a boundary layer must exist on one of the meridional walls to close the mass flux. By considering conservation of vorticity, it can be shown that the boundary layer must exist on the western wall (Stommel, 1965). With bottom friction of the Ekman sort, the dominant balance in the western boundary layer is

$$\beta v = \beta\psi_x \approx r\psi_{xx} ,$$

with the associated boundary layer structure

$$\psi \approx \exp(-\beta x/r) .$$

If the boundary layer thickness (r/β) $<< L$, the total solution can then be written:

$$\psi(x, y) = \frac{\pi\tau_o}{H\beta\rho_o}\left[\frac{(L-x)}{L} - e^{-\beta x/r}\right]\sin(\pi y/L) , \qquad (11.32)$$

in which the Sverdrup (interior) and boundary layer circulations are combined. This is the *Stommel model* of the wind-driven circulation (Stommel, 1948). Analogous analytic models incorporating the effects of lateral friction and weak nonlinearity have been examined by Munk (1950) and Charney (1955a). The earliest numerical extensions of this work were due to Bryan (1963).

11.4.2 Box models of the thermohaline circulation

The simple models for the wind-driven circulation described in the previous section assume that buoyancy variations play little or only a passive role in the dynamics. Despite this drastic assumption, they are quite successful in representing the basic patterns of the circulation and provide the underpinning of much of the theory of the ocean general circulation. For climate applications, however, it is the spatial and temporal variations of the temperature and salinity distributions of the ocean and its capacity for heat storage and transport that are of primary concern. The determination of the thermohaline (joint effects of heat and salt on buoyancy) driven circulation and the reciprocal effects of the circulation on the distribution of water mass properties are difficult problems for several reasons. First and foremost is the essential nonlinearity of the system. The models of the previous section could be obtained through a systematic scale analysis and linearization of the governing equations. In considering the thermohaline circulation, the advection of heat and salt by the circulation is central to the problem and cannot be neglected. A further complication arises from the difference between the form of the surface forcing for temperature and salinity, as discussed in Sec. 11.2.3. As a result of the different mathematical structure of the boundary conditions on heat and salt (often referred to in the literature as mixed boundary conditions) the problem cannot be reduced to one in a single buoyancy variable. Additional complications in modeling the thermohaline circulation arise from the nonlinear equation of state for sea water and the presence of double-diffusive phenomena.

An alternative to the formal mathematical derivation of simplified models from the full equations of motion is to pose a conceptual model or simple physical analog for the system or processes being considered. This approach has been fruitfully exploited in developing our understanding of the dynamics of the thermohaline circulation. The consequences of the difference in the nature of the feedbacks between surface temperature and salinity and their respective surface forcings were first explored by Stommel (1961) using a very simple model consisting of two well-mixed reservoirs connected by pipes. In particular, he showed that the thermohaline circulation may have multiple equilibria for a given surface forcing distribution.

These ideas are illustrated using a slightly modified version of the model (Fig. 11.2), as described by Marotzke (1989). The two reservoirs are taken to represent equatorial and polar regions of the ocean. In light of the discussion

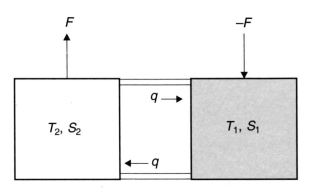

Fig. 11.2 Schematic of the two-box model of the thermohaline circulation. Box 1 represents high-latitude conditions, box 2 represents low-latitude conditions.

of surface boundary conditions above, it is assumed that the temperatures remain very close to the imposed atmospheric values T_1 and T_2 ($T_1 < T_2$) and are simply held fixed. Salinity, on the other hand, is forced by a flux of moisture F through the atmosphere from the low-latitude to the high-latitude box. The sense of the salinity forcing is to make the low-latitude box saltier and more dense and the high-latitude box fresher and less dense. The resulting torque thus opposes that of the imposed temperature differences. Conservation of salt in the boxes is given by:

$$V \frac{dS_1}{dt} = -FS_0 + |q|(S_2 - S_1) \tag{11.33}$$

$$V \frac{dS_2}{dt} = FS_0 + |q|(S_1 - S_2) \, , \tag{11.34}$$

where V is the volume of the boxes (assumed equal), S_0 is a (constant) reference salinity, and q is the rate of volume exchange in the pipes. The absolute value in the advective term arises because the same exchange is affected irrespective of the direction of the flow. The flow is driven by the pressure difference between the boxes (linearly proportional to the density difference) and retarded by friction in the pipes, and is assumed to be in instantaneous balance. This simplified dynamics is modeled using a resistivity k^{-1} to obtain:

$$q = -\frac{k}{\rho_o}(\rho_2 - \rho_1) \, . \tag{11.35}$$

The system is closed with a linear equation of state:

$$\rho = \rho_o(1 - \alpha T + \beta S) \tag{11.36}$$

where α is the thermal expansion coefficient and β is the haline contraction coefficient.

From (11.33) and (11.34) the total salt content of the system is conserved, and from (11.35) only the difference between the temperature and salinity of the two boxes enters the dynamics. A transformation is made to the following nondimensional variables

$$\delta \equiv \frac{\beta(S_2 - S_1)}{\alpha(T_2 - T_1)} \tag{11.37}$$

$$r \equiv \frac{q}{k\alpha(T_2 - T_1)} \tag{11.38}$$

$$E \equiv \frac{\beta F S_0}{kV\alpha^2(T_2 - T_1)^2} . \tag{11.39}$$

Time is also nondimensionalized by $\frac{2k\alpha(T_2 - T_1)}{V}$. The quantity δ represents the relative contributions of salinity and temperature to the buoyancy difference between the boxes, r represents the strength of the flow relative to the purely thermally driven system, and E represents the strength of the salinity forcing relative to advection. Subtracting (11.33) from (11.34) and substituting (11.37–11.39) gives

$$\dot{\delta} = E - |r|\delta , \tag{11.40}$$

and (11.35) becomes

$$r = 1 - \delta . \tag{11.41}$$

The three steady state solutions of (11.40–11.41) are given by

$$\delta_1 = \frac{1}{2}(1 - \sqrt{1 - 4E}) \tag{11.42a}$$

$$\delta_2 = \frac{1}{2}(1 + \sqrt{1 - 4E}) \tag{11.42b}$$

$$\delta_3 = \frac{1}{2}(1 + \sqrt{1 + 4E}) \tag{11.42c}$$

and are shown in Fig. 11.3. The system can support multiple equilibria for $E < \frac{1}{4}$; there are three solutions for a given value of E. The solution δ_2 can be shown to be unstable and hence not realizable in the real climate or a numerical model.

Solution δ_1 represented by the solid portion of the curve in Fig. 11.3, has $0 \le \delta \le 0.5$; i.e., the contribution of temperature to the buoyancy difference dominates that of salinity. The flow is relatively strong ($0.5 \le r \le 1$) with "sinking" in the cold box, deep flow to low latitudes, "upwelling" in the warm box, and surface flow back to high latitudes. This corresponds to the configuration of the thermohaline circulation of the North Atlantic under present climate conditions.

Solution δ_3 , represented by the dash-dot portion of the curve in Fig. 11.3, has $\delta \ge 1$; i.e., the contribution of salinity to the buoyancy difference dominates over that of temperature. In this regime, the flow is reversed from the previous case ($r < 0$) with "sinking" in the warm box, deep flow from low to high latitudes, "upwelling" in the cold box, and surface flow from high back to low latitudes. There is evidence that such a circulation with warm

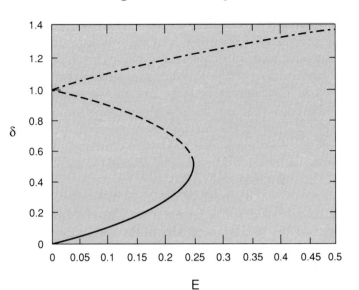

Fig. 11.3 Solution to the two-box model. The solid portion of the curve corresponds to Eq. (11.42*a*), the dashed portion to (11.42*b*), and the dash-dot portion to (11.42*c*).

salty deep water may have existed in the geologic past. In the forcing regime, where multiple equilibria are supported, the salinity dominated "inverse" circulation is weaker than the thermally dominated solution. This can be understood physically in terms of the relationship between residence time of water parcels in the boxes and the effect of the flux boundary condition on salinity. The longer a parcel remains in one box or the other, the more its salinity will be changed by the constant input or removal of fresh water through the surface. The slower circulation thus allows the salinity difference to build up and eventually dominate over temperature.

Welander (1986) has extended this system by adding a third box representing a second polar region. He shows that the number of stable equilibria increases to four, corresponding to combinations of the solutions of two independent two-box models: a symmetric solution with sinking in the polar boxes and upwelling in the equatorial box, a symmetric solution with sinking in the equatorial box and upwelling in the polar boxes, and two asymmetric solutions with sinking in one of the polar boxes and upwelling in the other polar box. The existence of multiple equilibria has also been found in more complex ocean models and coupled ocean–atmosphere models as described further in the following sections and in Chapter 17.

11.4.3 *Two-dimensional meridional plane models*

Box models provide a means of exploring basic physical processes active in the thermohaline circulation, and have been particularly useful in marine biogeochemistry (see Chapter 16). However, they have many limitations. The process of deep water formation in limited areas through convective overturning and subsequent upwelling through the stratified thermocline are key features of the thermohaline circulation. These cannot be easily modeled in the context of large well-mixed reservoirs. Another class of simplified models that can address these issues are two-dimensional models for the circulation in the meridional–vertical plane. The primary motivation for using two-dimensional models is economy. The adjustment times for the thermohaline circulation can be several thousand years, making studies with full three-dimensional GCMs very expensive.

Zonal mean diagnostics and models of the zonal mean circulation are quite common in atmospheric studies but have not been used extensively in ocean modeling. The reason lies in the complexity of the geometry of ocean basins. Beginning with a simplified momentum balance of geostrophy supplemented by vertical viscosity, zonal averaging yields

$$-f[v] = -\frac{\Phi(L) - \Phi(0)}{L} + \nu_v \frac{\partial^2 [u]}{\partial z^2} \qquad (11.43)$$

$$f[u] = -\frac{\partial [\Phi]}{\partial y} + \nu_v \frac{\partial^2 [v]}{\partial z^2} , \qquad (11.44)$$

where

$$[(\,)] = \frac{1}{L} \int_0^L (\,) dx \qquad (11.45)$$

and L is the width of the basin. The fundamental difficulty arises from the term $\frac{\Phi(L) - \Phi(0)}{L}$. Unlike the cyclic atmospheric case, the zonal pressure gradient term in a bounded basin does not vanish on zonal averaging and cannot be determined directly from the equations. An ad hoc parameterization is necessary to close the system. Marotzke et al. (1988) accomplish this by neglecting rotation. The zonal momentum equation then becomes decoupled from the rest of the system, reducing the dynamics to the simple relation

$$\frac{\partial [\Phi]}{\partial y} = \nu_v \frac{\partial^2 [v]}{\partial z^2} . \qquad (11.46)$$

The vertical component of the flow can be obtained from the zonally averaged continuity equation. Wright and Stocker (1991) propose a more elaborate parameterization that relates the zonal pressure gradient to the meridional gradient through a series of plausible assumptions. For further discussion on the formulation of this class of models the reader is referred to the above references or Thual and McWilliams (1991). The latter provides a careful examination of the relationship between the mathematical structure of the solutions of a hierarchy of box models and two-dimensional models of this type.

The growing interest in the role of the thermohaline circulation in the

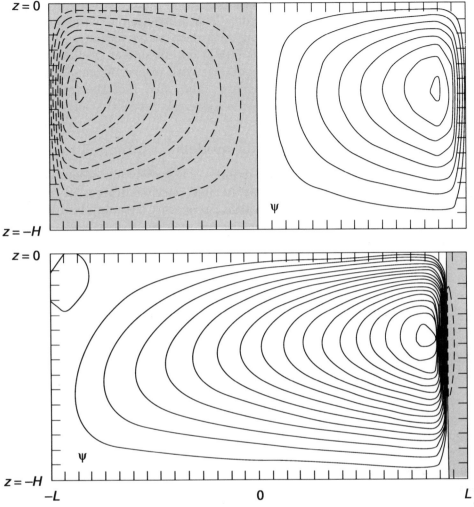

Fig. 11.4 Solutions of the two-dimensional model of Marotzke et al. (1988). The top panel
shows the symmetric solution with downwelling at each pole. The bottom panel
shows the pole-to-pole solution with downwelling at only one pole, and upwelling
throughout the interior.

climate system (Broecker and Denton, 1989) has led to a revival in the
use of this class of models despite their ad hoc nature. In particular,
they have been used to explore the question of the existence of multiple
equilibria in the thermohaline circulation resulting from the use of mixed
boundary conditions, and the conditions under which the ocean might switch
from one of these states to another. Marotzke et al. (1988) demonstrated
that a model forced with symmetric (about the centerline of the model)
surface temperature and symmetric surface salt flux had both symmetric
solutions and asymmetric pole-to-pole equilibrium solutions (Fig. 11.4). Of
particular interest is their result that vanishingly small perturbations in

high-latitude surface salinity (such as might occur from continental runoff during deglaciation) can lead to a jump from a solution with deep water formation in both hemispheres to one with deep water formation in only one hemisphere. Wright and Stocker (1991) obtain similar results and go on to examine the stability of these equilibria with respect to small changes in the atmospheric fresh water transport. They find that the transitions between equilibria are characterized by a hysteresis type behavior.

11.5 Ocean general circulation models

As discussed in the introduction, a hierarchy of ocean GCMs (OGCMs) have been developed that include varying degrees of spatial coverage, resolution, geographical realism, processes, etc. In this section, discussion is restricted to three-dimensional models that include active thermodynamics and hence are most directly applicable to climate studies. Until recently, only primitive equation models have fallen into this category. The features discussed focus on the role of various modeling choices in establishing the three-dimensional structure of the circulation, and in mechanisms responsible for ocean heat transport, whose importance was outlined in Chapter 4.

11.5.1 Idealized geometry models

Models with idealized basin geometry have been used extensively in ocean modeling and have played a major role in the development of new modeling methodologies, in establishing model sensitivity to forcing specification and parameterizations, and in exploring the role of various processes in the ocean circulation. Using a simplified geometry offers a number of advantages. In addition to the simplification in the shape of the basin itself, the distribution of winds and buoyancy forcing are generally chosen as simple functions of latitude, thereby reducing the number of parameters in the problem. The simple geometry and resulting flow configuration often make it easier to isolate specific phenomena or processes and make linkages with established theories of the general circulation. By taking advantage of the symmetries and uniformity of the problem, it is often much more economical to do extended integrations with this type of model. A disadvantage is that it is not possible to make direct comparisons with observed oceanographic fields.

The most common geographic idealization is a basin consisting of a sector of the sphere bounded by two meridians (usually taken to be about 60° apart) with a flat bottom. Frequently a symmetry condition is imposed at the equator. The first three-dimensional ocean general circulation model experiments (Bryan and Cox, 1967) as well as the first coupled ocean–atmosphere model experiments (Manabe and Bryan, 1969) were carried out in this configuration.

The first and perhaps most significant application of this class of models has been the elucidation of the relative roles of wind and buoyancy forcing in driving the ocean general circulation. The use of numerical models in this line of investigation is especially significant owing to the slow progress on this problem by analytical methods. Early theories of the ocean thermocline described the circulation in terms of self-similar solutions to the governing nonlinear partial differential equations (Veronis, 1973). While these solutions captured the general structure of the circulation, there were restrictions on the type of the surface boundary conditions that could be applied. Further, theories with quite different assumptions about the underlying physics gave nearly indistinguishable solutions. This was a disadvantage in terms of their ability to identify and isolate important processes. Finally, analogous to the Sverdrup theory of the wind-driven circulation, these were all inviscid solutions and hence could not explicitly account for transport in a western boundary layer. The pioneering calculations of Bryan and Cox (1967) provided a first look at the three-dimensional structure of the combined wind- and thermohaline-driven circulation. Partly in response to a revival of interest in the theory of the thermocline (Abarbanel and Young, 1987), a later series of experiments (Bryan and Sarmiento, 1985) refined and extended these early studies.

When forced with a zonally uniform meridional buoyancy gradient, but in the absence of wind forcing, the circulation in the simple sector models consists of a single anticyclonic gyre in the upper layers and a cyclonic gyre at depth (Fig. 11.5). There is essentially no vertically integrated transport. The surface and deep ocean are connected by sinking in the northeast corner of the basin and weak upwelling throughout the interior. The meridional flow is intensified in surface and deep western boundary currents, and the interior surface flow is nearly zonal as determined by thermal wind balance and the imposed surface buoyancy distribution. The circulation in the meridional–vertical plane (Fig. 11.6) shows a narrow region of sinking at the northern wall and nearly uniform upwelling throughout the interior. The meridional heat transport (Fig. 11.7) reaches a maximum at about the middle of the basin. When wind stress is applied, the surface circulation shows the familiar pattern of anticyclonic circulation in the subtropics and cyclonic circulation in the subpolar region (Fig. 11.5). This pattern is also reflected in the vertically averaged circulation. In addition to the meridional cell apparent in the thermally driven case, several shallow cells appear in the zonally averaged circulation with the addition of wind forcing (Fig. 11.6). These have downward motion in the subtropical gyre as a result of convergent Ekman transport and upwelling at the equator and in high latitudes in response to divergent Ekman transport. These shallow wind-driven overturning cells have a strong influence on the poleward heat transport since the temperature difference between the surface and lower return branches can be large. The latitude of maximum poleward

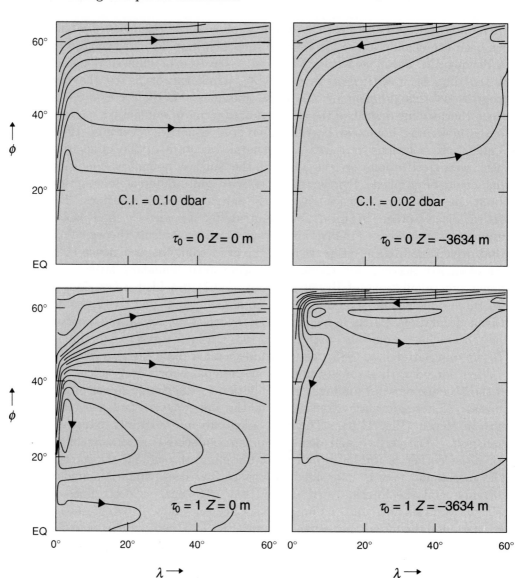

Fig. 11.5 Pressure for purely thermally driven (top) and wind- and thermally driven (bottom) cases of Cox and Bryan (1984) at the surface (left), and near the bottom (right). The flow is approximately parallel to the contours of pressure.

heat transport moves equatorward due to the large contribution of the tropical Ekman overturning cell (Fig. 11.7). This is roughly consistent with observations, which indicate that the maximum in the oceanic poleward heat transport occurs between 20° and 30° latitude (see Chapter 4, Figs. 4.18 and 4.20), whereas the atmosphere has a maximum poleward heat transport near 45° latitude.

Internal parameters can also affect the structure of the circulation. In

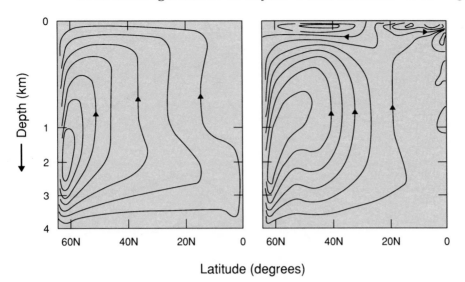

Fig. 11.6 Streamfunction for the zonally averaged meridional mass transport from the experiments of Cox and Bryan (1984) with no wind (left), and normal wind (right). Note the the vertical scale is expanded in the upper 1 km.

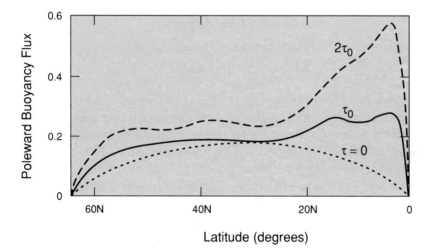

Fig. 11.7 Poleward heat transport (petawatts) for the no-wind (dotted), normal wind (solid), and doubled wind (dashed) cases of Cox and Bryan (1984).

particular, the magnitude of the vertical eddy diffusivity, κ_v, plays an important role in determining the relative contributions of wind and thermal forcing in the three-dimensional circulation. With increasing vertical diffusivity the strength of the meridional overturning and poleward heat transport both become larger. Examination of the solutions (F. Bryan, 1987) shows that the circulation becomes increasingly like the purely thermally driven case described above as the vertical diffusivity increases.

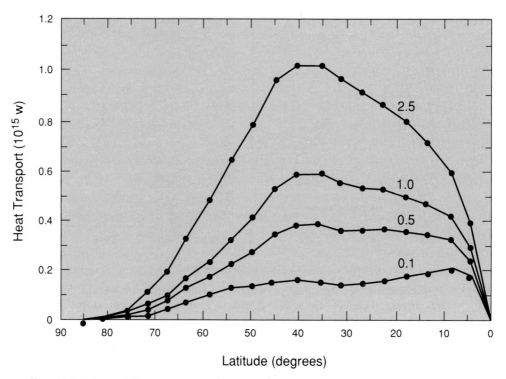

Fig. 11.8 Poleward heat transport (petawatts) for several cases with increasing vertical diffusivity from F. Bryan (1987).

This is apparent in the heat transport (Fig. 11.8) which has its maximum at increasingly higher latitudes as the vertical diffusivity becomes larger. Both the sensitivity to wind stress and to vertical diffusivity can be understood in terms of simple scaling relationships presented in F. Bryan (1987) and Bryan and Sarmiento (1985). These papers give a much fuller description of these issues and other aspects of the circulation in sector geometry models.

The single hemisphere sector models capture a number of basic features of the combined wind- and thermohaline-driven circulation. Inevitably, there are features of the real ocean that are left out however. One such feature is the role of a circumpolar current. In an illuminating study, Cox (1989) examined this question and others pertaining to the global distribution of water masses with a model comprised of several interconnected sectors approximating the true global geography. Figure 11.9 shows the meridional overturning in the "Atlantic" sector with a circumpolar channel open and closed. The basin extends to slightly higher latitude in the Southern Hemisphere, hence the deep water formed there is slightly colder and denser than that formed in the Northern Hemisphere. As a result, in the case with the channel closed, the southern deep water source dominates the meridional circulation and the northern overturning cell is confined to the upper 1,000 m. The meridional flow is in approximate geostrophic balance

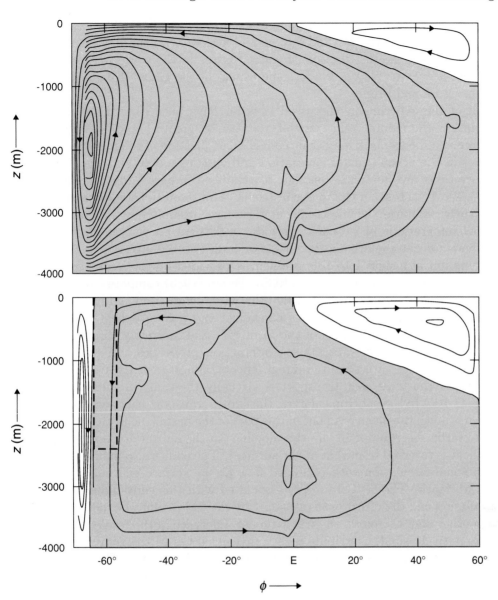

Fig. 11.9 Streamfunction for the zonally averaged meridional mass transport from the experiments of Cox (1989) with Drake Passage closed (top) and open (bottom). Neither case includes wind forcing.

with the east–west pressure gradient across the basin. Within the the area of the open channel, the east–west pressure gradient vanishes in the zonal mean, so the southern meridional cell is greatly diminished. The northern cell strengthens and deepens. Based on these results, Cox suggests that the global ocean may be relatively insensitive to perturbations in the rate of Antarctic Bottom Water formation, since these will be largely trapped on the poleward side of the circumpolar current. Additional experiments have examined the role of wind forcing and a Mediterranean-like salt source in determining the distribution of water masses in the world ocean. Each of these is seen to play a role in some aspect of the global distributions of temperature and salinity.

All of the experiments discussed thus far have been carried out at low resolution (horizontal grid sizes of 100 km or greater) and with steady forcing and, hence, lack mesoscale eddies. The question of the importance of eddies in the ocean general circulation and climate is a crucial one. Ocean observations are not dense enough to reliably compute eddy correlations such as $\overline{v'T'}$ across an entire ocean basin. The resolution requirements to explicitly compute the mesoscale motions are prohibitively expensive for global integrations of extended periods, and there are as yet no reliable methods for parameterizing the effects of eddies on the larger-scale flow.

Some initial insight into this problem is provided by the high-resolution sector model experiments of Cox (1985). Bryan (1986) compares the results of this eddy-resolving model and an otherwise identical non-eddy-resolving sector model calculation. Eddies can affect the poleward heat transport directly through correlations of the form $\overline{v'T'}$ and indirectly through their effect on the mean flow or air–sea interaction properties. In this model experiment, the direct and indirect effects almost exactly cancel each other and the total heat transport is virtually the same in the two cases (Fig. 11.10). In the subtropical gyre, the eddies transport heat equatorward, which is down the mean gradient in this part of the basin (the isotherms slope up towards the equator in the thermocline, so the subsurface temperature gradient is reversed from that at the surface). The eddies also induce a mean meridional circulation with poleward flow at the surface and equatorward flow at depth. The heat transport associated with this eddy-induced mean circulation and the direct heat transport by the eddies almost exactly cancel each other and, therefore, there is little net difference in the total transport. The relatively small contribution of transient eddies to the total poleward heat transport is in strong contrast to the atmospheric case. This is not to say that the eddies play no role in the circulation. Indeed, in the experiments of Cox (1985), the distribution of tracers and potential vorticity are very different in the eddy-resolving and non-eddy-resolving cases. This may have important implications for the ocean's capacity to absorb and sequester trace gases, for example.

11.5.2 *Basin-scale models*

To compare OGCM results with observations or to investigate processes associated with specific geographic or topographic features, it is necessary to go beyond idealized geometry models to models that include realistic representations of coastlines, surface forcing, etc. Ideally,

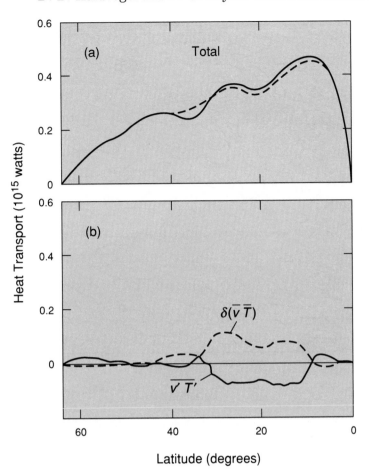

Fig. 11.10 (a) Comparison of the poleward heat transport in eddy-resolving (solid) and non-eddy-resolving (dashed) sector models of Cox (1985). (b) Contribution to the mean heat flux from transient eddies (solid) and enhanced mean flow (dashed) in the eddy-resolving model.

global models would be used in these cases, but, frequently, the cost of running a global model at the desired resolution would be prohibitive. Models of an individual basin can be run at lower cost and at higher resolution than would be possible for the world ocean, allowing for more realistic representations of complex geometry and physical processes. While reducing the computational cost, they do introduce the complication of open boundaries. Nearly the entire world ocean has been examined basin-by-basin using this approach.

The Atlantic is the most thoroughly observed ocean and includes a diverse range of physical processes, making it an ideal context for the application and testing of OGCMs. In this section, a brief review is given of some results of the Atlantic basin calculation of F. Bryan and Holland (1989) as a

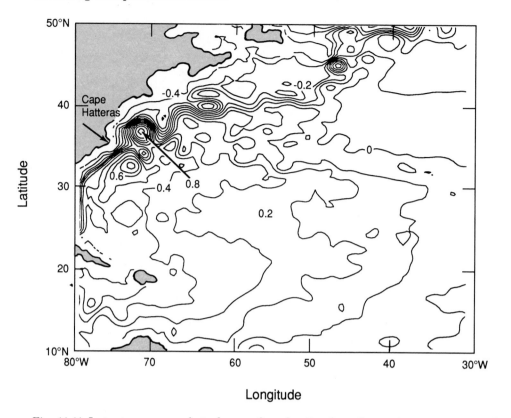

Fig. 11.11 Instantaneous snapshot of sea surface elevation from the northwestern portion of the eddy-resolving North Atlantic model of F. Bryan and Holland (1989). Note the overshoot of the Gulf Stream north of Cape Hatteras.

representative example of this class of model. The model extends from 15°S to 65°N with a horizontal resolution of $\frac{1}{3}^{\circ}$ latitude by $\frac{2}{5}^{\circ}$ longitude and 30 levels in the vertical, sufficient to marginally resolve mesoscale eddies. The model was forced with seasonally varying climatological winds, a Haney (1971) type surface heat flux parameterization, and a linear damping of surface salinities towards observed values. The open boundaries at the north and south were treated by damping the model-predicted temperatures and salinities towards observed values on a time scale of a few weeks within narrow buffer zones adjacent to the boundaries. No flow was allowed to pass through the boundaries however.

The mean circulation and the eddy variability are in qualitative agreement with observations. A discrepancy with the observed circulation occurs in the vicinity of Cape Hatteras, where the simulated Gulf Stream follows the coast too far north before separating and turning east (Fig. 11.11). This problem is common in both high- and low-resolution models, and models with and without active thermodynamics. Owing to the large thermal gradients in the Gulf Stream, small errors in the position of the current axis can lead to

402

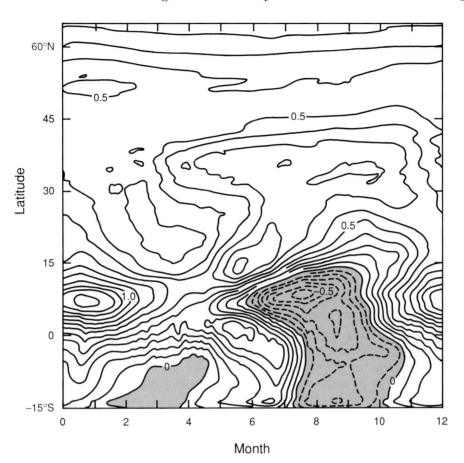

Fig. 11.12 Mean annual cycle of northward heat transport in the eddy-resolving North
Atlantic model of F. Bryan and Holland (1989).

large local errors in the predicted sea surface temperature and surface heat
flux. While the magnitude of the eddy variability is close to the observed
values near the Gulf Stream axis, the values in the eastern portion of the
basin are too low by approximately an order of magnitude. Thus, it may
be necessary to increase resolution even further to accurately portray the
observed dynamics.

In the previous section, the long-term mean poleward heat transport by
the ocean was discussed. Another important influence of the ocean on the
climate system is through seasonal variations in heat transport and heat
storage. The annual cycle of heat transport from this simulation is shown
in Fig. 11.12. Particularly striking is the large seasonal cycle in the tropics.
During late northern fall and winter there is northward heat transport
associated with northward surface flow in the Brazil Current. In spring
and summer, the flow separates from the coast near 7°N and flows eastward
in the North Equatorial Counter Current. The heat transport during this

period becomes equatorward. This seasonal variation in heat transport is largely balanced by seasonal heat storage resulting from deepening and shallowing of the thermocline between the equator and 10°N. While this simulation includes mesoscale eddies, the results are consistent with those of the sector model calculations described above in that the eddies make little direct contribution to the poleward heat transport.

Passive tracers can provide powerful diagnostics of ocean circulation both from observations and in models. In the simulation of F. Bryan and Holland (1989) an idealized tracer called "age" was included that satisfied the equation

$$\frac{dA}{dt} = \kappa_h \nabla^2 A + \kappa_v \frac{\partial^2 A}{\partial z^2} + 1$$

with the surface boundary condition

$$A(z = 0) = 0 \ .$$

In equilibrium, the age would measure the time scale at which any point in the interior communicates with the sea surface, referred to as the ventilation time scale. The tracer equation was integrated for only ten years in this experiment, so only the near surface waters are in equilibrium and the maximum ages are less than ten years. The tracer distribution can still be used to diagnose the rates and pathways of thermocline ventilation however. The distribution of age at a level in the middle of the thermocline is shown in Fig. 11.13. See Plate 11 for another view of this. The ages are near zero in the northern part of the basin, where wintertime convection provides rapid communication with the sea surface. A strong front extends from the coast of West Africa across the basin. This delimits the "ventilated" and "shadow" zones of thermocline theory (Pedlosky, 1987b). As mentioned in the discussion of sector models, eddies can strongly affect the transport of tracers. Figure 11.14 compares the horizontally averaged ages in the eddy-resolving experiment with a companion non-eddy-resolving experiment (horizontal resolution of 1° latitude by 1.2° longitude). Throughout the thermocline the ages are lower in the eddy-resolving case. This indicates that the eddies are inducing a more rapid vertical penetration of tracers into the ocean. This is consistent with the results of Cox (1985), who found that passive tracers were more rapidly mixed along isopycnal surfaces in the eddy-resolving sector model than in the non-eddy-resolving model.

11.5.3 *Global models*

The number of global OGCM experiments is still relatively small. Most of these have been carried out as a preliminary step in constructing coupled ocean–atmosphere models. In this section, some of the issues that become important when considering the global ocean modeling problem and a few results from global models are introduced.

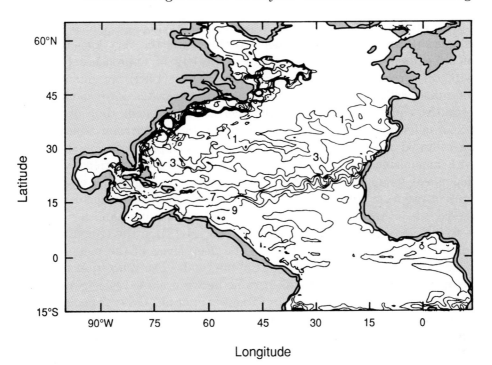

Fig. 11.13 Instantaneous snapshot of age at 295 m depth at year 10 of the integration of the eddy-resolving North Atlantic model of F. Bryan and Holland (1989).

Level Mean Ages

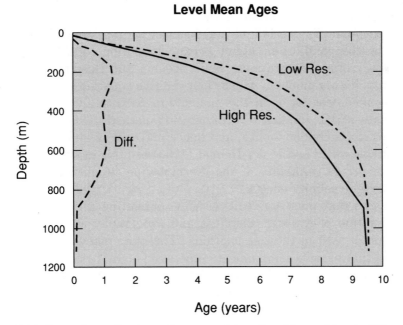

Fig. 11.14 Comparison of horizontally averaged ages for eddy-resolving (solid) and non-eddy-resolving (dash-dot) model of F. Bryan and Holland (1989).

One of the most challenging aspects of carrying out global ocean or coupled ocean–atmosphere model experiments is the very long time scale of adjustment for the deep ocean. A variety of approaches to dealing with this problem have been developed. Bryan (1984) presents a method for numerically altering the time scales of the faster wave modes of the primitive-equation system and the thermal adjustment time of the deep ocean, allowing longer time steps to be taken and thermal adjustment to be achieved more rapidly. This technique is appropriate for simulations with annual mean forcing, but its applicability to calculations with time-dependent forcing or resolved eddies is restricted. Toggweiler et al. (1989) present results of a global ocean simulation using this technique in which the model was integrated for 3,500 years to full equilibrium. Another approach is to introduce additional terms, similar to (11.19), in the temperature and salinity equations at all levels or in the deep levels of the model. This is referred to as the "robust-diagnostic" method. The relaxation time scale below the surface is usually taken to lie between a few years (Semtner and Chervin, 1988) and a few decades (Toggweiler et al., 1989). The adjustment time scale for the deep ocean is then comparable to the relaxation time scale, rather than the longer advective and diffusive time scales of the natural system. In addition, the model-predicted temperatures and salinities will be constrained to remain close to the observed values. By comparing the simulated radiocarbon distributions in otherwise identical prognostic and robust-diagnostic world ocean models, Toggweiler et al. (1989) show that while this technique does reduce errors in the temperature and salinity fields, it degrades the simulation of the convective overturning and deep circulation patterns. This technique may be most appropriate if the primary interest is in the equilibrium upper ocean circulation, or as a method for initializing climate integrations. Maier-Reimer and Hasselmann (1987) present results from a model in which the primitive equations are simplified by dropping nonlinear terms in the momentum balance and that uses an unconditionally stable, implicit time integration method. This permits time steps of approximately one month, and hence makes very long integrations possible. Mikolajewicz and Maier-Reimer (1990) use this model to examine very long-time-scale variability of the thermohaline circulation associated with high latitude salinity effects.

Another issue that must be dealt with in configuring a global model is the approximation of realistic coastlines and topography at the generally coarse resolution used in these simulations. There are many narrow straits and channels that provide communication between ocean basins that cannot be represented in coarse-resolution models. Even the Drake Passage, the narrowest point of the circumpolar ocean, is difficult to represent at a horizontal grid resolution of 4° or 5° typically used in global models. The contribution of Mediterranean Water to the salinity budget of the North Atlantic may be crucial in maintaining North Atlantic Deep Water formation (see Chapter 4, Sec. 4.7), but the Strait of Gibraltar cannot be resolved even at the highest resolutions being used in eddy-resolving models. In these cases it is often necessary to artificially open or deepen passages until a reasonable flow is obtained.

A common problem in coarse-resolution global models is that the mid-latitude thermocline tends to be too deep and diffuse. Toggweiler et al. (1989) report that horizontally averaged temperatures in their global model are up to 4°C warmer than observed. This is likely to be due to excessively large values of the vertical diffusivity in the main thermocline. The warmer thermocline leads to dynamic heights that are systematically too large in the subtropical gyres. Since the strength of the Antarctic Circumpolar Current is proportional to the difference in dynamic heights on its equatorward and poleward edges, many models overpredict the strength of the circumpolar transport partly as a result of this error in the temperature and salinity fields.

A quantity of primary interest for climate studies is the poleward heat transport by the ocean (see Chapter 4). Bryan (1982) provides a review of observational and model estimates of ocean heat transport (also see Figs. 4.18 and 4.20). A recent example of a model prediction of ocean heat transport is that of Semtner and Chervin (1988) shown in Fig. 11.15. Most of the features can be identified in other global models despite considerable differences in resolution, forcing and parameter choices. It is traditional to break the total heat transport down into components associated with "overturning" motion in the meridional-vertical plane, "gyre" motion in the horizontal plane, and parameterized eddy diffusion. The overturning component dominates in the tropics of both hemispheres. This is primarily due to the shallow wind forced overturning cell associated with equatorial upwelling. The Pacific Ocean makes the largest contribution to this component due to the width of the basin and the steadiness of the trade winds in the Pacific compared with the Atlantic or Indian Oceans. The overturning component remains the dominant term in the midlatitudes of the Northern Hemisphere. This is primarily due to the overturning cell associated with the formation of North Atlantic Deep Water. In subpolar latitudes the gyre component and diffusion component can become equally important. In the Southern Hemisphere, the heat transport associated with the overturning component is equatorward over a range of latitudes between 35° and 50°S. This has no counterpart in the Northern Hemisphere. This feature is associated with strong equatorward Ekman transports under the midlatitude westerlies of the circumpolar belt. The total transport remains poleward, however, at all latitudes of both hemispheres. The contribution of the gyre component tends to be larger in the Southern Hemisphere due to north–south meandering of the Antarctic Circumpolar Current.

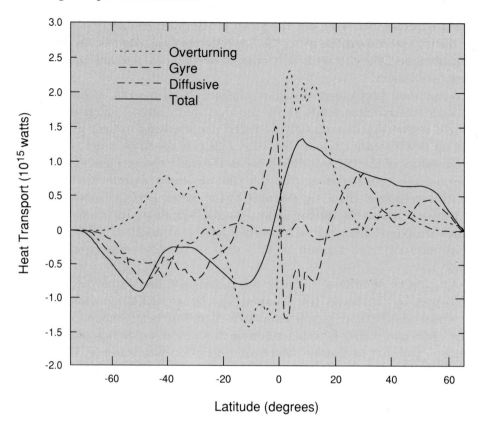

Fig. 11.15 Northward heat transport in the global ocean model of Semtner and Chervin (1988).

11.6 Future directions

It is likely that the oceanic component of any realistic global climate system model will contribute significantly to, and perhaps dominate, the computational resources required to carry out future climate simulations. It has been estimated, for example, that a truly eddy-resolving (1/8 × 1/8 degree, say) global integration of the Geophysical Fluid Dynamics Laboratory (GFDL) primitive equation model will require nearly a full year of CPU time on computers projected to be available in 1995 to complete a single 100-year simulation. This estimate is for the ocean model only; integration of the full climate system model would, of course, be more costly still. Moreover, it has been assumed in these estimates that our ocean models will keep pace with hardware improvement, in particular, the development of increasingly parallel computer architectures. We are optimistic that this will prove to be true. Nonetheless, ocean modelers are, and will continue to be, sorely pressed to produce global ocean models efficient enough for routine inclusion in global climate models.

What is to be done? The answer almost certainly lies in making substantial and simultaneous improvements in the physical, computational and algorithmic design of OGCMs.

11.6.1 Simplifying the equations of motion

Despite the fact that the HPE make only a few (sensible) physical approximations, in many applications it has proven desirable to further approximate the equations of motion prior to seeking their numerical solution. This is, in part, due to the very completeness of the HPE physics. In particular, the HPE include within them the physics necessary to support a variety of phenomena, some of which (e.g., surface and internal gravity waves) may not be of particular importance to the dynamics of the large-scale ocean circulation. Nonetheless, their appearance in the equations places sizable constraints on the operational form of numerical HPE models. (As noted in Chapter 9, for example, the usable time step in a numerical model is typically inversely proportional to the magnitude of admissible wave and/or current speeds.) Additional physical approximations may therefore be made which are valid for specific types of oceanic motion, and which lead to a substantial increase in computational efficiency and ease of interpretation.

A variety of reduced systems of equations is presently being explored. The derivation of such reduced systems typically proceeds along one of two lines. The first approach, which has already been applied in a simple form in (11.4.1), is to scale the equations of motion in a fashion consistent with the phenomena of interest, and then to eliminate higher-order terms, possibly in a way dictated by additional criteria such as the retention of global invariants. An alternative approach is to decrease the size (but not totally eliminate) certain terms in the full equations of motion – e.g., those responsible for the propagation of sound and gravity waves – thereby relaxing the restrictive time-stepping constraints that these high-frequency phenomena would normally impose.

A well-known, and historically important, reduced equation which can be derived in the former fashion is the *quasi-geostrophic (QG) potential vorticity equation* (Pedlosky, 1987a). The QG approximation has now a long history in ocean modeling. Perhaps most notably, models based upon QG dynamics have been instrumental in exploring the effects and importance of mesoscale eddies.

Unfortunately, the dynamical premises underlying the QG equations make them unsuitable to certain quite fundamental climate-related problems – e.g., the nature and maintenance of the oceanic thermohaline circulation. Hence, efforts are underway to identify suitable alternatives to quasi-geostrophy. One such family of models, the *balanced models*, are intermediate between quasi-geostrophy and the primitive equations in terms of physical content, complexity, and computational efficiency. Initial applications of the linear balance equations to the wind-driven ocean circulation have shown systematic improvement over analogous QG solutions, even for moderately small Rossby number (McWilliams et al., 1990).

For time scales in excess of a pendulum day and for horizontal scales comparable to the Earth's radius, scaling of the equations of motion yields the planetary geostrophic equations (PGE). The PGE also have the great computational virtue of explicitly filtering out high-frequency waves, leaving only the nondispersive baroclinic Rossby waves appropriate to the problem of planetary-scale baroclinic adjustment. However, the approximations inherent in their derivation rule out accurate inclusion of mesoscale eddy dynamics. Nonetheless, they are being used with great success to study buoyancy driven oceanic flows on planetary scales and their interaction with wind forcing (Colin de Verdière, 1989).

11.6.2 *What about those eddies?*

The removal of various high-frequency wave phenomena from OGCMs is both computationally desirable (since larger time steps may be used) and physically sensible (since in most oceanic situations the dynamical linkages between these waves and the lower-frequency oceanic motions are weak). An even greater gain in efficiency would be made, however, if mesoscale eddy processes could be ignored, since this would presumably allow a larger grid spacing (in at least the two horizontal coordinates) as well as a larger time step.

Though computationally desirable, the removal of explicit eddy effects is problematic on physical grounds. Given the preponderance of energy contained within the mesoscale space/time continuum, the nonlinear coupling between the eddies and the time-mean, larger-scale oceanic circulation is quite strong. It is now well known, for instance, that the eddies contribute in a substantial way to both the structure and intensity of oceanic current systems (Holland, 1985). There are suggestions that they may do so in a manner which leaves certain important climate-related parameters relatively unaffected (e.g., the net poleward heat transport). Nonetheless, it is difficult at this point to argue with the belief that eddy effects must be included in the oceanic component of a coupled climate model.

If such processes must be included, then either they must be explicitly (and adequately) resolved, or they must be implicitly (but adequately) parameterized. In the former instance, horizontal grids may need to be as fine as 10 km to capture the full inertial effects of the eddies. Unfortunately, the resulting computational penalty would be severe – one which would leave us far short of our goal of carrying out lengthy, routine, global ocean simulations for some time to come.

In the latter case, the need is to develop a sufficient parameterization – or closure, in the terminology of Sec. 11.2 – for the equations of motion. A few of the important effects of eddies are now understood though: how they drive deep mean flows, how they act to shape the mean potential vorticity field in the oceanic interior, and how they are responsible for systematic form stresses in the neighborhood of continental margins and other steep topographic features (e.g., see Young, 1987). Despite these theoretical efforts, however, suitable eddy parameterizations will be difficult to develop.

11.6.3 Numerical algorithms

Apart from the (long-term) prospect of discovering a suitable parameterization scheme for the ocean's mesoscale eddies, the economies associated with the model improvements thus far discussed would still leave us well short of a routinely affordable global ocean model by probably a factor of 100. Fortunately, the speed of supercomputers has been undergoing rapid and systematic improvement at a rate approaching an order of magnitude per decade. This issue is discussed in more detail in Chapter 9. If modelers of the global ocean, indeed of the entire coupled climate system, wish to take advantage of the projected factor of ten improvement in computer capability to be available in the 1990s, they will have to be clever in their development of new models – or redesign of existing models – to effectively utilize these new parallel architectures. An important step in this direction has been taken by Semtner and Chervin (1988), whose parallelized version of the GFDL model has enabled the first marginally eddy-resolving, fully baroclinic, decadal, global ocean simulations.

The impressive gains being made in supercomputer speed and capacity, however, often overshadow the fact that novel numerical algorithms in scientific computing have reduced execution times by at least as large a factor as have the improvements in hardware. Unfortunately, ocean modelers have been slow in taking advantage of novel algorithmic approaches to its problems. Among those developments which hold most promise for OGCMs are the greater utilization of higher-order approximation techniques such as spectral, spectral-finite-element, and spectral multi-grid methods (e.g., Haidvogel et al., 1991b); adaptive and/or composite mesh techniques (Spall and Holland, 1991); and more accurate numerical advection algorithms (Rood, 1987).

11.6.4 Optimal combination of models and data

The development of increasingly efficient models of the global ocean circulation is a great challenge; however, it may not be the single greatest constraint on future progress. In order to be able to apply and to validate the increasingly realistic models of the future, in principle, an increasingly complete observational description of the global ocean circulation itself, including its variability, and the exchanges of momentum, heat and moisture with the atmosphere will be required. Already noted above are reasons why this will remain a dilemma for oceanographers: datasets are difficult to collect and are very sparse and inhomogeneous (see also Chapter 4). The situation will no doubt improve considerably during the decade of the 1990s, as data from anticipated space-borne sensors (e.g., altimeters, scatterometers, etc.) become available. Even so, the great bulk of the ocean will continue to be incompletely (even poorly) observed, unless perhaps efficient acoustic means of observing the oceanic interior can be developed (Munk and Forbes, 1989).

One idea which is being intensively explored in regional and basin-scale ocean models, and which may ultimately prove practical for application to global models as well, is the application of optimal and suboptimal methods for the combination of prognostic models with data. Several alternate approaches are under evaluation. One family of approaches, broadly practiced in meteorology, is data assimilation – the systematic combination of dynamical (usually model-produced) and observational information to produce increasingly accurate representations (usually predictions) of large- and meso-scale motions (Ghil and Malanotte-Rizzoli, 1991). Assimilative techniques from the atmospheric sciences, appropriately adapted for the oceanic case, are being used with simulated datasets to explore the impact of assimilation on the steady and transient ocean response. Of particular interest is the determination of which space/time scales of oceanic motion are effectively "constrained" by assimilation of data of different kinds.

Another complementary approach to the combined analysis of oceanic models and observations is through the formulation of an optimization or inverse problem. Formulated in this way, poorly known model parameters, along with some estimate of how well their values are constrained by the data, can be explicitly calculated. The computational penalties associated with the optimization approach can be severe – amounting to the inversion of matrices whose dimensions are the square of the number of unknowns in the problem. Nonetheless, efficient implementations of the optimization approach are possible (e.g., based on the adjoint equations techniques). Applied to simple ocean models, the resulting formulation allows the optimal estimation of parameters, such as the bottom friction and forcing functions such as the wind stress, and provides a particularly attractive approach to the interpretation of transient tracer datasets (Wunsch, 1989).

12

Sea ice models

William D. Hibler, III and Gregory M. Flato

12.1. Introduction

The sea ice cover in the polar regions is a thin, variable layer of ice formed from freezing sea water and accumulated snowfall. This ice cover drifts in response to winds and currents, and grows and melts with the changing seasons. The seasonal variability of the ice cover in both hemispheres is discussed in Chapter 4 (see Fig. 4.12). Due to continual deformation, growth and melt, the Arctic ice cover contains a variety of thicknesses ranging from open water to pressure ridges tens of meters thick, although the mean thickness in the Arctic is generally 3–4 m. In the Antarctic, the relatively unimpeded motion and large oceanic heat flux lead to an ice cover with very few thick ridges and a mean thickness of 1–2 m.

When sea water freezes under quiescent conditions, much of the dissolved salt is expelled leaving newly formed (first-year) ice with a salinity of about 5–10 ppt. The remaining salt is trapped in small (0.1–1 mm) brine pockets giving sea ice mechanical, optical, electromagnetic, and thermodynamic properties much different from freshwater ice. Some of the salt in these brine pockets drains or is flushed out by melt water in subsequent summers and hence multi-year sea ice in the Arctic has a salinity of only 2–4 ppt. The turbulent conditions in the stormy southern ocean, together with an unconstrained ice margin, disrupt the orderly ice growth described above and instead create an ice cover composed primarily of small ice particles which form in the supercooled upper layer of the ocean. These particles agglomerate to form 'frazil ice' which makes up over half of the Antarctic ice volume. For a much fuller discussion of the physical properties of sea ice in both the Arctic and Antarctic see, e.g., Weeks and Ackley (1986) and Gow and Tucker (1990).

Due to the almost constant motion of sea ice, the dynamics and thermodynamics are closely related. One aspect of this coupling arises because of the strong dependence of growth rate on ice thickness – thin ice grows much faster than thick ice because it offers less resistance to the upward conduction of heat. In winter, local deformation, due to nonuniform ice motion, causes openings (leads) to appear in the ice cover which

413

dramatically enhances the thermodynamic growth of new ice. Ridging, another result of deformation, causes thin ice to be piled up into thick, roughly triangular features which are likely to survive the following summer's melt. Another aspect of dynamic–thermodynamic coupling involves the absorption of shortwave radiation by open water (which has a much lower albedo than ice) and the subsequent motion of the ice over this warmer water, enhancing the overall melt rate (Maykut and Perovich, 1987).

The presence of this ice cover can play a major role in high-latitude climate sensitivity. In particular, the ice cover exerts a strong control over the heat and moisture fluxes into the atmosphere and thus influences polar atmospheric circulation and precipitation patterns. The strong albedo contrast between ice and open water[1] and the positive feedback of sea to air heat transfer with air temperature are the principal reasons for the enhanced sensitivity to CO_2 warming in the Arctic exhibited by global circulation models (e.g., Manabe and Stouffer, 1980; Manabe et al., 1990 and see Chapter 20, Sec. 20.4). Horizontal transport of fresh water in the form of sea ice may play an equally important role in controlling high-latitude ocean circulation. Ice freezing in one location, drifting, and then melting in another creates a significant surface salt flux in some regions of the ocean, hence affecting the baroclinic component of the ocean circulation. A possible process in the large-scale behavior of the climate system involves the southward flow of ice through Fram Strait which subsequently melts in the Greenland and Norwegian Seas and maintains the strong stratification in this region. The strength of this stratification modifies the rate of deep convective overturning which, in turn, affects the ice edge location, and is a source of deep water for the world's oceans (see Chapter 4, Sec. 4.7). Simulating the large-scale transport of ice is thus one of the most important aspects of sea ice modeling in the context of the climate system.

Sea ice modeling involves a circulation modeling problem somewhat similar to that in the atmosphere and the ocean. However, a major difference is that the ice may be considered to be a two-dimensional continuum with body forces exerted by the atmosphere and ocean. In this continuum the ice thickness characteristics form the main material property that dictates the dynamic and thermodynamic properties of the ice.

A coupled dynamic–thermodynamic sea ice model relevant to climate studies may be divided into four components: a *momentum balance* describing sea ice drift which includes air and water stresses, Coriolis force, internal ice stress, inertial forces and ocean currents; *ice thickness distribution* equations which describe the evolution of the ice thickness characteristics caused by thermodynamic and dynamic effects; an *ice rheology* which relates the ice stress to the ice deformation and ice thickness

---

[1] The albedo of sea ice ranges from about 0.3 to 0.7 whereas the albedo of open water is about 0.1 to 0.2.

distribution; and a *thermodynamic model* which specifies the growth and decay rates of various ice thicknesses in the environment of a sea ice cover of variable thickness. The thermodynamic model must also include some type of oceanic boundary layer to take into account the heat storage in the ocean, as well as some parameterization of the atmospheric boundary layer for the purposes of determining the incoming long- and shortwave radiation and the surface sensible and latent heat fluxes.

In this modeling overview we will describe these four components of a sea ice model and examine some of the sensitivity characteristics of coupled dynamic–thermodynamic sea ice models relevant to climate studies. We first discuss sea ice dynamics which incorporates the ice momentum balance and sea ice rheology. As part of this ice rheology discussion we also discuss simpler sea ice dynamics models usable in numerical investigations of climate. More complete reviews of some of the subjects covered in this chapter can be found in the book edited by Untersteiner (1986).

12.2 Sea ice dynamics

Sea ice moves in response to wind and water currents and the internal stress in the ice. Considering sea ice to be a two-dimensional continuum, it obeys the normal Euler equations of motion

$$m\frac{D\mathbf{v}}{Dt} = \nabla \cdot \boldsymbol{\sigma} + \mathbf{X} \tag{12.1}$$

where $\mathbf{v}$ is the ice velocity, m the ice mass per unit area, $\boldsymbol{\sigma}$ is the second-order internal stress tensor in the sea ice due to ice interaction, $\mathbf{X}$ is the total body force on the ice and the substantial derivative of the ice velocity is

$$\frac{D\mathbf{v}}{Dt} = \frac{\partial \mathbf{v}}{\partial t} + (\mathbf{v} \cdot \nabla)\mathbf{v}.$$

Since sea ice is a two-dimensional continuum, the interfacial stresses between the ice, atmosphere and ocean are part of the body forces $\mathbf{X}$ in Euler's equation. Additionally there is a gravitational force due to the tilt of the ocean surface. Explicitly writing out these terms

$$m\frac{D\mathbf{v}}{Dt} = -mf\mathbf{k} \times \mathbf{v} + \boldsymbol{\tau}_a + \boldsymbol{\tau}_w + \nabla \cdot \boldsymbol{\sigma} - mg\nabla h \tag{12.2}$$

where, in addition to the symbols defined in (12.1), $\mathbf{k}$ is a unit vector normal to the surface, f is the Coriolis parameter, $\boldsymbol{\tau}_a$ and $\boldsymbol{\tau}_w$ are the forces due to air and water stresses, and h is the height of the sea surface. In this formulation, $\boldsymbol{\tau}_w$ is assumed to include frictional drag due to the relative movement between the ice and the underlying ocean.

In this momentum balance, the air and water stresses are normally determined from idealized boundary layers, assuming constant turning angles (McPhee, 1979; Brown, 1980; Leavitt, 1980):

$$\boldsymbol{\tau}_a = c'_a(\mathbf{V}_g \cos \varphi_a + \mathbf{k} \times \mathbf{V}_g \sin \varphi_a) \tag{12.3}$$

$$\boldsymbol{\tau}_w = c'_w[(\mathbf{V}_w - \mathbf{v}) \cos \varphi_w + \mathbf{k} \times (\mathbf{V}_w - \mathbf{v}) \sin \varphi_w)] \tag{12.4}$$

where $\mathbf{V}_g$ is the geostrophic wind, $\mathbf{V}_w$ the geostrophic ocean current, c'_a and c'_w air and water drag coefficients, and φ_a and φ_w air and water turning angles. In practice, both the currents and the sea surface tilt are estimated from geostrophic considerations by setting h equal to the dynamic height and computing currents by $\mathbf{V}_w = \frac{g}{f}\mathbf{k} \times \nabla h$. In general, both c'_a and c'_w are nonlinear functions of the winds and currents. The two most commonly used formulations are (1) linear, where c'_a and c'_w are taken to be constant, and (2) quadratic, where

$$c'_a = \rho_a c_a |\mathbf{V}_g| \tag{12.5}$$

and

$$c'_w = \rho_w c_w |\mathbf{V}_w - \mathbf{v}| \tag{12.6}$$

with c_a and c_w being dimensionless drag coefficients (with typical values of 0.0012 and 0.0055 respectively; McPhee, 1980) and turning angles φ_a and φ_w of about 25° in the Arctic or −25° in the Antarctic. In the linear models, a linear drag coefficient of $\rho(fK_v)^{1/2}$ (where ρ is the density, K_v is the vertical eddy viscosity) is obtained from classical Ekman layer theory with a turning angle of 45°. In a real turbulent boundary layer the vertical eddy viscosity varies with depth and velocity. For this reason, quadratic drag laws with turning angles different from 45° give a better fit to observations (McPhee, 1980) than the simple linear Ekman theory. However, it should be noted that the best fit to observations occurs with drag laws where the turning angle varies with the surface stress.

Basically, in the momentum balance (12.2), the essential causes of the ice motion can be thought of as the geostrophic wind above the atmospheric boundary layer and the ocean current beneath the oceanic boundary layer. These forces are transmitted to the ice via simple integral boundary layers. In addition, a steady ocean current introduces a tilt (∇h) of the sea surface height (or alternatively a pressure on the ice) which also affects the ice motion.

The dominant terms in the momentum balance are the air and water stresses and the ice interaction. The Coriolis term, tilt and steady current terms are about an order of magnitude smaller. Also the inertia term is rather small as it would take about an hour for ice to come to a steady state drift state after the sudden imposition of a wind field. A typical force balance is shown in Fig. 12.1, where the ice interaction has been deduced by residual calculations with all the other terms being measured. That the ice interaction is the same order of magnitude as the wind and water drag has been verified by observation in enclosed areas, such as the Bay of Bothnia, where the ice often does not move at all even with significant winds (Lepparanta, 1980; Omstedt and Sahlberg, 1977).

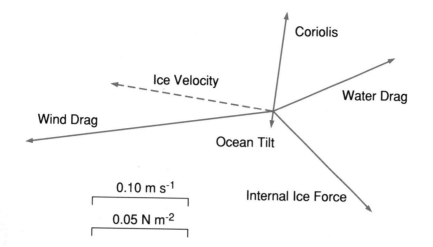

Fig. 12.1 An estimate of the force balance on sea ice for winter conditions based on wind and water stress measurements. In this balance the force due to internal ice stress is determined as a residual and the dashed line shows the ice velocity.

a.   Free drift analysis

To understand the nature of ice drift it is instructive to examine a special case of the momentum balance in which there is no ice interaction. This case is usually referred to as free drift. To further simplify matters, only steady free drift will be considered, so that the inertial terms can also be removed. Such inertial effects do play a role in short-term prediction of ice trajectories but are not critical for the longer-term steady ice drift relevant to climate.

Following McPhee (1980) and Thorndike and Colony (1982), it is convenient to use complex notation to analyze this case. Writing all vectors in complex form, i.e. $\mathbf{A} = Ae^{i\delta}$, with δ the counterclockwise turning angle from the real axis, and noting that the force on the ice due to sea surface tilt is given by $-imf\mathbf{V}_w$, the steady state momentum balance can be written in the form

$$\rho_a c_a \mathbf{V}_g V_g e^{i\varphi_a} = imf\mathbf{V} + \rho_w c_w \mathbf{V}Ve^{i\varphi_w} \qquad (12.7)$$

where $\mathbf{V}$ is the ice velocity relative to the steady currents:

$$\mathbf{V} = \mathbf{v} - \mathbf{V}_w. \qquad (12.8)$$

Letting δ be the clockwise angle from geostrophic wind to ice drift direction, (12.7) can be rewritten

$$V_g^2 e^{i\delta} = V^2 e^{-i\varphi_a} \frac{\rho_w c_w}{\rho_a c_a} \left(e^{i\varphi_w} + \frac{mf}{\rho_w c_w V} e^{i\pi/2} \right). \qquad (12.9)$$

A solution of this equation may easily be obtained by assuming a value for the ice velocity and then solving for the geostrophic wind. Figure 12.2

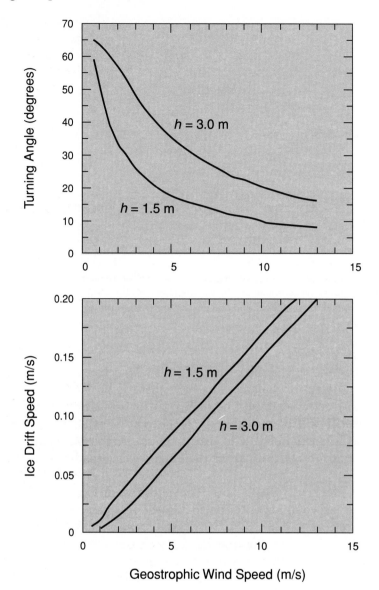

Fig. 12.2 The clockwise turning angle from the geostrophic wind to the ice velocity (above), and the ice drift speed versus the geostrophic wind speed (below). These curves were calculated assuming free drift and quadratic boundary layers. Curves for different ice thicknesses are shown.

shows a solution to the free drift equations assuming drag coefficients and Coriolis values similar to those employed by Hibler (1979): $\rho_w = 10^3$ kg m^{-3}, $\rho_a = 1.3$ kg m^{-3}, $c_w = 0.0055$, $c_a = 0.0012$, $f = 1.46 \times 10^{-4}$ s^{-1}, $\varphi_a = \varphi_w = 25°$). The results basically show that the ice drifts at a fixed fraction of the wind speed with a drift direction slightly to the right of the wind. As the ice mass increases, the ice tends to drift a little slower and slightly further to the right of the wind. An interesting feature of this plot is the very nearly linear relation between the wind speed and ice drift for all but very small speeds, despite the fact that boundary layer drag coefficients are nonlinear. This situation arises because both boundary layers are quadratic, and hence, except at low speeds, largely balance one another. However, the turning angle does show somewhat greater sensitivity to wind speed. Also note that although V and V_g are linearly related, the best linear fit does not go through the origin, which would be the case for linear wind and water drag.

Although the free drift model produces velocity fields which correlate well with observed drift, it is a very poor model for simulations longer than a few weeks. The reason is the lack of resistance to buildup near boundaries (coastlines) which quickly leads to unrealistically large ice thicknesses when advection is included. Including an ice interaction term, as described in the following section, changes the details of the ice velocity field such that this unbounded thickness buildup is avoided.

b.   Ice interaction and ice rheology for climatic studies

The most obvious effect of ice stress is to change the ice drift direction and magnitude from the free drift case. In many cases the net effect of the ice interaction is to cause a force opposing the wind stress roughly in a manner shown in Fig. 12.1. As a consequence, to achieve the same ice velocity under these circumstances a larger wind stress more nearly parallel to the ice velocity is required. This feature is consistent with observations by Thorndike et al. (1975) which show ice motion to be very highly correlated with geostrophic winds with the ice drift rate decreasing somewhat in winter for the same geostrophic wind. However, examination of drift statistics shows higher winds producing an almost discontinuous shear near the coast. Moreover, observations show that while ice drift against the shore increases the ice thickness, the buildup is not unlimited.

As noted by Hibler (1979), both these characteristics of discontinuous ice drift and a limit on near shore ice buildup may be explained by nonlinear plastic ice rheologies. These rheologies have yield stresses that are relatively independent of strain rate. Hence far from shore even though the ice is interacting strongly, there may be very low stress gradients since the stresses are relatively constant. Also this fixed yield stress will cause a discontinuous slippage at coastal points and prevent the ice from building up without bound. Without such a nonlinear rheology it is very difficult to obtain these features.

For climatic purposes, probably the most important feature of the ice interaction is some type of resistance to ice buildup without drastically affecting the ice drift. While this can be accomplished by a nonlinear plastic rheology including a shear strength, reasonable results may also be obtained by considering the ice to have resistance only to compression and no resistance to dilation or shearing. This *cavitating fluid* model has been compared by Flato and Hibler (1990) to a more complete plastic rheology with good success.

Formally, ice rheology may be examined by noting that for a two-dimensional isotropic continuum, a general constitutive law may be written in the form

$$\sigma_{ij} = 2\eta\dot{\varepsilon}_{ij} + \left([\zeta - \eta]\dot{\varepsilon}_{kk} - \frac{P}{2} \right) \delta_{ij} \qquad (12.10)$$

where σ_{ij} is the two-dimensional stress tensor and $\dot{\varepsilon}_{ij}$ the two-dimensional strain rate tensor, $\delta_{ij} = 1$ for $i = j$, or $\delta_{ij} = 0$ for $i \neq j$, and η and ζ, the shear and bulk viscosities, and P, the internal ice pressure, are, in general, functions of the two invariants of the strain rate tensor. These invariants can be taken as the principal values of the strain rate tensor or, alternatively, as the divergence rate and the maximum shear. For the cavitating fluid model there is no shear term and so the stress tensor is given by

$$\sigma_{ij} = -P\delta_{ij}. \qquad (12.11)$$

An exact iterative solution of the cavitating fluid has been presented by Flato and Hibler (1992) where formulation in a spherical grid has also been carried out. However, a particularly simple approximate solution to the cavitating fluid applicable to climate modeling was presented earlier by Flato and Hibler (1990). The basic idea in this model is to iteratively modify a free drift velocity field in a momentum conserving manner until there is either no ice convergence, or the ice pressure reaches its maximum value.

An example of the use of the cavitating fluid model is shown in Fig. 12.3 where the cavitating fluid correction is applied to the free drift velocity field shown in part (a) of the figure. Two types of corrections are applied: in part (b) a totally incompressible sea ice drift field is shown, while in part (c) a maximum allowable two-dimensional pressure of 5.53×10^4 N m^{-1} is assumed. As can be seen, including a maximum yield pressure modifies the velocity field somewhat inasmuch as some convergence is allowed. The main characteristic of both fields however, is that the cavitating fluid approximation does not damp out the ice velocity field but rather modifies it to prevent convergence.

A more complete sea ice rheology, the viscous-plastic elliptical yield curve rheology, was presented by Hibler (1979). In this scheme plastic failure and rate independent flow are assumed when the stresses reach the yield curve values represented by an ellipse in principal stress space. Here the compressive strength is given by the length of the ellipse while the shear strength is given by its width. Stress states inside the yield curve correspond to slow viscous creep deformation. This model has been widely used and produces realistic thickness and velocity patterns; however, its relative complexity and the dramatic slowdown in ice drift it produces when

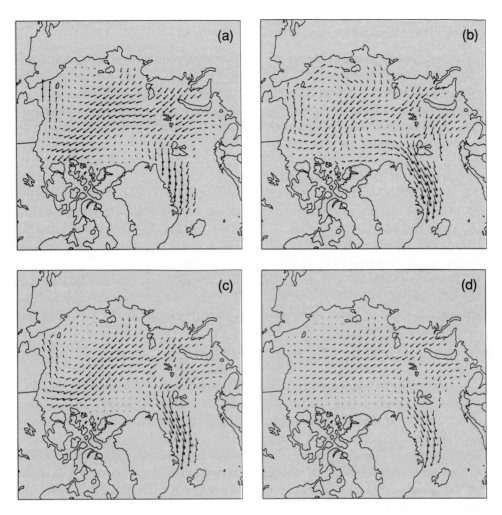

Fig. 12.3 Average ice velocity fields calculated using forcing from March 1983. (a) Free drift. (b) Incompressible cavitating fluid. (c) Cavitating fluid model with realistic compressive strength. (d) Viscous-plastic, elliptical yield curve model with shear and compressive strength. A vector one grid cell long is approximately 0.1 m s^{-1}.

the wind fields are temporally smoothed (Flato and Hibler, 1990) make it somewhat less desirable for long-term climate studies. As a comparison, the velocity field calculated using this model with a constant ice strength of $5.53 \times 10^4 \text{ N m}^{-1}$ is shown in Fig. 12.3(d). Readily apparent here is the less robust velocity field which results from the increased resistance afforded by the shear strength. For a more complete comparison of a variety of plastic rheologies for the Arctic Basin, the interested reader is referred to Ip et al. (1991) where several nonlinear rheologies are compared.

12.3 Ice thickness distribution models

A key coupling between sea ice thermodynamics and ice dynamics is the ice thickness distribution. The basic idea here is that the deformation causes pressure ridging and open water creation. When combined with ice transport, these factors change the spatial and temporal growth patterns of the sea ice and, when coupled with mechanical properties of ice, can modify its response to climatic change.

a. Multi-level ice thickness distribution

A theory of ice thickness distribution may be formulated by postulating an areal ice thickness distribution function and developing equations for the dynamic and thermodynamic evolution of this distribution. Following Thorndike et al. (1975), $g(H)dH$ is defined to be the fraction of area (in a region centered at position x, y at time t) covered by ice with thickness between H and $H + dH$. This distribution evolves in response to deformation, advection, growth and decay. Neglecting lateral melting effects, it is easy to derive the following governing equation for the thickness distribution:

$$\frac{\partial g}{\partial t} + \nabla \cdot (\mathbf{v}g) + \frac{\partial (f_g g)}{\partial H} = \Psi, \tag{12.12}$$

where f_g is the vertical growth (or decay) rate of ice of thickness H and Ψ is a redistribution function (depending on H and g) that describes the creation of open water and the transfer of ice from one thickness to another by rafting and ridging. Except for the last two terms, (12.12) is a normal continuity equation for g. The last term on the left-hand side can also be considered a continuity requirement in thickness space since it represents a transfer of ice from one thickness category to another by the growth rates. An important feature of this theory is that it presents an "Eulerian" description in thickness space. In particular, growth occurs by rearranging the relative areal magnitudes of different thickness categories.

This multi-level ice thickness distribution theory represents a very precise way of handling the thermodynamic evolution of a continuum comprised of a number of ice thicknesses. However, the price paid for this precision is the introduction of a complex mechanical redistributor.

In particular, to describe the redistribution one must specify what portion of the ice distribution is removed by ridging, how the ridged ice is redistributed over the thick end of the thickness distribution, and how much ridging and open water creation occur for an arbitrary two-dimensional strain field, including shearing as well as convergence or divergence. However, in selecting a redistributor one can be guided by the conservation conditions on Ψ:

$$\int_0^\infty \Psi dH = \nabla \cdot \mathbf{v} \tag{12.13}$$

$$\int_0^\infty H\Psi dH = 0 . \tag{12.14}$$

Equation (12.13) follows from the constraint that Ψ renormalizes the g distribution to unity due to changes in area. Equation (12.14) follows from conservation of mass and basically states that Ψ does not create or destroy ice but merely changes its distribution. An additional assumption in (12.14) is that the ice mass is related in a fixed manner to the thickness.

An additional consistency condition can be imposed if one asserts that all the energy lost in deformation goes into pressure ridging and that the energy dissipated in pressure ridges is proportional to the change in potential energy (Rothrock, 1975). With these assumptions

$$C \int_0^\infty H^2 \Psi(H) dH = \sigma_{ij}\dot{\epsilon}_{ij} \tag{12.15}$$

where, assuming a plastic rheology, C is a constant related to the densities of water and ice and the frictional energy losses during ridging.

A redistribution function which satisfies these constraints may be constructed (for an explicit form see Coon et al., 1974 or Hibler, 1980) by allowing open water to be created under divergence and ridging to occur under convergence. Within this formalism ridging occurs by the transfer of thin ice to thicker categories. In this transfer some assumptions must be made about the amount of ridging and hence open water created under pure shear or more generally under an arbitrary deformation state. This "energetic consistency" condition will affect the thermodynamic growth via open water fractions and hence the total ice created.

b. Two-level ice thickness distribution

Many features of the thickness distribution may be approximated by a two-level sea ice model (Hibler, 1979) where the ice thickness distribution is approximated by two categories: thick and thin. Within this two-level approach the ice cover is broken down into an area A (often called the compactness), which is covered by thick ice, and a remaining area $1 - A$, which is covered by thin ice, which, for computational convenience, is always taken to be of zero thickness (i.e., open water). The idea is to have the open water approximately represent the combined fraction of both open water and thin ice up to some cutoff thickness H_0. The remainder of the ice is distributed arbitrarily. However, since the thin ice mass is normally small, the mean thickness of the remaining "thick" ice is approximately equal to H/A.

For the overall mean thickness H and compactness A, the following continuity equations are used:

$$\partial H/\partial t = -\partial(uH)/\partial x - \partial(vH)/\partial y + S_h \tag{12.16}$$

$$\partial A/\partial t = -\partial(uA)/\partial x - \partial(vA)/\partial y + S_A \tag{12.17}$$

where $A < 1$, u, v are the components of the ice velocity vector, and S_h and S_A are thermodynamic terms. While (12.16) is a simple continuity equation for ice mass (characterized by the mean thickness H), with a thermodynamic source term, (12.17) is somewhat more complex. By including the restriction that $A < 1$, a mechanical sink term for the areal fraction of ice has been added to a simple continuity equation for ice concentration. This sink term turns on when $A = 1$ (i.e., no open water left) and under converging conditions removes enough ice area through ridging to prevent further increases in A. Although the sink term does not change the ice mass, it can cause the "thick" ice thickness to increase by allowing H to increase while A does not. Note that however the ridging problem is formally treated in this two-level model it does not affect the conservation of ice mass, which is explicitly guaranteed by (12.16).

The thermodynamic terms in (12.16) and (12.17) represent the total ice growth (S_h) and the rate at which ice-covered area is created by melting or freezing (S_A). It is the parameterization of the S_A term that is particularly difficult to do precisely within the two-level model and generally represents one of the weaknesses of the model.

In a formulation developed by Hibler (1979), the thermodynamic terms in the continuity equations (12.16) and (12.17) are given by

$$S_h = f_g\left(\frac{H}{A}\right) A + (1 - A)f_g(0) \tag{12.18}$$

$$S_A = \frac{f_g(0)}{H_0}(1 - A)\delta_1 + \frac{A}{2H}S_h\delta_2 \tag{12.19}$$

where $\quad \delta_1 = 1$ if $f_g(0) \geq 0$, $\delta_1 = 0$ if $f_g(0) < 0$

$\delta_2 = 0$ if $S_h > 0$, $\delta_2 = 1$ if $S_h < 0$

with $f_g(H)$ the growth rate of ice, and H_0 a fixed demarcation thickness between thin and thick ice. Details of determining $f_g(H)$ are given later.

The S_h term specifies the net ice growth or melt. Within the two-level approximation, S_h is given by the sum of the ice grown on open water plus the additional growth over the portion of the cell covered by thick ice. To approximate the growth and decay rate of this thick ice, its mean growth rate is estimated to be that of ice of constant thickness H/A. For melting conditions the same sum over open water and thick ice is used. A critical assumption here is that the heat absorbed by open water will horizontally mix and melt additional ice until the mixed layer returns to freezing.

The S_A term characterizes the way in which growth and decay change the relative areal extents of thin and thick ice. The basic physical notion is that the areal fraction of thin ice will decrease rapidly under freezing conditions, and increase slowly under melting conditions. To parameterize the freezing effect, the fraction of open water $(1 - A)$ is allowed to decay exponentially with a time constant of $H_0/f_g(0)$, which gives the first term in equation (12.19). The constant H_0 is chosen to be small compared to mean ice thicknesses but large enough so that heat fluxes through H_0-thick ice are substantially less than through open water. In Arctic simulations performed by Hibler (1979) $H_0 = 0.5$ m was used. However, for Antarctic simulations (Hibler and Ackley, 1983) H_0 was set equal to 1.0 m. This larger value is based on field observations reported in Ackley et al. (1980) and Gow et al. (1982), which suggest that frazil ice formation may well prolong ice production in open water regions. The second term in equation (12.19) accounts for melting. Its magnitude is derived by assuming that the thick ice is uniformly distributed between 0 and $2H/A$ in thickness, and all melts at the same rate.

To allow either the multi-level or two-level sea ice model to be integrated over a seasonal cycle, it is necessary to include some type of oceanic boundary layer or ocean model. The simplest approach (used for example by Hibler and Walsh, 1982) is to include a motionless fixed depth mixed layer (usually 30 meters in depth). With this model, any heat remaining after all the ice is melted is used to warm the mixed layer above freezing. Under growth conditions, on the other hand, the mixed layer is cooled to freezing before the ice forms. Another approach that treats vertical penetrative convection processes much better is to include some type of one-dimensional mixed layer. Such an approach, using a Kraus–Turner-like mixed layer, was carried out by Lemke et al. (1990) for the Weddell Sea. A third approach is to utilize a complete oceanic circulation model which also allows lateral heat transport in the ocean. The latter approach was used by Hibler and Bryan (1984) in a numerical investigation of the circulation of an ice-covered Arctic Ocean. In the Hibler and Bryan study a two-level dynamic–thermodynamic sea ice model was coupled to a fixed level baroclinic ocean circulation model, with the upper ocean layer 30 m thick and, as in the motionless case, it was not allowed to drop below freezing if ice was present. Inclusion of some type of penetrative convection in a similar ice–ocean circulation model is an item of high priority for the future.

The two-level model, combined with the simple cavitating fluid rheology, an oceanic boundary layer, and a thermodynamic model (such as discussed in the following section), allows most of the relevant sea ice processes to be included in global climate models. Examples of the spatial patterns of ice thickness buildup produced by such a model are shown in Fig. 12.4 for both the Arctic and Antarctic. The general features of the modeled thickness patterns agree quite well with observations except that, for reasons discussed below, the thickness in the Arctic is somewhat less than that observed. It is interesting to note here that, in agreement with observations, ice in the Antarctic is significantly thinner than in the Arctic. This result is only obtained in a model that includes realistic ice dynamics. Also shown in Fig. 12.4 are the observed 15% concentration contours to indicate ice edge location.

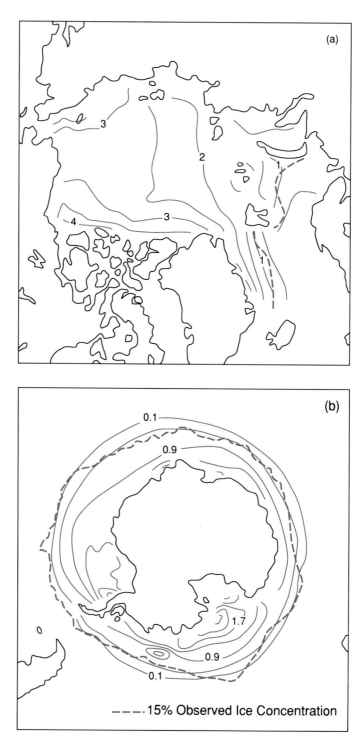

Fig. 12.4 Mean ice thickness contours in meters calculated by the cavitating fluid model (Flato and Hibler, 1991) for: (a) the Arctic at the end of March 1983; (b) the Antarctic at the end of July based on climatological forcing (C. Geiger, pers. comm.). The dashed lines indicate observed 15% ice concentration contour.

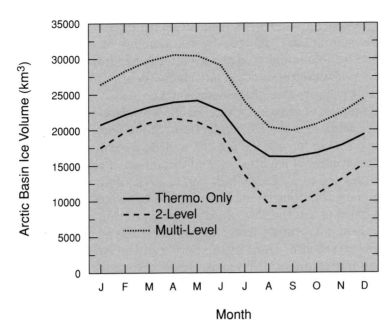

Fig. 12.5 Average seasonal cycle (1979–85) of ice volume in the Arctic Basin calculated by the multi-level, two-level and thermodynamic-only models (based on Flato, 1991). Note that 20,000 km³ represents a mean ice thickness of ~2.9 m.

The main relevance of the variable thickness distribution to climate modeling is its more precise treatment of the growth of ice. A comparison of the two-level and multi-level approaches to modeling the dynamic thermodynamic evolution of an ice cover can be found in Flato (1991). However, the ice thickness is considerably greater with the multi-level model owing to its more realistic treatment of the thickness distribution. It has been shown (Maykut, 1986) that the overall heat transfer to the atmosphere, and hence ice growth, depends to a large extent on the details of the thin end of the ice thickness distribution. An illustration of the differences in the seasonal cycle of ice volume in the Arctic Basin obtained by a simple thermodynamics-only model, the two-level model, and the multi-level model is shown in Fig. 12.5. The shortcomings of the two-level model discussed above are clearly evident. Unfortunately the multi-level model approach, while much more accurate in this regard, is quite complex and computationally demanding. It may be that a relatively simple scheme like that proposed by Walsh et al. (1985), in which the mean thickness is arbitrarily decomposed into a number of thickness categories, may improve the two-level scheme sufficiently for use in large-scale climate studies.

12.4 Sea ice thermodynamic models

Sea ice grows and decays in response to longwave and shortwave radiation forcing, air temperature and humidity via turbulent sensible and latent heat exchanges, and heat conduction through the ice. The heat conduction is significantly affected by the amount of snow cover on the ice and by the brine remaining in the ice after it has frozen. These internal brine pockets cause the thermodynamic characteristics of sea ice to be very different than freshwater ice of the same thickness.

A complete model of sea ice thermodynamics needs to include all the heat budget components and the ice conduction. Many features of the thermal processes responsible for sea ice growth and decay can be identified from semi-empirical studies of ice breakup and formation of relatively motionless lake ice and sea ice (e.g., Langleben, 1971, 1972; Zubov, 1943). Overall, the two dominant components of the surface heat budget relevant to sea ice growth and freezing are the shortwave radiation during melting conditions and sensible heat loss during freezing. Observations of fast ice (relatively motionless ice attached to land) at the border of the Arctic Ocean indicate that once the initial stages of breakup (snow cover melt and formation of melt ponds) have passed, the remaining decay of a stationary ice cover is almost entirely due to the shortwave radiation incident on the ice surface (Langleben, 1972). However, at the initial stages of breakup and decay, the sensible heat flux plays an important role in promoting rapid melting of the surface snow cover. This melting forms melt ponds (Langleben, 1971), which reduce the albedo and greatly enhance the rate of ice melt. After only a few weeks, drainage canals and vertical melt holes develop, and the characteristic appearance of a summer ice cover evolves, with leads and surrounding smooth hummocks. Once these melt ponds have formed, the remaining decay is dominated by the radiation absorbed by the water. Zubov (1943), for example, has constructed an empirical rule for the breakup of fast ice, using only the radiation absorbed by the water in melt ponds and assuming this energy is expended solely in decreasing the horizontal dimensions of the floes. The model has been improved by Langleben (1972), who showed that including the radiation absorption by the ice can reduce the decay time up to 30%, in better agreement with observations. Recent work by Maykut and Perovich (1987) has improved further on this by considering the transfer of heat between the ocean and ice in much more detail.

As is clear from this discussion, the decay of sea ice in nature is rather critically affected by the amount of open water which, because of its low albedo, can absorb much more radiation than the ice. On motionless ice sheets, the open water is present only through melt ponds or holes. However, in an actual dynamic variable-thickness ice cover, as occurs almost everywhere in the polar regions, the growth and decay of sea ice can be substantially affected by spatial thickness variations. Perhaps the most obvious example is the effect of open water on the adjacent pack ice. During melting conditions the radiation absorbed by leads can contribute to lateral melting by ablation at the edges of ice floes. This heat can also be carried under the ice where it will contribute to bottom ablation. If the ice cover is sufficiently disintegrated, some of the heat can be stored in the mixed layer, thus delaying autumn freeze-up. Thin ice can also affect adjacent heat exchanges by modifying one or both of the planetary boundary layers.

It is also probable that thick ice in the form of pressure ridges behaves differently than level ice. Koerner's (1973) ablation observations indicate that the upper surfaces of first-year ridges ablate much more rapidly than level ice. Rigby and Hanson (1976) observed up to 2 m of summer melt on the bottom of an old 10 m thick ridge. They also observed losses of 4 m from a 15 m first-year ridge. Basically, the deep keels of pressure ridges allow ablation at the sides as well as the bottom (with perhaps substantial internal melting in porous, first-year ridges) although significant mechanical erosion may also take place.

a. *Empirical analytic sea ice growth models*

Considerable insight into the growth of sea ice may be obtained by examining empirical models where all the heat budget components (except conduction) are parameterized by the air temperature.

To show how these models work, we first introduce the concept of a degree day, $\theta(t)$ where:

$$\theta(t) = \int_0^t (T_f - T_a)dt \tag{12.20}$$

and T_f is the freezing temperature of sea water, and T_a is the air temperature. The bottom temperature of the ice T_B is usually taken as being the same as T_f. A good empirical approximation to ice thickness, H, as a function of degree days based on data analysis, is given by Anderson (1961):

$$H^2 + 5.1H = 6.7\theta \tag{12.21}$$

where H is in cm and θ has units of °C day. To obtain a more physical basis for this type of law, consider a steady state conduction through ice of thickness H. Choosing the sign convection that fluxes going *into* a surface are positive and *away* from a surface negative, we obtain for the growth on the bottom surface of the ice:

$$\rho_i L \frac{dH}{dt} = -\frac{\kappa_i}{H}(T_0 - T_B) \tag{12.22}$$

where ρ_i is the ice density, L is the latent heat of fusion, κ_i is the thermal conductivity of ice, and T_0 is the surface temperature. In other words, the heat released by growth on the bottom of the ice cover is balanced by conduction through the ice.

The simplest approximation at this point is the "Stefan" approximation where the surface temperature of the ice is taken to be the air temperature, i.e., $T_0 = T_a$, which is not a bad approximation. Integrating over time, the right-hand side of (12.22) becomes proportional to degree days

$$\rho_i L \frac{H^2}{2} = \kappa_i \theta \, . \tag{12.23}$$

This equation provides a physical basis for the squared term in the empirical Anderson equation.

However, this formulation is unrealistic for thin ice since under those circumstances T_0 can be much warmer than T_a. To account for this, some type of average transfer coefficient C_t is assumed that describes both sensible and latent heat exchanges. Because there is no latent heat change at the top surface of the ice under growth conditions, this surface flux must be equal to the conductive flux, yielding

$$\frac{\kappa_i}{H}(T_0 - T_B) = C_t(T_a - T_0) \, . \tag{12.24}$$

Solving for T_0 and substituting into (12.22)

$$\frac{-\kappa_i C_t}{(\kappa_i + C_t H)}(T_a - T_B) = \rho_i L \frac{dH}{dt} \, . \tag{12.25}$$

Integrating this equation gives

$$H^2 + \frac{2\kappa_i H}{C_t} = \frac{2\kappa_i}{\rho_i L}\theta \tag{12.26}$$

which is then the same form as the Anderson empirical formula.

The effect of a snow cover can also be taken into account within this formulation, by noting that with a layer of snow on top of the ice, assuming there is no melting or freezing at the snow/ice interface, the steady state conductance $\frac{\kappa_i}{H}$ may be replaced by the thermal conductance of the combined snow/ice slab,

$$\gamma = \frac{\kappa_i}{H + \frac{\kappa_i}{\kappa_s} H_s} \tag{12.27}$$

where H_s is the snow depth and κ_s is the thermal conductivity of snow (see Maykut, 1978, for details). Substituting this term back into the surface heat loss and proceeding as before gives:

$$H^2 + \left(\frac{2\kappa_i}{\kappa_s} H_s + \frac{2\kappa_i}{C_t}\right) H = \frac{2\kappa_i}{\rho_i L}\theta \, . \tag{12.28}$$

With the numerical values (see Maykut, 1986) $C_t = 209$ J cm^{-1} day^{-1} K^{-1}, $\kappa_i = 1758$ J cm^{-1} day^{-1} K^{-1}, $\rho_i L = 272$ J cm^{-3}, $\kappa_s = \kappa_i/6.5$, and with H in cm, the above formula becomes

$$H^2 + 16.8H = 12.9\theta \tag{12.29}$$

for the no snow case, and

$$H^2 + (13.1H_s + 16.8)H = 12.9\theta \tag{12.30}$$

for the snow case.

Apart from the fact that neither of these analytic results really fit the

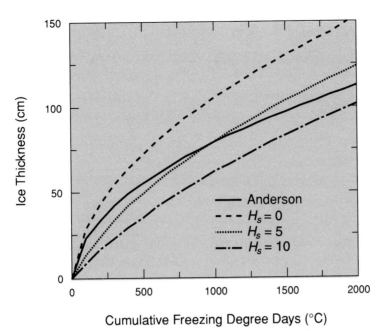

Fig. 12.6 Ice thickness versus freezing degree days calculated by the Anderson Eq. (12.27) and Eq. (12.31) with varying snow thicknesses.

Anderson empirical results, there are a number of useful conclusions that can be drawn from these formulae. Taking the derivative of (12.30) with respect to θ gives

$$\frac{dH}{d\theta} = \frac{12.9}{2H + 13.1H_s + 16.8}. \qquad (12.31)$$

From this result it is evident that the growth slows down drastically once the ice gets thicker than about 20 cm.

It is also clear that the effect of snow is very critical for thin ice and that it will greatly slow down the growth rate. Finally, a third interesting point is that the thicknesses are amazingly independent of the amount of heat transfer to the atmosphere as the latter only becomes critical for very thin ice. The square term in (12.30), which does not involve the rate of heat transfer to the atmosphere, becomes dominant for thicker ice. Thus many of the detailed heat loss mechanisms, such as wind speed, etc., are not highly critical to the final equilibrium thickness of the ice, which is basically a function of the degree days.

b. Full heat budget thermodynamic models

As shown in Fig. 12.6, the simple analytical ice growth model does not agree particularly well with Anderson's empirical expression. The main reason for this is lack of oceanic heat flux, which is difficult to include in the above formulation, and the fact that snowfall is really important only

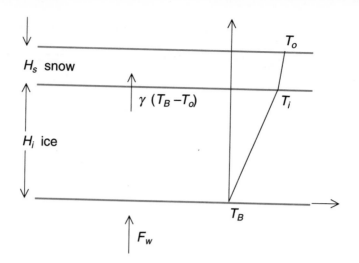

Fig. 12.7 Sketch of combined snow and ice system used in a more complete thermodynamic
sea ice model.

in the early part of the freezing season.

Because of this, what is usually done is to make use of an equilibrium
ice/snow system (i.e., neglecting heat capacity of snow and ice) in
conjunction with a complete surface heat budget. One can solve this system
iteratively and come up with an ice growth rate. Some variation of this
procedure is used in most climate studies involving sea ice. The basic idea
is illustrated in Fig. 12.7 with the steady-state temperature profile shown.
Assuming no melting at the snow ice interface, by continuity, the amount
of heat going through this interface must be the same in the snow and the
ice so that

$$(T_i - T_0)\frac{\kappa_s}{H_s} = \frac{\kappa_i}{H}(T_B - T_i) \tag{12.32}$$

where, as depicted in Fig. 12.7, T_i is the surface temperature of the ice and
T_0 the surface temperature of the snow. This equation allows a solution for
T_i in terms of T_B and T_0. Substituting the resulting expression into the
conductive flux through the ice gives

$$(T_i - T_0)\frac{\kappa_s}{H_s} = \gamma(T_B - T_0) \tag{12.33}$$

where γ is defined in (12.27).

As before the sign convention is such that fluxes into the surface are
considered positive. The complete heat budget equation then becomes

$$(1 - \alpha)F_s + F_l + \rho_a C_p C_H V_g(T_a - T_0) + \rho_a L_v C_E(q_a(T_a) - q_s(T_0))$$

$$-\epsilon\sigma T_0^4 + \gamma(T_B - T_0) = 0 \tag{12.34}$$

where α is the surface albedo, ρ_a is the density of air, C_p is the specific heat
of air, V_g the wind speed, q_a the specific humidity of the air, q_s the specific
humidity at the temperature of the surface, and F_s and F_l the incoming
shortwave and longwave radiation terms. The constants C_H and C_E are

432

bulk sensible and latent heat transfer coefficients, L_v is the latent heat of vaporization, and ϵ is the surface emissivity. The equation is usually solved iteratively (e.g., Appendix B of Hibler, 1980, for details and numerical values of various constants) for the ice surface temperature. The conduction of heat through the ice is used to estimate ice growth using

$$\gamma(T_B - T_0) - F_w = \rho_i L \frac{dH}{dt} \tag{12.35}$$

where F_w is the oceanic heat flux. In the case that the calculated surface temperature of the ice is above melting, it is then set equal to melting and the imbalance of surface flux used to melt ice.

To give some feeling for the relative role of the different components, Table 12.1 (taken from Hibler, 1980) gives a breakdown of the various heat budget components near the North Pole taken from the solution for such a balance equation without snow cover. As can be seen, the qualitative statements made earlier about the dominance of sensible heat fluxes in winter and the dominance of the shortwave fluxes in summer are borne out by the more detailed heat budget calculations.

Table 12.1a. Calculated heat budget components near the North Pole for different ice thicknesses in units of centimeters of ice per day. (1 cm of ice per day is equivalent to 34.95 W m^{-2}.)

Heat budget component	18 February				10 June				26 June			
	Open water	Ice thickness (m) 0.5	1.0	4.0	Open water	Ice thickness (m) 0.5	1.0	4.0	Open water	Ice thickness (m) 0.5	1.0	4.0
Net shortwave radiation	0	0	0	0	7.5	2.1	2.1	2.1	7.5	3.2	3.2	3.2
Net longwave radiation	−3.0	−0.4	−0.1	0.1	−0.9	−0.7	−0.8	−0.8	−0.6	−0.9	−0.9	−0.9
Sensible heat flux	−12.3	−2.5	−1.5	−0.5	−0.9	−0.8	−0.7	−0.7	0	−0.1	−0.1	−0.1
Latent heat flux	−3.1	−0.2	−0.1	0	−0.6	−0.6	−0.5	−0.5	0	−0.2	−0.2	−0.2
Conductive flux		3.1	1.7	0.4		0	−0.1	−0.1		−0.3	−0.1	0
Growth rate	18.4	3.1	1.7	0.4	−5.1	0	−0.1	−0.1	−6.9	−2.0	−2.0	−2.0

Table 12.1b. Central Arctic atmospheric temperatures, dew points and wind speeds.

	18 February	10 June	26 June
Air temperature (K)	241	269	272
Dew point temperatures (K)	236	268	271

c. *Effect of internal brine pockets*

Apart from the absence of dynamics, the global thermodynamic models mentioned above are still somewhat simplified, the main simplification being that no internal melting due to brine pockets in the ice has been considered. In sea ice the density, specific heat, and thermal conductivity are all functions of salinity and temperature (the dependence on temperature is indirectly also due to the salinity). These dependencies are caused by salt trapped in brine pockets that are in phase equilibrium with the surrounding ice. The equilibrium is maintained by volume changes in the brine pockets. A rise in temperature causes the ice surrounding the pocket to melt, diluting the brine and raising its freezing point to the new temperature. Because of the latent heat involved in this internal melting, the brine pockets act as a thermal reservoir, retarding the heating and cooling of the ice. Since the brine has a smaller conductivity and a greater specific heat than ice, these parameters change with temperature.

To take these variations into account Maykut and Untersteiner (1971) developed a time-dependent thermodynamic model for level multi-year sea ice and carried out a variety of calculations that yielded considerable insight into the growth and decay of sea ice. Basically they allowed the product of the density and the heat capacity, $\rho\nu$, and the thermal conductivity κ_i to be functions of the salinity S and temperature T. These parameters were then used in a time-dependent thermal diffusion equation which also allowed for shortwave radiation penetration into the ice.

A simplified model proposed by Semtner (1976) was that sea ice could be thought of as a matrix of brine pockets surrounded by ice where melting can be accomplished internally by enlarging the brine pockets rather than externally by decreasing the thickness. As a consequence, for the same forcing, sea ice can have a substantially greater equilibrium thickness than freshwater ice. In this model the snow and ice conductivities are fixed and the salinity profile does not have to be specified. To account for internal melting, an amount of penetrating radiation is stored in a heat reservoir without causing ablation. Energy from this reservoir is used to keep the temperature near the top of the ice from dropping below freezing in the fall. Using this "brine damping" concept, Semtner was able to reproduce many aspects of Maykut and Untersteiner's results within a few percent.

For an even simpler diagnostic model, Semtner proposed that a portion of the penetrating radiation I_0 be reflected away and the remainder applied as a surface energy flux. In addition, to compensate for the lack of internal melting, the conductivity is increased to allow greater winter freezing. In the simplest model, linear equilibrium temperature gradients are assumed in both the snow and ice. Since no heat is lost at the snow–ice interface, the heat flux is uniform in both snow and ice.

The results of both Semtner's prognostic and diagnostic models are compared to Maykut and Untersteiner's results in Fig. 12.8. This figure

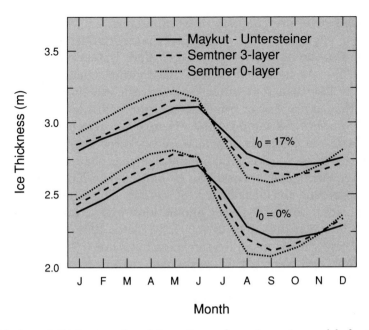

Fig. 12.8 Annual thickness cycles of three thermodynamic sea ice models for the cases of 0 and 17% penetrating radiation. (Redrafted from Semtner, 1976).

also shows the importance of the assumed penetration of radiation which causes internal melting. By allowing no radiation to penetrate, the internal melting is mitigated and the radiation instead is used to melt the ice, causing a reduced thickness. Note that while the diagnostic model reproduces the mean thickness well, the amplitude and phase of the seasonal variation of thickness are somewhat different from the prognostic models. This simplest diagnostic model has taken on special significance for numerical simulations of sea ice because almost continual ridging and deformation make it difficult to record the thermal history of a fixed ice thickness.

Finally, because of the problems with an excessive seasonal cycle in the simplest Semtner model, it may be useful to employ some type of brine damping in sea ice models that are used in climate studies. However, the difficulty here is that when a full dynamical model is used, the advection of ice makes it difficult to record the internal temperature characteristics of the ice. An approach that improves the summer thermodynamic response of the ice cover is to include a brine pocket storage term in an equilibrium one-level model. Such an approach was carried out by Flato (1991) in a numerical investigation of a multi-level thickness sea ice model. While his approach does not take into account the heat capacity of the ice, it does create a more realistic summer cycle in that it retards spring melting and autumn freezing. However, when used with a model involving full ice dynamics (see Flato, 1991) there are fewer effects of the brine damping since much of the growth and decay effects are dominated by the ice deformation with its attendant open water creating and pressure ridging.

The role of snow cover in modifying the thermodynamic growth of sea ice was also investigated by Maykut and Untersteiner (1971) and Semtner (1976). As indicated earlier, the thermal conductivity of snow is considerably less than that of ice; however, the albedo of snow is much higher than that of ice. Lower conductivity implies less ice growth whereas higher albedo implies less summer melt. These conflicting processes combine to produce little change in mean annual ice thickness for snowfall rates less than about 80 cm yr^{-1}. Since this is much higher than the snowfall rate for most of the Arctic, snow cover can be expected to have a rather small effect on the amount and distribution of sea ice. This notion is confirmed by recent dynamic–thermodynamic simulations with a multi-level sea ice model which included the effect of snowfall (Flato, 1991); here, a realistic snow cover decreased the mean ice thickness by about 0.5 m.

Precipitation, and snowfall in particular, is important in long-term simulations of Arctic ocean circulation. Although it appears to be relatively small on an annual basis, the fresh water supplied by precipitation is important in maintaining the stratification and hence the baroclinic circulation of the ocean (Ranelli and Hibler, 1991). A further complication is that snowfall which accumulates on the ice, and drifts with the ice, melts in a different location than it fell. This alters the apparent timing and spatial pattern of the precipitation received by the ocean and may affect the details of the circulation.

12.5 Summary

Sea ice plays a number of important roles in the climate system, yet the realistic treatment of sea ice in large-scale climate models is only now being addressed. On a regional scale, sea ice controls the heat and moisture fluxes between the ocean and atmosphere and may thus influence atmospheric circulation and precipitation patterns. Most important here is the coupling between thermodynamic processes (local energy balance and ice growth) and dynamic processes which continuously modify the state of the ice cover – destroying thin ice and creating thick ridges and open water leads. On a global scale, the high albedo of sea ice and its large spatial extent contribute to the much-studied ice-albedo feedback effect. Also on the global scale, ice transport southward from the Arctic provides an important source of fresh water to the Greenland and Norwegian Seas impacting the deep convective overturning and bottom water formation in this region. This process not only affects the distribution of heat carried by Atlantic currents, but also may influence the global-scale thermohaline circulation. These large-scale consequences of the polar ice cover can only be simulated in a climate model that contains a realistic sea ice dynamics component.

13

Land ice and climate

Cornelis J. van der Veen

13.1 Introduction

Land ice sheets are of concern for the climate system because they control the level of oceans and because of their physical interactions with the atmosphere. In simulations with GCMs, the large ice sheets of Antarctica and Greenland are usually represented as fixed lower boundary conditions. This is a valid first approach because the time scales for change in geometry of these ice sheets are much longer than the integration times of GCMs. Conversely, detailed models of land ice are largely concerned with understanding the dynamics of current systems, using measured atmospheric inputs as prescribed upper boundary conditions. A fully coupled GCM/ice-sheet model is required to study the interactions of large ice sheets with the atmosphere and oceans during periods of transition, such as from glacial to interglacial conditions. Such a model has yet to be developed, and common practice is to study these interactions with simple ice-sheet models coupled to simple climate models.

Terrestrial ice accounts for about 1.5% of all water present on the Earth and in its atmosphere, compared to 94% stored in the world's oceans. Nevertheless, small changes in the volume of grounded ice can have large effects on global sea level, as illustrated in Table 13.1. This table follows the division of land ice into mountain glaciers and small ice caps, the Greenland ice sheet, and the Antarctic ice sheet, reflecting the different climate characteristics and response times of each component. It should be stressed that the numbers given in Table 13.1 contain large uncertainties.

A glacier or ice sheet is nourished by accumulation, mainly in the form of snow, while mass is lost through surface melting in the ablation zone, and iceberg calving at the margin. Redistribution of mass from the accumulation area to the ablation zone is achieved through ice flow. In combination, these processes strive towards a steady state in which the shape of the glacier is constant (Fig. 13.1). Ice masses rarely achieve a steady state, however, because atmospheric boundary conditions, and therefore accumulation and ablation, are continuously changing. In addition, internal instabilities, such as periodic surges, may prevent the glacier from reaching a steady state.

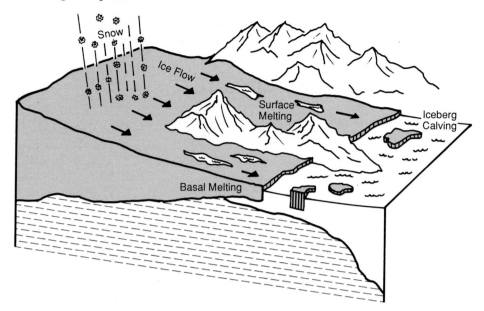

Fig. 13.1. Schematic representation of a glacier. The floating ice front shown here is characteristic of a marine ice sheet; tidewater glaciers usually have grounded calving fronts.

Table 13.1. Some characteristics of terrestrial ice. (From: IPCC, 1990a.)

	Antarctica (grounded ice)	Greenland	Glaciers (small ice caps)
Area (10^6 km^2)	11.97	1.68	0.55
Volume (10^6 km^3 ice)	29.33	2.95	0.11
Mean thickness (m)	2,488	1,575	200
Mean elevation (m)	2,000	2,080	–
Equivalent sea level (m)	65	7	0.35
Accumulation (10^{12} kg yr^{-1})	2,200	535	–
Ablation (10^{12} kg yr^{-1})	<10	280	–
Calving (10^{12} kg yr^{-1})	2,200	255	–
Mean equilibrium–line altitude (m)	–	950	0 – 6,300
Mass turnover time (yr)	~15,000	~5,000	50 – 1,000

The current state of balance of the components of terrestrial ice is poorly known, owing to a lack of comprehensive data. Worldwide, mountain glaciers have receded since about the mid-nineteenth century, with occasional interruptions of the retreat. The Greenland ice sheet appears to be thickening, or in near equilibrium in the interior regions, but may be thinning in the coastal areas. Estimates of the mass balance of the Antarctic ice sheet suggest that it is positive, although error limits allow for a slightly negative balance (cf., Van der Veen, 1991, for a more extensive overview).

Information on the history of ice sheets and climate can be obtained from ice cores because annual ice layers similar to those shown in the Quelccaya ice cap in Peru (see Plate 12) are preserved within the ice. Jouzel et al. (1989) discuss results from three deep Antarctic ice cores extending back to the last glacial period (65–15 ka B.P.). These data suggest that the central parts of Antarctica were thinner (up to a few 100 m) during the last glacial maximum than the present ice sheet. This thinning is consistent with lower accumulation during the glacial period. In the coastal areas, the ice sheet was considerably thicker during glacial times than presently, probably because of the close proximity to the sea and consequently stronger interaction with sea level and temperature.

During the late Wisconsinan (about 11 ka B.P.), the western margin of the Greenland ice sheet was up to 200 km farther advanced than the present-day ice edge (Reeh, 1989). Reeh (1985) argues that, in spite of the greater geographical extent, the ice sheet may have been as thin as, or even thinner than at present, due to decreased accumulation and the fact that the ice sheet in the late Wisconsinan consisted of much softer ice. The margin of the ice sheet in west Greenland reached its present position between 8 and 6 ka B.P. The recession is likely to have continued beyond the present extent until the onset of colder neoglacial conditions beginning around 3 ka B.P. The neoglacial changes in the marginal positions are poorly documented, but it appears that the neoglacial re-advance ended about 300 to 100 yr ago, and was followed by a widespread recession (Weidick, 1985).

Glaciers respond dynamically to changes in climate. In first approximation, this process of adjustment can be described by diffusive kinematic wave theory (e.g., Hutter, 1983, Chapter 6), which predicts that the time scale for adjustment depends strongly on the size of the glacier. For example, small mountain glaciers respond almost instantaneously to changes in climate, whereas central East Antarctica may be reacting to changes that occurred over the past 200,000 years; the response time of the Greenland ice sheet lies somewhere in between these extremes (1,000–10,000 yr) (Whillans, 1981). Because of this large range in response times, identifying probable causes of observed changes is often difficult, not in the least because feedback mechanisms (such as the coupling between surface elevation and snowfall or melting, or the glacio-isostatic response of the Earth's crust to the ice load; Oerlemans and Van der Veen, 1984), may strongly amplify the glacier's response.

For further reading, Paterson (1981) is an excellent introduction to glaciology, while Hutter (1983) provides a thorough theoretical description of glacier flow (suitable for readers with a strong mathematical background). Numerical aspects and the interactions between ice sheets and climate are discussed in Oerlemans and Van der Veen (1984). Also of interest are *Annals* of *Glaciology*, Vol. 11, (1988; *Antarctic Glaciology*), Vol. 12, (1989; *Ice Dynamics*) and Vol. 14, (1990; *Ice and Climate*).

13.2 Ice sheets and climate

In Table 13.2 some data on the global energy balance are given. From these numbers it can be seen that rapid deglaciation does not affect the global energy balance very much. For example, if a major Antarctic ice surge were to occur, in which one-tenth of the current ice volume is discharged into the oceans, the amount of energy needed to melt this excess ice corresponds to a decrease in mean ocean temperature of about 0.16 K. Locally, of course, the energy balance may be affected more strongly, but atmospheric motions and ocean currents can easily counteract the local loss of energy.

A more important coupling between ice sheets and climate is through the albedo feedback. Because snow and ice have a very high albedo (about 0.6 to 0.9), an expanding ice cover results in more solar radiation being reflected back into space. This leads to a reduction in global temperatures, allowing the ice cover to expand further. By using a simple (zero-dimensional) energy balance model (see Chapters 1 and 10) in which the planetary albedo increases with growing areal ice cover (which, in turn, is linked to the global temperature), it can be readily shown that the behavior of the model climate is described by a cusp catastrophe. That is, for a range of model parameters, three equilibria exist, namely a completely ice-free Earth, an Earth that is completely covered with ice, and an intermediate state. The latter solution is unstable, and small perturbations will cause the climate to move to either one of the other two (stable) equilibria (see Chapter 10).

In reality, the albedo feedback is more complicated, because the growth and decay of polar ice caps cannot be simply linked to variations in global temperature. A more appropriate way to include ice-sheet dynamics is by

Table 13.2. Some energy data of the global climate system. (From: Oerlemans and Van der Veen, 1984.)

Total precipitation	5.1×10^{14}	$m^3 \ yr^{-1}$
Associated release of latent heat	1.3×10^{24}	$J \ yr^{-1}$
Absorbed solar radiation	3.8×10^{24}	$J \ yr^{-1}$
Present ice volume	3.2×10^{16}	m^3
Energy needed to melt it	9.9×10^{24}	J
Ice-age ice volume	8.1×10^{16}	m^3
Return to interglacial requires	1.5×10^{25}	J
Energy needed to warm the ocean by 1 K	5.8×10^{24}	J
Energy needed to warm the atmosphere by 1 K	5.1×10^{21}	J

using the model of Weertman (1976). In this model, the ice is assumed to have a perfect plastic rheology, with a yield stress of about 1 bar. A perfectly plastic material does not deform until the deforming stress reaches a critical value called the yield stress, at which point the rate of deformation becomes very large. For glacier flow, the driving stress due to gravity is directly linked to the thickness and surface slope of the ice sheet. Thus, in the present context, the assumption of perfect plasticity implies that the ice sheet adjusts its geometry such that the driving stress equals the yield stress at all times. In that case, the profile of the glacier is parabolic:

$$h(x) = [c(L - |x|)]^{\frac{1}{2}}$$

where h is the elevation of the ice-sheet surface above sea level, L the half-width of the glacier, x the distance from the ice-sheet center and c a scaling constant defining the height-to-width ratio of the ice sheet (Oerlemans and Van der Veen, 1984). Growth and decay of the ice sheet can be determined from the total ice volume added to, or removed from, the glacier through surface accumulation and ablation. Thus, if V represents ice volume per unit width, and M the net accumulation minus ablation (expressed in meters of ice per unit time and unit width), the change in V is

$$\frac{dV}{dt} = \int_0^L M(x)dx.$$

Changes in the areal extent (i.e., in L) can be readily related to changes in ice volume through this assumption of a parabolic ice-sheet profile. Further refinements can be added to the model, for example by limiting the maximum areal extent of the glacier (on the Northern Hemisphere, the ice sheet cannot expand into the Arctic Ocean), or by including the glacio-isostatic response of the Earth crust to the ice load (Ghil and LeTreut, 1981).

Inclusion of glacier dynamics in the crude way described above, is sufficient to produce free oscillations of the climate system (depending, of course, on the choice of model parameters). These oscillations, with a typical period of 100 ka, are primarily caused by the slow response of the ice sheet to climate forcing resulting in a phase lag between ice extent and global temperature. Similar oscillations have been found in numerical models in which the evolution of the glacier is solved explicitly on a horizontal grid (e.g., Oerlemans and Van der Veen, 1984).

Simple coupled models such as those discussed above, are helpful tools in identifying important feedback mechanisms and investigating the most general features of ice sheet–climate interaction. The next step is to study the most important mechanisms with more realistic climate and ice-sheet models.

441

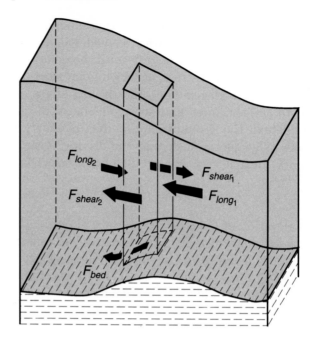

Fig. 13.2. Resistive forces acting on a glacier. Only the resistive forces in the mean direction of ice flow are shown; these are longitudinal pushes and pulls from up- and down-glacier ice (F_{long}), side shear (F_{shear}), and drag at the glacier base (F_{bed}). The sum of these resistive forces equals the driving force, directed in the direction of decreasing surface elevation.

13.3   Modeling glacier flow

Glacier flow is driven by gravity, causing ice to flow in the direction of decreasing surface elevation; this action is described by the driving stress. Acceleration terms in Newton's second law of motion can be neglected, and the equations of motion reduce to equations of static force balance. This means that, given the driving stress (which can be calculated from the glacier geometry), the resistive stresses that oppose this action can be estimated. This resistance to flow is due to drag at the glacier base and surrounding rock walls or slower moving ice, and gradients in longitudinal forces (Fig. 13.2). Depending upon the application of the model, one or more of these resistive stresses may be neglected. For example, in the laminar (or lamellar) flow theory, the driving stress is entirely balanced by basal drag, and a column of ice deforms by simple shear.

Stresses are linked to strain rates through the constitutive relation (Glen's flow law; Paterson, 1981), so that knowledge of the stresses in a glacier allows calculation of the deformational component of the ice velocity. Other contributions to the total ice velocity may be associated with sliding of the ice over the bed, and deformation of the underlying bed (Fig. 13.3).

442

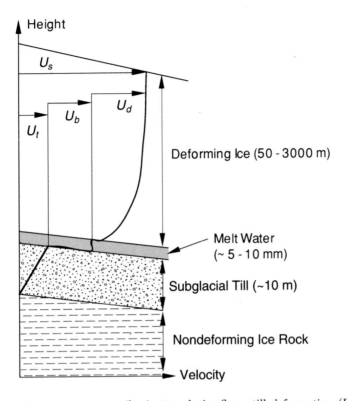

Fig. 13.3. Illustrating processes contributing to glacier flow: till deformation (U_t), basal sliding (U_b) and ice deformation (U_d). The vertical axis is not to scale; characteristic thicknesses for each layer are indicated on the right. The contribution of each process to the observable surface velocity ($U_s = U_t + U_b + U_d$) varies strongly among different glaciers. The linear increase in velocity in the till layer is based on a linear relation deformation rate and stress which has not been established unambiguously yet.

When the basal ice temperature is at the pressure-melting point, a lubricating water layer can form, allowing the ice to slide over its bed. Sliding velocities can be an order of magnitude larger than the deformational velocity (up to several hundreds of meters per year). Many theories have been proposed to calculate the sliding velocity, but the issue is far from being settled (cf., Paterson, 1981). Most sliding relationships used in numerical models are loosely based on the results of the empirical studies of Budd et al. (1979), linking the sliding velocity to the drag at the glacier base, and the effective pressure at the bed, being the weight of the ice above minus the subglacial water pressure.

The discovery of a layer of deformable till under Ice stream B, West Antarctica (Blankenship et al., 1986) has led to the hypothesis of continuous mobile drift (e.g., Alley et al., 1987) to explain the high velocities observed on this ice stream (up to 800 m yr^{-1}). The ice stream is taken to be underlain by a continuous layer of deforming debris, with the shear stress that controls glacial flow being transmitted through this debris. If the relation between this shear stress and the deformation rate of the subglacial till is known, the ice velocity due to deformation of the till layer can be calculated.

With the ice velocities known, the continuity equation, expressing conservation of mass (or volume if the ice density is taken constant), can be solved. To this effect, a numerical grid covering the glacier is defined and velocities at gridpoints calculated. The divergence of flow at each gridpoint, together with the local accumulation and melting rate, then yields the rate of thickness change at that point. By marching the model forward in time, evolution of the glacier can be studied.

13.4 Modeling the surface mass balance

The link between climate and glacier is forged through the net surface mass balance (accumulation minus ablation). Atmospheric temperature itself is also a link because, in part, this affects the temperature of the ice, and hence the rate of deformation. However, diffusion and advection of heat within the main body of a glacier is a very slow process. On the time scales of interest in this book (up to a few hundred years), the direct interaction between glacier flow and surface temperature, may be neglected.

The most satisfactory approach for estimating ablation is to explicitly calculate the contribution of surface melting to the surface mass balance. This can be done by using an energy balance model for the upper ice layers as described by, among others, Greuell and Oerlemans (1986). In this model, temperatures within the upper snow layers (to a depth of about 25 m below the surface) and melting rate are calculated from meteorological input parameters (cf., Fig. 1.2). Sensitivity experiments show that ablation is very sensitive to variations in atmospheric temperature, humidity, cloudiness, and wind speed.

The model of Greuell and Oerlemans only considers steady-state glaciers, and does not include evolution of the glacier geometry. In time-evolving glacier models, common practice is to parameterize melt rates in terms of meteorological variables. For mountain glaciers, this may be a dubious procedure. Many statistical studies have been conducted to find a relation between the surface mass balance and meteorological data (e.g., Letréguilly, 1988), but the disparity among the relations derived for individual glaciers makes it unclear whether any is suitable for general application. For the Greenland ice sheet, however, parameterization of melt rate and surface (ice) temperature appears to give reasonable agreement between observed and modeled ablation rates (Reeh, 1991).

In applying results of statistical studies, one should keep in mind that, in most cases, correlations are based on incomplete records. For example, for the Greenland ice sheet, only five air temperature records that cover an entire annual cycle, are available for interior stations, with the record length between one and three years (Reeh, 1991). Furthermore, it is not immediately clear whether such correlations are also valid under different climatic conditions (such as after a global warming).

Accumulation on a glacier is not any easier to model than ablation. An important factor is the air temperature, determining the form of precipitation (snow or rain), as well as the amount of water vapor that can be contained in the atmosphere. For example, in central East Antarctica precipitation rates are very low because the cold air contains very little water vapor. Another important factor is the surface orography. Due to the forced uplift of moist air, mountain ranges in general experience higher precipitation than their surroundings. It can thus be expected that precipitation rates will reach a maximum near the edge of an ice sheet, where the surface slope is largest.

To include a satisfactory calculation of precipitation, a balance equation for precipitable water should be used. This was done by Sanberg and Oerlemans (1983), who studied the effect of upslope precipitation on the evolution of the Fenno-Scandinavian ice sheet. However, in most numerical models, accumulation is simply linked to the surface elevation, and perhaps distance to the ice-sheet margin (cf., Oerlemans and Van der Veen, 1984).

13.5 Response of mountain glaciers to climate changes

Essential to the credibility of any numerical model is the ability to reproduce observed geometries or variations in geometry. Given the number of adjustable parameters in an ice-flow model, almost any (realistic) glacier profile can be reproduced. A more significant test is whether the model can simulate observed glacier changes. This has been attempted for a number of glaciers, but in general, results of even the "best" simulations are rather disappointing, and poor agreement between observed and modeled glacier advance and retreat exists (Oerlemans, 1989b).

The poor performance of the model cannot be ascribed to inadequacies of the flow model. Varying model parameters within realistic ranges suggests that these uncertainties cannot explain the discrepancies between modeled and observed historic glacier variations. Rather, the unsuccessful simulations are attributed to an inadequate reconstruction of past net surface mass balance.

Because local conditions, such as the shape of the valley, routes of major weather patterns, etc., play an important role in determining the local surface climate of a glacier, a generally applicable model does not exist. Instead, the response of each individual glacier to changes in climate needs to be addressed separately, at least when specific predictions are sought. Generalizations, such as in Oerlemans (1989a), in which the contribution of

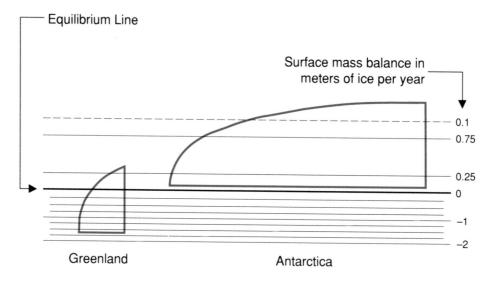

Fig. 13.4 Generalized surface mass balance field with the present-day ice sheets of Antarctica and Greenland superimposed (From: Oerlemans and Van der Veen, 1984.)

mountain glaciers and small ice caps to the sea-level rise as observed during the past century, is parameterized as a function of the global-mean surface air temperature, need to be viewed with skepticism, especially when past trends (valid for a small temperature increase) are extrapolated over much larger temperature ranges.

13.6 Response of the polar ice sheets to climate changes

The mass budgets of the ice sheets in Greenland and Antarctica are very different because of the large differences in surface climate (Fig. 13.4). In general, the surface mass balance (snowfall minus melting, averaged over one year) increases with elevation until a maximum is reached. At higher altitudes, a steady decrease occurs, more or less in proportion to the saturation water vapor pressure. This occurs largely because as air becomes colder, it can hold less water vapor and in a very cold climate, the amount of water vapor in the air becomes the limiting factor for precipitation. A large part of the surface of the Greenland ice sheet lies below the equilibrium line indicating that surface melting contributes significantly to ice loss. It is estimated that for this ice sheet, surface melting and iceberg calving at the edge are equally important loss mechanisms (Reeh, 1989). In Antarctica, on the other hand, surface melting is negligible and ice loss is achieved through calving at the front. The inland ice flows towards the margins, gradually concentrating in rapidly moving ice streams and outlet glaciers (such as Byrd Glacier, see Plate 13). These large rivers of ice (up to 500 km long and 50 km wide) drain in the ice shelves (Ross and Ronne-Filchner, and numerous smaller ones). At the seaward edge of these shelves, ice breaks off to produce icebergs.

Fig. 13.4 can be used to infer qualitatively the response of both ice sheets to a climatic warming. In first approximation, a warmer climate can be represented by an upward shift of the surface mass balance field. This means that melting on the Greenland ice sheet will become more important, whereas in Antarctica the surface mass balance should increase. In other words, the Greenland ice sheet may be expected to decrease in size while the Antarctic ice sheet may grow after a climate warming.

To this date, no modeling studies have addressed the issue of how the Greenland ice sheet may react to an altered climate. Only estimates of the sensitivity of the surface mass balance to climate changes have been made. For example, Ambach and Kuhn (in Oerlemans, 1989b) use a linear perturbation theory to expand the surface mass and energy budget at the equilibrium line (the line where the annual snowfall equals annual ablation), which allows calculation of the change in the equilibrium-line altitude for small changes in atmospheric temperature, precipitation and radiation. By extrapolating this elevation change to the entire ice sheet, an estimate can be made of the change in ablation and accumulation areas, and the effect on the ice sheet. What remains to be done, however, is to apply these results to a numerical model of the Greenland ice sheet, to more accurately assess the effects on ice flow, and hence the changes in ice volume.

Only two studies have been conducted to assess the response of the current Antarctic ice sheet to a climatic warming, using a two-dimensional numerical model (Oerlemans, 1982; Budd et al., 1987). Oerlemans argues that a climate warming may lead to increased snowfall in central Antarctica. Melting near the ice-sheet edge is also expected to increase, but this effect turns out to be smaller. In his warming experiments, Oerlemans prescribes a linear increase in annual-mean sea-level temperature (determining melt rates in the coastal regions) and precipitation rate, from the present values at the start of the integration, to the prescribed higher values after 100 yr, after which they are kept constant. The results of four warming scenarios are given in Fig. 13.5. In the experiments with a large temperature increase (+6 K), the ice volume increases initially, but after about 80 yr (when 80% of the temperature rise is effected), coastal melting causes the ice sheet to become smaller. The period of ice loss has only a short duration, and subsequently, the ice sheet starts to grow again. This behavior is associated with the dynamical response of the ice sheet. If melting near the edge becomes large, while farther inland the precipitation increases, the surface slope is no longer in equilibrium with the mass-balance distribution. The ice sheet responds by increasing the discharge of ice toward the edge, compensating the surface lowering there due to surface melting. Thus, the temporary decrease in ice volume reflects the fact that the ice sheet needs

447

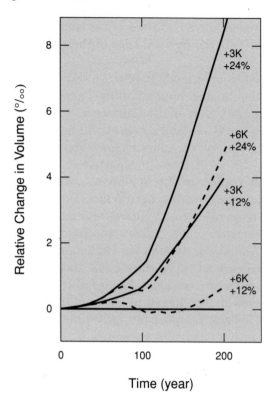

Fig. 13.5. Effect of increased sea surface temperature and precipitation rate on the Antarctic ice volume. The change in volume is given in percent of the present-day ice volume. Labels indicate the temperature change and increase in precipitation rate 100 years after the onset of warming. (From: Oerlemans, 1982.)

some time to react dynamically to the increased mass-balance gradient.

Whereas Oerlemans (1982) only considers the effect of increased surface accumulation and melting, Budd et al. (1987) believe that these effects will be over-shadowed by increased thinning rates near the grounding line (where the land ice flows into the floating ice shelves) due to higher temperatures of the sea water circulating under the ice shelves. Because, at present, it is computationally too time consuming to carry out large numbers of response calculations for the entire Antarctic ice sheet at sufficiently high resolution to resolve the major individual glaciers and ice streams (i.e., a grid spacing of 10 km or less; Oerlemans used a 100 × 100 km grid), Budd et al. use a hierarchy of numerical models. A flowband model is used to study the response of a cross-sectional profile across West Antarctica. Results of a high-resolution (20 km grid) two-dimensional model for West Antarctica, and a coarse-resolution (100 km grid) two-dimensional model for the entire Antarctic, are then used to extend the flowband changes to changes in the whole Antarctic ice sheet. It is not clear to what extent thinning rates at the grounding line could increase, but Budd et al. estimate a maximum of 1 cm yr^{-1}. Even for such a high thinning rate, the sea-level rise caused by the decrease in Antarctic ice volume, is only about 60 cm during the first century. A new steady state is reached after some 10,000 yr, with sea level about 4.5 m higher than the present-day stand.

13.7 Possible instability of the West Antarctic ice sheet

The possibility of a collapse of the marine-based West Antarctic ice sheet as a result of warmer ocean temperatures (and enhanced basal melting under the peripheral floating ice shelves) has received much attention outside the glaciological literature. The initial thesis was that, if melting rates under the ice shelves increase, ice rises (places where the shelves have run aground) may disappear or become less effective, while the restraining forces due to shear between the shelves and the enclosing mountains will become less as the shelves thin. This results in a reduced back-pressure on the grounded parts, so that discharge of inland ice may increase (Mercer, 1978). A positive feedback between thinning at the grounding line, and further reduction in backstress, may exist, and the entire ice sheet may collapse within a short time.

Very few studies have been undertaken to study this supposedly catastrophic behavior using numerical models that cover an entire flowband, and most of these contain dubious assumptions. However, both Van der Veen (1987) and Alley et al. (1987) conclude that the inland ice may be (much) more stable to changes at the grounding line, than suggested earlier. Admittedly, these models have only been applied to idealized geometries, and further modeling efforts and field measurements are needed to settle this issue once and for all.

13.8 Concluding remarks

The aim of most of the present-day numerical models is to obtain a better understanding of the mechanisms playing a role in glacier evolution, to explain past glacier changes, and to assess the range of possibilities for future changes. Most of these models include glacier dynamics in a simplified manner and theoreticians may argue that ice flow is much more complex, and should be treated mathematically more rigorously. This may be true for some purposes. However, for including land ice in climate models, one needs the simplest possible model with the minimum necessary degrees of freedom. Given the uncertainties involved in formulating the appropriate surface boundary conditions (e.g., past climate, or future warming trends) as well as the basal conditions (e.g., basal sliding, deforming bed) and the physics controlling ice flow (e.g., the constitutive relation for glacier ice), a mathematically more comprehensive approach is often not warranted.

Modeling and parameterization

When discussing the response of glaciers to climatic changes, it should be borne in mind that a constant climate does not necessarily imply a steady-state glacier. For example, many mountain glaciers exhibit surge-type behavior in which extended periods of normal motion are followed by brief periods of comparatively fast motion (Raymond, 1987). A number of numerical models have been developed that are capable of simulating glacier surges (Budd, 1975), as well as similar oscillations in large ice sheets (Oerlemans and Van der Veen, 1984), without having to resort to periodic climatic forcing. Essential in these models is that a positive feedback exists between the sliding velocity, and quantities from which this velocity is calculated (such as the amount of subglacial water).

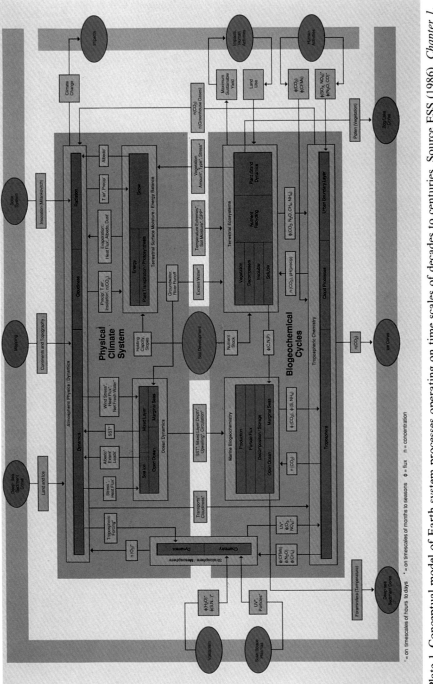

Plate 1 Conceptual model of Earth system processes operating on time scales of decades to centuries. Source ESS (1986). *Chapter 1.*

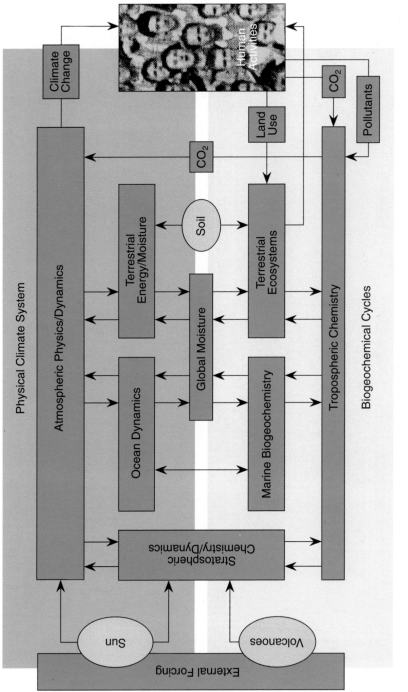

Plate 2 Schematic model of the fluid and biological Earth that shows global change in a time scale of decades to centuries. A notable feature is the presence of human activity as a major inducer of change; humanity must also live with the results of change from both anthropogenic and natural factors. Source ESS (1986). *Chapter 2.*

Plate 3 Deforestation in Rondonia, a part of the Amazon, has been attributed to the paving of the Cuiba–Porto Velho highway (BR-364) in 1984, which brought on a surge of migration. Source Malingreau (1988). *Chapter 2*.

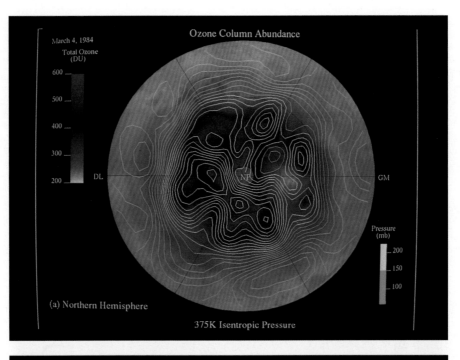

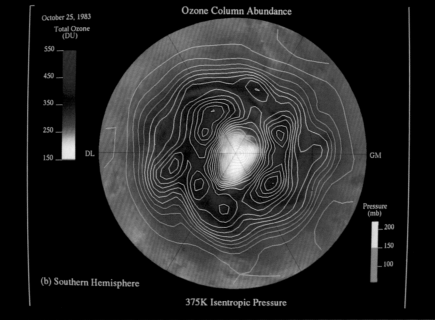

Plate 4 Distribution of ozone column abundance, as observed by the Total Ozone
Measurement Spectrophotometer (TOMS), flown aboard Nimbus-7 (color) and
pressure distribution (contours) on the 375 K isentropic surface, which is
approximately coincident with the tropopause. (a) NH on 4 March 1984. (b) SH
on 25 October 1983. Reprinted from Salby (in press) by permission of Academic
Press. *Chapter 3.*

Plate 5 Visible radiance at 1200 UTC on 4 March 1984, measured aboard the geostationary satellite GOES-E. Reprinted from Salby (in press) by permission of Academic Press. *Chapter 3.*

C2 5–YR MEAN ANNUAL CLOUD AMOUNT

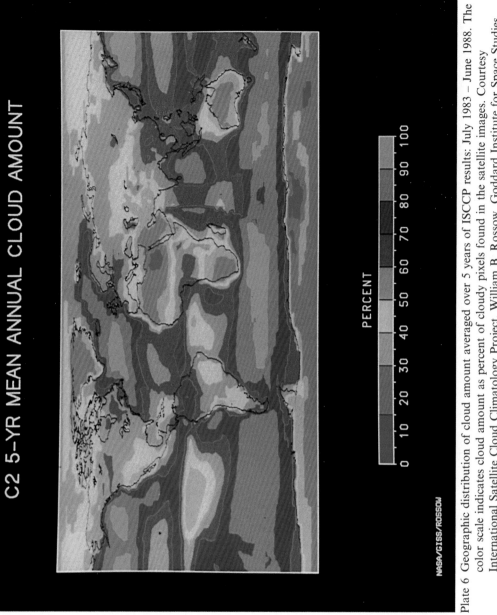

NASA/GISS/ROSSOW

PERCENT

0 10 20 30 40 50 60 70 80 90 100

Plate 6 Geographic distribution of cloud amount averaged over 5 years of ISCCP results: July 1983 – June 1988. The color scale indicates cloud amount as percent of cloudy pixels found in the satellite images. Courtesy International Satellite Cloud Climatology Project, William B. Rossow, Goddard Institute for Space Studies, NASA (see Rossow and Schiffer, 1991). *Chapters 3 and 5.*

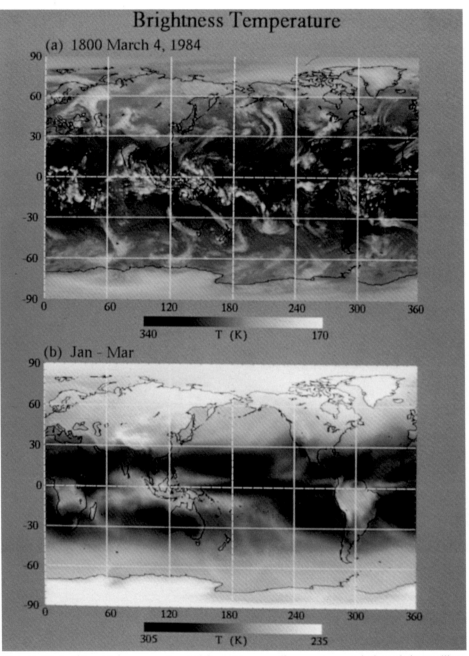

Plate 7 Global cloud image, constructed from 11 μm radiances measured aboard six satellites simultaneously viewing the Earth. (a) Instantaneous cloud field at 1800 UTC on 4 March 1984. (b) January–March mean cloud field. Bright regions correspond to cold (high) clouds, dark regions to warm low surface. Reprinted from Salby (in press) by permission of Academic Press. *Chapter 3.*

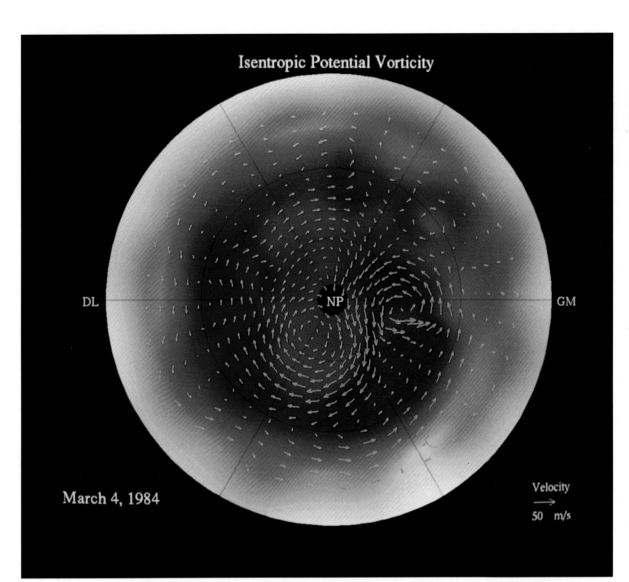

Plate 8 Potential vorticity and motion on the 850 K isentropic surface for 4 March 1984.
Reprinted from Salby (in press) by permission of Academic Press. *Chapter 3.*

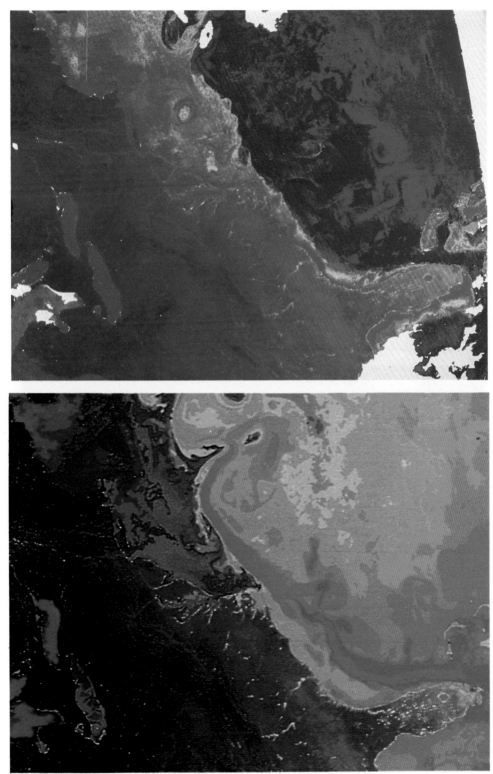

Plate 9 Satellite observations of sea surface temperature (left) and ocean color (right) demonstrate the correlation between marine productivity and physical oceanography. Note especially the contrast across the Gulf Stream with warm water associated with low productivity. Source: ESS (1986). *Chapters 8 and 16.*

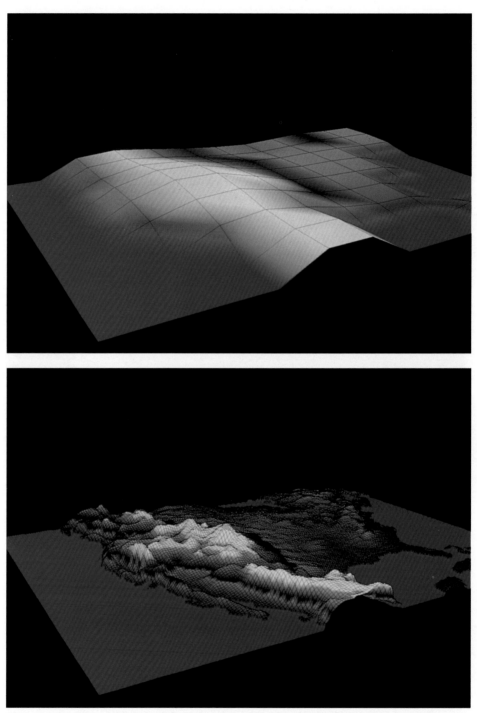

Plate 10 Schematic depiction of the topography over North America as represented in most coarse-resolution (480 km grid) atmospheric general circulation models used for climate simulation in the past (above) and in high-resolution (60 km grid) global operational forecast models (below). At the higher resolution a factor of 500 times the computing power is required. Courtesy Thomas Bettge, NCAR. *Chapters 9 and 23.*

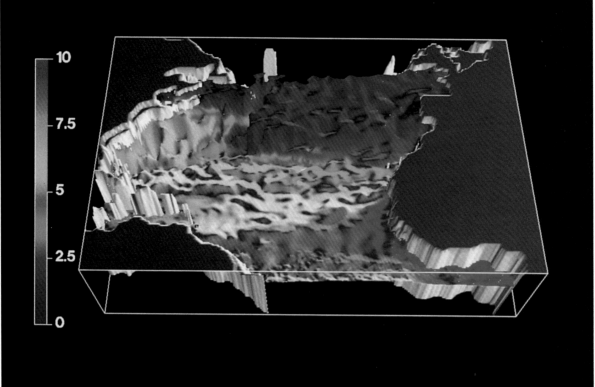

Plate 11 Ventilation time scales in the North Atlantic thermocline. A measure of the time scale at which the interior ocean communicates with the atmosphere, called age, is mapped as color (units on the color scale correspond to years) on a surface of constant potential temperature (15°C). Near the outcrop (where the temperature surface intersects the sea surface) the age is near zero, indicating that water in this region has recently been in contact with the atmosphere. There is a sharp gradient emanating from the coast of West Africa. This delineates the boundary between the "ventilated" and "shadow" zones of thermocline theory. Note also the east–west asymmetry near the central latitude of the subtropical gyre. Young, newly ventilated water is being carried south and into the thermocline over most of the interior of the subtropical gyre. On the western boundary, the Gulf Stream is carrying the older water north towards the outcrop where it will eventually be entrained into the surface mixed layer and be re-ventilated. Courtesy Frank Bryan, NCAR. *Chapter 11.*

Plate 12 Annual ice layers in the Quelccaya ice cap, Peru. Courtesy Lonnie G. Thompson, Ohio State University. *Chapters 13 and 18*.

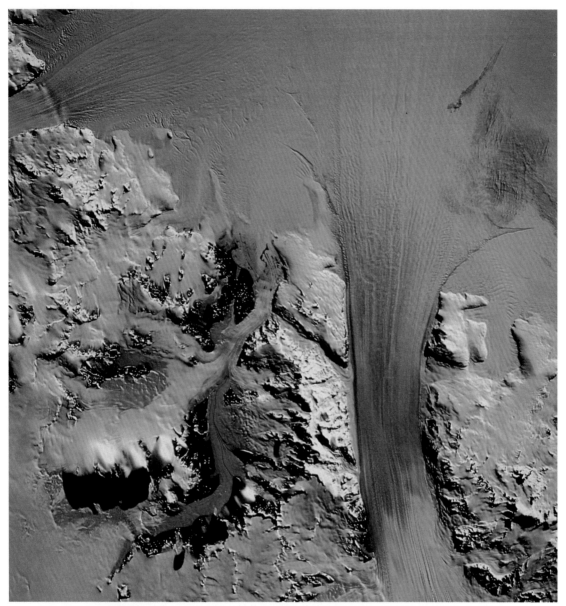

Plate 13 Byrd Glacier, Antarctica, flowing through the Transantarctic Mountains into the Ross Ice Shelf, as seen from space (Landsat image). USGS, Flagstaff Image Processing Facility, courtesy Baerbel K. Lucchitta. *Chapter 13.*

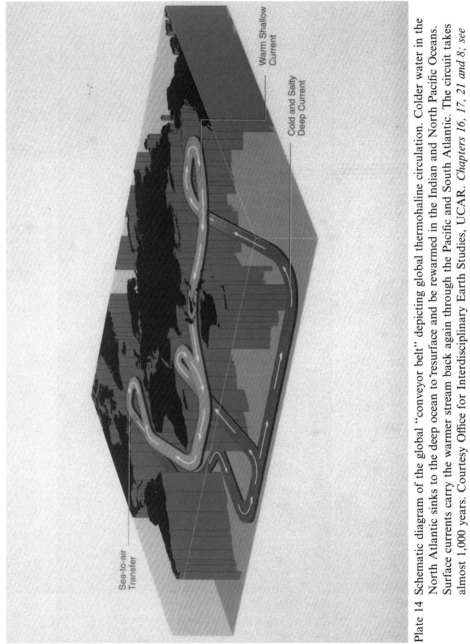

Warm Shallow
Current

Cold and Salty
Deep Current

Sea-to-air
Transfer

Plate 14 Schematic diagram of the global "conveyor belt" depicting global thermohaline circulation. Colder water in the North Atlantic sinks to the deep ocean to "resurface and be rewarmed in the Indian and North Pacific Oceans. Surface currents carry the warmer stream back again through the Pacific and South Atlantic. The circuit takes almost 1,000 years. Courtesy Office for Interdisciplinary Earth Studies, UCAR. *Chapters 16, 17, 21 and 8; see also Fig. 17.12.*

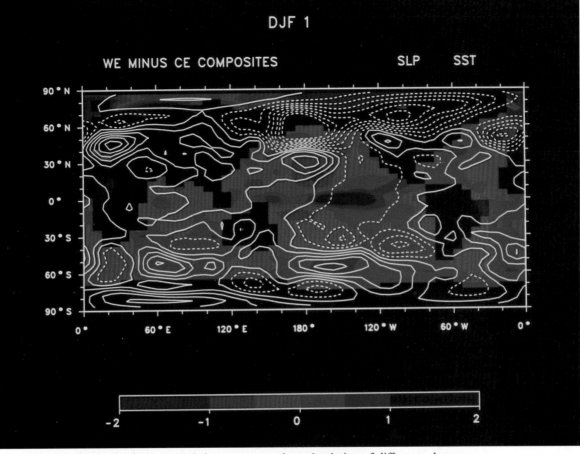

DJF 1

WE MINUS CE COMPOSITES   SLP   SST

Plate 15 Composite from a coupled ocean–atmosphere simulation of differences between warm events (WE or El Niño events) and cold events (CE or La Niña events) for December–January–February (DJF) during the mature phase of the events. The sea surface temperature is relatively higher during warm events in red areas and relatively lower in blue areas. Sea level pressure anomalies are superposed in white, with negative values dashed. The contour interval is 1.0 mb. Courtesy Gerald Meehl, NCAR. *Chapter 17.*

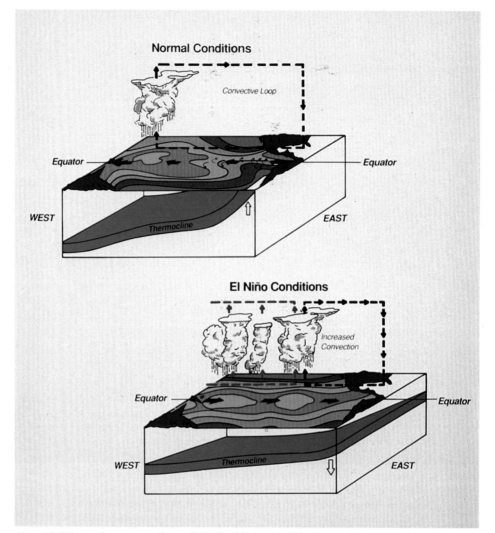

Plate 16 Schematic cross sections of the Pacific Basin with Australia at lower left and the Americas on the right depicting normal and El Niño conditions. Total sea surface temperatures are multicolored, with water exceeding 29°C in gold and a contour interval of 1°C. Regions of convection and overturning in the atmosphere are indicated and the thermocline is shown in blue. Changes in surface currents along the equator are indicated by the black arrows. Courtesy UCAR. *Chapter 18.*

14

Biophysical models of land surface processes

Piers J. Sellers

14.1 Introduction

The terrestrial biosphere and the global atmosphere are coupled to each other over a wide range of time scales through the exchanges of radiation, heat, water, CO_2, and trace constituents, see Fig. 14.1 and Chapters 5, 6, and 10. It is easier to realistically model the short time-scale processes (radiation and energy exchange), where physical mechanisms predominate, than the longer time-scale processes (trace

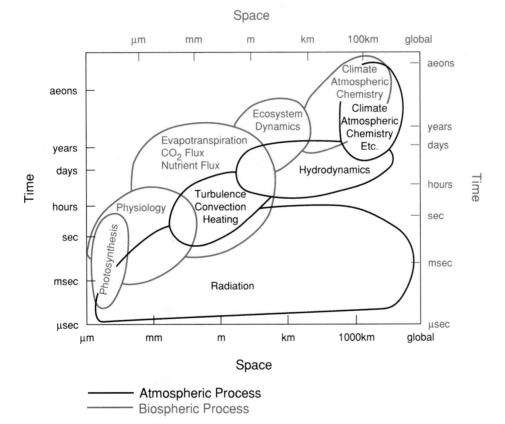

Fig. 14.1 Time–space scale diagrams of important processes as perceived by: (a) atmospheric scientists, and (b) biologists.

constituent exchange and ecological succession), where diverse biological processes can strongly influence the time-evolution of the system.

Figure 14.1 shows time–space scale domains as viewed by (a) atmospheric scientists and (b) biologists. The two communities have had their greatest scientific successes in different parts of the scale domain: atmospheric scientists have improved GCMs over the last three decades to the point where they are routinely applied for weather forecasting and climate simulation. The combination of supercomputers and efficient computational schemes has resulted in the development of realistic representations of atmospheric hydrodynamics over the medium range of time and space scales. Biologists, on the other hand, have made considerable progress in constructing rigorous models of vegetation photosynthesis (e.g., Farquhar et al., 1980; and see Chapter 6) that are based on sound biochemical and biophysical principles. However, the rigor and reliability of biological models tend to degrade as time and space scales increase, so that at the scales of ecological succession, the models are highly empirical and of uncertain reliability. Complete ecosystems, where organisms and environment interact with each other over an enormous range of possible interconnections, are very complex and do not generally lend themselves to simple parameterization. Nevertheless understanding some of the biospheric mechanisms that are potentially important to the study of climate change will require some description of the time-evolution of terrestrial ecosystems. The 'scale gap' between the experiences of the biological and atmospheric communities must be addressed (though, as we shall see, not necessarily by direct model coupling) if we are to construct an accurate picture of the complete climate system.

The simplest initial approach to modeling land processes is to break the problem down into components that can be treated more or less as separate items. Figure 14.2 shows the complete spectrum of interactions broken out into three coupled feedback loops, the surface components of which can be described as biophysics (or land surface climatology), biogeochemistry (or nutrient cycling), and community composition and structure (ecosystem dynamics), see Hall et al. (1988). The biophysics loop is the fastest, most strongly interacting, and most 'physical' of the three; the surface properties of albedo, roughness, and the biophysically controlled rates of photosynthesis and transpiration act as surface boundary conditions to the near-surface atmospheric forcings to give the radiation, heat, water vapor and momentum fluxes between the two systems, see Chapter 10. (If respiration and allocation can be realistically described, total CO_2 flux can also be calculated.) These short time-scale fluxes influence the local hydrology, carbon transfer, and nutrient cycle rates (through temperature and moisture availability), which, in turn, feed into the biogeochemistry loop. The biogeochemistry loop is influenced by climatic forcings on the time scale of months to years and by inputs from the biophysics system (see

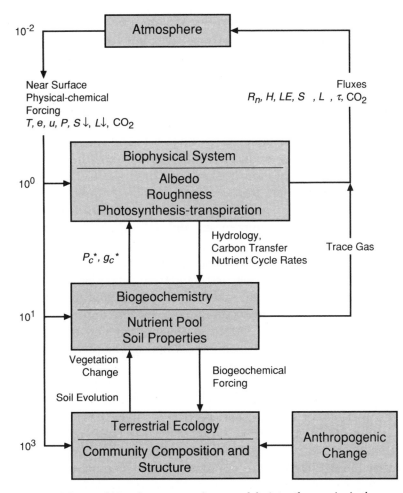

Fig. 14.2 Breakdown of biosphere–atmosphere models into three principal groups roughly
separated by time scale (in years, at left) of exchange process: biophysical models,
biogeochemical models, and terrestrial ecosystem models.

Fig. 14.2). The nutrient cycle rates and pool size, and the soil properties
and hydrology, in large part determine the trace gas fluxes from the system
and also regulate the area-averaged canopy photosynthetic capacity, P_c^*,
and maximum stomatal conductance, g_c^*, both of which are functions of
canopy nitrogen content among other things. P_c^* and g_c^* regulate the
photosynthesis–transpiration system and partially determine the fluxes of
water and CO_2 from the vegetated land surface.

The biogeochemical system feeds directly into the community composition
and structure (CCS) through its influence on soil properties and nutrient
status. The CCS, which can be loosely thought of as the ecological status of
the system (species composition, biomass, etc.), is also influenced directly
by physical climate forcings. For example, the southern edge of the North
American mid-continental boreal forest is largely controlled by summer soil

moisture deficits and fire frequency, see Chapter 6. The CCS, in turn, exerts direct influence on the biochemical and biophysical components of the system through the contemporary vegetation–soil system and P_c^* and g_c^*, respectively. Few attempts have been made to model more than two of these feedback loops at a time, and for good reason: the time and space scales that must be addressed to model any two of them in the context of a realistic coupled model are almost completely incompatible. With a few exceptions, what has been done up to now is model one of the loops in detail, while prescribing the boundary forcings from the other systems. For example, simple biophysical models have been linked to GCMs to calculate the surface fluxes of radiation, heat, water and momentum between the land surface and the atmosphere, and to date, all such schemes have prescribed the global fields of albedo, roughness, and P_c^*, g_c^*, either implicitly or explicitly. The bulk of this chapter is devoted to a discussion of this class of models: the basic concepts involved; the history of their development; model components; model testing and validation; and the future research and development directions. The other two classes of biosphere models are briefly reviewed in a subsequent section, mainly in the context of developing linkages with climate models. More detailed discussion of the different components of these other models may be found in Chapters 6 and 8.

14.2 The development and shortcomings of early models

This section reviews the basic concepts which formed the basis of the first land surface parameterizations for GCMs (LSPs), the history of their development and implementation, and the weight of evidence which suggests that biophysically based approaches are now required.

14.2.1 *Basic concepts*

Atmospheric GCMs require time series of the fluxes of radiation, sensible heat, latent heat, and momentum across the air–land interface to serve as lower boundary conditions. These may be specified by calculations which involve the surface parameters of albedo, roughness length, and moisture availability. In this section, the basic (minimal) equation set traditionally used to calculate these fluxes is reviewed.

For a surface with a specified albedo and assuming a thermal emissivity of unity, the net radiation may be defined as follows:

$$R_n = S(1 - \alpha) + L_{wd} - \epsilon \sigma T_s^4 , \qquad (14.1)$$

where R_n = net radiation; S = total solar radiation incident on the surface; α = broad-band surface albedo; L_{wd} = net downwelling longwave radiation; T_s = surface temperature; ϵ = surface emissivity; and σ = Stefan–Boltzmann constant.

The shear stress exerted on the atmosphere by the surface may be given by

$$\tau = \rho \frac{u_r}{r_{a_m}} ,$$

(14.2)

where τ = shear stress, kg m^{-1} s^{-2}; ρ = density of air, kg m^{-3}; u_r = wind speed at reference height above surface, m s^{-1}; and r_{a_m} = aerodynamic resistance for momentum transfer, s m^{-1}. τ is the rate of momentum transfer from the atmosphere to surface induced by the deceleration force exerted on the horizontal wind flow, u_r, as it moves over a rough surface. Here, the resistance analogue for a flux is used, whereby:

Flux = Potential difference/Resistance.

In this case, the potential difference is the difference between the wind speed at a height z_r above the surface $(u = u_r)$ and at the surface $(u = 0)$. In a GCM application, z_r would be located within the lowest atmospheric layer of the model. The resistance may be calculated using an eddy diffusion concept, whereby a 'conductance' to momentum transfer is inverted and integrated between the surface and the reference height. Under neutral conditions (strong ventilation relative to convection driven transport), r_{a_m} is given by:

$$r_{a_m} = \frac{1}{u_r} \left[\frac{1}{k} \ln \left(\frac{z_r}{z_0} \right) \right]^2 ,$$

(14.3)

where k = von Karman's constant, 0.41; z_0 = roughness length, m; and z_r = reference height, m.

In nature, the roughness length z_0 in (14.3) is a function of the height and density of the vegetation. Rough surfaces, e.g., forests and shrublands, are more strongly coupled to the atmosphere via turbulent transfer than are smooth surfaces, such as deserts and grasslands. Note that r_{a_m} is equivalent to $1/C_D u_r$, where C_D is the surface drag coefficient.

The net radiation R_n, as given by (14.1), from the surface to the atmosphere, is partitioned into four components at the Earth's surface,

$$R_n = H + LE + G + P,$$

(14.4)

where H = sensible heat flux (conduction to the air), W m^{-2}; LE = latent heat flux (energy release to the air by evaporation), W m^{-2}; L = heat of vaporization, J kg^{-1}; E = moisture flux into the air, kg m^{-2} s^{-1}; G = heat flux into the ground, W m^{-2}; and P = energy used for photosynthesis, W m^{-2}.

P amounts to less than 1% of absorbed insolation. Over a day or longer, the ground heat flux, G, averages out to be a relatively small term, seldom more than about 10% of R_n. H and LE normally make up the bulk of R_n, and the magnitude of the ratio of sensible to latent heat release into the atmosphere (the Bowen ratio) can be an important determinant of the local and regional climate. The sensible heat flux heats the air above the surface locally in both space and time – it forces the development of the planetary boundary layer (PBL) in the daytime hours. The latent heat flux may have a small local heating effect, but generally the released water vapor is transported into the free atmosphere, moved downstream by the winds, and condenses to form clouds and precipitation some distance in time and space from its point of release. Since these two modes of heat release have such different effects on the internal heating and radiative properties of the atmosphere (condensing water vapor forms clouds), the accurate specification of the terms in the surface energy budget (14.4) is needed for the realistic simulation of atmospheric circulation, see also Chapter 10.

The partitioning of R_n into H, LE and G has been the most problematical aspect of constructing realistic LSPs. Generally, a surface temperature is specified so that the sensible heat flux can be calculated by:

$$H = \left(\frac{T_s - T_r}{r_a}\right)\rho c_p,$$
(14.5)

where T_r = air temperature at reference height, z_r; c_p = specific heat of air, J kg^{-1} K^{-1}; r_a = aerodynamic resistance for heat and water vapor, s m^{-1} = $1/C_H u_r$; and C_H = heat transfer coefficient.

Traditionally, most LSPs have employed a similar equation to (14.5) to describe the transfer of water vapor from the surface to the air:

$$LE = \beta\left(\frac{e_*(T_s) - e_r}{r_a}\right)\frac{\rho c_p}{\gamma},$$
(14.6a)

where β = 'Beta function', the index of moisture availability; $e_*(T_s)$ = saturated vapor pressure at temperature T_s, Pa; e_r = vapor pressure of air at height z_r, Pa; and γ = psychrometric constant, Pa K^{-1}. The moisture availability function, β, is formulated to vary from 1 (freely available soil moisture) to 0 (dry soil conditions).

In (14.6a), the evaporative source is assumed to be a saturated reservoir of moisture at surface temperature, T_s. For the case of a lake or a recently wetted vegetation canopy, this is perfectly reasonable, as the surface moisture source does indeed consist of free water in direct contact with the atmosphere in exactly the same way as envisaged for sensible heat transfer. However, most land surfaces which have any significant soil moisture are covered by vegetation, and the outer surface of the plant leaves, in most cases, consists of a waxy cuticle which retains internal moisture to prevent desiccation. Most of the water evaporated from a vegetated surface to the atmosphere passes through tiny pores in leaf surfaces, the stomates, which cover a small fraction, around 1%, of the total leaf surface. These pores are under active physiological control (see later sections) and impose a significant resistance to the passage of water vapor from the saturated interior of the leaves to the external air. The water vapor must, therefore, traverse an additional 'stomatal' resistance on its way out of the vegetation to the air layer in contact with the leaf surface. A gradient of vapor pressure is required to drive this transfer and so the vapor pressure at the leaf surface, e_s, is usually considerably lower than $e_*(T_s)$. However, e_s represents the surface 'skin' source term for the aerodynamic transfer of moisture away from the vegetation and corresponds to T_s for sensible heat transfer. A total area-averaged surface resistance term, which integrates the action of the vegetation stomates and the soil evaporation term, if any, should therefore be used to describe the evaporation from the land surface. In place of (14.6a), we should write:

$$LE = \left(\frac{e_*(T_s) - e_r}{r_a + r_s} \right) \frac{\rho c_p}{\gamma}, \qquad (14.6b)$$

where r_s = surface resistance, s m^{-1}.

It has been argued that the effect of a nonzero r_s in (14.6b) can be mimicked by simply manipulating the value of β. A simple calculation set out in Sato et al. (1989) demonstrated that this is not the case but rather that the concept of the β-function of (14.6a) is fundamentally in error for the calculation of land surface evaporation rates, see the next section.

The ground heat flux, G, is usually modeled with diffusion–conduction models, which can be generalized by:

$$G = f\left(\frac{dT_s}{dt} \right). \qquad (14.7)$$

Given values of the atmospheric forcing variables $S^\downarrow$, $L^\downarrow$, u_r, T_r, e_r, and surface parameters α, z_0 and β or r_s, (14.1) through (14.7) can be solved to give values of R_n, τ, LE, H, G, and T_s. Normally T_s is used as a prognostic variable and is time-stepped in parallel with the physics calculations of the atmospheric GCM.

From the above, it can be seen that the essential surface parameters are the surface albedo (radiative transfer), the roughness length (momentum transfer), and the surface resistance or moisture availability term (partitioning of energy).

14.2.2   History of model development and implementation

Until the mid-1980s, the land surface properties that regulate the exchange of radiation, momentum and heat with the atmosphere were regarded as separable parameters which could be independently prescribed as boundary conditions within an atmospheric GCM. Although this approach was not very realistic, it lent itself to some simple sensitivity studies in which the influence of the albedo (radiative transfer), the surface roughness (momentum transfer), and the surface hydrological parameterization or prescription (sensible and latent heat transfer) were investigated separately.

The surface albedo is known to vary from about 12% (tropical rain forest) to around 35% (Sahara desert) over the snowfree land (Shuttleworth et al., 1984b; Matthews, 1985). In most GCMs, the land albedo fields are simply prescribed as fixed fields. Charney et al. (1977) conducted a series of GCM

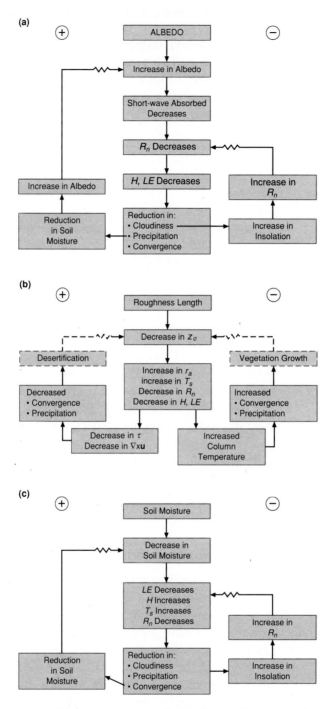

Fig. 14.3 Possible consequences of large-scale changes in the key surface parameters: albedo, roughness length, and soil moisture. The left-hand side of each figure refers to a positive feedback loop, where consequences tend to reinforce the initial perturbation, while the right-hand side shows negative feedback effects. For illustration, the parameter changes associated with desertification or deforestation have been shown; that is, reductions in roughness length and soil moisture availability and an increase in albedo: (a) albedo; and (b) roughness length; (c) soil moisture.

sensitivity experiments centered on albedo changes in the Sahel, which indicated that an increase in the albedo of the region would lead to a decrease in the surface evaporation rate and a reduction in the precipitation rate in the same area. This study was followed by similar ones on the Sahel/Sahara region (Sud and Fennessy, 1982; Chervin, 1979), which produced the same result qualitatively, if not quantitatively, see Fig. 14.3(a).

Fields of surface roughness, z_0, have traditionally been dealt with in an even simpler way. For many models, one value of z_0 is specified for all land surfaces and seasons. Specifying a suitable value for z_0 is made more difficult by differences between the aerodynamic resistance appropriate for water vapor and heat versus momentum.

The two transfer processes operate differently at the small and medium scales. At the leaf or plant scale, momentum is exchanged by both viscous (or skin drag) and bluff body (or pressure/form drag) interactions; heat and water vapor, on the other hand, are exchanged only via molecular diffusion through the laminar boundary layer overlying the leaf surface (skin drag). Generally speaking, the exchange of momentum is thus two to four times as efficient as for heat or vapor. At larger scales, that is, the grid-area average scale, momentum is also extracted from the atmosphere via pressure drag interactions with small-scale topography and gravity-wave drag exerted by mountains (McFarlane, 1987).

Thus, models that use one value of z_0 to describe all these exchange processes inadequately treat momentum and scalar exchanges (heat, water vapor, CO_2, etc.). Sud et al. (1988) investigated the effects of drastically reducing the roughness length in the desert regions in a GCM simulation and found that the curl of the surface stress, low-level convergence, and convective precipitation all decreased in those regions, see Fig. 14.3(b).

Most of the past sensitivity studies investigating soil moisture effects partitioned the absorbed energy, R_n, into ground heat flux, sensible heat flux, and latent heat flux (G, H, LE) terms using (14.5), (14.6a) and (14.7), where β was defined using the Budyko (1974) bucket model in which each land surface gridpoint was modeled as a bucket with a 150 mm rainfall equivalent capacity. The bucket was filled by rainfall and emptied by evaporation, where evaporation rate was given by (14.6a) and β was defined as a function of the amount of water in the bucket.

Sato et al. (1989a) presented a brief analysis, which assessed the errors involved in using (14.6b) in place of (14.6a). They considered an area of land exposed to a time-invariant forcing of T_r, e_r, u_r, S, and L_{wd}, and calculated the energy budget for the surface, assuming a thick vegetation cover (tropical forest), ample soil moisture and negligible heat storage $(G \rightarrow 0)$.

The analysis demonstrated that for a tropical forest site, the bucket model would calculate an unrealistic initial evaporation rate almost five times greater than that given by the biophysical model. To reduce the bucket model evaporation rate to that of the biophysical model it would be necessary to reduce β to 0.091, which implies a soil wetness on the order of 10% in most bucket models, clearly below the range of reasonable 'tuning' for a humid region.

This analysis is extreme, as it takes no account of the feedback from the excessive evaporation rate on the atmospheric forcing. When Sato et al. (1989a) introduced the bucket model into the GCM, the huge latent heat fluxes generated from the humid continental regions led to the simulation of an unrealistically cold, humid, and stable lower troposphere within the first few time steps. This, in turn, suppressed further evaporation from the surface to lower (but still unrealistically high) rates.

There is ample evidence to suggest that the bulk of GCM simulation studies carried out with β-function formulations have systematically overestimated the land surface evaporation rate for humid areas. In the case of weather forecast models, this overestimation leads to excessive convection and precipitation over the continents, particularly in summer, and an overestimation of spring and early summer evaporation rates, possibly leading to a complete drying out of the soil in the continental interiors by the mid-to-late summer. This 'boom and bust' behavior of the simulated annual hydrologic cycle is clearly at odds with reality (see Rasmusson, 1968). Figure 14.3(c) summarizes the consequences of changes in soil moisture on regional climate, see also Sato et al. (1989a).

14.2.3 *Shortcomings of the early modeling approaches*

As we have seen from the last section, the early modeling efforts suffered from flaws which made them unrealistic and, inevitably, inaccurate:

(a) *Independent specification of surface attributes*: Surface albedo, roughness length, and moisture availability were frequently specified as completely independent, unrelated characteristics of the land surface.

(b) *Unrealistic specification of albedo*: Albedo fields were often specified from data with no spatially integrating theory applied to check the reasonableness of the prescription. Generally, the specified albedos were too low (see Dorman and Sellers, 1989).

(c) *Unrealistic specification of roughness length, confusion of momentum and heat/vapor transfer pathways*: Unrealistic values of z_0, sometimes simply one number for all land surfaces, have frequently been used in GCMs. In one instance, z_0 values of 50 m or more were used in an attempt to increase the drag over mountain ranges – this same z_0 was used to calculate the transfer of heat and water vapor from the surface (Betts, personal communication). In all GCMs, one value of z_0 was used for heat, water, and momentum transfer up until the mid-1980s.

(d) *Unrealistic specification of soil hydrology*: The 'bucket' model traditionally had a reservoir with a total water-holding capacity of

150 mm, worldwide. Clearly, in nature, soil moisture capacities vary from virtually nil (desert, mountains) to several meters (tropical and some temperate forests).

(e) *Unrealistic description of the evapotranspiration process*: The use of the β-function, with a range from zero to unity, is probably the most damaging of all the flaws in the early models. By failing to account for surface resistance, this formulation systematically overestimates surface evaporation rates in the humid regions.

14.3   The development of biophysical models

The models of Dickinson (1984) and Sellers et al. (1986) were the first attempts at a biophysically realistic modeling approach for LSPs, where the emphasis is on modeling the soil–vegetation complex itself and thereby specifying the surface attributes of albedo, roughness and evapotranspiration rate as mutually consistent surface properties.   While the models of Dickinson (1984) and Sellers et al. (1986) have concentrated on global scale processes, Avissar and Pielke (1989) and Noilhan and Planton (1989) have extended the approach to mesoscale applications.

These new treatments have their origins in the small-scale models developed by micrometeorologists, forest scientists, and agronomists; see for example Waggoner and Reifsnyder (1968). In the case of the Simple Biosphere (SiB), a deliberate effort was made to base the radiative, momentum, and heat/mass transfer properties of the vegetated surface on a set of directly measurable surface parameters, some of which are involved in all three submodels (see Table 14.1).

The next three sections describe the three submodel components – radiative transfer, turbulent transfer, and heat/water vapor exchange. Examples of the application of specific methods or techniques are drawn from the Biosphere–Atmosphere Transfer Scheme (BATS) of Dickinson (1984) and the Simple Biosphere Model (SiB) of Sellers et al. (1986).

14.3.1   *Radiative transfer*

In the context of global modeling, an accurate and realistic description of the interception, absorption, and scattering of radiation by vegetation canopies and soil is important for three reasons:

(i) *Radiation budget*: The calculation of the absorbed radiative energy is an essential precursor to quantifying the surface heat fluxes [see (14.1)].

(ii) *Biophysical function*: Photosynthesis and transpiration are driven by absorption and utilization of photosynthetically active radiation (PAR), 0.4–0.7 μm, by the chlorophyll in plant leaves.

461

Table 14.1. Vegetation and soil parameters used by 3 submodels of SiB.
Note linkages between submodels through shared parameters.

Parameters	Radiative transfer	Momentum transfer	Stomatal resistance
Leaf area index	x	x	x
Leaf angle distribution function	x	x	x
Leaf optical properties	x		x
Soil optical propeties	x		
Green leaf fraction	x		x
Cover fraction	x	x	x
Canopy top and base heights		x	
Leaf dimension		x	
Leaf physiological properties, root physiology, morphology			x
Soil physical properties			x

(iii) *Remote sensing of the surface condition*: Satellite remote sensing is the only practical means of specifying the biophysical state of the land surface semi-continuously in time and over synoptic spatial scales. The correct interpretation of satellite data depends on the application of radiative transfer models to identify the contributions of the atmosphere, vegetation, and soil to the radiance as measured by the satellite sensor at the top of the atmosphere.

Biophysical models require points (i) and (ii) to be satisfactorily addressed or else they simply will not work properly. Ultimately, part (iii) will become important as we try to force the surface boundary conditions using satellite data.

The right-hand side of (14.1) contains the terms that must be obtained from the canopy radiative transfer description. The insolation, S, and the downward longwave flux, L_{wd}, represent forcings provided by the GCM or in off-line studies by micrometeorological observations. The parameters or terms that must be provided by the LSP are albedo, α, and surface emission of longwave radiation, which in (14.1) is lumped under the $\epsilon\sigma T_s^4$ term.

The calculation of the surface albedo is the more difficult task. Green leaves have spectral properties which vary sharply over the shortwave spectrum: they are relatively dark in the visible region, where chlorophyll absorbs these more energetic photons to drive photosynthesis, and almost perfect scatterers in the near-infrared region. At a minimum, a modeling treatment of shortwave radiation interaction with vegetated surfaces requires splitting the insolation into at least these two wavebands. Conventionally, the division is made at around 0.7 μm, where the leaf scattering coefficient sharply increases to divide the broadband visible (0.4–0.7 μm) and near-infrared (0.7–4.0 μm) regions.

Good reviews of different methods for modeling shortwave radiative transfer within vegetation canopies can be found in Asrar (1989). The more rigorous models include those of Kimes (1984) and Myneni et al. (1989): these models use either finite element (vector) or Monte Carlo methods, whereby the fate of scattered radiation is traced numerically. Although these techniques permit the inclusion of some intricate details of radiative transfer (e.g., nonisotropic scattering by leaf elements, penumbral effects, etc.), they tend to be computationally expensive and too cumbersome for operational use; climate models require manipulable, robust, computationally cheap formulations.

The two-stream approximation represents the next level down in sophistication: it is reasonably realistic but can be manipulated to yield an analytical solution for a given set of soil and (plane-parallel) vegetation characteristics. Dickinson (1983) reviewed the work of Meador and Weaver (1980) and others in which the two-stream method was used to describe shortwave radiation fluxes in the atmosphere. Dickinson (1983) adapted the same equations to describe radiative transfer in vegetative canopies.

In the model, a beam of radiation incident on a homogeneous, plane-parallel vegetation canopy is attenuated down through the foliage, according to the exponential extinction law, that is,

$$F_L = F_0 e^{-KL_z}. \tag{14.8}$$

where F_{L_z} = direct beam flux at level L_z (leaf area index) in the canopy; F_0 = direct beam flux incident on the canopy; K = optical depth of the direct beam per unit leaf area = $G_{(\mu)}/\mu$; $G_{(\mu)}$ = relative projected area of leaves in direction $\cos^{-1}\mu$; and μ = cosine of zenith angle of incident radiation.

In nature, the direct beam radiation intercepted by the leaves according to (14.8) may be scattered into an infinite number of possible radiances (vectors). The two-stream model treats all of these scattered radiances in a grossly simplified way by considering only two integrated fluxes, one upward and one downward, for the diffuse radiation. The upscatter and downscatter parameters for the direct beam contribution to these two diffuse fluxes can be determined by analysis (see Dickinson, 1983).

There are four possible fates for upward diffuse flux as it passes up through the canopy: it can pass through the gaps between leaves; it can be absorbed by leaves; or it can be rescattered by a leaf into either an upward or downward diffuse flux. Calculation of the fraction of radiation passing through gaps between leaves and absorbed by leaves is straightforward. The treatment of scattering is a little more complicated, however. The diffuse upscatter parameter may be inferred from the analysis of Norman and Jarvis (1975) where the leaf is treated as an inclined plane with isotropic forward and backscattering properties. A flat leaf will reflect downward fluxes into the upward direction only and will transmit in the downward direction only. As the leaf-angle inclination increases, a greater fraction of the downward diffuse flux may be reflected into the downward (hemispherically integrated) direction and transmitted into the upward (hemispherically integrated) direction.

All of these possible fates of the diffuse radiation fluxes and the contribution of the direct incident beam to them may be given by:

$$-\left(\frac{dI^\uparrow}{dL_z}\right) + \frac{1}{\bar{\mu}}[1 - (1 - \beta)\omega]I^\uparrow - \frac{\omega}{\bar{\mu}}\beta I^\downarrow = \omega K \beta_0 e^{-KL_z} \; , \qquad (14.9a)$$

$$\left(\frac{dI^\downarrow}{dL_z}\right) + \frac{1}{\bar{\mu}}[1 - (1 - \beta)\omega]I^\downarrow - \frac{\omega}{\bar{\mu}}\beta I^\uparrow = \omega K(1 - \beta_0)e^{-KL_z} \; , \qquad (14.9b)$$

where $I^\uparrow$, $I^\downarrow$ = integrated upward, downward diffuse fluxes, respectively; $\bar{\mu}$ = average inverse diffuse optical depth per unit leaf area; β = diffuse upscatter parameter and β_0 = direct beam upscatter parameter; ω = the leaf scattering coefficient.

The different contributions to and/or fates of upward diffuse flux, $I^\uparrow$, can be seen in (14.9a): the second term describes the proportion of the upwelling diffuse flux that is intercepted by leaves and rescattered back into the upward direction to contribute to $I^\uparrow$, while a proportion of flux, $(1 - \omega)/\bar{\mu}$, is absorbed; the third term defines the proportion of the downward diffuse flux, $I^\downarrow$, that is scattered into the upward direction; the term on the right-hand side is, of course, the contribution from the interception and scattering of the direct solar beam. Equation (14.9b) essentially does the same thing but for the downward diffuse flux.

Equation (14.9) may be solved in the $\omega \to 0$ limit (single-scattering approximation and semi-infinite canopy) to yield an upward diffuse flux at the top of the canopy, which may be taken as the single-scattering albedo. It can also be solved analytically to yield expressions for the diffuse fluxes within the canopy and the canopy reflectance given boundary conditions for the incident radiation flux and the soil surface reflectance.

The direct radiation terms on the right-hand sides of (14.9) may be dropped to solve for the case of incident diffuse fluxes. If this calculation is performed for the diffuse and direct beam components of visible and near-infrared fluxes, i.e., four components in all, a weighted average of the resulting reflectances will provide an estimate of the albedo term as used in (14.1).

The exact value of the calculated albedo will therefore depend upon the value of the constants in (14.9), which in turn depend on:

(1) the scattering coefficients of the leaves and soil;

(2) the leaf area index;

(3) the leaf angle distribution function;

(4) the proportions of different components of the incident radiation; and

(5) the angle of the direct-beam incident radiation.

This method can be used to calculate not only the total albedo of the vegetated surface but also the amount and kind of the radiation absorbed by different parts of the canopy and the soil surface, see Sellers (1985).

In theory, the treatment of longwave emission by the vegetation and soil could also be covered using the two-stream approximation. In practice, almost all researchers assume that natural surfaces (both soil and vegetation) have emissivities of unity, which greatly simplifies the treatment.

Such simplified modeling schemes form the basis of the radiative transfer packages in SiB and BATS. Given the constraints of supercomputer storage and (computational) time, it is unlikely that more sophisticated radiative transfer treatments will be used *online* in GCMs in the foreseeable future.

The two-stream model appears to be an effective method for calculating the fate of incident shortwave fluxes, see Dorman and Sellers (1989). However, the approach is not suitable for application over heterogeneous or 'clumpy' canopies, such as coniferous forests. For those applications, three-dimensional models are required, particularly when calculating radiances for the purpose of satellite data interpretation. Three-dimensional models could be used offline from GCMs to calculate radiance fields and albedos, thus completing the link between the radiative transfer and biophysical properties used in LSPs and the global data sets provided by satellite remote sensing.

Lastly, the effects of snow on surface radiative properties are dealt with fairly simply in most models. Usually, a snow cover fraction is calculated as a function of the depth of the snow and the surface albedos adjusted by simple area weighting.

14.3.2 *Turbulent transfer*

A description of near-surface turbulent transfer is necessary for an LSP for two reasons:

(i) *Momentum transfer:* The drag force on the lowest layer of the atmosphere may be important for the calculation of the curl of the surface stress (see Sud et al., 1988).

(ii) *Heat and mass transfer:* The transport of heat, water vapor, CO_2, and trace gases away from the surface is partly regulated by the aerodynamic resistance pathway. When the surface is wet, the free water evaporation rate (interception loss) is determined entirely by the available energy, aerodynamic resistance, and vapor pressure deficit of the air.

The turbulent transport of momentum and scalars between the atmosphere and the surface is a complex process that does not lend itself to simple modeling (see Raupach and Thom, 1981). Most of the more sophisticated treatments are based on higher-order closure models, such as those described in Raupach and Thom (1981), Yamada (1982), Shaw and Pereira (1982), and Meyers and Paw U (1986). Other treatments make use of Markov-chain numerical simulations, which trace the paths of hypothetical scalars released from the canopy–soil surfaces into the airflow, and eddy-simulation techniques which take a finite element approach to modeling the windflow above and around objects on the surface.

These treatments can be far more realistic than the simpler approaches, but are practically unsuited for LSP–GCM applications because of their computational expense and storage requirements. Additionally, as the sensitivity analysis of Sato et al. (1989a) showed, the aerodynamic resistance pathway plays only a secondary role in determining the amount and ratio of the surface heat fluxes. In the example given in Sato et al. (1989a), the surface resistance was roughly an order of magnitude greater than the aerodynamic resistance, so that the latent/sensible heat flux proportions are insensitive to large relative errors in r_a. All the same, a physically based, reasonably accurate method of calculating aerodynamic resistances is highly desirable for LSPs. The commonly used approach (SiB, BATS) is based on eddy-diffusion concepts which have been shown to be numerically adequate, even if they fall short of complete physical realism.

In SiB, the canopy is represented as a layer of porous material sandwiched in between two constant stress layers. Here the task is to:

(i) calculate the *total* effect of the canopy–soil complex on momentum transfer, that is, derive a single value of the roughness length, z_0; and

(ii) calculate the aerodynamic resistances which control the fluxes of water vapor and heat from the canopy–soil surfaces to the free atmosphere.

A vegetation canopy overlying a ground-cover soil-surface gives three such resistances: r_d, the resistance between soil surface and the well-mixed air of the canopy air space (CAS); r_b, the combined boundary layer resistances between all the canopy leaf surfaces and the CAS; and r_a, the aerodynamic resistance between the CAS and the lower atmosphere (see Fig. 14.4).

For neutral conditions, the value of r_a for momentum should be the same as those for heat and water vapor, as the same eddies in the atmosphere transport all three quantities. However, r_b and r_d are obviously different for momentum than for passive scalars because of their different modes of near-surface transport (viscous and form drag forces for momentum, viscous forces only for passive scalars).

Using the simple structure shown in Fig. 14.4, we may write the equations for momentum transfer following the methods described in Sellers et al. (1986). First of all, it is assumed that some distance above the surface,

466

Atmospheric Boundary Layer

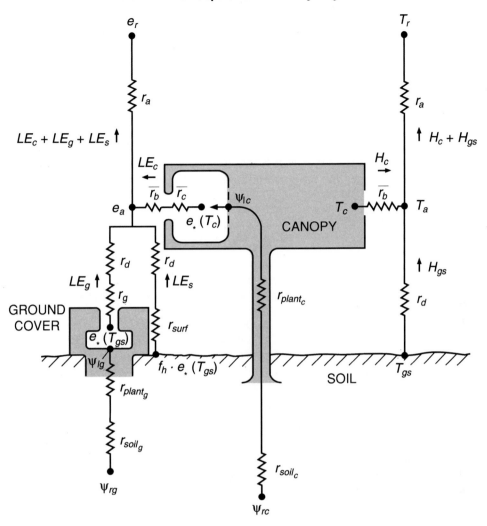

Fig. 14.4 Framework of the Simple Biosphere (SiB). The transfer pathways for latent and sensible heat flux are shown on the left- and right-hand sides of the diagram, respectively. The treatment of radiation and intercepted water has been omitted for clarity.

the log-linear wind profile (see Monteith, 1973) holds true. However, it has been demonstrated that extrapolation of the log-linear profile yields underestimates of the turbulent transfer coefficient close to the top of plant canopies. It is therefore assumed that the log-linear profile, often used to describe the variation of wind speed with height within the constant stress layer near the surface, is valid only above a certain transition height, z_m, which is taken to be a function of the canopy morphology and below which an empirical adjustment to the profile must be made (see Garratt, 1978; Raupach and Thom, 1981).

Under neutral conditions, equations for the transfer of momentum above and within the canopy may be written as follows:

Above the canopy transition layer $(z > z_m)$

$$\tau = \rho u_*^2 = \rho \left[\frac{ku}{ln\left(\frac{z-d}{z_0}\right)} \right]^2 , \qquad (14.10)$$

where τ = shear stress, kg m^{-1} s^{-2}; ρ = air density, kg m^{-3}; u_* = friction velocity, m s^{-1}; k = von Karman's constant = 0.41; d = zero plane displacement height, m; z_0 = roughness length, m; and z_m = transition height, m.

Within the canopy $(z_1 < z < z_2)$

$$\frac{\partial \tau}{\partial z} = \rho \frac{C_d \overline{L_d}}{P_s} u^2 , \qquad (14.11)$$

where u = wind speed, m s^{-1}; C_d = leaf drag coefficient; $\overline{L_d}$ = area-averaged stem and leaf area density, m^2 m^{-3}; and P_s = leaf shelter factor. Also,

$$\tau = \rho K_m \frac{\partial u}{\partial z} , \qquad (14.12)$$

where K_m = momentum transfer coefficient, m^{2}s^{-1}, and

$$d = \frac{\int\limits_{z_1}^{z_2} u^2 z \, dz}{\int\limits_{z_1}^{z_2} u^2 \, dz + \frac{\tau}{\rho}\Big|_{z_1} \frac{P_s}{C_d \overline{L_d}}} . \qquad (14.13)$$

The derivations of (14.10), (14.11) and (14.12) can be found in Monteith (1973); they are commonly used to describe the absorption of momentum by a rough surface. Equation (14.13) was first suggested by Thom (1971), where d, the zero plane displacement height, is defined as the moment height for momentum absorption. The term, $\tau/\rho|_{z_1} (P_s/C_d\overline{L_d})$, has been added to the original form, to take account of the momentum absorbed by the ground.

The momentum flux conditions at the height of the canopy top, z_2, must be related to those above the transition height, z_m, so that continuous profiles of K_m and u extend from above z_m to the soil surface. Garratt (1978) and Raupach and Thom (1981) have noted that estimates of the momentum flux coefficient at $z = z_2$ were 1.5–2.0 times larger than a simple downward extrapolation of (14.10) would indicate. For the time being, we refer to the ratio of the actual value of K_m to its extrapolated log-linear estimate as G_1.

To obtain the characteristics of the canopy, z_0 and d, and to estimate the resistances, $\overline{r_b}$, r_d, and r_a, the equation set (14.10)–(14.13) has to be closed by adding an equation that describes the variation of K_m within the canopy air space: commonly, it is assumed that K_m is proportional to u, see Denmead (1976) and Legg and Long (1975). This leaves five equations with five unknowns, u, K_m, z_0, d, and $\tau/\rho|_{z_1}$.

Boundary conditions must be imposed. The upper boundary condition is $u = u_r$, $z = z_r$, while the lower one may be specified in terms of the shear stress at ground level. Equations (14.10)–(14.13) may be solved for any given set of values of z_2, z_1, $C_d\overline{L_d}$, P_s, and $\tau/\rho|_{z_1}$.

The parameters that are the hardest to quantify are the shelter factor, P_s, reported to be between 1 and 4, depending on vegetation density and the value of G_1. The Appendices of Sellers et al. (1986) and (1989) describe procedures for solving the equation set for vertically homogeneous and vertically varying profiles of leaf area density, respectively. In Sellers et al. (1989), the value of G_1 was optimized so as to provide the best match between the values of z_0 and d derived from this method and those derived with the second-order model of Shaw and Pereira (1982) for a canopy of identical structure: This procedure yielded a value of $G_1 = 1.449$, which compares well with the reported observations of Garratt (1978) of $G_1 = 1.5$–2.0.

Figure 14.5 shows the resulting variation of z_0 with canopy density, taken from Sellers et al. (1989). Note how z_0 first of all increases with canopy density and then decreases as the canopy becomes even denser and less pervious to the airflow. Also, as canopy density increases, the value of d increases monotonically. Both of these results are in qualitative agreement with the findings of Shaw and Pereira (1982).

Whatever the arrangement of canopy density with depth, the solution of the equation set yields profiles of K_m and u down through and below the canopy (see Appendix of Sellers et al., 1989). These may now be used to obtain the values of r_a, r_b, and r_d shown in Fig. 14.4.

The boundary layer resistance r_b for a single leaf has been determined experimentally for many species (Goudriaan, 1977) and commonly yields an expression of the form

$$r_{b_i} = \frac{C_s}{L_i\sqrt{u_i}}, \tag{14.14}$$

where C_s = transfer coefficient; L_i = leaf area of ith leaf, m^2; u_i = local wind speed, m s^{-1}; and r_{b_i} = boundary layer resistance of ith leaf, s m^{-1}.

A bulk boundary layer resistance may be assigned to a group of leaves by integrating (14.14) over the depth of the canopy if the individual resistances, r_{b_i}, are assumed to act in parallel and the wind profile is defined. This yields:

$$r_b = \frac{C_1}{\sqrt{u_2}}. \tag{14.15}$$

The water vapor and sensible heat source height, h_a, may be defined as the center of action of r_b in the canopy, in much the same way as was defined with momentum [see Eq. (14.13) and Appendix of Sellers et al. (1989)]. For

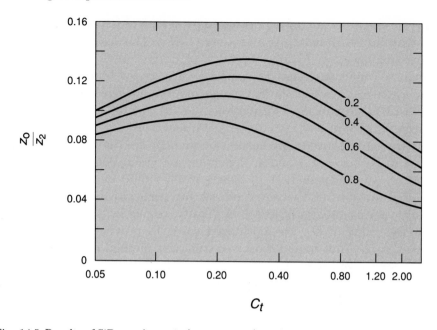

Fig. 14.5 Results of SiB aerodynamic (preprocessor) model with parameters forced to match the results of Shaw and Pereira (1982). Variation of the roughness length, z_0, with canopy drag coefficient, C_t, and height of maximum foliage density, z_c. Canopy drag coefficient is given by: $C_t = (C_d/P_s)L_T$. Height of maximum foliage density, z_c, divided by canopy height, z_2, is marked on graph.

dense vegetation canopies, h_a is approximately equal to d.

The neutral value of the aerodynamic resistance to the transfer of heat and water vapor from the ground cover and the soil surface to h_a may be written as

$$r_d = \int_0^{h_a} \frac{1}{K_s}\, dz = \frac{C_2}{u_2}\,, \tag{14.16}$$

where C_2 = surface dependent constant.

The aerodynamic resistance between h_a and z_r may be defined as

$$r_a = \int_{h_a}^{z_r} \frac{1}{K_s}\, dz = \int_{h_a}^{z_2} \frac{1}{K_s}\, dz + \int_{z_2}^{z_r} \frac{1}{K_s}\, dz\,, \tag{14.17}$$

where K_s = heat/water vapor transfer coefficient, m^2 s^{-1}.

Under neutral conditions, these calculations of z_0, d and u allow the integration of the terms on the right-hand side of (14.17) and:

$$r_a = \frac{C_3}{u_r}\,, \tag{14.18}$$

where C_3 = surface dependent constant; and u_r = wind speed at reference height, m s^{-1}.

The net result is that, when given a specification of some basic canopy parameters (z_2, z_1, L_d, P_s, C_d, and C_s), estimates of z_0, d and C_1, C_2, C_3 (r_b, r_d, and r_a) for neutral conditions may be derived. Note that the difference between the momentum and heat transfer properties of a densely vegetated area is related here to the difference between C_d in (14.11) and C_s in (14.14). Normally, C_d is three or four times as big as C_s for individual leaves and is probably even greater when leaves are arranged in clusters, and incorporation of this difference between these transfer parameters has the net result of taking the 'bluff-body' effect of Stewart and Thom (1973) explicitly into account in this simple turbulent transfer description. The parameters z_0, d, C_1, C_2, and C_3 need be calculated only once for any given vegetation–soil configuration, as they are assumed to be independent of wind speed.

The effects of nonneutrality on the values of the aerodynamic resistances can be fairly severe; for example, r_a may vary by an order of magnitude, depending on the strength of the sensible heat flux when the wind speeds are low. Almost all LSPs use formulations based on the work of Paulson (1970) to calculate the dependence of r_a on u_r and H, although many make use of simpler parameterizations that are curve-fitted to the transcendental solutions to the Paulson (1970) equations [see, for example, Deardorff (1972)].

An inspection of (14.5) will show that the sensible heat flux and r_a are highly interdependent through a nonlinear function if r_a itself is made a transcendental function of H. For a steady-state condition, i.e., no time variation in T_s, T_r, and u_r, an iterative technique can be used to derive the values of r_a and H. However, time variation in these parameters, for whatever reason, greatly complicates the solution procedure and can be a source of numerical noise.

Nonneutral corrections to r_b and r_d are less drastic, as these are primarily determined by mechanical turbulence. Simple empirical expressions based on experiments and theory over heated plates (see Monteith, 1973) were used in the SiB model (see Sellers et al., 1986).

It is useful to bear in mind the values of r_a, r_b, and r_d under normal conditions. For a tropical forest canopy, where z_r is 10 m to 40 m above the treetops and u_r is 5 m s^{-1}: r_a is on the order of 5–10 s m^{-1}, $r_b = 5$–15 s m^{-1}, and $r_d = 200$–500 s m^{-1}, under neutral conditions. For the same wind speed, r_a can vary from over 100 s m^{-1} (stable) to less than 2 s m^{-1}. As the vegetation gets shorter and sparser, r_a increases, r_b increases, and r_d decreases.

14.3.3 Transport of heat and water vapor

Given the calculation of the aerodynamic resistances, the calculation of sensible heat fluxes is fairly straightforward, see Fig. 14.4

and (14.5). The treatment of evapotranspiration is not so simple. The main problem is that there are different sources of water vapor on the land surface, all of which behave in markedly different ways. These include: (i) lakes; (ii) soil evaporation and soil surface interception loss (puddles); (iii) canopy interception loss, where rainwater or snow intercepted by the canopy is re-evaporated directly back to the atmosphere; (iv) vegetation transpiration, that is, water extracted from the root zone by plants and transpired through the stomates; and (v) evaporation from snow and ice. These are discussed in turn.

As yet, no LSP takes account of lake evaporation, even though in parts of Canada and Russia, lakes may occupy a significant fraction of a given grid area. To calculate lake evaporation correctly would require the inclusion of an extra temperature for the lake surface, with some account taken of the effects of lake volume and mixing. As a result, the process of lake evaporation must be considered as a separate energy balance calculation from the land portion of the grid area, the only communication with adjacent land areas being through the common overlying atmosphere.

Evaporation from the soil surface can be a significant proportion of the total evapotranspiration loss. Most LSPs consider a source of water vapor within a thin upper layer of the soil, say about 2 cm thick, which communicates with the atmosphere via an empirical soil surface resistance, r_{soil}. Thus,

$$LE_s = \left[\frac{h_s e_*(T_{gs}) - e_a}{r_{soil} + r_d} \right] \frac{\rho c_p}{\gamma}. \tag{14.19}$$

where E_s = soil evaporation rate, h_s = relative humidity of upper soil layer; $e_*(T_{gs})$ = saturated vapor pressure at soil surface temperature, T_{gs}; and e_a = vapor pressure of the canopy air space, see figure 14.4.

The relative humidity term is usually taken from the standard relationship:

$$h_s = \exp\left(\frac{\psi g}{R T_{gs}} \right), \tag{14.20}$$

where ψ = soil moisture tension or matrix potential in top layer, m; and R = gas constant.

From (14.20), the top layer of the soil behaves as an almost saturated source of water vapor until the soil moisture tension drops below -10 to -20 m (equivalent to about 50% of saturation for an 'average' loamy soil), after which it drops off rapidly. Even with this effect, many researchers, including Sun (1982) and Camillo and Gurney (1986) found that the r_{soil} term is necessary to prevent the calculation of excessive soil evaporation rates. This term can be thought of as the resistance to transfer of the water vapor through pores in the soil surface.

Canopy interception loss can be a significant proportion of the total evapotranspiration from tall vegetation in a humid region (see Shuttleworth et al., 1984a, and Sato et al., 1989a). For example, in the Amazon basin, it is thought that around 30% of the total evapotranspiration loss comes from this source (see Salati, 1987, and Franken et al., 1982).

Observations indicate that a leafy canopy is capable of holding 0.1 mm rainfall per unit leaf area index. If a linear relationship is assumed between the amount of water stored on a canopy and the wetted leaf area,

$$LE_I = \frac{(e_* (T_s) - e_a) \, \rho c_p}{2r_b} \cdot W_c \, , \tag{14.21}$$

where E_I = canopy interception loss rate, kg m^{-2} s^{-1}; W_c = fractional wetted canopy area = M_c/S_c; M_c = amount of water stored on the canopy, mm; and S_c = maximum value of M_c, mm $\simeq 0.1 \, L_T$, mm, and L_T = total leaf area index.

The '2' in the denominator of (14.21) is there because it is assumed that the interception loss occurs from only one side of the leaf, whereas sensible heat loss can take place from both sides. The instantaneous interception loss rate can be very high, so that sometimes the latent heat flux can exceed the available energy, R_n, and the extra energy is supplied by (negative) sensible heat fluxes drawn from the atmosphere to the surface (see Thom and Oliver, 1977, and Shuttleworth et al., 1984a).

The transpiration of water from vegetation is a complex process which truly requires a biophysical approach for a realistic representation within an LSP. There is now ample evidence to show that the stomatal function, and hence the canopy transpiration resistance, r_c, is primarily dependent on the photosynthetic rate and water-stress status of the vegetation.

Figure 14.6 is a schematic of a section of a C_3 plant leaf. The chloroplast captures PAR photons using the antenna complex of the chlorophyll molecule and uses this energy to split water molecules and combine the resulting hydrogen atoms with CO_2 molecules to produce glucose, releasing oxygen molecules to the atmosphere. CO_2 enters the leaf and water vapor escapes from the leaf via the stomata which are under active physiological control and the apertures are on the order of 10 μm wide.

$$6H_2O + 6CO_2 = C_6H_{12}O_6 + 6O_2 \, . \tag{14.22}$$

| From soil | From atmosphere | Glucose | Returned to atmosphere |

At low light intensities, the rate of reaction is limited by the incident PAR, which limits the number of electrons available to split the water molecules. At high (saturating) levels of PAR, the photosynthetic rate is limited by the amount and cycling rate of the carboxylasing enzyme (Rubisco) in the leaf. This enzyme is key to the biochemical cycle which actually combines the simpler molecules, CO_2 and H_2O, to make glucose. Farquhar and Sharkey (1982) and others have proposed convincing models of this process, which reproduces the photosynthetic performance of unstressed leaves.

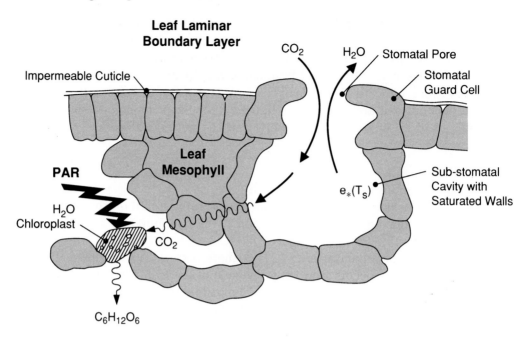

Fig. 14.6 Schematic of a leaf cross section, showing links between stomatal gas exchange (CO_2 and H_2O) and photosynthesis.

The stomatal function of unstressed leaves follows the photosynthetic rate (see Collatz et al., 1991). It is believed that under normal conditions, the stomata are controlled to maximize the role of carbon assimilation for a minimal loss of leaf water. However, there are some additional effects that are induced as part of a leaf's response to stress; these include a progressive stomatal closure as the external air dries (presumably to prevent desiccation) and a similar closure response to decreasing soil moisture. While the exact mechanisms involved in sensing stress and inducing stomatal closure are incompletely understood, the responses are sufficiently well behaved to allow the construction of robust semi-empirical models.

The BATS and SiB models used the work of Jarvis (1976) as a basis for formulating a leaf-level model of stomatal function independent of any leaf photosynthetic description. The model of Jarvis (1976) may be manipulated to give

$$g_s = \frac{1}{r_s} = [f(\mathbf{F} \cdot \mathbf{n})][f(T) \cdot f(\delta_e) \cdot f(\psi_l)] , \qquad (14.23)$$

where g_s = leaf stomatal conductance, m s^{-1}; r_s = leaf stomatal resistance, s m^{-1}; $\mathbf{F}$ = vector flux of PAR, W m^{-2}; $\mathbf{n}$ = vector of leaf normal; and $f(T)$, $f(\delta_e)$, $f(\psi_l)$ = adjustment factors to account for the effects of temperature, leaf water potential, and vapor pressure deficit stresses.

474

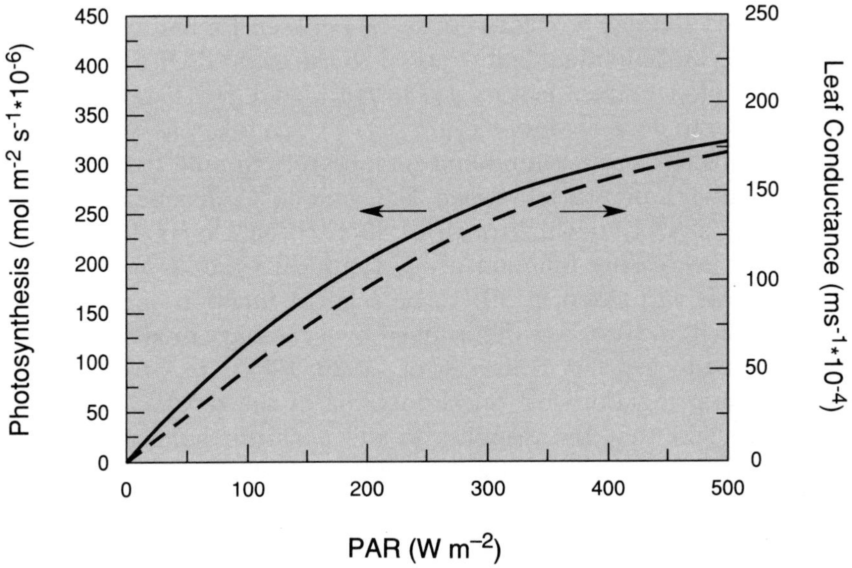

Fig. 14.7 Simulations for a single leaf using the Farquhar et al. (1980) and Collatz et al. (1991) stomatal–photosynthesis model (see Sellers et al., 1992). Solid line – photosynthesis; dashed line – stomatal conductance.

The formulation of (14.23) gives a saturating curve such as that shown in Fig. 14.7. The stress factors, $f(x)$, vary from unity, under optimal conditions, to zero when photosynthesis and transpiration are totally suppressed by adverse environmental conditions (see Jarvis, 1976, or Sellers et al., 1989).

Equation (14.23) may be integrated over leaf angle and canopy depth to provide an estimate of canopy conductance; this has been done in full for some regular leaf angle distribution functions (see Sellers, 1985, Table 4). The integration procedure can be simplified by using the Goudriaan (1977) formulation, which defines the average leaf projection $\overline{G_{(\mu)}}$ in a given zenith direction μ for a given leaf angle distribution function. The use of $\overline{G_{(\mu)}}$ obviates the need for integrating the leaf biophysical functions over the range of the leaf angles in the canopy and does not lead to serious errors (see Sellers, 1985). The canopy integral form of (14.23) is then

$$g_c = \frac{1}{r_c} = f(\Sigma) \int_0^{L_T} g_s\left(\vec{F}, \mu, \overline{G_{(\mu)}}\right)\overline{G_{(\mu)}}dL , \qquad (14.24)$$

where g_c = canopy conductance, m s^{-1}; and $f(\Sigma) = f(T) \cdot f(\delta_e) \cdot f(\psi_l)$. The combined stress factor $f(\Sigma)$ is assumed to be more or less constant throughout the canopy and can therefore be moved outside the integral.

The integration of (14.24), in practice, is done in different ways in different models (see Dickinson et al., 1991). In BATS, a two-layer numerical scheme is used, while in SiB an analytical integration of (14.24) is performed. For a fully developed canopy, g_c comes out to be between 1.5 and three times the value of g_s for an individual leaf exposed to the same PAR flux.

The forms of the stress factors $f(T)$, $f(\delta_e)$, and $f(\psi_l)$ were determined from curve fits to data by Jarvis (1976). $f(T)$ was taken as a smooth curve varying from zero at some minimum temperature around freezing to unity at an optimum temperature between 25°C and 35°C, depending on species, and dropping sharply down to zero at 45°C to 55°C. $f(\delta_e)$ was taken to be a linearly decreasing function of δ_e, the local vapor pressure deficit in the CAS. $f(\psi_l)$ was taken in SiB to be a linear function of the leaf water potential, which, in turn, was determined by a catenary model of water flow from root zone to leaf (see Sellers et al., 1986, 1989). In BATS, $f(\psi_l)$ was taken to be a straightforward linear function of the soil moisture content. It is now thought that the response to soil moisture is controlled by root hormones rather than any hydraulic relationship between water potential in the leaves and root zone, which suggests that the simpler approach used in BATS is actually more realistic. In any case, it seems that empirical functions will continue to be used for this portion of any biophysical model.

Lately, there have been some developments that promise to greatly simplify the modeling of canopy conductance. Farquhar et al. (1980) presented the enzyme kinetics–electron transport model, which calculates the leaf photosynthetic rate as a function of PAR, leaf temperature, leaf internal CO_2 concentration, C_i, and leaf carboxylase concentration, V_{max}. Now C_i is regulated by the leaf photosynthetic rate and the stomatal conductance.

$$A_n = A - R_d = (C_s - C_i)\frac{g_s}{1.6} , \qquad (14.25)$$

where A_n = net leaf assimilation rate, μmol m^{-2} s^{-1}; A = leaf assimilation rate, μmol m^{-2} s^{-1}; R_d = leaf respiration rate, μmol m^{-2} s^{-1}; C_s = CO_2 concentration on leaf surface, mol mol^{-1}; C_i = CO_2 concentration in leaf interior, mol mol^{-1}; and 1.6 = factor to account for different diffusivities of CO_2 and H_2O.

In (14.25), the units of g_s are expressed in mol m^{-2} s^{-1}. Collatz et al. (1991) analyzed a large set of data for C_3 plants and proposed a 'robust semi-empirical model' which relates leaf stomatal conductance to photosynthetic rate and environmental parameters:

$$g_s = \frac{mA_n}{C_s}h_s + b , \qquad (14.26)$$

where m, b = constants 9 and 0.01, respectively; and h_s = relative humidity on leaf surface.

Equation (14.26) replaces all the terms in (14.23) except for $f(\psi_l)$. Sellers et al. (1992) have developed methods of integrating the Farquhar et al. (1980) and Collatz et al. (1991) combined photosynthesis–transpiration model to estimate consistent canopy photosynthetic rates and conductances.

These calculations also tie in closely with those spectral properties of canopies which may be amenable to remote sensing, so that it is likely that the transpiration component of LSPs may be partially forced by satellite data in the near future.

Whatever the method for calculating $g_c(\equiv 1/r_c)$, the canopy transpiration rate is defined by

$$LE_c = \left(\frac{e_*(T_c) - e_a}{2r_b + r_c}\right)\frac{\rho c_p}{\gamma}(1 - W_c) . \tag{14.27}$$

In (14.27), transpiration is assumed to occur from only one side of the leaves, hence $2r_b$, and only from the dry foliage, hence $(1 - W_c)$.

Evaporation from snow and ice is treated in a similar way to saturated soil evaporation or canopy interception loss in most models, except that the temperature of the evaporating surface is constrained to be less than or equal to freezing point and the latent heat of vaporization, L, is augmented to take account of the additional phase change from ice to water.

All of the above discussion deals with the transport of sensible heat and water vapor from (or from just within) the surface to the free atmosphere. However, the conduction and storage of heat within the surface substrate (vegetation, soil, rocks, soil water, snow, etc.) and the vertical transfers of the liquid and/or solid phases of water are important determinants of the behavior of LSPs at longer time scales (hours to weeks).

There are many sophisticated models of heat and water transport in soils, some of which deal with multiphase transfer of water and coupled heat/water vapor exchanges [see de Vries (1975) for a general review and Camillo and Schmugge (1981) for a specific well-described example].

The coupled schemes, while realistic, are demanding in terms of computer time, parameter specifications, and number of prognostic variables and are prone to numerical problems. Almost all LSPs, therefore, make use of isothermal treatments, where heat and water transport in the soil are decoupled. The force-restore method of Deardorff (1977) is widely used: this is essentially a two-layer model where the heat capacity of the upper (thin) and the lower (thick) layers are fixed to be appropriate to the diurnal and annual heating cycles, respectively. This simple methodology is capable of reproducing the magnitude and diurnal pattern of the ground heat flux to an acceptable accuracy.

The transfers of liquid and solid water are handled in more idiosyncratic ways in LSPs. BATS and SiB both incorporate at least one vegetation interception store and the throughfall of water through gaps in the canopy joins with the canopy drainage, or excess interception capacity, from the canopy to arrive as an 'effective rainfall' rate on the soil surface. If this exceeds the local infiltration rate, a proportion is immediately lost to 'surface runoff,' while the rest enters the top of the soil column. Interestingly, this is the only area where SiB makes any concession to large-scale spatial heterogeneity: large-scale frontal rainfall is assumed to be spread uniformly over a grid-area, but convective rainfall, which is obviously more intensely concentrated in space and time, is applied nonuniformly, so that roughly 80% of the precipitation falls within 10% of the area (see Sato et al., 1989b). This local concentration of the precipitation will exceed the local interception capacity of the vegetation, on the order of 0.5 mm, and the local infiltration capacity of the soil, thus giving rise to a much higher area-averaged surface runoff rate than if the same amount of water were distributed uniformly over the area. Incorporation or neglect of this effect can have significant effects on the simulated hydrological cycle (see Sato et al., 1989a; Nobre et al., 1991).

For transfers of water within the soil profile, all LSPs make use of the same set of diffusion equations:

$$Q = K\left(\frac{d\psi}{dz} + 1\right) \tag{14.28a}$$

$$K = K_s W^{2B+3} \tag{14.28b}$$

$$\psi = \psi_s W^{-B} \tag{14.28c}$$

where Q = vertical transfer of water, m s^{-1}; ψ = soil moisture potential, m; K = soil hydraulic conductivity, m s^{-1}; K_s = value of K at saturation, m s^{-1}; W = soil wetness fraction; ψ_s = value of ψ at saturation, m; and B = empirical constant.

The parameters K_s, ψ_s, and B, which essentially determine the waterholding properties of the soil and its ability to supply the vegetation and soil surface with liquid water for evapotranspiration, depend on soil type, more exactly on soil texture (see Clapp and Hornberger, 1978). Sandy soils have low values of B (around 4) and relatively high values of K_s and ψ_s; this results in rapid drainage of the soil profile and poor retention of soil moisture. Clay soils, on the other hand, are associated with low conductivities and high soil moisture tensions (low values of K_s and high values of B – around 8) and a relatively large proportion of 'tightly held' soil moisture that is unavailable for transpiration.

Three- or four-layer models are used in an attempt to simulate the vertical distribution of soil moisture in most contemporary LSPs. Given the natural variability of soil properties and the effects of topography and the spatial variability of rainfall rates, vegetation cover, etc., it is hardly surprising that the soil hydrological components of LSPs are regarded as conceptual efforts, at best.

14.3.4 *Model assembly and operation*

The preceding three sections have outlined the equations used to describe the instantaneous exchange rates for a biophysical model at a given moment. Linking the equations together to provide a realistic and numerically well-behaved representation of exchanges between the surface and the atmosphere *globally* represents another class of problem.

Parameter sets for the model must be defined. Currently, BATS and SiB specify around a dozen vegetation types worldwide, each of which is allocated soil and vegetation parameters, some of which vary seasonally. The definition of the global distribution of vegetation types varies quite widely between models.

The linkage of individual model grid-areas with the lowest layer of the GCM and the time-integration schemes used also vary widely from model to model. All these schemes are essentially grappling with the same problem, that is, how to deal with the highly nonlinear interactions between the models' 'fast-varying' prognostic variables (atmospheric temperatures and humidities; surface temperatures), surface–atmosphere heat fluxes, and the aerodynamic resistances which link the two systems.

In most LSP–GCMs, the sequence of calculations is broadly similar (see Fig. 14.8). The radiative transfer calculations are carried out first: these depend on the slowly varying properties of the surface and the atmosphere, and therefore require no complex numerical scheme for their implementation. Once the net radiation of the surface has been determined, the heat fluxes and aerodynamic resistances for the time step are calculated. For example, in SiB, an implicit backwards method is used to calculate the tendencies of the temperature and humidity of the lowest layer and the temperatures of the vegetation and ground (see Sato et al., 1989b), which, in turn, imply the fluxes of heat and water vapor. Often, iteration is used in order to make the aerodynamic resistances consistent with the heat fluxes acting over the time step.

The operation of an LSP within a GCM can be more or less considered as a set of independent, parallel grid-area calculations, similar to those performed for a series of stand-alone micrometeorological models. Implicit in this is the assumption of scale-independence of all the processes involved in the model formulation, an assumption which is clearly not valid for any of the submodels.

For the treatment of grid-scale radiative transfer, the most severe shortcomings of current models center around the natural heterogeneity of the surface radiation field due to clouds, topography, and surface cover. Provided the *time-averaged* downward radiation flux is correct, the effect of unrealistic cloud distributions and geometries is probably not too serious. The effects due to topography and cover variations, including snow, are likely to be more serious. For example, if snow is assumed to be uniformly spread over the landscape, the whole grid area will remain highly reflective and cold until the last gram of snow has melted, after which it will be instantaneously replaced by a warmer, darker surface. A few attempts at modeling these effects have been made, but none widely adopted.

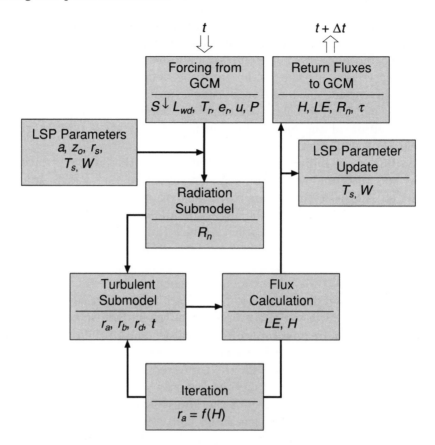

Fig. 14.8 Order of calculation for different submodels in a biophysical LSP: Radiative transfer is dealt with first and then turbulent transfer properties (resistances) are defined, followed by heat flux calculations. In some models, iteration is applied to the flux resistance calculation.

Turbulent transfer of heat and moisture is probably described adequately (for the purposes of LSPs) using the methods described in Sec. 14.3.2. As most transfer rates are dependent on the logarithm of z_0, large relative errors can be made here with relatively little damage to the calculation of heat fluxes. The transfer of momentum flux is significantly enhanced over heterogeneous terrain, where the topography exerts bluff-body effects on the low-level airflow. André and Blondin (1986) and Carson (1986) have proposed treatments which take some account of the surface heterogeneity of z_0 and subgrid-scale topography, respectively.

Lastly, the surface variability of hydrological and biophysical processes and parameters represents the toughest problem for parameterization: the small-scale variability of soil properties is notoriously extreme, typically two orders of magnitude within a few meters, and combined with topographical variations can give rise to extremely heterogeneous soil moisture fields. All of this is overlaid by variations in vegetation form and function and the total effect is to produce an apparently intractable system of complex variability. In nature, however, the real situation may not be so bad for the following reasons:

(i) Variations in soil properties, topography, soil moisture and vegetation are highly correlated. The final expression of soil moisture availability may be in the vegetation type and vigor.

(ii) Recent analyses have shown that an area-average index of vegetation cover, which should be amenable to remote sensing, corresponds closely to the area-integral of vegetation photosynthetic capacity and conductance, and thus evapotranspiration. The validity of these and other area-averaging schemes will be tested using data from large-scale field experiments (e.g., André et al., 1988, and Sellers et al., 1988).

14.3.5 Model testing and validation

There are several levels of model testing and validation: (i) offline tests; (ii) parameter set validation; (iii) climate (LSP–GCM) simulation. Each of these is discussed briefly below.

(i) *Offline tests:* This involves driving the model with atmospheric forcings provided by field studies conducted in vegetated areas throughout the world and comparing the calculated fluxes of sensible and latent heat, net radiation, and momentum with equivalent observations. Sellers and Dorman (1987) reviewed the results of several of these offline tests which used data collected at arable sites and a coniferous forest site in western Europe and the U.S.A. The tests indicated that the SiB model behaved realistically and accurately for these vegetation types and highlighted the sensitivity of the calculated partition of energy to the surface resistance specification.

Sellers et al. (1989) conducted an investigation into the energy partition of an Amazonian tropical forest site using the data of Shuttleworth et al. (1984a,b) in combination with SiB. These data consisted of two years worth of micrometeorological forcing data measured above the canopy, together with some extended periods, each of up to three weeks duration, of hourly flux measurements obtained with eddy correlation equipment. Figure 14.9 shows a comparison between the observed and simulated heat fluxes over the forest. The complete data set was also used in another mode: some of the model parameters were allowed to change to produce the best match between observed and calculated fluxes using an optimization procedure. This methodology has some considerable promise, as it can be used to calibrate some of the physiological parameters representative of an area of up to 1 km^2, as opposed to the normal species-specific definition of parameters from the literature. Several tests have shown that the latter approach can lead to large systematic errors in the calculation of the heat fluxes.

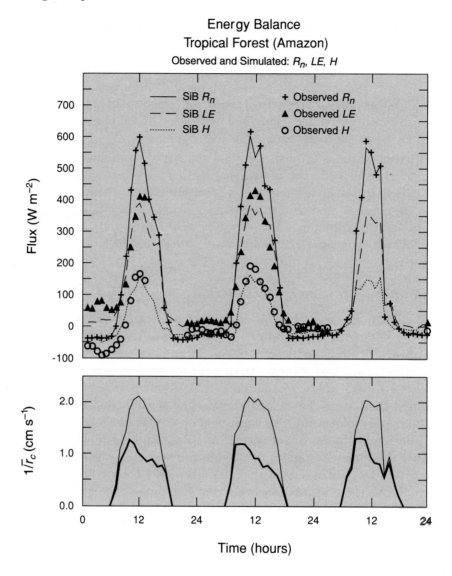

Fig. 14.9 Results of three days of SiB simulation compared with observations from a tropical forest site in Brazil; data from Shuttleworth et al. (1984). Time series of observed and calculated net radiation, latent heat flux, and sensible heat flux. The 'strapped' longwave radiation scheme was used, whereby observed and calculated net radiation fluxes are forced to agree.

(ii) *Parameter set validation:* Dorman and Sellers (1989) reported on an early version of the SiB parameter set (version 1G). Briefly, the world's vegetation formations are 'lumped' into 12 types or biomes, each of which has a parameter set associated with it. These were then used to generate fields of albedo, roughness length, and stomatal resistance. Satellite data may be used at some point in the future to check and modify these fields, but currently vegetation 'climatologies' of this type are used to define the global phenology and hence time variation of these fields.

(iii) *Climate (LSP–GCM) simulation:* Sato et al. (1989a) describe experiments conducted with a modified version of the U.S. National Meteorological Center (NMC) GCM as used by the Center for Ocean–Land–Atmosphere Interactions (COLA) at the University of Maryland. Two versions of the GCM are currently available: the first (Ctl–GCM) containing the original bucket hydrology, and the second (SiB–GCM) with the SiB formulation. In the work of Sato et al. (1989a), the two GCMs, Ctl–GCM and SiB–GCM, were integrated for 50 days from 0000 UTC 15 June 1986, and 30 days from 0000 UTC 15 December 1985. The simulated fields were compared to investigate the impact of including a biophysically based surface parameterization in the GCM. Particular attention was paid to the differences in surface hydrology and land surface energy budgets produced by the two models over a range of spatial scales. There were four steps in the analysis of the paired runs.

First, analyses of surface and near-surface processes for representative gridpoints within each model were compared with equivalent field measurements. In particular, the 30-day means of the simulated hourly net radiation, sensible heat and latent heat fluxes for a tropical forest, a coniferous forest, and an agricultural site were compared with observations. In all cases, the SiB–GCM produced substantially lower evapotranspiration rates and higher sensible heat fluxes than the Ctl–GCM. The SiB–GCM results were also in much better agreement with the available observations and indicated that the conventional 'bucket' hydrological model is responsible for yielding excessively high evaporation rates, as expected from the analysis described in Sec. 14.2.2. As a result, the Ctl–GCM simulation yields air temperatures that are too low and exhibit a markedly reduced diurnal cycle relative to the observations, whereas the SiB–GCM results match observations reasonably well. This difference is due to the different sensible heat fluxes generated by each model for this point; the SiB–GCM with its lower evaporation rates contributes to a much higher direct heating of the lower atmosphere and to a more dynamic growth of the atmospheric boundary layer during the day.

Second, the simulated time series of the energy and moisture budgets of four large regions (the Amazon basin, central and eastern United States, Asia, and the Sahara) were compared (see Table 14.2). The results shown in Table 14.2 reinforce the supposition that the main impact of including a biophysical model in a GCM is manifested in the calculated partition of the available energy at the surface. Although there were substantial differences between the albedo fields of SiB–GCM and Ctl–GCM, the available energy calculated by the two models differed by less than 5% in most cases, with SiB–GCM almost always having the lower values on account of its

Table 14.2. The mean surface energy budget as calculated by (a) Ctl–GCM, and (b) SiB–GCM for four regions: Amazon Basin, central and eastern United States, Sahara Desert and Asia, reproduced from Sato et al. (1989a). The means are for the 50-day (June 15 – August 4) summer simulations in all cases. The symbols used in the columns represent the following fluxes (W m^{-2}): S = insolation; $(1 - \alpha)S$ = net shortwave (absorbed by the surface); L_n = net longwave (absorbed by the surface); R_n = net radiation, available energy; LE_T = transpired latent heat flux (SiB only); LE_1 = interception loss + soil evaporation latent heat flux (SiB only); LE = total latent heat flux (sum of LE_T and LE_1 in SiB); H = sensible heat flux; and G = ground heat flux.

(a) Mean (50-day) surface energy budget calculated by Ctl–GCM

Region	S	$(1 - \alpha)S$	L_n	R_n	LE_T	LE_1	LE	H	G
Amazon	211.1	194.8	-31.5	163.3	–	–	161.8	4.6	-3.1
Central & eastern U.S.	260.2	231.8	-34.9	196.9	–	–	176.4	18.4	2.1
Sahara Desert	289.6	217.4	-92.4	125.0	–	–	3.1	117.1	4.8
Asia	281.9	236.6	-50.1	186.5	–	–	134.5	50.9	1.1

(b) Mean (50-day) surface energy budget calculated by SiB–GCM

Region	S	$(1 - \alpha)S$	L_n	R_n	LE_T	LE_1	LE	H	G
Amazon	210.9	184.7	-30.6	154.1	94.2	35.7	129.9	27.2	-3.0
Central & eastern U.S.	261.8	220.2	-39.9	180.3	114.9	20.1	135.0	41.1	4.2
Sahara Desert	292.2	202.0	-83.1	118.9	3.36	0.04	3.4	113.8	1.7
Asia	283.5	223.4	-54.0	169.4	93.2	18.0	111.2	57.4	0.8

higher albedos and generally higher surface temperatures. However, the partitioning of the available energy as calculated by the two models differed significantly, with SiB–GCM having evaporation rates which were typically 75% or less than the equivalent Ctl–GCM figures. This result follows directly from the stomatal resistance calculation of SiB, which is entirely absent from Ctl–GCM; as a result, Ctl–GCM typically produces latent heat fluxes which approach the flux of available energy (see Table 14.2).

Figure 14.10 shows the water budget of North America as simulated by Ctl–GCM and SiB–GCM for the June–August period. This region is calculated to have a very low net convergence of water vapor during the summer months by both models, a result which is supported by the data analysis of Rasmusson (1968). It follows that the precipitation and evapotranspiration rates are tightly coupled for this season and region.

N. America 6/15 – 8/4

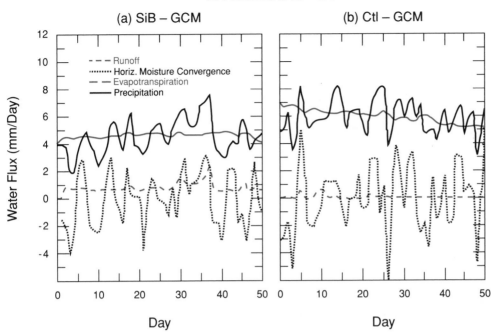

Fig. 14.10 (a) The daily mean moisture budget, as calculated by SiB–GCM for the eastern and central U.S. Region, June–August. The solid line is for precipitation, the coarse dashed line is for evaporation, the fine dashed line is for runoff, and the dotted line is for horizontal moisture convergence in the atmosphere. (b) Same as (a) but for Ctl–GCM.

The SiB–GCM simulation produces an evaporation rate that is about 1.5 mm day^{-1} less than that produced by Ctl–GCM, and this same difference is reflected in the precipitation rates. The SiB–GCM evapotranspiration rate of 4.5 mm day^{-1} and matching precipitation rate are probably too high; Rasmusson (1968) estimated the rainfall to be about 3.1 mm day^{-1} in July. The Ctl–GCM values of 6.0 mm day^{-1} are clearly far too high.

Third, the global fields of latent heat flux (evapotranspiration), sensible heat flux, and precipitation generated by SiB–GCM and Ctl–GCM were compared. Figure 14.11 shows the *difference* fields for the surface latent heat flux and precipitation, as compiled from the first 30 days of the June–July simulation. The shaded areas in both cases correspond to differences between the SiB–GCM and Ctl–GCM simulations of around 1 mm day^{-1}. The striking result from these analyses is the spatial coherence of the two fields, which indicates that the recycling of evaporated water into regional precipitation may play a strong role in continental climatology.

Fourth and last, the impacts of the two land surface parameterization schemes (SiB–GCM and Ctl–GCM) on the large-scale atmospheric

SiB—GCM Minus Ctl—GCM

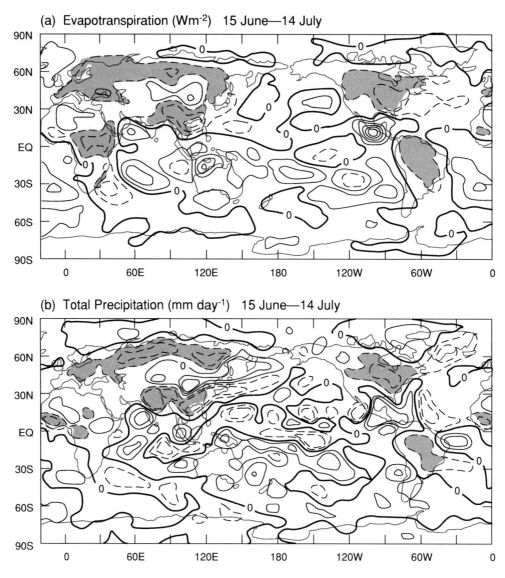

(a) Evapotranspiration (Wm⁻²)   15 June—14 July

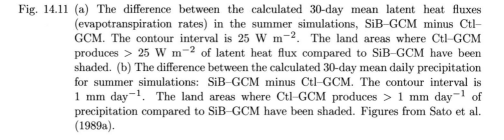

(b) Total Precipitation (mm day⁻¹)   15 June—14 July

Fig. 14.11 (a) The difference between the calculated 30-day mean latent heat fluxes (evapotranspiration rates) in the summer simulations, SiB–GCM minus Ctl–GCM. The contour interval is 25 W m^{-2}. The land areas where Ctl–GCM produces > 25 W m^{-2} of latent heat flux compared to SiB–GCM have been shaded. (b) The difference between the calculated 30-day mean daily precipitation for summer simulations: SiB–GCM minus Ctl–GCM. The contour interval is 1 mm day^{-1}. The land areas where Ctl–GCM produces > 1 mm day^{-1} of precipitation compared to SiB–GCM have been shaded. Figures from Sato et al. (1989a).

circulation were investigated. Among other effects, the 200 mb tropospheric jet in the Northern Hemisphere in SiB–GCM appears to be weaker than that of Ctl–GCM and analyses. Also, the jet stream shifted slightly northward in the June–July SiB–GCM simulation.

To summarize, Ctl–GCM was found to give rise to excessive evapotranspiration rates over the vegetated land surface, exceeding equivalent calculations by SiB–GCM by up to 2 mm day^{-1}. Generally, this difference was directly reflected in the regional precipitation rates. Additionally, SiB–GCM typically gave rise to 25–50 W m^{-2} more of sensible heat flux over the vegetated land, as compared to Ctl–GCM, and the resulting SiB–GCM atmospheric boundary layer was simulated to have a considerably greater diurnal amplitude, consistent with observations. These results are primarily related to the explicit calculation of biophysical surface resistances in SiB–GCM, which were completely absent in Ctl–GCM.

In sum, biophysical models can improve the realism of the LSP–GCM performance and, in some cases, markedly improve the simulation of near-surface climatology and continental hydrology. In spite of their shortcomings, biophysically based LSPs are likely to become the tool of choice for climate studies, weather forecast models, and large-scale perturbation or sensitivity studies.

14.4   The future of biophysical models

The present-day biophysical models discussed in this chapter are the direct descendants of micrometeorological models developed over the last thirty years. It is likely that the capability and design of LSPs will diversify from now on, as specific needs and applications are identified and developed. Three particular applications – weather forecast models, climate models, and biogeochemistry/ecology models – are discussed in more detail below. Contemporary biophysical models will probably provide the basis for the next generation of parameterizations required for these applications.

14.4.1   *Simpler models for forecast applications*

Shukla et al. (personal communication) showed that the incorporation of a biophysical model (SiB) into a GCM could improve the medium-range forecast. Also, work is being conducted on mesoscale studies, particularly on the effects of sharp boundaries between surface cover types on the mesoscale circulation, see Avissar and Pielke (1989). It is likely that a less detailed model may be adequate for these tasks, possibly initialized with satellite data. Xue et al. (1991) proposed a simplification of SiB which essentially reproduces SiB results, but with fewer parameters and a lower computational burden. Much of this economy is achieved by moving some of the calculations offline and curve-fitting simple parameterizations to the results; for example, the albedo calculation is replaced by a polynomial with a dependence on vegetation type and solar angle. We might expect further simplifications where the phenology and other vegetation attributes are directly fixed from satellite data (see Sellers et al., 1991), obviating the need for retaining large parameter sets.

14.4.2   *More complex models for climate and carbon cycle studies*

Biophysical models have been used to investigate the impact of large-scale deforestation on climate (Dickinson and Henderson-Sellers, 1988; Nobre et al., 1991; Shukla et al., 1990, Lean and Warrilow, 1989). However, the mechanisms involved in global change involve more than just the components of the physical climate system. It is likely that the biota may play a significant role in the Earth system response to increasing atmospheric CO_2 (see Tans et al., 1990; Post et al., 1990; Heimann et al., 1990). The analysis of Tans et al. (1990) presents a case for the existence of a large terrestrial sink of atmospheric CO_2 in the Northern Hemisphere, while the analysis by Heimann et al. (1990) indicates that the annual amplitude of atmospheric CO_2 concentration, dependent on the seasonal patterns of respiration and photosynthesis in the terrestrial biota, has increased over the last decade, which suggests a 'spin-up' of the terrestrial carbon cycle.

Realistic models of the Earth system are clearly essential for the exploration of the consequences of global change. For the case of increasing atmospheric CO_2, a combined treatment of the carbon cycle and the physical climate system offers the best hope for producing credible simulations. In a previous section, the combination of photosynthesis and stomatal function models was discussed – this step clearly represents a solid start for dealing with the carbon fixation part of the problem. The models describing allocation and respiration are less rigorous and probably not robust enough for global-scale applications, although some encouraging work has been carried out for particular ecosystems (see Parton et al., 1987). It is possible that a similar model to that of Parton et al. (1987), based on conceptual carbon stores of different availabilities connected via respiration fluxes with environmentally dependent time constants, could be run with a modified LSP–GCM and closure assumptions used to optimize the values of the various rate constants and store sizes. One method may be to operate an LSP–GCM over some period of the historical record where ice core data indicate that the biosphere was in equilibrium with regard to carbon flux, derive the rate constants, and then apply the model to the present-day estimates. This approach would require an interactive physical and biological ocean, among other things.

Whatever the methods for model calibration and validation, it is safe to assume that at least some of the LSP–GCMs applied to the study of the 'greenhouse' problem will involve coupled photosynthesis–transpiration physiological models and an overall greater level of complexity.

14.4.3 Links to biogeochemistry and terrestrial ecology

Figures 14.1 and 14.2 illustrate the increasing complexity and difficulty in the assembly of rigorous models as we move to larger spatial scales and longer time scales, specifically in dealing with problems associated with biogeochemistry (trace gas exchange, nutrient cycling) and ecology (succession, competition, invasion). Generally speaking, the number of variables goes up – ecosystem dynamics can involve physiology, seed dispersal, and fire frequency as comparably important processes – and the exactness or the 'physical directness' of controlling relationships becomes less clear.

At some point, a break must occur between the direct simulation of physical climate system processes, i.e., GCM operation, and biological processes. We have seen how the energy, heat, water, and carbon cycles may be modeled using directly coupled models of the atmosphere and biosphere. It is conceivable that the same approach may be extended to the modeling of biogeochemistry, given the assumption of quasi-static biota. However, it is probably sensible and economic not to explore the question of coupled climate and ecological change using directly coupled GCMs and terrestrial ecosystem models for the following reasons:

(i) The time constants associated with the two kinds of models are incompatible; and

(ii) Most advanced ecosystem models are semi-stochastic (see, for example, Shugart et al., 1973). Thus, for a given climate change scenario, a large number of ecosystem response cases would have to be combined into an ensemble, in order to extract a reliable signal of the most likely total ecosystem response.

A more practicable way of tackling the problem might be to extract the essential climatic forcing statistics from paired GCM control and climate change runs and apply these as a climatic upper boundary forcing to the ecosystem models, which may be two-dimensional to take account for the effects of regional gradients, subgrid-scale variability, etc. These offline forced models may be run repetitively and economically to determine the time-evolution of ecosystem status. If the changes are sufficiently large that they might feed back into the climate, then these results could be used to adjust the GCM boundary condition and the iterative process repeated. In this scenario, the ecological and LSP–GCM do *not* communicate with each other directly, but via a pair of modules which transform the output of one into input for the other and vice versa, see Fig. 14.12.

489

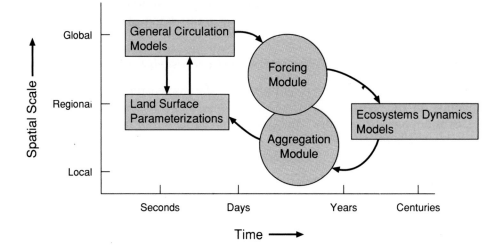

Fig. 14.12 Relationships between land surface parameterizations, atmospheric general circulation models, and ecosystem dynamics models show disparity in time and space scales. For more effective global change research, two new 'communication' techniques or 'modules' must be developed to interpret results between model classes.

14.5 Summary

The last ten years have seen some major advances in the coupling of atmospheric GCMs with biophysically based models of the land surface. These land surface parameterizations (LSPs) have been shown to produce more realistic simulations of the surface net radiation and surface fluxes of sensible and latent heat than pre-existing conceptual models. Research conducted since 1988 has explored the utility of these coupled LSP–GCMs for forecast work, climate simulation, and large-scale perturbation experiments.

The next few years should see the development and implementation of a number of offshoots from the work conducted in the 1980s. First, simplified biophysical models, probably partially initialized and forced using satellite data, may be used routinely in forecast applications. Second, LSP–GCMs may be combined with realistic photosynthesis and respiration models to study the complete biospheric response to increasing CO_2. Last, novel techniques may be developed to allow communication between LSP–GCMs and biogeochemistry and ecological models.

15

Chemistry–transport models

Guy P. Brasseur and Sasha Madronich

15.1 Introduction

Past, present, and future trends in atmospheric composition are the result of the concerted action of complex physical, chemical, and biological processes. Atmospheric chemistry models have as a primary objective the explanation and quantification of these trends, and so they must integrate the various individual processes and their interactions through simplified mathematical descriptions. The development of such models is necessarily an iterative procedure requiring sensitivity studies of model components as well as validation against observations, with further model refinement as necessary. Thus, it must be emphasized that a second major objective of the models is the identification of those observable quantities which most sensitively test the model accuracy, followed by a critical assessment of the model–measurements comparisons. Ultimately, models will be used to extrapolate trends into future decades, for different assumed scenarios of natural and anthropogenic perturbations. Over such longer time scales, the nonlinearity of the entire biogeochemical system is likely to be expressed even more strongly, so that the reliability of the predictions will depend on how well the coupled processes are represented in the model.

The desired output of the models generally includes both temporal and spatial distributions of chemical constituents. Cyclic temporal dependences can provide a fairly stringent test for model validity through comparison with measurements – examples include the diurnal cycles of HO_x and NO_x, the seasonal cycles of CO_2 and tropospheric O_3, and the 11-year solar cycle of stratospheric O_3. Models must also explain the origin of long-term trends which have been observed for many atmospheric constituents, e.g., CO_2, CH_4, tropospheric and stratospheric O_3, N_2O, and CFCs; see Chapter 7. Such trends are not simply related to changes in emissions, but may have significant contributions from nonlinear atmospheric interactions such as the oxidizing capacity of the troposphere, or the complex chemistry of stratospheric CFCs, HO_x, and NO_x. Spatial distributions provide an additional opportunity for model validation, particularly with the prediction of altitude profiles and latitudinal distributions, since trends in these dimensions are generally quite persistent. Full three-dimensional distributions may be required to study the relationships between some source and receptor regions, and the possible biospheric feedbacks – for example, in the assessment of damage induced by acid precipitation and oxidants to forested regions far from the primary emission sources. Inversely, comparisons of model-predicted and measured spatial distributions may be used to infer the location and magnitude of emission sources, especially when such sources are difficult to determine directly because of geographic location (e.g., large-scale oceanic and remote continental sources). Furthermore, the spatial distributions determine the rate of mixing of some reactive compounds, which, if remaining unmixed, would have an entirely different effect on atmospheric composition. Such may be the case, for example, for tropospheric ozone production by mixing NO_x-rich air from combustion sources with hydrocarbon-rich air from forest emissions.

This chapter provides an overview of chemistry–transport models. The necessary model components are discussed in Sec. 15.2. Section 15.3 describes the formulation of various commonly used models, and the simplifications made depending on the scientific questions to be addressed and on the computer resources available. Some examples of calculated trace gas distributions obtained by some of these models are also shown here. Section 15.4 addresses the time integration of the chemical source and sink terms, a particularly difficult problem due to the different time scales associated with different chemical constituents (the "stiffness" problem). Section 15.5 addresses the transport (advection and diffusion) problem, and outlines some basic solution methods.

The design and development of chemistry–transport models span many different disciplines. Our discussion here is necessarily schematic, and we have attempted only to identify the major issues, and to provide some initial overview and guidance for further study. Among the many texts and monographs which can be recommended are those of Seinfeld (1986) and Brasseur and Solomon (1986) for excellent overviews of tropospheric and stratospheric chemistry, respectively; Press et al. (1986) for time integration of the chemical rate equations; and Richtmeyer and Morton (1967), Mesinger and Arakawa (1976), Oran and Boris (1987), and Rood (1987) for practical solutions to the advection and diffusion equations.

15.2   Components of chemistry–transport models

The mass conservation law is the central element of models which describe the transport and chemical transformations of atmospheric chemical compounds. For each compound i, continuity can be equivalently expressed by the flux form

$$\frac{\partial \rho_i}{\partial t} + \nabla \cdot (\rho_i \mathbf{v}) = S_i \qquad (15.1a)$$

or by the advective form

$$\frac{dy_i}{dt} = \frac{S_i}{\rho_a} \qquad (15.1b)$$

with

$$\frac{d}{dt} = \frac{\partial}{\partial t} + \mathbf{v} \cdot \nabla .$$

In these expressions, t is time, $\mathbf{v}$ is the wind three-dimensional velocity, ρ_a the air mass (or number) density, ρ_i the mass (or number) density of constituent i, y_i its mass (or volume) mixing ratio, and S_i are source terms (expressed in mass or in number of particles per unit volume and time), that include both positive (actual sources) and negative (sinks) contributions.

The source terms may be a function of the concentration of other compounds, so that in practice, each continuity equation $(15.1a–b)$ represents a system of N coupled equations, where N is the number of transported species. The equations are solved subject to initial conditions and boundary constraints, with a number of variables prescribed externally to the model.

Initial conditions may be set with preliminary estimates of the solutions or with observed (assimilated) values from which the model is integrated forward in time. As the adopted initial conditions, in general, do not represent equilibrium values, the model may exhibit an initial transient behavior which might not be desirable, lasting for an integration period corresponding to the longest time constant in the system. Boundary conditions account for the influence of the external environment on the system in the spatial domain under consideration. Several types of boundary conditions exist and lead to different mathematical formulations. Perhaps the simplest condition at the external boundary is to specify a value of the concentration (Dirichlet condition) provided, for example, by observations. Another type of condition, which might be more representative of the physics of the problem, is provided by a flux derived, for example, from emission inventories or observed deposition velocities at the surface. In this case, the formulation involves the normal gradient on the boundary (Neumann condition). It should be emphasized that boundary values in chemistry–transport models are not easily established since very often they are related to complex processes at the interfaces between the atmosphere and the ocean, the atmosphere and the continental biosphere, the troposphere and the stratosphere, the atmosphere and the extraterrestrial space, etc.

Values of the coefficients representing the rate at which chemical and photochemical reactions proceed need to be specified in the formulation of the source terms. These coefficients (reaction coefficients and absorption cross sections) are usually adopted from laboratory investigations, while the spectrum of the solar actinic flux is provided by observational data. The intensity of the direct and diffuse solar flux is calculated by an appropriate radiative transfer code. In some models, where the temperature structure is derived consistently with the calculated distribution of trace species, a radiative code for terrestrial radiation is included in the model.

The transport of trace gases in the atmosphere depends on dynamical variables such as advection by the general circulation, convection, mixing associated with wave breaking, etc. The fields of these variables may be interpolated on climatological or episodical basis from compilations of meteorological observations, or may be taken from general circulation models. If the chemistry–transport code is run interactively with a general circulation model, the integration is said to be performed "online." In contrast, when the dynamical variables used are taken from the output of a dynamical model run independently, the model integration is said to be run "offline." Note that some environmental conditions which affect the physical and chemical processes in the model may be prescribed as input (e.g., fields of temperature, pressure, humidity, clouds, etc.), or calculated by the model as a function of the changing atmospheric composition (e.g., optical depth and its dependence on the overhead ozone abundance).

Numerical methods designed to accurately simulate advective processes in transport models will be discussed in Sec. 15.5.1. It should be noted that other physical processes such as dynamical exchanges between the boundary layer and the free troposphere, transfer between the boundary layer and the surface (e.g., wet and dry deposition), deep convection in tropical storms, mass exchanges in frontal systems, troposphere–stratosphere exchanges (e.g., tropopause folding events), etc., are not necessarily well represented by advective schemes since the spatial scales involved are usually smaller than the typical resolution of many models. These subgrid processes can modify substantially the fate of chemical species, so that accurate parameterizations need to be implemented in the transport codes. These may be expressed by diffusive terms added to the advective terms in the continuity equation (see Sec. 15.5.2) or by more complex formulations. Because the physical processes, which are intended to be represented by these empirical or semi-empirical parameterizations, are not necessarily well understood, the accuracy of the model simulations is severely limited by the parameterizations.

Numerous other parameterizations may be used in chemistry–transport models. These include, among others, biogenic emissions expressed as a function of temperature, radiation and soil composition; cloud and aerosol chemical and radiative processes; dry deposition for different soil types and boundary layer meteorology; and wet deposition in terms of precipitation rates.

Two broad categories of chemistry–transport models can be distinguished: (1) *Eulerian* models, in which the atmospheric variables (e.g., trace gas concentrations) are derived at specified points of a grid covering the entire model domain; and (2) *Lagrangian* approaches, in which the variables are calculated following specified air parcels which are displaced by predetermined wind velocity fields.

15.3 Hierarchy of chemistry–transport models

The numerical solution of the continuity equations (Eq. 15.1) for even a relatively small number of chemical constituents in a realistic, three-dimensional global atmosphere is usually extremely computation-intensive. Spatial discretization is a primary consideration, since even a relatively coarse latitude–longitude–altitude grid can become intractable with only 20–40 chemical species, especially if integrated for decades or longer. Thus, many smaller models have been developed for specific applications. Regional three-dimensional models can use much higher resolution and process detail (including more chemical species) over a limited geographical domain, as long as suitable lateral boundary conditions are prescribed (e.g., from the output of a larger-domain, coarser resolution model). Two-dimensional models are often used for the stratosphere, where longitudinal mixing is rapid, so that the spatial variability is largely confined to the latitude and altitude coordinates. One-dimensional models, in which latitude and longitude are fixed and only the altitude distribution is of interest, have been used for both stratospheric and tropospheric calculations, as they offer the advantage of rapid calculation even with fairly detailed chemistry, and can thus be used for sensitivity studies over long simulation times. Zero-dimensional models which follow a fixed air parcel along a trajectory are best suited for studying complex chemical systems, or for situations where chemical equilibrium is achieved on time scales much shorter than those for transport. This hierarchy of models, from the process-intensive box models to the transport-intensive three-dimensional models, also provides a systematic route for the development and evaluation of simplified parameterizations of individual processes. Chapter 10 provides a comprehensive introduction to the hierarchy of atmospheric models.

15.3.1 *Zero-dimensional models*

Zero-dimensional (or box) models consider the chemical transformations occurring in a single, independent, well-mixed air parcel, thus neglecting entirely the coupling between chemistry and transport. In this approximation, the continuity equation (15.1) for each chemical constituent i reduces to:

$$\frac{\partial y_i}{\partial t} = \frac{S_i}{\rho_a} \qquad (15.2)$$

and the integration of the N simultaneous equations is carried forward in time from the appropriate initial concentration values. The source terms usually represent chemical reaction rates with many different time scales, so that even this simple formulation can cause numerical difficulties, and must be handled with methods appropriate to "stiff" systems of differential equations (see Sec. 15.4.2).

Box models may be interpreted as following a specific air parcel along a trajectory in the absence of significant mixing with other air parcels, or, alternatively, a small parcel embedded in a large well-mixed air mass. Time-dependent environmental parameters, such as temperature, pressure, humidity, and photolytic radiation, can be prescribed and included through their effect on the chemical reaction rates. Because only a single air parcel is considered, box models are ideal for studying complex chemical systems, such as the tropospheric photooxidation of natural and anthropogenic hydrocarbons, which may require consideration of hundreds and even thousands of reactions – a task currently too large for multi-dimensional models. They are also applicable when the quantities of interest are in rapid local equilibrium with other, slowly varying quantities, which are either held fixed or varied in a prescribed way (e.g., the diurnal photostationary partitioning of the OH/HO_2 and NO/NO_2 systems). In general, box models are better suited for studying chemical relationships between species, rather than their individual budgets.

Some of the limitations of box models can be overcome by adding simple parameterizations to the right-hand side of (15.2), for example emission fluxes from the surface as zero-order sources, or surface deposition, rainout, and dilution with surrounding air masses as first order sinks. However, such zero-dimensional parameterizations must be viewed as approximate at best, because their spatial inhomogeneity properly requires multi-dimensional treatment. Models with multiple boxes may also be constructed, but because each box is treated as an independent chemical system with no mixing from adjacent boxes, this is equivalent to solving several box models separately.

As an example of a zero-dimensional model, consider a mixture of hydrocarbons, ozone, and nitrogen oxides typical of a remote U.S. continental region, as shown in Table 15.1. The degradation of the hydrocarbons is driven mainly by highly reactive radicals, especially the hydroxyl radicals (OH), the hydroperoxy radicals (HO_2), and various organic peroxy radicals (RO_2). As shown in Fig. 15.1, these radicals are produced mostly during daylight hours by solar ultraviolet radiation which breaks the bonds of more stable molecules (such as ozone, NO_2, and formaldehyde). The initial attack on the hydrocarbons is followed by the production and the destruction of intermediate partly oxygenated organic compounds, and ultimately by the production of simple carbon-containing species such as CO and CO_2. Figure 15.2 shows the time evolution of the various species comprising the carbon budget. In three days, about 20 ppbC (parts per billion of carbon) of the initial hydrocarbon has reacted, yielding not only CO ($\sim$10 ppbC) and CO_2 ($\sim$3.5 ppbC), but also a

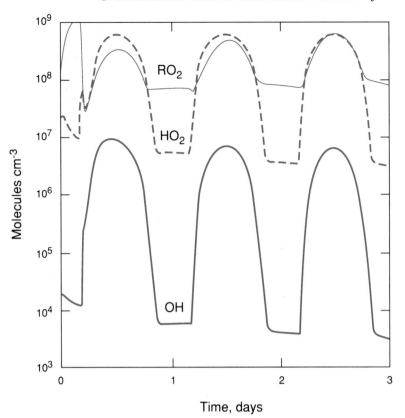

Fig. 15.1. Time evolution of hydroxyl (OH), hydroperoxy radicals (HO$_2$) and organic peroxy radicals (RO$_2$) in a continental air parcel.

Table 15.1. Initial conditions for zero-dimensional model. Conditions assume the summer solstice at 40°N, 3 km above sea level, temperature 295 K and 50% relative humidity. Hydrocarbon measurements are for Niwot Ridge, Colorado (from Greenberg and Zimmerman, 1984).

Species	Concentration (ppbv)
NO$_2$	1
O$_3$	20
CO	150
CH$_4$	1,700
C$_2$H$_6$	2.24
C$_2$H$_4$	0.46
C$_2$H$_2$	0.70
C$_3$H$_8$	1.27
C$_3$H$_6$	0.16
n-C$_4$H$_{10}$	0.51
n-C$_5$H$_{12}$	0.19
benzene	0.24
toluene	0.14
isoprene	0.63

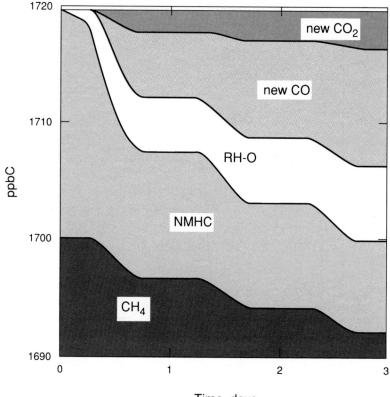

Fig. 15.2. Transformations of organic carbon in a continental air parcel. Differently shaded areas represent concentrations (in units of parts per billion of carbon atoms) for methane (CH_4), nonmethane hydrocarbons (NMHC), the sum of all oxygenated organic molecules (RH–O), and carbon monoxide (CO) and dioxide (CO_2) produced from the oxidation of the organic compounds. Note that the sum is conserved at 1,719.7 ppbC.

significant amount of complex oxygenated organic compounds ($\sim$6.3 ppbC). Detailed analysis of the zero-dimensional model results has identified these oxygenated organics as aldehydes, ketones, organic hydroperoxides, alcohols, organic acids, and organic nitrates. Some of these compounds have already been detected experimentally in continental as well as marine tropospheric air.

15.3.2 One-dimensional models

Much of the pioneering work in atmospheric chemistry has been performed with one-dimensional models that treat a vertical column of the atmosphere. In these models, transport is averaged over longitude and latitude. Because of this averaging, the mean vertical advection does not represent the net vertical motion of chemically reactive species. Instead, this vertical transport is controlled by the species concentration gradient, and therefore is more accurately parameterized as a vertical diffusion process. The one-dimensional continuity equation then becomes:

$$\frac{\partial y_i}{\partial t} = \frac{1}{\rho_a} \frac{\partial}{\partial z}\left(K_z \rho_a \frac{\partial y_i}{\partial z}\right) + \frac{S_i}{\rho_a} \tag{15.3}$$

where y_i, S_i, ρ_a, and t are defined as in (15.1), z is the altitude, and K_z the so-called eddy diffusion coefficient. Values of K_z are derived empirically from measurements of long-lived species and are of the order of 10 m^2 s^{-1} in the troposphere, 0.5 m^2 s^{-1} in the lower stratosphere, and 100 m^2 s^{-1} in the mesosphere.

The one-dimensional formulation is clearly a poor description of dynamical exchanges but provides an inexpensive tool to evaluate the relative role of chemical reactions as a function of altitude and to perform long-term integrations.

15.3.3 Two-dimensional models

Two-dimensional models are commonly used to study the behavior of chemical compounds in the atmosphere as a function of latitude and altitude. These models provide useful information on the zonally averaged distributions, the seasonal evolutions, and the budgets of these species. The relative role of chemistry, radiation, and transport can be assessed if these processes are treated interactively in the model.

Two-dimensional formulations are usually adopted for studies dealing with layers above the tropopause because, in the middle atmosphere, the variations with longitude of the key variables are substantially weaker, while in the troposphere dynamical and chemical processes are very inhomogeneous. Even in the stratosphere, the fluxes of mass and energy cannot be described only by the zonally averaged meridional and vertical winds, because transport is achieved in large part by three-dimensional planetary waves. The contribution by the eddies must be provided by a closure relation, in which the eddy transport of a quasi-conservative tracer is expressed, for example (Reed and German, 1965), as a function of the gradient of its mixing ratio (Eulerian models based on linear theory). The validity of this assumption has been questioned (Mahlman, 1975).

An alternative formulation, based on the transformed Eulerian mean circulation (TEM) (Andrews and McIntyre, 1976; Boyd, 1976; Dunkerton, 1978) is now commonly used. In this approach, the transformed circulation in the meridional plane is defined by

$$\bar{v}^* = \bar{v} - \frac{1}{\rho_a} \frac{\partial}{\partial z}\left[\frac{\rho_a \overline{v'\theta'} \cos\phi}{\partial \bar{\theta}/\partial z}\right] \tag{15.4a}$$

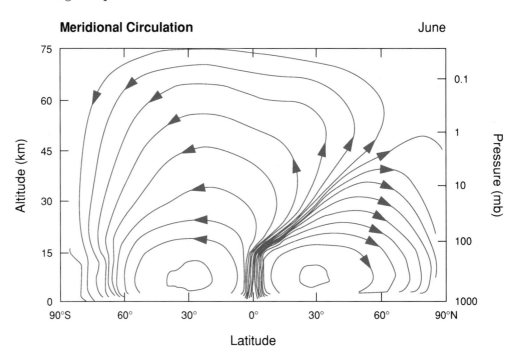

Meridional Circulation   June

Fig. 15.3. Mass streamfunction of the transformed Eulerian mean (TEM) meridional
circulation calculated for June conditions in the two-dimensional chemical–
radiative–dynamical model of Brasseur et al. (1990).

$$\overline{w}^* = \overline{w} + \frac{1}{a\cos\phi}\frac{\partial}{\partial\phi}\left[\frac{\overline{v'\theta'}\cos\phi}{\partial\overline{\theta}/\partial z}\right] \qquad (15.4b)$$

where $(\overline{v}, \overline{w})$ is the classic Eulerian zonal mean meridional circulation,
(v', w') the deviation from the mean wind, θ is the potential temperature, ϕ
the latitude, a the Earth's radius, and the overbar denotes zonal averages.
This circulation accounts for essentially all the mass and energy transport
and, under most circumstances, is very similar to the Lagrangian transport
of air parcels. In these models, eddy diffusivity accounts only for transient
and dissipative effects. The TEM circulation calculated for June is shown
in Fig. 15.3; see also Chapter 3, Sec. 3.10. The temperature is derived from
an energy conservation equation in which the diabatic heating is computed
with a radiative transfer code. The distribution of chemical constituents
is obtained from the continuity/transport equation, which, in the TEM
framework, is expressed by

$$\frac{\partial\overline{y}_i}{\partial t} + \overline{v}^*\frac{1}{a}\frac{\partial\overline{y}_i}{\partial\phi} + \overline{w}^*\frac{\partial\overline{y}_i}{\partial z} = \overline{S}_i + \overline{F}_i \qquad (15.5)$$

where $\overline{F}_i$ represents the contribution of small-scale eddy diffusion.

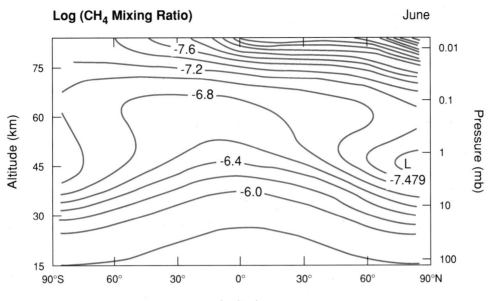

Log (CH$_4$ Mixing Ratio)   June

Fig. 15.4. Meridional distribution (June conditions) of the zonally averaged mixing ratio of methane (two-dimensional model of Brasseur et al., 1990). Note the logarithmic -6.0 corresponds to a mixing ratio of 10^{-6} (or 1 ppmv).

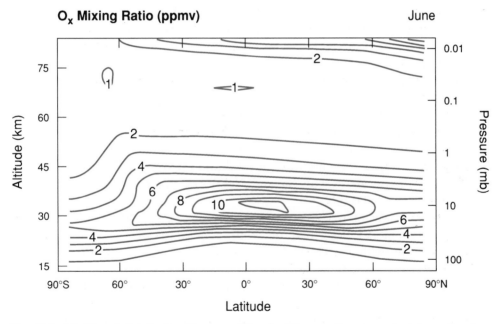

O$_x$ Mixing Ratio (ppmv)   June

Fig. 15.5. Meridional distribution (June conditions) of the zonally averaged mixing ratio of odd oxygen (ozone plus atomic oxygen), expressed in parts per million by volume (two-dimensional model of Brasseur et al., 1990).

Figures 15.4 and 15.5 show the two-dimensional distribution of methane and odd oxygen ($O_3 + O$) in the stratosphere, as simulated in a two-dimensional model.

501

15.3.4 *Three-dimensional models*

In principle, the most realistic representation of the transport of trace species is provided by three-dimensional models. These models are developed to study the evolution of chemical compounds either on the regional or global scale. In the first case, the horizontal resolution is relatively high (typically 10 to 100 km in the horizontal directions) and the integration is performed typically over 1 to 5 days. This approach is usually adopted to study the fate of chemical compounds during specific meteorological events. In the second case, the horizontal resolution is typically 100 to 500 km and the integration is performed over several months or years. Because of limitations in computer resources, the chemistry implemented in these models (especially the global models) is necessarily simplified. As one of the goals of three-dimensional models is to reproduce the transport of atmospheric compounds, advection must be simulated by the most accurate numerical schemes. Examples of global chemistry–transport models are provided by the studies of nitrogen oxides of Levy and Moxim (1989) and Penner et al. (1991).

The formulation adopted by Levy and Moxim (1989) is Eulerian. The model has 11 terrain-following levels and a horizontal grid size of approximately 265 km or 2.4° latitude. The transport is simulated off-line from a parent general circulation model with a time step of 26 minutes. Weather events such as frontal passages, cyclones, etc., are resolved. The continuity equation includes parameterized horizontal and vertical subgrid diffusion and an explicit turbulent mixing contribution in the boundary layer. The model has been used to simulate the global spread and deposition of reactive nitrogen emitted by fossil fuel burning. Reactive nitrogen is transported as a simple tracer with no explicit chemistry. Figure 15.6 shows the yearly averaged mixing ratio of total odd nitrogen (NO_y, which includes all nitrogen species except N_2 and N_2O) calculated by the model in the surface layer at 990 mb.

In their study of reactive nitrogen, Penner et al. (1991) have used the Lagrangian model described by Walton et al. (1988). The winds and precipitation fields are taken from the NCAR Community Climate Model. Basic chemical reactions for NO, NO_2 and HNO_3 are included but the concentration of the other compounds is specified. A set of 50,000 constant-mass air parcels with an average size of 320 km × 320 km and an average depth of 100 mb is advected by the specified winds. Monthly mean mixing ratios at any Eulerian location represent the average of a large sample of parcel mixing ratios as these parcels are advected past gridpoints. The model uses a split-operator method to compute the separate effects of advection, chemistry, interparcel mixing, convective motions, precipitation, etc. The surface mixing ratio of NO_x (NO + NO_2) calculated for July is shown in Fig. 15.7.

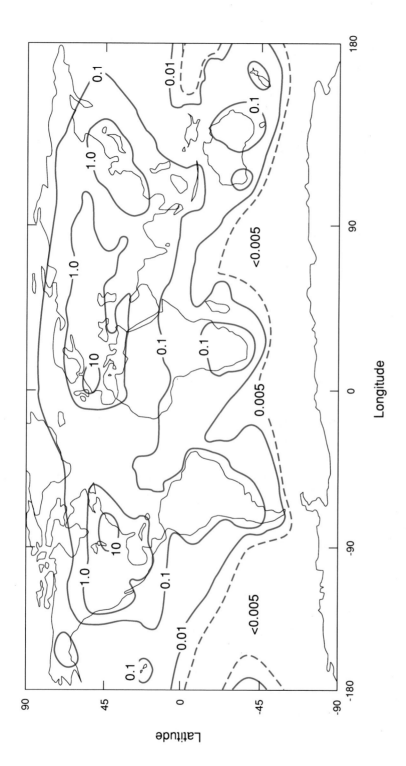

Fig. 15.6. Latitude–longitude distribution of the yearly average of the odd nitrogen mixing ratio (expressed in parts per billion by volume) in the surface layer at 990 mb; model of Levy and Moxim (1989).

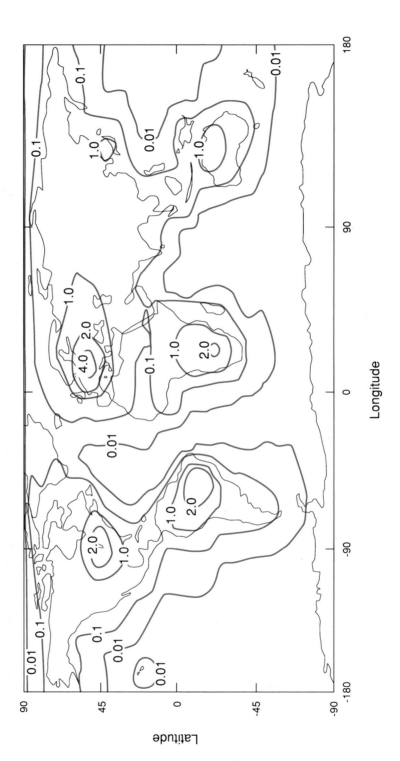

Fig. 15.7. Latitude–longitude distribution of the NO_x ($NO + NO_2$) surface mixing ratio calculated for July conditions. Values are expressed in parts per billion by volume; model of Penner et al. (1991).

15.4 Chemical systems

15.4.1 Formulation of chemical equations

The continuity equations for trace constituents are often solved through a time splitting technique in which chemistry and transport at each time step are treated sequentially. In this case, the effect of chemistry is expressed by the nonlinear vectorial differential equation

$$\frac{\partial}{\partial t}\mathbf{y} = \mathbf{S}(t, \mathbf{y}) \tag{15.6a}$$

subject to the initial condition

$$\mathbf{y}(t_o) = \mathbf{y}_o \tag{15.6b}$$

where $\mathbf{y}$ is a vector whose elements $(y_1, \ldots, y_N)$ are the concentrations of the N chemical compounds, and the function $\mathbf{S}(t, \mathbf{y})$ represents the chemical source and sink terms accounting for a variety of processes such as chemical transformations, condensation, evaporation, local injection, etc. For constituent i, taking into account only zero-order (e.g., injection), first-order (e.g., photolysis), and second-order (e.g., chemical bimolecular reactions) processes, the source term can be expressed as

$$S_i = a_i + \sum_j b_{i,j} y_j + \sum_{j,k} c_{i,jk} y_j y_k$$

where a_i, $b_{i,j}$ and $c_{i,jk}$ may vary with time and location through temperature, pressure, and solar intensity.

15.4.2 Time integration

A large number of numerical methods are available to solve nonlinear systems (15.6a–b) and are introduced in Chapter 9. The performance of these methods is generally expressed in terms of accuracy, stability, and computational efficiency. In addition, in the case of chemical systems, important criteria for the success of a numerical technique are its ability to conserve the number of atoms, and to provide positive concentration values at all time steps.

Because the rates of the different reactions in a given chemical system can vary by many orders of magnitude, the lifetime of the species can range from microseconds (e.g., hydrogen atoms) to centuries (e.g., nitrous oxide molecules). The lifetimes of various species can be calculated from the Jacobian matrix $\partial\mathbf{S}/\partial\mathbf{y}$, and the system is said to be "stiff" if the ratio between the largest and smallest eigenvalues of this matrix is significantly larger than unity. For this reason, direct solution of (15.6) by the explicit Euler forward scheme,

$$\mathbf{y}_{n+1} = \mathbf{y}_n + \Delta t \mathbf{S}(t_n, \mathbf{y}_n) \tag{15.7}$$

is rarely practical since it is stable only if the time step $\Delta t = t_{n+1} - t_n$ is smaller than

$$\Delta t \leq \frac{2}{\max_j |\mathrm{Re}(\lambda_j)|} \qquad j\epsilon[1, N]$$

where λ_j are the eigenvalues of the Jacobian matrix; see Chapter 9 for simple examples.

The implicit (or backward) Euler scheme,

$$\mathbf{y}_{n+1} = \mathbf{y}_n + \Delta t \mathbf{S}(t_{n+1}, \mathbf{y}_{n+1}) \tag{15.8}$$

is unconditionally stable, although the solution is not straightforward if $\mathbf{S}(t,\mathbf{y})$ includes nonlinear terms, and an iteration procedure must be used. However, stability does not guarantee accuracy and, in most applications, convergence to the true result is achieved only if the time step Δt is sufficiently small. For very stiff systems, extremely small steps may be required, which introduces a computational constraint that might be as severe as that associated with the explicit method. It is worth noting that, in the case of linear systems, the implicit method converges to the true steady-state solution for large step sizes. Thus, if excessive time steps are used, stability is maintained but the time evolution is not accurately represented.

Semi-implicit schemes such as

$$\mathbf{y}_{n+1} = \mathbf{y}_n + \Delta t [\xi \mathbf{S}(t_n, \mathbf{y}_n) + (1 - \xi) \mathbf{S}(t_{n+1}, \mathbf{y}_{n+1})] \tag{15.9}$$

may be used to optimize stability and accuracy. The trapezoidal formula, which corresponds to $\xi = 0.5$, is unconditionally stable for linear forcing, and, at sufficiently small time steps, provides results which are more accurate than those obtained with the pure implicit technique. Again, if the chemical forcing $\mathbf{S}(t,\mathbf{y})$ is not linear, an iteration technique has to be used to solve (15.9) for $\mathbf{y}_{n+1}$.

It is important to note that for closed chemical systems, (15.7), (15.8) and (15.9) conserve exactly the total number of atoms of each type. This feature can be shown by multiplying the equation for each chemical constituent by the number of atoms per constituent and by summing these equations; the sum of the production and destruction terms will necessarily be zero.

Many classic numerical methods such as Runge–Kutta, Burlirsch–Stoer, and predictor–corrector (see for example, Press et al., 1986), do not provide stable and accurate solutions for stiff systems, except when extremely small time steps are used. If very accurate solutions are required, multistep methods can be used. The solution of (15.6) in this case is expressed by

$$\sum_{j=-(k-1)}^{1} \alpha_j \mathbf{y}_{n+j} = \Delta t \sum_{j=-(k-1)}^{1} \beta_j \mathbf{S}(t_{n+j}, \mathbf{y}_{n+j}) \tag{15.10}$$

where α_j and β_j are constants chosen such that $\alpha_1 = 1$. In the implicit case (β_1 different from zero), the solution is obtained by iteration when the forcing function $\mathbf{S}(t, \mathbf{y})$ is nonlinear.

Gear (1967, 1971a, 1971b) has developed a multistep method which is "stiffly stable" (see Gear, 1969 for definition) and is appropriate for solving chemical systems. The method is stable for relatively large time steps and is defined by Eq. (15.10) in which the constants $\beta_j = 0$ for all j except $j = 1$. The order of the method (k) and the time step are variable and chosen to minimize computational burden for a specified accuracy. The Gear solver is probably one of the most accurate, robust methods available for nonlinear chemical systems. The scheme, however, is computationally expensive (due to the numerous iterations and matrix manipulations) and has large memory requirements, since the solutions at k previous integration times need to be stored. Hence, the method is impractical for multi-dimensional chemistry–transport models but is adequate for box models with detailed chemistry.

The more recent algorithm of Kaps and Rentrop (1979) (see also Press and Teukoloky, 1989), also adapted for stiff problems, is a generalization of the Runge–Kutta scheme and is simpler to implement that the Gear method. It seeks a solution of the form

$$\mathbf{y}_{n+1} = \mathbf{y}_n + \sum_{i=1}^{Q} c_i \mathbf{k}_i \tag{15.11}$$

where the corrections $\mathbf{k}_i$ are found by solving Q linear equations

$$(\mathbf{I} - \gamma \Delta t \mathbf{J})\mathbf{k}_i = \Delta t \mathbf{S} \left(\mathbf{y}_n + \sum_{j=1}^{i-1} \alpha_{ij} \mathbf{k}_j \right) + \Delta t \mathbf{J} \sum_{j=1}^{i-1} \gamma_{ij} \mathbf{k}_j \tag{15.12}$$

where $\mathbf{J}$ is again the Jacobian matrix, $\mathbf{I}$ the identity matrix, and $\gamma, \alpha_{ij}, \gamma_{ij}$ are fixed constants independent of the problem. The scheme is simply a Runge–Kutta scheme if $\gamma = \gamma_{ij} = 0$. An automatic step size adjustment algorithm is usually associated with the method.

"Linearization" techniques provide an alternative to the above schemes. If the source function can be expressed by

$$S(\mathbf{y}) = A + B\mathbf{y} + C(\mathbf{y})\mathbf{y} \tag{15.13}$$

Eq. (15.6) can be approximated

$$\mathbf{y}_{n+1} = \mathbf{y}_n + \Delta t [A + B\mathbf{y}_{n+1} + C(\mathbf{y}_n)\mathbf{y}_{n+1}]. \tag{15.14}$$

The solution is then obtained by simple matrix inversion. The method is unconditionally stable, and inherent conservation of properties in the system of equations depends on the specific form of $C(\mathbf{y})$. By expressing the quadratic terms for the species i and j, say $k\, y(i)y(j)$, as

$$k\, y(i)y(j) \approx \frac{k}{2}[y_n(i)y_{n+1}(j) + y_{n+1}(i)y_n(j)] \tag{15.15}$$

the resultant matrix $C(\mathbf{y})$ maintains the conservation properties of the

507

system (e.g., conservation of the number of atoms). This approximation scheme, called the semi-implicit symmetric method (Ramaroson et al., 1991) is second-order accurate for the nonlinear terms. It can, however, exhibit some oscillatory behavior if the adopted time step is too large.

Another practical approach (Hesstvedt et al., 1978) is to express the continuity equation in the form

$$\frac{dy_i}{dt} = P_i - L_i y_i \tag{15.16}$$

where P_i and $L_i y_i$ are the photochemical production and loss rates. The solution is provided by the following rules: if the lifetime of the species $\tau_i = 1/L_i < 10$ percent of the time step, Δt, steady state conditions are assumed; for lifetimes greater than 100 times the time step, an explicit Euler forward method is used; for intermediate lifetimes $\Delta t/10 < \tau < 100\Delta t$, the P_i and L_i terms in (15.16) are assumed to be constant over time interval Δt and the analytical solution

$$y_i^{n+1} = (P_i/L_i)^n + [y_i^n - (P_i/L_i)^n] \exp\left(-L_i^n \Delta t\right) \tag{15.17}$$

is used. An iterative procedure improves the accuracy of the solution. This method is inherently nonconservative, although for some chemical systems rather accurate solutions are obtained.

When computer capabilities are not sufficient for using complex and expensive methods, or when only limited accuracy (e.g., 1–5 %) is required, a "family grouping" technique can be used to reduce the stiffness. In this method (see e.g., Turco and Whitten, 1974), chemical species which are strongly coupled through fast chemical and photochemical conversions are grouped into approximately conservative families. An illustration is provided by a simplified chemical scheme for odd nitrogen compounds in the stratosphere. Nitric oxide (NO) is efficiently converted into nitrogen dioxide (NO_2) by ozone (O_3), and NO_2 is rapidly decomposed into NO, for example, by photolysis:

$$NO + O_3 \rightarrow NO_2 + O_2$$

$$NO_2 + h\nu \rightarrow NO + O$$

These processes are significantly faster than the mechanisms responsible for the "external" production

$$N_2O + O(^1D) \rightarrow 2\ NO$$

and destruction

$$NO_2 + OH \rightarrow HNO_3$$

of the NO_x family (defined here as $NO_x = NO + NO_2$); the chemical lifetime of NO_x is then substantially larger than that of NO or NO_2 taken individually.

Equations for the concentration of a total family can be obtained by summing with an appropriate weighting factor the equations of the individual members of the family. This weighting coefficient accounts for the fact that members may be formed by association of other members (e.g., N_2O_5 is composed of NO_2 and NO_3). The definition of a chemical family is not unique and the weighting factors may change accordingly. Because chemical families are nearly conserved on time scales much larger than the lifetime of individual members, the equations governing the fate of families are significantly less stiff than those associated with the individual members. The chemistry–transport equations for families often can be solved by using classic (e.g., explicit) numerical schemes with relatively large time steps. The concentration of the individual members is then derived by applying photochemical equilibrium conditions. This latter assumption provides usually rather accurate results in the sunlit atmosphere but may not be appropriate in dark regions. The concentration of fast-reacting species can also be obtained by solving for each constituent a linearized chemical equation. Because, in some linearization techniques, the number of atoms is not necessarily conserved, the obtained concentration may have to be adjusted so that their weighted sum corresponds to the concentration of the total family. This correction can also be applied to a subset of the family (Turco and Whitten, 1974).

The chemical families method is generally stable and computationally inexpensive. It is therefore often adopted for multi-dimensional chemistry–transport models. However, because families have to be defined on a case by case basis, no general solver can be used; the family grouping technique is therefore more difficult to program than iterative methods.

15.5 Tracer transport

15.5.1 Advection schemes

Advection transport is a dominant component of the evolution of trace species and appears explicitly in almost all chemistry–transport models. Although the continuous prognostic equation describing advection of a species in a predetermined flow field assumes the simple form of a first-order partial differential equation, in general its explicit analytic solution is unavailable, and numerical techniques (advection schemes) must be invoked in order to provide an approximate prognosis. Among a variety of existing methods there is no single, universal technique that can be recommended for all applications, and the choice of a method always compromises issues of overall accuracy, efficiency, and complexity of computations. Modern chemistry–transport models optimize these issues by using advection schemes which are too complex to describe here in any detail. The reader interested in currently used practical methods is referred to the monographs by e.g., Oran and Boris (1987) and Rood (1987). Fortunately, most of these advanced techniques are based on various improvements of only a few elementary methods, the knowledge of which then becomes a prerequisite to the understanding of the available options. This section extends the concepts introduced in Chapter 9 and illustrates some of these elementary advection algorithms and their properties.

For illustration, we consider the elementary one-dimensional advection equation

$$\frac{\partial y}{\partial t} + c\frac{\partial y}{\partial x} = 0 \tag{15.18}$$

where c is the velocity (assumed for simplicity to be constant in time and space), y is the advected quantity (e.g., concentration of a species), x and t the spatial and temporal variables.

The approximate solution to (15.18) can be obtained at time t_{n+1} and location x_j by employing one of the basic finite difference schemes. The classical example of an advection scheme is the forward in time and centered in space (FTCS) approximation,

$$y_j^{n+1} = y_j^n - \frac{c\Delta t}{2\Delta x}\left(y_{j+1}^n - y_{j-1}^n\right) \tag{15.19}$$

where $\Delta t = t_{n+1} - t_n$ is the time step and $\Delta x = x_{j+1} - x_j$ is the space interval between gridpoints. This formulation is explicit and therefore simple to resolve; however it is unconditionally unstable and therefore not useful for practical purposes. Despite its impracticality, the FTCS is important as it illustrates certain fundamental concepts inherent to finite difference approximations. For instance, expanding y_j^{n+1}, y_{j+1}^n, and y_{j-1}^n in the second-order Taylor sum about y_j^n, and noting that (15.18) implies

$$\frac{\partial^2 y}{\partial t^2} = c^2\frac{\partial^2 y}{\partial x^2}, \tag{15.20}$$

shows that the FTCS scheme actually approximates the advection equation (15.18) with the additional negative diffusion term

$$-c^2\frac{\Delta t}{2}\frac{\partial^2 y}{\partial x^2}$$

appearing on the right-hand side of the equation. This negative diffusion simulated by the FTCS scheme is responsible for the unconditional instability of the numerical solutions.

Many finite difference approximations to (15.18) may be viewed as FTCS schemes with additional terms to correct the inherent negative diffusion. For example, the Lax scheme, where y_j^n in (15.19) is replaced by $\frac{1}{2}(y_{j+1}^n + y_{j-1}^n)$, i.e.,

$$y_j^{n+1} = \frac{1}{2}(y_{j+1}^n + y_{j-1}^n) - \frac{c\Delta t}{2\Delta x}(y_{j+1}^n - y_{j-1}^n), \tag{15.21}$$

may be interpreted as the FTCS scheme with the centered-in-space approximation to

$$\frac{(\Delta x)^2}{2\Delta t}\frac{\partial^2 y}{\partial x^2}$$

added to the right-hand side of (15.19). The positive definiteness of the effective diffusion coefficient, and therefore the stability of the resulting scheme, is ensured providing the Courant–Friedrichs–Lewy (CFL) number (often called the Courant number) satisfies the condition

$$\frac{|c|\Delta t}{\Delta x} \leq 1 \qquad (15.22)$$

(see Chapter 9, Sec. 9.2.1). Although the Lax scheme achieves stable solutions, it is only first-order accurate and suffers from excessive numerical dissipation, a property that is not desired in chemistry–transport applications.

A numerical scheme which is often used to model tracer advection is the upwind differencing method (also called the donor cell method), where

$$y_j^{n+1} = y_j^n - \frac{c\Delta t}{\Delta x}\left(y_j^n - y_{j-1}^n\right) \text{ for } c > 0 \qquad (15.23a)$$

$$= y_j^n - \frac{c\Delta t}{\Delta x}\left(y_{j+1}^n - y_j^n\right) \text{ for } c < 0 \qquad (15.23b)$$

This scheme, which is stable for a Courant number less than unity, preserves the sign of the initial condition but has the disadvantage of being dissipative. Although implicit diffusion of the upwind scheme is less than that of the Lax scheme, it is still excessive for most chemistry–transport applications. The method can, however, be improved by introducing an appropriate antidiffusive velocity (Smolarkiewicz, 1984), which reduces the numerical diffusion produced by the scheme itself. The resulting scheme is considerably more complex than the upwind scheme, but does offer second-order accuracy for arbitrary flows and preserves the sign of the initial conditions.

Both the Lax and the upwind schemes are first-order accurate and lead to smooth but overly diffusive solutions. Second- and higher-order methods improve the accuracy of approximations, but lose the smoothness of the first-order schemes. For example, the second order in time leapfrog method,

$$y_j^{n+1} = y_j^{n-1} - \frac{c\Delta t}{\Delta x}\left(y_{j+1}^n - y_{j-1}^n\right) \qquad (15.24)$$

is stable when the CFL condition is fulfilled, and exhibits no implicit dissipation since the Taylor expansion of (15.24) contains no even-order terms. Disadvantages include excessive dispersion (due to the odd terms of the Taylor expansion), the memory requirements for storing values of y at two previous times (t_n and t_{n-1}) to compute values at t_{n+1}, and the appearance of so-called computational modes related to the fact that odd and even mesh points are completely decoupled (mesh drifting).

The one-step Lax–Wendroff method (also known as the Leith (1965) or Crowley (1968) scheme)

$$y_j^{n+1} = y_j^n - \frac{c\Delta t}{2\Delta x}\left(y_{j+1}^n - y_{j-1}^n\right) + \frac{1}{2}\left(\frac{c\Delta t}{\Delta x}\right)^2\left(y_{j+1}^n - 2y_j^n + y_{j-1}^n\right) \qquad (15.25)$$

may be viewed as an elaboration of the FTCS scheme where the third term on the right-hand side of (15.25) compensates (to second order) the negative diffusion implicit in the FTCS scheme. The method is stable within the CFL condition, and free of the computational modes characteristic of the leapfrog scheme; however it requires considerable efforts for generalization to multi-dimensional, arbitrarily variable flows, in contrast with the leapfrog scheme where such generalizations are straightforward.

In the two-step Lax-Wendroff method, which also avoids mesh drifting and produces only limited dissipation, the advected variable at time t_{n+1} and mesh point x_j is calculated from

$$y_j^{n+1} = y_j^n - \frac{c\Delta t}{\Delta x} \left(y_{j+\frac{1}{2}}^{n+\frac{1}{2}} - y_{j-\frac{1}{2}}^{n+\frac{1}{2}} \right) \tag{15.26a}$$

where the values at intermediate time steps and mesh points are calculated by the Lax scheme, e.g.,

$$y_{j+\frac{1}{2}}^{n+\frac{1}{2}} = \frac{1}{2} \left(y_{j+1}^n + y_j^n \right) - \frac{c\Delta t}{\Delta x} \left(y_{j+1}^n - y_j^n \right). \tag{15.26b}$$

The stability criterion is again given by the CFL condition.

Figure 15.8 shows the performance of some of the advection schemes described above, for the one-dimensional, constant velocity transport of an initially rectangular distribution of tracer (Rood, 1987). The FTCS scheme (Fig. 15.8a) is clearly unstable, with large oscillations. The solutions obtained with the upwind method (Fig. 15.8b) are positive and free of oscillations, but are strongly diffusive. The leapfrog method (Fig. 15.8c) is not diffusive, but produces occasional negative values and small-scale waves which, if left unfiltered, will rapidly dominate the solution. The one-step Lax–Wendroff method (Fig. 15.8d) produces moderate diffusion and smaller waves, although negative values are still possible. The problem of unphysical negative mixing ratios is generally addressed by applying numerical filters to the solution.

One complaint against traditional numerical methods for advection transport is that they may lead to negative values in the positive-definite scalar fields (e.g., chemical tracers) or, in more general terms, to spurious numerical over- and undershoots in regions of steep gradients in these variables. Such numerical noise, which may be acceptable where advection is the only concern, may lead to large errors in nonlinear interactions between the two advected scalars (e.g., where a small negative value multiplies a large positive value – as may happen in chemical reactions). Traditional techniques that are free of these problems are only first-order accurate and exhibit a tendency to overly diffuse numerical solutions.

In the last two decades, advanced finite difference methods for solving the transport problem have been designed that are essentially free of the spurious oscillations characteristic of higher-order methods yet do not suffer from the excessive implicit diffusion characteristic of low-order methods (see

Rectangle

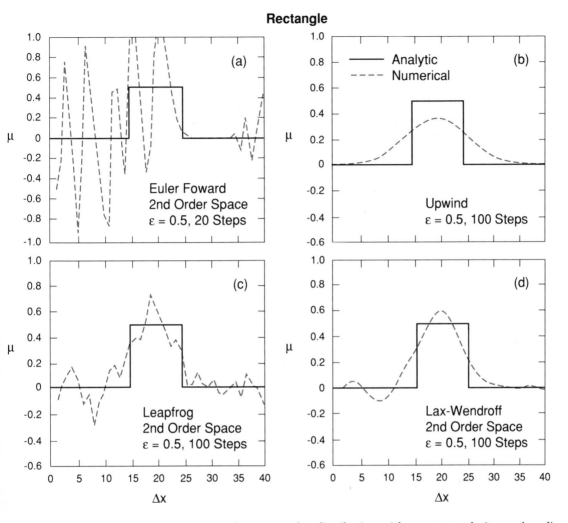

Fig. 15.8. Advection in one dimension of a rectangular distribution with constant velocity, and cyclic boundary conditions. The adopted Courant number is 0.5. (a) Euler forward FTCS scheme after 20 time steps; (b) upwind scheme after 100 time steps; (c) leapfrog time differences with second-order centered spatial differences after 100 time steps; (d) one-step Lax–Wendroff scheme after 100 time steps. (From Rood, 1987.)

Smolarkiewicz and Grabowski, 1990, for a review and discussion).

Although these essentially nonlinear algorithms are far more complex than the traditional linear schemes, they offer strong (linear and nonlinear) stability, maintain steep gradients of the transported fields, and preserve the monotonicity and/or sign of the advected variable. These properties make them potentially attractive tools for applications which combine transport with chemical reactions between the advected variables. The attractiveness of one scheme versus another depends primarily on its ability to be mass conservative, gradient preserving and positive definite, with good phase error characteristic, to limit numerical diffusion and to be computationally efficient. Some schemes, rather than providing a solution at different gridpoints (or a constant value inside a grid box), assume a distribution of the tracer mixing ratio inside these grid boxes. The schemes of Crowley (1968) and Tremback et al. (1987), for example, can be regarded as defining this distribution by a polynomial determined from information in neighboring boxes. Tremback et al. (1987) and Bott (1989) have shown that numerical diffusion is reduced by using higher-order polynomial fits (at the expense of computer time). The scheme presented by Prather (1986) represents the solution inside a grid box by a second-order polynomial. Numerical diffusion is small since, in addition to the zero-order moment (average tracer mixing ratio in the cell) the first-order (slope) and second-order (curvature) moments are also transported. The method is one of the most accurate but it is very expensive as it requires storing 11 variables for each grid cell. A fixer can be used to ensure positivity of the solution.

Another difficulty with traditional numerical methods for advection transport is that they may require much smaller time steps than those required to model the relevant physical scales of interest. The CFL stability condition (Eq. 15.22, necessary for the convergence of numerical integrations) represents a potential disadvantage of a majority of otherwise attractive transport schemes. In the last decade considerable efforts have been invested in semi-Lagrangian techniques which circumvent the CFL restriction and replace it with the much weaker condition characteristic of methods for ordinary differential equations (see e.g., Williamson and Rasch, 1989 and Staniforth and Côté, 1991)

The basic idea of a semi-Lagrangian approach is simple. It originates with a Lagrangian form of (15.1)

$$\frac{dy}{dt} = \frac{\mathbf{S}}{\rho_a}$$ (15.27)

and the particular interpretation of its integral

$$y(x_i, t) = y(x_o, t_o) + \int_T \frac{\mathbf{S}}{\rho_a} dt$$ (15.28)

where (x_i, t) and (x_o, t_o) are the two points connected by a parcel's trajectory T. Assuming that (x_i, t) represents points of a regular mesh, one computes backward-in-time the trajectories arriving at (x_i, t) and finds their departure points (x_o, t_o). Since the departure points do not necessarily coincide with the points of the mesh, information therein is not immediately available, and one has to invoke an interpolation procedure to determine $y(x_o, t_o)$. In essence, similar action must be taken with respect to the source term, with details depending upon a selected method of the integral's approximation and a particular problem at hand. The backward-in-time integration of the trajectories and the nonconservative form of the governing equations makes the approach distinct from a class of mixed arbitrary-Lagrangian-Eulerian (ALE) methods (Hirt et al., 1974) popular in the area of computational fluid dynamics. Advantages of semi-Lagrangian techniques include a circumvention of the CFL stability condition (typical of Eulerian methods) and the convenience of a regular-mesh discretization (contrasting with purely Lagrangian techniques). A variety of semi-Lagrangian methods have been designed exploiting different interpolation techniques and different temporal discretizations (see Staniforth and Côté, 1991, for a review).

As the semi-Lagrangian approach originates with nonconservative formulations of the continuous advection equation, it leads to algorithms which, although accurate and computationally efficient, usually do not satisfy the exact conservation constraint characteristic of flux-form Eulerian schemes. Although the semi-Lagrangian methods conserve transported variables with accuracy only to the truncation error, they appear to be attractive tools for those applications where overall accuracy and computational efficiency are the focus.

15.5.2 Diffusion schemes

In many cases, chemistry–transport equations include diffusive terms which are added to the advection and chemical terms discussed earlier. For example, in 3D models these diffusive terms account for subgrid mixing, while in 1D models they represent the globally averaged vertical transport. Diffusive terms are also introduced in zonally averaged 2D models as parameterizations of wave breaking. For illustration, consider the one-dimensional case given by

$$\frac{\partial y}{\partial t} = \frac{\partial}{\partial x}\left(D\frac{\partial y}{\partial x}\right) \tag{15.29}$$

where $D \geq 0$ is the diffusion coefficient. If, in a first case, D is assumed to be constant, Eq. (15.29) can be differenced and replaced by the FTCS expression

$$\frac{y_j^{n+1} - y_j^n}{\Delta t} = \frac{D}{(\Delta x)^2}\left[y_{j+1}^n - 2y_j^n + y_{j-1}^n\right] . \tag{15.30}$$

The stability criterion for this scheme is

$$\frac{2D\Delta t}{(\Delta x)^2} \leq 1 \tag{15.31}$$

so that the number of time steps required to notice the effect of diffusion on a spatial scale significantly larger than the cell width Δx is usually prohibitive.

When a fully implicit scheme is used,

$$\frac{y_j^{n+1} - y_j^n}{\Delta t} = \frac{D}{(\Delta x)^2}\left[y_{j+1}^{n+1} - 2y_j^{n+1} + y_{j-1}^{n+1}\right] \tag{15.32}$$

the solution is unconditionally stable; a set of simultaneous linear equations has, however, to be solved at each time step t_{n+1}

$$\alpha y_{j-1}^{n+1} + \beta y_j^{n+1} + \gamma y_{j+1}^n = y_j^n \quad (j = 1 \text{ to } J) \qquad (15.33)$$

where $\alpha = \gamma = -D\Delta t/(\Delta x)^2$, $\beta = 1 - 2D\Delta t/(\Delta x)^2$, and J is the number of mesh points. Supplemented by appropriate boundary conditions at $j = 0$ and $j = J+1$, this tridiagonal system can easily be solved. As with implicit advection methods for large time steps Δt, this does not reproduce the details of the small-scale evolution from the initial condition but does yield the correct equilibrium solution for $\Delta t \to \infty$.

Stable solutions with greater accuracy can be obtained using the Crank–Nicholson scheme, in which the right-hand side of (15.29) is approximated by the average of the explicit and implicit formulations.

The conclusions obtained for the simple formulations discussed above can easily be generalized to cases for which the diffusion coefficient D is a function of variable x. In this case, the right-hand side of (15.29) can be differenced as

$$\frac{\partial}{\partial x}\left(D\frac{\partial y}{\partial x}\right) \approx \left[D_{j+\frac{1}{2}}(y_{j+1} - y_j) - D_{j-\frac{1}{2}}(y_j - y_{j-1})\right]/(\Delta x)^2 \quad (15.34a)$$

$$\equiv D'' \text{ say}, \qquad (15.34b)$$

where $D_{j+\frac{1}{2}}$ is the diffusion coefficient at point $x_{j+\frac{1}{2}}$ and D'' is defined by the expression on the right-hand side. The explicit FTCS method is stable for

$$\Delta t \leq \min\left[\frac{(\Delta x)^2}{2D_{j+\frac{1}{2}}}\right] \qquad (15.35)$$

while the implicit technique again is unconditionally stable.

When multi-dimensional diffusion is considered, a time splitting method (also called method of fractional steps) may be an appropriate way to proceed. In the alternating-direction implicit (ADI) method, which is second-order accurate in time and space and is unconditionally stable, each time step is divided in K steps of size $\Delta t/K$, where K represents the spatial dimensionality of the problem. The idea is, at each substep, to treat a different dimension implicitly, while the others are treated explicitly. Consider the three-dimensional diffusion equation

$$\frac{\partial y}{\partial t} = \frac{\partial}{\partial x_1}\left(D_1\frac{\partial y}{\partial x_1}\right) + \frac{\partial}{\partial x_2}\left(D_2\frac{\partial y}{\partial x_2}\right) + \frac{\partial}{\partial x_3}\left(D_3\frac{\partial y}{\partial x_3}\right). \qquad (15.36)$$

The solution at time t_{n+1} is obtained in three successive steps

$$\frac{y_j^{n+\frac{1}{3}} - y_j^n}{\Delta t/3} = [D_1'']^{n+\frac{1}{3}} + [D_2'']^n + [D_3'']^n \qquad (15.37a)$$

$$\frac{y_j^{n+\frac{2}{3}} - y_j^{n+\frac{1}{3}}}{\Delta t/3} = [D_1'']^{n+\frac{1}{3}} + [D_2'']^{n+\frac{2}{3}} + [D_3'']^{n+\frac{1}{3}} \qquad (15.37b)$$

$$\frac{y_j^{n+1} - y_j^{n+\frac{2}{3}}}{\Delta t/3} = [D_1'']^{n+\frac{2}{3}} + [D_2'']^{n+\frac{2}{3}} + [D_3'']^{n+1}, \qquad (15.37c)$$

where $[D_i'']^n$ is evaluated from Eq. (15.34) at time n. At each of the $K(= 3)$ substeps, a tridiagonal system needs to be solved.

Operator splitting techniques are often profitably used in chemistry–transport models, for example, when combined advective–diffusion–chemical equations are to be solved. In this case, each physical process can be treated separately and appropriate numerical techniques can be used for each of these processes.

15.5.3   Final remarks

The numerical techniques described above are *finite difference methods* in which the continuity/transport equations are discretized on a finite number of gridpoints (or grid boxes). Alternative approaches are possible and may become common in the future. In one of them, the *finite element method*, an approximate solution is expressed by the sum of independent polynomials (shape functions) defined on finite spatial elements, and obtained by minimizing the resulting error weighted and integrated over the entire domain (Zienkiewicz, 1977). The advantages of the method are that it can easily be adopted for a multi-dimensional domain with irregular boundaries, and that the size and the shape of the finite elements can be arbitrarily chosen and dynamically modified to account for the presence of sharp gradients in certain areas of the computational domain. The disadvantage is the large memory size and computer time generally required. The use of well-chosen shape functions can, however, substantially reduce the integration time.

16

Biogeochemical ocean models

Jorge L. Sarmiento

16.1 Introduction

The ocean plays a major role in determining the atmospheric concentration of two of the anthropogenically perturbed greenhouse gases, CO_2 and N_2O (Table 16.1 and see Chapter 8). It is presently absorbing an estimated 18–40% of the CO_2 that is being released to the atmosphere by human activities, as well as serving as a source of between 15 and 46% of the total known N_2O sources to the atmosphere. In addition, the ocean is a repository for 95% of the estimated pre-industrial combined atmosphere–ocean–terrestrial biota and soils inventory of CO_2 (Fig. 6.6). The estimated annual exchange rate of the ocean with the atmosphere is 45 times larger than the net uptake of anthropogenic CO_2. A mass of CO_2 equivalent to that of the entire atmosphere cycles through the ocean once every 6 to 7 years. In addition to the impact of the ocean on greenhouse gases, it has been suggested that oceanogenic dimethyl sulfide (DMS) is a significant source

Table 16.1. Relative importance of oceanic source or sink for anthropogenic greenhouse gases (adapted from IPCC, 1990a). The error ranges given are those due to uncertainty in the oceanic source or sink only. Uptake of CFCs by the oceans is negligible due to their low solubility in water. No estimate for the magnitude of this uptake is given in the IPCC report.

Greenhouse gas	Contribution to change in radiative forcing from 1980 to 1990	Oceanic source relative to total sources	Estimated error
CO_2	55%	−29%	±11%
CH_4	15%	+ 2%	1–4%
N_2O	6%	+31%	15–46%
CFCs	24%	–	–

for cloud condensation nuclei in remote marine areas, and that changes in its supply rate may play a significant role in climate feedback (see Chapters 7 and 8).

Other than water vapor, the greenhouse gases have only minor oceanic sources and sinks. Oceanic uptake of chlorofluorocarbons (CFCs) is small because of their low solubilities and inertness towards chemical and biological processes. The flux of chlorofluorocarbons into the ocean can be determined straightforwardly from their solubilities, the gas exchange rate, and a model of ocean transport. Atmospheric methane has a minor oceanic source of order 2% of the total, as determined from observations of air–sea gradients. Because of the smallness of this source, and our poor knowledge of the oceanic processes which determine it, this chapter will not directly address the modeling of methane.

The first ocean model of fossil CO_2 uptake was developed more than 30 years ago (Revelle and Suess, 1957). This model consisted of one atmospheric and two oceanic boxes, with exchange rates between them obtained from calibration with radiocarbon observations. It showed that a large fraction of the fossil fuel CO_2 released into the atmosphere was being absorbed by the ocean. This suggestion of a substantial oceanic fossil fuel CO_2 sink has given rise to numerous subsequent studies.

Another major incentive to studies of the carbon cycle has been the discovery of reduced atmospheric CO_2 levels during the last ice age (e.g., see Fig. 21.3). A number of box models have been developed which explain these observations as resulting from a rearrangement of total carbon and alkalinity distributions in the ocean (e.g., review by Sarmiento et al., 1988). This continues to be an area of active research. One of the hypotheses for how the total carbon and alkalinity were redistributed is that an increased supply of iron by dust transport enhanced biological uptake in regions where the iron supply appears to be inadequate for biological needs at present (Martin, 1990). There have been recent suggestions of iron fertilization by humans to force a change in the distribution of total carbon and alkalinity in the ocean in such a way as to enhance the oceanic uptake of fossil CO_2 (e.g., Martin et al., 1990). Box models of the oceanic carbon cycle have been developed to estimate how much additional fossil CO_2 might be taken up (Peng and Broecker, 1991; Joos et al., 1991).

By contrast, the importance of N_2O and DMS to climate and the role of the oceans in determining their atmospheric concentrations have only recently begun to receive significant recognition. Ocean model studies of N_2O and DMS are essentially nonexistent, and the issues that need to be resolved in developing models of these substances can only be hinted at. The bulk of this chapter is therefore, of necessity, focused on CO_2.

The oceanic role in control of atmospheric CO_2 involves a large variety of processes. Chapter 21 briefly addresses the weathering, metamorphism, crustal degassing, and sedimentary deposition processes that control atmospheric CO_2 on time scales of millions of years. The biogeochemical cycling models that have been developed to deal with these time scales treat the ocean as a one-box repository in which the oceanic processes per se are not modeled. This chapter focuses on processes internal to the ocean which might lead to changes in atmospheric CO_2 on time scales of order hundreds to a few hundred thousand years. It also covers the oceanic response to the anthropogenic CO_2 transient. The latter problem is a relatively straightforward one in that anthropogenic CO_2 enters the ocean essentially as a passive tracer, i.e., one that is not involved in biological processes. Most of this chapter is thus dedicated to a discussion of the more complex biological processes introduced in Chapter 8 which, together with chemical and physical processes, determine the distribution of CO_2 in the ocean, and through this have a major impact on atmospheric CO_2. The following section introduces some basic modeling concepts by use of a series of box models which illustrate various important aspects of the carbon cycle. After this there is a discussion of recent attempts to develop more realistic models of the important processes making use of three-dimensional ocean circulation models, and then a discussion of models of anthropogenic CO_2 uptake.

16.2 Basic concepts

The conservation equation for tracers is

$$\frac{\partial C}{\partial t} = -\mathbf{v} \cdot \nabla C + \nabla \cdot (D\nabla C) + \mathrm{SMS} \qquad (16.1)$$

with SMS (sources minus sinks) representing chemical and biological transformations. This equation is comparable to Eq. (15.1) discussed in Chapter 15. Oceanic tracers are generally characterized as either conservative, with SMS $= 0$, or nonconservative, with a nonzero SMS. Salinity is an example of a conservative tracer. It is subject only to physical transport and the concentrating and diluting effects of evaporation and precipitation at the surface. Because of the salinity effects on density, and thus its importance in determining the oceanic pressure field, the modeling of salinity is a familiar problem to physical oceanographers (see Chapter 11). CO_2, on the other hand, is one of many nonconservative chemicals that undergo complex and often poorly understood chemical and biological transformations in the ocean interior. Biogeochemical modelers can draw on the expertise of physical oceanographers to account for the effect of transport, evaporation, and precipitation on the distribution of the nonconservative chemicals. Specification of the SMS terms, as well as boundary conditions such as gas exchange and sediment–water fluxes, all discussed in Chapter 8, are the major new problems that must be dealt with in modeling nonconservative tracers.

What are the nonconservative SMS processes affecting CO_2? The interaction between atmospheric CO_2 and oceanic carbon occurs only with oceanic carbon that exists in the form of CO_2. Within the ocean, however, CO_2 undergoes rapid chemical equilibration between the three forms of dissolved inorganic carbon (DIC) consisting of CO_2, HCO_3^-, and CO_3^{2-} [See (8.20)]. As discussed in Chapter 8, the distribution of DIC between these three chemical forms depends on the equilibrium constants involving the carbon system. The solubility and equilibrium constants are sensitive to the temperature (T) and salinity (S). The system of equilibrium equations involving the carbon system has two more unknowns than equations. Solution to the set of equations thus requires specifying two of the carbon system variables, usually alkalinity [Alk, defined by (8.22) and (8.23)] and DIC. Thus

$$CO_2 = f(T, S, \text{Alk}, \text{DIC})$$
$$\text{SMS}_{CO_2} = f(T, S, \text{SMS}_{\text{Alk}}, \text{SMS}_{\text{DIC}}). \tag{16.2}$$

The outstanding feature of the vertical oceanic DIC and Alk distributions in Fig. 8.5 is the fact that they have lower concentrations at the surface and higher concentrations at depth. The lower surface concentrations have been explained as resulting primarily from the biological pump, which is a consequence of uptake of these substances by production of organic matter and $CaCO_3$ at the surface, coupled with the remineralization of these substances at depth (Sec. 8.3.2 and 8.3.3). The other contributor to the vertical gradient of total carbon is the solubility pump, a consequence of the higher solubility of CO_2 in the cold waters which fill the deep ocean (Sec. 8.3.4).

The influence of the biological and solubility pumps on the oceanic total carbon and alkalinity distributions, and on the atmospheric CO_2 partial pressure, can be most readily illustrated by consideration of a set of simple box models. O_2 will be used initially rather than CO_2 in order to avoid the additional complexity of dealing with the fact that carbon exists as bicarbonate and carbonate ions as well as CO_2. Like CO_2, O_2 is more soluble in cold than in warm water and will thus experience the solubility pump. O_2 is also affected by the biological pump (Eq. 8.32), although the effect is opposite to that of CO_2 in that O_2 is released when CO_2 is consumed, and vice versa.

16.2.1 Two-box model

Consider first the simple two-box model shown in Fig. 16.1. This box model has been used with considerable success for studying the cycling of chemicals in the ocean (e.g., Broecker and Peng, 1982). It includes the effect of the biological pump through the organic matter flux term, P, but is missing the solubility pump in that the deep waters do not have a separate

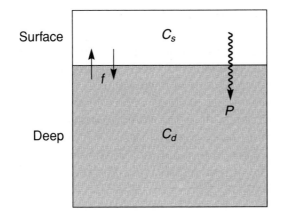

Fig. 16.1 A two-box model of the oceans. f represents physical exchange between the surface and deep box, C_s and C_d represent average surface and deep concentrations, and P represents the particulate rain of phosphate in organic matter. Taken from Broecker and Peng, 1982.

surface source of cold water. The oxygen conservation equation for the deep box is

$$\frac{dO_{2_d}}{dt} = f(O_{2_s} - O_{2_d}) - r_{O_2/P}P, \qquad (16.3)$$

where the subscripts d and s refer to deep and surface values, respectively. P, which represents the biological pump in this model, is the downward flux of phosphorus in organic matter, all of which is assumed to be metabolized in the deep ocean. $r_{O_2/P}$ is the Redfield ratio of oxygen released to phosphorus consumed in the formation of organic matter, for which a value of 172 will be used (Takahashi et al., 1985). From the surface box conservation equation for phosphorus, one can obtain $P = f(PO_{4_d}^{3-} - PO_{4_s}^{3-})$. Substituting this into (16.3) gives, in a steady state

$$O_{2_d} = O_{2_s} - r_{O_2/P}(PO_{4_d}^{3-} - PO_{4_s}^{3-}) \qquad (16.4)$$

from which one can find, using approximate global average values for oxygen and phosphate

$$O_{2_d} = 240 - 172(2.2 - 0.0) \ \mu\text{mol kg}^{-1} = -136 \ \mu\text{mol kg}^{-1}.$$

The observed value of deep oxygen is approximately 166 μmol kg^{-1}. What is wrong?

Part of the answer is that the solubility pump is missing. The waters which fill the deep ocean form in cold high-latitude regions with oxygen saturation concentrations of order 340 μmol kg^{-1}. This will give an increase of $340 - 240 = 100 \ \mu$mol kg^{-1} in the deep oxygen content, resulting in a modified O_{2_d} of $-136 + 100 = -36 \ \mu$mol kg^{-1}, still too low.

The other reason for the deep ocean oxygen problem is that a two-box model cannot correctly represent all aspects of the biological pump. There are limited regions of the surface ocean where nutrients do not go to 0 (see Fig. 8.6). These are small enough in areal extent so they do not have a significant impact on the global average surface nutrient content. However, the largest of the high surface nutrient regions is around Antarctica, which is one of the major source regions of deep water (see Sec. 4.7). In addition, the other major source region of deep waters in the North Atlantic also has higher surface nutrients than the global average. The water sinking into the deep ocean from these regions carries a high load of "preformed" nutrients, thus much of the excess nutrient in the deep ocean arrives there by transport, not by the organic matter rain P. Nutrients delivered by this manner will not consume oxygen.

16.2.2 High-latitude outcrop three-box model

How can the model in Fig. 16.1 be improved to take the high-latitude processes into consideration? The next level in complexity is to add a single high-latitude surface box (Fig. 16.2). If one puts down the deep ocean oxygen equation for the three-box ocean model, then substitutes in the values for $\mathrm{SMS}_{\mathrm{PO}_4^{3-}} = P_h + P_l$ obtained from the phosphorus conservation equations for the surface low- (l) and high- (h), latitude boxes (assuming, as before, that low-latitude phosphate concentration is 0), one arrives at the following expression

$$O_{2_d} = O_{2_h} - r_{O_2/P}(\mathrm{PO}_{4_d}^{3-} - \mathrm{PO}_{4_h}^{3-}). \tag{16.5}$$

The key difference in this equation as compared with (16.4) is that the surface concentrations of oxygen and phosphate are now those characteristic of the deep water formation regions rather than for global average surface water. The solubility pump has been included to account for the fact that the water filling the abyss is formed in regions of very low temperatures, and thus higher O_2 solubility; and the biological pump has been modified to account for the observation that biological uptake in the high latitudes, P_h, is not high enough to totally deplete surface high-latitude nutrients, PO_4^{3-}. Substituting in appropriate estimates for the parameters and concentrations in (16.5) gives the far more satisfactory result:

$$O_{2_d} = 340 - 172(2.2 - 1.2) = 168 \ \mu\mathrm{mol \ kg}^{-1}.$$

Atmospheric CO_2 can be solved for in this model by adding conservation equations for Alk and DIC, by adding an atmospheric box, and by including the air–sea exchange of CO_2. The model shows that atmospheric CO_2 and low-latitude surface CO_2 are determined primarily by the high-latitude surface ocean CO_2 concentration. This is because of the rapid exchange between this box and the vast CO_2 reservoir of the deep ocean. The high-latitude surface box is the window through which the relatively small reservoirs of the atmosphere and low-latitude surface ocean see the deep ocean and are fixed by it. Equations for surface high-latitude concentrations

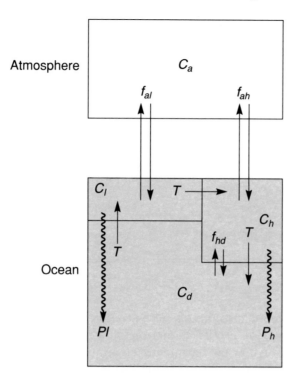

Fig. 16.2 A three-box model of the oceans similar to the two-box model in Fig. 16.1, only the surface ocean has now been separated into a high-latitude and low-latitude component indicated by the subscripts h and l, respectively. f_{hd} is exchange between the high-latitude and deep boxes. T represents thermohaline overturning. The Ps are as defined in Fig. 16.1. Taken from Sarmiento and Toggweiler (1984).

of Alk and DIC, on which the CO_2 depends, can be obtained by substituting total carbon and alkalinity in place of oxygen in (16.5), and rearranging the equation. The sign in front of the final term in (16.5) must be changed, since carbon and alkalinity are consumed when oxygen is produced, and vice versa:

$$\mathrm{DIC}_h = \mathrm{DIC}_d - r_{\mathrm{DIC/P}}(\mathrm{PO}_{4_d}^{3-} - \mathrm{PO}_{4_h}^{3-})$$
$$\mathrm{Alk}_h = \mathrm{Alk}_d - r_{\mathrm{Alk/P}}(\mathrm{PO}_{4_d}^{3-} - \mathrm{PO}_{4_h}^{3-}).$$

(16.6)

A powerful result of this model is that it shows that large changes in high-latitude DIC and Alk, and thus atmospheric CO_2, can occur by the simple mechanism of changing the preformed nutrient level, $\mathrm{PO}_{4_h}^{3-}$. Table 16.2 summarizes results obtained with this model, and other similar models, by modifying $\mathrm{PO}_{4_h}^{3-}$ over a range from 0 to an upper limit where it is equal to the deep ocean concentration. The concentration of $\mathrm{PO}_{4_h}^{3-}$ depends on the rate of nutrient supply from the deep ocean by circulation, relative to the rate of nutrient removal by high-latitude productivity P_h (see Fig. 16.2).

Table 16.2. Box model scenarios of atmosphere–ocean CO_2 distribution. Full operation of the biological pump corresponds to a complete removal of surface nutrients. The partial operation of the biological pump describes the present situation without the anthropogenic CO_2 perturbation.

Solubility pump	Biological pump	Atmospheric pCO_2(ppm)	Source
No	No	720	Volk and Hoffert, 1985
Yes	No	450 to 530	Wenk, 1985; Bacastow and Maier-Reimer, 1990; Baes and Killough, 1986
Yes	Partial	280	Neftel et al., 1985
Yes	Full	165	Sarmiento et al., 1988

Changes in the relative magnitudes of these processes have been postulated to have played a key role in the reduction of atmospheric CO_2 during the last ice age (see review by Sarmiento et al., 1988), although recent measurements do not support all the predictions made by this model, and a satisfactory explanation for the ice-age CO_2 reductions remains to be found. This mechanism also serves as the basis for the suggestion that atmospheric CO_2 might be reduced by fertilizing the high-latitude regions with iron, as discussed in the introduction. The effect of shutting down both the solubility and biological pumps is also shown in Table 16.2. The solubility pump is shut down by forcing the entire ocean to the global average temperature.

16.2.3 Three-box vertical model

An important aspect of ocean carbon chemistry, not addressed by the two-box and three-box ocean models, is the fact that remineralization of organic matter and $CaCO_3$, which jointly determine the total carbon and alkalinity distributions, occur at different depths (Sec. 8.3.3). A simple model consisting of three stacked boxes makes it possible to gain some insight into how changes in regeneration depth of these two substances, and the effect of mixing in enhancing or reducing the effect of differential regeneration depth on the vertical profiles, might affect atmospheric CO_2 (Sarmiento et al., 1988).

Figure 16.3 shows the three-box vertical model of the ocean. The high-latitude outcrop box of Fig. 16.2 has been removed in order to keep the model simple. The deep ocean is separated into a "thermocline" box where most of the organic matter produced by organisms regenerates, and a deep box where most of the $CaCO_3$ regenerates. A solution for the surface DIC

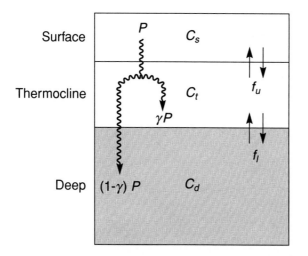

Fig. 16.3 A three-box vertical model of the oceans. Symbols are as defined in Fig. 16.1. γ represents the fraction of the surface particle production which is regenerated in the middle "thermocline" box, with the remainder being regenerated in the deep box. Taken from Sarmiento et al. (1988).

and Alk in terms of PO_4^{3-} is obtained as above:

$$DIC_s = DIC_d - r_{C/P} \cdot PO_{4_d} \cdot \left[\frac{1 + \frac{f_u}{f_l} \cdot (1 - \gamma_{DIC})}{1 + \frac{f_u}{f_l} \cdot (1 - \gamma_{PO_4})} \right]$$

$$Alk_s = Alk_d - r_{Alk/P} \cdot PO_{4_d} \cdot \left[\frac{1 + \frac{f_u}{f_l} \cdot (1 - \gamma_{Alk})}{1 + \frac{f_u}{f_l} \cdot (1 - \gamma_{PO_4})} \right] .$$

$$(16.7)$$

Equation (16.7) is identical to (16.4) and (16.6) except for the term in the brackets involving f_u/f_l, the ratio of the upper exchange rate to the lower exchange rate, and the γs, which are the proportion of the regeneration of a given substance that occurs in the thermocline box. The magnitude of the γs, which tends to be small for Alk, and large for DIC and PO_4^{3-}, is such that Alk is more sensitive than DIC to changes in the exchange rate ratio, f_u/f_l. In simplest terms, these equations change atmospheric CO_2 essentially by the extent to which they sequester Alk in the deep ocean. Recall from Sec. 8.3.1 that CO_2 varies inversely with Alk, thus an increase in the deep ocean sequestering efficiency, or a more vigorous mixing between the upper and thermocline boxes, relative to mixing between the thermocline and deep boxes, will lead to an increase in atmospheric CO_2.

Figure 16.4 shows the atmospheric CO_2 predicted by this model for various values of the two exchange rate terms, and with $\gamma_{PO_4^{3-}} = 1$, $\gamma_{DIC} = 0.8$, and $\gamma_{Alk} = 0$. The atmospheric CO_2 can be driven up to greater than 20,000 ppm before the sequestering of Alk in the deep ocean is great enough to drive the surface Alk to 0. Thus, this mechanism has a far greater range on the high

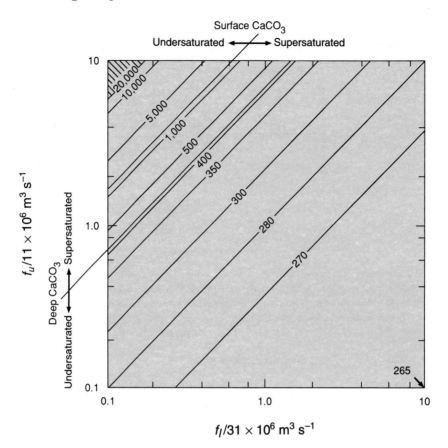

Fig. 16.4 Best values for f_u and f_l of 11×10^6 m^3 s^{-1} and 31×10^6 m^3 s^{-1}, respectively, were obtained for the model in Fig. 16.3 by forcing the model to fit the pre-industrial value of pCO$_2$ (280 ppm) and the deep ocean pre-bomb radiocarbon average of -160‰. These values were then varied over an order of magnitude in each direction, as shown. As the atmospheric pCO$_2$ increases, the deep ocean goes from being undersaturated to being saturated, which would sequester CaCO$_3$ even more efficiently by burying it in the sediments, thus driving the model even more rapidly in the direction of higher pCO$_2$ values. Eventually, however, the upper ocean becomes undersaturated, at which point calcareous organisms would have difficulty surviving. Taken from Sarmiento et al. (1988).

end than does the high-latitude outcrop mechanism, but it has very little flexibility on the lower end (of order 15 ppm) because the present level of deep ocean sequestering of Alk is quite small. However, it would probably be difficult with this mechanism to drive the atmospheric CO$_2$ any higher than of order 1,100 ppm, because at this point the surface ocean becomes undersaturated in CaCO$_3$ and the organisms which use CaCO$_3$ in their shells would have difficulty surviving. It is of interest to consider the possibility that a stagnation of the deep ocean might well be associated with extinction events of organisms that use CaCO$_3$ to build their shells, and conversely, that the development of such organisms would require a relatively rapid exchange of the deep ocean with the upper ocean.

One can readily infer from the foregoing results that a change in the ratio of photosynthetic production of organic carbon to production of $CaCO_3$, which was assumed to be 4 in the above study (see Sec. 8.3.3), could have a significant impact on atmospheric CO_2. Other related issues which could be addressed by box models such as these are the potential impact on CO_2 of changes in the Redfield ratio of organic matter, or the depth of regeneration of different components of organic matter. At present we do not have a good idea of what controls these processes and what changes in them, if any, might be reasonable to expect in response to climate or other changes in the environment.

16.3 Three-dimensional simulations of the natural carbon cycle

Powerful as they are in demonstrating the role of the solubility and biological pumps in controlling atmospheric CO_2, box models such as those discussed in the previous section cannot realistically simulate many important processes affecting the ocean carbon cycle, such as the role of the global thermohaline conveyor belt in explaining the high nutrient concentrations of the deep Pacific, and the effect of seasonality and biological processes on surface ocean CO_2. In addition, such models are not able to directly predict the effect of climate change on the carbon cycle. This requires more realistic ocean circulation models coupled to atmospheric models, as well as realistic models of the biological processes that drive the biological pump. This section uses two recent studies to illustrate the modeling strategies that are being developed to understand processes such as these.

The box model discussions of the previous section demonstrated that the modeling of DIC and Alk, from which the CO_2 is determined [Eq. (16.2)], could be greatly simplified by tying them to the cycling of phosphorus (or, alternatively, nitrogen) through the use of Redfield ratios. Both DIC and Alk are affected by the cycling of organic matter represented by (8.32), and by the cycling of $CaCO_3$, as discussed in Sec. 8.3.3. The aim of this brief discussion is to show exactly how DIC and Alk are linked to phosphorus (P) and $CaCO_3$. This is simple for DIC. From (8.32), and from the 1:1 ratio of C to $CaCO_3$, one obtains

$$\text{SMS}_{\text{DIC}} = r_{\text{C/P}}\text{SMS}_{\text{P}} + \text{SMS}_{\text{CaCO}_3} . \tag{16.8}$$

In order to do this for Alk one must go back to its definition as given by (8.22)

$$\text{Alk} \approx ([\text{Na}^+] + [\text{K}^+] + 2[\text{Mg}^{2+}] - [\text{Cl}^-] - 2[\text{SO}_4^{2-}] - [\text{Br}^-])$$
$$+ (2[\text{Ca}^{2+}] + [\text{NH}_4^+] - [\text{NO}_3^-]) . \tag{16.9}$$

529

All the chemicals in the first set of parentheses on the right-hand side of (16.9) are conservative, i.e., they behave like salinity and can thus be ignored for purposes of this discussion. The second set of parentheses contains those chemicals which are nonconservative, including Ca^{2+}, which is affected by the biological formation and chemical dissolution of $CaCO_3$. The contribution of ammonium to alkalinity is minor, and will be ignored in most of what follows. Thus

$$SMS_{Alk} = 2 \cdot SMS_{CaCO_3} - SMS_N \qquad (16.10)$$

where SMS_N is sources minus sinks of nitrate. Nitrate cycling is also a function of phosphorus cycling according to (8.32), i.e., $SMS_N = r_{N/P} SMS_P$. Inserting this relationship into (16.10) gives $SMS_{Alk} = 2 \cdot SMS_{CaCO_3} - r_{N/P} SMS_P$. The modified form of (16.2) thus gives SMS for CO_2 in terms of $CaCO_3$ and P:

$$SMS_{CO_2} = f(T, S, SMS_{CaCO_3}, SMS_P). \qquad (16.11)$$

A further simplication that has been used in most models to date is to assume that the cycling of $CaCO_3$ is also linked to that of PO_4^{3-} through a simple Redfield ratio relationship. Thus, the problem of predicting the sources minus sinks of CO_2 is reduced to the problem of predicting the PO_4^{3-} distribution, or alternatively, the NO_3^- distribution.

The following sections describe two recent simulations of the sources minus sinks of nutrients which are representative of the state-of-the-art in this field. The first study focuses on the large-scale distribution of phosphate, addressing primarily the pathways followed by organic matter from the time it leaves the surface ocean until it is regenerated at depth. The paradigm that had existed until recently is that the majority of this transport occurred by rapidly sinking particles of organic matter which were regenerated below where they were formed at the surface. This study shows that such a mechanism leads to unrealistic phosphate distributions, with a sizable trapping of nutrients occurring under regions of high surface production. A more realistic prediction requires that a large portion of the organic matter leave the surface in a form that can be transported laterally by ocean currents, away from the regions of production. The results also illustrate the functioning of the thermohaline conveyor belt (cf. Fig. 17.12 and Plate 14) and its role in leading to the increase in ocean concentrations that occur along the pathway of the deep waters that leave the North Atlantic and travel all the way to the deep North Pacific (Fig. 8.12).

The model used for the phosphate study has nonseasonal annual mean forcing, and thus is not able to simulate the seasonality of biological cycling in the surface ocean discussed in Sec. 8.3.2. In addition, the biological pump is highly parameterized, with essentially no direct representation of the actual biological processes that drive the pump. The second study is based on a seasonally forced model of new production in the surface ocean in which the biological processes are characterized by an ecosystem model.

It illustrates the seasonal cycle in detail but, more importantly, points the way for future models that are based on a more realistic characterization of the biological pump.

16.3.1 Simulation of new production and regeneration

Najjar (1990) has carried out a series of studies of new production and regeneration of phosphorus in the global ocean general circulation model (GCM) of Toggweiler et al. (1989a). He predicts new production, the net biological uptake of inorganic matter in the surface ocean which must be balanced by an export of organic matter (see Sec. 8.3.2), by using a simple scheme which makes use of the observed nutrient distribution. The SMS term at the surface takes the form $SMS_P = -\gamma \cdot (PO_4^{3-} - PO_{4_{obs.}}^{3-})$, which predicts new production by damping the model predicted PO_4^{3-} field towards the observed $PO_{4_{obs.}}^{3-}$ field with a rate constant $\gamma = 1/30$ days (Sarmiento et al., 1988; Najjar et al., 1992). The tendency of ocean transport will be to bring subsurface high PO_4^{3-} waters into the euphotic zone. The damping term will force these high values back towards the observed low values at the surface. The removal of PO_4^{3-} that this requires is the new production.

Once the new production is determined, the organic matter thus formed must be removed from the surface and regenerated at depth. Najjar has two scenarios for how new production is removed from the surface ocean. He originally began with the assumptions that the flux of organic matter out of the surface was in the form of vertically sinking particulate matter, and that the regeneration dynamics of the particles could be represented by the sediment trap work summarized by (8.34). However, this approach led to numerous difficulties. Beneath high productivity shallow upwelling regions along the equator, the regeneration of sinking particles resulted in a trapping of nutrients which led to excessively high concentrations (Fig. 16.5b). In addition, the average concentration in the upper 2,000 to 3,000 m was too low and the concentrations in the deep waters were too high (Fig. 16.6). It is difficult to separate problems with the ocean circulation model from problems with the parameterization of the regeneration process (Najjar et al., 1992), but one explanation for the low intermediate depth concentrations and high deep concentrations may be that a large fraction of the particulate organic matter (POM) is sinking all the way to the sediment–water interface before being regenerated. The low regeneration in the water column would give low concentrations there. The high regeneration at the sediment–water interface would contribute to high concentrations in the deep waters. A way was needed to export the organic matter away from the high production regions so that it could be regenerated elsewhere; and to regenerate a higher portion of the organic matter in the upper water column.

The observations of dissolved organic matter (DOM) by Suzuki et al.

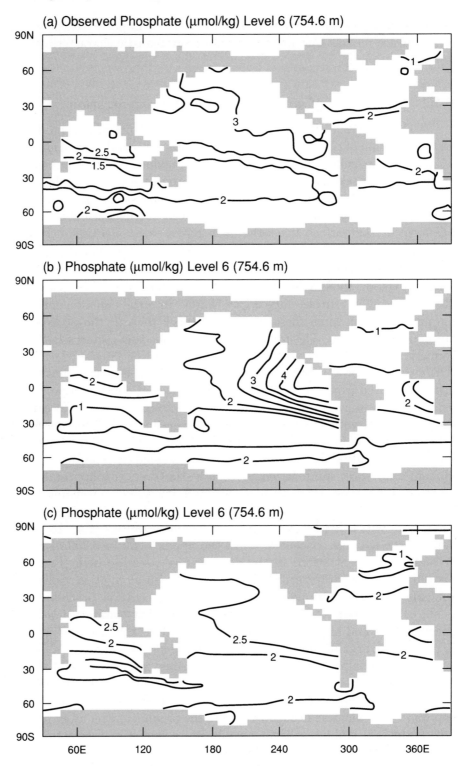

(a) Observed Phosphate (μmol/kg) Level 6 (754.6 m)

(b) Phosphate (μmol/kg) Level 6 (754.6 m)

(c) Phosphate (μmol/kg) Level 6 (754.6 m)

Fig. 16.5 Results from the model study of water column regeneration by Najjar (1990).
(a) The upper panel shows an objective map of phosphate observations at
295 m. (b) The middle panel shows results from a simulation in which all
new production is assumed to be regenerated immediately below where it occurs
(following the sediment trap based scaling of Martin et al., 1987). Note the high
concentrations under the high production regions in the equatorial Pacific and
Atlantic. This trapping continues to great depth in the water column. Elsewhere
the concentrations tend to be too low. (c) The bottom panel shows results from
a simulation in which 80% of the new production is put into dissolved organic
matter which is then regenerated by decaying with a time scale that increases
exponentially with a scale depth of 750 m, following roughly the observed bacterial
biomass as summarized by Cho and Azam (1988). The concentrations are now
more evenly distributed than in the middle panel, with concentrations under the
high production regions reduced dramatically.

(1985), and Sugimura and Suzuki (1988), and the interpretations of these
observations by Toggweiler (1989), suggested that DOM might provide a
way out of this problem. In a second parameterization of the removal
of organic matter from the surface and its regeneration at depth, Najjar
et al. (1992) split the new production, putting 20% into particles and 80%
into DOM. Unlike sinking particles, DOM is advected away from high-
productivity regions, eliminating the trapping of nutrients. The particles
were regenerated according to (8.34) and the DOM according to SMS =
$k_e \exp[-(z - z_e)/z^*]$, with a 750 m scaling depth based on a rough fit to
the bacterial biomass data of Cho and Azam (1988). Bacteria are the main
consumers of DOM. k_e is adjusted each time step so as to conserve the DOM
inventory at its initial guess value. The initial guess is based on observations
of the relationship of dissolved organic to dissolved inorganic matter (Suzuki
et al., 1985; Sugimura and Suzuki, 1988). Najjar et al.'s simulations show
dramatic improvement when 80% of the new production goes into DOM
(Figs. 16.5 and 16.6). Indeed, the only area of major discrepancy between
the model including DOM and the observations is in the deep Atlantic.
There the major problem is that the ocean circulation model is predicting
a too-shallow formation of nutrient poor North Atlantic Deep Water. Thus
the deep ocean is filled with high nutrient Antarctic Bottom Water.

The major conclusion from this study is that there is a need in the model
for a large part of the organic matter produced at the surface to leave the
surface in a long-lived form that can be transported by ocean circulation.
This has been simulated as DOM. A role for small nonsinking particulate
organic matter can be dismissed on the basis that the population of particles
is too low to give a large enough supply to explain the observed nutrient
fields. The model gives a DOM regeneration time scale of 30 years at
the base of the euphotic zone, increasing with depth to time scales well
in excess of the ocean circulation time scale of 1,000 years in the deep
ocean. Application of this model to the carbon cycle is being undertaken at

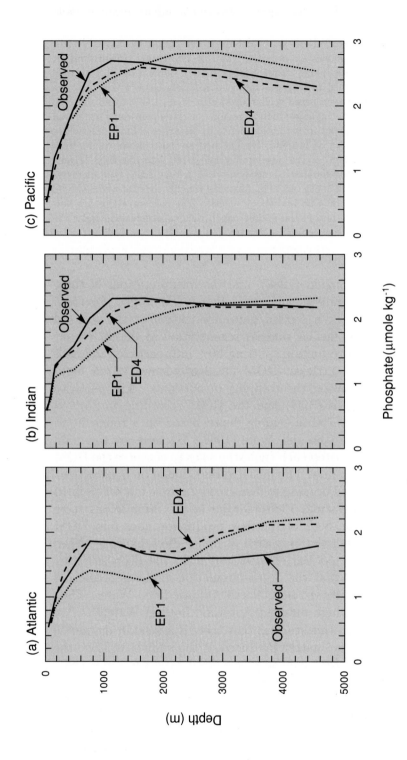

Fig. 16.6 Horizontally averaged vertical profiles of phosphate from Najjar (1990). Experiment EP1 (the particle only model) is as in Fig. 16.5(b) and ED4 (the model including dissolved organic matter) is as in Fig. 16.5(c). The dramatic improvement of the dissolved organic matter simulation is due primarily to the fact that the particle only model exports too high a fraction of the organic matter to the deep ocean before it is regenerated. The regeneration of the dissolved organic matter simulation occurs higher in the water column due to the fact that concentrations of dissolved organic matter are higher there, and the decay time scale shorter (see Fig. 16.5(c) caption).

the present time. Similar three-dimensional ocean GCM studies by Maier-Reimer and Hasselmann (1987), and by Bacastow and Maier-Reimer (1990) already include the carbon cycle, and have been used as a basis for studying the uptake of anthropogenic CO_2.

16.3.2 Euphotic zone food web model

The simplified approach of the Najjar et al. (1992) model for dealing with new production has a number of drawbacks. For one thing, the seasonal resolution of oceanic nutrient data is very limited. In addition, a model which depends on present nutrient data has no predictive value for situations in which the oceanic nutrient field might be expected to change. Dealing with such conditions requires the development of models which predict how biological processes respond to the physical environment. The following discussion of process models is based primarily on Fasham et al. (1990), who, in addition to developing a model of their own, have summarized the important literature in this area.

The first step in developing a euphotic zone nutrient cycling process model is to characterize the photosynthetic uptake of dissolved inorganic nutrients. The influence on nutrient uptake of temperature and light, and the concentration of nutrients limiting to growth (N_i), is generally assumed to be separable into independent functions such that the SMS term for N_i can be represented by

$$\mathrm{SMS}_{N_i} = -Ph \cdot f(T, \text{ light}) \cdot \prod_{i=1}^{n} f_i(N_1, N_2, \ldots, N_{n-1}, N_n) + \text{ sources}.$$

$$(16.12)$$

This equation is a reformulation of the first term on the right-hand side of (8.33). Ph is the concentration of photosynthesizing organisms, $f(T, \text{light})$ is the temperature and light-dependent growth rate of Ph under conditions of abundant nutrient supply. It has units of t^{-1}. $\prod$ is the product of the $f(N_i)$s, which are dimensionless nutrient limitation factors for each limiting nutrient. n, the number of limiting nutrients, is generally taken as 1: NO_3^- or PO_4^{3-}; or 2: NO_3^- and NH_4^+ (nitrate uptake is inhibited in the presence of ammonium); although other substances such as trace metals are very likely of importance in some environments as well.

A poorly known and highly variable fraction of $\sim 70\%$ of the organic matter produced by photosynthesis is eventually recycled back to dissolved inorganic nutrients within the surface rather than being exported (Platt et al., 1989). The production of dissolved inorganic nutrients by recycling is represented by the "source" term in (16.12). This recycling involves complex food web interactions which include processes such as zooplankton grazing and bacterial degradation (Sec. 8.3.2b). A byproduct of these food web interactions is dissolved inorganic nutrients. One of the major challenges in euphotic zone modeling lies in how to model this recycling.

One approach to solving (16.12) is to treat Ph, and $f(T, \text{light})$ as a single parameter which is a function of light only, $g(\text{light})$. Bacastow and Maier-Reimer (1990) follow this approach, representing $Ph \cdot f(T, \text{light})$ by a nontemperature dependent adjustable parameter which has a constant value throughout the model, and a dimensionless light function which gives the annual mean incident light at each latitude relative to the equator. The nutrient limitation expression $f(N_i)$ is given as a Michaelis–Menten kinetic expression. Thus

$$\text{SMS}_{\text{PO}_4^{3-}} = g(\text{light}) \cdot \left(\frac{\text{PO}_4^{3-}}{K + \text{PO}_4^{3-}} \right).$$

No recycling source terms are included, and the population of phytoplankton, Ph, is not included explicitly in the model. Thus the effect of recycling and of processes that control the population of Ph, such as grazing, is implicit in the choice of $g(\text{light})$ and the Michaelis constant K.

The Michaelis–Menten kinetic expression is the standard approach used in dealing with enzyme catalyzed biological processes such as the uptake of nutrients in photosynthesis. However, the $g(\text{light})$ expression significantly underrepresents our knowledge of the processes it represents. $g(\text{light})$ in (16.12) can be readily modified to include the effect of temperature on the maximum photosynthetic rate, as well as the effect of season, depth attenuation, and cloudiness on the light supply which Bacastow and Maier-Reimer (1990) leave out (e.g., Fasham et al., 1990; Platt and Sathyendranath, 1991). However, it is difficult to see how one can develop a more realistic approach for dealing with the influence of Ph on the photosynthesis term short of some form of food web model which includes processes such as grazing of phytoplankton. More realistic models for the influence of regeneration on the SMS "source" term also require a food web model. Food webs are complex and highly variable in space and time so that no standard approach for dealing with them can be suggested. The most promising approaches to this problem involve simple process models which lump all photosynthesizers into one or a few "phytoplankton" categories. Similarly, grazing on live organisms, as well as consumption of organic matter, are represented by one or a few categories of "zooplankton" and "bacteria."

Fasham et al. (1990) have developed an ecosystem model with the aim of predicting the new production and how it relates to other processes occurring in the euphotic zone. Their seven-component food web model of nutrient cycling is shown in Fig. 8.7, which serves as the basis for the more detailed depiction given in Fig. 16.7. The ecosystem model was kept as simple as possible, including only one type of phytoplankton, zooplankton, and bacteria, but including four forms of nitrogen. Uptake of nitrate is, by definition, the new production (see Sec. 8.3.2*b*). The organisms produce

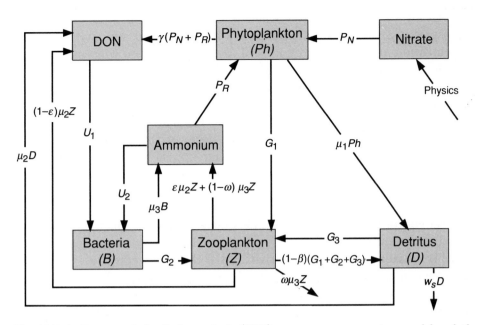

Fig. 16.7 A diagram of the Fasham et al. (1990) upper ocean ecosystem model and the parameters used in the results that are shown below. See Fasham et al. (1990), for a full discussion of the model and the parameters chosen.

and consume ammonium as well as dissolved organic nitrogen (DON) and particulate nitrogen. These latter processes give rise to an enhanced primary production based on ammonium uptake by phytoplankton, which is known as "regenerated" production. This ability of the model to predict both new production and regenerated production makes it a powerful tool for studying the impact of biology on oceanic nutrient distributions, as well as the relationship between observations of primary production and the quantity, new production, which is of importance to the carbon cycle. The details of the ecosystem model and its calibration with observations at Bermuda Station S are discussed by Fasham et al.

The Fasham et al. ecosystem model has been incorporated into a 2° horizontal resolution, 25 vertical level, seasonal ocean general circulation model of the Atlantic (Sarmiento, 1986); and initial biogeochemical output for the North Atlantic has been produced and tested against observations (Sarmiento et al., 1991). The behavior of the seven components of the ecosystem model (nitrate, phytoplankton, zooplankton, bacteria, ammonium, DON, and detritus) is determined by solution of equations of the form (16.1), with the biological SMS terms for the euphotic zone given by Fasham et al. (1990). New production (P_N in Fig. 16.7) is defined as the photosynthetic nitrate uptake, which is dependent on the phytoplankton and nitrate concentrations as well as light supply. Ammonium inhibition of nitrate uptake is included to account for the observation that ammonium is preferred by phytoplankton when it is present. P_R is defined as the photosynthetic uptake of ammonium, also dependent on phytoplankton concentration and light supply as well as the ammonium concentration. G represents zooplankton growth, and U represents bacterial uptake. G and U depend on the food supply concentration. The remaining terms all represent exudation and excretion.

The regeneration of organic matter that is transported below the euphotic zone is simulated by a simple first-order decay with a time scale of $1/10$ day. Any detritus that sinks out of the upper 123 m is regenerated instantaneously to ammonium following the sediment trap POM-based scaling of (8.34) with $a = 0.858$. The DOM pool represents only very short-lived highly labile organic matter, most of which is rapidly consumed by bacteria within the euphotic zone. This model does not include the longer-lived DOM required by the study discussed in the previous section. The model is initialized with an objective map of the observed nitrate field. The only other information that is needed is the light supply, which is specified following standard formulations with cloud cover from Levitus (personal communication).

The following brief discussion covers some salient results which trace the pathway of nitrogen through the food chain in the euphotic zone (upper 123 m in the model). Figure 16.8 shows the annual mean supply of nitrate to the euphotic zone at year 3. The net transport of nitrate to the surface is 64% by upwelling, 7% by vertical mixing, and 29% by convective overturning (implemented in the model by homogenizing adjacent layers that are unstable with respect to each other). Supply of nitrate by vertical transport (Fig. 16.8b) is positive almost everywhere. The only exception to this is the southern portion of the subtropical gyre. The little nitrate that is supplied to the euphotic zone in the southern portion of the subtropical gyre arrives there by horizontal transport (Fig. 16.8a) which comes from the tropics. The processes that give rise to this horizontal convergence are Ekman upwelling and divergence in the tropics, coupled with convergence and downwelling in the subtropical gyre (see Fig. 4.7). The upward flux of nitrate by deep wintertime mixing in the northern half of the subtropical gyre counterbalances the removal of nitrate by downwelling, but the southern half of the subtropical gyre does not experience deep wintertime mixing and thus has an overall removal of nitrate by vertical transport.

Nitrogen enters the food chain by photosynthetic uptake and is removed from the surface primarily as detrital organic matter formed by zooplankton egestion and mortality, and by phytoplankton mortality (Fig. 16.7). Figure 16.9 shows the primary production, and nitrate and ammonium uptake by phytoplankton in the model. The increase of the phytoplankton standing crop which occurs in spring as the water column restratifies and the solar insolation increases (i.e., the spring bloom) is evident in Fig. 16.9(b). A comparison of Fig. 16.9(b) with 16.9(d) shows that direct uptake of nitrate

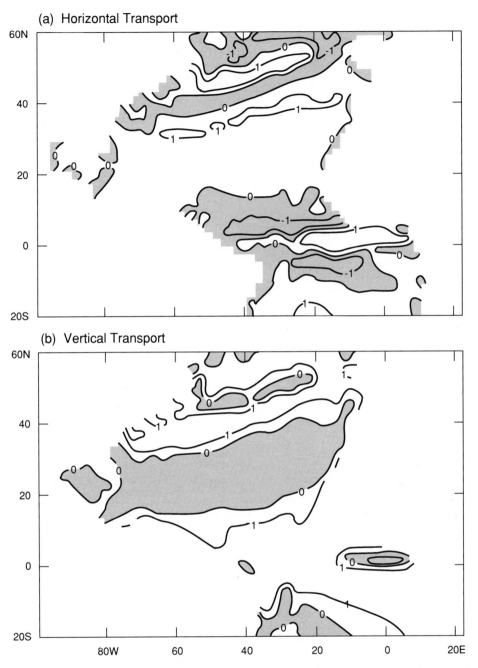

Fig. 16.8 Horizontal and vertical annual mean supply of nitrate to the top 123 m of the North Atlantic ecosystem model. The contours are 0 and ± 1.0 μmol-N m^{-2} day^{-1}. The shading indicates removal of nitrate. Note, in particular, that the vertical transport is removing nitrate from the southern half of the subtropical gyre, with supply being primarily by horizontal tranport. The net supply to the southern half of the subtropical gyre is extremely low, resulting in the low production which is evident in Fig. 16.9(a). Taken from Sarmiento et al. (in prep.).

539

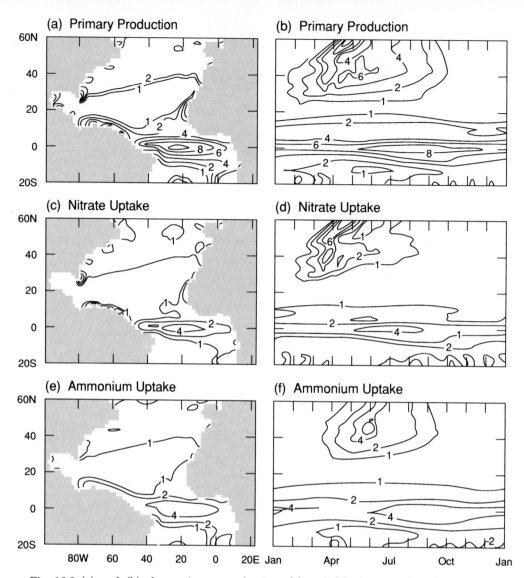

Fig. 16.9 (a) and (b) show primary production, (c) and (d) show uptake of nitrate by phytoplankton, and (e) and (f) show ammonium uptake by phytoplankton. Note that the units and contour intervals are different for all the plots. The left-hand panels show annual means over the top 123 m. The right-hand panels show zonal integrals over the top 123 m as a function of time. The development of the spring bloom shows as contours which slope upward to the right (e.g., in b and d), as the bloom begins first at low latitudes then progresses gradually to higher latitudes. Taken from Sarmiento et al. (in prep.).

(new production) fuels the early part of the bloom. Later on, however, the bloom is fueled primarily by ammonium uptake (Fig. 16.9f). The ammonium is produced by zooplankton and bacteria (see balances in Fig. 16.7), which do not develop until later in the bloom. The annual average ratio of new production resulting from nitrate uptake to the sum of new production and regenerated production resulting from ammonium uptake (the "f-ratio") is

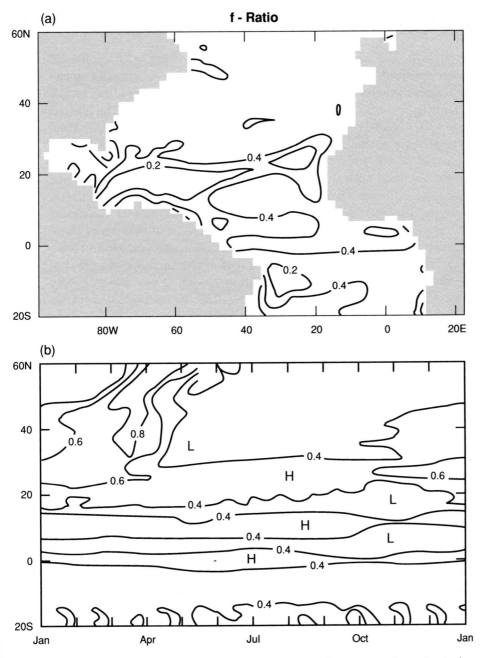

f - Ratio

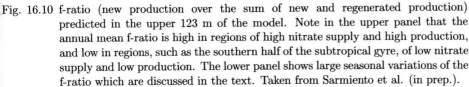

Fig. 16.10 f-ratio (new production over the sum of new and regenerated production)
predicted in the upper 123 m of the model. Note in the upper panel that the
annual mean f-ratio is high in regions of high nitrate supply and high production,
and low in regions, such as the southern half of the subtropical gyre, of low nitrate
supply and low production. The lower panel shows large seasonal variations of the
f-ratio which are discussed in the text. Taken from Sarmiento et al. (in prep.).

0.43, but Fig. 16.10 shows the high values during the early part of the
spring bloom (the time when new production is dominant) and abrupt
plunge afterwards (when regenerated production is dominant) that would

be expected from the above results. The seasonal changes in biological processes summarized in Figs. 16.9 and 16.10 are a major contributor to the seasonal changes in the various surface ocean properties depicted in Figs. 8.7 and 8.8. They have a major impact on oceanic CO_2, and thus on the air–sea exchange of CO_2. Models such as these are also being developed to simulate the net oceanic production and air–sea flux of DMS.

A comparison of the predictions made by this model with observations such as those obtained with the satellite Coastal Zone Color Scanner (color can be used to estimate chlorophyll, which is related to phytoplankton biomass; see e.g. Plate 9), show that there are many problems with the model (Sarmiento et al., in prep.). Interestingly, many of the most obvious of these problems, such as a too high pigment concentration in the equatorial band, and too low concentrations in the southern half of the subtropical gyre, are related to problems with the ocean circulation model. These have already been discussed in some detail by Sarmiento (1986) in connection with an analysis of the heat balance in the surface ocean which strongly suggests that the supply of cold (nutrient rich) waters to the surface is too high in the equatorial band, and too low in the southern half of the subtropical gyre. Overall, however, the model predictions are qualitatively very similar to the satellite observations and other observational constraints, suggesting that it may indeed be possible to construct reasonably realistic models of the effect of oceanic food webs on nutrient cycling by focusing on the processes rather than on the specific organisms.

16.4 Simulation of anthropogenic CO_2 uptake

The most important application of models of ocean biogeochemical cycling has been in estimating the oceanic uptake of anthropogenic CO_2. One way to model this is to make use of nutrient models such as those discussed in Sec. 16.3, and solving the carbon conservation equation using the Redfield ratio approach introduced at the beginning of Sec. 16.3. The fossil CO_2 perturbation is added to an atmospheric box and permitted to invade the ocean (e.g., Maier-Reimer and Hasselmann, 1987; and Bacastow and Maier-Reimer, 1990). An alternative approach, which has been the most common one used to date, is to assume that the natural cycle of carbon has not changed during the two centuries that the anthropogenic transient has been significant, and to model the CO_2 invasion as a perturbation. Siegenthaler et al. (1988) have made measurements of atmospheric CO_2 in ice cores from the tenth to the eighteenth centuries which support the assumption that the natural cycle of carbon is not likely to have led to significant atmospheric CO_2 changes on a time scale of two centuries. A recent application of the perturbation approach in box-modeling studies is that of Siegenthaler and Oeschger (1987). This section discusses the recent three-dimensional perturbation study of Sarmiento et al. (1992).

The perturbation approach to modeling oceanic uptake of anthropogenic CO_2 stems from the assumption that the processes affecting the distribution of pre-anthropogenic carbon, DIC_0, continue to operate unchanged, thus

$$\frac{\partial DIC_0}{\partial t} = 0 = -\mathbf{v} \cdot \nabla DIC_0 + \nabla \cdot (D\nabla DIC_0) + SMS_{DIC_0}.$$

This being the case, one can define a perturbation carbon distribution

$$\delta DIC = DIC - DIC_0$$

which satisfies the conservation equation

$$\frac{\partial (DIC - DIC_0)}{\partial t} = \frac{\partial \delta DIC}{\partial t} = -\mathbf{v} \cdot \nabla \delta DIC + \nabla \cdot (D\nabla \delta DIC). \qquad (16.13)$$

The SMS term drops out because perturbation DIC is affected only by transport processes, i.e., $SMS_{DIC} = SMS_{DIC_0}$. δDIC has an initial condition of 0 everywhere because the simulation is begun before anthropogenic effects became significant. Solution of (16.13) requires specification of transport of CO_2 due to air–sea gas exchange at the air–sea interface.

$$\nabla \cdot (D\nabla \delta DIC) = \nabla(f_{\delta CO_2}) \qquad\qquad \text{at } z = 0 \qquad (16.14)$$

where

$$f_{\delta CO_2} = k_g(1 - \gamma_{ice})(\delta pCO_{2_{atm}} - \delta pCO_{2_{oc}})$$
$$= -k_g(1 - \gamma_{ice})\Delta \delta pCO_2. \qquad (16.15)$$

$f_{\delta CO_2}$ is the flux of perturbation CO_2 into the ocean. k_g is the gas exchange coefficient, equal to k_w as defined by (8.8) multiplied by the CO_2 solubility. The value of k_w used is that of Broecker et al. (1985b). δpCO_2 is $pCO_2 - pCO_{2_0}$, with the subscript 0 indicating the concentration at the beginning of the simulation and subscripts atm and oc indicate atmosphere and oceanic values. γ_{ice} is the fraction of sea ice cover. It is possible in models for a single grid box to have only partial ice cover, i.e., for γ_{ice} to be other than 0 or 1. The assumption is made that gas exchange is not allowed through the ice. The fraction of sea ice can be obtained from models, such as those discussed in Chapter 12, or from observations.

In order to solve (16.13) and (16.14) we need to specify k_g, $\delta pCO_{2_{atm}}$, and $\delta pCO_{2_{oc}}$ while $\mathbf{v}$ and D are supplied by primitive equation ocean circulation models of the type discussed in Chapter 11. The results shown below were obtained with the model of Toggweiler et al. (1989a). $\delta pCO_{2_{atm}}$ is defined as

$$\delta pCO_{2_{atm}} = pCO_{2_{atm}} - pCO_{2_0}. \qquad (16.16)$$

$pCO_{2_{atm}}$ can either be modeled, or prescribed from observations such as the measurements of CO_2 made in trapped air bubbles from the Antarctic Siple ice core by Neftel et al. (1985) and Friedli et al. (1986), combined with atmospheric measurements made at Mauna Loa by Keeling et al. (1989) beginning in 1958 (Fig. 2.2). pCO_{2_0} must be specified and is set to a constant value of 280 ppm.

$\delta p\mathrm{CO}_{2_{oc}}$ is calculated from $\delta\mathrm{DIC}$, which, in turn, is obtained by solution of (16.13). The relation between $\delta p\mathrm{CO}_{2_{oc}}$ and $\delta\mathrm{DIC}$ can be approximated very well, without explicitly solving for the relatively complex carbonate equilibria each time step at each surface gridpoint, by noting that the ratio $\delta p\mathrm{CO}_{2_{oc}}/\delta\mathrm{DIC}$ is nearly a linear function of $\delta p\mathrm{CO}_{2_{oc}}$ for a given temperature (and constant alkalinity) over a range in excess of $\delta p\mathrm{CO}_{2_{oc}} = 0$ to 200 ppm. This approach is equivalent to the buffer factor approach used in one-dimensional models, but somewhat simpler for three-dimensional calculations, since the buffer factor approach would require DIC_0 to be specified over the whole ocean. Given this, we can write

$$\frac{\delta p\mathrm{CO}_{2_{oc}}}{\delta\mathrm{DIC}} = \frac{p\mathrm{CO}_{2_{oc}}}{\mathrm{DIC} - \mathrm{DIC}_0} = z_0 + z_1\delta p\mathrm{CO}_{2_{oc}} \qquad (16.17)$$

where z_0 and z_1 are functions of temperature obtained from the carbon system equations as explained by Sarmiento et al. (1991). $\delta p\mathrm{CO}_{2_{oc}}$ is in ppm and $\delta\mathrm{DIC}$ in $\mu\mathrm{mol\ kg}^{-1}$. Rearranging (16.17) gives

$$\delta p\mathrm{CO}_{2_{oc}} = \frac{z_0\delta\mathrm{DIC}}{1 - z_1\delta\mathrm{DIC}}. \qquad (16.18)$$

The final expression for the gas flux is obtained by substituting (16.16) and (16.18) into (16.14)

$$f_{\delta\mathrm{CO}_2} = k_g(1 - \gamma_{ice})\left[(p\mathrm{CO}_{2_{atm}} - p\mathrm{CO}_{2_0}) - \frac{z_0\delta\mathrm{DIC}}{(1 - z_1\delta\mathrm{DIC})}\right]. \qquad (16.19)$$

$\delta\mathrm{DIC}$ is calculated by solving (16.13) with an initial condition of 0 everywhere. Note the convention for reporting $p\mathrm{CO}_2$ is that they be given in dry air, thus a correction needs to be applied to account for the fact that air at the air–sea interface is essentially at 100% water vapor saturation.

Figure 16.11(b) summarizes the cumulative oceanic and atmospheric inventories predicted by an ocean GCM solution to (16.12) and (16.18) which is forced by the Siple ice core/Mauna Loa atmospheric CO_2 shown in Fig. 16.11(a) (Sarmiento et al., 1992). Already in the early nineteenth century a relatively large annual input to the atmosphere is necessary in order to explain the atmospheric CO_2 concentration increase at that time (cf. Fig. 16.11a). It is very likely that there were major CO_2 emissions from deforestation starting before 1800, a point which has been discussed in detail by Siegenthaler and Oeschger (1987). Note that the validity of this conclusion depends critically on the assumption that the Siple ice core record used in obtaining Fig. 16.11(a) truly reflects the pre-Mauna Loa CO_2 history.

More recently, i.e., nearly every year after 1941, the combined oceanic and atmospheric uptake of carbon by the model calculation shown in Fig. 16.11 is less than the fossil production as estimated by Rotty and Masters (1985) and Marland (1989; see Fig. 16.11b). A deforestation source must be accounted for as well. This model thus agrees with most previous

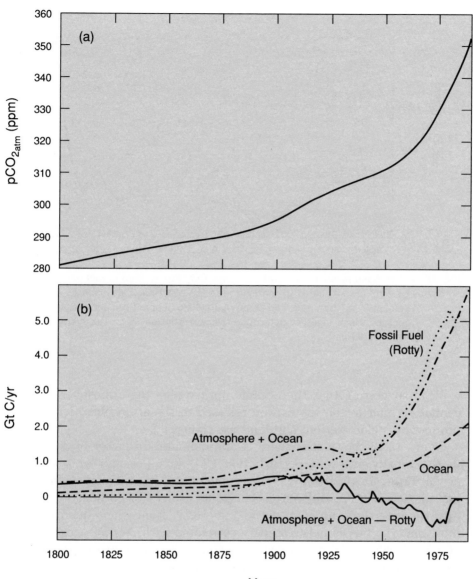

Fig. 16.11 (a) Spline fit to the Siple ice core/Mauna Loa atmospheric pCO_2 which is used to force the perturbation CO_2 simulation. (b) The annual change in carbon calculated by the ocean model, summed to the annual change in the atmospheric inventory. The atmospheric inventory is based on the Siple ice core and Mauna Loa CO_2 data as analyzed by Siegenthaler and Oeschger (1987). The fossil CO_2 production is taken from Rotty and Masters (1985) with updates by Marland (1989). The difference between the combined atmosphere/ocean change and the "Fossil" curve shows what the net release from sources other than those modeled would have to be if the ocean model were correct. Note, in particular, that the present model requires an additional source of carbon prior to ∼1940, and a sink after that. Taken from Sarmiento et al. (1992).

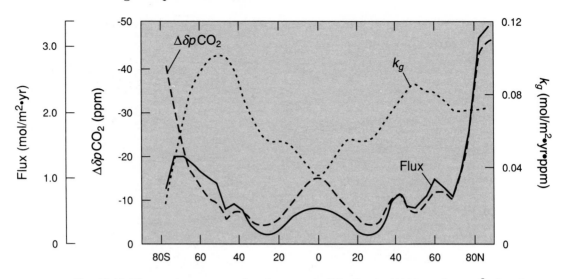

Fig. 16.12 The zonal mean annual anthropogenic CO_2 flux in 1986 in mol per m^2 of ice free ocean. Also included in the figure are the zonal mean gas transfer coefficient in mol m^{-2} yr^{-1} ppm^{-1}, and $\Delta\delta pCO_2$ for 1986 in ppm (also per unit area of ice free ocean) which together determine the perturbation flux. Taken from Sarmiento et al. (1991).

work in suggesting that the oceanic uptake plus the atmospheric increase cannot account for all the carbon released to the atmosphere by fossil fuel production, deforestation, and land use changes.

Figure 16.12 shows that there is approximately a factor of 30 variation in the zonally averaged perturbation CO_2 flux per unit area of ice-free ocean. From (16.15) the flux is the product of the gas transfer coefficient, k_g, and air–sea $\Delta\delta pCO_2$ difference. The zonal average of both these quantities over the ice-free ocean is shown in Fig. 16.12, from which one can readily infer that the flux is most strongly correlated with $\Delta\delta pCO_2$, except south of $\sim$70°S, where the gas transfer coefficient is exceptionally small due to the very low annual mean wind speeds extrapolated from the Esbensen and Kushnir (1981) data set that was used for this study. The pattern of $\Delta\delta pCO_2$ derives primarily from the different rates at which convective overturning and upwelling supply the surface with fresh water relatively uncontaminated with perturbation CO_2.

Of course δDIC does not just stay in the region where it is added to the ocean, but is redistributed horizontally. Figure 16.13 shows that δDIC is lost from the equatorial regions and subpolar gyres, with the loss accumulating primarily in the subtropical gyres as well as some gain in the Arctic. This meridional redistribution is controlled predominantly by upwelling of waters with a small excess CO_2 load in the equatorial region and subpolar gyre,

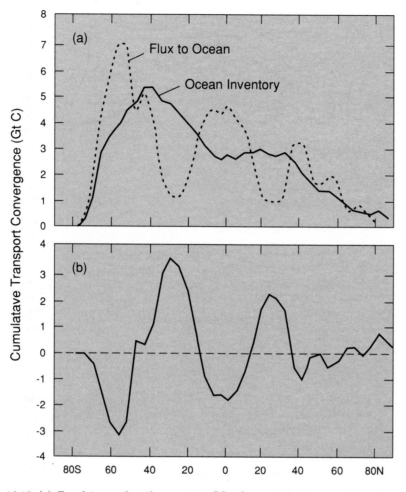

Fig. 16.13 (a) Zonal integral anthropogenic CO_2 flux and anthropogenic carbon standing crop in GtC per 4.5° latitude band in 1986 from model version A1.1. (b) is the zonally integrated carbon transport divergence in GtC yr^{-1} per 4.5° latitude band for the year 1986. Taken from Sarmiento et al. (1992).

accompanied by Ekman transport towards and downwelling within the subtropical gyre. The tendency towards shallower penetration of δDIC in the equatorial region and deeper penetration in the subtropics is clearly evident in the zonally averaged ocean basin profiles of Fig. 16.14. The regions of deepest penetration are the North Atlantic Deep Water formation regions, and around the Southern Ocean in the Atlantic and Indian Oceans. The deeper penetration in the southern Indian Ocean region is probably not a realistic feature, as discussed by Toggweiler et al. (1989b).

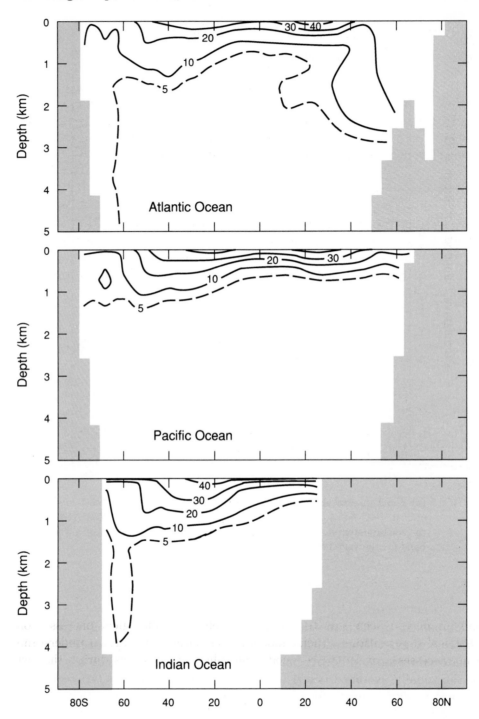

Fig. 16.14 Zonally averaged δDIC in the Atlantic, Pacific, and Indian Oceans. Taken from Sarmiento et al. (1992).

16.5 Future directions

16.5.1 *Anthropogenic CO₂ uptake*

Ocean circulation models, such as the one described in the previous section, provide the major constraint on estimates, such as those shown in Fig. 6.7, of how much anthropogenic CO_2 is entering the ocean at the present time. Improvements in those estimates will occur primarily as a result of improvements in the ocean circulation models. There is a modest uncertainty in the perturbation approach which arises from the assumption that pre-anthropogenic CO_2 was at 280 ppm everywhere in the surface ocean. However, models such as those of Maier-Reimer and Hasselmann (1988) and Bacastow and Maier-Reimer (1990), which have a more realistic simulation of pre-anthropogenic CO_2, do not differ significantly in the results they obtain. This suggests that it is not likely that improved chemical and biological models will make a significant difference in model estimates of anthropogenic CO_2 uptake, although it is important to demonstrate this.

The reason that models are so important in constraining anthropogenic CO_2 uptake is that direct measurements of the air–sea gas exchange, or of the change in total oceanic carbon inventory, are difficult to obtain with sufficient spatial and temporal resolution, and accuracy. Nevertheless, there is an active program to obtain such measurements at the present time, with preliminary results that pose a major challenge to the model uptake estimates. Tans et al. (1990) make use of oceanic air–sea CO_2 gas exchange estimates, as well as an atmospheric CO_2 transport model and atmospheric CO_2 observations, to conclude that oceanic uptake is less than half that predicted by most oceanic models. However, these estimates are being challenged on several grounds, including the failure of Tans et al. (1990) to take into account all aspects of the ocean carbon cycle such as the large flux into the ocean by rivers, with only a small sedimentary burial, as shown in Fig. 6.6. Clearly much work needs to be done to reconcile other constraints on the carbon cycle with those obtained by ocean models.

16.5.2 *The natural carbon cycle and other chemicals*

The most difficult challenge in the development of biogeochemical models of the oceans is to predict the effect of biological processes on chemical distributions. Such models are essential if we are to understand the processes that led to reduced atmospheric CO_2 levels during the last ice age, and if we are to predict the oceanic role in the impact of climate change on the future evolution of atmospheric CO_2, including the possibility of human intervention, such as the iron fertilization scenario mentioned in the introduction. The key issues identified by the models in Sec. 16.2 are: (1) what controls the surface nutrient concentration of high-latitude deep water formation regions, (2) how are the carbon and alkalinity related to the nutrients in organic matter (i.e., what is the Redfield ratio both in uptake and regeneration), and (3) what processes control the depth of regeneration of the organic and inorganic matter produced at the surface?

The successful modeling of euphotic zone processes requires above all that we first understand what factors determine the extent of nutrient removal that occurs in a given region of the ocean. The development of the food web model discussed in Sec. 16.3.2 is based on the premise that ecosystem interactions, in conjunction with physical processes, play a critical role in limiting the ability of biological processes to take up nutrients. If it could be demonstrated that the wind-borne supply of iron-containing dust particles is the key factor, then food web interaction models would not be so critical, and modeling efforts should rather be concentrated on atmospheric dust transport. The highest priority is for collaborative field and modeling work that can pin down the important control mechanisms. Euphotic zone food web models are also essential for predicting the production of DMS (e.g., Wolfe et al., 1991).

The whole strategy of linking carbon and alkalinity cycling to nutrient cycling through the Redfield ratios is open to question. It has proved difficult to constrain just what the Redfield ratios are. The analysis of observed dissolved inorganic property distributions is complicated because it is difficult in most places to clearly isolate that portion of the nutrient, total carbon, and alkalinity which is due to regeneration or uptake, as opposed to transport. The use of sediment traps to catch sinking particles produced in the surface continues to be controversial because of considerable uncertainty as to what portion of the sinking material is being caught. In addition, recent studies, such as those discussed in Sec. 16.3.1, suggest a major role for DOM in transport out of the surface to depth, but there are as yet few measurements available, and those that are available do not agree very well with each other.

As problematic as our understanding of euphotic processes is, it is considerably better than our understanding of water column regeneration processes, for which mechanistic models, such as the euphotic zone food web model discussed in Sec. 16.3.2, have not even been attempted as yet. Observational constraints are extremely difficult to obtain in the subsurface ocean, and models are thus far all highly parameterized simplifications such as those discussed in Sec. 16.3.1, based on indirect evidence such as observations of the dissolved nutrient distribution.

One of the ways in which ocean biogeochemical models may find important use is in the long-term monitoring of oceanic new production. There exist a large number of in situ measurements of radiocarbon uptake from which global maps of estimated primary production have been produced (e.g., Fig. 8.8 and Berger et al., 1989). There also exist global maps of surface chlorophyll concentration made with measurements obtained by the satellite Coastal Zone Color Scanner with intermittent spatial and temporal resolution during 1979–1986 (e.g., Lewis, 1989); see Plate 9 for an example in the North Atlantic. Ocean color instruments will be going up on new satellites in the near future. However, neither radiocarbon uptake nor ocean color measurements are directly related to new production. Radiocarbon uptake measures the total uptake of carbon, i.e., primary production, not the $\sim 30\%$ fraction which is exported. Similarly, the CZCS surface chlorophyll concentrations can be used to estimate the primary production by utilization of empirical vertical chlorophyll profile shapes (e.g., Platt et al., 1988), but not the new production.

Translating estimates of primary production into estimates of new production requires knowledge of the ratio of new production over primary production (known as the "f-ratio"). The extremely limited f-ratio data set has been used to construct simple empirical relationships (e.g., Eppley and Peterson, 1979; and Harrison et al., 1987) which have been applied to obtain estimates of new production from satellite data (e.g., Dugdale et al., 1989) as well as from radiocarbon uptake measurements. However, there is not yet enough information on f-ratios or new production to know how succesful these various approaches will be. Food web model studies of f-ratios show a highly complex behavior in space and time (e.g., Fig. 16.10) which suggests that the use of satellite color observation to estimate new production may ultimately require the development of food web models that can assimilate the data.

PART 4

COUPLINGS AND
INTERACTIONS

17

Global coupled models: atmosphere, ocean, sea ice

Gerald A. Meehl

17.1 Introduction

Recent years have witnessed a rapid increase in computer-model simulation of the global coupled ocean–atmosphere system, prompted by simultaneous advances in several areas. First, each successive generation of supercomputers has expanded the limits of computing power and has allowed longer runs with more complicated models. Second, better understanding of observed phenomena, themselves the products of air–sea coupling (e.g., El Niño–Southern Oscillation, or ENSO), has provided insight into the modeling of the coupled system. And third, the evolution of computationally efficient computer models of atmosphere and ocean has enabled researchers to perform the multidecadal integrations necessary to study global climate-related problems. These models are, in effect, a synthesis as well as a measure of our current understanding of the coupled system.

Global coupled models of the Earth's climate make up a series of tools with which researchers can study a wide spectrum of problems (Meehl, 1984; 1990a). The term "coupled model" has multiple meanings in the hierarchy of formulations developed to study the ocean–atmosphere climate system. The concept of radiative equilibrium was first used in more simplified radiative convective models (Manabe and Strickler, 1964; see Chapter 10, Sec. 10.2.2*b*). These were early attempts to understand the thermodynamic aspects of climate equilibrium. Global models have been used to study both the dynamic and thermodynamic aspects of the climate problem.

Ocean representations in global coupled models include:

(1) a "swamp" ocean in which sea surface temperatures (SSTs) are computed from surface energy balance only, with no heat storage or ocean currents;

(2) a simple mixed-layer or "slab" ocean in which SSTs are computed from surface energy balance and heat storage in a fixed-depth mixed layer with no ocean currents;

(3) the ocean General Circulation Model (GCM) in which SSTs are computed from surface energy balance and heat storage and include contributions from ocean currents and upwelling.

The first two ocean representations are nondynamical.

To provide a context for descriptions of the more complicated and computationally intensive coupled ocean–atmosphere GCMs in this chapter, the nondynamical ocean formulations of global coupled models are briefly reviewed and the issues dealing with coupling strategies for the coupled GCMs are examined. One of the most serious problems confronting coupled GCMs is climate drift or systematic errors in the coupled simulation, and several alternatives for dealing with this problem are presented along with some of the interesting and intriguing products of global coupled simulations. These features are the product of coupled interaction between the ocean and atmosphere. They involve aspects of the global climate simulation and include ENSO as well as the relationship of coupled low-frequency variability to the thermohaline circulation in the ocean.

17.2 Hierarchy of global coupled models

17.2.1 *Swamp ocean*

Historically, in modeling the global coupled ocean–atmosphere system, researchers have had to deal with the fundamentally different response times of the two major components – the atmosphere and the ocean. The atmosphere can typically adjust to a new forcing in a week to a month or so, the upper ocean on the order of weeks to several months, the mid-ocean several years, and the lower ocean on the order of hundreds of years. The computational burden of making long integrations with the coupled ocean–atmosphere system (order 10 to 100 years) has always played a major role in modeling. Any attempt at modeling the global system has to be tempered by the type of problem to be studied and the computing resources available. Consequently, some problems can be studied with ocean formulations that require less computer time when coupled to an atmospheric GCM.

The desired result is a simulation of interaction between atmosphere and ocean (i.e., the ocean surface temperature can change in response to atmospheric forcing, and the atmosphere, in turn, is affected by the new ocean surface temperature). Several alternatives have been developed that respond rapidly to altered atmospheric forcing and, therefore, require less computer time. The first is a simple interactive ocean temperature formulation where the ocean acts only as a wet surface. In this formulation, the SST responds to surface energy balance from the atmosphere, but it has no heat storage and there are no currents in the ocean that affect SST. Such a formulation is called a "swamp" ocean, since the ocean surface acts like a simple wet surface (Fig. 17.1a).

The SST is calculated with a swamp ocean by the surface energy balance

$$S + F^{\downarrow} - F^{\uparrow} - H - LE = 0 \,, \tag{17.1}$$

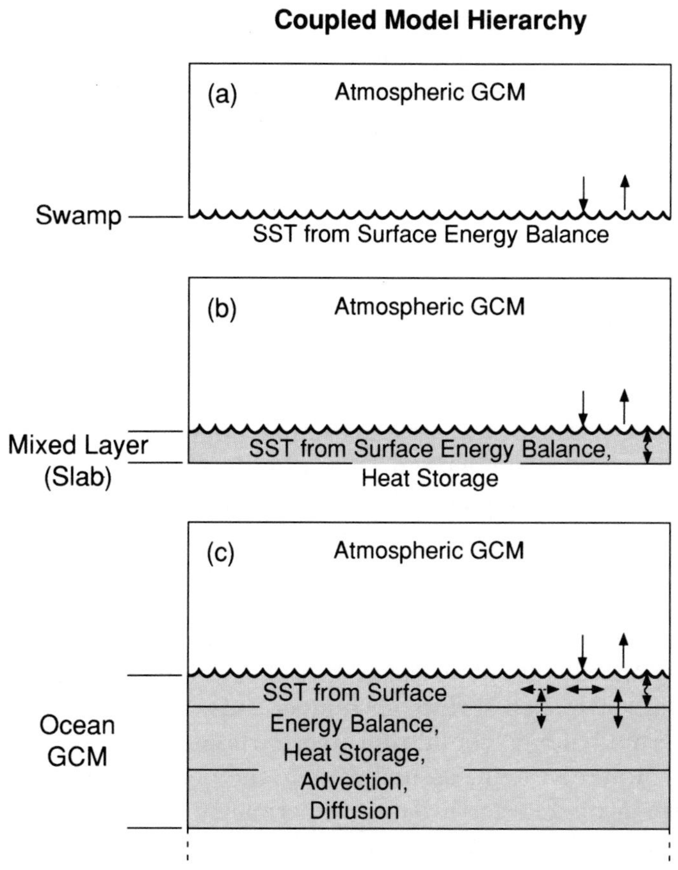

Coupled Model Hierarchy

Fig. 17.1 Schematic of global coupled model hierarchy: (a) atmospheric GCM coupled to a swamp ocean; (b) atmospheric GCM coupled to a simple mixed layer or slab ocean; and (c) atmospheric GCM coupled to an ocean GCM.

where S is absorbed solar flux, $F^{\downarrow}$ downward infrared flux, $F^{\uparrow}$ upward infrared flux, H sensible heat flux, and LE latent heat flux (latent heat of evaporation L multiplied by the evaporation E). Several of these terms are functions of surface temperature, and SST can be calculated iteratively from the surface energy balance. Chapter 10 gives details on the formulation of these terms.

Because the swamp ocean has no heat storage, only annual mean forcing from the sun can be applied. That is, no seasonal cycle of solar insolation is allowed, since in the winter hemisphere the sea ice would freeze well into the midlatitudes without stored heat in the oceans to modulate the seasonal swing of SST. This class of model is used mainly for basic sensitivity studies. Not being allowed to run a seasonal cycle is a serious constraint, but the advantage is that running an atmospheric GCM with a swamp ocean is computationally cheap. Because the ocean is simply a surface energy balance, the response to atmospheric forcing is almost instantaneous, and integrations of several years yield information on basic sensitivities. For example, a swamp-ocean formulation coupled to an atmospheric GCM can show sensitivity to an external forcing, such as a decrease of solar constant (Wetherald and Manabe, 1975) or an increase of atmospheric carbon dioxide (Washington and Meehl, 1983).

17.2.2 Mixed-layer ocean

The next step in the global coupled model hierarchy is to couple an atmospheric GCM to an ocean formulation that includes a nominal thickness of water, usually on the order of 50 to 100 m (Fig. 17.1b). This is important because this layer or "slab" of water adds simple upper-ocean seasonal heat capacity and the coupled model can be run with a seasonal cycle. Such a coupled model crudely simulates the seasonal cycle of upper ocean heat storage (Meehl and Washington, 1985) and permits one more element of the coupled system – the seasonal cycle and its important effects on climate sensitivity – to be studied simplistically with a model that can be run to equilibrium. Equilibrium in this context means that the atmosphere and ocean reach a balanced state of no mean drift of the climate toward another state (e.g., globally averaged surface air temperatures remain at some mean value). In a thermodynamic equilibrium sense, these models can represent changes in the climate system, assuming that the ocean heat transport is the same in the control and perturbation experiments. (For example, if flux corrections are used, the corrections must remain the same in control and perturbation experiments to satisfy this requirement; see discussion below.) The equation producing the SST in such a coupled model is represented by

$$\rho c_p h \frac{\partial T}{\partial t} = S + F^{\downarrow} - F^{\uparrow} - H - LE \, , \qquad (17.2)$$

where T is SST, ρ density of sea water, and c_p specific heat of sea water. The term at the left is the heat storage in the mixed layer of depth h, and the collection of terms at the right is the surface energy balance noted in Eq. (17.1) for the swamp ocean.

This type of global coupled model is more expensive computationally because of the time lag before the simple mixed layer with its heat storage comes into equilibrium with the atmosphere. This time lag is usually on the order of 20 years or so, somewhat longer than could be expected for a simple 50-m deep ocean surface layer. Most of the approach to equilibrium occurs in five years or so, but the last 10–20% takes about another 15 years, apparently due to slow convergence of small imbalances in the ocean–atmosphere system (Washington and Meehl, 1984). More computer time must be expended than with the swamp-ocean formulation, but the simulation is somewhat more realistic, to a point. Since a major part of the real ocean is not included (ocean currents), we cannot hope to simulate all aspects of the observed SST distribution with such a coupled simulation. Because the role of horizontal ocean heat transport by ocean currents is to move heat from the tropics poleward (see Chapter 4), the exclusion of currents from the simulation yields SSTs warmer in the tropics than observed and colder than observed in high latitudes. Since upwelling in the ocean produces lower SSTs in certain tropical and eastern ocean subtropical areas, the exclusion of upwelling contributes to warmer tropical SSTs in upwelling regimes in the tropical eastern Pacific and off the west coasts of the subtropical continents (Fig. 17.2, from Meehl and Washington, 1985). Such errors in the SST simulation are sometimes corrected by including the so-called Q-flux (or heat flux) correction terms in this class of model. Heat flux correction can be thought of as a source of heat to represent the effects of heat transport by the ocean currents (e.g., Hansen et al., 1984). Issues involved with flux correction are discussed in Sec. 17.4 of this chapter.

In spite of problems with systematic SST errors in the mixed-layer coupled-model simulation, this class of model has been useful in studying global climate sensitivity and has formed a basis for assessments of the global response to increased atmospheric CO_2 (Schlesinger and Mitchell, 1987; IPCC, 1990a).

17.2.3 Dynamical ocean GCM

An atmospheric GCM coupled to an ocean GCM represents the most complicated (and computationally intensive) step in the global coupled model hierarchy. Inasmuch as the ocean GCM includes all the elements of the swamp (surface energy balance) and mixed layer (surface energy balance and heat storage), plus the potential realism of ocean currents, upwelling, and subgrid-scale mixing processes to contribute to the SST and sea ice distribution, the coupled ocean–atmosphere GCM is the most powerful tool available to study global climate (Fig. 17.1c). Several problems exist in this class of coupled model, however, that were not factors in the swamp or mixed-layer models.

Foremost is computational expense. With the entire depth of the ocean involved, the spin-up of such a model with a tremendously slow response time is a significant problem. That is, how does the coupled model reach the point where sensitivity experiments can begin (where the atmosphere and ocean are in some state of internal consistency with each other) and how much computer resource must be expended? In a coupled GCM, the model, conceivably, must be run for hundreds of years just for the response of the middle and lower ocean to approach consistency with the surface waters and the atmosphere above. Additionally, small imbalances can lead to slow drift of sea ice extent (e.g., a slow global cooling can be associated with slowly expanding sea ice; conversely, if the ice-albedo scheme is using albedos that

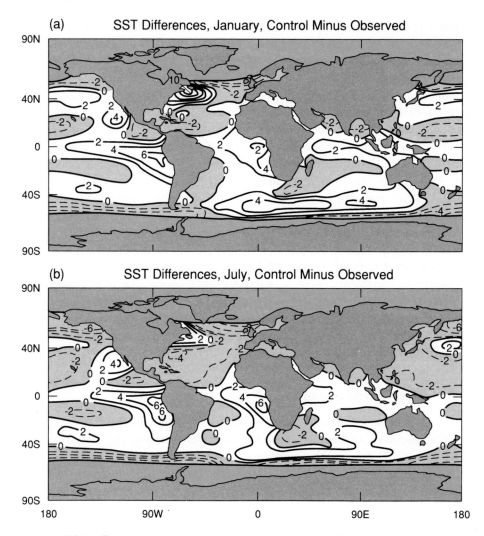

Fig. 17.2 SST differences, model minus Alexander and Mobley (1976) mean observed SSTs, (°C): (a) January; and (b) July. The model SSTs are generated by the simple mixed-layer ocean with no dynamics and with a constant 50-m depth coupled to an atmospheric GCM. Light colored areas are negative (cooler in the model compared to the observed). The dark color indicates regions covered by at least 0.2 m of sea ice in the model.

are too high, this can cause a cooling trend by itself). Climate-sensitivity experiments face the same problem. Because the entire depth of the ocean is involved, the ultimate response time of the model is only as short as the slowest responding part of the system, namely the deep ocean. Thus the model is not only more time-consuming because of the extra equations that must be included for the dynamic ocean and its multilevels, but the time required for spin-up and integration of useful experiments is much longer.

The issues of computational expense and need for long integrations also affect the choice of model resolution. To allow longer integrations, the typical tradeoff is to run relatively coarse-resolution coupled GCMs in computationally feasible experiments.

An ocean GCM includes equations for the calculation of ocean currents, temperature, and salinity (Chapters 4 and 11). The SST equation from an ocean GCM with finite-depth layers (ocean models with isopycnal surfaces are formulated somewhat differently; see Chapter 11) includes heat storage in the surface layer as before from the mixed-layer model; the effects of ocean currents and upwelling; the surface energy balance; and vertical and horizontal eddy diffusivity from subgrid-scale processes. (See Chapter 11 for more details of diffusion parameterizations as well as other mixing schemes and their rationalization.) The predictive equation for salinity is very similar to that for temperature, but without the radiative damping terms (see Chapter 11 and a complete discussion in Washington and Parkinson, 1986).

Coupled GCMs represent many possibilities for studying processes and sensitivities not resolved by the simpler and more computationally efficient alternatives in the hierarchy – the swamp and mixed-layer ocean formulations. Systematic errors become even more important, however. These are a consequence of physical processes not represented or poorly represented in the models, as in the swamp and mixed-layer formulations discussed previously. For example, the large horizontal heat diffusion required by a coarse grid ocean model to maintain computational stability can result in high-latitude SSTs that are too warm. The sea ice then melts accordingly and a systematic error in the ocean results in SST as well as in sea ice errors. In addition, as a result of limitations in parameterizations, unresolved space scales of certain processes, or inaccurate formulations, errors can be introduced that can be manifested in the coupled model as errors in the simulation. Systematic errors are discussed in Sec. 17.4.

17.3 Coupling strategies

17.3.1 *Coupling interface*

In order for elements to communicate with each other in a global coupled model, certain parameters are passed back and forth between the atmosphere and ocean GCMs. To couple with an ocean model, the atmospheric GCM must communicate or "hook up" with it. The parameters that must be passed in common between model elements constitute the coupling interface between atmosphere, ocean, and sea ice models (Fig. 17.3; the coupling interface between the atmosphere and land surface is discussed in Chapter 14). Since component models are almost always developed separately, the interface between models is crucial. For example, not only must the fields (a model field is a two-dimensional array of values

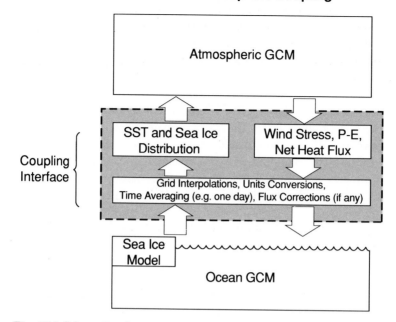

Fig. 17.3 Schematic of ocean–atmosphere coupling interface.

of some variable) be compatible, but the units must be shared in common or converted, interpolations between incompatible grids must be performed, and each component model must be set up to receive the expected quantities from the other models. Additionally, the frequency of communication between the models must be structured, appropriate averaging for time interval communication must be performed, and flux corrections may be applied.

17.3.2 *Atmosphere communicates with ocean and sea ice*

To begin with, the atmospheric model and its communication with the ocean and sea ice models is illustrated. More details on the structure and function of atmospheric GCMs appear in Chapters 3 and 10. Here, only communication at the air and sea ice interface is discussed. Consider an atmospheric model that communicates with the ocean and sea ice models once each model day. That is, the atmospheric model takes SSTs and sea ice distributions from the ocean GCM and sea ice model from the previous model day and holds them fixed for one atmospheric model day. If the atmospheric model has a 40-min time step, there are 36 iterations per model day. For each of the 36 iterations, the atmospheric model uses the same values of SST and sea ice averaged for the previous day from the ocean and sea ice models.

At the end of that model day, the fields required by the ocean and sea ice models are averaged over the period of that day. These values usually include the surface wind stress, net freshwater flux expressed as precipitation minus evaporation $(P - E)$, and, in some cases, runoff from continents, and net heat flux into the ocean. The equations for these quantities may be given as

$$\tau_x = C_D \rho_1 V_1 u_1 , \tag{17.3}$$

$$\tau_y = C_D \rho_1 V_1 v_1 , \tag{17.4}$$

$$F_f = P - E , \tag{17.5}$$

$$H_{net} = [S + F^\downarrow - F^\uparrow - H - LE] , \tag{17.6}$$

where u and v are the zonal and meridional components of wind from the lowest atmospheric model layer near the surface, the subscript 1 denotes the lowest model level near the surface, and the wind speed is

$$V_1 = (u_1^2 + v_1^2)^{\frac{1}{2}} . \tag{17.7}$$

C_D is the drag coefficient that is usually empirically determined or sometimes calculated as a function of surface wind speed and stability (see Fig. 4.6) and ρ is the density. Evaporation in $P - E$ also appears in the net surface heat flux as part of the latent heat flux term LE. To compute the net surface heat flux, the functions of SST use the value of SST from the previous day from the ocean model. Other fields may also be passed from the atmosphere into the ocean depending on the exact formulation of the ocean and sea ice models (e.g., snowfall from atmosphere passed to sea ice model).

17.3.3 *Ocean and sea ice communicate with atmosphere*

After the atmospheric model has produced the fields required by the ocean model for its calculations, the ocean model in this example runs for one model day with the values of wind stress, net freshwater flux expressed as precipitation minus evaporation (and runoff from the continents, if included), and net heat flux from the atmosphere held fixed for that entire model day (left side of Fig. 17.3). If the ocean uses a 40-min time step, for example, there are 36 iterations per model day. For each of the 36 iterations then, the values from the atmosphere are held constant and the ocean adjusts to that forcing to produce new values for all the ocean fields (details for ocean models are in Chapter 11).

Of most interest for the actual coupling are the parameters at the air–sea interface. From the ocean, therefore, the quantities "seen" by the atmosphere are SST and sea ice distribution. We noted earlier how SST could be calculated by an ocean GCM. The forcing from the atmosphere causes the SST, currents, and salinity to adjust in the ocean GCM, and similar forcing from the atmosphere and ocean causes adjustments in the sea ice model. In the ocean, there is usually a heat flux from the ocean to the bottom of the sea ice, as well as heat flux through the sea ice.

Usually, sea ice models in global coupled models are relatively simple because of computational restraints. Evidence is mounting, nevertheless, that different types of sea ice formulations can alter global climate sensitivity (e.g., Meehl and Washington, 1990). Early global coupled models almost exclusively included simple thermodynamic sea ice models (e.g., Semtner, 1976). These formulations only use information on temperature (from atmosphere above and ocean below) at each gridpoint for formation or melting of ice. No ice movement is allowed, nor is there any other kind of ice dynamics that could be important (Chapter 12). More complicated and realistic sea ice models have existed for some time (Hibler, 1979; Parkinson and Washington, 1979), but they consumed a prohibitive amount of computer time. As more streamlined sea ice models become available (e.g., Flato and Hibler, 1990), more realistic assessments of the role of sea ice in global climate will be possible. These models can include thermodynamic effects from the earlier simple sea ice models, ice motion, effects of leads, and some simple ice dynamics. More details on sea ice formulations are given in Chapter 12.

For the present example of forcing from an atmospheric GCM to a sea ice model (Fig. 17.3), the ocean model provides temperature and surface current data to an ice model, and that model computes new ice distributions based on the ocean and atmospheric data held constant for that model day.

After this model day has been run for the ocean and sea ice models, values of SST and sea ice distribution are averaged for that day and fed back to the atmospheric model. The cycle then starts over as the atmospheric model uses those values of SST and sea ice for one model day to compute its response to that altered forcing.

This illustration uses a model communication rate of once each model day. This is somewhat arbitrary, but it is based on a common technique in global coarse-grid coupled models of the late 1980s.

17.3.4   The problem of frequency of communication

One of the most fundamental issues in coupling the atmosphere and ocean is the different time scale of response of the respective media, the atmosphere being relatively fast, the ocean relatively slow. The historical struggle with this concept in relation to computer resources and degree of sophistication is informative and will provide the context for a discussion of state-of-the-art coupled modeling today.

In the late 1960s and 1970s, computing constraints necessitated the development of elaborate asynchronous coupling schemes, especially if the inclusion of an annual cycle was to be allowed. Asynchronous coupling involves running atmosphere and ocean for disproportionate amounts of time. That is, the atmospheric model takes more computer time but responds more rapidly to changes from the ocean. Conversely, the ocean model takes much less computer time and responds much more slowly. Asynchronous coupling takes advantage of these characteristics so that the atmospheric model is run, for example, for one year and the forcing for that year is averaged in some way to include the seasonal cycle (e.g., seasonal averages) and then the ocean is run with that forcing for, say, 10 years or more to allow it to respond to that one year of atmospheric forcing.

The first published description of an experiment with coupling an atmospheric GCM to an ocean GCM emerged from the Geophysical Fluid Dynamics Laboratory (GFDL) in the late 1960s (Manabe and Bryan, 1969). They set up the model with a very simple sector configuration partitioned between land and sea that used a coarse grid ($5° \times 5°$ latitude–longitude) for atmosphere and ocean. The atmospheric model had nine levels, the ocean five. The models were run asynchronously and the atmosphere was run only for one model year, while the ocean was run for 100 model years with annual mean solar forcing. The ocean and atmosphere communicated once every model day in the atmosphere. This pioneering calculation took 1,200 hours on a UNIVAC 1108, a huge block of computer time even by today's standards. In comparison, present coupled ocean–atmosphere GCMs can be run synchronously (atmosphere and ocean communicate at least once each model day) for over 100 model years for the same amount of time on a modern supercomputer.

Even with the simple land–sea configuration and other limitations of that first integration at GFDL, a respectable simulation of atmosphere and ocean temperatures was produced (Fig. 17.4b). Systematic errors in this simulation are still evident in today's coarse-grid coupled GCM simulations. For example, the tropical SSTs were too low, high-latitude SSTs too high, and the tropical troposphere in the atmosphere too cold. Yet, this marked a beginning and demonstrated that two independent media – the atmosphere and the ocean – could be coupled, interact, and produce a climate not far removed from that observed.

The researchers at GFDL produced an improved coupled GCM in 1975 (Manabe et al., 1975; Bryan et al., 1975). This coupled model had a $5° \times 5°$ coarse resolution in the atmosphere and ocean, 9 levels in the atmosphere, 12 levels in the ocean, and realistic geography. It was run asynchronously, as in their earlier experiment, with annual mean solar forcing. The atmosphere was integrated for 310 model days and the ocean for 272 model years, with communication between the two every 1/4 model day in the atmosphere.

Surface temperatures from observations (Fig. 17.5a), the coupled model (Fig. 17.5b), and the atmospheric model coupled to a simple energy balance or swamp ocean (Fig. 17.5c), show typical systematic errors for

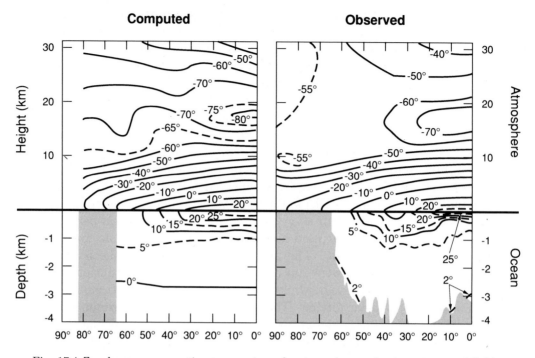

Fig. 17.4 Zonal mean cross-section temperatures for atmosphere and ocean; computed (left) and observed (right) from an early GFDL ocean–atmosphere model calculation (Manabe and Bryan, 1969).

this class of global coupled GCM. Namely, there is very little east–west temperature gradient in the tropical Pacific, temperatures are too high around Antarctica, and there is lack of a well-developed Gulf Stream associated with temperatures that are too low in the North Atlantic (Fig. 17.5b). Yet, the improvements in temperature simulation with a dynamical ocean are evident over a swamp ocean with no ocean currents (Fig. 17.5c). The swamp ocean has even lower temperatures in the North Pacific and North Atlantic, partly a consequence of lack of ocean heat transport and, perhaps, lack of a solar seasonal cycle. There is no hint of a tropical eastern Pacific cold tongue present in the dynamical model.

GFDL built on previous work and used the 1975 atmosphere-and-ocean-model configuration to attempt a seasonal-cycle integration through an atmospheric and oceanic data library (Manabe et al., 1979). As the atmosphere–ocean model was run asynchronously, the data library was updated and the new boundary values passed to the appropriate model (Fig. 17.6a). In this manner, the atmospheric model was run for 4.2 model years and the ocean for 1,200 model years.

About this time, the National Center for Atmospheric Research (NCAR) published its first results from a coupled GCM (Washington et al., 1980). The NCAR model was run asynchronously with a seasonal cycle by

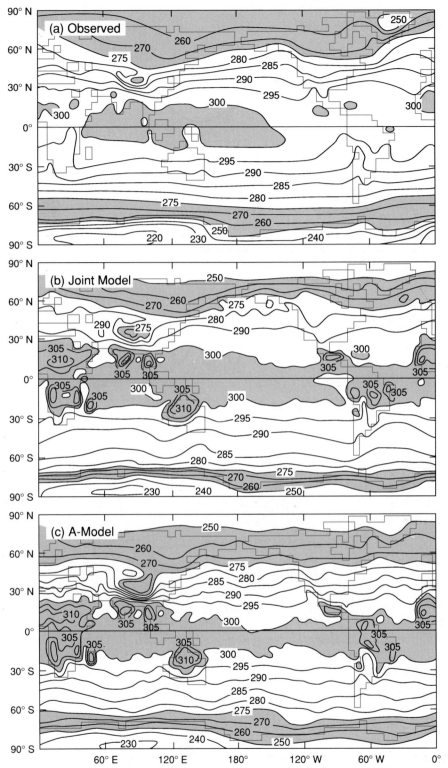

Fig. 17.5 Annual mean surface temperature (K) for: (a) observations; (b) coupled ocean–atmosphere model; and (c) atmospheric model coupled to a swamp-ocean formulation (after Manabe et al., 1975).

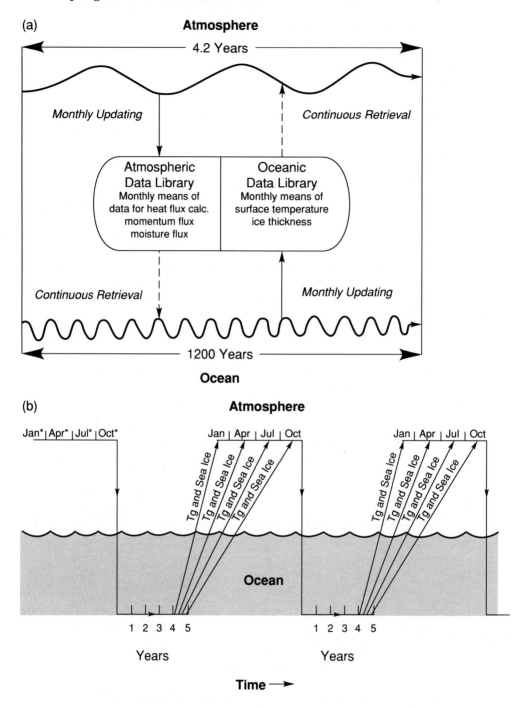

Fig. 17.6 Asynchronous coupling schemes for seasonal-cycle integrations for (a) GFDL coupled model (Manabe et al., 1979); and (b) NCAR coupled model (Washington et al., 1980).

integrating four months from the atmospheric model – January, April, July, and October. The first two harmonics were fitted to the forcing quantities from the atmospheric model (net heat flux, wind stress, and net freshwater flux) and were used to run the ocean model with a full seasonal cycle for five model years. Then the values of SST and sea ice for the four months needed by the atmospheric model were saved from the last year of the five-year ocean model integration and the atmospheric model was run again for four months, and so on for four cycles (Fig. 17.6b).

Inasmuch as the NCAR and GFDL coarse-grid models were similar in many ways and, in fact, used ocean models with common origins, it was not surprising that their climate simulations were similar. The coupled models did not run away to a totally foreign climate state, and they simulated many first-order aspects of the observed system. The systematic errors were similar – tropical oceans tended to be too cold (owing in part to coarse vertical resolution and equatorial upwelling that is too strong), the high-latitude southern oceans too warm (owing in part to coarse horizontal resolution requiring too strong heat diffusion to keep the model stable), and the North Atlantic too cold associated with an underdeveloped Gulf Stream (Fig. 17.7). Other errors in the atmospheric models were present arising in part from shortcomings in the convection and cloud schemes.

These cases illustrate the historical issue of time scale of response and asynchronous versus synchronous coupling. As global coupled modeling moved into the 1980s, problems with asynchronous coupling and deleterious effects on climate sensitivity (Dickinson, 1981) prompted efforts to have the component models communicate at least once each model day (synchronous coupling) during sensitivity experiments. This is now the minimum frequency of communication for synchronous coupling. Asynchronous methods are still used in the preceding spin-up phase of model coupling.

Communication frequency of once each model day is straightforward if the atmospheric model does not have a diurnal cycle. If a diurnal cycle is included, more frequent communication between atmosphere and ocean, on the order of once each model hour, could be required (Gates et al., 1985; Schlesinger et al., 1985). This increases the computational expenditure because the inclusion of a diurnal cycle in the atmospheric model requires that the full radiation subroutines be called much more frequently (Chapter 10).

For this reason, most late-1980s global coupled models did not include a diurnal cycle – but not because the diurnal cycle was unimportant. In fact, a diurnal cycle can significantly affect simulated climate (Wilson and Mitchell, 1986). However, longer runs to study basic climate sensitivity are possible without a diurnal cycle, and most first-order aspects of climate are simulated without this cycle. As coupled modeling moves into the 1990s, more models are being run with a diurnal cycle and more frequent communication between atmosphere and ocean. As always, the tradeoff

(a) GFDL

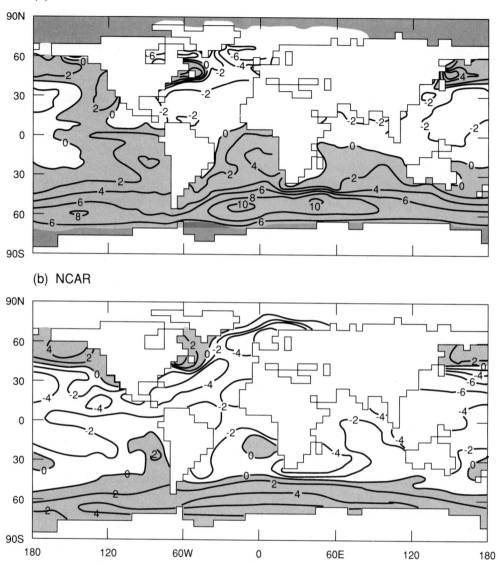

(b) NCAR

Fig. 17.7 Annual mean SST differences, computed minus observed (°C) for: (a) GFDL coupled model (Manabe et al., 1979); and (b) NCAR coupled model (Washington et al., 1980).

between realistic detail and computational efficiency to allow long runs is the central element of climate-modeling strategy.

17.3.5 Spin-up problem

As the time scales of the response between ocean and atmosphere are central to global coupled models, so too is this concept important to

570

Spin-Up Methodology

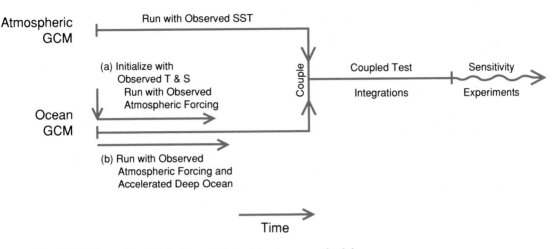

Fig. 17.8 Schematic of global coupled model spin-up methodology.

the spin-up of each component model. Spin-up is more of a factor, in computational expense, for the slower-responding ocean GCM. For example, the response time of the upper ocean is on the order of 10 years or less, the deep Atlantic Ocean about 100 years, and the deep Pacific about 1,000 years.

There are two basic spin-up strategies for the ocean (Fig. 17.8). The first involves accelerating the time step of the lower ocean so that, in effect, the deep ocean runs equivalently longer than the upper ocean (Bryan et al., 1975; Bryan and Lewis, 1979). This so-called asynchronous spin-up of the ocean (Fig. 17.8b) has the advantage that the entire ocean can be brought into a fairly stable equilibrium (Manabe et al., 1991). The disadvantage is that, due to the different time scales, mixing between upper and lower ocean can prolong an anomalous transient forcing from the upper ocean into a long-lived response in the lower ocean as it is accelerated.

The second spin-up strategy uses some kind of initialization technique (Fig. 17.8a) to force at the outset the three-dimensional structure of the ocean into agreement with observed data. For example, either two-dimensional (latitude–depth) or three-dimensional observed ocean temperatures and salinities can start the ocean model and force the model variables to be close to the observed variables (e.g., Washington and Meehl, 1989). One of the advantages of this technique is that the ocean model starts in a state that approximates the observed ocean and avoids some of the possible systematic errors that may be introduced due to time-scale problems associated with the accelerated procedure described above. The presumption is that, for most of the depth of the ocean, the time scales of the drift away from initial values will be on the order of 100 years or longer. For climate problems that can be studied on time scales of 100 years or less, therefore, such slow drifts in the ocean are much smaller than the signals being studied in the coupled model. The disadvantage is that since the model ocean will want to drift slowly towards its own climate, the drift, no matter how small, represents imbalances in the model system and reflects model deficiencies that may affect the results of the coupled simulation. Additionally, evidence from the paleoclimatic record (see Chapter 21) indicates that the observed ocean–atmosphere system is probably not in equilibrium. Therefore, ocean observations, incomplete as they are, can only be thought of as a snapshot of the ocean as it evolves on much longer time scales.

17.4 Climate drift in the coupled simulation

Each imperfect component model has its own set of systematic errors. Systematic errors are defined as stable errors in the model simulation that result from model deficiencies in the component models alone, additive errors from the component models after they are coupled, or errors that are produced as the result of the coupled interactions between imperfect component models. Sometimes systematic errors defined in this way are referred to as climate drift. In the present context, climate drift will not apply to the rapid evolution of the coupled model toward relatively stable systematic error patterns on the order of several years or less but, rather, to the slow drift of the coupled system on longer time scales (order 100 years) from an initialized state in the ocean or from small imbalances between model components. If the accelerated spin-up procedure has been used, by definition there is little climate drift in the system. But then there is the possibility of greater systematic error in the ocean-component model.

Because of these systematic errors, a philosophical decision faces coupled modelers. On the one hand, the modeler can decide that the systematic errors, while serious in terms of the control integration, are instructive, convey accurately the deficiencies of the coupled model, and provide an informative context within which to interpret results from sensitivity experiments (Washington and Meehl, 1989). The advantage of this approach is that the coupled model is internally consistent, nothing is being masked or hidden concerning model deficiencies, and sensitivity experiments can be interpreted directly based on the coupled model response. The disadvantage is that the systematic errors in the control integration may unrealistically affect the results of the sensitivity experiments.

On the other hand, the modeler can decide that the systematic errors present too big a bias in the control run and affect the results of sensitivity experiments to an unacceptable degree. The alternative is to somehow "fix" the errors in the control run to provide a more realistic basic state for sensitivity experiments. Such a technique has been devised and is variously called "flux correction" (Sausen et al., 1988) or "flux adjustment" (Manabe et al., 1991). The advantage is that the control simulation is corrected to look more like the observed system, and this improved basic state may provide better climate-sensitivity estimates. Also, there is an implicit assumption that the sensitivity experiment is a small perturbation on the basic state. The disadvantage is that the system is altered such that the results from sensitivity experiments may not accurately convey what the coupled model would have produced without the corrections.

Any number of variables at the air–sea coupling interface can be adjusted with this method (Fig. 17.3), and there are several ways to perform flux correction. All are ad hoc and correct only values at the air–sea interface. One way of performing the flux correction for a variable such as net heat flux is to first run the ocean model with observed values of wind stress and SST. The net heat flux forcing H_o for the ocean is obtained by

$$H_o = \gamma(T_{obs} - T_{comp}) \,, \tag{17.8}$$

where γ is a so-called Haney coupling coefficient (Haney, 1971; and detailed by Meehl et al., 1982, for an ocean model); see also Sec. 11.2.3. This coefficient introduces a time lag in the correction of the SST computed by the ocean GCM (T_{comp}) to the observed SST (T_{obs}). After running the ocean model for some time, values of H_o (a measure of the net surface heat flux required to correct the computed SST to the observed value) are saved as a function of time and space.

The next step is to run the atmospheric model with observed values of SST. At the end of this integration the net heat flux from the atmospheric model [H_a, the imbalance arising from Eq. (17.1)] is also saved as a function of time and space. The heat flux correction H_c in the coupled model, then, is simply the difference

$$H_c = H_o - H_a \tag{17.9}$$

and is added to the net heat flux going from atmosphere to ocean (Fig. 17.3) at each time step as a function of time and space.

For example, net surface heat fluxes from several model configurations can show where the fully coupled dynamical GCM will have problems (Fig. 17.9, from Meehl, 1989). The coarse-grid (5° latitude by 5° longitude) ocean model, run separately and forced with observed wind stress and temperature, shows large positive (downward into the ocean) heat flux H_o values in the equatorial Indian and Pacific Oceans (Fig. 17.9b). Meanwhile, the atmospheric model run separately with observed SSTs (Fig. 17.9c) shows smaller net heat flux H_a values in the equatorial tropics closer to the observed values (Fig. 17.9a). The net large positive heat flux H_c indicates that the ocean model wants to produce colder-than-observed SSTs. That is, the surface forcing is not positive and large enough to keep the model-calculated SSTs as warm as the observed SSTs. This is evidence of a systematic error in the ocean model. In this case, the coarse vertical

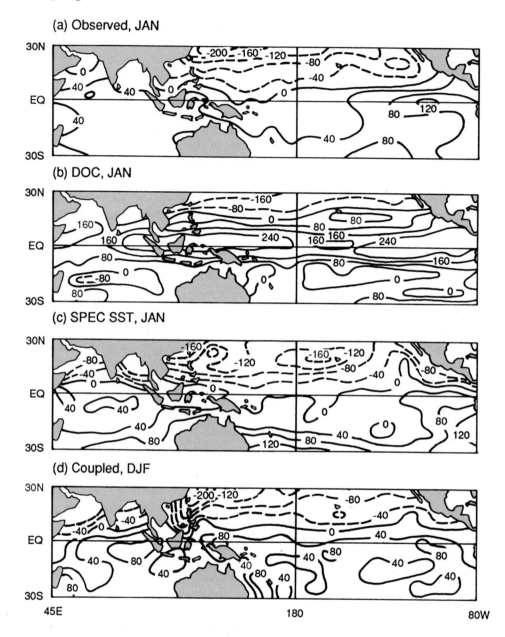

Fig. 17.9 Net heat flux at the surface (W m^{-2}; heat flux into the ocean is positive, negative values are dashed) for January: (a) observed (Esbensen and Kushnir, 1981); (b) decoupled ocean (DOC) (ocean model run with observed wind stress and surface air temperature forcing); (c) specified (SPEC) SST (atmospheric model run with observed SSTs and sea ice distributions); and (d) DJF values for coupled (atmospheric model coupled to the ocean model).

resolution (50 m surface layer, 450 m second layer) dictates that if divergence is occurring in the surface layer due to the easterly trade winds, water from the deep cold second layer is brought to the surface, producing colder-than-observed SSTs. However, in other areas, the main origin of the systematic errors may come from the atmospheric model, perhaps associated with poor simulation of stratus cloud regimes, errors in wind stress, etc.

After each model component is run separately for some time, they are coupled and the entire model is usually run for some years as the different components come into adjustment with each other (Fig. 17.8). Because the atmospheric model shows less systematic error of net surface heat flux in the equatorial tropics than the ocean model with its large cooling tendency there, the coupled simulation could be expected to be dominated by the systematic errors in the ocean model. This is the case in the coupled simulation. There are greater-than-observed positive heat flux values in the equatorial tropics (Fig. 17.9d) and colder-than-observed SSTs (not shown). The alternative would be to apply H_c as the flux correction to add heat flux to that coming from the atmosphere in the coupled model and heat the equatorial SSTs to correct for the systematic cold error there in the ocean model.

In an example of flux correction for other variables, the atmospheric model is first run with observed values of SST as before. Values of wind stress and precipitation minus evaporation ($P - E$ or freshwater flux) calculated by the atmospheric model are differenced from observed values of wind stress and equivalent freshwater fluxes derived from observed salinity as a function of time and space. The residuals are saved as the correction terms for the atmospheric model. When the models are coupled, the correction terms are added to wind stress and $P - E$ values from the atmospheric model at each time step. This effectively removes a large part of the systematic error from the atmospheric model, and the coupled simulation is presumably closer to observed values.

Flux-correction methods are designed to remove most of the tendency for the coupled model to "drift" toward its own climate replete with systematic errors. Yet, *the correction terms are only added at each time step*, and the coupled model can still exhibit slow drift (Cubasch, 1989). There is also the problem of changing surface conditions. For example, if all the sea ice at a gridpoint melts, the flux correction for the newly exposed ocean point must somehow be dealt with.

To demonstrate how flux correction can work in a coupled model, a preliminary simulation from a coupled model without freshwater-flux correction shows unrealistically low values of salinity in the North Atlantic compared with the observed (Fig. 17.10a,b). After a freshwater-flux correction is computed and the model has been spun back up with observed values of salinity, however, the freshwater-flux adjustment is able to maintain higher values of salinity in the North Atlantic (Fig. 17.10c), as observed. But

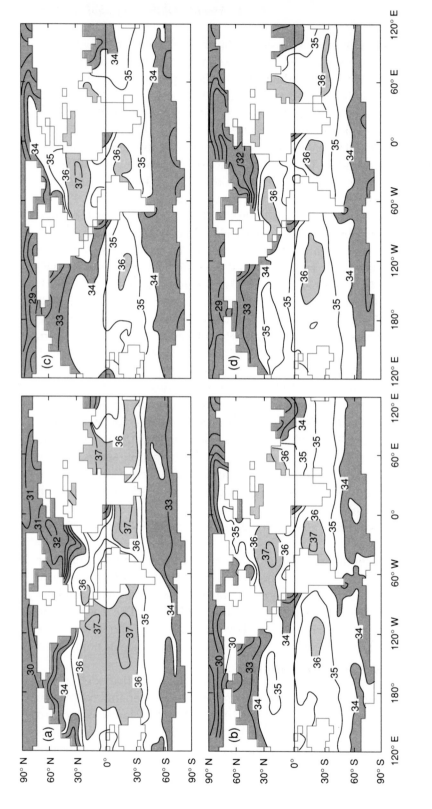

Fig. 17.10 Salinity distributions for: (a) preliminary experiment from coupled ocean–atmosphere model without freshwater-flux correction; (b) observations; (c) experiment with freshwater-flux correction added after coupled model spun back up with observed salinity distribution; and (d) experiment with freshwater-flux correction simply added to preliminary experiment in (a) (Manabe and Stouffer, 1988).

simply applying the freshwater-flux adjustment to the preliminary experiment (Fig. 17.10a) does not return the salinity values to the observed state (Fig. 17.10d). This result indicates that the coupled model used in that experiment has two stable climate states, one with a strong thermohaline circulation and one without. This issue will be discussed further in Sec. 17.5.

Systematic errors in the present generation of coupled models are not insignificant and the issue of how to handle them is controversial. If the flux-correction technique is used, often the correction terms are sizable (Latif et al., 1988), sometimes reaching factors of five or more times the values being corrected (Manabe et al., 1991). On the other hand, simulations without any flux correction suffer from considerable systematic errors in the coupled simulation (Washington and Meehl, 1989). For example, SSTs in some areas of the tropics in that simulation are too low by about 4°C.

The problem of systematic errors in model simulations is not new. The numerical weather prediction modeling community has faced it for over 20 years. Pressure to make more accurate short-term weather forecasts has often prompted schemes to correct systematic errors in weather forecast models. The usual approach by the numerical weather forecasting community is to not spend time and resources developing better and more elaborate flux-correction techniques, but rather to devote the majority of resources to improving the component models to eliminate as much as possible the sources of the systematic errors. By the same means, the present flux-correction dilemma will become less of an issue in coupled modeling as the component-model simulations are improved and have fewer systematic errors.

17.5 Products of coupled interaction

In spite of the difficulties with systematic errors, a number of useful results have been produced with the present generation of global coarse-grid coupled ocean–atmosphere GCMs. This section will not dwell on many of the details of the results from coupled models (these are described more thoroughly in Chapters 18 and 19), but a few unique results that are a product of coupled interaction in present-day global coupled models are mentioned here.

17.5.1 El Niño–Southern Oscillation

The limited-domain coupled models of ENSO discussed in Chapter 18 are explicitly designed to simulate ENSO phenomena. But most global coupled GCMs were conceived to study long-term climate-sensitivity problems, often involving increased CO_2 or trace gases. These models are coarse-grid and meant for multidecadal integrations. It comes as somewhat of a surprise, then, when certain aspects of ENSO phenomena appear in

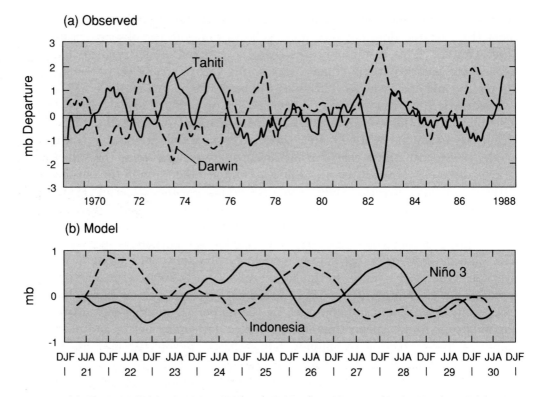

Fig. 17.11 (a) Observed SLP anomalies at Tahiti (solid line) and Darwin (dashed line); and (b) area-averaged SLP anomalies for the NINO3 area in the eastern Pacific (90°W to 150°W, 10°N to 10°S) and an area over Indonesia in the western Pacific (110°E to 155°W, 10°N to 13°S) (Meehl, 1990b).

some of the simulations. Correlations of sea level pressure (SLP) at a model gridpoint near Darwin, Australia, with all other gridpoints from a 23-year integration with one of these global coupled GCMs resemble the observed Southern Oscillation (Sperber et al., 1987).

This oscillation of SLP anomalies between eastern and western Pacific in another of these coupled models is also similar to that observed (Fig. 17.11, Meehl, 1990b). Patterns of coupled anomalies of SST and SLP are global in the coupled model as they are in the observations (Plate 15). Differences of composite warm and cold events in that model show a similar seasonal timing of SST anomalies to observed events, with the establishment of warm anomalies in the eastern Pacific in northern spring and slow movement of the anomalies westward. Associated with the SST anomalies are westerly wind-stress anomalies just to the west of the SST anomalies in both the computed and observed quantities in the eastern Pacific. The observed composites of SST and wind anomalies also reveal signals of positive SST anomalies and westerly wind-stress anomalies moving from the western Pacific eastward that are not represented in the coupled model. This suggests that the various flaws and limitations of the coupled model in the western Pacific allow only a subset of the interactive processes involved with ENSO to appear in the present generation of coarse-grid coupled models (Philander et al., 1989; Meehl, 1990b).

17.5.2 Low-frequency variability and the thermohaline circulation

With the introduction of interactive dynamical oceans, coupled GCMs have shown that some features of low-frequency variability (decadal and longer time scales) are simulated in certain respects. Some of this variability may be the product of the alteration of the thermohaline circulation in the ocean by the ocean–atmosphere coupling. The importance of the thermohaline circulation in global ocean poleward heat transport is discussed in Chapter 4, Sec. 4.7. A more detailed discussion of low-frequency variability appears in Chapter 19. Here, some general features of low-frequency variability and thermohaline circulation in global coupled GCMs are discussed.

Some coupled model simulations have indicated that changes of forcing from the surface of the ocean in a coupled GCM can alter or even break down the thermohaline circulation (Bryan and Spelman, 1985; F. Bryan, 1986; see Chapter 11, Sec. 11.4.2). The latter phenomenon has sometimes been referred to as the "thermohaline catastrophe." The thermohaline circulation, driven by density differentials in the ocean, acts as a giant conveyor belt (Gordon, 1986; Broecker, 1987, Fig. 17.12) transporting heat and salt great distances (see also Plate 14) and, it is thought, modulating global climate. In the Atlantic, the thermohaline circulation is manifested as warm saline water transported by the Gulf Stream and North Atlantic current systems to high latitudes in the North Atlantic in the surface layers. There, the saline water cools, becomes denser, sinks, and returns southward at depth to form a large-scale meridional cell in the ocean [Fig. 17.13(a) from Bryan and Spelman, 1985]. If surface forcing is altered, for example, from inherent variability in the atmosphere or by warming and freshening the surface ocean layer from increased CO_2 in the atmosphere (i.e., the atmosphere is warm and there is more precipitation, especially in winter), the water at high latitudes is no longer cold or saline enough to sink and the entire thermohaline circulation can be weakened or even break down entirely [Fig. 17.13(b) from Bryan and Spelman, 1985].

Consequently, there is some evidence that the coupled climate system may be able to exist in more than one stable state (i.e., there may be multiple equilibria), and the thermohaline circulation is an essential component of this phenomenon (see Figs. 17.10 and 21.6, Manabe and Stouffer, 1988). This has major implications on long time scales in the climate system and may be a factor in the changes from glacial to interglacial climates

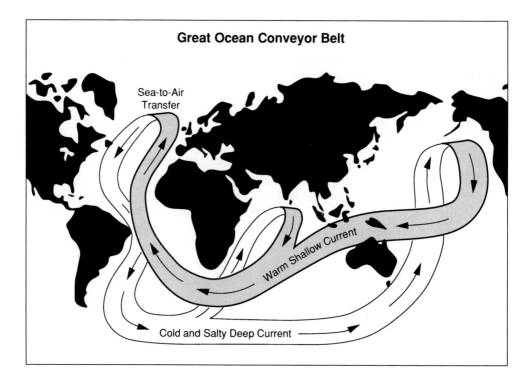

Fig. 17.12 Schematic diagram of the global "conveyor belt" depicting global thermohaline circulation (after Broecker, 1987).

(Chapter 21, Sec. 21.2) The implication of this model experiment is that the coupled system can exist stably both with and without the thermohaline circulation with significantly altered atmospheric and oceanic characteristics in the two states.

Additional studies of the consequences of changes of the thermohaline circulation in coupled models are just beginning. Certainly, since global coupled GCMs can simulate some aspects of this circulation, atmospheric forcing and response to its changes can be analyzed. The so-called "conveyor belt" (Fig. 17.12) that connects the thermohaline circulation in the Atlantic to the global oceans has been simulated in some respects with a high-resolution global ocean model with observed forcing (Semtner and Chervin, 1992). As computing capability expands over the next 5 to 10 years, such high-resolution ocean models will be coupled to atmospheric GCMs to provide even better tools to study the effects of an altered thermohaline circulation on low-frequency variability, global climate sensitivity, and climate change.

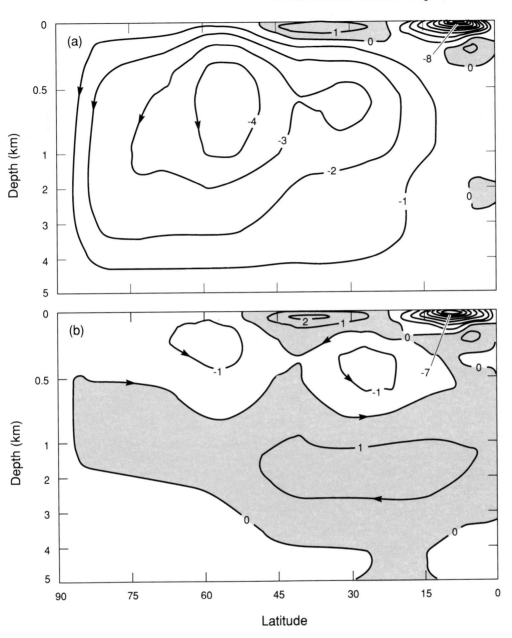

Fig. 17.13 Mean meridional circulation (10^{12} g s^{-1}) in the ocean from a sector ocean–atmospheric coupled GCM; (a) equilibrium solution for control case; and (b) average circulation for years 21–30 after atmospheric CO_2 is instantaneously quadrupled showing the breakdown of the thermohaline circulation (after Bryan and Spelman, 1985).

18

Tropical Pacific ENSO models: ENSO as a mode of the coupled system

Mark A. Cane

18.1 Introduction

The most dramatic, most energetic, and best defined pattern of interannual variability is the global set of climatic anomalies referred to as ENSO, an acronym derived from its oceanographic component, El Niño, and its atmospheric component, the Southern Oscillation. Occurring irregularly, but about every four years on average, ENSO warm events directly affect the climate of more than half the planet, often exacting a heavy toll in human life and economic well-being.

The devastating consequences of the 1982/83 ENSO warm event, the most extreme in at least a century, are graphically illustrated in Canby (1984). In Australia the worst drought ever recorded spawned firestorms that incinerated whole towns. The same drought conditions resulted in the burning of millions of acres of rainforest in Borneo. There were also severe droughts in the Nordeste region of Brazil, in the Sahel, and in Southern Africa. Normally arid regions of Peru and Ecuador were inundated by more than 3 m of rain, causing great loss of life and destruction of the transportation infrastructure. Drastic changes in the tropical Pacific Ocean resulted in mass mortality of fish and bird life. All in all, US $8 billion in damage and the loss of 2,000 lives have been attributed to the 1982/83 ENSO event.

While the consequences of the warm events have long been appreciated (some historical remarks are given below), awareness of the importance of the cold extremes of the ENSO cycle has developed quite recently. The linking of the cold event of 1988 to the drought in North America (Trenberth et al., 1988), was but one of the global consequences (cf. Ropelewski and Halpert, 1987).

El Niño historically refers to a massive warming of the coastal waters off Peru and Ecuador. It is accompanied by torrential rainfall, often resulting in catastrophic flooding. Widespread mortality of fish and guano birds further damages the local economy. El Niño events have been documented back to 1726 (Quinn and Neal, 1978) and there is other evidence indicating occurrences for at least a millennium prior to that (Quinn et al., 1987).

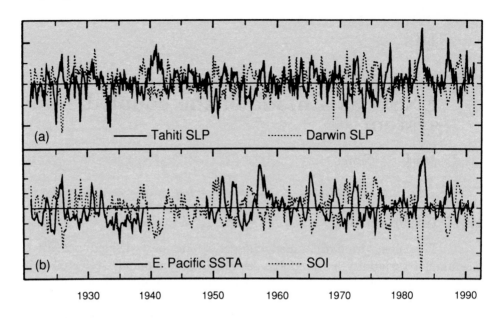

Fig. 18.1 (a) Sea level pressure (SLP) at Darwin, Australia (dashed) and Tahiti (solid). Each curve is normalized by its standard deviation. Note the signature of the Southern Oscillation: when SLP is high in the western Pacific (Darwin) it tends to be low in the east (Tahiti) and vice versa. (b) The dashed line is a Southern Oscillation Index (SOI): the normalized difference of sea level pressure between Darwin, and Tahiti. The solid line (E. Pacific SSTA) is an index of El Niño: sea surface temperature anomalies along the west coast of South America (ship track 6 in Rasmusson and Carpenter, 1982).

The atmospheric component of ENSO, the Southern Oscillation, is a more recent discovery. Although the term is sometimes used to refer to the global complex of climatic variations, the Southern Oscillation is specifically an oscillation in surface pressure (and thus atmospheric mass) between the southeastern tropical Pacific (the locus of the South Pacific high) and Australian–Indonesian regions (the Indonesian low). See Fig. 18.1(a).

A convenient Southern Oscillation Index (SOI) is the pressure difference between Tahiti and Darwin. A time series of the SOI is shown in Fig. 18.1(b). Also shown is a time series of sea surface temperature (SST) along a ship track off the coast of South America. The very strong connection between the two is immediately apparent: when the waters of the eastern tropical Pacific are abnormally warm (an El Niño event) sea level pressure drops in the eastern Pacific and rises in the west (low SOI). The reduction in the east–west pressure gradient is accompanied by a weakening of the usually reliable low-latitude easterly trades.

The ocean warming is not confined to the coast. At its peak, it covers a band from 10°S to 10°N extending more than a quarter of the way around the Earth (Fig. 18.2). Typically, early signs of the warming appear late in

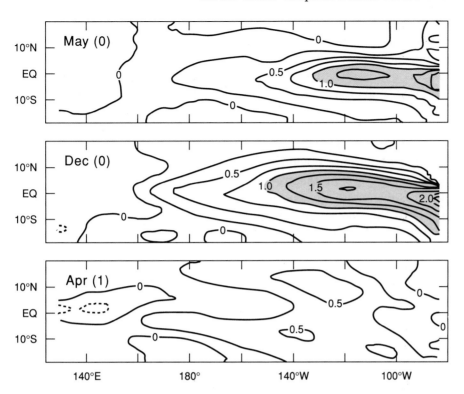

Fig. 18.2 Sea surface temperature anomalies (°C) for a composite El Niño (Rasmusson and Carpenter, 1982), constructed by averaging over 6 events (1951, 1953, 1957, 1965, 1969, 1972; cf. Fig. 18.1). Shown are maps for May and December of the El Niño year and April of the following year.

the boreal spring or summer, building to a peak at the end of the calendar year. By the following summer the warm event is usually over. This massive alteration in its lower thermal boundary condition engenders a substantial change in atmospheric heating, with global consequences of the sort noted above. The strongest and most reliable effects are in the tropics (e.g., a poor Indian Monsoon), especially the tropical Pacific, but there are also consistent influences in midlatitudes, referred to as teleconnections (Glantz et al., 1991).

ENSO has a number of special relations to the issue of greenhouse warming motivating so much of the current interest in the climate system. The ultimate realizations of all climate fluctuations, even those, like the march of the seasons – or greenhouse warming – which are externally forced, depend on complex internal feedbacks within the climate system. ENSO is the premier example of variability stemming entirely from the internal workings of the climate system. Generated by the interaction of ocean and atmosphere, it embodies physics essential to longer-term climate variability.

Thus, attempts to model and predict ENSO test our ability to model and predict longer time-scale climate variations, and do it in a context with more observational data to verify against. Many of the issues in climate modeling, such as the difficulties in coupling complex ocean and atmosphere models, arise in ENSO simulations and are already being addressed. Beyond that, ENSO is an important aspect of the climate system which must be included in any comprehensive climate simulation: the absence of ENSO would alter even the mean climate. Conversely, if the mean climate changes, it will be important to predict how the interannual variability such as ENSO will change. There are obviously important human consequences from an increase or decrease in the frequency or intensity of drought and flood. Finally, ENSO enters the controversy about detection of the greenhouse signal above the "noise" of the natural variability in the climate system. Accounting for ENSO reduces the noise, thus making the signal more detectable (e.g., Jones, 1988). Moreover, ENSO would likely influence the shape of the "fingerprint" of greenhouse warming.

The ENSO modeling experience offers valuable lessons for more comprehensive climate system modeling efforts. It is arguably the most advanced example of modeling two-way interactions between the atmosphere and ocean. It is one of the few examples of successful prediction of climate variations by objective means, and the only verified example where the prediction has been carried out with a dynamical model, now the common method in weather prediction.

ENSO modeling began from a base of useful hypotheses as to the mechanisms governing ENSO, hypotheses developed from analyses of data too meager to be conclusive. This background and our subsequent picture of the workings of ENSO are reviewed in the next section. A variety of models followed (see Sec. 18.4), including relatively simple models providing important analytical tools, and also "intermediate" models, which, while less complex than the comprehensive GCMs, achieved recognizable simulations of ENSO. Their results validated some of the earlier ideas and allowed further theoretical advances (Sec. 18.3). They also demonstrated some predictive skill (Sec. 18.5), which is both valuable in its own right and also increases confidence in their general correctness. Section 18.6 considers efforts along the more difficult path of coupling GCMs. A final section discusses implications of the ENSO modeling experience.

This chapter is organized with some attention to the history of ENSO research in the belief that this best reveals its lessons for this next stage in climate modeling. The ENSO literature is vast and the account here is necessarily sketchy. Good brief reviews, though already somewhat dated, are Rasmusson (1985) and Cane (1986). A more recent general review is Enfield (1989). McCreary and Anderson (1991) survey the myriad coupled models which have been used to investigate ENSO physics. A thorough treatment with emphasis on the physics of ENSO is the recent monograph by Philander (1990). The collection edited by Glantz et al. (1991) provides a more succinct account of relevant physics amid its emphasis on the global impacts of ENSO.

18.2 Our understanding of the mechanisms of ENSO: Pt I

The seminal figure in delineating the Southern Oscillation and showing its worldwide associations was Sir Gilbert Walker, Director General of the Observatory in India. Walker assumed his post in 1904, shortly after the famine resulting from the monsoon failure in 1899 (an El Niño year); his goal was to predict the monsoon fluctuations, an activity begun by his predecessors after the disastrous monsoon of 1877 (also an El Niño year).

Walker was already aware of work dating from just before the turn of the century which described the sea level pressure swing between South America and India–Australia. Over the next three decades he added correlates from all over the globe to this primary SO signal: rainfall in India and in the central equatorial Pacific; temperatures in southeastern Africa, southwestern Canada, the southeastern United States; etc. This work, being purely empirical and based on records too short to establish its reliability, was long regarded with considerable skepticism. However, in recent years Walker's correlations have been found to hold up when examined with more than 50 years of new data (see especially Horel and Wallace, 1981, and Ropelewski and Halpert, 1987).

Oddly, Walker missed the connection to El Niño. It only became widely appreciated after the revival of interest in the SO in the 1960s, principally through the work of Jacob Bjerknes and others (Rasmusson, 1985). In a series of papers (see especially Bjerknes, 1969), Bjerknes did more than point out the empirical relation between the two; he also proposed an explanation which requires a two-way coupling between the atmosphere and ocean in the tropical Pacific. It is fair to say that the Bjerknes hypothesis underlies all subsequent progress in understanding and modeling ENSO, though successful models of ENSO would not be possible without the rapid enhancement in our understanding of the tropical oceans and atmosphere since his time. For example, the model of Zebiak and Cane (1987), discussed below, was a conscious attempt to "model" the Bjerknes hypothesis which exploits these advances to translate his ideas into suitable equations.

Bjerknes' ideas developed from observations of large-scale anomalies in the atmosphere and tropical Pacific Ocean during 1957–58, the International Geophysical Year. A major El Niño occurred in those years, bringing with it all the atmospheric changes connected to a low SOI. It is implausible that a warming confined to coastal waters off South America could cause global changes in the atmosphere, but the 1957 data showed that the rise in SST extended along the equator out to the dateline (cf. Fig. 18.2). The essential elements of the hypothesis are depicted in Plate 16.

Bjerknes began by recognizing a striking fact about the normal state of the tropical Pacific (see Fig. 4.1 for mean SST maps): the SSTs at the eastern end are remarkably cold for such low latitudes (cf. especially Bjerknes, 1969, from which we quote freely). Since the western Pacific is very warm, there is a large SST gradient along the equator in the Pacific. As a result, there is a direct thermal circulation in the atmosphere along the equator: the relatively cold, dry air above the cold waters of the eastern equatorial Pacific flows westward along the surface toward the warm west Pacific. "There, after having been heated and supplied with moisture from the warm waters, the equatorial air can take part in large-scale, moist-adiabatic ascent." Some of the ascending air joins the poleward flow at upper levels associated with the Hadley circulation, and some returns to the east to sink over the eastern Pacific. There is a zonal surface pressure gradient associated with this equatorial circulation cell, high in the east and low in the west. Bjerknes named this the "Walker Circulation", because he felt that fluctuations in this circulation initiated pulses in Walker's Southern Oscillation. It can have such global consequences because "it operates a large tapping of potential energy by combining the large-scale rise of moist air and descent of colder dry air."

Bjerknes also pointed out that even while the surface winds are being driven westward along the equator by the zonal SST gradient, they are acting to create the cold ocean temperatures in the east responsible for that gradient. In a search for the mechanisms responsible, differences in surface heating can quickly be ruled out: while heat flux estimates are quite uncertain, there is no doubt that more heat is going into the equatorial Pacific at the east than at the west. The causes of the unusually cold SSTs are to be found in three features of wind-driven ocean dynamics.

(i) *Horizontal advection.* The easterly winds drive westward currents along the equator, advecting the cold waters from the South American coast.

(ii) *Equatorial upwelling.* The Coriolis force turns ocean currents to the right in the Northern Hemisphere and to the left in the Southern Hemisphere. Consequently, the surface flow at the equator is deflected poleward, and the poleward flow must be fed by waters which upwell along the equator, waters that are colder than the surface; see Chapter 4, Sec. 4.4.

(iii) *Upward thermocline displacement.* The tropical ocean can usefully be viewed as a two-layer fluid, consisting of a shallow, warm ocean layer above the layer of cold abyssal waters. In the real ocean the two are separated by the thermocline, a narrow (50–100 m) region of strong temperature change (10°C or more); see Fig. 4.10 for the seasonal thermocline development in midlatitudes. The easterlies along the equator (cf. Fig. 4.5) push the waters of the warm upper layer to the west and polewards, pulling the thermocline to the surface in the east.

As a result, the water upwelled there is colder than it would be if the upper layer waters were more evenly distributed with longitude.

The limitations of ocean theory and observation in his time made it impossible for Bjerknes to decide which of these three factors is most important (cf. Bjerknes, 1969). Even today there is considerable uncertainty as to their relative roles, although it seems to be the case that none are negligible (e.g., Seager et al., 1988).

Thus, the oceanic and atmospheric circulations over the tropical Pacific are mutually maintained by what Bjerknes (1969) referred to as a "chain reaction"; "an intensifying Walker Circulation also provides for an increase of the east–west temperature contrast that is the cause of the Walker Circulation in the first place." He also noted that the interaction could operate in the opposite sense: a decrease of the equatorial easterlies diminishes the supply of cold waters to the eastern equatorial Pacific (by any of the three mechanisms); the lessened east–west temperature contrast causes the Walker Circulation to slow down.

Bjerknes thus provided an explanation for the association of the low phase of the Southern Oscillation with El Niño as well as the association of the high phase with the normal cold state of the eastern Pacific. In each phase a positive feedback operates; in other words, an instability of the coupled ocean–atmosphere system (Philander et al., 1984). However, Bjerknes could not account for the turnabout from one state to the other. An explanation of the oscillation had to await two decades of research.

Wyrtki (1975, 1979) seized on the point that during El Niño the ocean response is dynamical rather than thermodynamic (i.e., due to variations in surface heat flux). He shifted attention from SST to sea level. SST variations are readily apparent only in the eastern part of the ocean, and, as noted above, even after one recognizes that SST changes are dynamically caused, it is a far more complex response than sea level and therefore more difficult to decipher. By collecting and charting sea level data, Wyrtki was able to show that the oceanic changes during El Niño are basinwide. He also showed that the initial changes in the wind were in the central and western Pacific, far from the locale of the SST changes. Finally, he suggested that the signal could propagate eastward from the area of the wind change to the South American coast through the equatorial wave guide in the form of equatorial Kelvin waves (see below). These ideas were amplified by a number of investigators and verified in a set of numerical experiments (Busalacchi and O'Brien, 1981; Busalacchi et al., 1983). In these experiments, a landmark application of numerical modeling to ENSO research, a linear shallow water model was driven by nearly two decades of monthly surface wind-stress fields. The wind-forced model thermocline anomalies showed a significant correlation with sea level observations.

Two additional studies from the early 1980s stand out as foundation stones for ENSO modeling. Gill (1980) showed that a single vertical mode linear model could capture the major features of the tropical atmosphere's response to the anomalous heating associated with variations in tropical SSTs. Rasmusson and Carpenter (1982) synthesized the incomplete and often perplexing observational fragments into a coherent picture of the evolution of the "canonical" ENSO warm event. Though undoubtedly a simplification, this invaluable distillation provided the first specific target for models to emulate.

18.3 Our understanding of the mechanisms of ENSO: Pt II

The Bjerknes–Wyrtki theory did not explain the perpetual turnabout from warm to cold states and back again. The explanation turns on some special features of the equatorial ocean dynamics that govern the variations in the upper tropical ocean over time scales of a few years or so. In this context, it is adequate to regard the ocean as a two-layer system: an active upper layer separated by the thermocline from a deep abyssal layer of greater density. Motions in the lower layer are quite slow, and may be disregarded relative to the active layer above. Making the useful assumption that linear physics govern the evolution of vertically integrated characteristics of this upper layer such as thermocline displacements and upper layer transports, the governing equations reduce to the shallow water equations of tidal theory.

The total response may be analyzed into a sum of free and forced waves. For the time and space scales relevant to El Niño, only two types of wave motions matter: long Rossby waves and equatorial Kelvin waves. The latter are strongly trapped to the equator and owe their existence to the vanishing of the Coriolis force there. Except within a few degrees of the equator, the geostrophic balance between Coriolis and pressure gradient forces dominates the dynamics of the ocean. This balance is characteristic of Rossby waves and strongly constrains their propagation speeds and their amplitude in response to wind driving. The Kelvin wave propagates eastward and is the fastest of the low-frequency ocean motions. It can cross the Pacific in less than three months whereas the fastest Rossby wave is three times slower.

Long Rossby waves propagate energy westward, while Kelvin waves travel to the east. When a Kelvin wave hits the eastern boundary, its reflection is made up of an infinite sum of Rossby waves, which collectively act to extend the equatorial wave guide up and down the eastern boundary to high latitudes. However, the faster Rossby waves at low latitudes carry much of the mass and energy brought east by the Kelvin waves back toward the west. The reflection process at the west is more efficient: all of the mass flux which the Rossby waves carry into the boundary is collected by boundary currents and brought equatorward, where it is returned eastward in the form of Kelvin waves.

The particular properties of equatorial waves mentioned above are essential to the El Niño phenomenon. Only at low latitudes can low-frequency waves cross the ocean in times matched to the seasonal variations in the winds. A given wind change generates a stronger response at the equator than at higher latitudes, and equatorial waves are less susceptible to the destructive influences of friction and mean currents. Finally, as will be seen, the asymmetries in the waves and their reflections are essential to the ENSO cycle. While the explanation for the perpetual oscillations of the ENSO cycle is inherent in the physics just described, it emerged only after the development of the numerical models discussed in Sec. 18.4.

An early version (Cane and Zebiak, 1985; Cane et al, 1986) emphasized the recharging of the equatorial "reservoir" of warm water as a necessary precondition for the initiation of a warm event. On the basis of his analysis of sea level data, Wyrtki (1985) developed a very similar hypothesis. The aftermath of a warm event leaves the thermocline along the equator shallower than normal (i.e., equatorial heat content is low and SST is cold; this is the "La Niña" phase). Over the next few years the equatorial warm water reservoir is gradually refilled. Once there is enough warm water in the equatorial band, the rapid (for the ocean) equatorial Kelvin waves allowed by linear equatorial ocean dynamics can move enough of the warm water to the eastern end of the equator to initiate the next event.

The theories of Suarez and Schopf (1988) and Battisti and Hirst (1989), which also have linear equatorial ocean dynamics at their core, provide a much clearer picture of how the ENSO cycle operates. More recent work along similar lines has expanded our understanding of this mechanism (Schopf and Suarez, 1990; Graham and White, 1988; Cane et al., 1990; Munnich et al., 1991; also see Cane and Zebiak, 1987), and a complete paradigm for the ENSO cycle can now be presented.

As in nature, let the main wind changes be in the central equatorial ocean while the SST changes are concentrated in the east. Then the surface wind amplitude, which depends on the east–west temperature gradient, varies with this eastern temperature. Further simplify by assuming that the eastern SSTs are principally controlled by thermocline depth variations. These variations are driven by the changes in the surface wind stress according to the linear shallow water equations on an equatorial beta plane. If the eastern SSTs are warm (thermocline high) then the wind anomaly will be westerly, forcing a Kelvin wave packet in the ocean to further depress the thermocline in the east thus enhancing this state.

However, this excess of warm water must be compensated somewhere by a region of colder water (shallower than normal thermocline). Equatorial dynamics dictates that this be in the form of equatorial Rossby wave packets, which must propagate westward from the wind forcing region. When they reach the western boundary they are reflected as "cold" equatorial Kelvin waves, which propagate eastward across the ocean to reduce the SST there.

Thus the original warm signal is invariably accompanied by a cold signal – but with a delay. This delayed oscillator mechanism accounts for the turnabout from warm to cold states.

To further appreciate the role of equatorial waves in sustaining the ENSO oscillation consider the state of affairs when the eastern thermocline and SST anomalies are near zero; for example, at the termination of a warm event. Then the wind anomaly must be near zero as well, so there is no direct driving to evolve the coupled system to its next phase. However, the previous warm event necessarily left a residue of cold Rossby waves in the western ocean, which eventually reflect at the west into a Kelvin wave which will reduce the SST in the east. The wind then becomes easterly and the cycle continues.

This paradigm may be distilled into a very simple system such as a single ordinary differential equation with a delay (Suarez and Schopf, 1988; Battisti and Hirst, 1989) or a recurrence relation in a single variable (Munnich et al., 1991). Perhaps the simplest version is that of Battisti and Hirst (1989):

$$\frac{\partial T}{\partial t} = -bT(t-\tau) + cT(t).\tag{18.1}$$

They derived this equation as well as values for the parameters b, c, and τ, from Battisti's (1988) version of the ZC numerical model (Zebiak and Cane, 1987). Here T is the SST anomaly in the eastern equatorial Pacific, and c is the sum of all the processes that induce local changes in T, including horizontal advection, thermal damping, anomalous upwelling and changes in the local subsurface thermal structure (including local wave effects). The b term accounts for the effect of Kelvin waves generated at the western boundary as the reflection of Rossby waves; τ is the delay associated with this reflection process. Growing, oscillating solutions to (18.1) – ENSO modes – exist when $b\tau > \exp(c\tau - 1)$, a relation which holds for the parameters characteristic of the numerical model (see the Appendix to Battisti and Hirst, 1989).

Though other mechanisms can give rise to unstable oscillations in coupled tropical models (e.g., Hirst, 1986; 1988; Neelin, 1991; and see Sec. 18.6), it is generally accepted that this paradigm accounts for the behavior of the numerical models discussed above, as well as that in the higher-resolution coupled GCM which exhibits an ENSO-like oscillation (see Sec. 18.6). It is more difficult to establish conclusively that it operates in nature. It is consistent with the refill idea described above, which is supported by data (Wyrtki, 1985, and the additional time series available in the *Climate Diagnostics Bulletin* of NOAA). The role for western boundary reflection is further supported by the semi-empirical studies of Zebiak (1989) and Graham and White (1990). Finally, the ZC coupled model, in which this mechanism is clearly operative, has demonstrated the ability to predict warm events a year or more in advance (Sec. 18.5).

While the restriction of these models to the tropical Pacific region serves to bolster Bjerknes' emphasis on this region, it also renders them incapable of simulating the global consequences of ENSO. What is perhaps more troubling is the inability of the paradigm to account for the changes in the western equatorial Pacific preceding the warming in the east (see the discussion in Cane et al., 1990). More generally, the SO is observed to exhibit behavior distinct from El Niño (viz Fig. 18.1), and this too is not reproduced. These tropical Pacific omissions suggest that connections important to the ENSO cycle may have been overlooked.

The observed ENSO cycle is not regular, and some of the models share this feature (e.g., Figs. 18.3, 18.11). Nonetheless, the cause of the observed aperiodicity remains an unsettled issue. The results from Battisti's (1988) model and the experiments of Schopf and Suarez (1988) suggest that it is solely due to noise; that is, atmospheric or oceanic fluctuations distinct from the ENSO cycle. On the other hand, the low-order ENSO model of Münnich et al. (1991) produces aperiodicity, doing so rather readily if a seasonal modulation is included.

Experiments and analysis with ENSO models have demonstrated very strong sensitivities to rather small changes in parameter values. In the anomaly models some of these changes are equivalent to changes in the mean background state. Since a greenhouse warming will alter this state, the implication of such sensitivity is that the characteristics of ENSO will be changed. There have been a few experiments to explore this possibility (Zebiak and Cane, 1991), but inferences must be highly tentative in deference to our limited confidence in the ENSO models and to the great uncertainties as to the nature of greenhouse induced changes. This area of research is likely to become quite active as climate modeling progresses.

18.4 Modeling of ENSO as a coupled system

Though other views were (and still are) available, by the early 1980s a basis for modeling ENSO had emerged which allows much of the daunting complexity of the full atmosphere–ocean system to be ignored. The Bjerknes–Wyrtki scenario is at the core of it, which means that though the consequences are global, the essential interactions between atmosphere and ocean take place in the equatorial Pacific. The crucial variations of SST result from ocean dynamics, not variations in heat exchange with the atmosphere. Furthermore, these dynamics are essentially linear and act remotely: equatorial Kelvin waves carry the message of a wind change in the central and western equatorial Pacific eastward to effect a change in SST in the eastern Pacific. The role of the surface heat exchange is to drive the circulation of the tropical atmosphere, including the surface wind stress so crucial to the coupling. This atmospheric response can be largely captured by a steady-state linear model.

Building on these ideas, a number of highly idealized coupled models were developed in the early 1980s which added significantly to our developing understanding of ENSO. For example, the stability analysis of Hirst (1986) provided concepts for interpreting the results of the more complex, and thus less intellectually tractable, models (also see Neelin, 1990). It is impossible to do justice to this work here; Philander (1990), and McCreary and Anderson (1991) provide good accounts with many references.

The first coupled model to generate results which could be said to simulate ENSO was that of Cane and Zebiak (1985; a complete description, including the values of all parameters, is given in Zebiak and Cane, 1987). It is not a comprehensive model (i.e., a GCM). Rather, it was explicitly designed to capture the physics essential to the Bjerknes–Wyrtki hypothesis. It is described in some detail to illustrate this modeling method.

Before attempting to couple them, the ocean and atmosphere components of the ZC coupled model were developed and tested independently. Both use approximations to the full primitive equations (Chapter 3, Sec. 3.6.2) suitable only for the long time and large space scales characteristic of ENSO. Both are models for the anomalies about a monthly mean state specified from climatological observations (specifically the Climate Analysis Center data set; see Rasmusson and Carpenter, 1982). This anomaly model approach bypasses the very elaborate and difficult problem of simulating the mean climate well enough to allow realistic ENSO simulations; as we will see below, even our most sophisticated GCMs have yet to solve this problem.

18.4.1 *The atmosphere model*

Our working hypothesis demands of the atmospheric model "only" that, if SST anomalies characteristic of ENSO are given, it produces the surface wind field needed to drive the tropical Pacific Ocean.

Observations show that the tropical anomalies are dominated by a simple vertical structure with a universal form, namely, a reversal of polarity between the lower and upper troposphere (e.g., regions of low-level convergence lie below regions of upper-level divergence). Linear dynamical models with a single degree of freedom in the vertical have proven surprisingly adept at reproducing the horizontal structure of the atmosphere though the physical interpretation of these models is uncertain (e.g., Neelin, 1989). The ZC atmospheric dynamics is of this Gill (1980) type in which the horizontal structure of this dominant vertical mode is described by steady-state, linear shallow-water equations on an equatorial beta plane:

$$\epsilon u_a - \beta y u_a = -p_x \,, \tag{18.2}$$

$$\epsilon v_a + \beta y u_a = -p_y \,, \tag{18.3}$$

$$\epsilon p + c_a^2 [(u_a)_x + (v_a)_y] = Q \,, \tag{18.4}$$

where (u_a, v_a) are zonal and meridional surface wind anomaly components, p is the anomaly pressure (divided by density), $\beta = df/dy$ at the equator $(y = 0)$, with f being the Coriolis parameter, and c_a is a constant related to the vertical structure. The model employs a linear dissipation for both momentum (u_a, v_a) and mass p in the form of Newtonian cooling and Rayleigh friction with common damping time ϵ^{-1}. It has been shown in many places (e.g., Zebiak, 1986; Weare, 1986) that, with the anomalous heating Q specified, the simple dynamics of (18.2)–(18.4) is able to simulate the tropical surface wind anomalies reasonably well (at least as well as atmospheric GCMs, in fact). The model is often used to explain GCM results (e.g., Shukla and Wallace, 1983; Philander, 1990). However, this is true only if the damping time is assumed to be 1 or 2 days, values which are too short to be readily justified in some interpretations of these models (but see Neelin, 1989).

The heating anomaly $Q = Q_s + Q_1$, where Q_s expresses the dependence on local evaporation and Q_1 on low-level moisture convergence:

$$Q_s = (\alpha T) \exp[\overline{T} - 30°C)/16.7°C], \qquad (18.5a)$$

$$Q_1 = \beta_*[M(\overline{c} + c) - M(\overline{c})], \qquad (18.5b)$$

$$c \equiv -(u_a)_x - (v_a)_y; \qquad (18.5c)$$

α and β_* are constants and the function $M(x)$ is defined by $M(x) = x$ for $x > 0$ and $M(x) = 0$ otherwise. Overbars denote mean quantities, specified from data in this model.

Several observational studies (e.g., Cornejo-Garrido and Stone, 1977) as well as GCM calculations have demonstrated the important contribution of moisture convergence to the overall tropical heat balance. The convergence feedback is incorporated into the model using an iterative procedure in which the heating at each iteration depends on the convergence field from the previous iteration. The scheme is analyzed in detail in Zebiak (1986). The feedback is nonlinear because the moisture related heating is operative only when the total wind field is convergent, and this depends not only on the calculated anomalous convergence, but also the specified mean convergence.

Zebiak (1986) shows that this model simulates the (Rasmusson and Carpenter composite) ENSO surface wind anomalies when driven by observed ENSO SST anomalies (also see Weare, 1986). However, this is true only in the deep tropics, where the atmospheric structure assumed in the model matches the observed structure and the dynamics are nearly linear. The model is unable to simulate winds at higher latitudes, where the simplifying assumptions no longer hold. Thus it would be unsuitable for studying teleconnections with midlatitudes, for example. Furthermore, if our working hypothesis were wrong and ENSO depended in an essential way on midlatitude influences, then one would expect the model to be unable to simulate the ENSO cycle.

18.4.2 The ocean model

According to our working hypothesis, only the tropical Pacific need be considered, and the details of coastal geometry are seemingly irrelevant. Thus the model ocean basin is taken as rectangular, extending from 124°E to 80°W, and 29°S to 29°N. The Wyrtki additions to the Bjerknes hypothesis imply that the model must be able to capture the large-scale changes in thermocline depth during the ENSO cycle. A number of studies, beginning with Busalacchi and O'Brien (1981) demonstrate that linear dynamics in a reduced gravity model is sufficient. Alternative derivations of these equations are discussed in Cane (1984) and references therein.

In this type of model the ocean below the thermocline is taken to be motionless; nonlinearities are neglected so the mean motions in the warm layer above the thermocline are governed by the equations:

$$u_t - \beta y v = -g' h_x + \tau_{s_x}/\rho H - ru, \tag{18.6}$$

$$\beta y u = -g' h_y + \tau_{s_y}/\rho H - ru, \tag{18.7}$$

$$h_t + H(u_x + v_y) = -rh, \tag{18.8}$$

where $\mathbf{v} = (u, v)$ is the horizontal velocity in the layer, h is the deviation of the layer thickness from its mean value H, r is a Rayleigh friction parameter, τ_s is the surface windstress, and $g' = g\Delta\rho/\rho$, with $\Delta\rho$ a characteristic density difference between the layers.

Our hypothesis demands of the ocean model that, when driven by the anomalous winds, it produce tropical SST anomalies, especially those in the eastern equatorial Pacific. The reduced gravity model with its single active layer is not sufficient, because SST is determined by the dynamics and physics of the much thinner surface layer as well as by surface heating. Including this layer means splitting the active upper ocean into two layers. Numbering these layers 1 and 2 we now have

$$\mathbf{v} = H^{-1}(H_1\mathbf{v}_1 + H_2\mathbf{v}_2). \tag{18.9}$$

The velocity shear $\mathbf{v}_s = \mathbf{v}_1 - \mathbf{v}_2$ is determined from

$$r_s u_s - \beta y v_s = \tau_{s_x}/\rho H_1, \tag{18.10}$$

$$r_s v_s + \beta y u_s = \tau_{s_y}/\rho H_1, \tag{18.11}$$

Equations 18.10 and 18.11 are essentially the equations for the Ekman layer (see Chapter 4, Sec. 4.4), but with a linear Rayleigh damping r_s added. For the value of the damping time used by ZC ($r_s^{-1} = 2$ days) this term is only significant within a few degrees of the equator. Nonetheless, it is too short to be justified as friction, and is best interpreted as a proxy for the nonlinear terms (especially $w\mathbf{v}_z$) omitted from the equations.

Vertical gradients within the surface layer are small, so the SST is the same as the temperature of this ocean mixed layer. The governing equation for the temperature anomaly in this layer is

$$\frac{\partial T}{\partial t} = -\overline{\mathbf{v}}_1 \cdot \nabla T - \mathbf{v}_1 \cdot \nabla(\overline{T} + T) - \{M(\overline{w}_s + w_s) - M(\overline{w}_s)\}\overline{T}_z$$
$$- M(\overline{w}_s + w_s)T_z - \alpha_s T \tag{18.12}$$

where $\mathbf{v}_1$ is the surface layer current anomaly, and the anomalous upwelling is

$$w_s = H_1[(u_1)_x + (v_1)_y].$$

The function M [cf. (18.5b)] accounts for the fact that surface temperature is affected by vertical advection only in the presence of upwelling. The anomalous temperature gradient, T_z, is defined by

$$T_z = (T - T_e)/H_1\,, \tag{18.13}$$

where H_1 is the surface layer depth, and T_e measures the anomalous temperature of the waters entrained into the surface layer. This water, at the base of the mixed layer, is a mixture of the surface water and the unperturbed waters below. Hence

$$T_e = (1 - \gamma)T + \gamma T_d\,, \tag{18.14}$$

where $\gamma \ (= 0.5)$ is a mixing parameter and $T_d(\overline{h}, h)$ relates the subsurface temperature anomaly to the mean and anomalous thermocline depths:

$$T_d = T_\pm\{\tanh[b_\pm(\overline{h} + h)] - \tanh[b_\pm\overline{h}]\}\,, \tag{18.15}$$

where the constants T_+, b_+ apply for $h > 0$ and the constants T_-, b_- apply for $h < 0$. The variable h is obtained from the model dynamics. Finally, the term $-\alpha_s T$ crudely accounts for anomalous heat exchange with the atmosphere.

Though (18.12) is an equation for the temperature anomaly, it is nonlinear and no term is omitted. In principle, all the mean quantities could be specified from data, but since the current data are so meager, mean currents were computed by running the model forced with climatological winds. The surface layer depth is taken as a constant 50 m. This is an acceptable simplification near the equator, though a bit questionable in the far eastern equatorial Pacific where the mixed layer may be only 25 m deep. It would seem to be inappropriate to higher latitudes, though the results of Seager (1989) indicate it is not a major source of error in modeling SST.

Equation (18.15) is essentially a curve fit to the equatorial vertical temperature profile (in addition to Zebiak and Cane, 1987; see Seager et al., 1988). Ocean GCMs commonly have trouble maintaining the thermocline; here it is specified. By making the temperature beneath the mixed layer, T_d, a function of thermocline depth, we are assuming that the subsurface thermal structure moves up and down without changing shape. This is consistent with observations of low-frequency changes. Finally, the anomalous surface heating is simply taken as a constant times T. It would be easy to use a more elaborate formula based more closely on the standard bulk formulae (cf. Seager et al., 1988), but this so-called Newtonian cooling form was chosen in accord with the idea that for ENSO the surface heat exchange acts only as a damping on the ocean, tending to destroy anomalies, not create them.

This model, far simpler than a GCM, was constructed with careful attention to the processes influencing SST, the oceanic variable essential in the coupled ENSO problem. When forced by (composite) ENSO wind anomalies, it is able to simulate the observed (composite) ENSO SST anomalies (Cane, 1986). It is generally accurate in the eastern Pacific and tends to understate the anomalies in the central and western Pacific. The specified value for thermal damping is reasonable for the eastern Pacific, but too large elsewhere. The model does poorly with the upwelling signal along the coast of South America, probably as a consequence of its coarse resolution and crude representation of coastal geometry. The overly simple representation of subsurface thermal structure is an additional source of error. These shortcomings notwithstanding, the model's overall performance for low-latitude SST anomalies is at least comparable to that of any ocean GCM. (A variant of this model performs well for climatological and total SST; cf. Seager et al., 1988; Seager, 1989.)

18.4.3 *Coupled model results*

A numerical experiment with the coupled model was initiated with an imposed 2 m s^{-1} westerly wind anomaly of four months duration. There was no external forcing thereafter: aside from the model physics, evolution of anomalies in SST, winds, etc. depends only on this initial condition and on the monthly mean climatological fields specified in the component ocean and atmosphere models. Furthermore, because of the damping in the model, the initial conditions are largely forgotten within a decade. A time series of model SST anomaly averaged over the eastern equatorial Pacific is shown in Fig. 18.3. There are peaks of varying amplitude occurring at irregular intervals but typically three to four years apart. They tend to be phase-locked to the annual cycle, with major events reaching maximum amplitude at the end of the calendar year and decaying rapidly thereafter. All of these features are characteristic of observed El Niño events. The amplitude of model events is similar to observed ones. The model appears to be somewhat more regular than nature; the high-frequency fluctuations present only in the real atmosphere and ocean may account for the broader natural spectrum (but see below).

Figure 18.4 depicts the evolution of SST during a typical El Niño event; it may be compared with Fig. 18.2. In December of the preceding year, (not shown), there was no discernible anomaly; by March of the El Niño

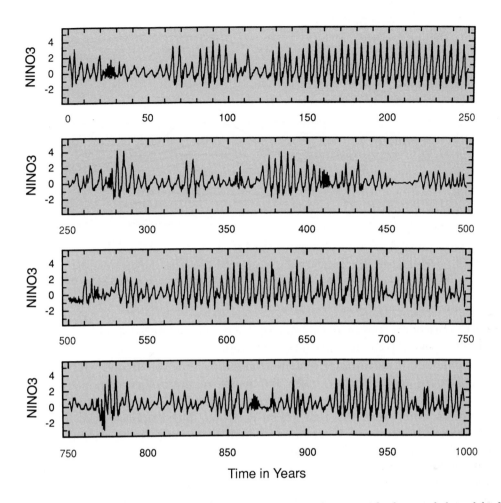

Fig. 18.3 Time series of NINO3 from 1,024–year simulations with the coupled model of Zebiak and Cane (1987). NINO3 denotes SST anomalies averaged over the region 5°N to 5°S, 90°W to 150°W. It is a widely used index of ENSO events, since warm anomalies in the NINO3 area are characteristic of El Niño (viz. Fig. 18.2).

year there is a small but systematic warming in the eastern Pacific; by December the anomaly extends to the dateline, with a maximum at about 135°W. Figures 18.5 and 18.6 show the first four EOFs[1] of model and data, respectively. The four model EOFs account for a higher percentage of the variance than the four observed EOFs. Since all the higher EOFs have only

---

[1] EOFs – empirical orthogonal functions – also referred to as Principal Components, provide an objective means of extracting the dominant structures of natural variability; a basic text is Morrison, 1976, Chapter 8.

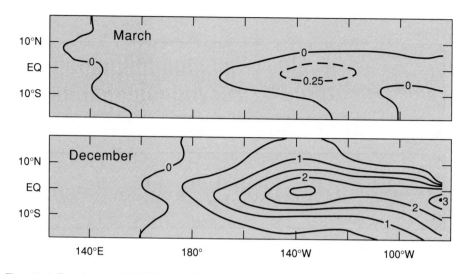

Fig. 18.4 Coupled model SST anomalies for March and December during a coupled model
El Niño event (note that the contour interval for March is 0.25°C and that for
December is 0.5°C). (From Cane and Zebiak, 1985).

a small part of the variance, this remainder may be regarded as noise, and it
is to be expected that the full, natural system will be noisier than the ENSO
specific model. The variance distribution shows that the second, third and
fourth modes are relatively more significant in nature, corresponding to
the fact that nature exhibits more propagation within an event, and more
differences from event to event. The correspondence between the model
and observed structures is obvious enough to permit the assertion that
the model is realistic, though there are important discrepancies. That the
model understates the strength of the signal at the South American coast
is likely due to its coarse resolution, which precludes a decent simulation of
coastal upwelling. The model's systematic understatement of the variability
westward of about 160°W is a shortcoming with more serious consequences.

Figure 18.7 illustrates the evolution of zonal wind along the equator.
The prominent feature is the band of westerly anomalies in the central
Pacific. It resembles the typical ENSO anomaly (Cane, 1983) but lacks
the observed eastward progression in its early stages. What is missing is
the initial anomaly west of the dateline. The model atmosphere has no SST
anomaly to respond to because the ocean model has too little variability
there. Thereafter, the spatial and temporal patterns are generally realistic
until the year following the event, year 32. The model westerly anomalies
persist several months longer than is typical of El Niño events. The same
is true for SST and other fields, and is characterisic of model events. A
possible cause is the model's inability to produce the easterly anomalies
in the far western Pacific which appear during the termination phase of
observed events. As is the case even when observed SST anomalies are

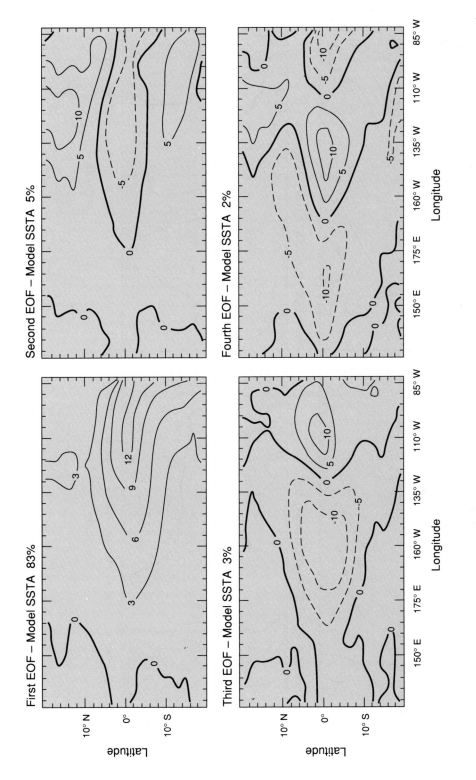

Fig. 18.5 First four EOFs of the ZC model SST anomalies, calculated from monthly values of the 1,024-year simulation shown in Fig. 18.3.

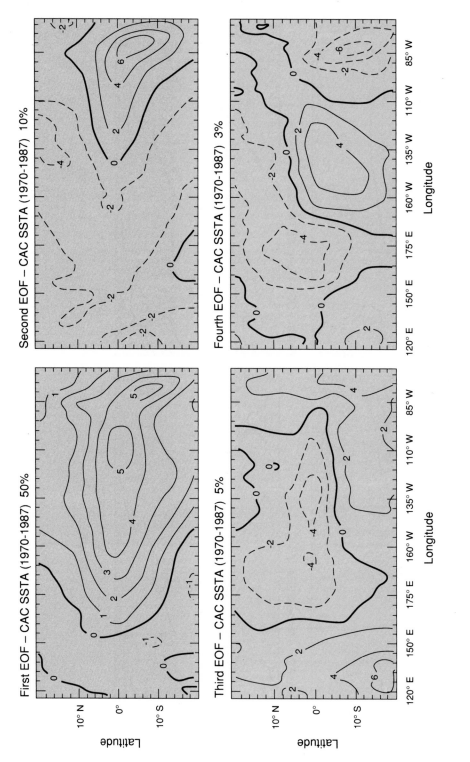

Fig. 18.6 First four EOFs of observed SST anomalies, calculated from monthly values between the years 1970 and 1987, inclusive (CAC analysis).

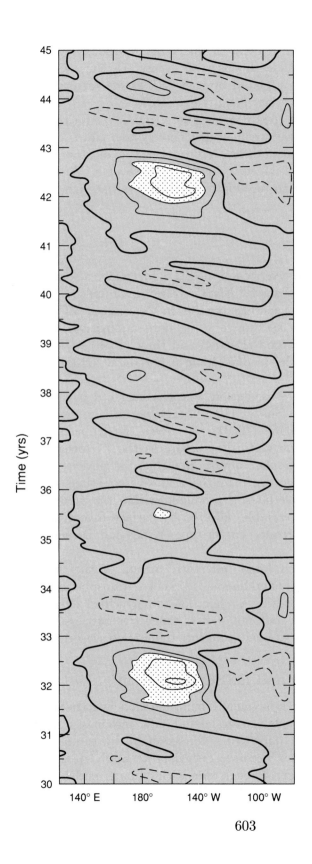

Fig. 18.7 Time–longitude sections for 16 years of the ZC coupled model integration showing the forcing for the gravest mode oceanic Kelvin wave, a measure of zonal wind anomalies along the equator. Positive (westerly) anomalies are indicated with solid lines, and negative (easterly) anomalies are indicated with dashed lines. Large westerly anomalies (>0.15 dynes cm^{-2}) are stippled. (From Cane and Zebiak 1985.)

603

specified, the model winds are poorest in the Asian Monsoon region and in the far eastern Pacific. As with observed events, El Niño anomalies disappear quite rapidly, to be replaced by cold SST in the eastern Pacific and stronger than normal easterlies along the equator.

In summary, the coupled ocean–atmosphere model, which greatly simplifies the physics of the real ocean–atmosphere system and is certainly unrealistic in many details, nonetheless reproduces the most prominent spatial and temporal features of the evolution of El Niño events. Mechanisms internal to the model allow it to terminate events and initiate new ones in a never ending ENSO cycle.

18.4.4   *Other models*

Battisti (1988) coded a version of the ZC model which also captures many of the features of the observed ENSO. However, it fails to exhibit the aperiodic behavior which generally obtains in the original ZC model. The Battisti version behaves as though it is more dissipative, but the reasons for this are not understood. Battisti and Hirst (1989) were guided by experiments with this model in developing a paradigm for the ENSO mechanism (see Sec. 18.3).

A coupled ocean–atmosphere oscillation was also achieved in the more complex model of Schopf and Suarez (1988). The ocean component was documented in Schopf and Cane (1983). It covers the tropical Pacific only, and has almost the same vertical structure as the ZC ocean, with two active layers above a motionless abyss. However, the physics is more comprehensive. Dynamics is governed by the primitive equations and the temperature of the lower active layer is predicted within the model. The depth of the surface layer is variable, affected by surface divergence and by mixed layer physics of the Kraus–Turner type (Kraus and Turner, 1967; Schopf and Cane, 1983).

The atmospheric model is a finite difference version of the two-level primitive equation model of Held and Suarez (1978). Its atmospheric physics parameterizes radiative processes by a relaxation to a specified zonal mean temperature. It includes dry, but not moist, convective processes. The coupled run is forced by mean annual radiation; there is no seasonal cycle.

18.5   Model verification and ENSO prediction

The issue of verification arises in all simulation modeling. In Sec. 18.4 some qualitative comparisons between model and data were made. In order to know if a model has been improved – for example, by tuning model parameters – it sooner or later becomes necessary to apply quantitative measures of model performance. Though rare with climate models, this is routine with weather forecasting models. Not coincidently, these are the most sophisticated and realistic models in the climate sciences.

It would be best to compare sequences of model states with sequences of observations. However, the aperiodicity of ENSO implies its predictability is inherently limited, so even a perfect model would not reproduce the time evolution of the observed cycle for very long. The same problem arises for all climate models. Short sequences on the order of a year may be verified; this is the prediction problem discussed below.

Another test is to ask whether the model's characteristic variability is "realistic." The EOF analysis of Sec. 18.4 (Fig. 18.5 vs Fig. 18.6), which compares typical model anomaly patterns with those in the observational data, is one test of this sort that could readily be made quantitative. Zebiak and Cane (1991) assayed an answer by comparing a number of statistics from model generated eastern Pacific (NINO3) SST time series with those calculated from observed time series. In this study 243 100-year long model runs were made in which each of five model parameters were given one of three values: unchanged from the values in the standard run (i.e., the run depicted in Fig. 18.3; cf. Zebiak and Cane, 1987 for a description of model parameters), increased by 5%, or decreased by 5%. The large number of runs allows one to conclude with (statistical) confidence that these modest changes in parameter values altered the characteristic amplitude and frequency of events, as well as the degree of irregularity of the oscillations. In most cases, aperiodic oscillations with an average period near four years still occur. One may make subjective judgments as to the relative reality of different parameter settings, but the short length of the observed NINO3 record used (18 years) allowed only a small number of cases with obviously pathological behaviors to be confidently labeled "unrealistic."

Zhao et al. (1991) were able to reach a more definite conclusion by comparing coupled model results with a 500 year record of accumulation from the Quelccaya ice cap in Peru shown in Plate 12. Significantly better agreement with the statistical characteristics of this record was obtained by decreasing the oceanic equivalent depth and increasing the sharpness of the thermocline while reducing its overall amplitude. (As above, changes are 5% variations from the standard run.) Obviously, this approach to model verification is limited by the paucity of long records germane to ENSO.

Its social and economic benefit makes prediction a principal goal of climate modeling. In addition to being the most socially significant application of climate modeling, attempts at prediction provide an objective and quantitative measure of our collective scientific understanding and modeling skill – perhaps the best measure we have.

Dynamical models are the method of choice in weather forecasting. Thus far ZC is the only dynamical model used for forecasting ENSO. (Procedures and results are described in Cane et al., 1986; and Cane, 1991)

Figure 18.8 illustrates the forecast performance at various leads in terms of the eastern equatorial Pacific SST index, NINO3. (Since the initial

605

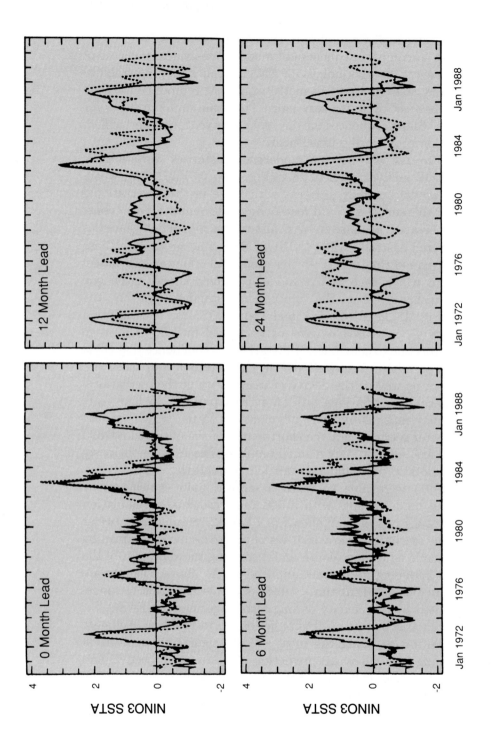

Fig. 18.8 Model forecasts (dashed line) and observed (solid line) NINO3 SST anomalies (°C) from 1970 to 1989. The forecasts are at the various lead times indicated; a zero-month lead forecast is actually the initial condition generated by the ocean model forced by observed winds. (From Cane, 1990.)

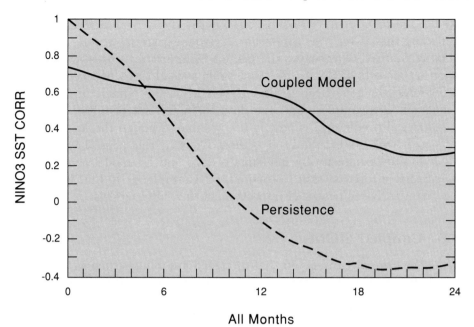

All Months

Fig. 18.9 A summary of the forecasting skill of the ZC model based procedure of Cane et al. (1986). Shown is the correlation coefficient of forecast and observed NINO3 SST anomalies at lead times for 0 to 24 months. The correlation of a persistence forecast with observed values is also shown. A value of 0.5 marks a forecast with the same error variance as a climatological forecast.

conditions do not use observed SST, the initial model SST may also be compared with data.) The major events are captured at leads out to a year, with few false events other than the consistent tendency to persist the 1982/83 event. Even the two-year lead captures much of the observed behavior. At all leads the model is better at forecasting the major departures from normal (i.e., the events) than at tracking the smaller- amplitude higher-frequency fluctuations. The impression that the three-month lead is only slightly worse than the initial conditions, and that the falloff in skill is slow from six months to a year, is confirmed in Fig. 18.9. This summary masks a marked seasonal dependence, with model skill being greatest for boreal fall and winter (viz. Cane, 1991, Fig. 11.5).

A number of statistical (Barnett et al., 1988; Xu and von Storch, 1990) and statistical-dynamical (Inoue and O'Brien, 1984; Latif et al., 1991) schemes have also been used to forecast ENSO. By revealing important connections in the data, statistical models have contributed substantially to our understanding of ENSO (e.g., Barnett et al., 1988). As with the dynamical model, the skill is in capturing the major events, not the smaller fluctuations, supporting the view of ENSO as a large-scale, low-frequency cycle. Generally, the statistical models do better than the dynamical model at short leads (less than five months), and less well at longer leads. This may be due to poor initial conditions for the dynamical model; if so, it is interesting that it recovers somewhat if given enough lead time.

The skill of ENSO forecasts, though not high, has already established that important aspects of ENSO are predictable at lead times of several seasons. Useful forecasts have been achieved even while our understanding of the phenomena is incomplete and the data for a full set of initial conditions (or statistical predictors) is largely unavailable. Though there are intrinsic limits to ENSO predictability, it is most likely at this primitive stage that skill is limited primarily by inadequate data and by model shortcomings. An increase in forecast skill (measured, for example, as in Fig. 18.9) would be a quantitative indicator of improvement in ENSO models.

18.6 Coupled GCMs

We have already noted that intermediate models such as ZC are not suitable for studies of midlatitudes. They cannot be used for predicting the global consequences of ENSO. Even within the tropics, their incomplete physics causes difficulties. These limits point to the desirability of a comprehensive ENSO model – a coupled ocean–atmosphere GCM (CGCM). Beyond the ENSO perspective, there is the need for a comprehensive climate system model which, while able to simulate ENSO and other natural climate variability, also contains the physics to respond correctly to changes in greenhouse gases. Not only is a detailed review of all the ENSO CGCMs impossible here, but the rapid ongoing progress ensures that it would soon be dated. A recent compendium of ENSO results from 17 coupled models (Neelin et al., 1992) may be consulted for the status of CGCMs as of early 1991.

The successes of the less comprehensive models establish that the physics needed to produce a reasonable ENSO cycle is known and can be modeled. Our ability to construct an ENSO CGCM is thus demonstrated in principle. Practice is another matter. In the less elaborate parameterizations of the simpler models, the effects of changes in parameter values are more apparent; i.e., these models are easier to tune. Their relative simplicity makes them easier (if not always easy) to understand. That these models are fast enough to run many times per GCM experiment compounds these advantages. Striving for a complete test of our understanding of the climate system, the GCMs try to assume as little as possible. The intermediate models with their limited goals are willing to specify a great deal. For example, in the ZC model the mean climatology is taken from data, as is the subsurface temperature structure.

CGCM ENSO work generally has not followed the path of numerical weather prediction, which progressed over time from the equivalent barotropic vorticity equation to increasingly more complex models. Instead, existing "state-of-the-art" atmospheric and oceanic GCMs have been coupled together. For subtle reasons, this approach has proven more frustrating than anticipated.

It is usual practice before coupling to test the component models independently by specifying climatological forcing: i.e., SST for the atmosphere GCM; surface wind stress, salt and heat flux for the ocean GCM. One then checks to see if they produce a reasonable climatology (see, for example, Chapter 17). In some cases they have also been forced with anomalies observed in ENSO years (e.g., Chapter 19). Few, if any, of these climate models have been compared with observations in anything like the quantitative manner of weather prediction models. The hard lesson of experience has been that coupling atmosphere and ocean GCMs together most often causes their seemingly small flaws to feed back on each other, moving the model climate system far from reality, a result known as "climate drift."

A GCM may do a good job overall while doing poorly on a feature crucial for the ENSO interaction. For example, atmospheric GCMs (AGCMs), even those with excellent simulations of surface pressure, 500 mb height, etc., typically understate the strength of tropical surface wind stress. By a global measure, the model surface wind may be quite good, but the equatorial winds count disproportionately for ENSO. In a simple model one might just increase the surface drag coefficient and use the higher stress to drive the ocean. Atmospheric GCMs tend to counter an increased drag coefficient by reducing the surface wind, leaving the stress almost the same.

The conventional wisdom has it that the ocean GCMs are the weak link; AGCMs have a much longer and richer history. In fact, surface wind stress and surface heat flux, the only aspects of the atmosphere which matter in driving the ocean, are not well simulated by AGCMs. Among other things, both depend strongly on the parameterization of the planetary boundary layer and on convection, not the strongest features of these models; see Chapter 10. Since these surface fluxes of heat and momentum matter little for weather prediction or for any other atmospheric model run with specified SSTs, they have long been neglected. This has changed only recently with the heightened interest in modeling the coupled system, ENSO in particular. A recent comparison of the operational U.S. National Meteorological Center surface wind analysis with tropical Pacific buoy data (Reynolds et al., 1989) shows that despite great progress, even the best of AGCMs make errors with significant consequences for ENSO modeling. The heat flux is an even more difficult problem (cf. the discussion in Seager et al., 1988).

One strategy that recognizes the problems and computer time demands on the atmospheric side is to construct a hybrid by coupling a simple atmospheric model to an ocean GCM. Latif et al. (1991) constructed an empirical atmosphere model from statistical relations between SST and surface wind which was coupled to a GCM and used in prediction studies. Their results are at least comparable to the statistical prediction techniques.

Neelin (1990) used a hybrid model to good advantage to study the changes in behavior with changes in mean climatology and in the "coupling strength", i.e., the amplitude of the wind anomaly for a given SST anomaly. The model climatology can be readily adjusted by a flux correction technique. It was found that a mean state with an excessive cold tongue inhibited interannual variability, but a more realistic state allows it.

In all the true CGCMs the AGCM has been too coarse (usually R15 or T21; cf. Chapter 9) to resolve all the significant features of the tropical atmosphere, most notably the ITCZ. For many of the CGCMs, especially those intended for global climate studies rather than ENSO, the ocean GCM is also low resolution, typically 4×5 degrees of latitude and longitude. This is too coarse to simulate equatorial wave dynamics properly, and makes equatorial upwelling far too weak. In addition, coarse vertical resolution leads these models to erode the sharp equatorial thermocline. As a result thermocline depth changes have too weak an influence on SST. Thus the models are unable to capture essential features of equatorial oceans, features which were also central to the ENSO paradigm presented above. As a rule, the zonal SST gradient along the equator is too weak in these models. In some the minimum temperature is at the center of the basin or the gradient is reversed relative to observations. (Neelin et al., 1992 and Meehl, 1990b review most of these coarse ocean CGCMs.)

Nonetheless, a number of these coupled simulations exhibit interannual variations with the three- to four-year time scale characteristic of ENSO. Figure 18.10, from Neelin et al. (1992), is an example; see also Fig. 17.11. It is typical in that the anomalies are 1°C or less, much weaker than the observed ENSO. They invariably propagate westward. Neelin (1991) identifies them as "SST modes", instabilities involving a two-way coupling between SST and the atmosphere, but without a role for the wave dynamics central to the paradigm presented above.

A number of the half dozen or so CGCMs with high-resolution ocean components give reasonable climatologies (Neelin et al., 1992 describes all these CGCMs), but to date only one exhibits interannual variability with amplitudes comparable to the observed ENSO (Philander et al., 1991). The response of the atmospheric component, the GFDL R15 climate model, (cf. Chapter 19) to observed SST anomalies, has been described by Lau (1985). The ocean model climatology is described by Philander et al. (1987). There is no flux correction. Annually averaged solar insolation is applied; there is no seasonal cycle.

The evolution of SST for a 28-year run from this model is shown in Fig. 18.11. The mean SSTs are a bit warm, but are reasonably close to observed values. The anomalies tend to be stationary, with realistic amplitudes and patterns. However, model episodes tend to persist for about three years, roughly twice the length of observed events. The absence of an annual cycle may account for some loss of realism. The results of this

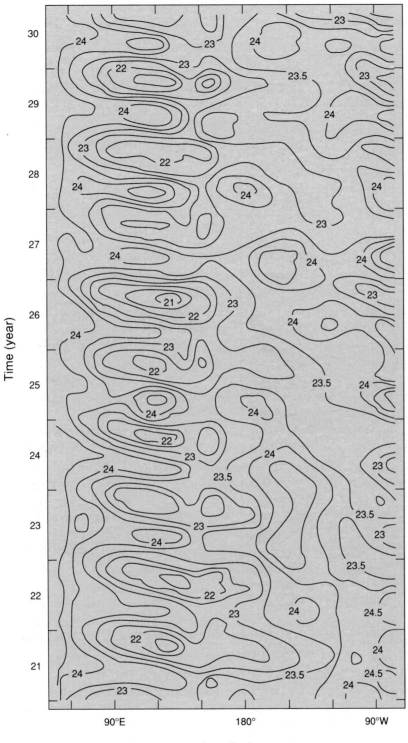

Fig. 18.10 Time–longitude plot of SST seasonal means for years 21–30 in the coupled model of Meehl (1990), 50°E to 80°W, averaged from 10°N to 10°S. Contour interval is 0.5°C.

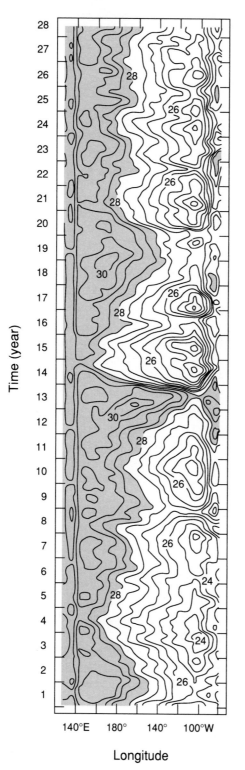

Fig. 18.11 Pacific SST along the equator over 28 years of simulation (without seasonal cycle) by the coupled model of Philander et al. (1991). Smoothing by a 13-month running mean has been applied. Contour interval 0.5°C, with shading over 28°C. (From Neelin et al., 1992.)

simulation, especially the temporal variations in heat content, suggest the equatorial wave mechanism discussed above as an appropriate explanation for the oscillation (cf. Neelin et al., 1991; Philander et al., 1992).

18.7 Discussion

ENSO modeling began with a fairly specific goal, to simulate the observed cycle. It could be built on a sound physical hypothesis and rapidly developing knowledge of the tropical atmosphere and ocean. Despite the apparent realism and immediate availability of GCMs, the earliest successes were achieved with less comprehensive models designed to capture the physics singled out by the early hypotheses. But modeling success did not have to wait for a complete theory. In fact, the models were instrumental in extending the theory.

One of these models proved able to forecast ENSO with some skill a year or more ahead. In view of the many features of the atmosphere and ocean it is unable to reproduce, this is surprising. Together with the statistical models' performance, it supports the concept that the interactions shaping this primary feature of natural variability in the climate system have large space and time scales (cf. Barnett et al., 1988). It can be pushed further, to suggest that the core interaction is too robust to be disturbed by the myriad factors ignored (smaller time and space scales, land processes, biology, etc.). Probably this is pushing too far: it is likely that some of those factors do influence the evolution of ENSO, so a prediction model unable to simulate them is handicapped to some degree.

However, it is not known to what degree this is true. It could be that other factors which influence ENSO are themselves unpredictable, so a model able to simulate them will have no greater forecast skill. (Midlatitude synoptic sytems are a well-known example of features unpredictable on the seasonal and longer lead times of interest for ENSO.) A corollary is that the existence of such influences would make ENSO less predictable by any model.

To generalize, when the predictability of climate features is inherently limited, it may be that nothing is to be gained by building a more inclusive model. The best to be done is a statistical description of possible features. A thorough simulation of such features may be unproductive, but some consideration of their influence (i.e., a parameterization) is still needed. This is absent in present intermediate models for ENSO.

Whatever the causes, the limited intrinsic predictability of ENSO phenomena complicates attempts to verify models. Short simulations (predictions) may be directly compared with data, but not long time behavior. Verifying statistics of model variability demands very long time series.

All of the same issues arise for climate system models. It is even more difficult to validate models intended to predict the consequences of changes that have no past precedent. Since changes in interannual variability are an important possible consequence of increases in greenhouse gases, a necessary, but hardly sufficient, requirement for a greenhouse climate model is an ability to simulate ENSO.

The relatively simple models first used for numerical weather prediction were consciously built on the basis of existing observational and theoretical understanding (e.g., equivalent barotropic structure; baroclinic instability rather than thermodynamics as the cause of synoptic systems). Computer time was an important constraint, but so was the limited understanding of how to make a complex model work. It took two decades before operational forecasts were made with GCM class models.

GCMs existed at the outset of ENSO modeling, so it was possible to try to skip ahead to a CGCM rather than building up to it slowly. However, it has proven far more difficult than anticipated to achieve realistic simulations of ENSO by coupling comprehensive state-of-the-art GCMs. Thus, it turned out in practice that ENSO modeling has recapitulated the history of numerical weather prediction. Limited computational resources dictate either insufficient spatial resolution, or experiments of very short duration. Limits to understanding are revealed by coupling: the coupled system stresses model weaknesses that were safely ignored when the components were run separately.

The difficulties do not diminish the desirability of a CGCM. The simpler models demonstrate that it can be done, and provide some valuable guidance on how to do it. The often exquisite parameter sensitivities of CGCMs notwithstanding, the demonstrated ability of at least one CGCM to generate realistic interannual variations is enormously encouraging. It seems altogether likely that satisfactory ENSO CGCMs will follow the first model successes in less than a decade, rapid progress in such a complex task.

PART 5

SENSITIVITY EXPERIMENTS
AND APPLICATIONS

19

Climate variability simulated in GCMs

Ngar-Cheung Lau

19.1 Introduction

One of the most striking characteristics of the Earth's climate system is its variability on a broad range of time scales. The temporal changes that immediately come to mind are those associated with the climatological seasonal and diurnal cycles. Superimposed on these periodic phenomena are many different types of irregular variations. Such departures from the long-term means are often referred to as "anomalies". We shall focus on the numerical simulation of such anomalous fluctuations in this chapter. The pertinent model results will be discussed mostly from a meteorological perspective. This approach is motivated by the important role of the atmospheric circulation in climate variability, and by the devotion of a substantial portion of our modeling efforts to atmospheric simulations. However, we must bear in mind that the atmospheric variability described here is intimately linked to the behavior of other components of the climate system. We shall consider only a subset of these interactions here, namely those involving the atmosphere and the underlying land and ocean surfaces. For a more in-depth treatment of the connections between various climatic components, the reader is referred to other relevant chapters in this book, particularly those in Parts 2 and 4.

As an illustration of the richness of the frequency spectrum for the atmospheric component of the climate system, typical time series of the observed daily and monthly fluctuations of surface air temperature are displayed in Fig. 19.1. The top panel of this diagram depicts day-to-day temperature changes during the winter of 1984/85 at 40°N, 75°W (incidentally, this particular location is selected for its proximity to Princeton, NJ). This time series indicates the frequent occurrence of temperature rises and falls on time scales of less than a week. A closer inspection of this panel reveals the occasional tendency for the rapid temperature changes to "ride" on more slow varying fluctuations. Examples of the latter, longer-term events include the warm spells from November 22 to December 4, and from December 8 to 24, as well as the persistent cold episode lasting from January 8 to February 10. The bottom panel of Fig. 19.1 describes the variations of the monthly averaged, Northern

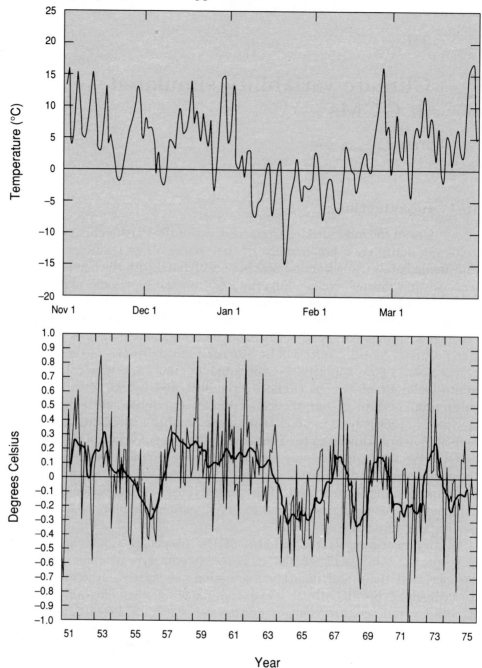

Fig. 19.1 Time series of daily mean 1,000 mb temperature at the gridpoint 40°N 75°W for the period November 1984–March 1985 (top panel), and monthly (light curve) and 12-month running mean (heavy curve) Northern Hemisphere average surface air temperature anomalies for the period 1951–1975 (bottom panel). Data sources: European Centre for Medium-Range Weather Forecasts analyses for top panel; and Yamamoto et al. (1975) and Manabe and Hahn (1981) for bottom panel.

Hemisphere mean temperature anomalies in a 25-year period. The averaged temperatures for individual months (light curve) exhibit notable departures from the climatological mean. The amplitudes of such perturbations are considerably lower than those associated with the daily and local changes portrayed in the top panel. In analogy with the daily temperature record, the monthly means displayed in the bottom panel are modulated by fluctuations with time scales of several years, as highlighted by the time series of the 12-month running means (heavy curve). Of particular interest is the prolonged warm period from 1957 to 1963, and the prominent temperature oscillations from 1965 to 1975 with typical periods of 2–4 years. Turning our attention now to the variability on even longer time scales, it is seen from global temperature variations (IPCC, 1990a) that the year-to-year perturbations are superposed on secular changes with even longer time scales. The warming trend from the beginning of the twentieth century to 1940, and the cooling trend in the 1960s are particularly evident (see Fig. 1.4).

The atmospheric variability in a particular portion of the frequency spectrum is often linked to a specific family of meteorological phenomena, which are, in turn, associated with a distinct set of dynamical and physical processes. In studying the behavior of the climate system, it is essential to have a clear understanding of the nature and origin of atmospheric variability on different time scales. As demonstrated in the preceding paragraph, the characteristic atmospheric time scales which are worth consideration include:

- *Periods of several days*

  Fluctuations on such short periods (referred by many meteorologists as the "synoptic" time scales) are often manifestations of the passage of transient weather systems. These circulation features possess distinct structural and propagation characteristics, and primarily owe their existence to internal atmospheric processes, such as hydrodynamic instability and convection.

- *Periods ranging from about 10 days to a season*

  This time range is particularly relevant to medium- and long-range weather forecasting. The atmospheric variability on this time scale is linked to both the inherent atmospheric variations and interactions with other parts of the climate system. From a meteorological standpoint, the latter interactions affect the atmosphere through alterations of the "boundary conditions" or "boundary forcing" at various interfaces. Some of the atmospheric phenomena in this category, such as persistent high pressure centers in the extratropics, exert a considerable influence on the trajectory and intensity of synoptic-scale systems for extended periods of time. The prevalence of such long-lived meteorological patterns over a given region is often associated with notable changes in the local climate on subseasonal time scales, such as the occurrence of droughts, floods, and temperature extremes (see examples in the top panel of Fig. 19.1).

- *Periods of several years*

  The roles of different processes operating at various interfaces with the atmosphere become more important for these time scales. In view of the long "memory" embedded in many maritime and land processes, the efforts in understanding interannual variability of the climate system have mostly concentrated on interactions of the atmospheric circulation with the underlying ocean and continental surfaces.

- *Periods of decades and beyond*

  Decadal and centennial climate variability may be attributed to the effects of processes with even longer characteristic time scales, such as those accompanying atmospheric interactions with the deep ocean or the cryosphere, secular changes in the concentration of chemical constituents in the atmosphere, and variations in the Earth's orbital parameters.

During the past decade, the modeling community has performed many investigations aimed at simulating the myriad phenomena associated with the entire range of time scales listed above. The primary tools for such endeavors are the current generation of General Circulation Models (GCMs) at various research centers. A general description of the atmospheric GCMs is given in Chapter 10. Integrations with these GCMs provide a controlled environment in which the contributions of specific internal and external processes to atmospheric variability can be diagnosed in a systematic manner. Most of these model variability studies have been conducted using one of the following experimental designs:

- *Atmospheric GCM runs with fixed, climatological boundary conditions*

  These integrations are particularly suited for studying the nature of model variability in the absence of certain factors which are considered to be external to the atmosphere, such as the influences of sea surface temperature anomalies. The variability in these experiments with climatological setting can mostly be attributed to dynamical and physical processes operating in the atmospheric interior.

- *Atmospheric GCM runs with perturbed boundary conditions*

  This class of experiments entails the prescription of anomalous conditions at certain interfaces between the atmosphere and other components of the climate system. These integrations allow for the boundary conditions to affect the atmosphere, but do not permit the atmosphere to change the boundary conditions. The sensitivity of the model atmosphere to the boundary conditions in question can be assessed by contrasting the output from such experiments with the output from "control" runs subjected to climatological forcing.

- *Coupled experiments with full interactions between the atmosphere and other subsystems*

  The variability of the atmosphere as well as other climatic components (e.g., ocean, cryosphere, ground hydrology) is modeled explicitly in these integrations. Two-way interactions (or "feedbacks") between the atmosphere and other components are incorporated in these coupled models.

Model studies of atmospheric variability performed to date have mostly focused on time scales of several years and less. The emphasis on the high-frequency end of the atmospheric spectrum is somewhat dictated by the availability of computer resources, which, until very recently, were only able to accommodate experiments with durations not longer than several decades. Phenomena and processes with time scales of a decade or more are therefore not sufficiently sampled in such simulations. Moreover, contemporary empirical and theoretical studies have fostered a rapid advance of our knowledge of atmospheric fluctuations with subseasonal and interannual time scales. On the other hand, our understanding of ultra-long period variations of the climate system is hindered by the scarcity of long and reliable observational records, as well as the lack of physical insights into the processes pertinent to these time scales.

The present chapter presents an overview of the different facets of model-simulated climate variability. Emphasis is placed on the baseline levels of variability of the atmosphere–ocean–land–cryosphere subsystem, in the absence of perturbations that might be ascribed to human activity, such as the increase of greenhouse gases. The following presentation is organized according to the typical time scales of the phenomena of interest, which range from periods shorter than a season (Sec. 19.2), to several years (Sec. 19.3), and to a decade and longer (Sec. 19.4). Since some processes and phenomena have a wide span of time scales, some overlap among the discussions in these individual sections is unavoidable.

For a more elaborate discussion of atmospheric and oceanic variability, the interested reader is encouraged to peruse the review articles in the monograph edited by Cattle (1987). Many of the lectures given at the NCAR Summer Colloquium (1987) on the dynamics of low-frequency phenomena in the atmosphere are also relevant to our subject matter. The notes from these presentations (available as a three-volume publication from NCAR) provide a comprehensive survey of the pertinent literature.

19.2 Variability on daily and monthly time scales

19.2.1 Phenomena with periods of several days

At first glance, the typical periods of these circulation features may appear to be too short to have any noticeable impact on the behavior of the climate system. Nonetheless, these short-lived phenomena do play a crucial role in the long-term balances of energy, momentum and water vapor in the atmospheric branch of the climate system. As will be elaborated later in this chapter, these high-frequency disturbances are also capable of influencing the more slowly varying flow patterns through mutual dynamical interactions, and by communicating the effects of anomalous boundary forcings to the quasi-stationary component of the atmospheric circulation. Moreover, the synoptic climatology (such as the seasonal dependence of precipitation amounts and temperature extremes) of many geographical sites is partially determined by the frequency of occurrence of these weather-scale features. The day-to-day (or "synoptic scale") fluctuations at a midlatitude site (such as those described in the top panel of Fig. 19.1) are mostly associated with migratory cyclones and anticyclones. The preferred trajectories of these disturbances are often referred to as "storm tracks" in the meteorological literature, and have been identified in the extratropical oceans of both the Northern and Southern Hemispheres (Blackmon, 1976; Trenberth, 1981). During the winter season, these zonally elongated sites of eddy activity are located downstream and slightly poleward of the quasi-stationary jet streams (narrow zones of high wind speed). Many observational studies of the structural and transport characteristics of the transient fluctuations residing in the storm tracks have been performed. These empirical results reveal many similarities with those associated with the life-cycle of unstable waves, as produced in simplified mechanistic models (e.g., Simmons and Hoskins, 1978). In particular, vigorous heat and momentum transports by the cyclone waves take place along the storm tracks, thus resulting in enhanced energy exchanges between the transient disturbances and the more slowly varying background flow.

The ability of the GCMs to reproduce the essential characteristics of the extratropical storm tracks has been the subject of many investigations. Most of these studies are concerned with the geographical distribution of temporal variance and covariance statistics based on time-filtered model data. As an illustration of the capability of the current generation of GCMs to simulate the storm track properties, the wintertime distributions of the root-mean-squares of the local height of the 500 mb surface are shown in Fig. 19.2. The height data have been filtered (Blackmon, 1976) to retain fluctuations with time scales between 2.5 and 6 days. Maxima in these charts indicate the occurrence of enhanced variability on subweekly time scales in the nearby regions. It is evident that the geographical locations of the observed sites of synoptic activity over the North Pacific and North Atlantic, as well as the typical eddy amplitude therein, are well simulated in the model atmosphere.

Synoptic-scale weather systems are found not only in the temperate latitudes. Summertime wavelike disturbances with periods of several days are also observed to be active in certain tropical zones (Lau and Lau, 1990). Such transient fluctuations are mostly westward moving, and exhibit well-

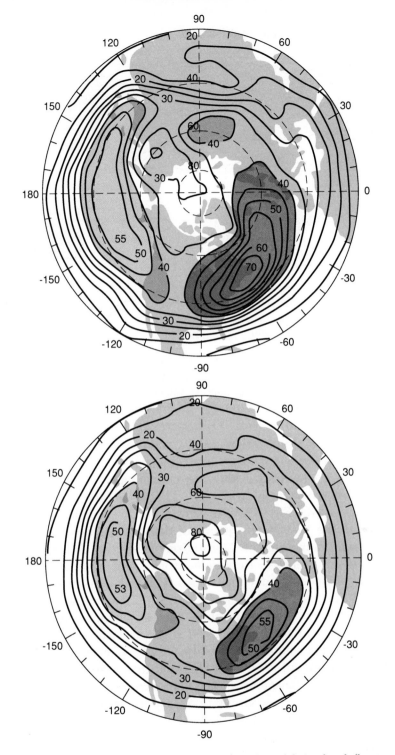

Fig. 19.2 Distributions of the observed (top panel) and model-simulated (bottom panel) root-mean-squares of wintertime 500 mb height data which have been filtered to retain fluctuations with periods between 2.5 and 6 days. Contour interval: 5 m. From Lau and Nath (1987).

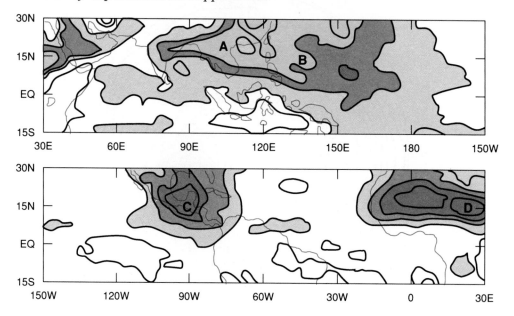

Fig. 19.3 Distribution of the model-simulated root-mean-squares of unfiltered 850 mb
vorticity during the northern summer. Contour interval: 4×10^{-6} s^{-1}. Light,
medium, and dense shading indicate values of 10–14×10^{-6} s^{-1}, 14–18×10^{-6}
s^{-1}, and $> 18 \times 10^{-6}$ s^{-1}, respectively. From Lau (1991).

defined wavelengths, propagation speeds, and spatial relationships with the
ambient large-scale circulation. They appear to be intimately associated
with deep moist convection, and are occasionally accompanied by the
formation of severe tropical vortices. The appearance of these tropical
features in a GCM (Fig. 19.3) is shown by the northern summertime
distribution of the root-mean-squares of daily fluctuations in the 850 mb
vorticity, which is a measure of the rotation of the flow. This pattern clearly
illustrates the occurrence in the model atmosphere of enhanced variability
over the Bay of Bengal/Indochina region, tropical western Pacific, Central
America and northern Africa (marked in Fig. 19.3 by the labels A, B, C
and D, respectively). The disturbances over the Asian monsoon regions in
sites A and B appear to be associated with the local semipermanent low
pressure zone during summer. The eddy activity over northern Africa bears
a strong relationship with the local climatological easterly jet stream. The
locations of various active sites in the model tropics, and their association
with the seasonally averaged circulation features, are in agreement with the
corresponding observational results.

19.2.2 *Phenomena with periods ranging from 10 days to a season*

The processes contributing to atmospheric variability on this time
scale may be classified under two broad categories – internal atmospheric

dynamics and forcing by anomalous boundary conditions. A comprehensive review of the nature of various internal and boundary mechanisms has been offered by Wallace and Blackmon (1983). The internal atmospheric processes relevant to this range include nonlinear interactions among different atmospheric wave components, and forcing by high-frequency transients (such as those described in the previous subsection). The second class of processes entails the forcing of the atmospheric circulation by the slowly varying changes in the properties of the land and ocean surfaces. Among the various interactive processes between the atmosphere and other components of the climate system, particular emphasis has been placed on the impact of sea surface temperature (SST) conditions on the atmospheric circulation. To date, the diagnosis of subseasonal variability has mostly been performed on GCM experiments with fixed, climatological SST conditions (see Sec. 19.1 of this chapter for a general discussion on experimental designs). A majority of the simulations with prescribed anomalous forcing is concerned with atmospheric changes on time scales of a season and longer. To simplify the presentation, we shall limit our attention to the role of internal atmospheric processes in this subsection, and defer the discussion on boundary forcing to the next section.

A detailed statistical analysis of the atmospheric variability simulated in a 15-year integration of a GCM (Manabe and Hahn, 1981) shows that, even in the absence of any anomalous SST forcing, the amplitudes of pressure and temperature fluctuations on monthly time scales in the model extratropics approach those of the observed atmosphere. The wintertime hemispheric distributions of model and observed root-mean-squares of 500 mb height values which have been filtered to retain fluctuations with periods between 10 days and a season (Fig. 19.4) reveal that the observed sites of enhanced variability on this time scale over the North Pacific and North Atlantic are mimicked by the model atmosphere, with the simulated amplitudes reaching to 70–80% of the observed values.

The output from Manabe and Hahn's experiment provides additional evidence that the model atmosphere with climatological SST forcing can reproduce the frequency dependence of the three-dimensional structure and temporal evolution of observed midlatitude phenomena with subseasonal time scales (Lau, 1981; Lau and Nath, 1987). In particular, some of the best-known recurrent circulation anomalies (often referred to as "teleconnection patterns") are discernible in the model atmosphere. The space-time evolution of model features with time scales of 10–30 days is similar to that associated with the dispersion of wave energy, with little or no phase propagation, whereas perturbations with periods longer than a month are organized in dipolar structures in the Northern Hemisphere. This distinction between the simulated eddy behavior in the two period ranges is in agreement with observational analyses (e.g., see Blackmon et al., 1984). The structural and energetic behavior of both observed and simulated

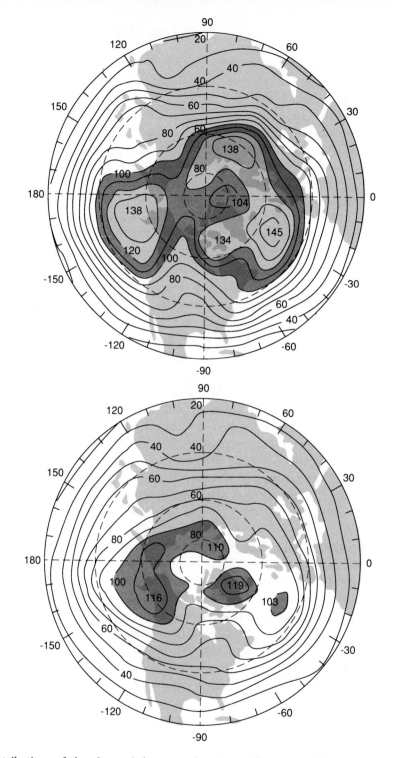

Fig. 19.4 Distributions of the observed (top panel) and model-simulated (bottom panel) root-mean-squares of wintertime 500 mb height data which have been filtered to retain fluctuations with periods between 10 days and a season. Contour interval: 10 m. From Lau and Nath (1987).

perturbations with monthly time scales tend to be elongated in the east–west direction, and thereby extract kinetic energy from the time-averaged jet streams (Wallace and Lau, 1985).

The largely statistical description in the preceding paragraphs is complemented by case-study approaches aimed at highlighting specific classes of model phenomena with weekly and monthly time scales. Of particular interest is the simulation of long-lived, quasi-stationary high pressure centers (often referred to as "blocking highs") in the extratropics. The climatology of blocking events appearing in a 1,200-day wintertime experiment with a GCM subjected to fixed boundary conditions (Blackmon et al., 1986) shows that the structural characteristics of the blocking highs in the model atmosphere are in agreement with the observations. For instance, the composite pattern of simulated blocking highs over the North Pacific bears a striking resemblance to its observed counterpart (Fig. 19.5). In the charts shown, the air flow is oriented approximately parallel to the contours. The set of parabola-shaped contours in the vicinity of 160–170°W represents a marked impediment to the normal west-to-east flow (hence the term "blocking"). Examples of model blocks occurring in conjunction with explosive cyclone formation or strong interactions with synoptic-scale eddies were also presented. The realistic simulation of the blocking episodes in a model environment with climatological boundary conditions led these investigators to suggest that internal atmospheric dynamics and physics could play an important role in the formation and maintenance of such phenomena. The heat and vorticity budgets as well as the transient eddy forcing of the model-generated blocking anticyclones were analyzed (Mullen, 1986). These results highlight the association of blocking events with systematic displacements of the tracks of synoptic-scale disturbances, and with well-defined changes in the heat and momentum transport properties of the high-frequency eddies. Many of the relationships between the simulated storm tracks and the slowly varying circulation are consistent with those pertaining to the observed atmosphere.

Another family of intraseasonal phenomena that has attracted considerable attention in the modeling community is the eastward propagating, planetary-scale "Madden–Julian" oscillation in the tropical troposphere, with a characteristic period of 40–50 days. This feature is associated with observed large-scale variations in pressure, wind, and convective activity near the equator (Madden and Julian, 1972). The origin of this oscillation has been linked to the interaction between moist convection and various wave modes which are preferentially excited in the deep tropics (e.g., Lau and Peng, 1987). A multitude of model diagnostic studies have demonstrated that some of the essential characteristics of the tropical intraseasonal oscillations are captured by GCMs subjected to nonvarying boundary forcing. An example of the model-simulated perturbations in precipitation (contours) and upper-level divergence (stippling) associated

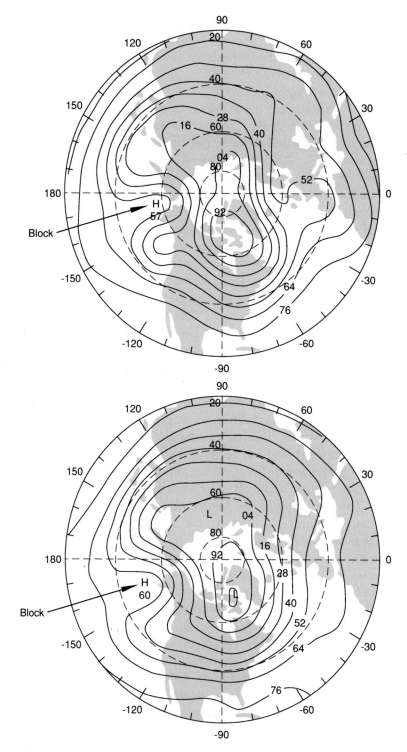

Fig. 19.5 Composite distribution of 500 mb height for observed (top panel) and model-simulated (bottom panel) blocking events over the North Pacific. Contour interval: 120 m. From Blackmon et al. (1986). Note the high pressure region (labelled as "Block") near 160–170°W.

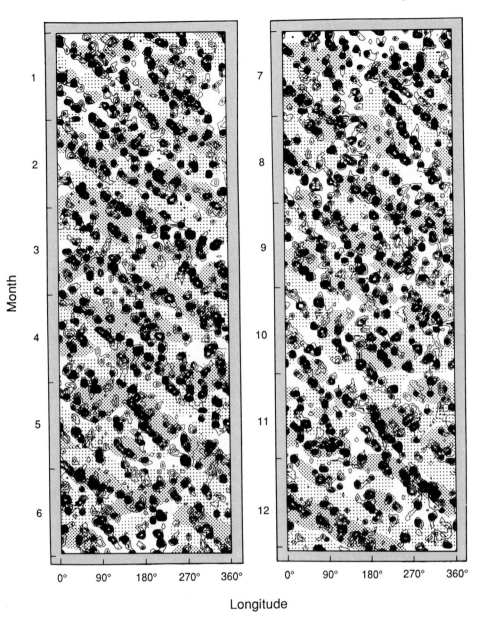

Fig. 19.6 Time–longitude distribution of model-simulated daily precipitation rate near the equator. Contour interval: 4 mm day^{-1}. Light and dense stippling indicate large-scale convergence and divergence near the tropopause, respectively. From Lau et al. (1988).

with this propagating phenomenon extracted from a 1,400-day integration using an idealized GCM with a zonally symmetric climate is displayed in Fig. 19.6. Note the tendency for most of the precipitation signals to migrate towards increasing longitude as one scans down the time axis. This diagram hence illustrates the prevalence of organized, eastward-traveling convective cells and divergence centers along the equatorial belt. Analogously, dry episodes are typically accompanied by upper-level convergence. The typical period of this oscillation in the model atmosphere is about 20–30 days. The notably shorter periods (as compared with the observations) seen here appear to be a deficiency shared by many GCMs. The phase speeds of these oscillations are seen to be sensitive to the details of model parameterizations of static stability and penetrative cumulus convection.

19.3 Variability on time scales ranging from months to several years

Simulations of atmospheric variability on time scales of several months to several years have mostly been concerned with the roles of interactive processes taking place at the air–sea and air–land interfaces. We shall therefore devote this section to describing phenomena associated with atmospheric interactions with the underlying boundaries. As pointed out in Sec. 19.2.2, conditions at the land and ocean surfaces may influence the atmospheric circulation on time scales even shorter than a season. Hence part of the following discussion has relevance to some of the subseasonal circulation features mentioned in Sec. 19.2.2.

19.3.1 *Phenomena associated with air–sea interaction in the tropics*

A substantial fraction of the modeling activity on the impact of tropical SST anomalies is focused on the atmosphere/ocean behavior during El Niño–Southern Oscillation (ENSO) episodes in the Pacific. These ENSO cycles have a typical time scale of several years, and are accompanied by coherent changes in many elements of the climate system. For instance, the interannual variations in Northern Hemisphere mean temperature previously noted in the bottom panel of Fig. 19.1 are rather well correlated with the timing of ENSO events in the same era. In view of the widespread influences of ENSO-related processes, secular changes in the frequency of occurrence of such episodes may complicate attempts to detect global warming due to increasing greenhouse gases (Trenberth, 1990). For a more detailed description of the important role of air–sea coupling in generating various ENSO phenomena, the reader is referred to Chapter 18.

Many simulations of the atmospheric phenomena related to ENSO have been conducted by prescribing a perturbed SST forcing at the lower boundary which is *spatially and temporally fixed*. Such experiments are hence primarily aimed at studying the *steady-state* response of the model atmosphere to a set of anomalous (but constant) boundary conditions. An additional degree of realism was built into the series of model runs analyzed by Lau (1985), in which the observed month-to-month changes

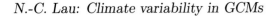

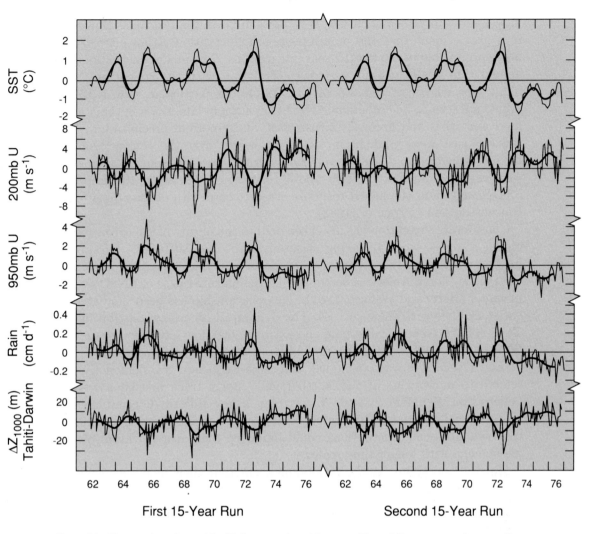

First 15-Year Run   Second 15-Year Run

Fig. 19.7 Time series of monthly (light curves) and lowpass filtered (heavy curves) anomalies of sea surface temperature, zonal wind at 200 and 950 mb, precipitation rate, and difference in 1,000 mb height between Tahiti and Darwin, for two independent 15-year model simulations. The tropical indices are computed using data for the central equatorial Pacific. From Lau (1985).

in the tropical Pacific SST field were specified, so that the lower boundary forcing of the atmospheric circulation varies in both space and time. The latter experimental design allows for a more detailed simulation of the *time-varying* atmospheric response in different phases of the ENSO cycle.

In Lau (1985), experiments were initiated using two independent sets of atmospheric conditions, but were subjected to the identical sequence of anomalous SST forcing observed during the 1962–1976 period. The time series of representative indices of the atmospheric circulation in a pair of 15-year integrations (Fig. 19.7) indicate that the interannual component of the model variability (heavy curves) is rather similar in the two runs. However, it is also evident that the month-to-month fluctuations (thin curves) are not reproducible in the two 15-year integrations. The model fluctuations on annual time scales exhibit a strong correlation with the concurrent SST input (see curves at the top of Fig. 19.7). In particular, warm ENSO events are seen to be accompanied by below-normal intensities in the upper-level westerlies and lower-level easterlies, above-normal precipitation, and reduced east–west gradient in 1,000 mb height over the central tropical Pacific, whereas cold events are characterized by atmospheric anomalies with reversed polarities. The above relationships between the imposed SST changes and the model response are in accord with observations (e.g., Rasmusson and Carpenter, 1982).

Considerable attention has also been paid to the nature of the *extratropical* model response to SST forcing situated in the low latitudes. Of special interest is the observed occurrence of a prominent teleconnection pattern in the Pacific/North American sector during the mature phase of ENSO events in winter (Horel and Wallace, 1981). This relationship has been simulated in various GCMs with some success. The dynamical processes contributing to the link between tropical forcing and extratropical flow have been examined by Held et al. (1989) using the model output from the experiments described in Lau (1985). It was concluded that high-frequency transient activities related to movements of the North Pacific storm track and equatorward penetration of planetary-scale waves play a key role in determining the extratropical response. This finding reinforces the earlier claim in Sec. 19.2.1 that synoptic-scale fluctuations could influence the nature of atmospheric phenomena with longer time scales.

Manabe and Hahn (1981) have pointed out that their model experiment with climatological SST forcing produces lower levels of variability in the tropics than observed. The variance statistics of their "control" experiment have been compared with those derived from the model run with prescribed SST changes (Lau, 1985). It is seen that the presence of changing SST boundary conditions leads to a significant enhancement in the amplitude of variability in the model tropics. Hence a large fraction of tropical variability may be attributed to air–sea interactions.

Some GCM experiments have also been performed to evaluate the role of SST anomalies in the occurrence of severe droughts. The prescription of a dipole-like SST anomaly in the tropical Atlantic produced drought-like conditions in northeastern Brazil (Moura and Shukla, 1981). Model evidence on the importance of worldwide SST anomalies in the incidence of prolonged dry periods in the Sahel region of Africa has been presented (Folland et al., 1986). The possible causes of the North American drought in the summer of 1988 have been analyzed (Trenberth et al., 1988). It was concluded in the latter study that this event was primarily brought about by an anomalous circulation pattern, which in turn was a forced response to the distribution of atmospheric heating associated with perturbed SST conditions in the tropical Pacific. The 30-day experiments conducted with operational models at various centers (Palmer and Brankovic, 1989; Kálnay et al., 1990) appear to lend some support to this proposed link between the cold phase of ENSO and the summer drought in 1988.

19.3.2 Phenomena associated with air–sea interaction in the extratropics

The current emphasis on SST anomalies in the equatorial zone has somewhat overshadowed the potential effects of *extratropical* oceanic conditions on atmospheric variability. The relationships between SST variations in the midlatitude oceans and short-term climate changes have long been recognized (e.g., Namias, 1969). The research interest in this issue has recently been revived by new observational and modeling results. These studies demonstrate that recurrent SST anomaly patterns in the North Pacific and North Atlantic are accompanied by well-defined circulation changes in the overlying atmosphere. The experiment analyzed by Pitcher et al. (1988) indicates that the atmospheric responses to prescribed midlatitude SST forcings exhibit a notable degree of nonlinearity, i.e., the response does not undergo a straightforward sign reversal when the polarity of the imposed SST anomaly is changed. This nonlinear behavior is also evident in similar model runs analyzed by Kushnir and Lau (1992). The relative contributions of SST anomalies in the tropics and the extratropics to atmospheric variability on interannual time scales were evaluated with a 30-year integration (Lau and Nath, 1990). The experimental design of this model run is similar to that used in Lau (1985), except that the domain of prescribed (and temporally varying) SST anomalies has been extended to all maritime gridpoints lying north of 40°S, so as to incorporate the boundary forcing in the North Pacific and North Atlantic as well as the tropical oceans. A global survey of the model output indicates that the seasonally averaged circulation during the northern winter exhibits strong covariability with the thermal conditions at two specific ocean sites in the extratropics, namely, the regions just south of Newfoundland and to the northwest of Hawaii. The distributions of regression coefficients of model-simulated 515 mb height versus the SST input at these two forcing locations (Fig. 19.8) depict the amplitude and polarity of typical pressure changes at various gridpoints corresponding to an SST increase of 1°C at the forcing site (indicated by a solid dot). The spatial pattern of the model response with respect to the SST forcing is in accord with the observational results. Lau and Nath also reported that the circulation in the Pacific/North American sector of the model atmosphere is more sensitive to in situ SST variations in the North Pacific than to remote forcing from the tropical ENSO region. The same conclusion was

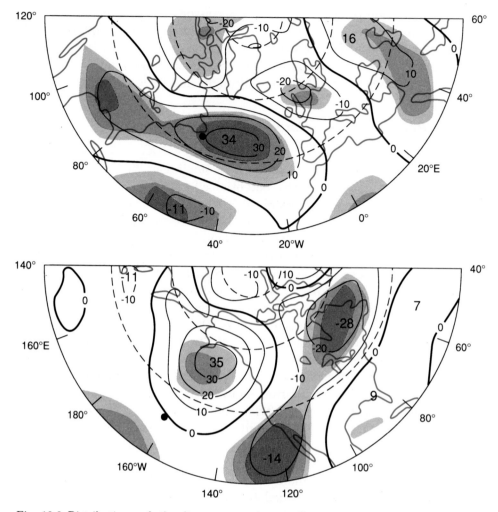

Fig. 19.8 Distributions of the linear regression coefficients between model-simulated, seasonally averaged 515 mb height at individual gridpoints and the imposed sea surface temperature anomalies at 45°N 56°W (top panel) and 31.5°N 161°W (bottom panel). The locations of the sea surface temperature forcing sites are indicated by solid dots. Significant regression values at the 90%, 95% and 99% levels are indicated by light, medium and dense shading, respectively. Contour interval: 10 m per °C. From Lau and Nath (1990).

reached earlier in the observational analysis by Wallace and Jiang (1987). A comprehensive treatment of climate variability must therefore take into account the effects of extratropical air–sea interaction.

Diagnosis of the model output by Palmer and Sun (1985) and Lau and Nath (1990) indicates that not only do the seasonally averaged flow patterns respond to the imposed oceanic forcing, the preferred sites of synoptic-scale disturbances are also noticeably altered due to relocations of the zones of strong temperature gradient at the lower boundary. The displacements of the storm tracks, in turn, exert a strong influence on the distributions of precipitation (and hence atmospheric heating) anomalies, as well as the spatial patterns of heat and moisture fluxes across the air–sea interface. Moreover, realignments of the storm tracks lead to notable changes in the dynamical interaction between the quasi-stationary circulation and the synoptic-scale eddies. Here is yet another example of the strong relationships between storm track dynamics and atmospheric behavior on lower frequencies.

In addition to the *simultaneous* relationships between midlatitude SST anomalies and the atmospheric circulation described above, there also exists some empirical evidence on the tendency for the atmospheric flow to *lead* the oceanic changes by one to several months (Davis, 1976; Wallace et al., 1990). Such results suggest that fluctuations in the extratropical oceans may actually be the result, rather than the initiator, of air–sea interactions. The asynchronous relationships have been investigated by the present author on the basis of an experiment in which an atmospheric GCM was coupled to a motionless mixed ocean layer of uniform thickness. The temperature of this ocean layer was allowed to vary according to the intensity of local heat fluxes across the air–sea boundary. Fluctuations of the latter fluxes are largely determined by the surface wind speed and air-to-sea temperature differences. Analysis of the model output indicates that even such a simple coupled system is capable of reproducing some of the observed spatial modes of SST and sea level pressure variability. Moreover, the simulated atmospheric changes typically lead the changes in the mixed layer temperature by approximately one month. It thus appears that, even in the absence of detailed ocean dynamics, the extratropical mixed layer in this simplified coupled model responds to local changes in the air–sea heat exchange in a rather realistic manner. These model results further substantiate the notion that the driving of the ocean by the atmospheric circulation plays a significant role in the extratropical variability of the coupled system. However, by virtue of the strong degree of temporal persistence of the oceanic fluctuations, a given SST anomaly produced several months earlier might last long enough to affect the current atmospheric state. It is therefore reasonable that a considerable degree of simultaneous correlation still remains between the variations in the ocean and the atmosphere, as demonstrated in the observational and modeling studies cited earlier in this subsection. Analysis of experiments with fully coupled ocean–atmosphere GCMs should be very helpful in advancing our understanding of the full array of feedback mechanisms involved in extratropical air–sea interaction.

19.3.3 Air–land interaction

One of the most conspicuous properties of the land surface that exhibits marked variability from winter to winter is the latitudinal extent of continental snow cover (see Chapter 5). Idealized GCM experiments confirm that changes in the snow cover alter the albedo and availability of ground water, and thereby exert a strong influence on the heat and soil moisture budgets of the land surface (Yeh et al., 1983). These model runs further illustrate that the removal of snow cover in late winter results in fluctuations in the climate system that persist well into the following spring and summer, with significant impacts on the thermal and dynamical structure of the atmosphere. Observational evidence on the relationships between seasonal snow cover and the atmospheric circulation (Walsh et al., 1982) shows that the extent of the snow line over North America exhibits a strong correlation with a characteristic teleconnection pattern at 700 mb (top panel of Fig. 19.9). Above-normal snow cover extent at a given longitude coincides with below-normal 700 mb heights directly aloft, and vice versa. The model output from the 15-year experiment of Manabe and Hahn (1981) analyzed by the present author in the same manner shows that the resulting correlation pattern (bottom panel of Fig. 19.9) bears a considerable resemblance to the observations. The lag-correlation statistics indicate that the changes in the atmospheric circulation tend to precede the variations in the snow cover in both model and observations. It thus appears that interactions between continental snow cover and the atmospheric circulation are properly captured by the GCM under investigation. In analogy with the discussion on extratropical air–sea interaction in Section 19.3.2, the nature of the simultaneous and lagged relationships between the atmospheric changes and the snow cover anomaly suggests that the atmospheric circulation could exert a considerable influence on the extent of continental snow cover. The persistence of the snow cover anomalies on monthly time scales could in turn allow for such anomalies to modify the present atmospheric flow.

Another property of the land surface affecting climate variability on interannual time scales is soil moisture. Diagnosis of GCM experiments with idealized geography indicates that soil moisture anomalies can persist for at least several months (Yeh et al., 1984). The length of this persistence time scale exhibits a latitudinal dependence, and is governed by the interplay of various hydrological processes such as evaporation, precipitation, runoff, and atmospheric moisture transport. It was also demonstrated that the soil moisture changes result in notable perturbations of the atmospheric circulation. Experimentation with a two-level GCM allowing for atmospheric interactions with the soil moisture content (Gordon and Hunt, 1987; Hunt and Gordon, 1988) reveals that, even in the absence of anomalous SST forcing, annual and multi-annual droughts do occur in selected geographical regions in the model climate. When compared with observed droughts, the simulated events tend to be more locally confined.

The summertime response of model atmospheres with realistic geography to imposed soil moisture perturbations demonstrate that the prescribed soil moisture changes have noticeable effects on the temperature and

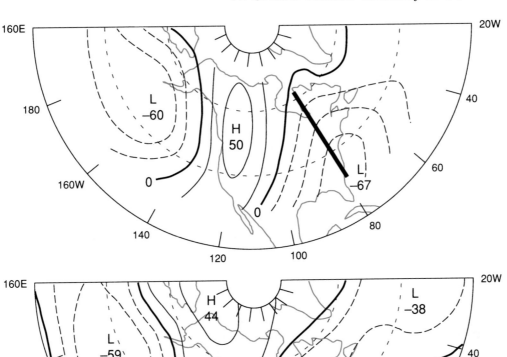

Fig. 19.9. Distributions of the correlation coefficients (in percent) between wintertime
700 mb height at individual gridpoints and the latitudinal extent of snow cover
near the 80°W meridian, as obtained from observations (top panel, from Walsh
et al., 1982), and from the author's analysis of the 15-year model simulation by
Manabe and Hahn (bottom panel).

precipitation in the vicinity of the anomaly area (see Rowntree and Bolton
(1983) for the European sector, and Rind (1982) and Oglesby and Erickson
(1989) for North America). In particular, Oglesby and Erickson (1989)
pointed out that reduced soil wetness may contribute to the initiation
and sustenance of warm and dry conditions over midlatitude continental
interiors, such as those observed over North America in the summer of 1988.

The nature of soil wetness variability and its influences on the near-
surface climate were interpreted in a stochastic framework analogous to

that outlined by Hasselmann (1976) in two model studies by Delworth and Manabe (1988, 1989). It was noted that the soil moisture parameterization scheme used in the experiments is a close mathematical analog to a first-order Markov process[1], in which random forcing with a white-noise spectrum (i.e., with variance uniformly spread over all frequencies) could yield a red-noise response (i.e., with more variance residing in lower frequencies). Evaporation serves as the damping term in the linear Markov process; whereas the white-noise forcing is supplied by the sum of rainfall and snowmelt. The soil moisture may hence be viewed as an integrator of random precipitation forcing from the atmosphere, thereby lending an increased low-frequency (red-noise) component to the variability of the ground hydrology. The model evidence presented by Delworth and Manabe confirms that the precipitation forcing is characterized by a white-noise spectrum, whereas the soil moisture response exhibits a distinct red-noise behavior, with half of the total variance residing at periods longer than seven months in the low latitudes, and as much as 13–20 months in the extratropics. It was shown that the persistence time scale of model soil wetness anomalies is inversely proportional to the rate of dissipation of soil wetness anomalies by evaporation. This dissipation rate is dependent on potential evaporation, so that the persistence time scale is longer when potential evaporation is weak (note that potential evaporation decreases poleward in relation to the poleward reduction of solar insolation). The occurrence of frequent runoff in regions where precipitation exceeds the mean potential evaporation rate can substantially reduce the persistence time scale of model soil wetness.

Some of these model findings have been confirmed by Vinnikov and Yeserkepova (1991) using soil moisture measurements taken at an extensive network of stations in the then Soviet Union. Delworth and Manabe have further evaluated the impact of soil processes on atmospheric variability by comparing the model statistics for a 50-year experiment which incorporates full interactions between the land surface and the atmosphere, with the corresponding statistics for a 25-year run in which such interactions are suppressed. It was concluded that soil processes act to increase the characteristic persistence time scale of the atmosphere. Due to influences of the soil wetness on the intensity of latent and sensible heat fluxes through the ground surface, the amplitude of the variability in temperature and relative humidity is also notably increased in the interactive run, particularly in the low latitudes and during the summer season.

19.4 Variability on decadal and centennial time scales

Thus far, only limited observational and modeling efforts have been devoted to climate variability on time scales of decades and longer.

---

[1] A concise discussion of Markov processes can be found in Chatfield (1984).

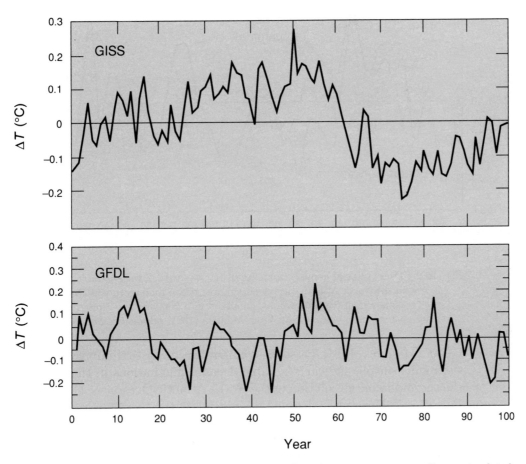

Fig. 19.10 Time series of annual global mean surface air temperature anomalies, as simulated in century-long integrations with coupled models at GISS (top panel, from Hansen and Lebedeff, 1987), and at GFDL (bottom panel, from Manabe et al., 1991).

Our understanding of such ultra-long period fluctuations therefore remains rather inadequate. It is, however, of interest to note the few available century-long integrations of coupled models with fixed external forcing do yield interdecadal variations of the global mean surface temperatures, with typical amplitudes of several tenths of a degree C (Fig. 19.10). In order to comprehend the origin of this low-frequency variability, one must invoke those processes in the climate system with very long time scales.

Since both the specific heat of sea water and the mass of the total water column are very large, the deep ocean has an enormous heat capacity, and exhibits fluctuations on time scales ranging from decades to centuries. Processes associated with the deep ocean circulation should therefore contribute significantly to very low-frequency variability of the climate system. The nature of atmosphere–ocean interactions on decadal time scales has recently been investigated using a 200-year integration of a coupled

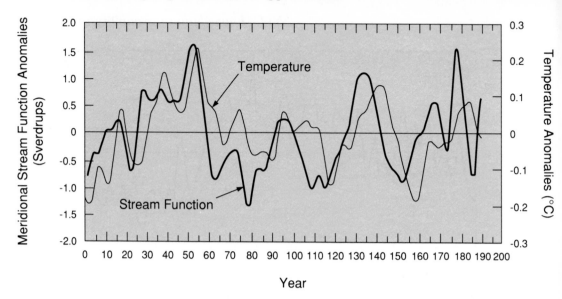

Fig. 19.11 Time series of annual anomalies of the intensity of the thermohaline circulation in the North Atlantic (heavy curve), and ocean temperature at 85 m depth in the same region (thin curve), as simulated in a 200-year experiment with a coupled atmosphere–ocean GCM. From Manabe (1991, personal communication).

GCM (Manabe, 1991, personal communication). Preliminary analysis of the model output indicates notable decadal fluctuations in the intensity of the thermohaline circulation in the North Atlantic Ocean (see heavy curve in Fig. 19.11). For a more detailed description of the thermohaline circulation, the reader is referred to Chapter 17. The changes in the thermohaline circulation are seen to be accompanied by pronounced variations in the subsurface ocean temperature (see thin curve in Fig. 19.11). There exists a distinct tendency for the fluctuations in the thermohaline circulation to lead the temperature anomalies by several years. These findings illustrate a potential role for the thermohaline circulation to modulate the North Atlantic climate on decadal time scales. Diagnosis of the same 200-year experiment also reveals substantial low-frequency variability in the Southern Ocean, with local SST anomalies persisting for several decades. These phenomena are apparently manifestations of long-period fluctuations of the circumpolar deep ocean circulation around Antarctica.

The amplitude of low-frequency variability simulated in the above-mentioned coupled experiment is shown by the distribution of the standard deviation of annually averaged SST, as calculated using data from the first 50 years of the integration (Fig. 19.12). This pattern indicates that enhanced variability is simulated in the North Atlantic, northwestern Pacific and the waters surrounding Antarctica. The maxima in the North Atlantic and the southern oceans coincide well with the sites of pronounced fluctuations in the thermohaline and the circumpolar deep

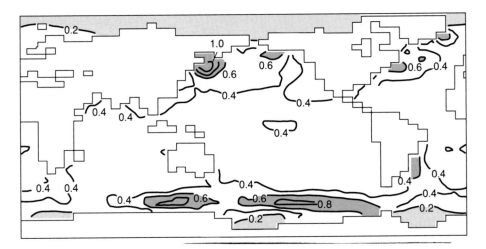

Fig. 19.12 Distribution of the standard deviation of annual mean sea surface temperature, as computed using the first 50 years of a 200-year experiment with a coupled atmosphere–ocean GCM. Contour interval: 0.25°C. From Manabe (1991, personal communication).

ocean circulations, respectively, as discussed in the preceding paragraph. The Northern Hemisphere, extratropical features in Fig. 19.12 are in fair agreement with the observations. The model variability in the tropical Pacific is noticeably weaker than that inferred from ship measurements. The latter discrepancy is primarily a consequence of the low resolution of the ocean GCM used in this experiment, which leads to ENSO events with reduced amplitudes.

In analogy with the discussion of air–land interaction in Sec. 19.3.3, the variability introduced by ocean–atmosphere coupling may also be viewed in a stochastic framework. In the present context, the principal damping mechanism is represented by deep mixing in the ocean interior, whereas the white-noise forcing may be supplied by randomly distributed freshwater fluxes from the atmosphere. This framework was tested using the output of a 4,000-year integration with an ocean model (Mikolajewicz and Maier-Reimer, 1990). Their results suggest that the ocean response to white-noise atmospheric forcing does exhibit a red-noise character for periods of 30 years and less. On even longer time scales, the model behavior is significantly different from that associated with a simple first-order Markov process. Instead, it was shown that the internal dynamics of the model drive a well-defined mode of ocean variability with a typical period of about 320 years. This very low-frequency oscillation is accompanied by coherent changes in the intensity of the basin-wide thermohaline circulation in the Atlantic, and in the mass transport through the Drake Passage.

19.5 Outlook

In this brief survey, it has been demonstrated that the GCMs serve as a powerful tool in expanding our knowledge of the nature of climate variability on a wide range of time scales. Properly designed experiments with these GCMs have provided useful insights on the individual roles and relative importance of different processes in various phenomena occurring within the climate system. With the advent of ever-increasing computational capacity, it will soon be feasible to conduct experiments with coupled systems with durations of a millennium (or even longer) on a routine basis. The model output from such integrations should provide unprecedented details on the ultra low-frequency behavior of the climate system. In fact, the rapidly increasing temporal and spatial coverage of the model-simulated data may soon surpass the corresponding coverage of any conceivable instrumental record of the observed climate system. Thus the array of model tools will in future constitute an unique resource for studying long-term climate change.

Climate modeling activities have always been conducted in conjunction with observational analyses of the real system. The observations provide the ultimate test of model fidelity. Moreover, observational evidence of certain behavior in the climate system often motivates the design and diagnosis of new model experiments. Conversely, model predictions of cause-and-effect relationships or new phenomena often suggest innovative interpretations and analysis procedures of the pertinent observations, or provide useful guidelines for gathering new empirical data. This synergism among model experimentation and real data analysis will hopefully be sustained in future studies of climate variability.

Notwithstanding the immense potential of the numerical models for climate research, it is necessary to bear in mind that a realistic simulation of the climate system is contingent upon the accurate model representation of the complex processes operating within a given component of the system, as well as the interactions among different components. Model development efforts aimed at incorporating the various interactive mechanisms mentioned here are still in a relatively primitive stage. These endeavors would benefit enormously from the bridging of our current research and educational activities across traditional academic disciplines, such as meteorology, oceanography, hydrology, geochemistry and glaciology. The salient relationships between various subsystems of the Earth's climate clearly call for such interdisciplinary approaches.

20

Climate-model responses to increased CO$_2$ and other greenhouse gases

Warren M. Washington

20.1 Introduction

The Swedish chemist Svante Arrhenius (1896) was the first researcher to express concern about the possible climatic effects of increased concentrations of greenhouse gases, such as carbon dioxide (CO$_2$). At the same time, the Industrial Revolution of the late 1890s saw a great increase in fossil fuel use, particularly coal, that has led to the observed increases in CO$_2$ in the atmosphere (Fig. 2.2). Arrhenius concluded that atmospheric CO$_2$ was important to the Earth's heat balance and that increases in these gases would lead to increased atmospheric temperature. Arrhenius estimated that an increase of 2.5 to 3 times the CO$_2$ concentration would result in a globally averaged temperature increase of 8–9°C, an effect not too different from that estimated by today's complex computer climate models.

Somewhat earlier, the Irish scientist John Tynall (1861), in his classic paper, *On the absorption and radiation of heat by gases and vapours, and on the physical connexion of radiation, absorption and conduction,* described the effect of increasing CO$_2$ and other gases, including water vapor, on the Earth's radiative balance. In a greenhouse analogy, solar radiation would be transmitted largely through greenhouse glass while significant terrestrial or infrared radiation would be trapped by the glass.

With simple models, Callendar (1938), Plass (1961), Mitchell (1961), and others tried to link CO$_2$ increases to temperature change. This chapter discusses model experiments of the response of the climate to increases in greenhouse gases, especially CO$_2$. Beyond the use of simple models to examine aspects of this problem (Secs. 20.5 and 20.6), there are two main kinds of experiments that have been performed with General Circulation Models (GCMs). The first are equilibrium experiments in which the amount of CO$_2$ in the model atmosphere is doubled and the differences in resulting climate examined (Sec. 20.7). The second are so-called transient experiments in which the amount of CO$_2$ is increased slowly in a realistic fashion to examine how the climate adjusts to the slowly increasing greenhouse effect (Sec. 20.8). In all these experiments, it is not just the direct changes in greenhouse radiative heating that are important, but also the myriad of complex feedback effects that must be dealt with and that make this a difficult problem (Secs. 20.3 and 20.4). For a review of climate-modeling studies of the greenhouse effect, we refer the reader to the following sources: Dickinson (1986), Schlesinger and Mitchell (1987), Mitchell (1989), Chapters 5 and 6 of the report of the Intergovernmental Panel on Climate Change (IPCC, 1990a), and the book, *Prospects for Future Climate* (MacCracken et al., 1990).

20.2 Radiative effects of increased greenhouse gases

As evidenced by their radiative physics, greenhouse gases are strong absorbers and emitters of thermal infrared radiation in the 5- to 100-μm range. A more complete description of the radiation aspects appears in Chapters 1, 7, and 10. One curious finding is that oxygen and nitrogen – the main constituent gases in the Earth's atmosphere – are not radiatively active in the infrared wavelength range. The structure of these particular diatomic molecules does not allow them to absorb or emit infrared radiation. Here we point out that triatomic molecules such as CO_2, water vapor (H_2O), nitrous oxide (N_2O), and ozone (O_3) are all strong absorbers and emitters in the infrared wavelength range. These effects are even more strongly manifested in molecules of more than three atoms, such as methane (CH_4) and chlorofluorocarbons (CFCs). For polyatomic molecules, the atoms can (1) vibrate toward and away from each other so as to absorb and emit radiational energy in specific wavelength ranges (vibrational energy), and (2) rotate about each other in a complex manner (rotational energy). These infrared absorption and emission properties can be calculated for simple molecules; for complex molecules, however, the absorption spectra are measured experimentally.

Absorption spectra in the 8- to 12-μm range are especially sensitive to greenhouse-gas concentrations, and climate models used in climate-change scenarios must provide a reasonably accurate treatment of these gases in this wavelength region. Climate models provide for water-vapor and CO_2-absorption parameterizations [see Chapter 10, Sec. 10.3.1, or Chapter 2 of IPCC (1990a)]. An increase of CO_2 alone does not yield a sizable climate effect. It is the water-vapor-increase feedback that amplifies the direct radiative effect of CO_2 or other greenhouse trace gases. Without increased evaporation and increased water-vapor feedback, the direct greenhouse effect from trace gases would be small.

While most climate-sensitivity experiments have involved models with only increased CO_2, it is true that other radiatively active gases are also increasing. In Table 7.1 of Chapter 7, the following percentages of yearly increases in the atmosphere are noted: CO_2 – 0.5%, O_3 – 1% in the troposphere, CH_4 – 0.9%, N_2O – 0.2%, and CFCs – 4%. It is not known with certainty if water vapor, the principal greenhouse gas, is increasing or decreasing in concentration. Accordingly, a widespread practice has been to substitute an "effective" CO_2 amount for the other greenhouse gases, although this has recently been questioned (see Sec. 20.7).

20.3 Water-vapor and cloud feedbacks

Some aspects of water-vapor and cloud feedback to the climate system are covered in Chapters 1 and 3. The major emphasis here is to review several feedback-sensitivity experiments in relation to increased greenhouse gases.

Möller (1963) demonstrated the critical role of water-vapor feedback on the climate system and was the first to point out the *large* effect of water-vapor feedback on surface temperature. He assumed that the climatological distribution of relative humidity remains nearly the same, although the surface temperature changes. In effect, when the surface temperature increases, absolute humidity increases since relative humidity is approximately the same. Downward infrared radiation also increases, but solar radiation absorption increases in the atmosphere and solar absorption decreases at the Earth's surface. In a simple model with many assumptions (some questionable), Möller found that, for a surface temperature near 15°C, a doubling of CO_2 would result in a 10°C temperature increase. The main point of this research was to show that water vapor, the most important greenhouse gas, amplifies the effect of other greenhouse gases.

Cloud feedback is a most difficult aspect of greenhouse-gas climate modeling. Clouds can cool or warm the climate system, depending on how they change. The most common schemes in GCMs tie cloudiness to a simple dependence upon relative humidity (Chapter 10, Sec. 10.3.4). Complex cloud physical processes involving cloud liquid-water and ice-water content, and cloud and ice-crystal size, are not usually included. The methods of treating GCM cloud microphysical processes have been highly simplified compared with those in some GCMs with detailed cloud submodels (e.g., Roeckner et al., 1987; Smith, 1990).

There are several possible effects of clouds. First, low-level clouds (often with large amounts of cloud water) could increase in areal extent, thus increasing the amount of solar radiation reflected. An increase of only a few percent could compensate for greenhouse warming. On the other hand, enhancement of the cirrus-cloud extent and amount (usually ice clouds) increases the trapping of infrared radiation and intensifies the greenhouse effect. A second possibility involves changing cloud heights. Higher clouds are usually colder, radiate less infrared radiation, and increase the greenhouse effect. A third prospect is that clouds may change their water content and thus their optical thickness, thereby changing their albedo. Ramanathan and Collins (1991) have shown from satellite data that, under the special circumstance of ocean temperatures in the range of 305 K, cumulus convection can lead to cirrus with sizable solar albedos. If

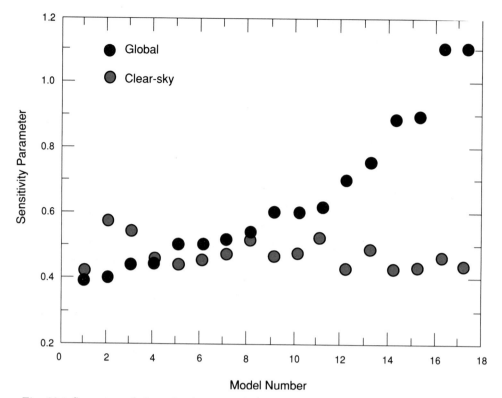

Fig. 20.1 Summary of clear-sky (open circles) and global-sensitivity parameters for both clear and cloudy sky (solid circles) for 17 GCMs.

a greenhouse warming leads to thicker clouds, it would likely result in a negative feedback and less greenhouse warming in the tropics. Experiments with climate models indicate that cloud feedback can cause less or greater greenhouse warming (see Mitchell et al., 1989, for one modeling study).

Although clouds can either enhance or decrease the amount, all of the models show that the greenhouse effect of clouds plus water vapor is always a warming, hence a net positive feedback. Cess et al. (1989) have shown a wide range of positive feedback values in an intercomparison study of climate models with prescribed hypothetical greenhouse-gas-caused ocean-temperature change. Figure 20.1 summarizes the sensitivity-feedback parameter from the 17 models; the sensitivity parameter is defined in Chapter 1 [Eq. (1.3)]. For clear sky, all models show positive feedback from increased water vapor, but for global cloudiness the sensitivity parameter ranges from a modest diminution of the global warming to a large amplification. When the sensitivity is large, then a sizable surface warming can result for doubled CO_2.

All of the climate model schemes for clouds are relatively crude when compared to nature. Many models show sizable cloud adjustments to increased CO_2 concentration, some tending to compensate or amplify in different regions of the atmosphere, so that a single globally averaged sensitivity number may be misleading. Figure 20.1 is shown only to give some idea of the range of model differences in cloud feedbacks to a specified ocean-surface temperature change (Cess et al., 1989). Because the cloud modeling and radiative components of each of these models are somewhat different, it is not surprising that the feedbacks can be different. A better grasp of how clouds work on time and spatial scales of GCMs is a high priority for improvement in climate modeling.

20.4   Snow-, sea, and land ice-albedo feedbacks

Most modeling studies have shown a large positive feedback of the climate system to albedo changes of snow and sea ice. The mechanism is thought to be simple. Surfaces covered by snow, sea ice or permanent land ice (e.g., Greenland) all have a high albedo. Any warming results in melting exposing bare ground or open sea water which will drastically change that region to low albedo. This will increase the absorption of solar radiation at the surface and thus enhance the warming. Greenhouse experiments that began with small quantities of snow and ice in the control experiment or unperturbed state will have less warming with increased greenhouse gases than model experiments with extensive snow and ice (e.g., Washington and Meehl, 1986; Ingram et al., 1989; Meehl and Washington, 1990). The dependence of the temperature of the control experiment on the amount of greenhouse warming is discussed in Sec. 20.7.

Cess et al. (1991) have shown that this view of the role of snow and sea ice albedo is too simple. In some models, although snow and sea ice decrease, clouds may increase so as to increase the planetary albedo. This is plausible because increased open water at the surface and thus in atmospheric water vapor can result in increases in low-level stratus clouds. Models, however, do not handle these interactions well. Sea ice has additional important feedback because of seasonal heat storage. For example, the high albedo of sea ice will lead to less heating in that region. Moreover, substantial amounts of heat are required to melt the sea ice. If sea ice disappears, then sensible heat can be stored in the upper ocean and retard freezing in the subsequent fall and winter. This feedback may further retard sea ice formation in subsequent seasons.

Even with our present knowledge of the key mechanisms involved in the feedbacks, the treatment of snow and sea ice processes in climate models needs to be improved substantially.

20.5   Energy balance climate-model estimates

Energy balance models (EBMs) were the first climate models to incorporate details of the important feedback processes involved in the greenhouse effect; see Chapter 1 for an overview and Chapter 10 for a description of the hierarchy of climate models. Used in the 1960s by Budyko (1974) and Sellers (1974), these models compute a surface temperature that balances heat fluxes. Depending on the model's details, terms are added to account for meridional (i.e., north–south) fluxes of energy by the atmosphere and ocean system.

The effect of increased greenhouse gases with such models is investigated by changing the infrared radiation aspects of the model. More infrared flux toward the surface caused by increased greenhouse gases will result in a warmer surface. Because the surface and lower atmosphere in EBMs are closely coupled, the system responds by increasing surface temperature. It is also possible to include the change of albedo of snow and ice surfaces in the zonally averaged EBMs as a result of warmer surface temperatures. This positive feedback can amplify the warming, especially in the higher latitudes.

Reasonable assumptions in the surface physical processes of an EBM yield a global warming of about 3°C for doubled CO_2 concentrations. In this type of model, ad hoc assumptions must be made about the role of dynamics in the redistribution of heat and feedback mechanisms, but such models have proven useful in understanding the physics of climate change.

20.6 Radiative-convective model estimates

The next level of model complexity in studies of the greenhouse effect are radiative-convective models (RCMs), first invented by Manabe and others at the Geophysical Fluid Dynamics Laboratory (GFDL) in the early 1960s. These models are one-dimensional in the vertical for which column radiative heating and cooling rates can be computed. See Chapter 10, Sec. 10.2.2*b* for details of such models in a broad context. The convective aspect of the model is shown in Fig. 20.2, where the heating by convection causes the atmosphere to warm and adjust to a moist adiabatic lapse rate. In adjusting to a new stable lapse rate, convection by dry or moist processes warms the upper troposphere and cools the lower troposphere. The relative humidity is specified and so too are the levels and radiative properties of the clouds.

Figure 20.3 from Manabe and Strickler (1964) shows the vertical distribution of the heating and cooling rates of various atmospheric gases, including water vapor, CO_2, and ozone. It is evident from Fig. 20.3 that these gases are the largest factors in radiatively heating and cooling the atmosphere. In the figure, L denotes longwave or infrared and S denotes solar radiation. The net radiative heating and cooling are near zero in the stratosphere, but in the troposphere the net radiative cooling is compensated by vertical transfers of heat from the surface by moist and dry convection.

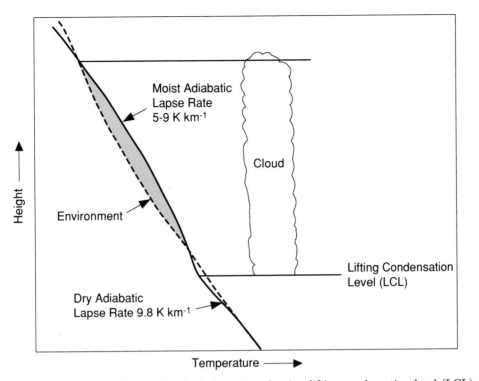

Fig. 20.2 Schematic of cumulus-cloud formation showing lifting condensation level (LCL), dry and moist adiabatic lapse rates, and lapse rate for air outside cloud (labeled "environment").

Solar and infrared heating and cooling rates are strongly affected by CO_2 concentrations.

Manabe and Wetherald (1967) were the first to use a RCM for CO_2 climate-change experiments. They set solar forcing to a global annual mean, assumed that the convective adjustment of temperature has a fixed critical lapse rate that cannot be exceeded, and specified relative humidity and cloudiness. The temperature structure in the stratosphere is based on radiative equilibrium between solar and infrared fluxes, whereas in the troposphere it is based on a balance of radiative fluxes and convective processes. Manabe and Wetherald show the distribution of temperature for three different concentrations of CO_2 (Fig. 20.4). At 150 ppm, the stratosphere warms and the troposphere cools; at 600 ppm, the stratosphere cools and the troposphere warms. Doubling the CO_2 results in a surface temperature increase of 2.9°C, and halving the concentration results in a surface temperature decrease of 2.4°C.

Because the RCMs are so highly parameterized, they should not be expected to give reliable estimates of greenhouse-gas-caused climate change, but they are instructional in the role of convective and radiative processes. With RCMs, Manabe and Wetherald (1967), Karol and Rozanov (1982), and

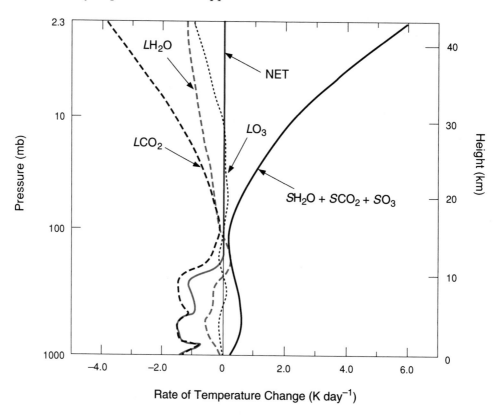

Fig. 20.3 Heating rates in atmosphere due to absorption of solar radiation (S) by atmospheric H_2O, CO_2, and O_3, and heating and cooling rates due to absorption and emission of longwave or infrared radiation (L) by H_2O, CO_2, and O_3. From Manabe and Strickler (1964).

Schlesinger (1985, 1988b) estimated surface temperature warming at ranges from $0.5°$ to $4.2°C$. These estimates vary for many reasons, one being how clouds and water vapor are handled in the model.

In his study of radiative and water-vapor feedback, Ramanathan (1981) used an RCM to aid our understanding of the basic mechanisms. His simple one-dimensional atmospheric RCM was coupled to a simple, uniformly mixed, 50-m layer of sea water. This type of model does not have all feedbacks of the real climate system, but it does include many fundamental processes, especially those dealing with water-vapor feedback. Figure 20.5 shows changes in surface temperature and flux of various quantities involved in the RCMs. Processes (1) and (2) are direct infrared radiative heating of the ground and the troposphere by a doubling of the Earth's CO_2 concentrations. The total additional heating of the troposphere is about $4 \ W \ m^{-2}$, of which $3 \ W \ m^{-2}$ in process (2) is heating the troposphere, and $1 \ W \ m^{-2}$ in process (1) is direct radiative heating of the Earth's surface.

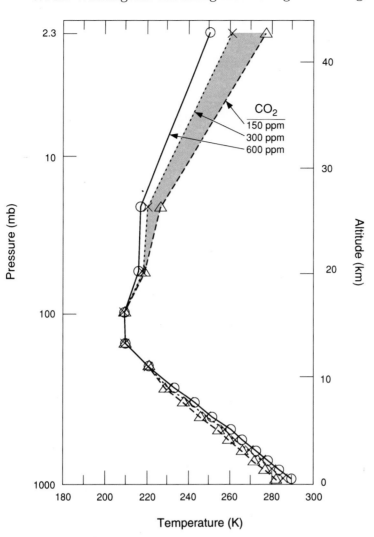

Fig. 20.4 Vertical distributions of temperature in a radiative-convective model for fixed relative humidity (FRH) and fixed cloud cover (FCL). Surface temperature change is 2.88°C for a CO_2 doubling from 150 to 300 ppm and 2.36°C for 300 to 600 ppm. From Manabe and Wetherald (1967).

Process (3) enhances the direct heating effect by a series of feedbacks. The ocean surface is warmed by direct heating which causes more evaporation and a larger absolute humidity (Δq) above the surface. This, in turn, increases latent-heat release and infrared flux back toward the ocean surface ($\Delta F^{\downarrow}$), causing increased surface temperature ($\Delta T_m > 0$). In a cyclic manner, temperature increases cause evaporation increases, etc. The process does not "run away" and the system settles down to a new equilibrium.

From their extensive survey of the literature, Schlesinger and Mitchell (1985) found the global surface temperature increases from doubled CO_2 in

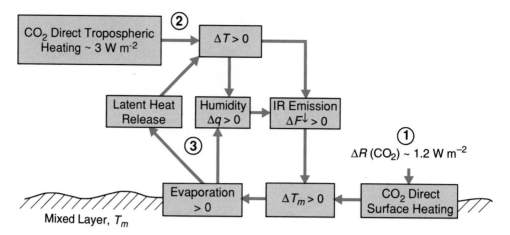

Fig. 20.5 Schematic illustration of ocean–atmospheric feedback processes by which CO_2 increase warms surface. All numbers correspond to hemispherically averaged conditions and apply to doubled CO_2. From Ramanathan (1981).

RCMs to be from 0.8° to 4°C. Although RCMs include more climate physical processes than EBMs, they are far from complete in feedback mechanisms. Most include a simple cloud mechanism. Treatment of snow and sea ice feedbacks is difficult since the models apply only to one particular location or for the globally averaged state.

20.7 Global coupled-model equilibrium estimates with simple oceans

The preferred tool for studying climate change is the three-dimensional GCM, because such a model can be compared directly with the observed climate system (particularly the atmosphere) and because GCMs make fewer model assumptions than EBMs or RCMs. As the name implies, GCMs compute explicitly the motions of the atmosphere or the ocean. An atmospheric GCM is usually coupled to an ocean GCM or to a simple ocean without dynamics (see Chapter 17 for more details on coupled atmosphere–ocean models). In the latter coupled models, the doubling of CO_2 is imposed in an instantaneous fashion. Since such a model cannot take into account the sequestering of heat in the mid- and lower oceans, it is not possible to investigate the effect of a gradual (or transient) increase of CO_2. The instantaneously doubled experiment is usually run to equilibrium. In contrast, when a full ocean GCM is coupled to the atmospheric GCM, it allows the transient response of the climate system to be examined (Sec. 20.8).

By itself, the conventional atmospheric GCM cannot be used for climate-change studies because it usually *specifies* ocean-surface temperatures and sea ice distributions. To simulate many of the feedback mechanisms, these two quantities must be *computed* as part of the solution. Manabe and Wetherald (1975) were pioneers in investigating climate change from increased CO_2 concentrations with a GCM. They considered two conditions for their climate-change sensitivity experiments: (1) that the model have a stable equilibrium climate, and (2) that external forcing (such as doubled CO_2 concentration) is not large enough to force the model climate from stable equilibrium into a markedly different state.

Because of the complexity of the climate system, these conditions are probably difficult to ensure in a mathematically rigorous sense, but they are reasonable criteria for any climate experiment. As an example of a drastically different climate state, picture a hypothetical ice-covered planet in which snow- and ice-albedo effects dominate so that surface temperature is not warm enough to melt snow and ice. The planet and its atmosphere remain so cold that the ice and snow albedos are unable to change and produce a warming. Such a state appears as a possible solution in simple models (see Sec. 10.2.1 and Fig. 10.1). Climate-sensitivity studies have indicated that the climate system is perturbed by increased CO_2, but it has not shifted into an entirely different state such as has occurred between the glacial and interglacial periods. Other climate-change mechanisms exist; for example, Broecker et al. (1985) suggest that a fluctuation in the transition from the glacial to the interglacial period may be caused by changes in the flow of the deep water of the North Atlantic (Chapter 17 describes this process in more detail). It may be possible for the climate system to have many stable equilibria. Therefore, the Manabe and Wetherald criteria (1) and (2) are useful for climate experiments, but they may not be sufficient.

Chapter 17 provides a review of the coupled atmosphere–ocean models in climate research. Manabe and Wetherald first used a simplified ocean and atmosphere. Because of limited computer resources, their model had simple geography, annually averaged solar forcing, and an ocean without heat storage or transport. Their term, "swamp ocean," meant that the ocean did not have heat capacity and was merely a wet surface with ocean temperature computed on the basis of the balance of fluxes of the Earth's surface (Chapter 17). They included a simple parameterization of sea ice based on the condition that sea ice would form whenever the ocean temperature dropped below the freezing point of sea water (approximately $-1.8°C$). The latitude–height zonally averaged mean distribution of temperature

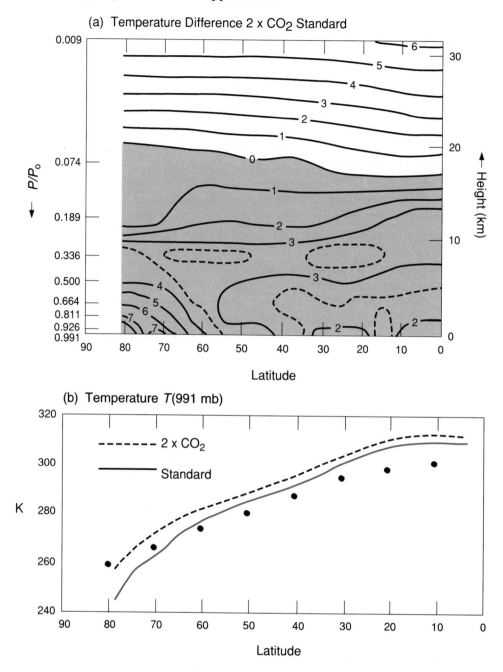

Fig. 20.6 (a) Latitude–height distribution of increase in zonal mean temperature (K) resulting from doubled CO_2 concentration; and (b) zonal mean temperature at lowest prognostic level (i.e., $\equiv$ 991 mb). Dots indicate observed distribution of zonal mean air-surface temperature (Oort and Rasmusson, 1971). Shading indicates decrease in temperature. From Manabe and Wetherald (1975).

change for doubled CO_2 is shown in Fig. 20.6(a). The overall significant pattern in Fig. 20.6(a) is for stratospheric cooling of about $-6°C$ at 30 km and tropospheric warming of up to $10°C$, with the largest warming in the polar region. The latitudinal distribution of surface temperature for the control $(1 \times CO_2)$, as well as the observed and doubled $(2 \times CO_2)$, is shown in Fig. 20.6(b). Part of the polar warming is caused by the snow- and sea ice-albedo feedback, and part can be attributed to a weaker surface temperature inversion. Some models show warming in the tropical troposphere caused by a stabler moist adiabatic lapse rate, the result of the cumulus convection process. Several climate-modeling groups have found that the horizontal temperature gradient decreases with increased meridional transport of energy, particularly latent heat from increased poleward transport of moisture.

The next level of ocean model used with a GCM in greenhouse-effect studies is the mixed-layer ocean. The distinct advantage of this type of model over the swamp model is that heat can be stored from one season to the next. Heat stored in the warming phase of the year, mostly summer, can be released in the subsequent winter. A great deal of the seasonal change of ocean temperature in mid- and high latitudes is explained by such a simple model, as shown in Chapter 17. The thickness of the mixed layer having uniform temperature with depth is typically 50–70 m. Several modeling studies of the changes in the simulated climate resulting from a doubling of CO_2 are part of the literature – Wetherald and Manabe (1986) with the GFDL model; Hansen et al. (1984) with the GISS model, Washington and Meehl (1984) with the NCAR model, Schlesinger and Zhao (1989) with the Oregon State University (OSU) model; Wilson and Mitchell (1987a,b) with the United Kingdom Meteorological Office (UKMO) model; and Boer et al. (1991) with the Canadian model. The characteristics of the models are quite different, and the range of globally averaged mean air-surface temperature is 2.8 to $5.2°C$ (Table 20.1).

Modeling studies have shown a dependence of temperature change upon basic-state temperature. MacCracken et al. (1990) describe a model

Table 20.1. Changes in surface global mean air temperature in different models for doubling of CO_2.

Model	$\Delta T_a(°C)$
Wetherald & Manabe	4.0
Hansen et al.	4.2
Washington & Meehl	3.5
Schlesinger & Zhao	2.8
Wilson and Mitchell	5.2
Boer et al.	3.5

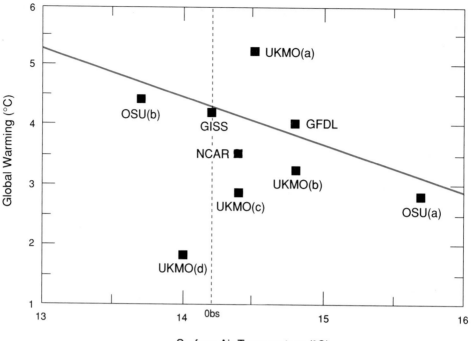

Fig. 20.7 Increases in global-mean air-surface temperature as simulated for CO_2 doubling by five GCMs versus their simulated $1 \times CO_2$ global-mean air-surface temperature. For some models, different approximations of physical processes give different results. "Obs" is observed global-mean air-surface temperature based on data of Jenne (1975). Adapted from Cess and Potter (1988), including results from Mitchell et al. (1989).

intercomparison of globally averaged warming as a function of globally averaged air-surface temperature (Fig. 20.7). Less greenhouse warming occurs when the surface temperature itself is warmer – with one exception. In experiments with a novel cloudiness parameterization, Mitchell and Ingram (1992) found a large change in sensitivity. Spelman and Manabe (1984), Washington and Meehl (1986), and Meehl and Washington (1990) found that the amount of global warming depended strongly upon sea ice extent. The thrust of Fig. 20.7 is that, all else remaining the same, the amount of global warming depends on global surface temperature, such that a warmer Earth usually means less warming from doubling CO_2. A great part of the explanation of this dependence is the feedback mechanism of water vapor, snow, sea ice, land ice, and cloud. This works multifold with many feedbacks, but one of the largest is the snow and sea ice albedo.

The latitude–height distribution of temperature change from several modeling groups is shown in Fig. 20.8 for December–January–February (DJF). The pattern of stratospheric cooling and tropospheric warming is

Zonal Mean Temperature Differences for DJF

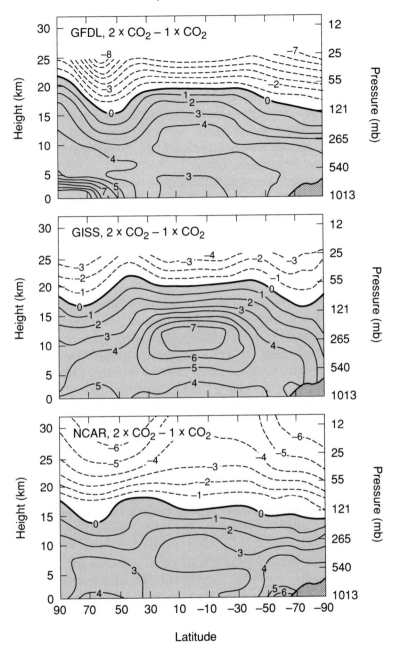

Fig. 20.8 DJF zonal mean air temperature differences (in °C) between $2{\times}CO_2$ experiments and control experiments ($1{\times}CO_2$) with GFDL, GISS, and NCAR models. Results are plotted as functions of latitude and height. From Schlesinger and Mitchell (1985).

quite similar, but the details are not (Schlesinger and Mitchell, 1985). This can be explained by the differences in the model and by how a simple mixed layer is included. The UKMO (Wilson and Mitchell, 1987a,b) and Canadian (Boer et al., 1991) groups have found this same characteristic pattern in their simulation of climate change from doubled CO_2. Two noteworthy features are (1) that the greatest surface warming occurs in the winter hemisphere, in part due to a weaker surface inversion and snow-ice-albedo feedback, and (2) that the greatest warming above the surface occurs in the upper tropical troposphere where moist water vapor and cloud convection cause greater warming. These seem to be associated with the incorporation of convective parameterization in the models.

One of the major advantages of GCMs over other modeling approaches is that the geographical distribution of the climatic variables can be studied. Chapter 19 describes the natural variability of the simulated climate from GCMs and compares the variability with the observed data. Despite the differences between the models and the observed, many major features of the climate system are simulated, thus demonstrating the models' credibility in climate-change experiments.

The geographical distribution of air-surface temperature from doubled CO_2 for DJF is shown in Fig. 20.9 for three of the five different models considered by MacCracken et al. (1990). All models exhibit a warming from about 2–3°C in the tropics to 20°C in the winter poleward regions. The greatest changes seem to be associated with variations in sea ice extent and the position of sea ice in the control experiment is a factor in the location of the warming maxima. The differences between the models are difficult to understand because their construction varies and the feedback mechanisms may be substantial. Barnett and Schlesinger (1987) and others have devised a new pattern-correlation method to make identification of a greenhouse "signal" more objective. In this chapter, only the geographic temperature change is discussed, but other changes, such as cloudiness, soil moisture, precipitation, winds, and sea level pressure, have been compared. All have ramifications for potential climate change (see review sources mentioned in Sec. 20.1).

In some experiments, the influence of other greenhouse gases has been considered by using an "effective" increase of CO_2 concentration. According to Wang et al. (1991), this may lead to improper greenhouse warming assessment, since several gases have radiative properties different from those of CO_2. In their study, they included CH_4, N_2O, CFC-11, and CFC-12. For example, $CFCl_3$ is optically thin while CO_2 is optically thick. The latitude–height heating-cooling distribution with only increased effective CO_2 concentration confines most of the heating to the lower troposphere, while the experiment with increased concentration of individual trace gases shows a large secondary maximum at about 12 km. The climate temperature-change patterns are quite different in the two sets

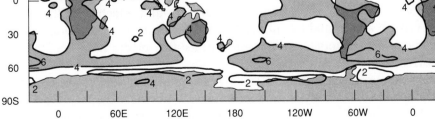

(a) OSU (Dec, Jan, Feb)

(b) UKMO (Dec, Jan, Feb)

(c) GISS (Dec, Jan, Feb)

Fig. 20.9 Geographical distribution of changes in air-surface temperature (°C), $2 \times CO_2$, for DJF simulated by (a) OSU model; (b) UKMO model; and (c) GISS model. Shading indicates temperature increases larger than 4°C.

of experiments with the individual greenhouse gases. The global surface warming is 4.2°C with "effective" CO_2, but 5.2°C with increased individual trace gases. This study strongly suggests that future, more definitive, studies should use individual trace-gas increases. These will also require reliable trend estimates of several radiatively active gases.

Most climate-change studies have been conducted with relatively coarse-resolution models because of the need for extended experiments. In his investigation of the effect of horizontal resolution on the greenhouse change, Senior (1991) carried out control and doubled CO_2 experiments with $5° \times 7.5°$ and $2.5° \times 3.75°$ (latitude/longitude) resolutions. Although there are some expected differences, the overall warming values and cloud-forcing patterns are quite similar.

20.8 Global coupled-model estimates with dynamical oceans

Even in their present primitive state of development and realism, coupled dynamical atmosphere and ocean models are shedding new light on the myriad of interactions within the climate system that can lead to regional changes. Some of the earliest research with such coupled dynamical models was conducted at GFDL by Bryan et al. (1982, 1988) and at OSU by Schlesinger and Jiang (1988). In his review of three-dimensional coupled-model development in Chapter 17, Meehl describes the problems encountered by researchers as various techniques were explored. Throughout most of general circulation modeling history, ocean temperatures have been specified or computed with simplified ocean physics. When ocean temperature is computed in a GCM, it can differ substantially from the observed distribution for many reasons, such as errors in the surface fluxes or improperly simulated ocean transport and diffusion processes. As Meehl points out, these errors can be corrected by various and somewhat arbitrary adjustments. Without adjustments, the model can drift (referred to as climate drift) into a state different from present climatology.

Over the last decade and with various climate models, many researchers have explored how the oceans can delay a greenhouse-gas-caused warming [e.g., Schlesinger et al. (1985), Washington and Meehl (1989), Stouffer et al. (1989), and Manabe et al. (1990, 1991)] with three-dimensional models. In some of these models, systematic surface-flux errors have been adjusted or corrected.

Because these coupled models have a complete ocean, it is possible to study more realistically the interactions with the oceans. One of the important aspects of the ocean in greenhouse warming is its enormous thermal capacity. Stouffer et al. (1989) showed a latitude–time distribution of surface temperature (Fig. 20.10). The transient surface warming takes much longer to reach sizable values (greater than 1°C) because of the ocean's

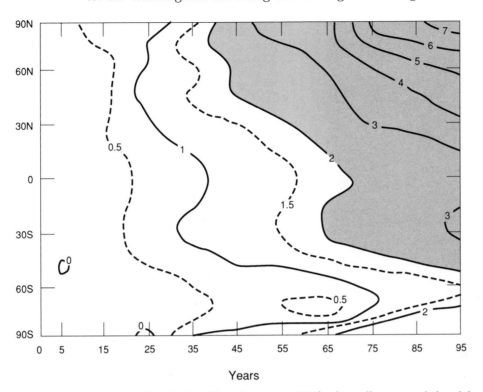

Fig. 20.10 Latitude–time distribution (Stouffer et al., 1989) of zonally averaged decadal-mean surface air temperature (°C) difference between 1% per year increase (compounded) of CO_2 concentration and a control with constant concentration.

capacity to sequester large amounts of heat. In their simulation, they show a greater warming in the polar region of the Northern Hemisphere and a delayed warming near the edge of Antarctica. The delayed ocean warming near Antarctica is caused by the dynamics of the atmosphere–ocean system that yields deep mixing and upwelling of colder water which slows surface warming. The overall effect of ocean thermodynamics, through sequestering of heat and ocean dynamics, is to lessen the global warming of surface temperature.

To examine the global and regional air-surface temperature changes, Meehl et al. (1992b) extended to nearly 50 years the coupled atmosphere–ocean climate-model experiment of Washington and Meehl (1989) with gradually increased CO_2. They calculated five-year seasonal means from year 26 to 45 to determine whether or not clear *regional* warming is evident for particular seasons. Five-year means rather than individual one-year seasonal means were used because the tropical Pacific El Niño and La Niña events produced by the coupled model (Meehl, 1990b) can profoundly affect mid- and high-latitude regional patterns on a time scale of individual years (Chapters 17 and 18).

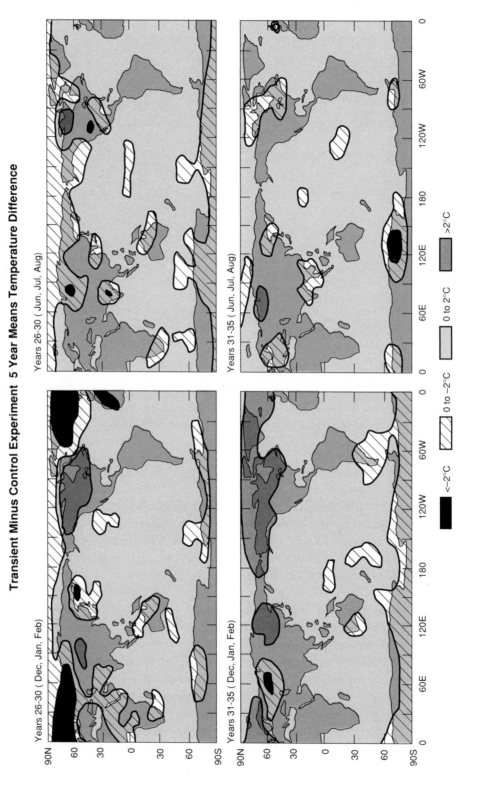

Transient Minus Control Experiment 5 Year Means Temperature Difference

Years 26-30 (Jun, Jul, Aug)

Years 31-35 (Jun, Jul, Aug)

Years 26-30 (Dec, Jan, Feb)

Years 31-35 (Dec, Jan, Feb)

Fig. 20.11 Distribution of geographical temperature difference of transient (1% per year) CO_2 concentration, increase minus control (330 ppm). Light shading is 0–2°C; heavy shading is 2°C; slants are 0 to −2°C; and solid black −2°C; DJF and JJA.

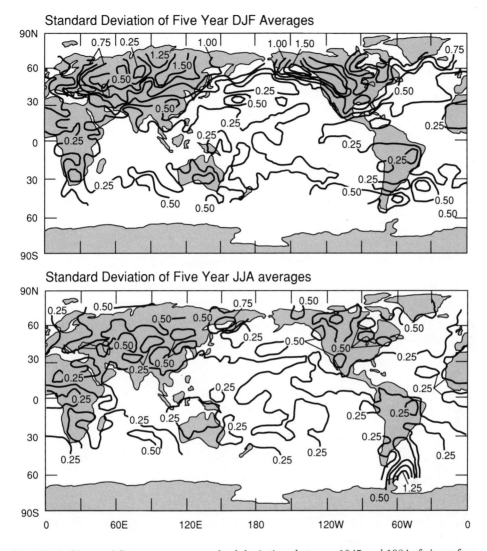

Fig. 20.12 Observed five-year mean standard deviations between 1945 and 1984 of air-surface temperatures for DJF and JJA.

In the following figures, the regional five-year differences between the increasing (a cumulative 1% per year) CO_2 experiment and the control with a constant concentration of CO_2 at 330 ppm are shown. The 1% per year roughly accounts for the combined heating effect of all greenhouse gases – such as CO_2, CH_4, CFCs, and N_2O. Figure 20.11 shows the seasonal differences. Winter, spring, and autumn have the greatest warming. Regional cooling exists, however, over parts of globe even with five-year mean patterns. Figure 20.12 presents the standard deviation of five-year mean DJF and JJA observed air-surface temperatures (1945 to 1984) (Meehl et al., 1992b). The observed standard deviations are roughly the same

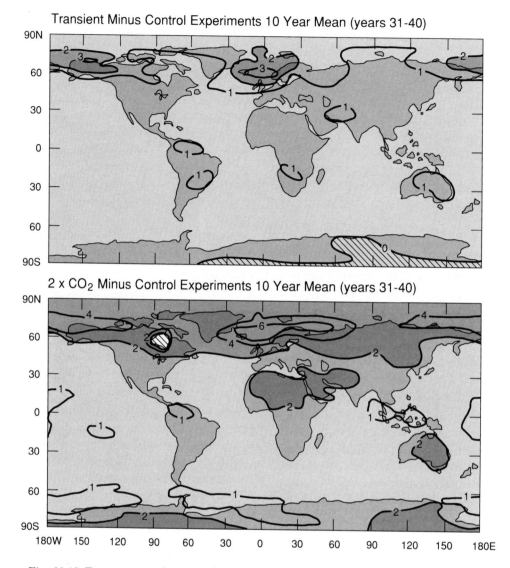

Fig. 20.13 Ten-year annual means of temperature difference (top) transient-minus-control; instantaneous doubling-minus-control (bottom).

magnitude as those from the model. Trenberth (1990) has shown large regional changes in the observed interdecadal surface temperatures over the Northern Hemisphere. All of these observed studies underscore the difficulty of identifying the "fingerprint" of greenhouse warming from short climatic records. The length of climatic record is also an issue in the model experiments.

In Fig. 20.13, a ten-year (31–40) annual mean temperature difference simulated in our transient CO_2 experiments exhibits general warming, especially at high northerly latitudes for this period. Also shown in

Fig. 20.13 are annual mean differences for years 31–40 from an experiment in which CO_2 is instantaneously doubled ($2\times CO_2$) at year 1. The $2\times CO_2$ difference from the control shows approximately the same regional pattern as the transient experiment. More consistent regional warming patterns are evident in the longer averaged *annual* means but not in the seasonal five-year means.

Figure 20.14 is a simulation of the temperature differences caused by increased CO_2 in a dynamical atmosphere–ocean GFDL GCM with runs of 100 years. In this set of experiments, low-frequency natural variability was observed in the control run (see Chapter 19, Fig. 19.10), and a similar pattern was found in the 1% (compounded) increase of CO_2 per year run. In Fig. 20.14, the time-dependent temperature response (1% per year at 70 years) (a) is compared with the temperature response due to doubling (instantaneous $2\times CO_2$) in a model after it reaches a new equilibrium (b), to give the ratio of the two (c). The time-dependent pattern has reached about 60 to 80% of the equilibrium response over most of the globe. The ratio is low in the North Atlantic and in the southern oceans (cf. Fig. 20.10), for reasons discussed earlier.

These results suggest that, to observe the regional greenhouse-effect signal, we may have to consider changes in seasonal means for periods substantially longer than five to ten years. Changes over shorter periods may reflect natural variability much more than the transient greenhouse buildup. Future climate models with increased resolution and improved treatment of climate processes, such as clouds, will have to contend with the large natural variability of multiyear or, perhaps, longer-than-decadal time scales of regional patterns. This will make establishment of clear and consistent regional greenhouse seasonal patterns more difficult than previously believed. In addition, these results hint that identification of regional greenhouse-gas-caused patterns in the observed data may also be difficult to verify on a seasonal basis. Model experiments that show a large increase in regional and global warming must contend with the fact that the "observed" warming is not large. This observational record constraint must be kept in mind. The IPCC attempted to give a "best" estimate of the global warming. Its consensus statement of 1°C warming by 2025 and 3°C before the end of the next century is likely to change as our knowledge of the climate system improves.

Stouffer et al. (1989) and Manabe et al. (1990) have shown that an unrealistic equilibrium state results in too weak a thermohaline circulation in the North Atlantic and too strong a halocline in that same region. The increasing CO_2 case has weaker circulation than the control, thus indicating that the thermohaline is weakened by greenhouse warming. In a coupled atmosphere–ocean model, Washington and Meehl (1989) found a similar weakening of the thermohaline circulation with greenhouse warming.

The El Niño–Southern Oscillation (ENSO) phenomenon, one of the most

(a) Years 60-80 of Time-Dependent Temperature Response

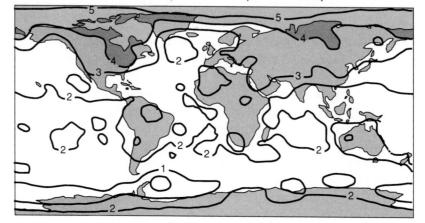

(b) Equilibrium Temperature Response

(c) Ratio of Time-Dependent Response to Equilibrium Response

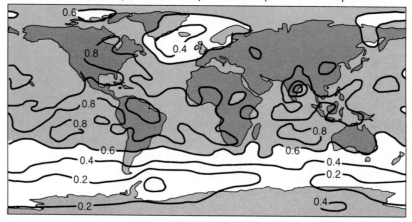

Fig. 20.14 (a) Time-dependent response of air-surface temperature (°C) in coupled ocean–atmosphere model to a 1% (compounded) yr^{-1} increase of atmospheric CO_2. Difference is shown between 1% yr^{-1} perturbation run and years 60–80 of control run when atmospheric CO_2 concentration approximately doubles; (b) equilibrium response of air-surface temperature (°C) in model to doubling of atmospheric CO_2; (c) ratio of time-dependent-to-equilibrium response shown above. From Manabe (1990, personal communication).

significant fluctuations in the climate system, can be simulated in coupled atmosphere–ocean GCMs (Chapters 17 and 18). Meehl et al. (1992a) have shown that the ENSO phenomenon in a doubled CO_2 experiment leads to increased low-level moisture convergence and increased precipitation over the warm ENSO SST temperature anomaly, and strengthens the east–west Walker circulation. The dry areas become drier and wet areas wetter. Although it is difficult to establish with certainty, the midlatitudes appear to have significant teleconnections with the tropics. If other studies substantiate these findings, increased greenhouse gases may affect one of the largest sources of interannual variability.

20.9 Future studies

To make intercomparison easier, future studies of possible climate-system changes need standard testing scenarios. The IPCC (1990a) report addressed the subjects of how concentrations of various greenhouse gases are increasing and how they are contributing to the heating rate of the troposphere. Estimates from 1980 to 2100 are made of the contribution to total heating by greenhouse gases. At present, as pointed out in Chapter 2, CO_2 is responsible for 55% of greenhouse warming, CFCs 11 and 12 for 17%, N_2O for 6%, and CH_4 for 15%. Proposed greenhouse-emission scenarios range from business-as-usual emissions (assuming no curtailment) to scaling back emissions (with various assumptions). International scientific organizations are attempting to agree upon emission rates so that standardized control and transient GCM experiments can be carried out by the climate-modeling groups.

The following are some findings from greenhouse-gas climate experiments:

1. The response of equilibrium global surface temperature to a doubling of CO_2 in a climate model with simple oceans is a global warming of 2–5°C. This results from a combination of factors, such as feedback of water vapor, clouds, and snow-ice-albedo changes. Observational records of surface temperature change indicate a warming of ~0.5°C over the past century but it cannot be unequivocally attributed to increasing greenhouse gases.

2. Climate-model estimates that take into account full ocean processes, albeit crudely, indicate a global surface warming in the 1–2°C range

from transient and doubled CO_2 experiments on time scales of the order of a century. Warming near the Earth's surface is greatest in mid- and high latitudes of the Northern Hemisphere, with significant regional variability. Modeling studies consistently show that the stratosphere will cool substantially. Precipitation amounts increase near the upper branch of the Hadley circulation and decrease in the subtropics. Soil-moisture content becomes near saturation in the winter hemisphere, but more drying occurs in the summer hemisphere because of increased evapotranspiration. The models generally show that snow and sea ice limits move poleward.

3. In modeling studies, warmer ocean-surface water forms and becomes less saline at high latitudes, thus weakening the thermohaline circulation and the formation of deep water, particularly in the North Atlantic (Chapter 11). These results are still tentative, since the component models (atmosphere–ocean–sea ice) do not represent all of the detailed physical processes, some of which may have significant negative feedback.

4. Although studies are limited, there is some indication that the manifestations of El Niño events (droughts and floods) may be more intense in a globally warmed climate. Because the ENSO events (see Chapters 17, 18 and 19) account for a large amount of the natural variability on monthly-to-decadal time scales, this may be an important change in the climate system in addition to the long-term mean changes. Present models are not sufficiently accurate for us to understand how hurricanes and typhoons may be affected except, possibly, through their strong correlation with warm sea surface temperatures. Since the ocean generally warms the tropics in the model experiments, it would not be unreasonable to expect that warmer ocean temperatures from increased greenhouse gases in regions prone to hurricanes and typhoons may lead to more frequent and more intense tropical storms. This, of course, has not been modeled in detail.

5. As outlined in this chapter, virtually every aspect of a climate model is subject to uncertainties (e.g., clouds, dynamics, moisture prediction, convective processes, boundary processes, ocean–sea ice processes, vegetation, and resolution effects) that limit the climate-change predictive skill of climate models. As climate models improve and simulate more faithfully the present climate and paleoclimates (see Chapter 21), we can begin to rely on them to simulate more quantitatively possible climate change from increasing greenhouse gases.

21

Modeling large climatic changes of the past

John E. Kutzbach

21.1 Introduction

The Earth's climate has always been changing and will no doubt continue to change. Climate has a past and a future. The future climate is unknown, but major efforts are underway to estimate the climatic changes that may be associated with increases of greenhouse gases (Chapter 20). Until fairly recently the climates of the past, although not unknown, have only been described qualitatively. This situation is now changing. New observational techniques are being used to obtain climatic records from soils, lake and ocean sediments and ice strata. These records are being obtained worldwide, so that near global coverage is possible. Increasingly accurate methods are available for radiometric dating of these environmental records. As a result, we are gaining detailed information about the evolution of the atmosphere and ocean, of shifts of continents and the rise and fall of mountains, and of the wax and wane of ice sheets, forests, lakes and deserts. Figure 21.1 provides an overview of the geological time scale along with notes on major geological events. This new information about past climates also helps place in perspective the magnitude of possible future changes of climate compared to changes of the past (Fig. 21.2).

This wealth of new knowledge has, in turn, stimulated efforts to simulate climates of the past with the aid of climate models. These modeling studies serve a number of purposes. First, modeling studies help identify potential causes of climatic change by testing the sensitivity of the model's climate to changes in external forcing or orographic and geographic boundary conditions. For example, the possible sensitivity of climate to changes of Earth's orbital parameters was suggested in the 1860s by Croll but the idea received serious consideration only when quantitative estimates of the climatic effects of the orbital changes were made in 1920 by Milankovitch using a zonal-average energy budget climate model. Second, modeling studies help us explore the internal mechanisms of climate change, such as the coupled interactions of the atmosphere, oceans, ice sheets and biosphere as the climate shifts between glacial and interglacial states. Third, comparisons of the results of simulations of past climate with observations of past climate help us evaluate the adequacy and accuracy of climate models.

669

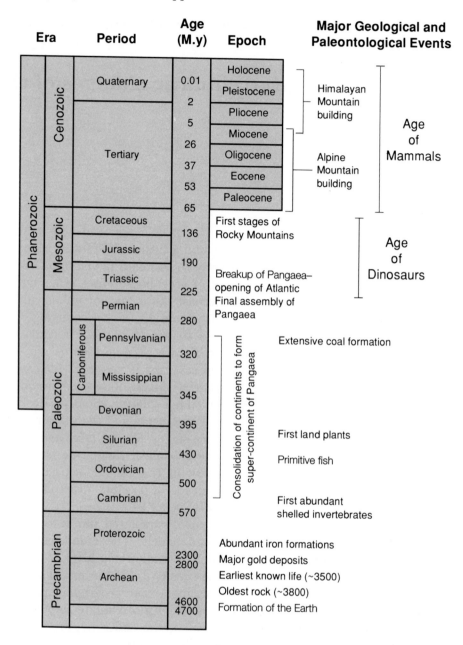

Fig. 21.1 The geological time scale (M.y is millions of years before present). (Crowley, 1983).

Schematic Comparison of Greenhouse Warming with Past Climates

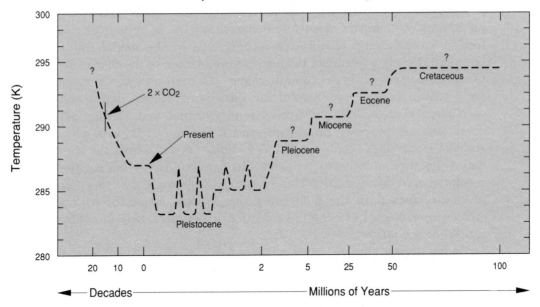

Fig. 21.2 Schematic comparison of possible future greenhouse warming with estimates of past changes in temperature. Pleistocene glacial–interglacial cycles are more numerous than shown. The characteristic amplitude of global temperature change during glacial–interglacial cycles is 3–4 K. Note that pre-Pleistocene changes are not well fixed in magnitude, but their relative warmth is approximately correct. Maximum warming in the Cretaceous is based on estimates by Barron and colleagues. Time intervals in between have been scaled accordingly. (Crowley, 1989).

This chapter outlines some of the challenging problems to be faced in understanding the causes and mechanisms of large climatic changes and gives examples of initial studies of these problems with climate models. It covers three main periods of Earth history: (1) the climatic changes of the past several glacial–interglacial cycles, where changes of Earth's orbital parameters change the distribution of solar radiation at the top of the atmosphere; (2) the climatic changes of the past several hundred million years, where plate movements and crustal movements cause important changes at the Earth's surface; and (3) the climatic changes of early Earth history, billions of years ago, where solar luminosity, Earth rotation rate, and atmospheric composition may all have differed very significantly from present-day conditions. Variability of climate on decadal and centennial time scales is described in Chapter 19.

671

All of the modeling studies described here are called climate sensitivity experiments, in which the response of a climate model (and up to now an incomplete climate model) to some known or hypothetical change in forcing is analyzed. Qualitative comparisons of the model results with observations help to evaluate the possible importance of the change in forcing for explaining the observed climatic change. Climate sensitivity experiments have proven to be very helpful for identifying causes and mechanisms of climate change. Alternatively, if all external and surface boundary conditions are properly prescribed, and if the model is an appropriately complete representation of the coupled climate system, then the studies are called climate simulation experiments. In simulation experiments, the model results should agree rather closely with the geologic evidence, provided of course, that both the model and the geologic estimates are accurate. Achieving this level of climate simulation experimentation remains largely a task for the future.

Studies of past climates use a broad range of models. AGCMs or AGCMs coupled to ocean mixed-layer models are being used to explore the full three-dimensional structure of past climates, including details of the hydrologic cycle. However, the extensive computing resources required for GCMs have precluded, until now, very long simulations. For similar reasons, extensive work with fully coupled AGCMs and OGCMs remains largely a task for the future. Simplified climate system models, such as energy budget models, (see Chapters 1 and 10) are often used to explore the long-term evolution of climate.

This chapter provides a brief introduction to climate sensitivity experiments pertinent to the study of past climates. Both observations and model studies of paleoclimates are treated extensively in Crowley and North (1991).

21.2 Climatic changes of the past several glacial–interglacial cycles

Observational studies have shown that variations of Earth's orbital parameters are pacemakers of glacial–interglacial cycles (Hays et al., 1976) and of wet–dry cycles in the tropics (Prell and Kutzbach, 1987). A significant fraction of the variance in time series of the estimated volume of glacial ice and indicators of temperate-latitude vegetation and tropical monsoons is phase-locked with the orbital cycles. Illustrations of this kind of phase-locking are in Fig. 21.3.

The discovery that large climatic changes are apparently paced by relatively small changes in Earth's orbital parameters presents a major opportunity and challenge: namely, to analyze and explain the processes and feedbacks that produce the observed large climatic response to the precisely

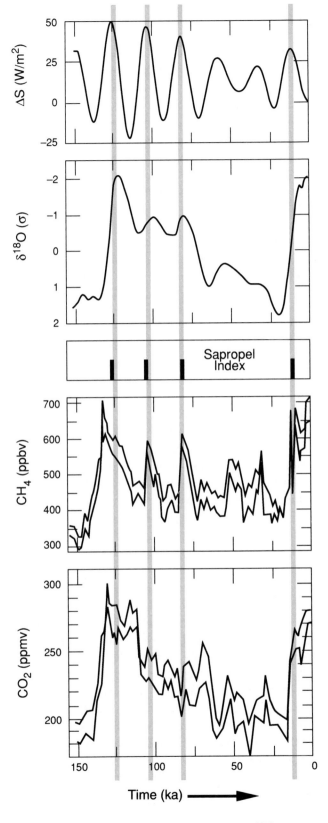

Fig. 21.3 Records of solar radiation changes due to orbital changes (top) and various climatic indicators (bottom) for the past 150,000 years (ka is thousands of years ago). ΔS refers to changes in average Northern Hemisphere summer radiation. $\delta^{18}O\,(\sigma)$ is an indicator of the volume of glacial ice on the continents (-2, the present minimum; $+2$, glacial maximum). The sapropel index from the Mediterranean is an indicator of the strength of the African Monsoon. (Prell and Kutzbach, 1987). The records of methane and carbon dioxide are from the Vostock (Antartica) ice core (Barnola et al., 1987). The vertical bars are added to show the timing of major radiation maxima.

673

known changes in external forcing that modify the seasonal and latitudinal distribution of insolation. If successful, we will have learned a great deal about the internal workings of the climate system. Variations of the Earth's orbital parameters include changes in the axial tilt (range: 22° to 24.5°; period: about 41,000 years), season of perihelion (range: all times of the year; period: about 22,000 years) and orbital eccentricity (range: 0 to 0.06; period: about 100,000 years).

A recent time of substantially altered orbital parameters was at 9 ka (9,000 years ago), when perihelion was in July and the axial tilt was greater than at present. The solar radiation in July of 9 ka, averaged over the Northern Hemisphere (NH), was about 7% (30 W m^{-2}) greater than it is at present. In January of 9 ka, solar radiation was correspondingly less than present (Fig. 21.4). Climate sensitivity experiments have helped to show how the orbitally caused changes in the seasonal cycle of solar radiation interact with the different thermal properties of land and ocean to cause large climatic changes. These climate sensitivity experiments were made initially with AGCMs using prescribed (modern) sea surface temperatures, and subsequently with AGCMs coupled to mixed-layer oceans (see Chapter 17, Sec. 17.2.2). The only coding changes required, relative to control (modern) simulations, are adjustments of the axial tilt, season of perihelion, and orbital eccentricity.

The change between the orbital configuration of 9 ka and today therefore provides us with an experiment, performed by nature, for studying the climatic response to an enhanced seasonal insolation cycle in the NH, and a reduced seasonal insolation cycle in the Southern Hemisphere (SH). The enhanced seasonal cycle strengthens NH monsoons (Kutzbach and Guetter, 1986; Fig. 21.5). Northern continents are warmer in summer (temperature increase of 2–4°C) and colder in winter; the temperature changes of the surrounding oceans are much less owing to their large heat capacity. In northern summer, the warming of the land relative to the ocean increases the land/ocean temperature contrast and produces an increased land/ocean pressure gradient (lower pressure over land relative to ocean) and a significantly expanded and intensified region of low pressure across North Africa and South Asia. Summer monsoon winds are strengthened and precipitation is increased for parts of North Africa and South Asia (Mitchell et al., 1988; COHMAP Members, 1988). These simulated changes in the hydrologic regime of past millennia are in qualitative agreement with geologic and paleobotanical observations of changes in tropical lake levels and vegetation (Kutzbach and Street-Perrott, 1985). For example, between 12 ka and 6 ka lakes and savanna vegetation existed about 1,000 km north of present limits in tropical North Africa. Northern continental interiors are warmer (in summer) and drier than present in the model experiment and this too agrees qualitatively with observations (Gallimore and Kutzbach, 1989). The increased insolation in northern summer, stemming both from

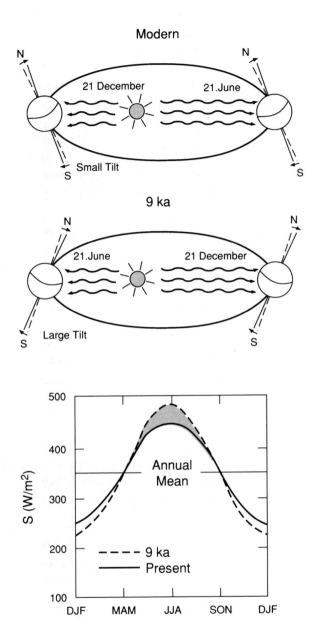

Fig. 21.4 Top: Changes in the Earth's orbit from the present configuration, where perihelion (minimum Earth–sun distance) is in northern winter, to the configuration for 9,000 years ago (9 ka), where perihelion was in northern summer and the axial tilt was 24° rather than $23\frac{1}{2}$°. The change in the season of perihelion is caused by the slow precession of Earth's rotational axis. Bottom: Changes in average Northern Hemisphere solar radiation through the annual cycle: present, solid line; 9,000 years ago, dashed line. (Modified from Kutzbach and Webb, 1991).

Surface Winds (July)

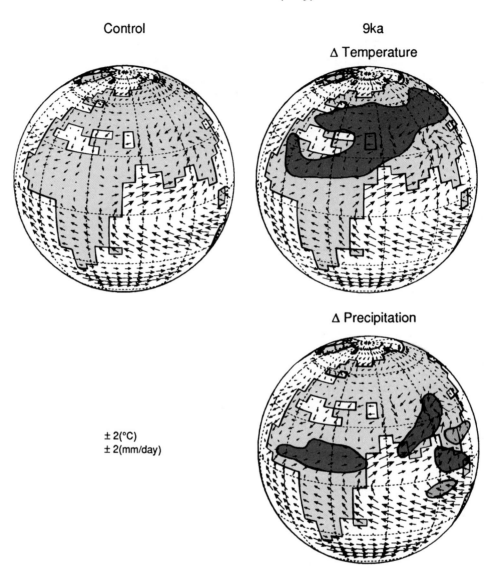

Fig. 21.5 Model simulation of July surface winds for a control experiment with present-day solar radiation (left) and an experiment with solar radiation set for 9,000 years ago (right). The right-hand panels also depict regions where surface temperature is increased by 2°C or more (dark areas) and where precipitation is increased/decreased by 2 mm day^{-1} or more (dark circled areas/light circled areas). In general, continental interiors were warmer than present and northern monsoon lands were wetter than present.

the summertime perihelion and the increased tilt, produces considerable melting of Arctic sea ice. Some observational evidence also suggests reduced sea ice cover around 9 ka.

Although many features of the observed climate around 9 ka are in qualitative agreement with the results of the climate sensitivity experiments using orbitally caused insolation changes, the agreement is far from perfect. Much needs to be learned about the response of soil moisture, runoff, vegetation, sea ice, and ocean currents to the changed insolation. In keeping with the sense of the definition of sensitivity experiments given earlier, the simulated climate sensitivity will change, perhaps significantly, as models of the climate system become more complete and accurate.

Turning to the question of the relationship between orbital changes and glacial–interglacial cycles, issues about climate sensitivity are more complicated. The geologic evidence provides the challenge of explaining not only the accumulation and wastage of several-kilometer-high ice sheets, but also substantial changes in monsoon circulations, vegetation zones, and atmospheric concentrations of carbon dioxide and methane (Fig. 21.3). Describing in detail how this system operates will require the use of fully coupled climate system models, including models of ice sheets, upper lithosphere adjustments to ice loading and biogeochemical cycles capable of simulating changes in atmospheric composition.

One aspect of the problem of explaining glacial–interglacial cycles has received considerable attention. Climate sensitivity experiments have shown that orbital conditions favoring cool northern summers might reduce temperatures sufficiently to prevent the melting of high-latitude snows. If snow were to persist through the summer months, glaciation could be initiated. Cool summers would be favored at times of large eccentricity, January perihelion, and minimum axial tilt. Experiments by North et al. (1983) using an energy budget climate model showed that for these orbital conditions summer temperatures could be several degrees Celsius lower than present in northern continental interiors where present-day summer temperatures are only a few degrees Celsius above freezing. When an ice-albedo feedback was included, the cooling was enhanced. While this result lends support to the hypothesis that certain orbital configurations favor initiation of glaciation, the accuracy of the ice-albedo feedback mechanism remains uncertain because the energy budget model has no explicit hydrologic cycle. Similar sensitivity experiments with AGCMs with explicit precipitation and snow cover parameterizations are therefore needed. One such AGCM experiment produced lowered temperatures in summer and year-round, and wetter conditions in high northern latitudes and especially Canada (Royer et al., 1983). On the other hand, Rind et al. (1989) found that temperature was not lowered sufficiently to maintain snow cover through the summer. Clearly, the sensitivity of climate to the insolation changes depends upon model parameterizations, resolution, and other prescribed (or interactive) boundary conditions (see Oglesby, 1990).

Leaving aside, for the moment, the question of the ultimate cause of glacial–interglacial cycles, it remains of great interest to explore the characteristics of glacial-age climates because conditions were so different from those existing now. Many features of glacial-age climates have been simulated with AGCMs (Gates, 1976) and with AGCMs coupled to mixed-layer oceans. The pioneering work of the CLIMAP project (CLIMAP Project Members, 1981) has been instrumental in providing estimates of ice-age conditions such as ice sheets, vegetation, and sea surface temperature. These estimates are used for lower boundary conditions or for model validation. Analysis of fossil air trapped in glacial ice and retrieved from ice cores in Antarctica and Greenland has shown that atmospheric concentration of CO_2 was about 200 ppmv during glacial times (Barnola et al., 1987), Fig. 21.3.

In the simulations of glacial climates with AGCMs, large anticyclones develop over the ice sheets. Temperatures are generally lower and precipitation rates are reduced. In the middle and upper troposphere, the presence of the large North American ice sheet causes the jet stream flow to split into two branches, a northern branch along the Arctic flank of the ice sheet and a southern branch located well south of the ice sheet border (Manabe and Broccoli, 1985). The northern branch of the split atmospheric flow brings cold Arctic air over the North Atlantic producing cold water and extensive sea ice cover that agreed with the marine geologic evidence. While many of the results of these experiments agree with paleoclimatic observations, many puzzles remain, such as the changed behavior of the ocean circulation and biogeochemical cycles, as manifested in the reduced atmospheric concentration of carbon dioxide.

Learning more about the role of the ocean in large climatic changes is of particular importance. For example, observational evidence indicates that North Atlantic Deep Water flow was significantly reduced during glacial times, that the upper waters of the oceans, and particularly the North Atlantic, were depleted in nutrients compared to today, and that the deep ocean was cooler (Boyle and Keigwin, 1987). If the vertically overturning thermohaline circulation of the Atlantic, described as a conveyor belt (Chapter 17, Fig. 17.12 and Plate 14), slows down or stops, the climate of the North Atlantic region would be much colder than present. Broecker et al. (1985a) and Broecker and Denton (1989) describe evidence for this bimodality, both on the long-term scale of glacial–interglacial cycles and on the abrupt "event" scale of centuries. These observational findings have been complemented by studies with ocean models showing possible bimodality in the ocean's thermohaline circulation (see Bryan, 1986, and references to earlier work therein, and Chapter 17, Sec. 17.5.2). Experiments with a coupled ocean–atmosphere GCM found two stable equilibria: in one the North Atlantic had a vigorous thermohaline circulation and a relatively

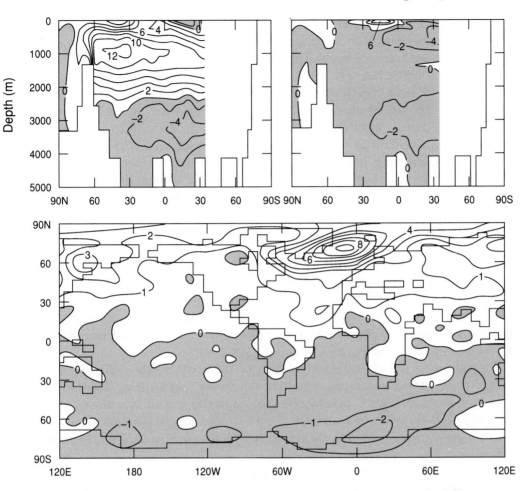

Fig. 21.6 Top: Streamfunction illustrating meridional circulation in the Atlantic Ocean. Units are in Sverdrups, 10^6 m^3 s^{-1}. In Experiment I (left) there is a strong thermohaline circulation whereas in Experiment II (right) it is absent. The streamfunction is not shown in the southern Atlantic which is not enclosed by coastal boundaries and freely exchanges water with other oceans. Bottom: Difference in surface air temperature (°C) between Experiments I and II. The surface air temperatures in the North Atlantic sector are significantly warmer in the simulation with strong thermohaline circulation and large poleward heat transport in the North Atlantic ocean. (Manabe and Stouffer, 1988).

warm climate in regions bordering the Atlantic; in the other there was no thermohaline circulation and the regional climate of adjacent lands was much cooler (Manabe and Stouffer, 1988) (Fig. 21.6). Birchfield (1989) has examined similar kinds of bimodality in coupled atmosphere–ocean box models (Chapter 11, Sec. 11.4.2). These oceanic changes may be linked to changes in biogeochemical cycling that could explain the glacial–interglacial differences of about 70 ppmv in the atmospheric concentration of CO_2 (pre-industrial value ~270 ppmv; glacial-age value ~200 ppmv). Many indicators of ocean climate show significant amplitude variability and consistent phase relationships with orbital cycles (Imbrie et al., 1989).

For large climatic changes, there may also be significant changes in the distribution of land vegetation and soil carbon (Prentice and Fung, 1990) and possible biosphere-albedo feedbacks (Street-Perrott et al., 1990).

In general, paleoclimate simulation experiments (in contrast to sensitivity experiments) are difficult to perform because models are inadequate and observations are very incomplete. For the period since the last glacial maximum, however, there is detailed information on the size of the wasting ice sheets, the atmospheric concentration of carbon dioxide, and the ocean surface temperature (as inferred from information in marine sediments). Using these geologic observations to prescribe surface boundary conditions of ice sheets, ocean temperature and atmospheric CO_2 levels, along with orbitally prescribed insolation, AGCMs have been used to simulate "snapshots" of the climate at 3,000-year intervals from 18 ka to the present (COHMAP Members, 1988). The simulated climate agrees qualitatively with many features of the observed climate. These initial attempts at simulating sequences of paleoclimates will no doubt be repeated in coming years using more complete models.

Embedded in the general deglaciation of the past 18,000 years are one or more very abrupt changes in climate (Broecker and Denton, 1989). For example, between 11,000 and 10,000 years ago the general warming trend was interrupted by a very significant return to cold conditions that persisted for a few centuries, and was then followed by an equally abrupt warming. The cause of such abrupt events is unknown. Perhaps the climate is sensitive to small changes of ice sheet height or shape or to melt-water discharge to the Gulf of Mexico or the North Atlantic. The possible causes of these and other abrupt events are under investigation with climate models (Rind et al., 1986; Overpeck et al., 1989; Oglesby et al., 1989).

The time-dependent behavior of the fully coupled climate system during glacial–interglacial cycles is being addressed with the aid of highly simplified models of the climate system. These models incorporate, often in heuristic fashion, the slow-response climate variables such as ice volume, bedrock depression, deep ocean temperature, and atmospheric concentration of carbon dioxide (Saltzman, 1985, 1990). The simulated time-dependent variations of one or more climate variables, such as global ice volume, are then compared with observations of ice volume variations inferred from oxygen isotope records. Some studies have focused on understanding the forced response of coupled climate systems to changes of orbital parameters while others have demonstrated that the coupled systems themselves exhibit free oscillatory behavior.

In part because of computing limitations, time-dependent climate models simulating the slow-response variables use very simplified treatments of the atmosphere. This limitation will certainly be relaxed as computing power increases. In the meantime, an intermediate class of atmospheric models, statistical-dynamical models, are efficient enough to be used in time-dependent integrations. For example, Berger et al. (1990) have performed a long time-dependent integration with a two-dimensional (latitude–height) zonal average atmospheric model coupled asynchronously to mixed-layer ocean and ice sheet models and forced by orbital changes. They simulate an ice-volume chronology that matches the observational record of the past 125,000 years quite well.

In summary, there is strong empirical evidence linking precisely known changes in Earth's orbital parameters to glacial–interglacial cycles in high northern latitudes and cycles of enhanced or weakened monsoons in middle and tropical latitudes. This advantage of knowing both the external forcing changes and the climatic response (from observations) provides a major opportunity to explore the internal workings of the climate system. Much has already been accomplished but many puzzles remain. The initial studies show encouraging agreement between experiments and geologic observations and underscore the notion that we can learn a great deal about the behavior of the climate system using both observations and models of the large climatic changes of the geologic past.

21.3 Climatic changes of the past several hundred million years

Plate movements (changing geography), crustal movements (changing orography), and associated changes in outgassing, weathering, and biogeochemical cycles must exert a strong "lower-boundary" forcing on Earth's climate on the scale of millions of years. To the extent that these "solid earth" processes are viewed as "external" to the climate system, these more distant geologic periods also provide extraordinary opportunites for understanding the full range of behavior of the climate system in response to external change. These ancient climates are very different from present and include periods when the Earth was warmer than present and likely experienced higher concentrations of atmospheric CO_2 than present.

Climate sensitivity experiments performed with energy budget climate models have been particularly useful, because of their computational efficiency, in exploring the first-order effects of the location and size of continents on climate (Crowley and North, 1991). The primary climatic response is due to the different thermal characteristics of land and ocean to the seasonal insolation cycle and the influence of the size of the land mass on the continentality of the climate. Over the past 100 million years, the gradual isolation and movement of Antarctica toward the South Pole and the gradual northward drift and separation of Greenland from adjacent land masses may have both lowered the average temperature and reduced summertime warmth.

AGCMs coupled to various kinds of simplified oceans are being used to explore the role of geography in shaping ancient climates. Kutzbach and Gallimore (1989) used an AGCM coupled to a mixed-layer ocean model to show that the supercontinent of Pangaea, as it existed over 200 million years ago, must have experienced strong monsoonal circulations (Fig. 21.7). In addition to simulating large extremes of summer and winter temperature, as reported with energy budget models (Crowley et al., 1989), the explicit hydrologic cycle of the AGCM produced regions of wet monsoonal climates along tropical coasts and of arid climates in the interior.

OGCMs, with basin geometry and surface forcing altered to correspond to estimated conditions for mid-Cretaceous (Barron and Peterson, 1990) and Pangaean time (Kutzbach et al., 1990), are also being used in ocean sensitivity experiments. These ocean modeling studies illustrate possible major alterations of the climate of the ocean over geologic time, including a major transition from relatively warm bottom water (10–15°C) around 50 million years ago to the cold bottom water of today.

Throughout this long period of geologic time there is some evidence that polar climates were warm and equable compared to present. This evidence presents a challenging puzzle because sensitivity experiments based upon the altered locations of continents and oceans have so far been unable to duplicate such conditions (Barron and Washington, 1984). One possibility is that the observational evidence has been misinterpreted. Another possibility is that other factors, such as significantly increased atmospheric concentration of carbon dioxide, may be involved. Thus attention is being given to modeling the carbon cycle through the geologic past (Berner, 1990) by incorporating processes such as volcanic outgassing, sedimentary burial of organic matter and carbonates, and continental weathering of silicates, carbonates and organic matter. Climate processes of atmosphere and ocean are not yet being included explicitly as variables but land/ocean fraction and continental runoff are prescribed, time-dependent forcings derived from geologic observations. The results show CO_2 concentrations at least ten times present levels between 600 and 350 million years ago, near present levels from 350 to 250 million years ago, and then another period of elevated CO_2 several times present levels, from 250 to 50 million years ago, followed by a gradual drop to the present concentration (Fig. 21.8).

A coupled ocean–atmosphere GCM with idealized geography has been used to explore equilibrium climates of atmosphere and ocean for a range of atmospheric CO_2 concentrations from eight times present levels to one-half times present levels (Manabe and Bryan, 1985). The results agree qualitatively with some trends in paleoclimatic observations over the past 50 million years (Fig. 21.2), suggesting that the observed long-term cooling of the continents and oceans over this interval may have been associated

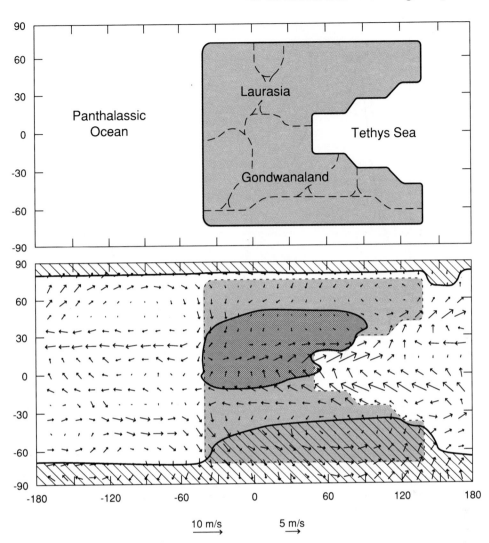

Fig. 21.7 Top: The idealized Pangaean continent with Laurasia in the Northern Hemisphere and Gondwanaland in the Southern Hemisphere. Panthalassa, the world ocean, and the Tethys Sea are indicated. Fine dashed lines indicate very approximately the outlines of modern land masses, but these outlines are only schematic. Bottom: Surface winds (arrows) and features of surface temperature (warmer than 30°C, shading, colder than 0°C, hatch) for June–July–August based on a climate model simulation with an atmospheric general circulation model coupled to a mixed-layer ocean model. (Kutzbach and Gallimore, 1989).

with a gradual decrease in atmospheric CO_2. The challenge for the future will be to study these problems with models that include realistic geography and orography, and that couple the atmosphere–ocean components of the Earth system with its biogeochemical components.

In the past several million years the overall plate movements have been

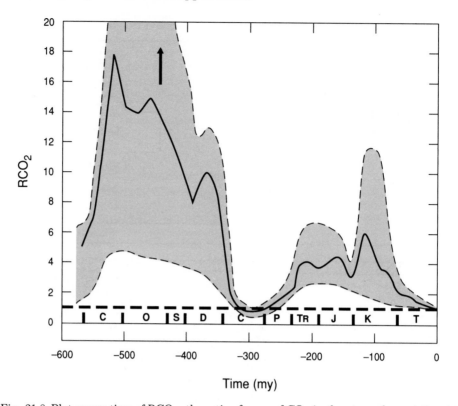

Fig. 21.8 Plot versus time of RCO_2, the ratio of mass of CO_2 in the atmosphere at time t to that in the present atmosphere, based on a carbon mass balance model. The time scale is in millions of years (my) and the main geological periods are indicated, see Fig. 21.1. The solid curve represents the best estimate of the various parameters that go into the model. Dashed lines show the envelope of approximate error based on sensitivity analysis. The vertical arrow denotes that early Paleozoic CO_2 levels may have been even higher than those shown; see text for discussion. (Berner, 1990).

relatively small and yet geologic observations suggest significant changes of climate toward cooler and generally drier conditions, with the cooling leading ultimately to the initiation of glacial–interglacial cycles around 2.5 million years ago (Fig. 21.2). Since orbital variations have been occurring throughout time, one interpretation is that some other factor (or factors) caused the climate to cool to the point where orbitally caused glacial cycles could occur. Mountain uplift, lowered CO_2 levels, and changes in ocean circulation are among the suggested factors.

Major worldwide uplift of plateaus and mountains has occurred in the past five to ten million years, with a possible doubling of heights in many regions (Ruddiman et al., 1989). In a series of climate sensitivity experiments with AGCMs for no-mountain, half-mountain (perhaps approximating the situation of five to ten million years ago) and full (present-day) mountains as prescribed lower boundary conditions, Ruddiman and Kutzbach (1989)

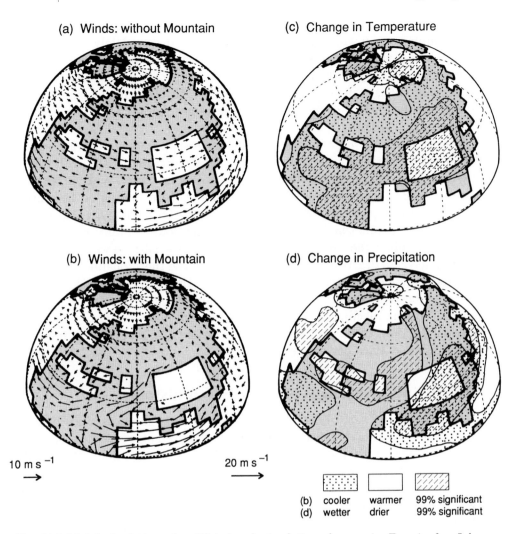

Fig. 21.9 Model simulation of uplift-induced circulation changes in Eurasia for July.
(a) Near-surface winds in "no-mountain" experiment. (b) Near-surface winds in
"mountain" experiment. (c) Change in surface temperature due to uplift; cooler
regions stippled; warmer regions blank. (d) Change in precipitation due to uplift;
wetter regions stippled, drier regions blank. Diagonal (broken-line) shading in (c)
and (d) is used for areas where temperature or precipitation changes are significant
at the 99% confidence level. (Ruddiman and Kutzbach, 1989).

show that many of the patterns of regional cooling and drying that have been
estimated from geologic observations can be explained by uplift (Fig. 21.9).
By comparing simulations with and without the Tibetan plateau, Manabe
and Broccoli (1990) show that the modern-day Asian deserts are simulated
correctly only with the Tibetan plateau present. These kinds of studies with
AGCMs give useful preliminary indications of the atmospheric changes to
be expected from changing orography but ignore the likely changes in ocean

temperature and circulation that would also occur. A climate model with a fully coupled ocean will be needed to simulate more completely the response of the climate system to uplift, and the model will also likely need to include interactive coupling with biogeochemical and vegetation processes. This is so because changes in weathering associated with uplift might be expected to change the carbon cycle and lower the atmospheric CO_2 concentration (Raymo et al., 1988).

Ocean gateways may open or shut in response to otherwise subtle horizontal or vertical crustal movements. Maier-Reimer et al. (1991) have used an OGCM forced by modern observed wind stress and surface air temperature to study the possible consequences of the closing of the central America isthmus that is believed to have occurred over the past ten million years. They found that with an open isthmus lower-salinity waters from the Pacific dilute the salinity of the North Atlantic surface waters and this dilution leads to drastically reduced strength of the thermohaline circulation cell and the poleward ocean heat transport in the North Atlantic.

Many questions remain concerning the causal factors that produced the general cooling of climate over the past several million years. All of the above mentioned factors (uplift, falling levels of CO_2, and changes in ocean gateways) may have played a role.

21.4 Climatic changes of early Earth history

Studies of early Earth history must include the possibility of very large differences in conditions compared to present. Models of solar evolution suggest that solar luminosity may have been about 75% of the modern value around 4–5 billion years ago. If the early atmosphere had the same composition as our present atmosphere, this so-called faint early Sun should have produced an ice-covered Earth, a possibility for which there is no evidence. In fact, there is evidence of relatively warm conditions, including running water and early forms of life. This faint-Sun paradox has led to investigations of the radiative effects of enhanced greenhouse gases on early Earth surface temperature (Kiehl and Dickinson, 1987). These studies are increasing our understanding of radiative and photochemical processes for concentrations of CO_2, methane, and other gases that differ considerably from present-day values. Of course, estimates of both the composition of the early Earth atmosphere and the climate are still subject to large uncertainties (Walker et al., 1983). Earth's rotation rate may also have been faster during early Earth history (Hunt, 1979). Even with the large uncertainties that exist concerning the actual climate of early Earth, studies of Earth's early climate history are underway using simple system models incorporating linked energy budget climate models, hydrologic cycle models and carbon cycle models (Marshall et al., 1988; Kuhn et al., 1989).

21.5 Conclusions

The climatic history of our Earth provides an increasingly data-rich environment for testing ideas about the causes and mechanisms of large climatic changes. Moreover, the possibility exists, in the future, to use modeling studies in combination with observational studies to assess the adequacy and accuracy of climate models.

This brief overview has illustrated some of these opportunities and some of the obstacles. Of necessity, most studies have used models of the climate system that fall well short of the desired level of breadth and detail. On the one hand, studies with fully coupled models are greatly simplified and include only a few variables (global ice volume, global deep ocean temperature, etc). On the other hand, detailed studies with individual system components (atmospheric or ocean GCMs) are likewise of limited value because they are not coupled to other important components of the climate system.

The climate sensitivity experiments that are used to infer the possible effects of past changes in orbital parameters, geography and orography, CO_2 levels, and solar irradiance are similar in methodology to the climate sensitivity experiments that are used to infer possible effects of future increases in greenhouse gases. The advantage in studying past climates is that the paleoclimatic observations help us assess the model's response. We know what happened (or we can find out).

As coupled climate models are developed, their accuracy must be evaluated. One test of the accuracy of coupled climate models will be the degree to which they can simulate the observed seasonal cycle. The recent few decades and centuries of historical records are useful for testing the accuracy of coupled climate systems on the scale of interannual variability. However, only the more distant paleoclimatic records of past millennia provide examples of large climatic changes of the order that might be associated with doubling or tripling of atmospheric concentrations of CO_2 and increases in other greenhouse gases over the next century. For example, the estimated global-average warming from the most recent glacial maximum (around 18 ka) to the present is about $4°C$ – roughly the size of the anticipated warming due to increased greenhouse gases over the next century. Because the future changes may occur much more rapidly than the deglacial warming that occupied a period of about ten thousand years, perhaps one hundred times more rapidly, abrupt as well as gradual changes in climate need to be studied. Another example of a period of the past that is now of substantial interest is the climate of several million years ago. This period had generally warmer conditions than present and perhaps elevated CO_2 levels. It may prove to be the most recent example of a climate substantially warmer than present (Fig. 21.2).

If we can construct realistic models of the coupled climate system and of the even broader Earth system, we will have many opportunities to use them – not only for addressing practical questions that we face in the next century, but also for working in an interdisciplinary mode with geologists, ecologists, archaeologists and paleontologists in solving puzzles about the Earth's past.

22

Changes in land use

Robert E. Dickinson

22.1 General scaling arguments

Changes at the land surface can affect surface energy balance and exchanges of water, heat, and momentum between Earth and atmosphere. Since humans have been and will be changing the land surfaces in myriad ways, it is necessary to develop analytical tools to simply examine whether a process may be significant or not. Different boundary conditions on a mesoscale (a few tens to hundreds of km) can alter mesoscale wind and precipitation patterns (Avissar and Pielke, 1989). However, such alterations would have to be massive before they would aggregate to significant changes on a continental or larger scale. Thus, for most purposes, it will be adequate to examine the possible effect of changes averaged to the scale of global model grids.

The basic driver of climate is solar input. About 240 W m^{-2} of solar energy is absorbed on the average by the terrestrial system. Absorbed over the whole system, a persistent 1 W m^{-2} change can eventually alter global average temperature by 0.5 to 1°C (cf. Chapter 1), which would be highly significant. Changes much smaller than this would probably be unimportant and certainly undetectable. Changes over the smallest scales resolved by global models must involve larger disruptions in energy exchange to merit comparable emphasis, since energy redistributions will tend to weaken their effect. We distinguish between those changes that affect the energy available in the land–atmosphere–ocean system, e.g., change in surface albedo, versus those that simply modify rates of exchange between surface and atmosphere, such as by altering rates of evapotranspiration.

Because of horizontal redistribution of energy, regional changes in net energy input will be several times less effective in changing temperatures over the area of application than would the same change applied globally. Another factor of two or more reduction in effectiveness can be expected if it is simply exchange rates being altered. Furthermore, regional changes will be less detectable against the background of natural climate fluctuations than would be global ones. Thus, we suggest that mean changes over a continental-scale region of less than 5 W m^{-2} in absorbed energy or 10 W m^{-2} in energy exchange rates will probably not be significant for most questions. Since about half of a change in surface albedo is masked by clouds and atmospheric attenuation, the latter figure can also be applied to changes in reflection of solar energy at the surface under clear-sky conditions. This value of 10 W m^{-2} is also a lower limit to likely future measurements of change by remote sensing, and suggests the limits of accuracy required in our model land surface parameterizations. Over continental-scale areas, we might lower these limits by a factor of two to be adequately conservative.

With the above analysis, we have reduced the question of what land changes might be important to what land changes might modify energy exchanges by more than the lower limits noted above. In particular, we find that all the modifications by urbanization, including heat released directly by fossil fuel combustion, are unimportant on the current spatial scales of climate models. The well-known *urban heat island* warming of up to a few degrees, a significant and real effect locally, does not contradict this statement, since the areas so modified are small compared to current scales of global climate models. However, modeling of the urban heat islands may be required to better interpret observational temperature records maintained in urban areas.

Our lower limit to a significant change corresponds to changes in surface albedo of 0.03 or more under summer or tropical conditions, and changes in evapotranspiration of 10 mm/month or more. Changes of less than 10 W m^{-2} in sensible heat fluxes could be important under warm wet conditions, where the total sensible flux may be only a few tens of W m^{-2} or less. The most direct effect of changing sensible heat fluxes is on the height of the daytime planetary boundary layer (PBL), see Chapter 10, Sec. 10.3.3.

The only changes in land cover likely to remain important in the context of the above analysis are those that involve large-scale modifications of the vegetation cover of a region. Studies of such change have, up to now, largely focused on tropical and semitropical regions, where the issue of forest conversion to other cover is referred to as *deforestation* and removal of relatively short vegetation in semiarid regions as *desertification*.

An additional issue that has received less attention is the effect of shifting the margins between tundra and taiga in northern latitudes on surface albedo by altering the masking of snow albedos in the spring by overlying forest cover (Peng et al., 1982).

22.2 Modeling sensitivity studies

Because of the complexity of the land system, and the relative lack of attention to modeling it as part of the climate system, we have little confidence in the details of any simulation results involving land processes and generally are unable to validate them. Furthermore, projection of actual future land covers is considerably more linked to unpredictable human economic decisions and less established than projections of increasing greenhouse gases. Thus, climate-modeling studies make no attempt at predictions but rather simply examine "what if" scenarios. These scenarios have usually been studied primarily to better understand the performance of the climate models and whatever improvements might be needed in their land parameterizations rather than to suggest possible future climates or what the effect of human activities might be. However, because these studies have also recognized the possible linkages of their conclusions to changes wrought by human activity, we shall also summarize such possible linkages here.

How can changing land as a boundary condition to the atmosphere affect distribution and amounts of precipitation? Speculations on this question extend well back into human history. However, observational studies have been too limited in coverage and involved too many uncontrolled variables for any convincing evidence addressing this question to be developed. Only with three-dimensional models of the atmospheric system have we been able to quantitatively examine this issue. Mintz has reviewed prior studies in *The sensitivity of numerically simulated climates to land-surface conditions* (1984). These studies have either considered the effect of changing surface albedo or soil wetness. Most simply, surface hydrological conditions have been specified in terms of a *wetness factor*, i.e., the surface is assumed to evaporate at a rate proportional to that of a wet surface, with the proportionality given by the prescribed wetness factor.

The persistence over several decades of drought in the Sahel region of Africa (semiarid zone south of the Sahara) has posed the question as to whether changing surface conditions could be contributing to this climate change. Charney (1975) hypothesized an important role for increases in surface albedo. He suggested that removal of vegetation in the region would expose the underlying sandy soil with a higher albedo than that of the overlying vegetation. With higher albedo, the surface would absorb less solar radiation and so it would provide less sensible heat to the overlying atmospheric column. This relative cooling of the overlying atmosphere would be dynamically compensated by adiabatic sinking and warming, in turn stabilizing and drying the atmospheric column. Hence precipitation would be reduced and the regrowth of vegetation suppressed. This process was referred to as a biophysical feedback because of the positive feedback (in the hypothesis) between vegetation and precipitation.

Charney et al. (1977) explored this hypothesis using simulations with the GISS GCM and with a fixed soil wetness. In their control-model simulations, all land surfaces have an albedo specified to be 0.14, except for deserts, whose albedo was taken to be 0.35. Some perturbation studies were made in which the albedo of three semiarid regions was changed to 0.35 and others where the albedo of three moist regions was changed to 0.35 (Fig. 22.1). The integrations were done for a 6-week period from mid-June to the end of July, and changes in precipitation and surface energy balance were analyzed for July.

691

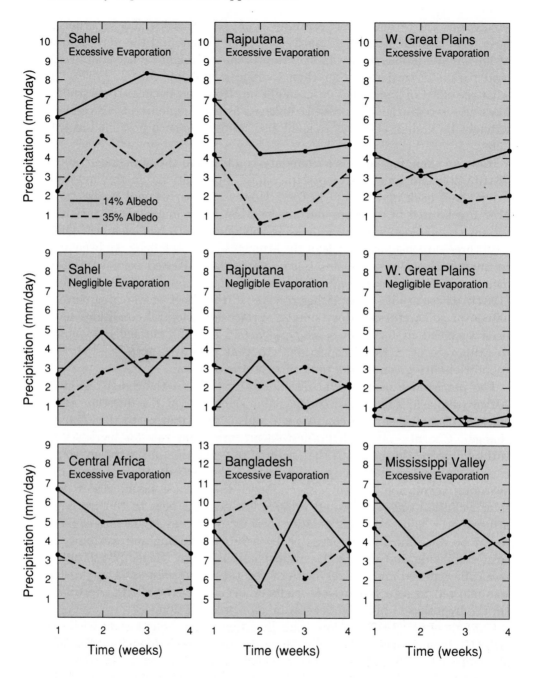

Fig. 22.1 Weekly averaged rainfall obtained by Charney et al. (1977) for various regions and for wet (excessive evaporation) or dry (negligible evaporation) surfaces for high versus low albedo.

Two procedures were used to introduce a prescribed surface wetness factor. Neither provided a reasonable description, for with one the land surface was specified as wet, wetness factor of nearly 1.0, and with the other as dry, wetness factor close to 0.0. The enhanced albedo in moist regions was introduced for only the wet case.

The numerical results obtained by Charney with the GISS model supported the general hypothesis that increasing surface albedo could reduce precipitation but did not support the detailed mechanisms he suggested, except possibly for the one dry Sahel case. Figure 22.1 shows that the precipitation decreased uniformly only for the semiarid cases with wet surface and for one of the three moist regions. However, the monthly average precipitation decreased in all but two cases: the case of Bangladesh, where it was essentially zero, and the dry Rajputana (northwestern India) case, where it increased slightly. Net surface radiation balance generally went down, but only in the dry cases was the solar absorption at the surface reduced by a magnitude consistent with the raised albedos. In all the wet cases, reductions in cloud cover compensated for the increased albedo, so that absorbed solar radiation was little changed, and indeed, contrary to the Charney hypothesis, for all of the semiarid cases, solar absorption increased.

Figure 22.2 shows for the excessive evaporation boundary condition the differences in energy balance between the high versus low albedo cases for the Sahel region, where the precipitation was reduced by the greatest amount. Reduction in net radiation was a direct result of a large increase in the net loss of thermal infrared radiation. Part of this increase is directly ascribable to reductions in cloudiness that allowed more emitted thermal radiation to escape. Also, since absorbed solar radiation increased (because of the reduced clouds) but latent heat fluxes decreased, the surface must have warmed up and emitted more blackbody radiation. (Changes in temperature or surface thermal emission were not recorded as part of the simulation history.) For the control albedo configurations, the differences in the precipitation between wet and dry cases were as pronounced as, or more pronounced than, the differences in precipitation between low and high albedo for the wet surface assumption. For example, for the Sahel region, either specifying the surface to be dry or of high albedo reduced monthly average precipitation from 7.4 to 4.0 mm day^{-1}.

In analyzing the dependence of precipitation on surface changes, it is useful to budget precipitation changes in terms of changes in surface fluxes of moisture and changes in the convergence of atmospheric moisture. The original Charney hypothesis required a reduction in atmospheric moisture convergence but did not include surface moisture fluxes. On the other hand, past literature has been replete with suggestions that precipitation might correlate with surface moisture fluxes. In all the GCM cases studied by Charney, increased albedo led to reduced surface evapotranspiration. In the moist surface semiarid regions, in addition, the moisture convergence in the

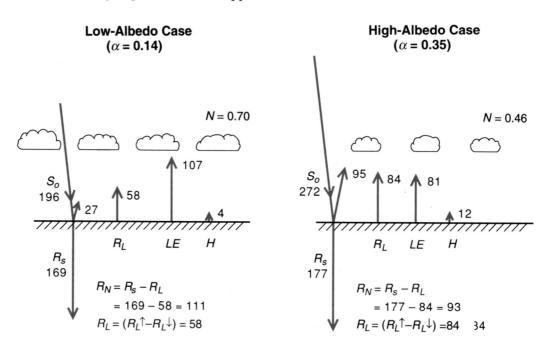

Fig. 22.2 The Sahel region energy budgets in W m^{-2} for Charney et al. (1977) as presented by Mintz (1984). The symbols used are N = cloud cover, S_o = incident solar radiation, R_s = absorbed solar radiation, R_L = net longwave, LE = latent heat flux, H = sensible heat flux.

atmospheric column was reduced by up to several times the reduction in surface evaporation. Ambiguities in the sign of the change in precipitation in the other cases resulted from differences in the sign of the change in moisture convergence.

In sum, the prototype studies by Charney of possible impacts of increasing surface albedos suggested that precipitation might be reduced much of the time, but that the actual atmospheric response could be quite complex and not intuitive, with interactive cloudiness having a major role in the final answer. Whether changes in moisture convergence would enhance or cancel the effects of decreased surface evapotranspiration in modifying rainfall could be quite dependent on details of regional climate and its simulation.

The changes of land properties that Charney assumed were much larger than might be supported by observations and did not represent vegetation in any manner except as a changing albedo. However, these studies helped to intensify interest in further exploration of how vegetation might interact with the climate system. More recently, modeling studies (e.g., Owen and Folland, 1988) have suggested that much of the year-to-year variations in

Sahel rainfall might be linked to variations in ocean surface temperatures. The potential role of land changes in general or as a feedback on the response to ocean forcing remains to be clarified. Charney's results also suggested that changes in evapotranspiration could be as important as or more important than changes in albedo in modifying atmospheric climate. This importance of evapotranspiration has become increasingly obvious with further studies.

Shukla and Mintz (1982) asked: "What if all continental surfaces were either completely dry or completely wet?" Their integrations were again carried out for the short period from mid-June to the end of July, and July averages were compared. The dry case differs from the wet case through its boundary conditions that imply an absence of surface water source and evaporative cooling. The surface warms up without this evaporative cooling. Precipitation rates over continents are greatly reduced (as shown in Fig. 22.3), either from this warming or from the reduced water source. Likewise, cloud cover is reduced, and increased solar radiation further warms the surface, giving the net balances shown in Fig. 22.4. The increased solar heating is more than half the loss of evaporative cooling, although the net increase of surface radiation is considerably smaller; the surface longwave emission has increased, and trapping by clouds has decreased. Overall, land temperatures north of 20°N in the summer hemisphere increased by 15–30°C. The relative contributions of changes in atmospheric convergence and evapotranspiration to the changed precipitation were not examined. However, tropical regions showed significant compensation for the interrupted supply of moisture from the surface, with rainfall over Bangladesh even increasing as a result of a shift in the region of maximum monsoon rainfall.

Several other sensitivity studies of the impact of land surface properties on the atmospheric circulation from modeling studies are discussed in Chapter 19, Sec. 19.3.3. Interactive soil moisture models (reviewed in Mintz, 1984) have been used to consider the effects of changing initial values of soil moisture on subsequent rates of precipitation. Large reductions of precipitation were realized in the cases with dryer initial soils for up to several months or more into the future. Thus, anomalies in soil moisture may require a considerably longer time to return to normal conditions than would be the case for unchanging climates because of the positive feedbacks between the soil moisture and precipitation.

Some recent GCM studies of global warming have examined differences over land between prescribed cloudiness simulations and ones with interactive clouds (Manabe and Wetherald, 1987). These have found considerable sensitivity to interactive cloudiness for global warming scenarios. Reductions in midcontinental soil moisture and surface temperature increases could be greatly amplified. Since such inclusion of interactive clouds in GCMs is very preliminary, the actual effect is uncertain.

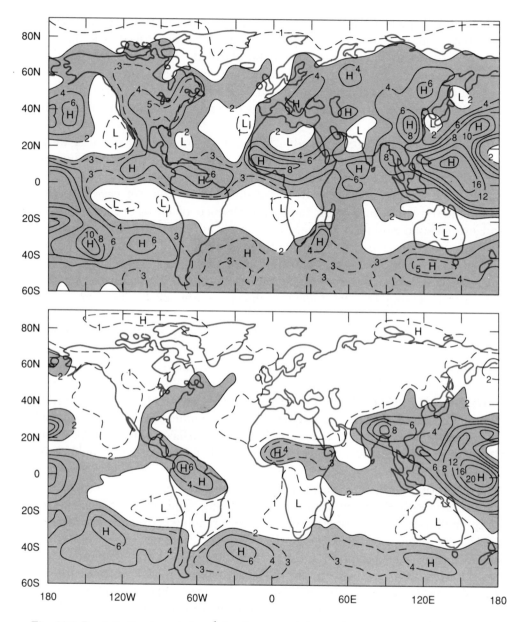

Fig. 22.3 Precipitation in mm day^{-1} for the wet and dry surface experiments of Shukla and Mintz (1982). Precipitation greater than 2 mm day^{-1} is shaded. Note for the dry case the extreme reduction of precipitation over the continental surfaces.

22.3 Climate effects of tropical deforestation

Tropical forests cover large areas in regions near the equator on several continents, where rainfall averages over 2,000 mm a year. The rainfall in these regions is overall a manifestation of the tropical rainbelt

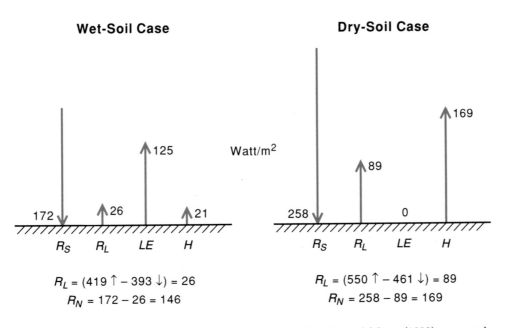

Fig. 22.4 Surface energy budgets for the experiments of Shukla and Mintz (1982), averaged over continental regions between 20° and 60°N. Symbols same as in Fig. 22.2.

and the distributions of oceans and continents. However, the question has long been raised as to whether precipitation over the tropical forests could be significantly modified by changing surface conditions, in particular by removal of the forests. This question becomes especially germane because of the ongoing shrinkage of tropical forests through extraction of timber resources and conversion to ranching, farming and agroforestry, as well as industrial and mining uses (see Chapter 2, Sec. 2.3 for a more complete discussion). The forests of the Amazon basin in South America, the largest single area of tropical forest, were until recently largely untouched, but are now being rapidly exploited with many possible implications for the global environment (e.g., Dickinson, 1987). The question of the climatic role of the Amazon forest is being examined by climate modelers.

The oceanic source of water for the Amazon basin is almost entirely from the Atlantic Ocean, transported by westward-moving trade winds. The basin is bounded by the high Andes mountains to the west and by highland plateaus to the north and south. Various scientists have examined the hydrological budget of the Amazon region through measurement of fluxes of water, gradients of isotopic tracers (^{18}O and deuterium), overall amounts of precipitation, and return of water to the Atlantic through the Amazon River. These studies have given a consistent picture that only about half of the precipitation over the basin is supplied by the Atlantic, with the rest originating from the evapotranspiration of the forest itself.

This result suggests that the forest may have a major role in providing its own rainfall. On the other hand, it is difficult to extrapolate from this inference any quantitative statement as to what changes might be expected with removal of the forest. In particular, there is no basis in the observational water-budget studies for inferring that other surfaces besides forest would evaporate less water, or for excluding an increase in the convergence of moisture from the ocean to compensate for changes in the surface supply of water.

Thus, it is necessary to use GCMs to explore this question. Earliest studies considered deforestation as simply increasing surface albedos. Recent studies (Dickinson and Henderson-Sellers, 1988; Lean and Warrilow, 1989; and Shukla et al., 1990), however, have attempted to include the dominant hydrological processes of the forests and a hypothetical degraded grassland that is assumed to replace the forest. It is unlikely that a short grassland would ever completely replace the Amazon forest or even most of it. Various kinds of scrubland and secondary forest are alternative replacements. However, grassland is assumed as the most extreme possibility already visible over large previously forested areas. This example is studied more to explore the functioning of vegetation in a climate model than to make any particularly relevant projection of future climates over the Amazon basin.

In a recent integration with the low-resolution (R15) NCAR Community Climate Model (CCM1) with the Biosphere Atmosphere Transfer Scheme (BATS) surface package, the effects of deforestation were simulated for a period of 3 years. Figure 22.5 shows fields for the forest and grass surface, averaged over all the forest points and by month over the 3-year period. Figure 22.6, for comparison, shows for a 12-month average the distribution of changes obtained by Shukla et al. (1990). Both evapotranspiration and atmospheric moisture convergence decrease so that runoff out of the basin is decreased.

Conversion from forest to grassland in recent simulations has reduced the net radiation available to a surface in two comparably important ways: (a) the models have generally assumed that the albedo of the grass would be higher than that of the forest, and (b) the models have assumed that the grass surface was much smoother than the forest. The latter change in all simulations to date has increased surface temperatures by more than they were decreased by the change in albedo. Hence the energy loss by upward thermal infrared radiation has increased. Both an albedo increase and a roughness decrease are plausible but difficult to quantify precisely. Recent studies have differed in their assumption about surface albedo change by about a factor of two. Indeed, since the albedos of vegetated surfaces depend on other factors besides just height, it is possible to find examples of some short vegetation with the same albedo as some stands of tall vegetation. Furthermore, surface roughness depends on other factors than just the height of the vegetation. A heterogeneous surface of tall mixed with short

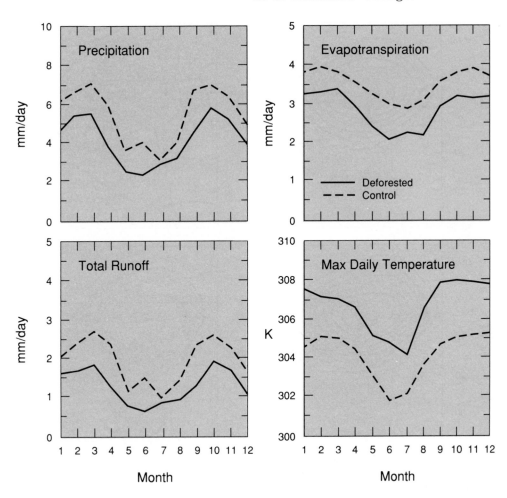

Fig. 22.5 Amazon-wide averages of monthly mean fields for control and deforested cases using CCM1 and BATS1e codes. Averages are over a 3-year simulation.

vegetation may have a larger roughness length than a completely forested surface. All these questions would have to be addressed more carefully in attempting to project actual changes, but can be ignored in present attempts to explore hypothetical scenarios to establish the possible role of vegetation in climate models.

Perhaps a more appropriate criticism of current studies might be that different authors have not only assumed different models for the role of vegetation but have made a wide range of assumptions about parameter changes. Thus, conclusions may be model-dependent in three ways: dependent on differences between GCMs, dependent on different land parameterizations, and dependent on assumed changes in model parameters. Such a rich variety of model differences may help establish which results are sufficiently robust to be model-independent. However, it increases

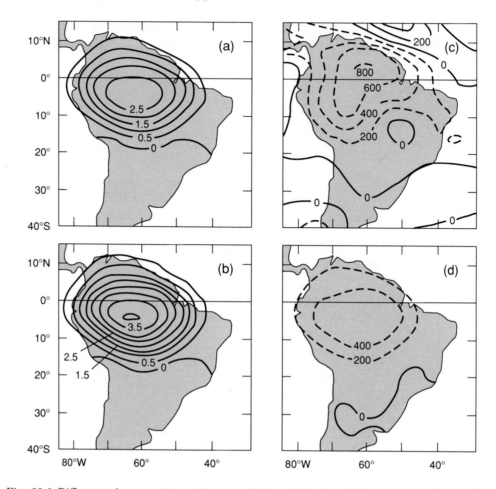

Fig. 22.6 Differences between 12-month means (1 Jan.–31 Dec.) of deforestation and control
cases (deforested – control) from Shukla et al. (1990) one year simulation: sector:
(a) surface temperature increase in °C; (b) deep soil temperature increase in °C;
(c) total precipitation changes (dashed lines indicate a decrease) in mm; and
(d) evapotranspiration decrease in mm.

the difficulty of intercomparing model results and determining significant
differences. Adoption of common numerical values for prescribed parameters
or differences and further study of the dependence of individual model results
on these differing parameters should help to remedy this situation.

22.4   Appropriate role of land use studies

Climate models that include descriptions of land processes are
marvelous research tools for exploring the mechanisms involved in the
interaction between land and atmospheric climate processes.   However,
they have not yet advanced to a point where they will reliably predict the

consequences of hypothetical changes of the land surface. Until it has been demonstrated that they can do that, studies of changes in land use will have to focus more on the goal of testing and improving this aspect of climate models than on opening windows to consequences of human activities.

PART 6

FUTURE PROSPECTS

23

Climate system modeling prospects

Lennart O. Bengtsson

23.1 Introduction and background

The mathematical modeling of the climate system, as described in this volume, has undergone a spectacular development. This can be demonstrated by comparing the first numerical experiments of the general circulation by Phillips (1956) using a two-layer quasi-geostrophic model on a β-plane to today's century long simulations of coupled ocean–atmospheric models, (Stouffer et al., 1989; Washington and Meehl, 1989; Cubasch et al., 1992).

The modeling of the climate system, so far, has mainly been concerned with modeling the different components of climate system separately and with the main emphasis on the atmosphere. Very recently, experiments have been undertaken towards a more comprehensive coupling of the different components into a complete system. The development of atmospheric modeling, in particular, has benefited from the synergism with numerical weather prediction.

23.1.1 Atmospheric GCM development

Numerical weather prediction and atmospheric general circulation modeling grew from similar origins in the early 1950s, but modeling for the two applications differed widely. Weather forecasting was concentrated on a time range of up to a few days ahead, over which period close attention to the climatological balance between resolved and parameterized processes was not of paramount importance. It was also subject to operational time constraints, which made it necessary to restrict the integration domain to limited areas. However, it became evident as the numerical predictions were extended into the medium range that the climatological balance of the forecast model became of paramount importance (Hollingsworth et al., 1980; Bengtsson and Simmons, 1983). As a consequence, short-term (30–90 days) climate simulations play an important role in the development of global forecast models.

Thus a clear distinction can no longer be drawn between numerical atmospheric models for climate studies and weather forecasting, and it may be anticipated that future improvements in model design will result not only from comparisons of the climate simulations produced by different models, but also from studies of the growth of systematic errors or *climate drift* (see Chapter 17) in extended forecast experiments.

Three areas have been identified of major importance in accelerating the development of atmospheric GCMs (AGCMs) in the 1980s:

(1) Improvements in the global observing system following the Global Weather Experiment (GWE) in 1979 and associated special observing programs to study specific processes or phenomena such as GATE (GARP Atlantic Tropical Experiment), ALPEX (Alpine Experiment), TOGA (Tropical Oceans Global Atmosphere) etc.

(2) The development of powerful supercomputers which has made it possible over a period of 10–15 years to improve the resolution of climate models by a factor of 3 to 5 (see Chapter 9 and Table 23.1).

(3) The synergism created by the parallel use of AGCMs for numerical forecasting and for climate simulation.

The usage of advanced AGCMs in daily forecasts has made it possible to validate the models against the widest possible range of initial conditions. Through a systematic use of ensemble predictions on different time scales, in combination with case studies using observations from special observing programs, a gradual improvement of the AGCMs has taken place. Figure 23.1 shows, for example, the improvements in the skill of the medium-range forecasting during the 1980s (cf., Fig. 9.2 for short-term forecasting). At the same time the systematic model error has been reduced by some 75%.

Another necessary condition for the improvement in the global forecasting models was the rapidly increased performance of supercomputers (Table 23.1). Comparing for example the first supercomputer, CRAY 1, developed by Cray Research in the late 1970s and the latest CRAY Y-16

Table 23.1. Computational requirements in the form of elapsed time for a range of Cray computers during the time period 1980–1992 for a range of spectral resolutions. Values are for a one-year simulation using a 19-level vertical resolution; they may be scaled linearly for alternative numbers of levels or alternative simulation periods. They are based on European Centre for Medium Range Weather Forecasts models optimized for high-resolution use and multi-tasked.

Year	1980	1984	1988	1992
Computer	Cray 1	Cray X-MP	Cray Y-MP	Cray Y-16
Model	1 Processor	2 Processors	8 Processors	16 Processors (est.)
T21	6	1	0.3	0.05
T42	50	8	2.2	0.4
T63	160	27	7.5	1.4
T106	800	125	36	7

Predictive Skill in Numerical Weather Prediction

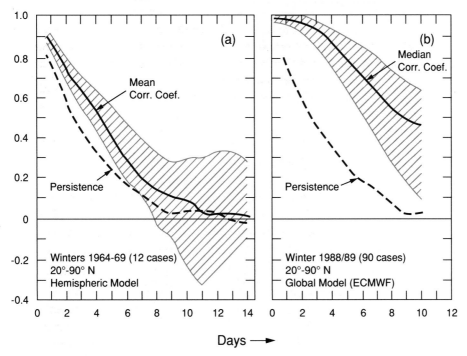

Fig. 23.1 The mean anomaly correction coefficients for a series of forecasts. Left shows the mean anomaly correction coefficients for a series of GFDL forecasts, Miyakoda et al., (1972). The dashed curve shows persistence. Right shows the same for the operational ECMWF model winter 1988–89.

to become available in 1992, the computing power has increased by a factor more than 100. This means that a T106[1] model in 1992 will require about the same elapsed time as a T21 model did in 1980! Although differences in timing can result from differences in technique and in the complexity of parameterization, the dominant sensitivity is to the resolution of the model.

So far, the climate-modeling community has been slow to increase the horizontal resolution of the climate models, and only recently has the resolution been extended to T42 or the equivalent gridpoint resolution. Instead the emphasis has been to integrate the models for longer time periods.

---

[1] See Chapter 9, Sec. 9.2.2 for an explanation of this notation.

23.1.2 The rest of the climate system

Climate modeling requires that due attention is paid to oceans and land surfaces, and to the chemical and biological processes and the way they interact with the dynamics and thermodynamics of the atmosphere and the oceans. As has been demonstrated in this volume, major achievements have taken place in recent years, including realistic simulations of the three-dimensional circulation of the oceans (Chapter 11). Sea ice modeling (Chapter 12) is also developing to become more realistic and physically based.

The modeling of atmospheric chemistry (Chapter 15) has progressed very well and there have been GCM studies including ozone and its three-dimensional variations. The sulfur cycle is being modeled in a similar way. More comprehensive chemical modeling, including up to 60 different chemical species, has been done over limited areas and for shorter time scales. Incomplete knowledge of the complex chemistry involved and how best to couple it with the physics and dynamics of the atmosphere model, as well as insufficient computer resources, are still preventing such schemes from being tested in global models over time scales relevant for climate.

Understanding the long-term variations of the greenhouse gases in the climate system (Chapter 20) does require a proper incorporation of relevant land surface processes, including realistic hydrology and relevant biology into climate models (Chapters 14 and 16). Similarly, biome models are used to present vegetation as a function of basic meteorological parameters (Chapter 6), and are being tested in response to climate change scenario experiments. These models are in their infancy and considerable efforts lie ahead before the biome models are interactively combined with the GCMs. On very long time scales, changes in the major ice sheets (Chapter 13) must be considered.

23.2 Climate models

A comprehensive validation of climate models in the context of simulations of climate change due to increases in greenhouse gases was recently carried out by the International Panel on Climate Change (IPCC, 1990a). Here we make use of some of the material from this report but discuss climate models in more general terms. So far, as shown in the IPCC report, climate models have for the most part consisted of AGCMs coupled to simple or rudimentary oceans and with many processes known to be important either left out or represented only very crudely. The following discussion therefore focuses mostly on the atmospheric part of the climate models. Other aspects of AGCMs are discussed in more detail in Chapter 10.

In selecting model variables in validating an AGCM, the following may be considered:

(a) *Variables important for the description of the general circulation.* Examples include sea level pressure, and wind and temperature through the depth of the atmosphere. This includes monthly averages, the daily variability (e.g., as portrayed by the eddy kinetic energy), seasonal transitions, and changes between characteristic circulation regimes (Großwetterlagen) on time scales from weeks to years. To put into simple terms: the objective should be that a sequence of simulated fields should not be distinguishable from a sequence of analyzed fields even by a trained meteorologist!

(b) *Variables critical for diabatic processes and the hydrological cycle.* These include boundary layer fluxes, friction, and energy dissipation in the free atmosphere. Except for the broad structure for inhabited parts over land, reliable observations of precipitation do not yet exist. Our present knowledge is based on energy balance considerations available as a climate average. Ensemble averages of diabatic tendencies and precipitation obtained from operational data assimilation systems (Arpe, 1991) have been found to be a useful complement pending reliable observed data.

(c) *Variables important for climate feedbacks.* Examples are clouds and their radiative effects, snow cover, soil moisture, and sea ice. If these are poorly simulated, errors will gradually occur in the description of the general circulation or, alternatively, the errors can be masked by ad hoc parameterizations which can be useful for a time but will, in the long term, be detrimental.

(d) *Variables of extreme events.* Examples of this are intense extratropical features; cyclones, cut-off lows, blocking anticyclones; frequency and distribution of tropical cyclones, and episodes of drought and excessive precipitation.

In carrying out simulations with AGCMs with specified SSTs and sea ice, the success of the simulation is often, at least in part, due to the constraints imposed at the lower boundary.

The sea level pressure is generally well simulated with models having a resolution of T42 or higher, or the equivalent. A slightly lower resolution may be sufficient with a careful tuning of horizontal diffusion (Laursen and Eliasen, 1989). The same can be said for the other large-scale variables. A remaining common deficiency is the general coldness of the simulated atmosphere. This is particularly the case in the polar upper troposphere and lower stratosphere in the summer hemispheres, with errors of more than 10°C.

The atmospheric momentum balance is delicate. There are clear

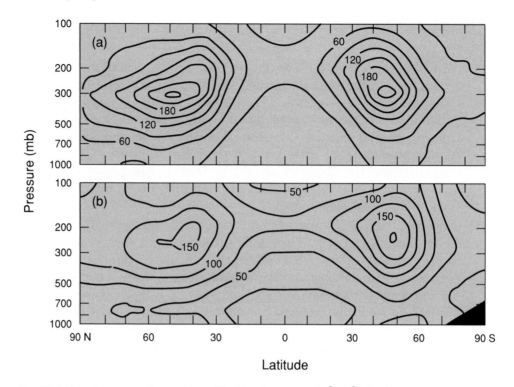

Fig. 23.2 Zonally averaged transient eddy kinetic energy (m^2 s^{-2}) for December–February for (a) observed (Trenberth and Olson, 1988) based on ECMWF analyses, (b) T30 version of the Canadian Climate Centre model (Boer and Lazare, 1988).

indications (Miller et al., 1989) that some of the aspects of the comparatively good simulations of low-resolution models ∼T20 in the Northern Hemisphere are due to an underestimation of the momentum flux, which, in turn, is compensated for by insufficient vertical transport of momentum in the westerlies. For a model with higher horizontal resolution, the poleward flux of horizontal momentum is more realistic, which means that there must also be a stronger downward transport in the westerlies. Parameterization of this, such as gravity wave drag (see Sec. 10.3.5), is consequently required for the higher resolution models.

Although the daily standard deviation of daily atmospheric variations is quite good, models still underestimate eddy kinetic energy, see Fig. 23.2. Present AGCMs have some ability to simulate blocking events of one to two week duration in the atmosphere (see Sec. 19.2.2 and Fig. 19.5). However, in studies of the performance of medium-range forecasting, Tibaldi and Molteni (1988) suggest that the frequency of blocking is significantly reduced after about a week. Recent evaluation of the spectral European Centre for Medium Range Weather Forecasts (ECMWF) model suggests that changes in the physical parameterization of radiation, convection, and reduction of vertical diffusion in the free atmosphere are leading to an increase in

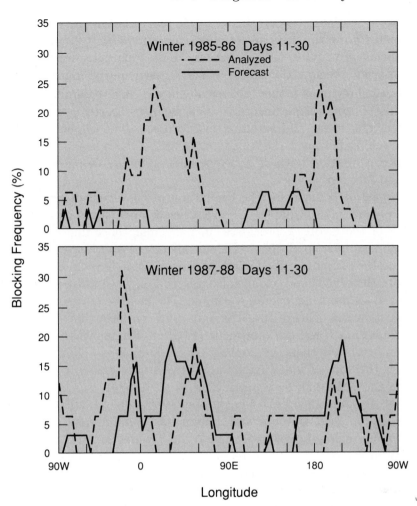

Fig. 23.3 The improvement in the prediction of blocking with the ECMWF model between the winter 1985–86 and the winter 1987–88. The dashed curve shows observed blocking according to Tibaldi and Molteni, (1988). The full lines show the prediction of blocking in the time range 11–30 days. Ten forecasts for each winter, resolution T63.

simulated eddy kinetic energy, and have also improved the prediction and simulation of blocking (see Fig. 23.3).

As far as the observations of precipitation can be relied upon, present "high-resolution" AGCMs produce a reasonably realistic simulation of the average global precipitation and its distribution. Further improved horizontal resolution, leading to a better resolution of orography (see Plate 10) and small-scale synoptic systems is likely to improve the situation further. Parameterization of subgrid-scale processes of moisture fluxes, in particular over orography and warm water pools in the tropics (Miller et al.,

1992), are essential for generating realistic precipitation patterns. Another yet unsolved problem is the role of mesoscale convective systems in creating more efficient precipitation mechanisms. This apparently leads to a delay in the diurnal precipitation cycle which is particularly common over the western Sahel region. Further improvements will require detailed diagnostic evaluation of model experiments verified against observations from in situ measurements, radar and satellite observations. The Global Energy and Water Cycle Experiment (GEWEX) later in the 1990s is expected to make a useful contribution towards a better understanding of precipitation and associated fluxes of moisture.

Fundamentally crucial for the evaluation of the effect of an increase of the greenhouse gases (Chapter 20) are a better understanding and a better handling of the feedback processes, in particular those due to changes in clouds and surface conditions. Clouds affect the shortwave radiation by increasing the planetary albedo and, by this process, have an overall cooling effect. On the other hand, clouds also absorb longwave radiation effectively and that gives rise to an overall warming. The balance between cooling and warming is rather delicate and it is only quite recently that data from the Earth Radiation Budget Experiment (ERBE) (Ramanathan et al., 1989) have demonstrated that the cooling effect dominates. Net global cloud forcing is the sum of longwave and shortwave forcing: $c = c_{lw} + c_{sw}$, in W m^{-2}, and according to Ramanathan et al. (1989), was -13.2 in April 1985, -16.4 in July 1985, -17.4 in October 1985 and -21.3 in January 1986. The values for July, October and January are preliminary. The positive values in any region do not exceed 25 W m^{-2} while negative values can be > 100 W m^{-2}.

Due to the great importance of the so-called cloud radiative forcing, an accurate simulation of clouds is very important. This will require a successful simulation not only of the total cloud cover and its geographical distribution, but also of the vertical distribution of clouds. Persistent boundary layer clouds, common over the cold ocean currents and over the Arctic in the summer have a strong cooling effect while deep convective systems and associated cirrus anvils have a warming effect. These cloud systems are notoriously difficult to parameterize and only recently have nontrivial representations of these types of clouds been introduced into AGCMs (see Chapter 10, Sec. 10.3.4).

Prediction of changes in the surface albedo due to snow and ice will require accurate handling and resolution of snow-producing weather systems, as well as sea ice (Chapters 12 and 13), and proper handling of the thermal processes in the soil leading to changes in albedo and snow melt. Vegetation must also be represented in a reasonably realistic manner since vegetation may shield the snow and thus decrease the albedo substantially (see Chapters 14 and 22).

The occurrence of extreme events is one important aspect of climate and, in some respects, more important than the mean climate. Many large-scale extreme events such as intense heat and cold, and prolonged wet and dry spells can be diagnosed from climate-model experiments (e.g., Mearns et al., 1984). The ability of climate models to simulate smaller-scale extreme events, such as tropical storms and small-scale extratropical storms and associated areas of intense weather, has so far been limited by the lack of resolution of the AGCMs. Bengtsson et al. (1982) demonstrated the capability of a GCM to generate hurricane-type vortices by analyzing the daily forecasts by the ECMWF model for one year. Studies by Dell'Osso and Bengtsson (1985) and Krishnamurti and Oosterhof (1989) have demonstrated the importance of resolution. Horizontal resolution of the order of at least 100 to 150 km is required to produce tropical storms by a numerical model with a reasonable accuracy. However, as has been noted by Haarsma et al. (1991) and Broccoli and Manabe (1990), even present climate models can simulate some of the structure that is characteristic of tropical cyclones. With respect to even finer features, it would be of interest to develop an analog to Model Output Statistics (MOS), which is used in numerical weather prediction to interpret large-scale numerical predictions into weather.

23.3 Predictability

Of the many systems that occur in nature some are convergent or stable, while others are divergent or unstable. By a stable system we mean a system which, if slightly disturbed, will converge towards a succession of states which would also have occurred if there had been no disturbance. By an unstable system we mean just the opposite, namely a system whose future state following a slight disturbance will diverge from what its future state would have been without any disturbance. Figure 23.4 shows an example of a simple stable and unstable system. These concepts are important for predictability; clearly the stable system may be potentially predictable while the unstable system is not. The atmosphere is typically unstable in this sense, and thus weather has limited deterministic predictability.

Lorenz (1982) introduced a new method to estimate atmospheric predictability based on the operational ECMWF weather forecast model. Forecasts from one to ten days in advance are issued daily at ECMWF. The feature which makes it possible to use these forecasts in an error growth study is the rather high quality of the one-day forecasts; thus the one-day forecasts for today's weather pattern may be regarded as equal to today's pattern plus a moderately small error. To find out how much this error grows in one day, when both patterns are governed by the equations of the model, it is sufficient to compare the two-day forecast for tomorrow with the one-day forecast for tomorrow. In the same way, we can determine the error growth during the next day by comparing the three-day with the two-day

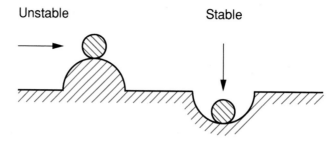

Fig. 23.4 Example of a simple stable and unstable system.

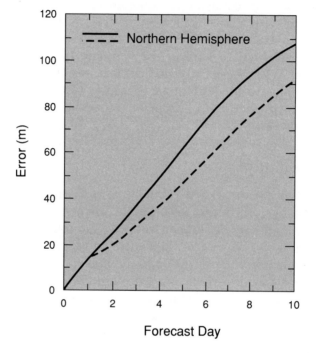

Fig. 23.5 Predictability curves for the ECMWF model winter 1989–90. The full curve shows
        the rms error growth for the 500 mb geopotential height, Northern Hemisphere
        extratropics. The dashed curve shows the estimated predictability for the same
        quantity according to Lorenz (1982).

forecast for the day after tomorrow and we may, in fact, continue the process
for nine days. ECMWF has regularly continued to produce these kinds of
predictability estimates and Fig. 23.5 shows the error growth for the 500 mb
geopotential height for the Northern Hemisphere for the winter 1989–90.
The doubling rate for small errors is around two days. Similar estimates for
other times of the year, as well as for the Southern Hemisphere, indicate
an overall doubling rate for small errors of 2–2.5 days. These results are
supported by studies of analog weather maps (Lorenz, 1969).

These error growth studies, based on global AGCMs, reveal only the growth of errors on the spatial scales which the models resolve (> 200 km). By omitting the smaller scales, the initial errors in those scales are effectively large enough that they undergo no further growth. Even if the smaller scales were resolved, the error growth in these scales would constitute only a minor part of the total growth, since these scales account for only a small part of the total variability. Nevertheless, the influence of errors in the smaller scales upon errors in the larger scales can be large. Lorenz (1969) has studied the propagation of errors towards large scales. Assuming the initial errors were confined to the smallest scales, those not resolved by our present observational network, it was found that the largest scales were predictable for a week or two. Assuming the existence of a mesoscale spectral gap (Lorenz, 1983) the predictability for the largest scales are extended by some four days.

In principle, the predictability of the ocean can be determined in the same way as for the atmosphere. The equivalence to the predictability of the weather is the predictability of the small-scale ocean eddies which are mostly generated along the strong ocean currents such as the Gulf Stream. Scale considerations suggest that the predictability of such eddies is of the order of months.

In principle, the prediction of climate is treated in the same way as the prediction of weather. While, in weather prediction, the concern is with predicting the change in the weather, that is the actual temporal evolution of the dependent parameters ($\mathbf{v}$, T, q, p_s, and associated quantities such as precipitation, clouds etc.), climate is concerned with the long-term statistical properties of the same quantities. There is no stringent definition of climate. Changes from week-to-week or month-to-month are seldom regarded as climate, year-to-year changes occasionally, decade-to-decade changes frequently, and century-to-century changes are usually considered as climate.

Due to the dominating role of the ocean–atmosphere coupling on time scales beyond weeks, climate prediction will implicitly include the prediction of the state of the ocean. Small-scale ocean eddies will not be considered.

As was originally suggested by Lorenz, it is useful to consider two different kinds of climate prediction. In the first of these, the concern is with the prediction of specific phenomena, such as the occurrence of El Niño, or when a drought in the Sahel region will begin or cease, or on the length and intensity of the Indian Monsoon. The difference between weather prediction and climate prediction is naturally diffuse. Lorenz has suggested denoting such predictions as *climate prediction of the first kind.*

Then there is another kind of climate prediction, which differs from the first, in that we are not particularly concerned with the chronological order in which atmospheric states occur. This type of climate prediction has to do with our ability to simulate the evolution of characteristic climate processes and is relevant when we wish to predict the kind of climate which may occur when there is a change in the external forcing, such as an increase in the atmospheric greenhouse gases. This kind of prediction is called *climate prediction of the second kind.*

Climate predictability of the first kind is strongly related to the predictability of the instantaneous weather. At middle and high latitudes, even the large-scale weather patterns and their time averages are internally driven by dynamically unstable synoptic systems and consequently are of a chaotic nature. The success, so far, in long-range weather forecasting (e.g., at NMC, ECMWF, and the U.K. Meteorological Office), both by empirical and numerical methods, has been very modest and of very limited use. There are nevertheless indications that cases with extreme SST anomalies, such as well-established El Niño phenomena, do affect the weather patterns in a detectable way in some extratropical areas such as over the extratropical Pacific and downstream areas over the western United States (see Chapters 18 and 19).

However, in the tropics the situation is different. This is mainly because of much weaker nonlinear coupling between different scales of motion. The general circulation in the tropics is dominated by the large-scale Hadley and Walker cells, which are driven by diabatic forcing mainly through the release of latent heat. The three centers of action (cf. Fig. 3.9), Indonesia, central Africa and northern South America are affected by SST anomalies, and numerous model experiments have demonstrated the fundamental role of SST anomalies in the equatorial Pacific and the tropical Atlantic (see Chapter 18). The quasi-linear character of tropical SST anomalies, coupled with the strong feedback between evapotranspiration over land and convective precipitation make it indeed possible, at least in principle, to extend climate predictions of the first kind to the order of perhaps a year or two.

The predictability of the second kind is intimately related to the issue of whether the climate is *transitive* or *intransitive*. If the dynamics of a system lead to a unique and stable set of infinitely long-term statistics, i.e., a unique climate in the infinite sense, the system is called *transitive*. If, instead, there are two or more physically possible climates, the system is called *intransitive*. Which of the climates may prevail is thus a matter of chance, since over a suitable period each may be initiated by an arbitrarily small random perturbation. Both transitive and intransitive systems occur in nature. Atmospheric systems on their own, except when drastically simplified, are probably not intransitive.

However, there are no means at present of determining whether the atmosphere–ocean–earth system is transitive or intransitive. Perhaps the most intriguing and one of the most challenging problems in climate research is the suggestion that the thermohaline circulation in the Atlantic Ocean is intransitive. Using an idealized model of the Atlantic Ocean, F. Bryan

(1986) has shown that with a given forcing of the circulation with freshwater fluxes, the ocean can develop at least two completely different stable sets of circulation, depending on the initial state. These aspects are discussed further in Chapters 17 (Sec. 17.5.2) and 21 (Sec. 21.2). Bryan's result was postulated on the basis of a careful examination of geological evidence by Broecker et al. (1985a) in relation to the return of glaciation between 10,000 and 11,000 years before present, known as the *Younger Dryas* event. The Younger Dryas event was preceded by a strong release of melt water into the North Atlantic, probably due to a deflection of the discharge of fresh water from the Mississippi to the St. Lawrence river. It has been suggested by Broecker et al. (1988) and Berger and Killingley (1982) that the reduction of deep water formation by the resulting strong stratification, caused by the injection of fresh water, could drastically reduce the meridional heat transport of the Atlantic and, consequently, provide a substantial cooling of the Arctic.

Maier-Reimer and Mikolajewicz (1989) have recently undertaken numerical experiments in order to study the sensitivity to the amount of discharged melt water in the circulation in the North Atlantic and the corresponding heat transport. The experiment demonstrated that there is a considerable sensitivity, in particular, to the discharge of the St. Lawrence river. Even comparatively minor enhancement of the flow, equivalent to about a doubling of the present 0.01 Sv (1 Sv = 1 Sverdrup = 10^6 m^3 s^{-1}) changed the circulation drastically *over a period of a few decades*. Only when the melt water was reduced to 0.007 Sv was the present circulation maintained. It appears, in fact, based on numerical experiments by Maier-Reimer (personal communication), that the present thermohaline circulation in the Atlantic Ocean is a less "common" state than the one with a very weak net heat transport. The regional effect is, of course, potentially severe, with enhanced ice coverage in the North Atlantic and a colder climate there as well as downstream over northern Europe (Fig. 21.6).

23.4 Improvements required

One of the difficulties in outlining a comprehensive strategy for climate modeling is the extreme complexity of the system, as it encompasses different subsystems with vastly different characteristic time scales. The large time-scale separation between atmosphere, ocean, land surface and biosphere, and cryosphere stretching from minutes and hours to hundreds of millennia requires the development of a hierarchy of models with different tradeoffs between spatial and temporal resolution and the length of integration.

This strategy is a basic component of which WCRP recognizes three streams of climate research:

(1) The first stream: the physical basis for long-range weather forecasting;

(2) The second stream: interannual variability; and

(3) The third stream: long-term climatic trends and climate sensitivity.

In Chapter 19 a discussion was given of natural variability on different time scales. Here we will consider climate scenarios on three time scales: up to a year, of the order of 100 years, and of the order of 10,000 years.

23.4.1 Long-range weather prediction/short-term climate forecasts

The modeling strategy for long-range prediction is straightforward. It is climate prediction of the first kind and will address integrations of not longer than a year or two, or as long as well-defined climate features, such as El Niño, can sensibly be predicted. The models required are the same as for long-range weather forecasting with an added ocean model and an improved treatment of land surface processes. Comprehensive models with the objective of semi-operational use are presently under development at major forecasting centers.

Experience from medium-range weather prediction suggests that these models shall have highest possible resolution with a minimum of 2° latitude–longitude in the atmosphere and 1° for oceans. The major problems will be to overcome the climate drift of the two subsystems; atmosphere and ocean, and to refine the coupling between them. The improvement of such complex systems will require a broad program of research, since most problems are interrelated.

Integration procedures need to be refined in order to reduce truncation errors, in particular, in the advection of quantities having sharp gradients such as moisture and liquid water; a Lagrangian advection is here an interesting possibility (see also Chapters 9 and 15).

Physical parameterization, in particular of clouds, boundary layer processes and surface fluxes, have apparent deficiencies. It is not unlikely that the combination of insufficient resolution of synoptic features and our inability to adequately parameterize organized subgrid-scale system is one of the causes of the systematic cooling of AGCMs in long integrations. Similarly, these deficiencies also reduce the exchange of heat, moisture and momentum between the atmosphere–oceans and the atmosphere–land surfaces.

Finally, individual processes cannot be studied in isolation; instead "packages" of comprehensive, physically consistent parameterization systems should be developed.

Of particular importance will be an ongoing validation of the climate predictions. For that reason it will be very important to run such forecasts on a routine basis to be assessed and evaluated by groups of scientists working in close cooperation with the modelers. Because of the time scales involved, verifications of the interannual changes, such as those associated with El Niño events, will be especially valuable.

23.4.2 Climate scenarios of order 100 years

On this time scale we are interested only in climate predictions of the second kind. The objective is to develop a model which is capable of reproducing the present climate and which can be used as a tool for climate change projections, such as changes caused by the greenhouse effect (see Chapter 20).

It is probably a reasonable assumption, at least initially, to exclude the changes in the major ice sheets and the biosphere, and to concentrate the modeling efforts on the atmosphere, the ocean, sea ice and land surface processes, and the coupling between the different subsystems.

Since we are not interested in the prediction of individual weather systems per se, but only in their realistic reproduction, there is no initialization problem. However, it does not seem possible to significantly reduce resolution because the generation and maintenance of quasi-stationary features in the atmosphere and air–sea interaction, for example, are strongly controlled by the transient synoptic systems. Resolution of orography and well-defined land–sea boundaries are equally important for the local and regional climate. A high resolution is also required in order to satisfy the need for regional and local climate change prediction.

Systematic errors and climate drift exist in individual subsystems. Typical errors in the atmosphere are the tendency for cooling and zonalization of the large-scale flow. Typical errors in the ocean models are problems in resolving the sharp temperature gradients, such as those existing along the Gulf Stream and the Kuroshio, or the diffusing away of the thermocline. In long integrations a larger climate drift in individual subsystems is kept under control by the stabilizing effect of the other subsystem, specified by observed data.

However, when the systems are coupled, one normally discovers that the coupled systems slowly drift into another state which is far from the observed climate. The drift clearly implies that the original subsystems contain errors which were too small to be detected, or which were compensated for in some way.

One part of the drift which can be compensated for is the part which is related to the simulation of the climate equilibrium state. To simulate the global mean temperature to within, say, $\delta T = \pm 2$ K (a little less than the difference in global mean temperature between the climate of the last ice age and today), then the accuracy required of the equivalent radiation fluxes, R, which are proportional to the fourth power of the absolute temperature, and ignoring feedback factors, implies $\delta R/R = 4\delta T/T = 8/288 = 3\%$. Accuracies of this order are difficult to achieve with present atmospheric and ocean GCMs. Typical model heat flux computations are accurate to within 10%, comparable with the accuracy of the heat flux observations. The problem is to find a method for coupling the individual subsystems together in such a way that they interact as if they were experiencing variations around their designed climate equilibrium state, rather than around a far-removed equilibrium state into which the coupled model has drifted.

A way to overcome this has been suggested by Sausen et al. (1988) and by Manabe and Stouffer, (1988), using a flux correction technique. The principle of the flux correction is given Chapter 17 (Sec. 17.4). This flux correction or flux adjustment technique requires no additional numerical experiments beyond the uncoupled mode experiments needed to verify the individual models for the subsystems. The basic approach is thus simple. Constant flux correction terms are applied to each of the subsystem models separately in order to cancel the drift arising from the mismatch of the fluxes computed by the separate models. Since the flux corrections are constant in time (but not in space) their influence in the dynamics of the variations in the system is presumed to be of secondary importance. However, the flux correction terms are additive, and the coupled model can still exhibit a drift. Moreover the flux correction terms cannot change during the course of a climate change experiment (i.e., it is effectively assumed that the model errors are the same for both the control and the perturbed climates). An example is here the effect the flux correction can have on the response to a change in the thermohaline circulation or to the change in the Arctic ice cover.

Clearly large flux correction terms (they are of the same order as the fluxes themselves) indicate serious problems with the component atmosphere and ocean GCMs and an urgent objective must be to reduce the flux correction terms to be significantly smaller values than at present.

Greenhouse simulations so far have used an "effective" CO_2 concentration in representing the other greenhouse gases. This is a gross simplification, as the major differences in the optical properties of the different greenhouse gases are that in some part of the atmosphere the thermal response has the same sign while in other parts, such as in the lower stratosphere, there is a compensatory effect (Chapter 20, Sec. 20.7). Secondly, CH_4, N_2O and CFCs interact chemically with stratospheric O_3 as well. Ozone plays an important part in determining the life-cycles of the greenhouse gases (see Chapter 7). Of importance also is a consideration of the sulfur cycle in view of its potential "negative greenhouse" effect due to its influence on the optical properties of clouds.

The biosphere has been only very crudely represented in climate models. A genuine problem has to do with the vast difference in scale between terrestrial ecosystems and the scales which can be represented in global climate models (Chapters 6 and 14, see Fig. 14.1). The biosphere and surface hydrology are important in two different ways. Firstly and directly they are important via evapotranspiration and albedo. This effect is significant. It is probably the reason for the very long periods of drought observed in exposed areas such as the Sahel region. There are also indications from numerical experiments that drastic changes in the biosphere, such as the destruction of the tropical forests in Amazonia, may have irreversible regional effects on the climate (Chapter 22).

The second important role of the biosphere, both on land and in the oceans, is the way the biosphere controls the carbon cycle (Chapters 8 and 16). Drastic rapid changes in the biosphere, both directly, such as from rapid destruction of ecosystems, or indirectly from, say, greenhouse warming, will lead to feedback processes which have to be considered at least in a crude way in climate modeling on this time scale.

23.4.3   *Climate scenarios of order 10,000 years*

Models addressing time scales of the order of 10,000 years and longer must include a complete representation of the climate system, including the cryosphere and the biosphere. It appears that an approach suggested by Hasselmann (1988) may be suitable on this time scale. We can look upon this as a mixture of a slow system consisting of the land ice, the deep ocean and the astronomical forcing, and a fast system consisting of the atmosphere, the land surfaces and the rest of the oceans.

An efficient integration technique can be established from the fact that one is not interested in the complete time history of the fast system, but only in establishing the *quasi-equilibrium statistics of the fast system as it responds to the more slowly evolving slow system.*

One method of achieving this is the "burst integration technique", in which the integration of the fast system is carried out intermittently during a sequence of finite time intervals, T_1, Fig. 23.6 (see also Chapter 21).

The integration period T_1 is chosen to be sufficiently long that the fast system has time to adjust to its statistically quasi-equilibrium state, but it is still significantly smaller than the interval T_2 between successive integrations. The statistical properties of the fast system computed over a period T_1 are assumed to apply over the entire intermittent interval T_2. The integration of the slow system is carried out continually, with a time step increment ΔT_s in the range $T_1 < \Delta T_s < T_2$. The method clearly requires a fairly pronounced time-scale separation between the fast and the slow system, and is justified only if the statistical properties of the fast system do not change significantly over the period T_2.

For a coupled system of the kind discussed here, T_1 could be chosen to be 10^2 years and T_2 to be 10^5 years.

So far there is little experience with this type of integration. The statistics of the fast system must be calculated to include the full stochastic forcing, including the low frequencies exerted by the fast system on the slow system.

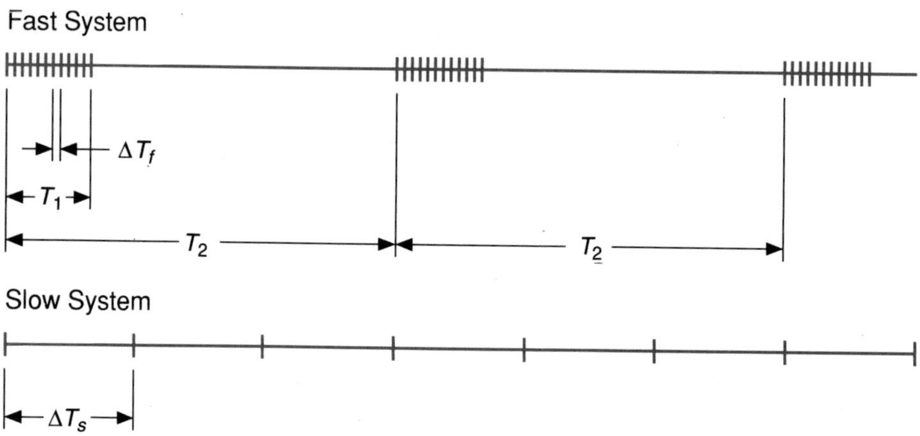

Fast System

Slow System

Fig. 23.6 Burst integration techniques. The fast system is integrated with time resolution ΔT_f for periods T_1, that are repeated at intervals T_2 The slow system is integrated continuously with resolution ΔT_s, where ΔT_s is in the range $T_1 < \Delta T_s < T_2$. After Hasselmann, (1988).

Errors in the statistical estimation of the mean fluxes and the low-frequency variability will generally lead to long-term fluctuations and drifts in the slow system, which may be difficult to diagnose and separate from the time fluctuations included by real fluxes.

23.5   Future prospects

In spite of the impressive documentation in this volume of the progress of climate modeling, there is no doubt that many of the major issues are still poorly understood and massive scientific and technical efforts are required to meet such a formidable challenge as the modeling and prediction of the complete climate system. A research program must formulate realistic step-wise objectives while in that process keeping the long-term goal of building a comprehensive system clear in its view.

One such objective will be to explore the natural variability and the predictability of the climate system on the time scale, say, from a season to a century. A thorough understanding of the natural variability is a necessary condition for separating the signal (e.g., increased greenhouse forcing) from the noise (natural variability).

Climate variability can be simulated by numerical experiments using coupled land–ocean–atmosphere models with the aim of reproducing the statistical character, space-time patterns and amplitudes of the observed variation, and to identify the mechanisms with which they are associated and assess the extent of their predictability (Chapter 19).

The development of models will depend on the Global Climate Observing System (GCOS) presently being considered. This system will be building on the World Weather Watch and on-going experimental programs such as TOGA and WOCE. Of particular importance will be enhanced observations of the upper ocean worldwide and on-going monitoring of significant aspects of the world ocean circulation at all depths. One central aspect will be to use the air–sea fluxes of momentum (wind stress), energy and water, obtained through data assimilation procedures, in order to drive the ocean dynamics and thermodynamics realistically and validate these fluxes with similar precision in model simulations of the coupled ocean–atmosphere system. The final thrust is to develop a capability for predicting variations in sea surface temperature, transport and storage of heat by the ocean, and other relevant properties that may be predictable.

The questions as to the role of the short-term dynamical variability of the system and the need to incorporate it in a long-term climate simulation is a central issue, since it will, to a large extent, determine the computational requirements. Intense air–sea interactions such as in the North Atlantic during the boreal winter are coupled to extreme synoptic events. Experience from numerical predictions suggest that horizontal resolution in the atmosphere of the order of 200 km is required to simulate such events reasonably accurately. At least a similar resolution is needed to handle orographically induced synoptic phenomena.

An equivalent problem exists in the oceans with the potential need to resolve ocean eddies. This will require a horizontal resolution of 5–10 km and computational resources far beyond what is presently available. Numerical experiments will be necessary to find out to what degree such small-scale eddies and vortices can be successfully parameterized.

A second primary objective will be to incorporate relevant chemical processes into the dynamical models in order to explore the subtle but important interaction between chemistry and atmospheric dynamics. Of particular importance is the modeling of stratospheric O_3 in order to reproduce the depletion at high latitudes in the spring (in particular the Antarctic ozone hole in the Austral spring) and the extent to which the globally observed reductions in recent years are due to longer-term natural fluctuations or anthropogenic effects (see Chapter 7, Secs. 7.3.3. 7.3.6 and 7.4.2). Other important aspects are to better understand the role of the sulfur cycle and whether its effect on cloud particles and hence the optical properties of clouds enhances cloud albedo and through this has an overall cooling effect. The other greenhouse gases except CO_2, notably CH_4, N_2O and the CFCs interact with stratospheric O_3. They also respond differently to the absorption of terrestrial radiation than CO_2 and should not therefore be incorporated as effective CO_2 in greenhouse scenario integrations.

A third objective is the carbon cycle and the need to understand and model its variability between glacial and interglacial periods, and to explain the constant atmospheric CO_2 for some 1,000 years prior to the industrial revolution. Thereafter it will be of major importance to understand the discrepancy between the emission of anthropogenic CO_2 and burning of vegetation and the estimated increase of carbon in the atmosphere and the ocean (see Sec. 7.3.2). At present, there is an amount of some 1.5 Gtonne per year not accounted for. The role of the biosphere both on land and in the oceans will have to be incorporated in such studies.

The numerical modeling of the climate system leading to a better understanding and to more reliable forecast or projections is not only one of the grand scientific challenges of our time, but also of greatest importance for the future of mankind. It will require major resources in manpower, computers and observational programs, as well as intense, sustained scientific effort over a very long period of time.

References

Abarbanel, H. D. I., and W. R. Young (Eds.), 1987: *General Circulation of the Ocean.* Springer-Verlag, 291 pp.

Aber, J. D., and J. M. Melillo, 1991: *Terrestrial Ecosystems.* Saunders College Pub., 427 pp.

Aber, J. D., J. M. Melillo and C. A. Federer, 1982: Predicting the effects of rotation length, harvest intensity and fertilization on fiber yield from northern hardwood forests in New England. *Forest Science,* **28**, 31–45.

Aber, J. D., J. M. Melillo and C. A. McClaugherty, 1990: Predicting long-term patterns of mass loss, nitrogen dynamics and soil organic matter formation from initial fine litter chemistry in temperate forest ecosystems. *Canadian J. Botany,* **68**, 2201–2208.

Aber, J. D., K. J. Nadelhoffer, P. Steudler and J. M. Melillo, 1989: Nitrogen saturation in northern forest ecosystems. *BioScience,* **39**, 378–386.

Acker, J. G., R. H. Byrne, S. Ben-Yaakov, R. A. Feely and P. R. Betzer, 1987: The effect of pressure on aragonite dissolution in seawater. *Geochim. et Cosmochim. Acta,* **51**, 2171–2175.

Ackley, S. F., A. J. Gow, K. R. Buck and K. M. Golden, 1980: Sea ice studies in the Weddell Sea aboard USCGC Polar Sea. *Antarctic J. United States,* **15**, 84–86.

Agarwal, A., and S. Narain, 1991: *Global Warming in an Unequal World: A Case of Environmental Colonialism.* Centre for Science and Environment, New Delhi, India.

AIP (American Institute of Physics), 1975: *Efficient Use of Energy.* New York, AIP.

AIP, 1985: *Energy Sources: Conservation and Renewables.* New York, AIP.

Alexander, R. C., and R. L. Mobley, 1976: Monthly average sea-surface temperatures and ice pack limits on a 1° global grid. *Mon. Wea. Rev.,* **104**, 143–148.

Alldredge, A. L., and M. W. Silver, 1988: Characteristics, dynamics and significance of marine snow. *Prog. Oceanogr.,* **20**, 41–82.

Alley, R. B., D. D. Blankenship, S. T. Rooney and C. R. Bentley, 1987: Till beneath Ice stream B 4. A coupled ice - till flow model. *J. Geophys. Res.,* **92B**, 8931–8940.

Anderson, D. L., 1961: Growth rate of sea ice. *J. Glaciol.,* **3**, 1170–1172.

Anderson, D. L. T., and A. E. Gill, 1975: Spin-up of a stratified ocean, with application to upwelling. *Deep-Sea Res.,* **22**, 583–596.

Anderson, D. L. T., and J. Willebrand (Eds.), 1989: *Oceanic Circulation Models: Combining Data and Dynamics.* Kluwer, 605 pp.

André, J.-C., and C. Blondin, 1986: On the more effective roughness length for use in numerical three-dimensional models. *Bound. Layer Met.,* **35**, 231–245.

References

André, J.-C., J.-P. Goutorbe, A. Perrier, F. Becker, P. Bessemoulin, P. Bougeault, Y. Brunet, W. Brutsaert, T. Carlson, R. Cuenca, J. Gash, J. Gelpe, P. Hildebrand, J.-P. Lagouarde, C. Lloyd, L. Mahrt, P. Mascart, C. Mazaudier, J. Noilhan, C. Ottle, M. Payen, T. Phulpin, R. Stull, J. Shuttleworth, T. Schmugge, O. Taconet, C. Tarrieu, R.-M. Thepenier, C. Valencogne, D. Vidal-Madjar and A. Weill, 1988: Evaporation from land surfaces: First results from HAPEX-MOBILHY special observing period. *Annales Geophysicae*, **6**, 477–492.

Andreae, M. O., 1990: Ocean–atmosphere interactions in the biogeochemical sulfur cycle. *J. Mar. Chem.*, **30**, 1–29.

Andrews, D. G., and M. E. McIntyre, 1976: Planetary waves in horizontal and vertical shear: The generalized Eliassen–Palm relation and the mean zonal acceleration. *J. Atmos. Sci.*, **33**, 2031–2048.

Anon., 1990: Tropical forests across three continents. *UNESCO Sources*, **19**, 6.

Anthes, R. A., 1977: A cumulus parameterization scheme utilizing a one-dimensional cloud model. *Mon. Wea. Rev.*, **105**, 270–286.

Anthes, R. A., 1984: Enhancement of convective precipitation by mesoscale variations in vegetative covering in semiarid regions. *J. Appl. Meteor.*, **23**, 541–54.

Arakawa, A., 1966: Computational design for long-term numerical intergrations of the equations of atmospheric motion. *J. Comput. Phys.*, **1**, 119–143.

Arakawa, A., 1972: Design of the UCLA general circulation model. Numerical Simulation of Weather and Climate, Dept. of Meteorology, University of California, Los Angeles, *Tech. Rept. 7*, 116 pp.

Arakawa, A., and V. R. Lamb, 1977: Computational design of the basics dynamical processes of the UCLA general circulation model. *Methods in Computational Physics* **17**, Academic Press, 174–265.

Arakawa, A., and W. H. Schubert, 1974: Interaction of a cumulus cloud ensemble with the large-scale environment, Part I. *J. Atmos. Sci.*, **31**, 674–701.

Arakawa, A., and M. J. Suarez, 1983: Vertical differencing of the primitive equations in sigma coordinates. *Mon. Wea. Rev.*, **111**, 34–45.

Armi, L., and D. M. Farmer, 1988: The flow of Mediterranean water through the Strait of Gibraltar. *Prog. Oceanogr.*, **21**, 105 pp.

Arpe, K., 1991: The hydrological cycle in the ECMWF short range forecasts. *Dyn. Atmos. Oceans*, **16**, 33-59.

Arrhenius, S., 1896: On the influence of carbonic acid in the air upon the temperature of the ground. *Philosoph. Mag.*, **41**, 237–275.

Arrhenius, S., 1908: *Worlds in the Making*. Harper and Brothers.

Aselman, I., and P. J. Crutzen, 1989: Global distribution of natural freshwater wetlands and rice paddies, their net primary productivity, seasonality and possible methane emissions. *Atmos. Chem.*, **8**, 307–358.

Asrar, G., 1989: *Theory and Applications of Optical Remote Sensing*. Wiley and Sons, 734 pp.

Aubreville, A., 1949: *Climate, Forests and Desertification in Tropical Africa*. Paris, France, Sociéte d'Editions Géographiques, Maritimes et Coloniales.

Avissar, R., and R. A. Pielke, 1989: A parameterization of heterogeneous land surfaces for atmospheric numerical models and its impact on regional meteorology. *Mon. Wea. Rev.*, **117**, 2113–2136.

Bacastow, R., and E. Maier-Reimer, 1990: Ocean-circulation model of the carbon cycle. *Clim. Dyn.*, **4**, 95–125.

Baes, C. F., and G. G. Killough, 1986: Chemical and biological processes in CO_2-ocean models. In *The Changing Carbon Cycle, A Global Analysis*, J. R. Trabalka and D. E. Reichle (Eds.), Springer-Verlag, 329–347.

Ball, J. T., I. E. Woodrow, and J. A. Berry, 1987: A model predicting stomatal conductance and its contribution to the control of photosynthesis under different environmental conditions. In *Progress in Photosynthesis Research*, Vol. IV, J. Biggens (Ed.), Martinus Nijhoff Pub., 221–228.

Barnett, T., and M. Schlesinger, 1987: Detecting changes in global climate induced by greenhouse gases. *J. Geophys. Res.*, **92**, 14772–14780.

Barnett, T., N. Graham, M. A. Cane, S. E. Zebiak, S. Dolan, J. J. O'Brien and D. Legler, 1988: On the prediction of the El Niño of 1986–1987. *Science*, **241**, 192–196.

Barnola, J. M., D. Raynaud, Y. S. Korotkevitch and C. Lorius, 1987: Vostok ice core: a 160,000 year record of atmospheric CO_2. *Nature*, **329**, 408–414.

Barron, E. J., and A. D. Hecht (Eds.), 1985: *Historical and Paleoclimatic Analysis and Modeling*. Wiley, 445 pp.

Barron, E. J., and W. H. Peterson, 1990: Mid-Cretaceous ocean circulation: results from model sensitivity studies. *Paleoceanography*, **5(3)**, 319–337.

Barron, E. J., and W. M. Washington, 1984: The role of geographic variables in explaining paleoclimates: results from Cretaceous climate model sensitivity studies. *J. Geophys. Res.*, **89**, 1267–1279.

Bartlett, K. B., P. M. Crill, R. L. Sass, R. C. Harriss and N. B. Dise, 1991: Methane emissions from tundra environments in the Yukon–Kuskowim Delta, Alaska. *J. Geophys. Res.*, (in press).

Battisti, D., 1988: The dynamics and thermodynamics of a warming event in a coupled tropical atmosphere/ocean model. *J. Atmos. Sci.*, **45**, 2889–2919.

Battisti, D. S., and A. Hirst, 1989: Interannual variability in the tropical atmosphere/ocean system: influence of the basic state and ocean geometry. *J. Atmos. Sci.*, **46**, 1687–1712.

Bazzaz, F. A., 1990: The response of natural ecosystems to the rising global CO_2 levels. *Ann. Rev. Ecol. Systematics*, **21**, 167–196.

Benedick, R. E., 1991: *Ozone Diplomacy: New Directions in Safeguarding the Planet*. Harvard University Press.

Bengtsson, L., and A. J. Simmons, 1983: Medium range weather prediction – operational experience at ECMWF. In *Large Scale Dynamical Processes in Atmosphere*, B. J. Hoskins and R. P. Pearce (Eds.), Academic Press, 337–363.

Bengtsson, L., H. Böttger and M. Kanamitsu, 1982: Simulation of hurricane-type vortices in a general circulation model. *Tellus*, **34**, 441–457.

Berger, A. H., H. Gallee, T. Fichefet, I. Marsiat and C. Tricot, 1990: Testing the astronomical theory with a coupled climate-ice-sheets model. *Global and Planetary Change*, **3**, 113–124.

Berger, W. H., and J. S. Killingley, 1982: The Worthington effect and the origin of the Younger Dryas. *J. Marine Res. Suppl.*, **40**, 27–38.

References

Berger, W. H., V. S. Smetacek and G. Wefer, 1989: Ocean productivity and paleoproductivity – an overview. In *Productivity of the Ocean: Present and Past*, W. H. Berger, V. S. Smetacek, and G. Wefer (Eds.), John Wiley and Sons, 1–34.

Berger, W. H., K. Fischer, C. Lai and G. Wu, 1987: Ocean productivity and organic carbon flux, Pt I: overview and maps of primary production and export production. SIO Ref. 87-30. Scripps Institute of Oceanography, Univ. of California, San Diego.

Berner, R. A., 1990: Atmospheric carbon dioxide levels over Phanerozoic time. *Science*, **249**, 1382–1392.

Betts, A. K., and M. Miller, 1986: A new convective adjustment scheme. Part II: Single column tests using GATE Wave, BOMEX, ATEX and arctic air-mass data sets. *Quart. J. Roy. Meteor. Soc.*, **112**, 693–709.

Birchfield, G. E., 1989: A coupled ocean–atmosphere climate model: temperature versus salinity effects on the thermohaline circulation. *Clim. Dyn.*, **4**, 57–71.

Bjerknes, J., 1969: Atmospheric teleconnections from the equatorial Pacific. *Mon. Wea. Rev.*, **97**, 163–172.

Bjerknes, V., 1904: Das problem von der wettervorhersage, betrachtet vom standpunkt der mechanik und der physik. *Meteor. Z.*, **21**, 1–7.

Blackmon, M. L., 1976: A climatological spectral study of the 500 mb geopotential height of the Northern Hemisphere. *J. Atmos. Sci.*, **33**, 1607–1623.

Blackmon, M. L., S. L. Mullen and G. T. Bates, 1986: The climatology of blocking events in a perpetual January simulation of a spectral general circulation model. *J. Atmos. Sci.*, **43**, 1379–1405.

Blackmon, M. L., Y.-H. Lee, J. M. Wallace and H.-H. Hsu, 1984: Time variation of 500 mb height fluctuations with long, intermediate and short time scales as deduced from lag-correlation statistics. *J. Atmos. Sci.*, **41**, 981–991.

Blankenship, D. D., C. R. Bentley, S. T. Rooney and R. B. Alley, 1986: Seismic measurements reveal a saturated porous layer beneath an active Antarctic ice stream. *Nature*, **322**, 54–57.

Bleakley, B. H., and J. M. Tiedje, 1982: Nitrous oxide production by organisms other than nitrifiers or denitrifiers. *Applied Environmental Microbiology*, **44**, 1342–1348.

Bleck, R., and L. T. Smith, 1990: A wind-driven isopycnic coordinate model of the north and equatorial Atlantic ocean. 1. Model development and supporting experiments. *J. Geophys. Res.*, **95**, 3273–3285.

Boer, G. J., and M. Lazare, 1988: Some results concerning the effect of horizontal resolution and gravity wave drag on simulated climate. *J. Climate*, **1**, 789–806.

Boer, G. J., N. McFarlane and M. Lazare, 1991: Greenhouse gas induced climate change simulated with the CCC second generation GCM. *Clim. Change*, submitted.

Botkin, D. B., J. F. Janak and J. R. Wallis, 1972: Some ecological consequences of a computer model of forest growth. *J. Ecol.*, **60**, 849–872.

Bott, A., 1989: A positive definite advection scheme obtained by nonlinear renormalization of the advective fluxes. *Mon. Wea. Rev.*, **117**, 1006-1051.

Bourke, R. H., and R. P. Garrett, 1987: Sea ice thickness distribution in the Arctic Ocean. *Cold Reg. Sci. Tech.*, **13**, 259-280.

Boville, B. A., 1985: The thermal balance of the NCAR Community Climate Model. *J. Atmos. Sci.*, **42**, 695–709.

Boville, B. A., 1991: Sensitivity of simulated climate to model resolution. *J. Climate*, **4**, 469–485.

Boville, B. A., and D. P. Baumhefner, 1990: Simulated forecast error and climate drift resulting from the omission of the upper stratosphere in numerical models. *Mon. Wea. Rev.*, **118**, 1517–1530.

Boville, B. A., and W. J. Randel, 1986: Observations and simulation of the variability of the stratosphere and troposphere in January. *J. Atmos. Sci.*, **43**, 3015–3034.

Bowditch, N., 1966: American practical navigator. U.S. Navy Hydrographic Office Pub. No. 9, U.S. Naval Oceanographic Office, 1966 corr. print.

Boyd, J. P., 1976: The noninteraction of waves with the zonally averaged flow on a spherical Earth and the interrelationships of eddy fluxes of energy, heat, and momentum. *J. Atmos. Sci.*, **33**, 2285–2291.

Boyle, E. A., and L. Keigwin, 1987: North Atlantic thermohaline circulation during the past 20,000 years linked to high-latitude surface temperature. *Nature*, **330**, 35-40.

Brasseur, G. P., and S. Solomon, 1986: *Aeronomy of the Middle Atmosphere*. Reidel, 441 pp.

Brasseur, G. P., M. H. Hitchman, S. Walters, M. Dymek, E. Falise and M. Pirre, 1990: An interactive chemical dynamical radiative two-dimensional model of the middle atmosphere. *J. Geophys. Res.*, **95**, 5639–5655.

Bremner, J. M., and A. M. Blackmer, 1978: Nitrous oxide: emission from soils during nitrification of fertilizer nitrogen. *Science*, **199**, 295–296.

Bremner, J. M., and A. M. Blackmer, 1981: Terrestrial nitrification as a source of atmospheric nitrous oxide. In *Denitrification, Nitrification and Atmospheric Nitrous Oxide*, C. C. Delwiche (Ed.), John Wiley and Sons, 151–170.

Brewer, A., 1949: Evidence for a world circulation provided by the measurements of helium and water vapour distributions in the stratosphere. *Quart. J. Roy. Meteor. Soc.*, **75**, 351–363.

Briegleb, B. P., 1992: Delta-Eddington approximation for solar radiation in the NCAR Community Climate Model. *J. Geophys. Res.*, **97**, 7603–7612.

Brocolli, A. J., and S. Manabe, 1990: Can existing climate models be used to study anthropogenic changes in tropical cyclone climate? *Geophys. Res. Lett.*, **17**, 1917–1920.

Broecker, H. C., and W. Siems, 1984: The role of bubbles for gas transfer from water to air at higher windspeeds. Experiments in the wind-wave facility in Hamburg. In *Gas Transfer at Water Surfaces*, W. Brutsaert and G. H. Jirka (Eds.), D. Riedel, 229–236.

Broecker, W. S., 1987: The biggest chill. *Natural History Magazine*, October, 74–82.

Broecker, W. S., and G. H. Denton, 1989: The role of ocean–atmosphere reorganizations in glacial cycles. *Geochim. Cosmochin. Acta* **53**, 2465–2501.

Broecker, W. S., and T.-H. Peng, 1982: *Tracers in the Sea*. Lamont-Doherty Geological Observatory, Palisades, N.Y., 690 pp.

Broecker, W. S., D. Peteet and D. Rind, 1985a: Does the ocean–atmosphere system have more than one stable mode of operation? *Nature*, **365**, 21–26.

Broecker, W. S., T.-H. Peng, G. Østlund and M. Stuiver, 1985b: The distribution of bomb radiocarbon in the ocean. *J. Geophys. Res.*, **90**, 6953–6970.

References

Broecker, W. S., M. Andree, W. Wolfi, H. Oeschger, G. Bonani, J. Kennet and D. Peteet, 1988: The chronology of the last deglaciation: implication to the course of the Younger Dryas event. *Paleoceanography*, **3**, 1–19.

Broomwell, G., and J. R. Heath, 1983: Classification categories and historical development of circuit switching topologies. *Computing Surveys*, **2**, 95–133.

Brown, R. A., 1980: Planetary boundary layer modeling for AIDJEX. In *Sea Ice Processes and Models*, R. S. Pritchard (Ed.), University of Washington Press, 387–410.

Browning, G. L., J. J. Hack and P. N. Swarztrauber 1989: A comparison of three numerical methods for solving differential equations on the sphere. *Mon. Wea. Rev.*, **117**, 1058–1075.

Bryan, F., 1986: High latitude salinity effects and interhemispheric thermohaline circulations. *Nature*, **323**, 301–304.

Bryan, F., 1987: Parameter sensitivity of primitive equation ocean general circulation models. *J. Phys. Oceanogr.*, **17**, 970–985.

Bryan, F., and W. R. Holland, 1989: A high resolution simulation of the wind- and thermohaline-driven circulation in the north Atlantic ocean. In *'Aha Huliko'a Parameterization of Small-Scale Processes*, P. Muller and D. Henderson (Eds.), Hawaii Institute of Geophysics, 354 pp.

Bryan, K., 1963: A numerical investigation of a nonlinear model of a wind-driven ocean. *J. Atmos. Sci.*, **20**, 594–606.

Bryan, K., 1982: Poleward heat transport by the ocean: Observations and models. *Ann. Rev. Earth Planet. Sci.*, **10**, 15–38.

Bryan, K., 1984: Accelerating the convergence to equilibrium of ocean climate models. *J. Phys. Oceanogr.*, **14**, 666–673.

Bryan, K., 1986: Poleward buoyancy transport in the ocean and mesoscale eddies. *J. Phys. Oceanogr.*, **16**, 927–933.

Bryan, K., 1989: The design of numerical models of the ocean circulation. In *Oceanic Circulation Models: Combining Data and Dynamics*, D. L. T. Anderson and J. Willebrand (Eds.), Kluwer, pp. 465–500.

Bryan, K., and M. D. Cox, 1967: A numerical investigation of the oceanic general circulation. *Tellus*, **19**, 54–80.

Bryan, K., and L. J. Lewis, 1979: A water mass model of the world ocean. *J. Geophys. Res.*, **84**, 2503–2517.

Bryan, K., and J. L. Sarmiento, 1985: Modeling ocean circulation. *Adv. Geophys.*, **28A**, 433–459.

Bryan, K., and M. J. Spelman, 1985: The ocean's response to a carbon dioxide-induced warming. *J. Geophys. Res.*, **90**, 11679–11688.

Bryan, K., S. Manabe and C. Pacanowski, 1975: A global ocean–atmosphere climate model. Part II. The oceanic circulation. *J. Phys. Oceanogr.*, **5**, 30–46.

Bryan, K., S. Manabe and M. J. Spelman, 1988: Interhemispheric asymmetry in the transient response of a coupled ocean–atmosphere model to a CO_2 forcing. *J. Phys. Oceanogr.*, **18**, 851–867.

Bryan, K., F. G. Komro, S. Manabe and M. J. Spelman, 1982: Transient climate response to increasing atmospheric carbon dioxide. *Science*, **215**, 56–58.

Budd, W. F., 1975: A first simple model for periodically self-surging glaciers. *J. Glaciol.*, **14**, 3–21.

Budd, W. F., P. L. Keage and N. A. Blundy, 1979: Empirical studies of ice sliding. *J. Glaciol.*, **23**, 157–170.

Budd, W. F., B. J. McInnes, D. Jenssen and I. N. Smith, 1987: Modelling the response of the West Antarctic Ice Sheet to a climatic warming. In *Dynamics of the West Antarctic Ice Sheet*, C. J. van der Veen and J. Oerlemans (Eds.), Reidel Publ. Co., 321–358.

Budyko, M. I., 1974: *Climate and Life.* English edition edited by D. H. Miller, International Geophysical Series, Vol. 18, Academic Press, 508 pp.

Bunker, A. F., 1976: Computations of surface energy flux and annual air–sea interaction cycles of the North Atlantic Ocean. *Mon. Wea. Rev.*, **104**, 1122–1140.

Bunker, A. F., 1980: Trends of variables and energy fluxes over the Atlantic Ocean. *Mon. Wea. Rev.*, **108**, 720–732.

Burke, I. C., T. G. F. Kittel, W. K. Lauenroth, P. Snook and C. M. Yonker, 1991: Regional analysis of the central Great Plains: sensitivity to climate variability. *BioScience*, in press.

Burke, I. C., D. S. Schimel, C. M. Yonker, W. J. Parton, L. A. Joyce and W. K. Lauenroth, 1990: Regional modeling of grassland biogeochemistry using GIS. *Landscape Ecology* **4**, 45–54

Busalacchi, A. J., and J. J. O'Brien, 1981: Interannual variability of the equatorial Pacific in the 1960s. *J. Geophys. Res.*, **86**, 10901–10907.

Busalacchi, A. J., K. Takeuchi and J. O'Brien, 1983: Interannual variability of the equatorial Pacific – revisited. *J. Geophys. Res.*, **88**, 7551–7562.

Butler, J. N., 1982: *Carbon Dioxide Equilibria and Their Applications.* Addison-Wesley, 259 pp.

Callendar, G. S., 1938: The artificial production of carbon dioxide and its influence on temperatures. *Quart. J. Roy. Meteor. Soc.*, **64**, 223–237.

Camillo, P. J., and R. J. Gurney, 1986: A resistance parameter for bare soil evaporation models. *Soil Science*, **141**, 95–105.

Camillo, P. J., and T. J. Schmugge, 1981: *A Computer Program for the Simulation of Heat and Moisture Flow in Soils.* NASA TM-82121, NASA/GSFC, Greenbelt, MD.

Canby, T., 1984: El Niño's ill wind. *Natl. Geogr.*, **165**, 144–183.

Cane, M. A., 1983: Oceanographic events during El Niño. *Science*, **222**, 1189–1194.

Cane, M. A., 1984: Modeling sea level during El Niño. *J. Phys. Oceanogr.*, **14**, 586–606.

Cane, M. A., 1986: El Niño. *Ann. Rev. Earth Planet Sci.*, **14**, 43–70.

Cane, M. A., 1991: Forecasting El Niño with a geophysical model. In *Teleconnections Linking Worldwide Climate Anomalies*, M. H. Glantz, R. W. Katz, and N. Nicholls (Eds.), Cambridge University Press, 345–370.

Cane, M. A., and S. Zebiak, 1985: A theory for El Niño and the Southern Oscillation. *Science*, **228**, 1085–1087.

Cane, M. A., and S. Zebiak, 1987: Deterministic prediction of El Niño events. In *Atmospheric and Oceanic Variability*, H. Cattle (Ed.), Roy. Meteorol. Soc./Amer. Meteorol. Soc. 153–182.

Cane, M. A., M. Münnich and S. Zebiak, 1990: A study of self-excited oscillations of the tropical ocean–atmosphere system. Part I: Linear analysis. *J. Atmos. Sci.*, **47**, 1562–1577.

References

Cane, M. A., S. E. Zebiak and S. C. Dolan, 1986: Experimental forecasts of El Niño. *Nature,* **321**, 827–832.

Carson, D. J., 1986: Issues concerning the evaluation of effective surface roughness of heterogeneous terrain. *ISLSCP, Proc. International Conference, Rome, Italy,* ESA SP–248, Paris, France, 77–82.

Cattle, H. (Ed.), 1987: *Atmospheric and Oceanic Variability.* Roy. Meteorol. Soc., 182 pp.

Cayan, D. R., 1990: Variability of latent and sensible heat fluxes over the oceans. Ph.D. Thesis, U. California San Diego.

CEES (Committee on Earth and Environmental Sciences), 1991: *Our Changing Planet: The FY 1992 U.S. Global Change Research Program.* CEES, U.S. Geological Survey.

Cess, R. D., and G. L. Potter, 1988: A methodology for understanding and intercomparing atmospheric climate feedback processes in general circulation models. *J. Geophys. Res.,* **93**, 8305–8314.

Cess, R. D., G. L. Potter, J. P. Blanchet, G. J. Boer, S. J. Ghan, J. T. Kiehl, H. Le Treut, Z.-X. Li, X.-Z. Liang, J.F.B. Mitchell, J.-J. Morcrette, D. A. Randall, M. R. Riches, E. Roeckner, U. Schlese, A. Slingo, K. E. Taylor, W. M. Washington, R. T. Wetherald and I. Yagai, 1989: Interpretation of cloud-climate feedback as produced by 14 atmospheric general circulation models. *Science,* **245**, 513–516.

Cess, R. D., G. L. Potter, M.-H. Zhang, J. P. Blanchet, G. J. Boer, S. Chalita, D. A. Daylich, A. D. Del Genio, V. Dymnikov, V. Galin, D. Jerrett, E. Keup, A. A. Lacis, H. Le Treut, X.-Z. Liang, J.-F. Mahfouf, B. J. McAvaney, V. P. Meleshko, J. F. B. Mitchell, J.-J. Morcrette, P. M. Norris, D. A. Randall, L. Rikus, E. Roeckner, J. F. Royer, U. Schlese, D. A. Sheinin, J. M. Slingo, A. P. Sokolov, K. E. Taylor, W. M. Washington, R. T. Wetherald and I. Yagai, 1991: Interpretation of snow-climate feedback as produced by 17 general circulation models. *Science,* **253**, 888–892.

Chamberlin, T. C., 1899: An attempt to frame a working hypothesis of the cause of glacial periods on an atmospheric basis. *J. Ecol.,* **7**, 545–561.

Chapman, S., 1930: A theory of upper atmospheric ozone. *R. Meteorol. Soc. Mem.,* **3**, 103–125.

Charlson, R. J., 1988: Have concentrations of tropospheric aerosols changed? In *The Changing Atmosphere,* F. S. Rowland and I.S.A. Isaksen (Eds.), Dahlem Workshop Reports, J. Wiley and Sons.

Charlson, R. J., J. E. Lovelock, M. O. Andreae and S. E. Warren, 1987: Oceanic phytoplankton, atmospheric sulfur, cloud albedo and climate. *Nature,* **326**, 655–661.

Charlson, R. J., J. Langner, H. Rodhe, C. B. Leovy and S. G. Warren, 1991: Perturbation of the Northern Hemisphere radiative balance by backscattering from anthropogenic sulfate aerosols. *Tellus,* **43A-B**, 152–163.

Charney, J. G., 1955a: The gulf stream as an inertial boundary layer. *Proc. Natl. Acad. Sci., USA,* **41**, 731–740.

Charney, J. G., 1955b: The use of the primitive equations of motion in numerical prediction. *Tellus,* **7**, 22–26.

Charney, J. G., 1975: Dynamics of deserts and drought in the Sahel. *Quart. J. Roy. Meteor. Soc.,* **101**, 193–202.

Charney, J. and P. Drazin, 1961: Propagation of planetary scale disturbances from the lower into the upper atmosphere. *J. Geophys. Res.,* **66**, 83–109.

Charney, J. G., R. Fjørtøft and J. von Neumann, 1950: Numerical integration of the barotropic vorticity equation. *Tellus*, **2**, 237–254.

Charney, J. G., P. H. Stone and W. J. Quirk, 1975: Drought in the Sahara: A biogeophysical feedback mechanism. *Science*, **187**, 434–435.

Charney, J. G., W. J. Quirk, S. H. Chow and J. Kornfield, 1977: A comparative study of the effects of albedo change on drought in semi-arid regions. *J. Atmos. Sci.*, **34**, 1366–1385.

Chatfield, C., 1984: *The Analysis of Time Series*. Chapman and Hall, 286 pp.

Chervin, R. M., 1979: Response of the NCAR general circulation model to changed land surface albedo. In *Report of the JOC Study Conference on Climate Models: Performance, Intercomparison and Sensitivity Studies, Vol. I*, GARP Publications Office, World Meteorological Organization, Geneva, 563–581.

Cho, B. C., and F. Azam, 1988: Major role of bacteria in biogeochemical fluxes in the ocean's interior. *Nature*, **332**, 441–443.

Cicerone, R. J., and R. S. Oremland, 1988: Biogeochemical aspects of atmospheric methane. *Global Biogeochem. Cycles*, **2**, 299–327.

Clapp, R. B., and G. M. Hornberger, 1978: Empirical equations for some soil hydraulic properties. *Water Resour. Res.*, **14.4**, 601–604.

CLIMAP Project Members, 1981: Seasonal reconstructions of the earth's surface at the last glacial maximum. Geological Society of America Map and Chart Series, MC-36.

Coakley, J. A., Jr., and R. D. Cess, 1985: Response of the NCAR Community Climate Model to the radiative forcing by the naturally occurring tropospheric aerosol. *J. Atmos. Sci.*, **42**, 1677–1692.

Cohen, Y., and L. I. Gordon, 1978: Nitrous oxide in the oxygen minimum of the eastern tropical North Pacific: evidence for its consumption during denitrification and possible mechanisms for its production. *Deep-Sea Res.*, **25**, 509–524.

Cohen, Y., and L. I. Gordon, 1979: Nitrous oxide production in the ocean. *J. Geophys. Res.*, **84**, 347–353.

COHMAP Members, 1988: Climatic changes of the last 18,000 years: Observations and Model Simulations. *Science*, **241**, 1043–1052.

Colin de Verdière, 1989: On the interaction of wind and buoyancy driven gyres. *J. Mar. Res.*, **47**, 595–633.

Collatz, G. J., C. Grivet, J. T. Ball and J. A. Berry, 1991: Physiological and environmental regulation of stomatal conductance, photosynthesis and transpiration: A model that includes a laminar boundary layer. *Agric. For. Meteorol.*, **54**, 107–136.

Cooley, J. W., and J. W. Tukey, 1965: An algorithm for the machine calculation of complex Fourier series. *Math. Comput.*, **19**, 297–301.

Coon, M. D., G. A. Maykut, R. S. Pritchard, D. A. Rothrock and A. S. Thorndike, 1974: Modeling the pack ice as an elastic-plastic material. *AIDJEX Bull.*, **24**, 1–105.

Cornejo-Garrido, A. G., and P. H. Stone, 1977: On the heat balance of the Walker Circulation. *J. Atmos. Sci*, **34**, 1155–1162.

Cornillon, P., and L. Stramma, 1985: The distribution of diurnal sea surface warming events in the western Sargasso Sea. *J. Geophys. Res.*, **90**, 11811–11815.

Courant, R., K. O. Friedrichs and H. Lewy, 1928: Über die partiellen differenzengleichungen der mathematischen physik. *Math. Annalen*, **100**, 67–108.

References

Court, A., 1974: The climate of the conterminous United States. Chapter 3 of *World Survey of Climatology*, Vol. II, R. A. Bryson and F. K. Hare (Eds.), 193–261.

Cox, M. D., 1985: An eddy resolving numerical model of the ventilated thermocline. *J. Phys. Oceanogr.*, **15**, 1312–1324.

Cox, M. D., 1989: An idealized model of the world ocean. Part I: The global-scale water masses. *J. Phys. Oceanogr.*, **19**, 1730–1752.

Cox, M. D., and K. Bryan, 1984: A numerical model of the ventilated thermocline. *J. Phys. Oceanogr.*, **14**, 674–687.

Cox, S., 1981: Radiation characteristics of clouds. In *Clouds, Their formation, Optical Properties, and Effects*, P. Hobbs and A. Deepak (Eds.), Academic Press, 241–280.

Craig, R., 1965: *The Upper Atmosphere, Meteorology and Physics*. Academic Press, 509 pp.

Crowley, T. J., 1983: The geological record of climatic change. *Rev. Geophys. Space Phys.*, **21**, 828–877.

Crowley, T. J., 1989: Paleoclimate perspectives on a greenhouse warming. In *Climate and Geo-Sciences*, A. Berger et al. (Eds.), Kluwer, 179–207.

Crowley, T. J., and G. R. North, 1991: *Paleoclimatology*. Oxford University Press, 339 pp.

Crowley, T. J., W. T. Hyde and D. A. Short, 1989: Seasonal cycle variations on the supercontinent of Pangea. *Geology*, **17**, 457–460.

Crowley, W. P., 1968: Numerical advection experiments. *Mon. Wea. Rev.*, **96**, 1–11.

Crutzen, P. J., 1987: The role of the tropics in atmospheric chemistry. In *The Geophysiology of Amazonia*, R. E. Dickinson (Ed.), John Wiley, 107–130.

Cubasch, U., 1989: Coupling a global atmospheric model with a global ocean model using the flux correction method. In *Aspects of Coupling Atmosphere and Ocean Models*, R. Sausen (Ed.), Meteorologisches Institut, Universität Hamburg, Report No. 6, 39–60.

Cubasch U., K. Hasselmann, H. Höck, E. Maier-Reimer, U. Mikolajewiez, B. D. Santer and R. Sausen, 1992: Time-dependent greenhouse warming computations with a coupled ocean–atmosphere model. *J. Climate*, accepted.

Dacey, J. W. H., and N. V. Blough, 1987: Hydroxide decomposition of dimethylsulfoniopropionate to form dimethyl sulfide. *Geophys. Res. Lett.*, **14**, 1246–1249.

Dacey, J. W. H., and S. G. Wakeham, 1986: Oceanic dimethyl sulfide: Production during zooplankton grazing on phytoplankton. *Science*, **233**, 1314–1316.

Davidson, E. A., J. M. Stark and M. K. Firestone, 1990: Microbial production and consumption of nitrate in an annual grassland. *Ecology*, **71**, 1968–1975.

Davis, M. B., 1981: Quarternary history and the stability of forest communities. In *Forest Succession*, D. C. West, H. H. Shugart, and D. B. Botkin (Eds.), Springer-Verlag, 132–153.

Davis, M. B., 1988: Ecological systems and dynamics. In *Toward an Understanding of Global Change*, National Academy Press.

Davis, R. E., 1976: Predictability of sea surface temperature and sea level pressure anomalies over the North Pacific *J. Phys. Oceanogr.*, **6**, 249–266.

Davis, R. E., R. E. DeSzoeke and P. P. Niiler, 1981: Variability in the upper ocean during MILE. Part II: Modeling the mixed layer response. *Deep-Sea Res.*, **28**, 1427–1451.

Deacon, E. L., 1977: Gas transfer to and across an air–water interface. *Tellus*, **29**, 363–374.

Deardorff, J. W., 1972: Parameterization of the planetary boundary layer for use in General Circulation Models. *Mon. Wea. Rev.*, **100**, 93–106.

Deardorff, J. W., 1977: Efficient prediction of a ground surface temperature and moisture with inclusion of a layer of vegetation. *J. Geophys. Res.*, **83**, 1889–1903.

Dell'Osso, L., and L. Bengtsson 1985: Prediction of a typhoon using a fine-mesh NWP model. *Tellus*, **37A**, 97–105.

Delworth, T. L., and S. Manabe, 1988: The influence of potential evaporation on the variabilities of simulated soil wetness and climate. *J. Climate*, **1**, 523–547.

Delworth, T. L., and S. Manabe, 1989: The influence of soil wetness on near-surface atmospheric variability. *J. Climate*, **2**, 1447–1462.

Denmead, O. T., 1976: Temperature cereals. In *Vegetation and the Atmosphere*, **2**, J. L. Monteith (Ed.), Academic Press, 1–31.

de Vries, D. A., 1975: Heat transfer in soils. In *Heat and Mass Transfer in the Biosphere*, Scripta Books.

Dickinson, R. E., 1973: Method of parameterization of infrared cooling between altitudes of 30 and 70 kilometers. *J. Geophys. Res.*, **78**, 4451–4457.

Dickinson, R. E., 1981: Convergence rate and stability of ocean–atmosphere coupling schemes with a zero-dimensional climate model. *J. Atmos. Sci.*, **38**, 2112–2121.

Dickinson, R. E., 1983: Land surface processes and climate – Surface albedos and energy balance. In *Advances in Geophysics*, **25**, Academic Press, 305–353.

Dickinson, R. E., 1984: Modeling evapotranspiration for three-dimensional global climate models. In *Climate Processes and Climate Sensitivity*, J. E. Hanson and T. Takahashi (Eds.), Geophysical Monograph 29, Maurice Ewing Volume 5, Amer. Geophys. Union, 58–72.

Dickinson, R. E., 1986: How will climate change? The climate system and modelling of future climate. Chapter 5 in *The Greenhouse Effect, Climatic Change, and Ecosystems*. B. Bolin, B. R. Döös, J. Jäger and R. A. Warrick (Eds.), Wiley, Chichester, 207–270.

Dickinson, R. E. (Ed.), 1987: *The Geophysiology of Amazonia. Vegetation and Climate Interactions.* John Wiley and Sons, 91–101.

Dickinson, R. E., and A. Henderson-Sellers, 1988: Modelling tropical deforestation: A study of GCM land-surface parameterizations. *Quart. J. Roy. Meteor. Soc.*, **114**, 439–462.

Dickinson, R. E., A. Henderson-Sellers, C. Rosenzweig and P. J. Sellers, 1991: Formulation of vegetation resistance in climate models. *Ag. Forest Meteor. J.*, **54**, 373–388.

Dickson, R. R., J. Meinke, S-A. Malmberg and A. J. Lee, 1988: The "Great Salinity Anomaly" in the northern North Atlantic, 1968–82. *Prog. Oceanogr.*, **20**, 103–151.

Dines, W. H., 1917: The heat balance of the atmosphere. *Quart. J. Roy. Meteor. Soc.*, **43**, 151–158.

Dines, W. H., 1929: Atmospheric and terrestrial radiation. *Quart. J. Roy. Meteor. Soc.*, **46**, 163–173.

Dobson, D., 1930: Observations of the amount of ozone in the Earth's atmosphere and its relation to other geophysical conditions – Part IV. *Proc. Roy. Soc. London, Sec. A*, **129**, 411–433.

DOE (Department of Energy), 1990: *Energy and Climate Change*. Report of the DOE Multi-Laboratory Climate Change Committee. Lewis Publishers, Inc.

Donner, L. J., H.-L. Kuo and E. J. Pitcher, 1982: The significance of thermodynamic forcing by cumulus convection in a general circulation model. *J. Atmos. Sci.*, **39**, 2159–2181.

References

Dorman, J. L., and P. J. Sellers, 1989: A global climatology of albedo, roughness length and stomatal resistance for atmospheric general circulation models as represented by the Simple Biosphere model (SiB). *J. Appl. Meteor.*, **28**, 833–855.

Driedonks, A. G. M., and H. Tennekes, 1981: Parameterization of the atmospheric boundary layer in large-scale models. *Bull. Amer. Meteor. Soc.*, **62**, 594–598.

Dugdale, R. C., A. Morel, A. Bricaud and F. P. Wilkerson, 1989: Modeling new production in upwelling centers: a case study of modeling new production from remotely sensed temperature and color. *J. Geophys. Res.*, **94**, 18119–18132.

Dunkerton, T. J., 1978: On the mean meridional mass motions of the stratosphere and mesosphere. *J. Atmos. Sci.*, **35**, 2325–2333.

Dupre, W., H. Enzer, S. Miller and D. Hillier, 1976: *Energy Perspectives 2.* U.S. Department of the Interior, Washington, D.C.

Dutton, J., 1976: *The Ceaseless Wind.* McGraw Hill, 579 pp.

Earth System Science, 1986: *A Program for Global Change.* National Aeronautics and Space Administration, Washington, D.C.

Eckholm, E. P., 1977: The other energy crisis. In *Desertification: Environmental Degradation in and around Arid Lands*, M. H. Glantz (Ed.), 39–56.

Ekman, V. W., 1905: On the influence of the earth's rotation on ocean currents. *Arch. Math. Astron. Phys.*, **2**, (11).

Eliasen, E., B. Machenhauer and E. Rasmussen, 1970: On a numerical method for integration of the hydrodynamical equations with a spectral representation of the horizontal fields, Report No. 2, Institut for Teoretisk Meteorologi, University of Copenhagen, 35 pp.

Emanuel, W. R., H. H. Shugart and M. P. Stevenson, 1985: Climatic change and the broad-scale distribution of terrestrial ecosystem complexes. *Clim. Change*, **7**, 29–43.

Emden, R., 1913: Uber strahlungsgleichgewicht und atmospharische strahlung: ein beitrag zur theorie der oberen inversion. *Sitz. Koniglich Bay. Akad. der Wiss.* 55–142.

Emerson, S. R., and D. Archer, 1990: Calcium carbonate preservation in the ocean. *Phil. Trans. Roy. Soc. Lond.*, **A331**, 29–40.

Energy Information Administration, 1989: *Monthly Energy Review.* U.S. Department of Energy, Washington, D.C.

Enfield, D. B., 1989: El Niño, past and present. *Rev. Geophys.*, **27**, 159–187.

Eppley, R., and B. Peterson, 1979: Particulate organic matter flux and planktonic new production in the deep ocean. *Nature*, **282**, 677–680.

Esbensen, S. K., and Y. Kushnir, 1981: The heat budget of the global ocean: An atlas based on estimates from surface marine observations. Climatic Res. Inst. Rept. No. 29, Dept. Atmos. Sci., Oregon State University, Corvallis. OR, 220 pp.

Evans, G., and J. S. Parslow, 1985: A model of annual plankton cycles. *Biological Oceanography*, **3**, 327–347.

Farman, J. C., B. G. Gardiner and J. D. Shanklin, 1985: Large losses of total ozone in Antarctica reveal seasonal ClO_x/NO_x interaction. *Nature*, **315**, 207–210.

Farquhar, G. D., and T. D. Sharkey, 1982: Stomatal conductance and photosynthesis. *Ann. Rev. Plant Physiol.*, **33**, 317–345.

Farquhar, G. D., S. von Caemerrer and J. A. Berry, 1980: A biochemical model of photosynthetic CO_2 fixation in leaves of C_3 species. *Planta*, **149**, 78–90.

Farrell, B. F., 1991: Equable climate dynamics. *J. Atmos. Sci.*, **47**, 2986–2995.

Fasham, M. J. R., H. W. Ducklow and S. M. McKelvie, 1990: A nitrogen-based model of plankton dynamics in the oceanic mixed layer. *J. Mar. Res.*, **48**, 591–639.

Fearnside, P. M., 1986a: Spatial concentration of deforestation in the Brazilian Amazon. *Ambio*, **15**, 74–81.

Fearnside, P. M., 1986b: *Human Carrying Capacity of the Brazilian Rainforest.* Columbia University Press.

Fearnside, P., 1990: The rate and extent of deforestation in Brazilian Amazonia. *Environ. Cons.*, **17**, 213–226.

Fels, S., 1985: Radiative-dynamical interactions in the middle atmosphere. *Adv. Geophys.*, **28A**, 277–300.

Ferenci, T., T. Strom, and J. R. Quayle, 1975: Oxidation of carbon monoxide and methane by psuedomonas methanic. *J. General Microbiol.*, **91**, 79–91.

Finlayson-Pitts, B. J., and J. N. Pitts, Jr., 1986: *Atmospheric Chemistry: Fundamentals and Experimental Techniques.* John Wiley and Sons.

Firestone, M. K., and E. Davidson, 1989: Microbial basis of NO and N_2O production and consumption. In *Exchange of trace gases between terrestrial ecosystems and the atmosphere.* M. O. Andreae and D. S. Schimel (Eds.), John Wiley and Sons, 7–21.

Flanagan, P. W., and K. Van Cleve, 1983: Nutrient cycling in relation to decomposition and organic matter quality in taiga ecosystems. *Canadian J. Forest Res.*, **13**, 795–817.

Flato, G. M., 1991: Numerical investigation of the dynamics of a variable thickness Arctic ice cover. Ph.D. thesis, Thayer School of Engineering, Dartmouth College, Hanover, NH.

Flato, G. M., and W. D. Hibler III, 1990: On a simple sea ice dynamics model for climate studies. *Adv. Geophys.*, **14**, 72–77.

Flato, G. M., and W. D. Hibler III, 1992: Modeling pack ice as a cavitating fluid. *J. Phys. Oceanogr.*, **22**, 626–651.

Fleming, E., S. Chandra, M. Schoeberl and J. Barnett, 1988: *Monthly-Mean Climatology of Temperature, Wind, Geopotential Height, and Pressure for 0–120 km.* NASA TM-100697. Available from NASA Goddard Space Flight Center, Greenbelt MD.

Flynn, M. J. 1972: Some computer organizations and their effectiveness. *IEEE Trans. Computers*, **21**, 948–960.

Folland, C. K., T. N. Palmer and D. E. Parker, 1986: Sahel rainfall and world sea temperatures, 1901–85. *Nature*, **320**, 602–607.

Fowler, S. W., and G. A. Knauer, 1986: Role of large particles in the transport of elements and organic compounds through the oceanic water column. *Prog. Oceanogr.*, **16**, 147–194.

Franken, W., P. R. Leopoldo, E. Matsui and M. N. G. Ribeiro, 1982: Interceptacao das precipitacoes em fioresta Amazonica de terra firme. *Acta Amazon.*, **12**, 15–22.

Friedli, H., H. Løtscher, H. Oeschger, U. Siegenthaler and B. Stauffer, 1986: Ice core record of the $^{13}C/^{12}C$ ratio of atmospheric carbon dioxide in the past two centuries. *Nature*, **324**, 237–238.

Fung, I. Y., D. E. Harrison and A. A. Lacis, 1984: On the net longwave radiation at the ocean surface. *Rev. Geophys. and Space Phys.*, **22**, 177–193.

References

Fung, I. Y., C. J. Tucker and K. C. Prentice, 1987: Application of advanced very high resolution radiometer vegetation index to study atmosphere–biosphere exchange of CO_2. *J. Geophys. Res.*, **92**, 2999–3015.

Gallimore, R. G., and J. E. Kutzbach, 1989: Effects of soil moisture on the sensitivity of a climate model to earth orbital forcing at 9000 yr BP. *Clim. Change*, **14**, 175–205.

Garcia, R. R., and S. Solomon, 1983: A numerical model of the zonally averaged dynamical and chemical structure of the middle atmosphere. *J. Geophys. Res.*, **88**, 1379–1400.

Garratt, J. R., 1978: Flux profile relations above tall vegetation. *Quart. J. Roy. Meteor. Soc.*, **104**, 199–211.

Gaspar, P., Y. Grégoris and J.-M. Lefevre, 1990: A simple eddy kinetic energy model for simulations of the oceanic vertical mixing: Tests at Station Papa and long-term upper ocean study site. *J. Geophys. Res.*, **95**, 16179–16193.

Gates, W. L., 1976: The numerical simulation of ice-age climate with a global general circulation model. *J. Atmos. Sci.*, **33**, 1844–1873.

Gates, W. L., 1985: The use of general circulation models in the analysis of the ecosystem impacts of climatic change. *Clim. Change*, **7**, 267–284.

Gates, W. L., Y.-J. Han and M. E. Schlesinger, 1985: The global climate simulated by a coupled atmosphere–ocean general circulation model: Preliminary results. In *Coupled Ocean–Atmosphere Models*, J. C. J. Nihoul (Ed.), Elsevier Oceanogr. Series, **40**, 131–151.

Gear, C. W., 1967: The numerical integration of ordinary differential equations. *Math. Comp.*, **21**, 146–156.

Gear, C. W., 1969: The automatic integration of stiff ordinary differential equations. In *Information Processing*, **68**, A. H. Morrel (Ed.), North Holland Pub. Co., 187–193.

Gear, C. W., 1971a: The simultaneous numerical solution of differential-algebraic systems. *IEEE Transactions on Circuit Theory*, **18**(1), 89–95.

Gear, C. W., 1971b: *Numerical Initial Value Problems in Ordinary Differential Equations.* Prentice Hall, 253 pp.

GEOSECS Atlas, 1981: Atlantic Expedition, The National Science Foundation, Washington, D.C., **2**.

German Bundestag (Ed.), 1989: *Protecting the Earth's Atmosphere: An International Challenge.* Report of the Enquete Commission. Bonn, German Bundestag.

Ghil, M., 1981: Internal climatic mechanisms participating in glaciation cycles. In *Climatic Variations*, A. Berger (Ed.), Reidel, 539–557.

Ghil, M., and H. LeTreut, 1981: A climate model with cryodynamics and geodynamics. *J. Geophys. Res.*, **86C**, 5262–5270.

Ghil, M., and P. Malanotte-Rizzoli, 1991: Data assimilation in meteorology and oceanography. *Adv. Geophys.*, accepted.

Gill, A. E., 1980: Some simple solutions for heat induced tropical circulation. *Quart. J. Roy. Meteor. Soc.*, **106**, 447–462.

Gill, A. E., 1982: *Atmosphere–Ocean Dynamics.* Academic Press, Intl. Geophys. Ser. 30, 662 pp.

Gilliland, R. L., and S. H. Schneider, 1984: Volcanic, CO_2 and solar forcing of Northern and Southern Hemisphere surface air temperatures. *Nature*, **310**, 38–41.

Giorgi, F., 1990: Simulation of regional climate using a limited area model nested in a general circulation model. *J. Climate*, **3**, 941–963.

Glantz, M. H., 1980: Man, state and the environment: An inquiry into whether solutions to desertification in the West African Sahel are known but not applied. *Can. J. Devel. Studies*, **1**, 75–97.

Glantz, M. H., and N. Orlovsky, 1983: Desertification: A review of the concept. *Desertification Control Bull.*, **9**, 15–21.

Glantz, M. H., and R. W. Katz, 1987: African drought and its impacts: Revived interest in a recurrent phenomenon. *Desertification Control Bull.*, **14**, 22–30.

Glantz, M. H., R. W. Katz and N. Nicholls, 1991: *Teleconnections Linking Worldwide Climate Anomalies*. Cambridge University Press, 353 pp.

Goody, R. M., 1964: *Atmospheric Radiation*. Oxford University Press, 519 pp.

Goody, R. M., and Y. L. Yung, 1989: *Atmospheric Radiation: Theoretical Basis*. 2nd edn, Oxford University Press. 528 pp.

Gordon, A., 1986: Interocean exchange of thermocline water. *J. Geophys. Res.*, **91**, 5037–5046.

Gordon, H. B., and B. G. Hunt, 1987: Interannual variability of the simulated hydrology in a climatic model-implications for drought. *Clim. Dyn.*, **1**, 113–130.

Goudriaan, J., 1977: *Crop Micrometeorology: A Simulation Study*. Wageningen Center for Agricultural Publishing and Documentation, 249 pp.

Gow, A. J., and W. B. Tucker III, 1990: Sea ice in the polar regions. In *Polar Oceanography, Part A: Physical Science*, W. O. Smith (Ed.), Academic Press, 47–122.

Gow, A. J., S. F. Ackley, W. F. Weeks and J. Govoni, 1982: Physical and structural characteristics of Antarctic sea ice. *Ann. Glaciol.*, **3**, 113–117.

Graham, N. E., and W. B. White, 1988: The El Niño cycle: A natural oscillator of the Pacific ocean–atmosphere. *Science*, **240**, 1293–1302.

Graham, N. E., and W. B. White, 1990: The role of the western boundary in the ENSO cycle: experiments with coupled models. *J. Phys. Oceanogr.*, **20**, 1935–1948.

Grainger, A., 1990: *The Threatening Desert: Controlling Desertification*. Earthscan Publications Ltd.

Greenberg, J. P., and P. R. Zimmerman, 1984: Nonmethane hydrocarbons in remote tropical, continental, and marine atmospheres. *J. Geophys. Res.*, **89**, 4767–4778.

Gregg, M. C., 1987: Diapycnal mixing in the thermocline: A review. *J. Geophys. Res.*, **92**, 5249–5286.

Greuell, W., and J. Oerlemans, 1986: Sensitivity studies with a mass balance model including temperature profile calculations inside the glacier. *Z. Gletscherk. Glazialgeol.*, **22**, 101–124.

Haarsma, R., J. F. B. Mitchell and C. A. Senior, 1991: Tropical cyclones in the warmer climate. *J. Climate*, submitted.

Hack, J. J., 1986: Peak vs. sustained performance in highly concurrent vector machines. *IEEE Computer*, **19**, 11–19.

Hack, J. J., 1989: On the promise of general–purpose parallel computing. *Parallel Computing*, **10**, 261–275.

Haidvogel, D. B., A. Beckmann and K. S. Hedstrom, 1991a: Dynamical simulations of filament formation and evolution in the coastal transition zone. *J. Geophys. Res.*, **96**, 15017–15040.

References

Haidvogel, D. B., J. Wilkin and R. Young, 1991b: A semi-spectral primitive equation ocean circulation model using vertical sigma and orthogonal curvilinear horizontal coordinates. *J. Comp. Phys.*, **94**, 151–185.

Hall, F. G., D. E. Strebel and P. J. Sellers, 1988: Linking knowledge among spatial and temporal scales: vegetation, atmosphere, climate and remote sensing. *Landscape Ecology*, **2**, 3–22.

Hall, M. M., and H. L. Bryden, 1982: Direct estimates and mechanisms of ocean heat transport. *Deep-Sea Res.*, **29**, 339–359.

Haltiner, G. J., and R. T. Williams, 1980: *Numerical Prediction and Dynamic Meteorology*, 2nd edn, John Wiley and Sons, 477 pp.

Hanel, R. A., B. J. Conrath, V. G. Kunde, C. Prabhakara, I. Revah, V. V. Salomonson and G. Wolford, 1972: The Nimbus 4 infrared spectroscopy experiment: 1. Calibrated thermal emission spectra. *J. Geophys. Res.*, **77**, 2629–2641.

Haney, R. L., 1971: Surface thermal boundary conditions for ocean general circulation models. *J. Phys. Oceanogr.*, **1**, 241–248.

Hansen, J., and S. Lebedeff, 1987: Global trends of measured surface air temperature. *J. Geophys. Res.*, **92**, 13345–13372.

Hansen, J., D. Johnson, A. Lacis, S. Lebedeff, P. Lee, D. Rind and G. Russell, 1981: Climate impact of increasing atmospheric carbon dioxide. *Science*, **213**, 957–966.

Hansen, J., A. Lacis, D. Rind, G. Russell, P. Stone, I. Fung, R. Ruedy and J. Lerner, 1984: Climate sensitivity: Analysis of feedback mechanisms. In *Climate Processes and Climate Sensitivity*, J. E. Hansen and T. Takahashi (Eds.), Maurice Ewing Series, **5**, AGU, 130–163.

Harrison, W. G., T. Platt and M. R. Lewis, 1987: f-ratio and its relationship to ambient nitrate concentration in coastal waters. *J. Plankton Res.*, **9**, 235–248.

Harriss, R. C., M. C. Garstang, S. C. Wofsy, S. M. Beck, R. J. Bendura, J.R.B. Coelho, J. W. Drewry, J. M. Hoell, P. A. Matson, R. J. McNeal, L.C.B. Molion, R. L. Navarro, V. Rabine and R. L. Snell, 1990: The Amazon boundary layer experiment: Wet season 1987. *J. Geophys. Res.*, **95**, 16721–16736.

Hartmann, D., V. Ramanathan, A. Berroir and G. Hunt, 1986: Earth radiation budget data and climate research. *Rev. Geophys.*, **24**, 439–468.

Harvey, L. D. D., 1988: A semi-analytic energy balance climate model with explicit sea ice and snow physics. *J. Climate*, **1**, 1065–1085.

Harvey, L. D. D., and S. H. Schneider, 1984: Sensitivity of internally generated climate oscillations to ocean model formulation. In *Milankovitch and Climate Change*, A. Berger, J. Imbrie, J. Hays, G. Kukla and B. Saltzman (Eds.), Reidel, 653–667.

Hasselmann, K., 1976: Stochastic climate models. Part I: Theory. *Tellus*, **28**, 6, 473–485.

Hasselmann, K., 1988: Some problems in the numerical simulation of climate variability using high-resolution coupled models. In *Physically Based Modelling and Simulation of Climate and Climate Change.*, NATO Advanced Study Institute (1986: Erice, Italy), M. E. Schlesinger (Ed.), Kluwer, 583–605.

Hays J. D., J. Imbrie and N. J. Shackleton, 1976: Variations in the Earth's orbit: Pacemaker of the ice ages. *Science*, **194**, 1121–1132.

Heimann, M., C. D. Keeling and C. J. Tucker, 1990: A three-dimensional model of atmospheric CO_2 transport based on observed winds: 3. In *Seasonal Cycle and Synoptic Time Scale Variations*, Geophysical Monograph 55, Amer. Geophys. Union, 277–303.

Held, I. M, S. W. Lyons and S. Nigam, 1989: Transients and the extratropical response to El Niño. *J. Atmos. Sci.*, **46**, 163–174.

Held, I., and M. Suarez, 1978: A two-level primitive equation atmospheric model designed for climatic sensitivity experiments. *J. Atmos. Sci.*, **35**, 206–229.

Hellerman, S., and M. Rosenstein, 1983: Normal monthly wind stress over the world ocean with error estimates. *J. Phys. Oceanogr.*, **13**, 1093–1104.

Henderson-Sellers, A., and K. McGuffie, 1987: *Climate Modelling Primer*. John Wiley and Sons, 217 pp.

Henderson-Sellers, A., R. E. Dickinson and M. F. Wilson, 1988: Tropical deforestation: important processes for climate models. *Clim. Change*, **13**, 43–69.

Henderson-Sellers, A., M. F. Wilson, G. Thomas and R. E. Dickinson, 1986: Current global land-surface data sets for use in climate-related studies, NCAR Tech. Note, NCAR/TN-272+STR, 110 pp.

Henderson-Sellers, B., and A. M. Davies, 1989: Thermal stratification modeling for oceans and lakes. *Ann. Rev. Numerical Fluid Mech. Heat Transfer*, **2**, 86–156.

Hesstvedt, E., Ö. Hov and I. S. A. Isaksen, 1978: Quasi-steady state approximations in air pollution modeling: Comparison of two numerical schemes for oxidant prediction. *Int. J. Chem. Kinetics*, **10**, 971–994.

Hibler, W. D., III, 1979: A dynamic thermodynamic sea ice model. *J. Phys. Oceanogr.*, **9**, 815–846.

Hibler, W. D., III, 1980: Modeling a variable thickness sea ice cover. *Mon. Wea. Rev.*, **108**, 1942–1973.

Hibler, W. D., III, and S. F. Ackley, 1983: Numerical simulation of the Weddell Sea pack ice. *J. Geophys. Res.*, **88**, 2873–2887.

Hibler, W. D., III, and K. Bryan, 1984: Ocean circulation: Its effect on seasonal sea ice simulations. *Science*, **224**, 489–492.

Hibler, W. D., III, and J. E. Walsh, 1982: On modeling seasonal and interannual fluctuations of arctic sea ice. *J. Phys. Oceanogr.*, **12**, 1514–1523.

Hillis, W. D., 1985: *The Connection Machine*, MIT Press, 190 pp.

Hirst, A., 1986: Unstable and damped equatorial modes in simple coupled ocean–atmosphere models. *J. Atmos. Sci.*, **43**, 606–630.

Hirst, A., 1988: Slow instabilities in tropical ocean basin – global atmosphere models. *J. Atmos. Sci.*, **45**, 830–852.

Hirt, C. W., A. A. Amsden and J. L. Cook, 1974: An arbitrary Lagrangian-Eulerian computing method for all flow speeds. *J. Comput. Phys.*, **14**, 227–253.

Hitchman, M. H., and G. Brasseur, 1988: Rossby wave activity in a two-dimensional model: closure for wave driving and meridional eddy diffusivity. *J. Geophys. Res.*, **93**, 9405–9417.

Hogg, N., 1991: On the transport of the Gulf Stream between Cape Hatteras and the Grand Banks. *Deep-Sea Res.*, accepted.

Holdridge, L. R., 1947: Determination of world plant formations from simple climatic data. *Science*, **105**, 367–368.

Holland, W. R., 1985: Simulation of mesoscale ocean variability in mid-latitude gyres. *Adv. Geophys.*, **28**, 479–523.

References

Hollingsworth, A. K., K., Arpe, M. Tiedke, M. Capaldo and H. Saavijärvi, 1980: The performance of a medium-range forecast model in winter – impact of physical parameterizations. *Mon. Wea. Rev.*, **108**, 1736–1773.

Holloway, G., 1989: Subgridscale representation. In *Oceanic Circulation Models: Combining Data and Dynamics*, D. L. T. Anderson and J. Willebrand (Eds.), Kluwer, 513–593.

Holton, J., 1979: *An Introduction to Dynamic Meteorology*. Academic Press, 391 pp.

Horel, J. D., and J. M. Wallace, 1981: Planetary scale atmospheric phenomena associated with the Southern Oscillation. *Mon. Wea. Rev.*, **109**, 813–829.

Hortal, M., and A. J. Simmons, 1991: Use of reduced Gaussian grids in spectral models. *Mon. Wea. Rev.*, **119**, 1057–1074.

Houghton, J., 1986: *The Physics of Atmospheres*. Cambridge University Press, 271 pp.

Houghton, R. A., J. E. Hobbie, J. M. Melillo, B. Moore, B. J. Peterson, G. R. Shaver and G. M. Woodwell, 1983: Changes in the carbon content of terrestrial biota and soils between 1860 and 1980: A net release of CO_2 to the atmosphere. *Ecological Monographs*, **53**, 235–262.

Houze, R. A., 1989: Observed structure of mesoscale convective systems and implications for large-scale heating. *Quart. J. Roy. Meteor. Soc.*, **115**, 425–461.

Hubbert, M. K., 1969: Energy resources. In *Resources and Man*, National Resource Council, W. H. Freeman and Company, 157–242.

Hubbert, M. K., 1972: *U.S. Energy Resources, A Review as of 1972*. U.S. Senate Committee on Interior and Insular Affairs. U.S. Government Printing Office, Washington, D.C.

Hunt, B. G., 1979: The effects of past variations of the earth's rotation rate on climate. *Nature*, **281**, 188–191.

Hunt, B. G., and H. B. Gordon, 1988: The problem of "naturally"-occurring drought. *Clim. Dyn.*, **3**, 19–33.

Hutter, K., 1983: *Theoretical Glaciology*. Reidel Publ. Co., 510 pp.

Imbrie, J., and K. P. Imbrie, 1979: *Ice Ages: Solving The Mystery*. Enslow, 224 pp.

Imbrie, J., A. McIntyre and A. Mix, 1989: Oceanic response to orbital forcing in the late Quaternary: Observational and experimental strategies. In *Climate and Geo-Sciences*, A. Berger et al. (Eds.), Kluwer, 121–164.

Ingram, W. J., C. A. Wilson and J.F.B. Mitchell, 1989: Modelling climate change: an assessment of sea-ice and surface albedo feedbacks. *J. Geophys. Res.*, **94**, 8609–8622.

Inoue, M., and J. O'Brien, 1984: A forecasting model for the onset of El Niño. *Mon. Wea. Rev.*, **112**, 2326–2337.

Ip, C. F., W. D. Hibler III and G. M. Flato, 1991: On the effect of rheology on seasonal sea ice simulations. *Annals of Glaciology*, **15**, 17–25.

IPCC, 1990a: *Climate Change: The IPCC Scientific Assessment*, J. T. Houghton, G. J. Jenkins, and J. J. Ephraums (Eds.), Cambridge University Press, 365 pp.

IPCC, 1990b: *Climate Change: The IPCC Impacts Assessment*. WMO/UNEP, Geneva.

IPCC, 1991: *Climate Change: The IPCC Response Strategies*. WMO/UNEP, Geneva.

Irabarne, J., and W. Godson, 1981: *Atmospheric Thermodynamics*. Reidel, 259 pp.

Isaksen, I. S. A. (Ed.), 1989: *Tropospheric Ozone: Regional and Global Scale Interactions*. Reidel.

Isaksen, I. S. A., and F. Stordal, 1986: Ozone perturbations by enhanced levels of CFCs, N_2O and CH_4: A two-dimensional diabatic circulation study including uncertainty estimates, *J. Geophys. Res.*, **91**, 5249–5263.

Isemer, H.-J., and L. Hasse, 1991: The scientific Beaufort equivalent scale: Effects on wind statistics and climatological air–sea flux estimates in the North Atlantic Ocean. *J. Climate*, **4**, 819–847.

ISSC (International Social Science Council), 1990: *A Framework for Research on the Human Dimensions of Global Environmental Change.* Paris, ISSC.

Iverson, R. L., F. L. Nearhoof and M. O. Andreae, 1989: Production of dimethylsulfonium propionate and dimethylsulfide in estuarine and coastal waters. *Limnol. Oceanogr.*, **34**, 53–67.

Jahne, B., K. O. Munnich, R. Bosinger, A. Dutzi, W. Huber and P. Libner, 1987: On the parameters influencing air–water gas exchange. *J. Geophys. Res.*, **92**, 1937–1949.

Jahnke, R. A., 1990: Ocean flux studies: a status report. *Rev. Geophys.*, **28**, 381–398.

Jahnke, R. A., and G. A. Jackson, 1987: Role of sea floor organisms in oxygen consumption in the deep North Pacific Ocean. *Nature*, **329**, 621–623.

Jarvis, P. G., 1976: The interpretation of the variations in leaf water potential and stomatal conductance found in canopies in the field. *Phil. Trans. Roy. Soc. London*, **Ser. B**, **273**, 593–610.

Jarvis, P. G., G. B. James and J. J. Landsberg, 1976: Coniferous forest. In *Vegetation and the Atmosphere*, Vol. 2, J. L. Monteith (Ed.), Academic Press, 171–240.

Jenkins, W. J., and P. B. Rhines, 1980: Tritium in the deep North Atlantic. *Nature*, **286**, 877–880.

Jenne, R., 1975: Data Sets for Meteorological Research. NCAR Technical Note, NCAR TN/1A-III, 194 pp.

Johnson, K. S., 1982: Carbon dioxide hydration and dehydration kinetics in seawater. *Limnol. Oceanogr.*, **27**, 849–855.

Jones, P., 1988: The influence of ENSO on global temperatures. *Climate Monitor*, **17**, 80–89.

Jones, R. D., and R. Y. Morita, 1983: Methane oxidation by *Nitrosococcus oceanus* and *Nitrosomonas europea*. *Applied Environmental Microbiology*, **45**, 401–410.

Joos, F., J. L. Sarmiento and U. Siegenthaler, 1991: Estimates of the effect of Southern Ocean iron fertilization on atmospheric CO_2 concentrations. *Nature*, **349**, 772–775.

Jouzel, J., G. Raibeck, J. P. Benoist, F.Yiou, C. Lorius, D. Raynaud, J. R. Petit, N. I. Barkov, Y. S. Korotkevitch and V. M. Kotlyakov, 1989: A comparison of deep Antarctic ice cores and their implications for climate between 65,000 and 15,000 years ago. *Quatern. Res.*, **31**, 135–150.

Junge, C. E., 1963: Sulfur in the atmosphere. *J. Geophys. Res.*, **68**, 3975–3976.

Kálnay, E., M. Kanamitsu and W. E. Baker, 1990: Global numerical weather prediction at the National Meteorological Center. *Bull. Amer. Meteor. Soc.*, **71**, 1410–1428.

Kálnay, E., R. Petersen, M. Kanamitsu and W. E. Baker, 1991: U.S. operational numerical weather prediction. *Rev. Geophys.*, 104–114.

Kálnay, E., M. Kanamitsu, J. Pfaendtner, J. Sela, M. Suarez, J. Stackpole, J. Tuccillo, L. Umscheid and D. Williamson, 1989: Rules for interchange of physical parameterizations. *Bull. Amer. Meteor. Soc.*, **70**, 620–622.

References

Kantha, L. H., A. Rosati and B. Galperin, 1989: Effects of rotation on vertical mixing and small-scale turbulence in stratified fluids. *J. Geophys. Res.*, **94**, 4843–4854.

Kaps, P., and P. Rentrop, 1979: Generalized Runge Kutta methods of order four with stepsize control for stiff ordinary differential equations. *Numer. Math.*, **33**, 55–68.

Karl, D. M., G. A. Knauer and J. H. Martin, 1988: Downward flux of particulate organic matter in the ocean: a particle decomposition paradox. *Nature*, **332**, 438–441.

Karol, I. L., and E. V. Rozanov, 1982: Radiative-convective climate models. *Izvestiya. Atmospheric Oceanic Physics*, **18**, 1179–1191.

Karoly, D. J., 1989: Northern Hemisphere temperature trends: A possible greenhouse gas effect? *Geophys. Res. Lett.*, **16**, 465–468.

Kasahara, A., 1974: Various vertical coordinate systems used for numerical weather prediction. *Mon. Wea. Rev.*, **102**, 504–522.

Kasahara, A., 1977: Numerical integration of the global barotropic primitive equations with Hough harmonic expansions. *J. Atmos. Sci.*, **34**, 687–701.

Kasting, J. F., O. B. Toon and J. B. Pollack, 1988: How climate evolved on the terrestrial planets. *Sci. Amer.*, February, 90–97.

Keeling, C. D., R. B. Bacastow, A. F. Carter, S. C. Piper, T. P. Whorf, M. Heimann, W. G. Mook and H. Roeloffzen, 1989: A three dimensional model of atmospheric CO_2 transport based on observed winds: 1. Analysis of observed data. In *Aspects of climate variability in the Pacific and the western Americas*, D. H. Peterson (Ed), *Geophysical Monogr.* 55, AGU, Washington, D.C. 165–236.

Keller, M. D., W. K. Bellows and R.R.L. Guillard, 1989: Dimethyl sulfide production in marine phytoplankton. In *Biogenic Sulfur in the Environment*, E. S. Saltzman and W. J. Cooper (Eds.), American Chemical Society, Washington, D.C., 167–182.

Khalil, M. A. K., and R. A. Rasmussen, 1988: Carbon monoxide in the Earth's atmosphere: Indications of a global increase, *Nature*, **332**, 242–245.

Kiehl, J. T., and R. E. Dickinson, 1987: A study of the radiative effects of enhanced atmospheric CO_2 and CH_4 on early earth surface temperatures. *J. Geophys. Res.*, **92**, 2991–2998.

Kiehl, J. T., and V. Ramanathan, 1983: CO_2 radiative parameterization used in climate models: Comparison with narrow band models and laboratory data. *J. Geophys. Res.*, **88**, 5191–5202.

Kiene, R. P., and T. S. Bates, 1990: Biological removal of dimethyl sulfide from sea water. *Nature*, **345**, 702–705.

Kimes, D. S., 1984: Modeling the directional reflectance from complete homogeneous vegetation canopies with various leaf orientation distributions. *J. Opt. Soc. Amer.*, **1**, 725–737.

Knowles, R., 1982: Denitrification. *Microbiological Rev.*, **46**, 526–532.

Ko, M. K. W., K. K. Tung, D. K. Weinstein and N. D. Sze, 1985: A zonal-mean model of stratospheric tracer transport in isentropic coordinates: numerical simulations for nitrous oxide and nitric acid. *J. Geophys. Res.*, **90**, 2313–2329.

Koerner, R. M., 1973: The mass balance of the sea ice of the Arctic Ocean. *J. Glaciol.*, **12**, 173–185.

Kogge, P. M., 1981: *The Architecture of Pipelined Computers.* McGraw-Hill, 334 pp.

Kraus, E., (Ed.), 1977: *Modeling and Prediction of the Upper Layers of the Ocean.* Pergamon Press.

Kraus, E., and J. Turner, 1967: A one-dimensional model of the seasonal thermocline. II: The general theory and its consequences. *Tellus,* **19,** 98–106.

Krenz, J. H., 1980: *Energy: From Opulence to Sufficiency.* Praeger Press.

Krishnamurti, T. N., and D. Oosterhof, 1989: Prediction of the life cycle of a super typhoon with a high-resolution global model. *Bull. Amer. Meteor. Soc.,* **70,** 1218–1230.

Krishnamurti, T. N., S.-L. Nam and R. Pasch, 1983: Cumulus parameterization and rainfall rates II. *Mon. Wea. Rev.,* **111,** 815–828.

Kuhn, W. R., J. C. G. Walker and H. G. Marshall, 1989: The effect on earth's surface temperature from variations in rotation rate continent formation, solar luminosity, and carbon dioxide. *J. Geophys. Res.,* **94,** 11129–11136.

Kuo, H.-L., 1974: Further studies of the parameterization of the influence of cumulus convection on large-scale flow. *J. Atmos. Sci.,* **31,** 1232–1240.

Kushnir, Y., and N.-C. Lau, 1992: The general circulation model response to a North Pacific SST anomaly: Dependence on time scale and pattern polarity. *J. Climate,* **5,** 271–283.

Kutzbach, J. E., and R. G. Gallimore, 1989: Pangaean Climates: Megamonsoons of the Megacontinent. *J. Geophys. Res.,* **94,** 3341–3358.

Kutzbach, J. E., and P. J. Guetter, 1986: The influence of changing orbital parameters and surface boundary conditions on climate simulations for the past 18,000 years. *J. Atmos. Sci.,* **43,** 1726–1759.

Kutzbach, J. E., and F. A. Street-Perrott, 1985: Milankovitch forcing of fluctuations in the level of tropical lakes from 18 to 0 kyr BP. *Nature,* **317,** 130–134.

Kutzbach, J. E., and T. Webb III, 1991: Late quaternary climatic and vegetational change in eastern North America: Concepts, models, and data. In *Quaternary Landscapes,* L.C.K. Shane and E. J. Cushing (Eds.), University of Minnesota Press, 175–218.

Kutzbach, J. E., P. J. Guetter and W. M. Washington, 1990: Simulated circulation of an idealized ocean for Pangaean time. *Paleoceanography,* **5(3),** 299–317.

Kvenvolden, K. A., 1988: Methane hydrates and global climate. *Global Biogeochem. Cycles,* **2,** 221–229.

Lachenbruch, A. H., J. H. Sass, L. A. Lawver, M. C. Brewer, B. V. Marshall, R. J. Munroe, J. P. Kenelly, Jr., S. P. Galamis Jr. and T. H. Moses, Jr., 1988: Temperature and depth of permafrost on the Arctic Slope of Alaska. In *Geology and Exploration of the National Petroleum Reserve in Alaska, 1974 to 1982,* G. Gryc (Ed.), USGS Professional Paper 1399, 645–656.

Lacis, A. A., and J. E. Hansen, 1974: A parameterization for the absorption of solar radiation in the earth's atmosphere. *J. Atmos. Sci.,* **31,** 118–133.

Lamb, H. H., 1977: *Climate: Present, Past and Future, Vol. 2, Climatic History and the Future.* Methuen.

Land, K. C., and S. H. Schneider, 1987: Forecasting in the social and natural sciences: an overview and analysis of isomorphisms. *Clim. Change,* **11,** 7–31.

Langleben, M. P., 1971: Albedo of melting sea ice bottomside features in the Denmark Strait. *J. Geophys. Res.,* **79,** 4505–4511.

Langleben, M. P., 1972: Decay of an annual cover of sea ice. *J. Glaciol.,* **11,** 337–344.

References

Large, W. G., and S. Pond, 1981: Open ocean momentum flux measurements in moderate to strong winds. *J. Phys. Oceanogr.*, **11**, 324–336.

Large, W. G., J. C. McWilliams and P. P. Niiler, 1986: Upper ocean thermal response to strong autumnal forcing of the Northeast Pacific. *J. Phys. Oceanogr.*, **16**, 1561–1579.

Latif, M., J. Biercamp and H. von Storch, 1988: The response of a coupled ocean–atmosphere general circulation model to wind bursts. *J. Atmos. Sci.*, **45**, 964–979.

Latif, M., M. Flügel, and J.-S. Xu, 1991: An investigation of short range climate predictability in the tropical Pacific. *J. Geophys. Res.*, **96**, 2661–2673.

Lau, K.-H., 1991: An observational study of tropical summertime synoptic scale disturbances. Ph.D. dissertation, Princeton University, 243 pp.

Lau, K.-H., and N.-C. Lau, 1990: Observed structure and propagation characteristics of tropical summertime synoptic scale disturbances. *Mon. Wea. Rev.*, **118**, 1888–1913.

Lau, K.-M., and L. Peng, 1987: Origin of low-frequency (intraseasonal) oscillations in the tropical atmosphere. Part I: Basic theory. *J. Atmos. Sci.*, **44**, 950–972.

Lau, N.-C., 1981: A diagnostic study of recurrent meteorological anomalies appearing in a 15-year simulation with a GFDL general circulation model. *Mon. Wea. Rev.*, **109**, 2287–2311.

Lau, N.-C., 1985: Modeling the seasonal dependence of the atmospheric response to observed El Niños in 1962–76. *Mon. Wea. Rev.*, **113**, 1970–1996.

Lau, N.-C., and M. J. Nath, 1987: Frequency dependence of the structure and temporal development of wintertime tropospheric fluctuations – comparison of a GCM simulation with observations. *Mon. Wea. Rev.*, **115**, 251–271.

Lau, N.-C., and M. J. Nath, 1990: A general circulation model study of the atmospheric response to extratropical SST anomalies observed in 1950–79. *J. Climate*, **3**, 965–989.

Lau, N.-C., I. M. Held and J. D. Neelin, 1988: The Madden–Julian Oscillation in an idealized general circulation model. *J. Atmos. Sci.*, **45**, 3810–3832.

Laursen, L., and E. Eliasen, 1989: On the effects of the damping mechanisms in an atmospheric general circulation model. *Tellus*, **41A**, 385–400.

Leaman, D. K., and F. A. Schott, 1991: Hydrographic structure of the convective regime of the Gulf of Lions: Winter 1987. *J. Phys. Oceanogr.*, **21**, 575–598.

Lean, J., and D. A. Warrilow, 1989: Simulation of the regional climatic impact of Amazon deforestation. *Nature*, **342**, 411–413.

Leavitt, E., 1980: Surface-based air stress measurements made during AIDJEX. In *Sea Ice Processes and Models*, R. S. Pritchard (Ed.), Univ. Washington Press, 419–429.

Leck, C., U. Larsson, L. E. Bågander, S. Johansson and S. Hadju, 1990: Dimethyl sulfide in the Baltic Sea: Annual variability in relation to biological activity. *J. Geophys. Res.*, **95**, 3353–3363.

Ledwell, J. J., 1984: The variation of the gas transfer coefficient with molecular diffusivity. In *Gas Transfer at Water Surfaces*, W. Brutsaert and G. H. Jirka, (Eds.), 293–302.

Leetmaa, A., and A. Bunker, 1978: Updated charts of the mean annual wind stress, convergences in the Ekman layers, and Sverdrup transports in the North Atlantic. *J. Mar. Res.*, **36**, 311–322.

Leetmaa, A., P. P. Niiler and H. Stommel, 1977: Does the Sverdrup relation account for the Mid-Atlantic circulation? *J. Mar. Res.*, **35**, 1–10.

Legg, B. J., and I. F. Long, 1975: Turbulent diffusion within a wheat canopy: II. *Quart. J. Roy. Meteor. Soc.*, **101**, 611–628.

Legrand, M., C. Feniet-Saigne, E. S. Saltzman, C. Germain, N. I. Barkov and V. N. Petrov, 1991: Ice-core record of oceanic emissions of dimethylsulfide during the last climate cycle. *Nature*, **350**, 144–146.

Leith, C. E., 1965: Numerical simulation of the Earth's atmosphere. In *Methods in Computational Physics*, **4**, B. Adler, S. Ferenbach, and M. Rotenberg (Eds.), Academic Press, 385 pp.

Lemke, P., W. B. Owens, and W. D. Hibler III, 1990: A coupled sea ice – mixed layer – pycnocline model for the Weddell Sea. *J. Geophys. Res.*, **95**, 9513–9526.

Lepparanta, M., 1980: On the drift and deformation of sea ice fields in the Bothnian Bay. Winter Navigation Research Board, Helsinki, Finland, Research Report No. 29.

Letréguilly, A., 1988: Relation between the mass balance of western Canadian mountain glaciers and meteorological data. *J. Glaciol.*, **34**, 11–18.

Levine, J. (Ed.), 1985: *The Photochemistry of Atmospheres: Earth, The Other Planets, and Comets.* Academic Press, 518 pp.

Levitus, S., 1982: Climatological atlas of the world ocean. NOAA Prof. Pap. 13, U.S. Govt. Print. Office, Washington, D.C.

Levitus, S., J. L. Reid, M. E. Conkright and R. G. Najjar, 1992: The distribution of phosphate, nitrate and silicate in the world oceans. *Prog. Oceanogr.*, in press.

Levy, H., II, and W. J. Moxim, 1989: Simulated global distribution and deposition of reactive nitrogen emitted by fossil fuel combustion. *Tellus*, **41B**, 256–271.

Lewis, M. R., 1989: The variegated ocean: a view from space. *New Scientist*, 7 October issue.

Lilly, D. K., 1972: Wave momentum flux – A GARP problem. *Bull. Amer. Meteor. Soc.*, **53**, 17–23.

Lindzen, R. S., 1981: Turbulence and stress owing to gravity wave and tidal breakdown. *J. Geophys. Res.*, **86**, 9707–9714.

Lindzen, R. S., and A. Y. Hou, 1988: Hadley circulations for zonally averaged heating centered off the equator. *J. Atmos. Sci.*, **45**, 2416–2427.

Liou, K., 1980: *An Introduction to Atmospheric Radiation.* Academic Press, 392 pp.

Liss, P. S., and L. Merlivat, 1986: Air–sea gas exchange rates: introduction and synthesis. In *The Role of Air–Sea Exchange in Geochemical Cycling*, P. Buat-Menard (Ed.), D. Reidel, 113–127.

Liu, W.T., K. B. Katsaros and J. A. Businger, 1979: Bulk parameterization of air–sea exchanges in heat and water vapor including the molecular constraints at the interface. *J. Atmos. Sci.*, **36**, 1722–1735.

London, J., 1980: Radiative energy sources and sinks in the stratosphere and mesosphere. *Proceedings of the NATO Advanced Study Institute on Atmospheric Ozone: Its Variations and Human Influences, Rep. FAA-EE-80-20*, A. C. Aikin (Ed.), 703–721.

Lorenz, E. N., 1968: Climate determinism. *Meteorol. Monogr.*, **8**, No. 30, 1–3.

Lorenz, E. N., 1969: The predictability of a flow which possesses many scales of motion. *Tellus*, **21**, 289–307.

Lorenz, E. N., 1982: Atmospheric predictability experiments with a large numerical model. *Tellus*, **34**, 505–513.

References

Lorenz, E. N., 1983: Estimates of atmospheric predictability at medium range. *Predictability of Fluid Motions. American Inst. Physics, Conf. Proc.*

Lovejoy, T. E., 1990: Environmental myopia. *Unesco Sources*, **19**, 3.

Lovins, A. B., L. H. Lovins, F. Krause and W. Bach, 1981: *Least-Cost Energy Solving the CO_2 Problem.* (Second edition 1989). Rocky Mountain Institute, Snowmass, CO.

Lutzenberger, J., 1987: Who is destroying the Amazon rainforest? *Ecologist*, **17**, 155–160.

MacCracken, M. C., A. D. Hecht, M. I. Budyko and Y. A. Izrael (Eds.), 1990: *Prospects for Future Climate*. A Special US/USSR Report on Climate and Climate Change, Lewis Publishers, Inc., Chelsea, MI, 270 pp.

Machenhauer, B., 1979: The spectral method, *Numerical Methods Used in Atmospheric Models*, GARP Publication Series 17, World Meteorological Organization, 121–275.

Madden, R. A., and P. R. Julian, 1972: Description of global-scale circulation cells in the tropics with a 40-50 day period. *J. Atmos. Sci.*, **29**, 1109–1123.

Mahlman, J. D., 1975: Some fundamental limitations of simplified transport models as implied by results from a three-dimensional general circulation/tracer model, in *Proc. Fourth Conf. Climatic Impact Assess. Prog.*, T. M. Hard, and A. Broderick (Eds.), Publ. DOT-TSC-OST-75-38, 132–146, Dept. of Transportation, Cambridge, Mass.

Maier-Reimer, E., and K. Hasselmann, 1988: Transport and storage of CO_2 in the ocean – an inorganic ocean-circulation cycle model. *Clim. Dyn.*, **2**, 63–90.

Maier-Reimer, E. and U. Mikolajewicz, 1989: Experiments with an OGCM on the cause of the Younger Dryas. In *Oceanography 1988*, A. Ayala–Castanares, W. Wooster, A. Yanes–Arancibia (Eds.), UNAM Press, Mexico, 87–100.

Maier-Reimer, E., U. Mikolajewicz and T. Crowley, 1991: Ocean general circulation model sensitivity experiment with an open Central American isthmus. *Paleoceanography*, **5(3)**, 349–366.

Malingreau, J.-P., 1984: *Remote Sensing and Forest Fire Monitoring in Indonesia.* Report on a Consultancy (20 October–11 November 1984). The Ford Foundation.

Malingreau, J.-P., 1988: Large-scale deforestation in the southeastern Amazon Basin of Brazil. *Ambio*, **17**, 49–55.

Manabe, S., and A. J. Broccoli, 1985: The influence of continental ice sheets on the climate of an ice age. *J. Geophys. Res.*, **90**, 2167–2190.

Manabe, S., and A. J. Broccoli, 1990: Mountains and arid climates of middle latitudes. *Science*, **247**, 192–195.

Manabe, S., and K. Bryan, 1969: Climate calculations with a combined ocean–atmosphere model. *J. Atmos. Sci.*, **26**, 786–789.

Manabe, S., and K. Bryan, 1985: CO_2-induced change in a coupled ocean–atmosphere model and its paleoclimatic implications. *J. Geophys. Res.*, **90**, 11689–11708.

Manabe, S., and D. G. Hahn, 1981: Simulation of atmospheric variability. *Mon. Wea. Rev.*, **109**, 2260–2286.

Manabe, S., and R. J. Stouffer, 1980: Sensitivity of a global climate model to an increase of CO_2 concentration in the atmosphere. *J. Geophys. Res.*, **85**, 5529–5554.

Manabe, S., and R. J. Stouffer, 1988: Two stable equilibria of a coupled ocean–atmosphere model. *J. Climate*, **1**, 841–866.

Manabe, S., and R. F. Strickler, 1964: On the thermal equilibrium of the atmosphere with a convective adjustment. *J. Atmos. Sci.*, **21**, 361–385.

Manabe, S., and R. T. Wetherald, 1967: Thermal equilibrium of the atmosphere with a given distribution of relative humidity. *J. Atmos. Sci.*, **24**, 241–259.

Manabe, S., and R. T. Wetherald, 1975: The effects of doubling the CO_2 concentration on the climate of a general circulation model. *J. Atmos. Sci.*, **37**, 3–15.

Manabe, S., and R. T. Wetherald, 1987: Large-scale changes of soil wetness induced by an increase in atmospheric carbon dioxide. *J. Atmos. Sci.*, **44**, 1211–1235.

Manabe, S., K. Bryan and M. J. Spelman, 1975: A global ocean–atmosphere climate model. Part I. The atmospheric circulation. *J. Phys. Oceanogr.*, **5**, 3–29.

Manabe, S., K. Bryan and M. J. Spelman, 1979: A global ocean–atmosphere climate model with seasonal variation for future studies of climate sensitivity. *Dyn. Atmos. Oceans*, **3**, 393–426.

Manabe, S., K. Bryan and M. J. Spelman, 1990: Transient response of a global ocean–atmosphere model to a doubling of atmospheric carbon dioxide. *J. Phys. Oceanogr.*, **20**, 722–749.

Manabe, S., J. Smagorinsky and R. F. Strickler, 1965: Simulated climatology of a general circulation model with a hydrologic cycle. *Mon. Wea. Rev.*, **93**, 769–798.

Manabe, S., R. J. Stouffer, M. J. Spelman and K. Bryan, 1991: Transient responses of a coupled ocean–atmosphere model to gradual changes of atmospheric CO_2. Part I. Annual mean response. *J. Climate*, **4**, 785–818.

Marland, G., 1989: Fossil fuels CO_2 emissions. CDIAC Communications, Winter 1989, Carbon Dioxide Inf. Anal. Center, Oak Ridge National Lab., Oak Ridge, TN, 1–3.

Marotzke, J., 1989: Instabilities and multiple steady states of the thermohaline circulation. In *Oceanic Circulation Models: Combining Data and Dynamics*, D. L. T. Anderson and J. Willebrand (Eds.), Kluwer, 501–511.

Marotzke, J., P. Welander and J. Willebrand, 1988: Instability and multiple steady states in a meridional-plane model of the thermohaline circulation. *Tellus*, **40A**, 162–172.

Marsden, R. F., L. A. Mysak and R. A. Myers, 1991: Evidence for stability enhancement of sea ice in the Greenland and Labrador Seas. *J. Geophys. Res.*, **96**, 4783–4789.

Marshall, H. G., J. C. G. Walker and W. R. Kuhn, 1988: Long-term climate change and the geochemical cycle of carbon. *J. Geophys. Res.*, **93**, 791–801.

Martens, C. S., C. A. Kelley, J. P. Chanton and W. Showers, 1991: Carbon and hydrogen isotopic characterization of methane from wetlands and lakes of the Yukon–Kuskowim Delta, western Alaska. *J. Geophys. Res.*, in press.

Martin, J. H., 1990: Glacial–interglacial CO_2 change: the iron hypothesis. *Paleoceanography*, **5**, 1–13.

Martin, J. H., S. E. Fitzwater and R. M. Gordon, 1990: Iron deficiency limits phytoplankton growth in Antarctic waters. *Global Biogeochem. Cycles*, **4**, 5–12.

Martin, J. H., G. A. Knauer, D. M. Karl and W. W. Broenkow, 1987: VERTEX: carbon cycling in the northeast Pacific. *Deep-Sea Res.*, **34**, 267–285.

Martin, P. J., 1985: Simulation of the mixed layer at OWS November and Papa with several models. *J. Geophys. Res.*, **90**, 903–916.

Mathews, E., and I. Fung, 1987: Methane emissions from natural wetlands: Global distribution, area and environmental characteristics of sources. *Global Biogeochem. Cycles*, **1**, 61–86.

References

Matson, M., C. F. Ropelewski and M. S. Varnadore, 1986: An atlas of satellite-derived Northern Hemisphere snow cover frequency. NOAA Atlas. U.S. Dept. Commerce, 75 pp.

Matson, P. A., and P. M. Vitousek, 1990: Ecosystem approach to a global nitrous oxide budget. *BioScience*, **40**, 667–672.

Matthews, E., 1985: Atlas of archived vegetation, land-use and seasonal albedo data sets. NASA Tech. Memo. 86199, 53 pp.

Matveev, L., 1967: *Physics of the Atmosphere.* Israel Program for Scientific Translations, Jerusalem, 699 pp.

Maykut, G. A., 1978: Energy exchange over young sea ice in the central Arctic. *J. Geophys. Res.*, **83**, 3646–3658.

Maykut, G. A., 1986: The surface heat and mass balance. In *Geophysics of Sea Ice*, N. Untersteiner (Ed.), Plenum Press, 395–464.

Maykut, G. A., and D. K. Perovich, 1987: The role of shortwave radiation in the summer decay of a sea ice cover. *J. Geophys. Res.*, **92**, 7032–7044.

Maykut, G. A., and N. Untersteiner, 1971: Some results from a time dependent, thermodynamic model of sea ice. *J. Geophys. Res.*, **76**, 1550–1575.

McCormick, M. P., P. Hamill, T. J. Pepin, W. P. Chu, T. J. Swissler and L. R. McMaster, 1979: Satellite studies of the stratospheric aerosol. *Bull. Amer. Meteor. Soc.*, **60**, 1038–1046.

McCreary, P. J., and D. L. T. Anderson, 1991: An overview of coupled ocean–atmosphere models of El Niño and the Southern Oscillation. *J. Geophys. Res.*, **96**, 3125–3150.

McFarlane, N. A., 1987: The effect of orographically excited gravity wave drag on the general circulation of the lower stratosphere and troposphere. *J. Atmos. Sci.*, **44**, 1775–1800.

McIntyre, M., and T. Palmer, 1983: Breaking planetary waves in the stratosphere. *Nature*, **305**, 593–600.

McLellan, T., M. E. Martin, J. D. Aber, J. M. Melillo, K. J. Nadelhoffer and B. Dewey, 1991: Comparison of wet chemical and near infrared reflectance measurements of carbon fraction chemistry and nitrogen concentration of forest foliage. *Canadian J. Forest Res.*, in press.

McNulty, S. G., J. D. Aber and R. D. Boone, 1991: Changes in forest floor and foliar chemistry of spruce-fir forests across New England. *Biogeochem.*, submitted.

McNulty, S. G., J. D. Aber, T. M. McLellan and S. M. Katt, 1990: Nitrogen cycling in high elevation forests of the northeastern U.S. in relation to nitrogen deposition. *Ambio*, **19**, 38–40.

McPhee, M. G., 1979: The effect of the oceanic boundary layer on the mean drift of pack ice: Application of a simple model. *J. Phys. Oceanogr.*, **9**, 388–400.

McPhee, M. G., 1980: An analysis of pack ice drift in summer. In *Sea Ice Processes and Models*, R. S. Pritchard (Ed.), Univ. Washington Press, 62–75.

McWilliams, J. C., N. J. Norton, P. R. Gent and D. B. Haidvogel, 1990: A linear balance model of wind-driven, midlatitude ocean circulation. *J. Phys. Oceanogr.*, **20**, 1349–1378.

Meador, W. E., and W. R. Weaver, 1980: Two-stream approximations to radiative transfer in planetary atmospheres: A unified description of existing methods and a new improvement. *J. Atmos. Sci.*, **37**, 630–643.

Mearns, L. O., R. W. Katz and S. H. Schneider, 1984: Extreme high-temperature events: changes in their probabilities with changes in the mean temperature. *J. Climate Appl. Meteor.*, **23**, 1601–1613.

Mearns, L. O., S. H. Schneider, S. L. Thompson and L. R. McDaniel, 1990: Analysis of climate variability in general circulation models: Comparison with observations and changes in variability in 2xCO$_2$ experiments. *J. Geophys. Res.*, **95**, 20469–20490.

Meehl, G. A., 1984: Modeling the earth's climate. *Clim. Change*, **6**, 259–286.

Meehl, G. A., 1989: The coupled ocean–atmosphere modeling problem in the tropical Pacific and Asian monsoon regions. *J. Climate*, **2**, 1146–1163.

Meehl, G. A., 1990a: Development of global coupled ocean–atmosphere general circulation models. *Clim. Dyn.*, **5**, 19–33.

Meehl, G. A., 1990b: Seasonal cycle forcing of El Niño–Southern Oscillation in a global coupled ocean–atmosphere GCM. *J. Climate*, **3**, 72–98.

Meehl, G. A., and W. M. Washington, 1985: Sea surface temperatures computed by a simple ocean mixed layer coupled to an atmospheric GCM. *J. Phys. Oceanogr.*, **15**, 92–104.

Meehl, G. A., and W. M. Washington, 1990: CO$_2$ climate sensitivity and snow–sea-ice albedo parameterization in an atmospheric GCM coupled to a mixed-layer ocean model. *Clim. Change*, **16**, 283–306.

Meehl, G. A., G. W. Branstator and W. M. Washington, 1992a: Tropical Pacific interannual variability and CO$_2$ climate change. *J. Climate*, (in press).

Meehl, G. A., W. M. Washington and T. R. Karl, 1992b: Low-frequency variability and CO$_2$ transient climate change. Part 1: Time-averaged differences. *Clim. Dyn.*, (in press).

Meehl, G. A., W. M. Washington and A. J. Semtner, Jr., 1982: Experiments with a global ocean model driven by observed atmospheric forcing. *J. Phys. Oceanogr.*, **12**, 301–312.

Meentemeyer, V., 1978: Macroclimate and lignin control of litter decomposition rates. *Ecology*, **59**, 465–472.

Melillo, J. M., J. D. Aber and J. M. Muratore, 1982: Nitrogen and lignin control of hardwood leaf litter decomposition dynamics. *Ecology*, **63**, 621–626.

Mellor, G. L., and T. Yamada, 1982: Development of a turbulence closure model for geophysical fluid problems. *Rev. Geophys. Space Phys.*, **20**, 851–875.

Menzel, D. W., and J. H. Ryther, 1960: The annual cycle of primary production in the Sargasso Sea off Bermuda. *Deep-Sea Res.*, **6**, 351–367.

Mercer, J. H., 1978: West Antarctic ice sheet and CO$_2$ greenhouse effect: a threat of disaster. *Nature*, **271**, 321–325.

Mesinger, R., and A. Arakawa, 1976: *Numerical Methods Used in Atmospheric Models*, **1**. GARP Publ. Ser. 17, **WMO**, Geneva, 64 pp.

Meyers, T., and K. T. Paw U, 1986: Testing of a higher-order closure model for modelling airflow within and above plant canopies. *Bound. Layer Meteor.*, **37**, 297–311.

Mikolajewicz, U., and E. Maier-Reimer, 1990: Internal secular variability in an ocean general circulation model. *Clim. Dyn.*, **4**, 145–156.

Miller M., T. Palmer and R. Swinbank, 1989: Parameterization and influence of subgridscale orography in general circulation and numerical prediction models. *Meteor. Atmos. Physics*, **40**, 84–109.

Miller, M. J., A. C. M. Beljaars and T. N. Palmer, 1992: The sensitivity of the ECMWF model to parameterization of evaporation from the tropical oceans. *J. Climate*, **5**, 418–434.

Mintz, Y., 1984. The sensitivity of numerically simulated climates to land-surface boundary conditions. In *The Global Climate*, J. T. Houghton (Ed.), Cambridge University Press, 79–105.

751

References

Mitchell, J. F. B., 1989: The "greenhouse effect" and climate change. *Rev. Geophys.*, **27**, 115–139.

Mitchell, J. F. B., and W. J. Ingram, 1992: Carbon dioxide and climate. Mechanisms of changes in cloud. *J. Climate*, **5**, 5–21.

Mitchell, J. F. B., N. S. Grahame and K. H. Needham, 1988: Climate simulations for 9000 years before present: Seasonal variations and the effect of the Laurentide ice sheet. *J. Geophys. Res.*, **93**, 8283–8303.

Mitchell, J. F. B., C. A. Senior and W. J. Ingram, 1989: CO_2 and climate: A missing feedback? *Nature*, **341**, 132–134.

Mitchell, J. M., Jr., 1961: Recent secular changes of global temperature. *Annals New York Acad. Sciences,* **95**, 235–250.

Miyakoda, K., G. D. Hembree, R. F. Strickler and I. Schulmann, 1972: Cumulative results of extended forecast experiments. I: Model performance for winter cases. *Mon. Wea. Rev.*, **100**, 836–855.

Molina, M. J., and F. S. Rowland, 1974: Stratospheric sink for chlorofluoromethanes: Chlorine atom catalyzed destruction of ozone. *Nature*, **249**, 810.

Möller, F., 1963: On the influence of changes in the CO_2 concentration in air on the radiation balance of the Earth's surface and on the climate. *J. Geophys. Res.*, **68**, 3877–3886.

Monteith, J. L., 1973: *Principles of Environmental Physics.* Edward Arnold Ltd., 242 pp.

Monteith, J. L. (Ed.), 1976: *Vegetation and the Atmosphere, I, II.* Academic Press.

Mooney, H. A., B. G. Drake, R. J. Luxmoore, W. C. Oechel and L. F. Pitelka, 1991: Predicting ecosystem response to elevated CO_2 concentrations. *BioScience*, **41**, 96–104.

Morrison, D. F., 1976: *Multivariate Statistical Methods.* McGraw-Hill, 415 pp.

Mosier, A., D. Schimel, D. Valentine, K. Bronson and W. Parton, 1991: Methane and nitrous oxide fluxes in native, fertilized and cultivated grasslands. *Nature*, **350**, 330–332.

Moura, A. D., and J. Shukla, 1981: On the dynamics of droughts in northeast Brazil: Observations, theory and numerical experiments with a general circulation model. *J. Atmos. Sci.*, **38**, 2653–2675.

Mullen, S. L., 1986: The local balances of vorticity and heat for blocking anticyclones in a spectral general circulation model. *J. Atmos. Sci.*, **43**, 1406–1441.

Munk, W. H., 1950: On the wind-driven ocean circulation. *J. Appl. Meteor.*, **7**, 79–93.

Munk, W. H., and A. M. G. Forbes, 1989: Global ocean warming: An acoustic measure? *J. Phys. Oceanogr.*, **19**, 1765–1778.

Münnich, M., M. Cane and S. Zebiak, 1991: A study of self-excited oscillations of the tropical ocean atmosphere system. Part II: Nonlinear cases. *J. Atmos. Sci.*, **48**, 1238-1248.

Myneni, R. B., G. Asrar and E. T. Kanemasu, 1989: The theory of photon transport in leaf canopies. In *Theory and Applications of Optical Remote Sensing*, G. Asrar (Ed.), Wiley and Sons, 142–204.

Nadelhoffer, K. J., and B. Fry, 1988: Controls of natural ^{15}N and ^{13}C abundances in forest soil organic matter. *Soil Science Soc. American J.*, **52**, 1633–1640.

Najjar, R. G., 1990: Simulations of the phosphorus and oxygen cycles in the world ocean using a general circulation model. Ph.D. Thesis, Princeton University, Princeton, NJ.

Najjar, R. G., J. L. Sarmiento, and J. R. Toggweiler, 1992: Downward transport and fate of organic matter in the ocean: Simulations with a general circulation model. *Global Biochem. Cycles*, **6**, 45–76.

Namias, J., 1969: Seasonal interactions between the North Pacific Ocean and the atmosphere during the 1960's. *Mon. Wea. Rev.*, **97**, 173–192.

Namias, J., 1972: Influence of Northern Hemisphere general circulation on drought in northeast Brazil. *Tellus*, **24**, 4, 336–343.

NCAR Summer Colloquium (1987): Dynamics of low frequency phenomena in the atmosphere. Volumes I to III. National Center for Atmospheric Research, Boulder, CO.

Neelin, J. D., 1989: On the interpretation of the Gill Model. *J. Atmos. Sci.*, **46**, 2466–2468.

Neelin, J. D., 1990: A hybrid coupled general circulation model for El Niño studies. *J. Atmos. Sci.*, **47**, 674–693.

Neelin, J. D., 1991: The slow sea surface temperature mode and the fast-wave limit: Analytic theory for tropical interannual oscillations and experiments in a hybrid coupled model. *J. Atmos. Sci.*, **48**, 584–606.

Neelin, J. D., N., M. Latif, M. Allaart, M. Cane, U. Cubasch, W. Gates, P. Gent, M. Ghil, C. Gordon, N. Lau, C. Mechoso, G. Meehl, J. Oberhuber, S. Philander, P. Schopf, K. Sperber, A. Sterl, T. Tokioka, J. Tribbia and S. Zebiak, 1992: Tropical air–sea interaction in general circulation models. *Climate Dyn.*, **7**, 73–104.

Neftel, A., E. Moor, H. Oeschgeer and B. Stauffer, 1985: Evidence from polar ice cores for the increase in atmospheric CO_2 in the past two centuries. *Nature*, **315**, 45–47.

Neftel, A., H. Oeschger, J. Schwander, B. Stauffer and R. Zumbrunn, 1982: Ice core sample measurements give atmospheric CO_2 content during the past 40,000 years. *Nature*, **295**, 220–223.

Nguyen, B. C., S. Belviso and N. Mihalopoulos, 1988: Dimethyl sulfide production during natural phytoplankton blooms. *Mar. Chem.*, **24**, 133–141.

Niiler, P. P., and W. S. Richardson, 1973: Seasonal variability of the Florida Current. *J. Mar. Res.*, **21**, 144–167.

Nobre, C., P. J. Sellers and J. Shukla, 1991: Amazonian deforestation and regional climate change. *J. Climate*, **4**, 957-988.

Noilhan, J., and S. Planton, 1989: A simple parameterization of land surface processes for meteorological models. *Mon. Wea. Rev.*, **117**, 536–549.

Nordhaus, W. D., and G. W. Yohe, 1983: *Future Paths of Energy and Carbon Dioxide Emissions.* Changing Climate, Report of the Carbon Dioxide Assessment Committee, National Research Council, National Academy Press, Washington, D.C., 496 pp.

Norman, J. M., and P. G. Jarvis, 1975: Photosynthesis in Sitka spruce [*Picea Sitchensis (Bong.) Carr*]: V. Radiation penetration theory and a test case. *J. Appl. Ecol.*, **12**, 839–878.

North, G. R., 1975: Theory of energy-balance climate models. *J. Atmos. Sci.*, **32**, 2033–2043.

North, G. R., R. F. Cahalan and J. A. Coakley, 1981: Energy balance climate models. *Rev. Geophys. Space Phys.*, **19**, 91–121.

North, G. R., J. G. Mengel and D. A. Short, 1983: A simple energy balance model resolving the seasons and the continents: Application to the astronomical theory of the Ice Ages. *J. Geophys. Res.*, **88**, 6576–6586.

NRC (National Research Council), 1990: *Research Strategies for the U.S. Global Change Research Program.* National Academy Press. Washington, D.C.

O'Brien, J. J. (Ed.), 1986: *Advanced Physical Oceanographic Numerical Modeling.* Reidel, 608 pp.

References

Oerlemans, J., 1982: Response of the Antarctic ice sheet to a climatic warming: a model study. *J. Climatol.* **2**, 1–11.

Oerlemans, J., 1989a: A projection of future sea level. *Clim. Change*, **15**, 151–174.

Oerlemans, J. (Ed.), 1989b: *Glacier fluctuations and climatic change.* Kluwer, 417 pp.

Oerlemans, J., and C. J. van der Veen, 1984: *Ice Sheets and Climate.* Reidel, 217 pp.

Oglesby, R. J., 1990: Sensitivity of glaciation to initial snowcover, CO_2, snow albedo, and ocean roughness in the NCAR CCM. *Clim. Dyn.*, **4**, 219–235.

Oglesby, R. J., and E. J. Erickson III, 1989: Soil moisture and the persistence of North American drought. *J. Climate*, **2**, 1362–1380.

Oglesby, R. J., K. A. Maasch and B. Saltzman, 1989: Glacial meltwater cooling of the Gulf of Mexico: GCM implications for Holocene and present-day climates. *Clim. Dyn.*, **3**, 115–133.

Olson, J. S., J. A. Watts and L. J. Allison, 1983: Carbon in live vegetation of major world ecosystems. DOE/NBB Report No. TR004, Oak Ridge National Laboratory, Oak Ridge, TN 37830, 152 pp.

Omstedt, A., and J. Sahlberg, 1977: Some results from a joint Swedish-Finnish sea ice experiment, March 1977. Winter Navigation Research Board, Norrkoping, Sweden, Research Report No. 26.

Oort, A. H., and E. M. Rasmusson, 1971: Atmospheric circulation statistics. NOAA Professional Paper No. 5, U.S. Department of Commerce, Washington, DC.

Oort, A. H., and T. Vonder Haar, 1976: On the observed annual cycle in the ocean atmosphere heat balance over the Northern Hemisphere. *J. Phys. Oceanogr.*, **6**, 781–800.

Oran, E. S., and J. P. Boris, 1987: *Numerical Simulation of Reactive Flow.* Elsevier, 601 pp.

Orszag, S. A., 1970: Transform method for calculation of vector coupled sums: Application to the spectral form of the vorticity equation. *J. Atmos. Sci.*, **27**, 890–895.

Otterman, J., 1985: Bidirectional and hemispheric reflectivities of a bright soil plane and sparse dark canopy. *Int. J. Remote Sens.*, **6**, 897–902.

Overpeck, J. T., L. C. Peterson, N. Kipp, J. Imbrie and D. Rind, 1989: Climate change in the circum-North Atlantic region during the last deglaciation. *Nature*, **338**, 553–557.

Owen, J. A., and C. K. Folland, 1988: Modelling the influence of sea-surface temperatures on tropical rainfall. In *Recent Climatic Change, A Regional Approach*, S. Gregory (Ed.), Bellhaven Press, 141–153.

Palmer, T. N., and C. Brankovic, 1989: The 1988 US drought linked to anomalous sea surface temperature. *Nature*, **338**, 54–57.

Palmer, T. N., and Z.-B. Sun, 1985: A modelling and observational study of the relationship between sea surface temperature in the northwest Atlantic and the atmospheric general circulation. *Quart. J. Roy. Meteor. Soc.*, **111**, 947–975.

Parkinson, C. L., and W. M. Washington, 1979: A large-scale numerical model of sea ice. *J. Geophys. Res.*, **84**, 311–337.

Parkinson, C. L., J. C. Comiso, H. J. Zwally, D. C. Cavalieri, P. Gloersen and W. J. Campbell, 1987: Arctic sea ice, 1973–1976: Satellite passive-microwave observations. NASA SP-489, Washington D.C. 296 pp.

Parsons, T. R., M. Takahashi and B. Hargrave, 1984: *Biological Oceanographic Processes.* Pergamon Press, 330 pp.

Parton, W. J., J. W. B. Stewart and C. V. Cole, 1988: Dynamics of C, N, P and S in grassland soils: A model. *Biogeochem.*, **5**, 109–132

Parton, W. J., D. S. Schimel, C. V. Cole and D. S. Ojima, 1987: Analysis of factors controlling soil organic matter levels in Great Plains grasslands. *Soil Sci. Soc. Amer. J.*, **51**, 1173–1179.

Pastor, J., and W. M. Post, 1986: Influence of climate, soil moisture, and succession on forest carbon and nitrogen cycles. *Biogeochem.*, **2**, 3–28.

Pastor, J., R. H. Gardner, V. H. Dale and W. M. Post, 1987: Successional changes in nitrogen availability as a potential factor contributing to spruce declines in boreal North America. *Canadian J. Forest Res.*, **17**, 1394–1400.

Paterson, W. S. B., 1981: *The Physics of Glaciers.* Pergamon Press, 2nd edn, 380 pp.

Paulson, C. A., 1970: Mathematical representation of wind speed and temperature profiles in the unstable atmospheric surface layer. *J. Appl. Meteor.*, **9**, 857–861.

Pedlosky, J., 1987a: *Geophysical Fluid Dynamics.* 2nd edn, Springer-Verlag, 710 pp.

Pedlosky, J., 1987b: Thermocline theories. In *General Circulation of the Ocean*, H. D. I. Abarbanel and W. R. Young (Eds.), Springer-Verlag, 710 pp.

Peng, L., M.-D. Chou and A. Arking, 1982: Climate studies with a multilayer energy balance model. Part I: Model description and sensitivity to the solar constant. *J. Atmos. Sci.*, **39**, 2639–2656.

Peng, T.-H., and W. S. Broecker, 1991: Dynamic limitations on the Antarctic iron fertilization strategy. *Nature*, **349**, 227–229.

Peng, T.-H., T. Takahashi, W. S. Broecker and J. Olafsson, 1987: Seasonal variability of carbon dioxide, nutrients and oxygen in the northern North Atlantic surface water: Observations and a model. *Tellus*, **39B**, 439–458.

Penner, J. E., C. S. Atherton, J. Dignon, S. J. Ghan, J. J. Walton and S. Hameed, 1991: Tropospheric nitrogen: A three-dimensional study of sources, distributions, and deposition. *J. Geophys. Res.*, **96**, 959–990.

Philander, S. G. H., 1990: *El Niño, La Niña, and the Southern Oscillation.* Academic Press, 293 pp.

Philander, S. G. H., W. Hurlin and A. Siegel, 1987: Simulation of the seasonal cycle of the tropical Pacific ocean. *J. Phys. Oceanogr.*, **17**, 1986–2002.

Philander, S. G. H., T. Yamagata and R. Pacanowski, 1984: Unstable air–sea interactions in the tropics. *J. Atmos. Sci.*, **41**, 604–613.

Philander, S. G. H., N.-C. Lau, R. C. Pacanowski and M. J. Nath, 1989: Two different simulations of Southern Oscillation and El Niño with coupled ocean–atmosphere general circulation models. *Phil. Trans. Roy. Soc. Lond.*, **A329**, 167–178.

Philander, S. G. H., R. Pacanowski, N.-C. Lau and M. J. Nath, 1992: A simulation of the Southern Oscillation with a global atmospheric GCM coupled to a high-resolution, tropical Pacific Ocean GCM. *J. Climate*, **5**, 308–329.

Phillips N. A., 1956: The general circulation of the atmosphere: A numerical experiment. *Quart. J. Roy. Meteor. Soc.*, **82**, 123–164.

Phillips, N. A., 1957a: A map projection system suitable for large-scale numerical weather prediction. *J. Meteor. Soc. Japan*, 75th Anniversary Volume, 262–267.

Phillips, N. A., 1957b: A coordinate system having some special advantages for numerical forecasting. *J. Meteor.*, **14**, 184–185.

References

Phillips, N. A., 1959: An example of nonlinear computational instability. In *The Atmosphere and Sea in Motion*, Rossby Memorial Volume, Rockefeller Institute Press, 501–504.

Phillips, N. A., 1962: Numerical integration of the hydrostatic system of equations with a modified version of the Eliassen finite-difference grid. *Proc. International Symposium on Numerical Weather Prediction*, Tokyo, Meteor. Soc. Japan, 109–120.

Pillsbury, T. E., 1912: The grandest and most mighty terrestrial phenomenon: The Gulf Stream. *The National Geographic Magazine*, **23**, 767–778.

Pinker, R. T., O. E. Thompson and T. F. Eck, 1980: The albedo of a tropical evergreen forest. *Quart. J. Roy. Meteor. Soc.*, **106**, 551–558.

Pitcher, E. J., M. L. Blackmon, G. T. Bates and S. Munoz, 1988: The effect of North Pacific sea surface temperature anomalies on the January climate of a general circulation model. *J. Atmos. Sci.*, **45**, 171–188.

Plass, G. N., 1961: Comments on the influence of carbon dioxide variations on the atmospheric heat balance by L. D. Kaplan. *Tellus*, **13**, 296–300.

Platt, T., and S. Sathyendranath, 1991: Biological production models as elements of coupled, atmosphere–ocean models for climate. *J. Geophys. Res.*, **96**, 2585–2592.

Platt, T., S. Sathyendranath, C. M. Caverhill and M. R. Lewis, 1988: Ocean primary production and available light: further algorithms for remote sensing. *Deep-Sea Res.*, **35**, 855–879.

Platt, T., W. G. Harrison, M. R. Lewis, W. K. W. Li, S. Sathyendranath, R. E. Smith and A. F. Vezina, 1989: Biological production in the oceans: the case for consensus. *Mar. Ecol. Prog. Ser.*, **52**, 77–88.

Platzman, G. W., 1979: The ENIAC computations of 1950 – gateway to numerical weather prediction. *Bull. Amer. Meteor. Soc.*, **60**, 302–312.

Pond, S. L., and G. L. Pickard, 1978: *Introductory Dynamic Oceanography*. Pergamon Press, 241 pp.

Post, W. M., J. Pastor, P. J. Zinke and A. G. Strangenberger, 1985: Global patterns of soil nitrogen storage. *Nature*, **317**, 613–616.

Post, W. M., T-H Peng, W. R. Emanuel, A. W. King, V. H. Dale and D. L. DeAngelis, 1990: The global carbon cycle. *American Scientist*, **78**, 310–326.

Prather, M. J., 1986: Numerical advection by conservation of second-order moments, *J. Geophys. Res.*, **91**, 6671–6681.

Prell, W. L., and J. E. Kutzbach, 1987: Monsoon variability over the past 150,000 years. *J. Geophys. Res.*, **92**, 8411–8425.

Prentice, K., and I. Y. Fung, 1990: The sensitivity of terrestrial carbon storage to climate change. *Nature*, **346**, 48–51.

Press, W. H., and S. A. Teukolsky, 1989: Integrating stiff ordinary differential equations. *Computers in Physics*, May/June, 88–91.

Press, W. H., B. P. Flannery, S. A. Teukolsky, and W. T. Vetterling, 1986: *Numerical Recipes*. Cambridge University Press, 818 pp.

Price, J. F., R. E. Weller and R. Pinkel, 1986: Diurnal cycling: Observations and models of the upper ocean response to diurnal heating, cooling and wind mixing. *J. Geophys. Res.*, **91**, 8411–8427.

Pruppacher, H. R., and J. D. Klett, 1978: *Microphysics of Clouds and Precipitation*. Reidel.

Quinn, W., and V. Neal, 1978: El Niño occurrences over the past four and a half centuries. *J. Geophys. Res.*, **92**, 14449–14461.

Quinn, W., D. Dopf, K. Short and R. W. Kuo-Yang, 1987: Historical trends and statistics of the Southern Oscillation, El Niño, and Indonesian droughts. *Fish. Bull.*, **76**, 663–678.

Ramanathan, V., 1981: The role of ocean–atmosphere interactions in the CO_2 climate problem. *J. Atmos. Sci.*, **38**, 918–930.

Ramanathan, V., and J. A. Coakley, 1978: Climate modeling through radiative-convective models. *Rev. Geophys. Space Phys.*, **16**, 465–489.

Ramanathan, V., and W. Collins, 1991: Thermodynamic regulation of ocean warming by cirrus clouds deduced from observations of the 1987 El Niño. *Nature*, **351**, 27–32.

Ramanathan, V., R. J. Cicerone, H. B. Singh and J. T. Kiehl, 1985: Trace gas trends and their potential role in climate change. *J. Geophys. Res.*, **90**, 5547–5566.

Ramanathan, V., E. J. Pitcher, R. C. Malone and M. L. Blackmon, 1983: The response of a spectral general circulation model to refinements in radiative processes. *J. Atmos. Sci.*, **40**, 605–630.

Ramanathan, V., R. D. Cess, E. F. Harrison, P. Minnis, B. R. Barkstrom, E. Ahmad and D. Hartmann, 1989: Cloud radiative forcing and climate: Results from the earth radiation budget experiment. *Science*, **243**, 57–63.

Ramaroson, R. A., M. Pirre, and D. Cariolle, 1991: A box model for on-line computations of diurnal variations in multidimensional models: Application to the one-dimensional case. *Ann. Geophys.*, submitted.

Randall, D. A., Harshvardhan, D. A. Dazlich, and T. G. Corsetti, 1989: Interaction among radiation, convection, and large-scale dynamics in a general circulation model. *J. Atmos. Sci.*, **46**, 1943–1970.

Ranelli, P. H., and W. D. Hibler III, 1991: Seasonal Arctic sea-ice simulations with a prognostic ice–ocean model. *Annals Glac.*, **15**, 45–53.

Rasch, P. J., and D. L. Williamson, 1991: Sensitivity of a general circulation model climate to the moisture transport formulation. *J. Geophys. Res.*, **96**, 123–137.

Rasmusson, E. M., 1968: Atmospheric water vapor transport and the water balance of North America. II: Large-scale water balance investigations. *Mon. Wea. Rev.*, **96**, 720–734.

Rasmusson, E. M., 1985: El Niño and variations in climate. *Am. Sci.*, **73**. 168–178.

Rasmusson, E. M., and T. H. Carpenter, 1982: Variations in tropical sea surface temperature and surface wind fields associated with the Southern Oscillation/El Niño. *Mon. Wea. Rev.*, **110**, 354–384.

Raupach, M. R., and A. S. Thom, 1981: Turbulence in and above plant canopies. *Ann. Rev. Fluid Mech.*, **13**, 97–129.

Raval, A., and V. Ramanathan, 1989: Observational determination of the greenhouse effect. *Nature*, **342**, 758.

Raymo, M. E., W. F. Ruddiman and P. N. Froelich, 1988: Influence of late Cenozoic mountain building on ocean geochemical cycles. *Geology*, **16**, 649–653.

Raymond, C. F., 1987: How do glaciers surge? A review. *J. Geophys. Res.*, **92B**, 9121–9134.

Redfield, A. C., B. H. Ketchum and F. A. Richards, 1963: The influence of organisms on the composition of seawater. In *The Sea*, Vol. 2, M. N. Hill (Ed.), Interscience, 26–77.

References

Reed, R. J., 1977: The development and status of modern weather prediction. *Bull. Amer. Meteor. Soc.*, **58**, 390–399.

Reed, R. J., and K. E. German, 1965: A contribution to the problem of stratospheric diffusion by large-scale mixing. *Mon. Wea. Rev.*, **93**, 313–321.

Reeh, N., 1985: Was the Greenland ice sheet thinner in the late Wisconsin than now? *Nature*, **317**, 797–799.

Reeh, N., 1989: Dynamics and climatic history of the Greenland ice sheet. In *Quaternary Geology of Canada and Greenland*, R. J. Fulton (Ed.), Geological Survey of Canada, Geology of Canada, No. 1, 793–822.

Reeh, N., 1991: Parameterization of melt rate and surface temperature on the Greenland ice sheet. *Polarforschung*, in press.

Reid, J. L., 1986: On the total geostrophic circulation of the South Pacific Ocean: Flow patterns, tracers and transports. *Prog. Oceanogr.*, **16**, 61 pp.

Reining, P., 1978: *Handbook on Desertification Indicators*. Washington, D.C., American Association for the Advancement of Science.

Revelle, R., and H. E. Suess, 1957: Carbon dioxide exchange between atmosphere and ocean and the question of an increase of atmospheric CO_2 during the past decades. *Tellus*, **9**, 18–27.

Reynolds, R., K. Arpe, C. Gordon, S. Hayes, A. Leetmaa and M. McPhaden, 1989: A comparison of tropical Pacific surface wind analyses. *J. Climate*, **2**, 105–111.

Rhines, P. B., 1977: The dynamics of unsteady currents. In *The Sea*, **VI**, E. Goldberg (Ed.), Wiley, 189–318.

Richardson, L. F., 1922: *Weather Prediction by Numerical Process*. Cambridge University Press, reprinted Dover, 1965, 236 pp.

Richardson, P. L., 1989: Worldwide ship drift distributions identify missing data. *J. Geophys. Res.*, **96**, 6169–6176.

Richtmeyer, R. D., and K. W. Morton, 1967: *Finite Difference Methods for Initial Value Problems*. Wiley-Interscience, 405 pp.

Rigby, F. A., and A. Hanson, 1976: Evolution of a large Arctic pressure ridge. *AIDJEX Bull.*, **34**, 43-71.

Rind, D., 1982: The influence of ground moisture conditions in North America on summer climate as modeled in the GISS GCM. *Mon. Wea. Rev.*, **110**, 1487–1494.

Rind, D., D. Peteet and G. Kukla, 1989: Can Milankovitch orbital variations initiate the growth of ice sheets in a general circulation model? *J. Geophys. Res.*, **94**, 12851–12871.

Rind, D., D. Peteet, W. Broecker, A. McIntyre and W. Ruddiman, 1986: The impact of cold North Atlantic sea surface temperatures on climate; Implications for the Younger Dryas cooling (11–10K). *Clim. Dyn.*, **1**, 3–34.

Robertson, G. P., 1982: Nitrification in forested ecosystems. *Phil. Trans. Roy. Soc. Lond.*, Series B, **296**, 445–457.

Robinson, A. R. (Ed.), 1983: *Eddies in Marine Science*. Springer-Verlag, 609 pp.

Roeckner, E., U. Schlese, J. Biercamp and P. Loewe, 1987: Cloud optical depth feedbacks and climate modelling. *Nature*, **329**, 138–139.

Røed, L. P., and C. K. Cooper, 1986: Open boundary conditions in numerical ocean models. In *Advanced Physical Oceanographic Numerical Modeling*, J. J. O'Brien (Ed.), Reidel, 608 pp.

Roemmich, D., and C. Wunsch, 1985: Two transatlantic sections: Meridional circulation and heat flow in the subtropical North Atlantic Ocean. *Deep-Sea Res.*, **32**, 619–664.

Rood, R. B., 1987: Numerical advection algorithms and their role in atmospheric transport and chemistry models. *Rev. Geophys.*, **25**, 71–100.

Ropelewski, C., and M. Halpert, 1987: Global and regional scale precipitation patterns associated with the El Niño/Southern Oscillation. *Mon. Wea. Rev.*, **114**, 2352–2362.

Rosenzweig, M. L., 1968: Net primary production of terrestrial communities: Prediction from climatological data. *Amer. Naturalist*, **102**, 67–74.

Rossow, W. B., and R. A. Schiffer, 1991: ISCCP cloud data products. *Bull. Amer. Meteor. Soc.*, **72**, 2–20.

Rothrock, D. A., 1975: The energetics of the plastic deformation of pack ice by ridging. *J. Geophys. Res.*, **80**, 4514–4519.

Rotty, R. M., and C. D. Masters, 1985: Carbon dioxide from fossil fuel combustion: trends, resources, and technological implications. In *Atmospheric Carbon Dioxide and the Global Carbon Cycle*, J. Trabalka (Ed.), DOE/ER-0239, U.S. Department of Energy, Washington, D.C., 63–80.

Rowntree, P. R., and J. R. Bolton, 1983: Simulation of the atmospheric response to soil moisture anomalies over Europe. *Quart. J. Roy. Meteor. Soc.*, **109**, 501–526.

Royer, J. F., M. Deque and P. Pestiaux, 1983: Orbital forcing of the inception of the Laurentide ice sheet. *Nature*, **304**, 43–46.

Ruddiman, W. F., and J. E. Kutzbach, 1989: Forcing of late Cenozoic Northern Hemisphere climate by plateau uplift in Southern Asia and the American West. *J. Geophys. Res.*, **94**, 18409–18427.

Ruddiman W. F., W. L. Prell and M. E. Raymo, 1989: Late Cenozoic uplift in Southern Asia and the American West: Rationale for General Circulation Modeling Experiments. *J. Geophys. Res.*, **94**, 18379–18391.

Ruiz Murrieta, J., and J. Saavedra Andaluz, 1990: Peru: Food from the forest. *UNESCO Sources*, **19**, 13.

Running, S. W., and J. C. Coughlan, 1988: A general model of forest ecosystem processes for regional applications: I. Hydrologic balance, canopy gas exchange and primary production processes. *Ecological Modeling*, **42**, 125–154.

Sadourny, R., A. Arakawa and Y. Mintz, 1968: Integration of the nondivergent barotropic vorticity equation with an icosahedral-hexagonal grid for the sphere. *Bull. Amer. Meteor. Soc.*, **96**, 351–356.

Salati, E., 1987: The forest and the hydrological cycle. In *The Geophysiology of Amazonia*, R. E. Dickinson (Ed.), Wiley and Sons, 273–296.

Salby, M., 1992: *Fundamentals of Atmospheric Physics*. Academic Press, (in press).

Salby, M., and R. Garcia, 1990: Dynamical perturbations to the ozone layer. *Physics Today*, **43**, 38–46.

Saltzman, B., 1978: A survey of statistical-dynamical models of the terrestrial climate. *Adv. Geophys.*, **20**, 183–303.

Saltzman, B., 1985: Paleoclimate Modeling. In *Paleoclimate analysis and modeling*, A. D. Hecht (Ed.), Wiley, 445 pp.

Saltzman, B., 1990: Three basic problems of paleoclimatic modeling: a personal perspective and review. *Clim. Dyn.*, **5**, 67–78.

References

Sanberg, J. A. M., and J. Oerlemans, 1983: Modelling of pleistocene European ice sheets: the effect of upslope precipitation. *Geologie en Mijnbouw*, **62**, 267–273.

Sarmiento, J. L., 1986: On the North and Tropical Atlantic heat balance, *J. Geophys. Res.*, **91**, 11677–11689.

Sarmiento, J. L. and J. R. Toggweiler, 1984: A new model for the role of the oceans in determining atmospheric pCO_2. *Nature*, **308**, 621–624.

Sarmiento, J. L., M. J. R. Fasham, R. D. Slater, J. R. Toggweiler and H. Ducklow, 1991: The role of biology in the chemistry of CO_2 in the ocean. *Chemistry of the Greenhouse Effect*, M. Farrell (Ed.), Lewis Publ., (in press).

Sarmiento, J. L., J. C. Orr and U. Siegenthaler, 1992: A perturbation simulation of CO_2 uptake in an ocean general circulation model . *J. Geophys. Res.*, **97**, 3621–3645.

Sarmiento, J. L., J. R. Toggweiler and R. Najjar, 1988: Ocean carbon cycle dynamics and atmospheric pCO_2. *Phil. Trans. Roy. Soc. Lond.*, **A 325**, 3–21.

Sarmiento, J. L., R. D. Slater, M. J. R. Fasham, H. W. Ducklow, J. R. Toggweiler and G. T. Evans, in prep. A seasonal three-dimensional model of nitrogen cycling in the North Atlantic euphotic zone.

Sass, R. L., F. M. Fisher, P. A. Harcombe and F. T. Turner, 1990: Methane production and emission in a Texas rice field. *Global Biogeochem. Cycles*, **4**, 47–68.

Sato, N., P. J. Sellers, D. A. Randall, E. K. Schneider, J. Shukla, J. L. Kinter III, Y.-Y. Hou and E. Albertazzi, 1989a: Effects of implementing the Simple Biosphere Model (SiB) in a general circulation model. *J. Atmos. Sci.*, **46**, 2757–2782.

Sato, N., P. J. Sellers, D. A. Randall, E. K. Schneider, J. Shukla, J. L. Kinter III, Y.-Y. Hou and E. Albertazzi, 1989b: Implementing the Simple Biosphere Model (SiB) in a general circulation Methodology: model and results. NASA Contractor Rep., Washington, D.C., 70 pp. plus references and figures.

Sausen R., R. K. Barthels and K. Hasselmann, 1988: Coupled ocean–atmosphere models with flux correction. *Clim. Dyn.*, **2**, 154–163.

Schimel, J. P., L. E. Jackson and M. K. Firestone, 1989: Spatial and temporal effects on plant–microbial competition for inorganic nitrogen in a California annual grassland. *Soil Biol. Biochem.*, **21**, 1059–1066.

Schimel, D. S., W. J. Parton, T. G. Kittel, D. G. Ojima and C. V. Cole, 1990: Grassland biogeochemistry: Links to atmospheric processes. *Climatic Change*, **17**, 13–25.

Schlesinger, M. E., 1985: Feedback analysis of results from energy balance and radiative-convective models. In *Projecting the Climatic Effects of Increasing Atmospheric Carbon Dioxide,* M. C. MacCracken and F. M. Luther (Eds.), U.S. Department of Energy, Washington, DC, 280–319.

Schlesinger, M. E. (Ed.), 1988a: *Physically Based Modelling and Simulation of Climate and Climatic Change – Part I.* Kluwer, 624 pp.

Schlesinger, M. E., 1988b: Quantitative analysis of feedbacks in climate modeling simulations of CO_2-induced warming. In *Physically Based Modelling and Simulation of Climate and Climatic Change,* M. E. Schlesinger (Ed.), NATO Advanced Study Institute Series, Kluwer, 653–736.

Schlesinger, M. E., and X. Jiang, 1988: The transport of CO_2-induced warming into the ocean: An analysis of simulations by the OSU coupled atmosphere–ocean general circulation model. *Clim. Dyn.*, **3**, 1–17.

Schlesinger, M. E., and J. F. B. Mitchell, 1985: Model projections of equilibrium climatic response to increased carbon dioxide. In *The Potential Climatic Effects of Increasing Carbon Dioxide*, M. C. MacCracken and F. M. Luther (Eds.), DOE/ER–0237, December 1985, Washington, DC, 81–147.

Schlesinger, M. E., and J. F. B. Mitchell, 1987: Climate model simulations of the equilibrium climatic response to increased carbon dioxide. *Rev. Geophys.*, **25**, 760–798.

Schlesinger, M. E., and Z.-C. Zhao, 1989: Seasonal climatic change induced by doubled CO_2 as simulated by the OSU atmospheric GCM/mixed-layer ocean model. *J. Climate*, **2**, 459–495.

Schlesinger, M. E., W. L. Gates and Y.-J. Han, 1985: The role of the ocean in CO_2-induced climate change: Preliminary results from the OSU coupled atmosphere–ocean general circulation model. In *Coupled Ocean–Atmosphere Models*, J.C.J. Nihoul (Ed.), Elsevier Oceanogr. Series, **40**, 447–478.

Schlesinger, W. H., 1991: *Biogeochemistry: An Analysis of Global Change*. Academic Press.

Schmitz, W. J., Jr., and N. G. Hogg, 1983: Exploratory observations of abyssal currents in the South Atlantic near Vema Channel. *J. Mar. Res.*, **41**, 487–510.

Schmitz, W. R. Jr., J. D. Thompson and J. R. Luyten (1992). The Sverdrup circulation for the Atlantic along 24°N. *J. Geophys. Res.* **97**(C5), 7251–7256.

Schneider, E. K., 1984: Response of the annual and zonal mean winds and temperatures to variations in the heat and momentum sources. *J. Atmos. Sci.*, **41**, 1093–1115.

Schneider, S. H., 1987: Climate Modeling. *Sci. Amer.*, **256 (5)**, 72–80.

Schneider, S. H., 1990: *Global Warming, Are We Entering the Greenhouse Century?* Vintage Books, 343 pp.

Schneider, S. H., and R. E. Dickinson, 1974: Climate modeling. *Rev. Geophys. Space Phys.*, **12**, 447–493.

Schneider, S. H., and T. Gal-Chen, 1973: Numerical experiments in climate stability. *J. Geophys. Res.*, **78**, 6182–6194.

Schneider, S. H., and R. Londer, 1984: *The Coevolution of Climate and Life*. Sierra Club Books, 563 pp.

Schneider, S. H., and C. Mass, 1975: Volcanic dust, sunspots, and temperature trends. *Science*, **190**, 741–746.

Schneider, S. H., and S. L. Thompson, 1981: Atmospheric CO_2 and climate: Importance of the transient response. *J. Geophys. Res.*, **86**, 3135–3147.

Schopf, P., and M. A. Cane, 1983: On equatorial dynamics, mixed-layer physics and sea surface temperature. *J. Phys. Oceanogr.*, **13**, 917–935.

Schopf, P., and M. Suarez, 1988: Vacillations in a coupled ocean–atmosphere model. *J. Atmos. Sci.*, **45**, 549–566.

Schopf, P., and M. Suarez, 1990: Ocean wave dynamics and the timescale of ENSO. *J. Phys. Oceanogr.*, **20**, 629–645.

Schulze, E. D., 1989: Air pollution and forest decline in a spruce (*Picea abies*) forest. *Science*, **244**, 776–783.

Schutz, C., and W. L. Gates, 1971: Global climatic data for surface, 800 mb: January. Rand, Santa Monica, **R–915** ARPA, 173 pp.

Schutz, C., and W. L. Gates, 1972: Global climatic data for surface, 800 mb, 400 mb: July. Rand, Santa Monica, **R–1029** ARPA, 180 pp.

References

Schutz, H., A. Holzapfel-Pschorn, R. Conrad, H. Rennenberg and W. Seiler, 1989: A 3-year continuous record on the influence of daytime, season and fertilizer treatment on methane emission rates of an Italian rice paddy. *J. Geophys. Res.*, **94**, 16405–16416.

Seager, R., 1989: Modeling tropical Pacific sea surface temperature: 1970-1987. *J. Phys. Oceanogr.*, **19**, 419–434.

Seager, R., S. Zebiak and M. A. Cane, 1988: A model of the tropical Pacific sea surface temperature climatology. *J. Geophys. Res.*, **93**, 1265–1280.

Sebacher, D. I., R. C. Harriss, K. B. Bartlett, S. M. Sebacher and S. S. Grice, 1986: Atmospheric methane sources: Alaskan tundra bogs, an alpine fen, and a subarctic boreal marsh. *Tellus*, **38**, 1–10.

Seinfeld, J. H., 1986: *Atmospheric Chemistry and Physics of Air Pollution*. John Wiley and Sons, 738 pp.

Sellers, P. J., 1985: Canopy reflectance, photosynthesis and transpiration. *Int. J. Remote Sens.*, **6**, 1335–1372.

Sellers, P. J., and J. L. Dorman, 1987: Testing the simple biosphere model (SiB) with point micrometeorological and biophysical data. *J. Climate Appl. Meteor.*, **26**, 622–651.

Sellers, P. J., Y. Mintz, Y. C. Sud and A. Dalcher, 1986: A simple biosphere model (SiB) for use within general circulation models. *J. Atmos. Sci.*, **43**, 505–531.

Sellers, P. J., J. A. Berry, G. J. Collatz, C. B. Field and F. G. Hall, 1992: Canopy reflectance, photosynthesis and transpiration. III: A reanalysis using improved leaf models and a new canopy integration scheme. *Rem. Sens. Env.*, (in press).

Sellers, P. J., F. G. Hall, G. Asrar, D. E. Strebel and R. E. Murphy, 1988: The First ISLSCP Field Experiment (FIFE). *Bull. Amer. Meteor. Soc.*, **69**, 22–27.

Sellers, P. J., W. J. Shuttleworth, J. L. Dorman, A. Dalcher and J. M. Roberts, 1989: Calibrating the Simple Biosphere Model (SiB) for Amazonia tropical forest using field and remote sensing data. Part I: Average calibration with field data. *J. Appl. Meteor.*, **28**, 727–759.

Sellers, W. D., 1974: A reassessment of the effect of CO_2 variations on a simple global climate model. *J. Appl. Meteor.*, **13**, 831–833.

Semtner, A. J., Jr., 1976: A model for the thermodynamic growth of sea ice in numerical investigations of climate. *J. Phys. Oceanogr.*, **6**, 379–389.

Semtner, A. J., Jr. and R. M. Chervin, 1988: A simulation of the global ocean circulation with resolved eddies. *J. Geophys. Res.*, **93**, 15502–15522.

Semtner, A. J., Jr., and R. M. Chervin, 1992: Ocean general circulation from a global eddy-resolving simulation. *J. Geophys. Res.*, **97**, 5493–5550.

Senior, C. A., 1991: The impact of resolution on the simulation of cloud and radiation. Chapter 4 in *Modelling changes in climate due to enhanced CO_2, the role of atmospheric dynamics, cloud and moisture*, by C. A. Senior, J. F. B. Mitchell, H. Le Treut, and Z-X Li. Climate Res. Tech. Note 13, Hadley Centre, Meteorol. Office, England, 9 pp and 7 figs.

Shackleton, N. J., and J. Imbrie, 1990: The $\delta^{18}O$ spectrum of oceanic deep water over a five-decade band. *Clim. Change*, **16**, 217–230.

Shaw, R. H., and A. R. Pereira, 1982: Aerodynamic roughness of a plant canopy: A numerical experiment. *Agric. Meteor.*, **26**, 51–65.

Shea, D. J., K. E. Trenberth and R. W. Reynolds, 1990: A global monthly sea surface temperature climatology. NCAR Technical Note NCAR/TN-345+STR, 167 pp.

Shoumatoff, A., 1990: *The World Is Burning: Murder in the Rain Forest*. Little, Brown and Co.

Shugart, H. H., and D. C. West, 1977: Development of an Appalachian deciduous forest succession model and its application to the impact of chestnut blight. *J. Environmental Management*, **5**, 161–179.

Shugart, H. H., T. R. Crow and J. M. Hett, 1973: Forest succession models: a rationale and methodology for modeling forest succession over large regions. *For. Sci.*, **19**, 203–212.

Shukla, J., and Y. Mintz, 1982: Influence of land-surface evapotranspiration on the earth's climate. *Science*, **215**, 1498–1501.

Shukla, J., and J. Wallace, 1983: Numerical simulation of the atmospheric response to equatorial Pacific sea surface temperature anomalies. *J. Atmos. Sci*, **40**, 1613–1630.

Shukla, J. C. Nobre and P. J. Sellers, 1990: Amazon deforestation and climate change. *Science*, **247**, 1322–1325.

Shuttleworth, W. J., J. H. C. Gash, C. R. Lloyd, C. J. Moore, J. Roberts, A. de O. Marques Filho, G. Fisch, V. de P. Silva Filho, M. de N. G. Ribeiro, L. C. B. Molion, L. D. de A. Sá, J. C. A. Nobre, O. M. R. Cabral, S. R. Patel and J. C. De Moraes, 1984a: Eddy correlation measurements of energy partition for Amazonian forest. *Quart. J. Roy. Meteor. Soc.*, **110**, 1143–1163.

Shuttleworth, W. J., J. H. C. Gash, C. R. Lloyd, C. J. Moore, J. Roberts, A. de O. Marques Filho, G. Fisch, V. de P. Silva Filho, M. de N. G. Ribeiro, L. C. B. Molion, L. D. de A. Sá, J. C. A. Nobre, O. M. R. Cabral, S. R. Patel and J. C. De Moraes, 1984b: Observations of radiation exchange above and below Amazonian forest. *Quart. J. Roy. Meteor. Soc.*, **110**, 1163–1169.

Siegenthaler, U., and H. Oeschger, 1987: Biospheric CO_2 emissions during the past 200 years reconstructed by deconvolution of ice core data. *Tellus*, **39B**, 140–154.

Siegenthaler, U., H. Friedli, H. Loetscher, E. Moor, A. Neftel, H. Oeschger and B. Stauffer, 1988: Stable-isotope ratios and concentration of CO_2 in air from polar ice cores. *Annals of Glaciology*, **10**, 1–6.

Silberman, I. S., 1954: Planetary waves in the atmosphere. *J. Meteor.*, **11**, 27–34.

Simmons, A. J., 1990: Studies of increased horizontal and vertical resolution. In *Ten Years of Medium Range Weather Forecasting*, Seminar Proc., ECMWF, Reading, England.

Simmons, A. J., and B. J. Hoskins, 1978: The life cycles of some nonlinear baroclinic waves. *J. Atmos. Sci.*, **35**, 414–432.

Simpson, G. C., 1928: Further studies in terrestrial radiation. *Mem. Roy. Meteorol. Soc.*, **III**, 1–26.

Simpson, J. J., and C. A. Paulson, 1979: Mid-ocean observations of atmospheric radiation. *Quart. J. Roy. Meteor. Soc.*, **105**, 487–502.

Slingo, J. M., 1987: The development and verification of a cloud prediction scheme for the ECMWF model. *Quart. J. Roy. Meteor. Soc.*, **113**, 899–927.

Slotnick, D. L., 1967: Unconventional systems. *Computer Design*, December, 49–52.

SMIC Report, 1971: *Study of Man's Impact on Climate*. MIT Press, 308 pp.

Smagorinsky, J., 1960: On the numerical prediction of large scale condensation by numerical models. *Geophys. Monogr.* **5**, 71–78.

Smagorinsky, J., S. Manabe and J. L. Holloway, 1965: Numerical results from a nine-level general circulation model of the atmosphere. *Mon. Wea. Rev.*, **93**, 727–768.

References

Smith, L. D., and T. H. vonder Haar, 1991: Clouds–radiation interactions in a general circulation model: impact upon the planetary radiation balance. *J. Geophys. Res.*, **96**, 893–914.

Smith, R. N. B., 1990: A scheme for predicting layer clouds and their water content in a general circulation model. *Quart. J. Roy. Meteor. Soc.*, **116**, 435–460.

Smolarkiewicz, P. K., 1984: A fully multidimensional positive definite advection transport algorithm with small implicit diffusion. *J. Comput. Phys.*, **54**, 325–362.

Smolarkiewicz, P. K., and W. W. Grabowski, 1990: The multidimensional positive definite advection transport algorithm: Nonoscillatory option. *J. Comput. Phys.*, **86**, 355–375.

Solar Energy Research Institute, 1981: *Solar Radiation Energy Resources Atlas of the United States.* U. S. Govt. Printing Ofc., Washington, D.C., SERI/SP-642-1037.

Solomon, A. M., 1986: Transient response of forests to CO_2-induced climate change: Simulation modeling experiments in eastern North America. *Oecologia*, **68**, 567–579.

Somerville, R., 1987: The predictability of weather and climate. *Clim. Change*, **11**, 239–246.

Spall, M. A., and W. R. Holland, 1991: A nested primitive equation model for oceanic applications. *J. Phys. Oceanogr.*, **21**, 205–220.

Spelman, M. J., and S. Manabe, 1984: Influence of oceanic heat transport upon the sensitivity of a model climate. *J. Geophys. Res.*, **89**, 571–586.

Sperber, K. R., S. Hameed, W. L. Gates and G. L. Potter, 1987: Southern Oscillations simulated in a global climate model. *Nature*, **329**, 140–142.

Spitzer, W. S., and W. J. Jenkins, 1989: Rates of vertical mixing, gas exchange and new production: Estimates from seasonal gas cycles in the upper ocean near Bermuda. *J. Mar. Res.*, **47**, 169–196.

Staniforth A., and J. Côté, 1991: Semi-Lagrangian integration schemes for atmospheric models – A review. *Mon. Wea. Rev.*, **119**, 2206-2223.

Starius, G., 1980: On composite mesh difference methods for hyperbolic difference equations. *Numer. Math.*, **35**, 241–255.

Stephens, G. L., and P. J. Webster, 1981: Clouds and climate: sensitivity of simple systems. *J. Atmos. Sci.*, **38**, 235–247.

Steudler, P. A., R. D. Bowden, J. M. Melillo and J. D. Aber, 1989: Influence of nitrogen fertilization on methane uptake in temperature forest soils. *Nature*, **341**, 314–316.

Stewart, J. B., and A. S. Thom, 1973: Energy budgets in a pine forest. *Quart. J. Roy. Meteor. Soc.*, **99**, 154–170.

Stoel, T. B. Jr., A. S. Miller and B. Milroy, 1980: *Fluorocarbon Regulation: An International Comparison.* Lexington Books, D. C. Heath and Co.

Stolarski, R. S., and R. J. Cicerone, 1974: Stratospheric chlorine: A possible sink for ozone. *Can. J. Chem.*, **52**, 1610–1615.

Stommel, H., 1948: The westward intensification of wind driven ocean currents. *Trans. Amer. Geophys. U.*, **29**, 202–236.

Stommel, H., 1961: Thermohaline convection with two stable regimes of flow. *Tellus*, **13**, 224–228.

Stommel, H., 1965: *The Gulf Stream.* University of California Press, 248 pp.

Stommel, H., and A. B. Arons, 1960: On the abyssal circulation of the worlds oceans. I. Stationary planetary flow patterns on a sphere. *Deep-Sea Res.*, **6**, 140–154.

Stommel, H., and E. Stommel, 1983: *Volcano Weather: The Story of 1816, The Year Without a Summer.* Seven Seas Press, 177 pp.

Stommel, H., P. P. Niiler and D. Anati, 1978: Dynamic topography and recirculation of the North Atlantic. *J. Mar. Res.,* **36,** 449–468.

Stouffer, R. J., S. Manabe and K. Bryan, 1989: Interhemispheric asymmetry in climate response to a gradual increase of CO_2. *Nature,* **342,** 660–662.

Street-Perrott, F. A., J. F. B. Mitchell, D. S. Marchand and J. S. Brunner, 1990: Milankovitch and albedo forcing of the tropical monsoons: a comparison of geological evidence and numerical simulations for 9000 yr BP. *Phil. Trans. Roy. Soc. Edinburgh, Earth Sci.,* **81,** 407–427.

Stull, R. B., 1988: *An Introduction to Planetary Boundary Layer Meteorology.* Kluwer, 666 pp.

Stumm, W., and J. J. Morgan, 1981: *Aquatic Chemistry.* John Wiley and Sons, 780 pp.

Suarez, M., and P. Schopf, 1988: A delayed action oscillator for ENSO. *J. Atmos. Sci.,* **45,** 3283–3287.

Suarez, M. J., A. Arakawa and D. A. Randall, 1983: The parameterization of the planetary boundary layer in the UCLA general circulation model: formulation and results. *Mon. Wea. Rev.,* **111,** 2224–2243.

Sud, Y. C., and M. J. Fennessy, 1982: A study of the influence of surface albedo on July circulation in semi-arid regions using the GLAS GCM. *J. Climate,* **2,** 105–125.

Sud, Y. C., J. Shukla and Y. Mintz, 1988: Influence of land-surface roughness on atmospheric circulation and precipitation: A sensitivity study with a General Circulation Model. *J. Appl. Meteor.,* **27,** 1036–1054.

Sugimura, Y., and Y. Suzuki, 1988: A high-temperature catalytic oxidation method for the determination of nonvolatile dissolved organic carbon in seawater by direct injection of a liquid sample. *Mar. Chem.,* **24,** 105–131.

Sun, S. F., 1982: Moisture and heat transport in a soil layer forced by atmospheric conditions. M.S. thesis, Dept. Civil Engineering, U. Connecticut, 72 pp.

Sundqvist, H., 1979: Vertical coordinates and related discretization. *Numerical Methods Used in Atmospheric Models,* GARP Publication Series No. 17, WMO, Geneva, 1–50.

Suzuki, Y. Y., Sugimura and T. Itoh, 1985: A catalytic oxidation method for the determination of total nitrogen dissolved in seawater. *Mar. Chem.,* **16,** 83–97.

Sverdrup, H. V., 1947: Wind-driven currents in a baroclinic ocean, with application to the equatorial currents of the eastern Pacific. *Proc. Nat. Acad. Sci. U.S.A.,* **33,** 318–326.

Taft, B. A., and M. McPhaden, 1988: Diurnal cycle of sea-surface temperature in the western tropical Pacific. Air–sea interaction in tropical western Pacific. *Proc. US-PRC Internat. TOGA Symposium,* China Ocean Press, Beijing.

Tai, C. K., 1990: Estimating the surface transport of meandering oceanic jet streams from satellite altimetry: Surface transport estimates for the Gulf Stream and Kuroshio Extension. *J. Phys. Oceanogr.,* **20,** 860–879.

Takacs, L. L., 1988: Effects of using *a posteriori* methods for the conservation of integral invariants. *Mon. Wea. Rev.,* **116,** 525–545.

Takahashi, T., W. S. Broecker and S. Langer, 1985: Redfield ratio estimates based on chemical data from isopycnal surfaces. *J. Geophys. Res.,* **90,** 6907–6924.

References

Talley, L. D., 1984: Meridional heat transport in the Pacific Ocean. *J. Phys. Oceanogr.*, **14**, 231–241.

Tans, P. P., I. Y. Fung and T. Takahashi, 1990: Observational constraints on the global atmospheric CO_2 budget. *Science*, **247**, 1431–1438.

Thom, A. S., 1971: Momentum absorption by vegetation. *Quart. J. Roy. Meteor. Soc.*, **97**, 414–418.

Thom, A. S., and H. R. Oliver, 1977: On Penman's equation for estimating regional evaporation. *Quart. J. Roy. Meteor. Soc.*, **103**, 345–357.

Thompson, P. D., 1978: The mathematics of meteorology. *Mathematics Today, Twelve Informal Essays*, L. A. Steen (Ed.), Springer-Verlag, 125–152.

Thompson, P. D., 1983: A history of numerical weather prediction in the United States. *Bull. Amer. Meteor. Soc.*, **64**, 755–769.

Thompson, S. L., and S. H. Schneider, 1979: A seasonal zonal energy balance climate model with an interactive lower layer. *J. Geophys. Res.*, **84**, 2401–2414.

Thorndike, A. S., and R. Colony, 1982: Sea ice motion in response to geostrophic winds. *J. Geophys. Res.*, **87**, 5845–5852.

Thorndike, A. S., D. A. Rothrock, G. A. Maykut and R. Colony, 1975: The thickness distribution of sea ice. *J. Geophys. Res.*, **80**, 4501–4513.

Thual, O., and J. C. McWilliams, 1991: The catastrophe structure of thermohaline convection in a two-dimensional fluid model and a comparison with low-order box models. *Geophys. Astrophys. Fluid Dyn.*, in press.

Tibaldi S., and F. Molteni, 1988: On the operational predictability of blocking. *ECMWF Seminar on the Nature and Prediction of Extra-tropical Weather Systems*, 7–11 Sept. 1987, ECMWF, 329–371.

Tiedtke, M., 1984: The effect of penetrative cumulus convection on the large-scale flow in a general circulation model. *Beitr. Phys. Atmos.*, **57**, 216–239.

Tiedtke, M., 1988: Parameterization of cumulus convection in large-scale models. In *Physically Based Modelling and Simulation of Climate and Climatic Change – Part I.*, M. E. Schlesinger (Ed.), Kluwer, 375–431.

Tiedtke, M., 1989: A comprehensive mass flux scheme for cumulus parameterization in large-scale models. *Mon. Wea. Rev.*, **117**, 1779–1800.

Tiedtke, M., W. A. Heckley and J. Slingo, 1988: Tropical forecasting at ECMWF: The influence of physical parameterization on the mean structure of forecasts and analysis. *Quart. J. Roy. Meteor. Soc.*, **114**, 639–664.

Tissue, D. L., and W. C. Oechel, 1987: Physiological response of *Eriphorum vaginatum* to field elevated CO_2 and temperature in the Alaskan tussock tundra. *Ecology*, **68**, 401–410.

Toggweiler, J. R., 1989: Is the downward dissolved organic matter (DOM) flux important in carbon transport? In *Productivity of the Ocean: Past and Present*, W. H. Berger, V. Smetacek, and D. Wefer (Eds.), Dahlem Workshop Report, John Wiley and Sons, 65–84.

Toggweiler, J. R., K. Dixon and K. Bryan, 1989a: Simulations of radiocarbon in a coarse-resolution world ocean model, I. Steady state prebomb distributions. *J. Geophys. Res.*, **94**, 8217–8242.

Toggweiler, J. R., K. Dixon and K. Bryan, 1989b: Simulations of radiocarbon in a coarse-resolution world ocean model, II. Distributions of bomb-produced ^{14}C. *J. Geophys. Res.*, **94**, 8243–8264.

Tremback, C. J., J. Powell, W. Cotton and R. A. Pielke, 1987: The foreward-in-time upstream advection scheme: Extension to higher orders. *Mon. Wea. Rev.*, **115**, 540–555.

Trenberth, K. E., 1979: Mean annual poleward energy transports by the ocean in the Southern Hemisphere. *Dyn. Atmos. Oceans*, **4**, 57–64.

Trenberth, K. E., 1981: Observed Southern Hemisphere eddy statistics at 500 mb: Frequency and spatial dependence. *J. Atmos. Sci.*, **38**, 2585–2605.

Trenberth, K. E., 1990: Recent observed interdecadal climate changes in the Northern Hemisphere. *Bull. Amer. Meteor. Soc.*, **71**, 988–993.

Trenberth, K. E., and J. G. Olson, 1988: ECMWF global analysis 1979–86: Circulation statistics and data evaluation. NCAR Technical Note, NCAR/TN-300+STR, 94 pp plus 12 fiche.

Trenberth, K. E., G. W. Branstator and P. A. Arkin, 1988: Origins of the 1988 North American drought. *Science*, **242**, 1640–1645.

Trenberth, K. E., W. G. Large and J. G. Olson, 1989: The effective drag coefficient for evaluating wind stress over the oceans. *J. Climate*, **2**, 1507–1516.

Trenberth, K. E., W. G. Large and J. G. Olson, 1990: The mean annual cycle in global ocean wind stress. *J. Phys. Oceanogr.*, **20**, 1742–1760.

Tucker, C. J., and L. D. Miller, 1977: Soil spectra contributions to grass canopy spectral reflectance. *Photogrammetric Engineering and Remote Sensing*, **43**, 721–726.

Tucker, C. J., J. R. G. Townshend and T. E. Goff, 1985: African land-cover classification using satellite data. *Science*, **227**, 369–375.

Tucker, C. J., I. Y. Fung, C. D. Keeling and R. H. Gammon, 1986: Relationship between atmospheric CO_2 variations and a satellite-derived vegetation index. *Nature*, **319**, 195–199.

Tully, J. P., and L. F. Giovando, 1963: Seasonal temperature structure in the eastern subarctic Pacific Ocean. *Roy. Soc. Can. Spec. Publ.*, **5**, M. J. Dunbar (Ed.), 10–36.

Turco, R. P., 1985: The photochemistry of the stratosphere. Chapter 3, in *The Photochemistry of Atmospheres: Earth, The Other Planets, and Comets*, J. Levine (Ed.), Academic Press, 77–128.

Turco, R. P., and R. C. Whitten, 1974: A comparison of several computational techniques for solving some common aeronomic problems. *J. Geophys. Res.*, **79**, 3179–3185.

Turco, R. P., R. C. Whitten and O. B. Toon, 1982: Stratospheric aerosols: Observation and theory. *Rev. Geophys. Space Phys.*, **20**, 233–279.

Turner, S., G. Malin and P. S. Liss, 1988: The seasonal variation of dimethyl sulfide and dimethylsulfoniopropionate concentrations in nearshore waters. *Limnol. Oceanogr.*, **33**, 364–375.

Twomey, S., 1977: *Atmospheric Aerosols*. Elsevier.

Twomey, S. A., M. Piepgrass and T. L. Wolfe, 1984: An assessment of the impact of pollution on global cloud albedo. *Tellus*, **36B**, 356–366.

Tynall, J., 1861: On the absorption and radiation of heat by gases and vapours, and on the physical connexion of radiation, absorption and conduction. *Philosophical Magazine and J. Science*, Series 4, **22 (146)**, 169–194 and **22 (147)**, 273–285.

U.N. Department of Economic and Social Affairs, 1976: *World Energy Supplies, 1950–1974*. United Nations, New York.

References

U.S. Congress, 1990: Senate Bill 169 of the 101st Congress. U.S. Government Printing Office, Washington, D.C.

U.S. Standard Atmosphere, 1976: National Oceanic and Atmospheric Administration, National Aeronautics and Space Administration, United States Air Force. U.S. Government Printing Office, NOAA-S/T 76-1562, Washington, D.C., 228 pp.

UNCOD (United Nations Conference on Desertification), 1977: *Desertification: Its Causes and Consequences*. Compiled and edited by Secretariat of United Nations Conference on Desertification, Nairobi, Kenya, 29 August–9 September 1977. Pergamon Press.

Untersteiner, N., 1986: *The Geophysics of Sea Ice*. Plenum Press, 1196 pp.

Vairavamurthy, A., M. O. Andreae and R. L. Iverson, 1985: Biosynthesis of dimethyl sulfide and dimethylpropiothetin by *Hymenomonas carterae* in relation to sulfur source and salinity variations. *Limnol. Oceanogr.*, **30**, 59–70.

Van der Veen, C. J., 1987: Longitudinal stresses and basal sliding: a comparative study. In *Dynamics of the West Antarctic Ice Sheet*, C. J. van der Veen and J. Oerlemans (Eds.), Reidel, 223–248.

Van der Veen, C. J., 1991: State of balance of the cryosphere. *Rev. Geophys.*, **29**, 433–455.

Veronis, G., 1973: Large scale ocean circulation. *Advances in Applied Mechanics*, Vol. 13, C.-S. Yih, (Ed.), Academic Press, 1–92.

Vinnichenko, N. K., 1970: The kinetic energy spectrum in the free atmosphere – 1 second to 5 years. *Tellus*, **22**, 158–166.

Vinnikov, K. Y., and I. B. Yeserkepova, 1991: Soil moisture: empirical data and model results. *J. Climate*, **4**, 66–79.

Volk, T., and M. I. Hoffert, 1985: Ocean carbon pumps: analysis of relative strengths and efficiencies in ocean-driven atmospheric pCO_2 changes. In *Carbon Dioxide and the Carbon Cycle, Archean to Present*, E. T. Sundquist and W. S. Broecker (Eds.), Geophys. Monogr. Ser., **32**, 99–110, AGU, Washington, D.C.

Vonder Haar, T. H., and A. H. Oort, 1973: New estimate of annual poleward energy transport by northern hemisphere oceans. *J. Phys. Oceanogr.*, **3**, 169–172.

Waggoner, P. E., and W. E. Reifsnyder, 1968: Simulation of temperature, humidity, and evaporation profiles in a leaf canopy. *J. Appl. Meteor.*, **7**, 400–409.

Walker, J.C.G., C. Klein, M. Schidlowski, J. W. Schopf, D. J. Stevenson and M. R. Walter, 1983: Environmental evolution of the Archean–early Proterozoic earth. In *Earth's Earliest Biosphere*, J. W. Schopf (Ed.), Princeton Univ. Press.

Wallace J., and P. Hobbs, 1977: *Atmospheric Science: An Introductory Survey*. Academic Press, NY, 467 pp.

Wallace, J. M., and M. L. Blackmon, 1983: Observations of low-frequency atmospheric variability. In *Large-Scale Dynamical Processes in the Atmosphere*, Academic Press, B. J. Hoskins and R. P. Pearce (Eds.), 55–94.

Wallace, J. M., and Q.-R. Jiang, 1987: On the observed structure of the interannual variability of the atmosphere/ocean climate system. In *Atmospheric and Oceanic Variability*. H. Cattle (Ed.), Roy Meteor. Soc., pp.17–43.

Wallace, J. M., and N.-C. Lau, 1985: On the role of barotropic energy conversions in the general circulation. *Adv. Geophys.*, **28A**, 33–74.

Wallace, J. M., C. Smith and Q.-R. Jiang, 1990: Spatial patterns of atmosphere/ocean interaction in the northern winter. *J. Climate*, **3**, 990–998.

Walsh, J. E., W. D. Hibler III and B. Ross, 1985: Numerical simulation of northern hemispheric sea ice variability, 1951–1980. *J. Geophys. Res.*, **90**, 4847–4865.

Walsh, J. E., D. R. Tucek and M. R. Peterson, 1982: Seasonal snow cover and short-term climatic fluctuations over the United States. *Mon. Wea. Rev.*, **110**, 1474–1485.

Walton, J. J., M. C. McCracken and S. J. Ghan, 1988: A global-scale Lagrangian trace species model of transport, transformation and removal processes, *J. Geophys. Res.*, **93**, 8339–8354.

Wang, W.-C., M. P. Dudek, X.-Z. Liang and J. T. Kiehl, 1991: Inadequacy of effective CO_2 as a proxy in simulating the greenhouse effect of other radiatively active gases. *Nature*, **350**, 573–577.

Wanninkhof, R., 1992: Relationship between wind speed and gas exchange over the ocean. *J. Geophys. Res.*, **97**, 7373–7382.

Wanninkhof, R., J. R. Ledwell and W. S. Broecker, 1987: Gas exchange on Mono Lake and Crowley Lake, California. *J. Geophys. Res.*, **92**, 14567–14580.

Waring, R. H., A. J. S. McDonald, S. Larsson, T. Ericsson, A. Wiren, E. Arwidsson, A. Ericsson and T. Tohammar, 1985: Differences in chemical composition of plants grown at constant relative growth rates with stable mineral nutrition. *Oecologia*, **66**, 127–137.

Warren, B. A., and C. Wunsch (Eds.), 1981: *Evolution of Physical Oceanography.* MIT Press, 623 pp.

Warren, S. G., and S. H. Schneider, 1979: Seasonal simulation as a test for uncertainties in the parameterizations of a Budyko–Sellers zonal climate model. *J. Atmos. Sci.*, **36**, 1377–1391.

Warren, S. G., C. J. Hahn, J. London, R. M. Chervin and R. L. Jenne, 1986: Global distribution of total cloud cover and cloud type amounts over land. U. S. Dept. Energy (DOE/ER/60085 H1) and National Center for Atmospheric Research (Technical Note NCAR/TN 273+STR), 229 pp.

Washington, W. M., and G. A. Meehl, 1983: General circulation model experiments on the climatic effects due to a doubling and quadrupling of carbon dioxide concentration. *J. Geophys. Res.*, **88**, 6600-6610.

Washington, W. M., and G. A. Meehl, 1984: Seasonal cycle experiment on the climate sensitivity due to a doubling of CO_2 with an atmospheric general circulation model coupled to a simple mixed-layer ocean model. *J. Geophys. Res.*, **89**, 9475–9503.

Washington, W. M., and G. A. Meehl, 1986: General circulation model CO_2 sensitivity experiments: Snow–sea-ice albedo parameterization and globally averaged surface air temperature. *Clim. Change*, **8**, 231–241.

Washington, W. M., and G. A. Meehl, 1989: Climate sensitivity due to increased CO_2: experiments with a coupled atmosphere and ocean general circulation model. *Clim. Dyn.*, **4**, 1–38.

Washington, W. M., and C. L. Parkinson, 1986: *An Introduction to Three-Dimensional Climate Modeling.* University Science Books and Oxford University Press, 422 pp.

Washington, W. M., A. J. Semtner, C. Parkinson and L. Morrison, 1976. On the development of a seasonal change sea-ice model. *J. Phys. Oceanogr.*, **61**, 679–685.

Washington, W. M., A. J. Semtner, Jr., G. A. Meehl, D. J. Knight and T. A. Mayer, 1980: A general circulation experiment with a coupled atmosphere, ocean and sea ice model. *J. Phys. Oceanogr.*, **10**, 1887–1908.

References

Watson, A. J., R. C. Upstill-Goddard and P. J. Liss, 1991: Air–sea gas exchange in rough and stormy seas measured by a dual-tracer technique. *Nature*, **349**, 145–147.

Watson, R. T., et al., 1988: *Report of the International Ozone Trends Panel*, Chapters 10 and 11, WMO Report No. 18.

Wayne, R. P., 1985: *Chemistry of Atmospheres*. Clarendon Press.

Weare, B., 1986: A comparison of shallow water model results for three estimates of a composite El Niño forcing. *J. Atmos. Sci.*, **43**, 162–170.

Weare, B., T. Strub and M. Samuel, 1981: Annual mean surface heat fluxes in the tropical Pacific Ocean. *J. Phys. Oceanogr.*, **11**, 705–717.

Webster, P., 1983: In *Large-Scale Dynamical Processes in the Atmosphere*, B. Hoskins and R. Pierce (Eds.), Academic Press, 235–276.

Webster, P. J., and G. L. Stephens, 1980: Tropical upper tropospheric extended cloud: Inferences from winter MONEX. *J. Atmos. Sci.*, **37**, 1521–1541.

Weeks, W. F., and S. F. Ackley, 1986: The growth, structure, and properties of sea ice. In *The Geophysics of Sea Ice*, N. Untersteiner (Ed.), Plenum Press, 9–164.

Weertman J., 1976: Milankovitch solar radiation variations and ice age ice sheet sizes. *Nature*, **261**, 17–20.

Weidick A., 1985: Review of glacier changes in West Greenland. *Z. Gletscherk. Glazialgeol.*, **21**, 301–309.

Weiss, R. F., J. L. Bullister, R. M. Gammon and M. J. Warner, 1985: Atmospheric chlorofluoromethanes in the deep equatorial Atlantic. *Nature*, **314**, 608–610.

Welander, P., 1986: Thermohaline effects in the ocean circulation and related simple models. In *Large-scale Transport Processes in Oceans and Atmosphere*, J. Willebrand and D. L. T. Anderson (Eds.), Reidel, 163–200.

Wenk, T., 1985: Einflüsse der Ozeanzirkulation und der marinen biologie auf die atmosphärische CO_2-konzentration. Ph.D. thesis, University of Bern, Switzerland.

Wessman, C. A., J. D. Aber, D. L. Peterson and J. M. Melillo, 1988a: Foliar analysis using near infrared spectroscopy. *Canadian J. Forest Res.*, **18**, 6–11.

Wessman, C. A., J. D. Aber, D. L. Peterson and J. M. Melillo, 1988b: Remote sensing of canopy chemistry and nitrogen cycling in temperate forest ecosystems. *Nature*, **335**, 154–156.

Wetherald, R. T., and S. Manabe, 1975: The effects of changing the solar constant on the climate of a general circulation model. *J. Atmos. Sci.*, **32**, 2044–2059.

Wetherald, R. T., and S. Manabe, 1986: An investigation of cloud cover change in response to thermal forcing. *Clim. Change*, **10**, 11–42.

Whillans I. M., 1981: Reaction of the accumulation zone portions of glaciers to climatic change. *J. Geophys. Res.*, **86C**, 4274–4282.

Whittaker, R. H., 1975: *Communities and ecosystems*. MacMillan, 385 pp.

Wigley, T. M. L., 1989: Possible climate change due to SO_2-derived cloud nuclei. *Nature*, **339**, 365–367.

Wigley, T. M. L., and S. C. B. Raper, 1990: Natural variability of the climate system and detection of the greenhouse effect. *Nature*, **344**, 324–327.

Williamson, D. L., 1968: Integration of the barotropic vorticity equation on a spherical geodesic grid. *Tellus*, **20**, 4, 642–653.

Williamson, D. L., 1971: A comparison of first- and second-order difference approximations over a spherical geodesic grid. *J. Comp. Phys.*, **7**, 301–309.

Williamson, D. L., 1979: Difference approximations for fluid flow on a sphere. Chapter 2 in *Numerical Methods used in Atmospheric Models*, GARP Publ. Ser. No. 17, WMO, Geneva, Switzerland, 51–120.

Williamson, D. L., 1988: The effect of vertical finite difference approximations on simulations with the NCAR Community Climate Model. *J. Climate*, **1**, 40–58.

Williamson, D. L., 1990: Semi-Lagrangian moisture transport in the NMC spectral model. *Tellus*, **42A**, 413–428.

Williamson, D. L., and P. J. Rasch, 1989: Two-dimensional semi-Lagrangian transport with shape-preserving interpolation. *Mon. Wea. Rev.*, **117**, 102–129.

Williamson, D. L., J. B. Drake, J. J. Hack, R. Jakob and P. N. Swarztrauber, 1992: A standard test set for numerical approximations to the shallow water equations on the sphere. *J. Comp. Phys.*, submitted.

Wilson, C. A., and J. F. B. Mitchell, 1986: Diurnal variation and cloud in a general circulation model. *Quart. J. Roy. Meteor. Soc.*, **112**, 347–369.

Wilson, C. A., and J. F. B. Mitchell, 1987a: A doubled CO_2 climate sensitivity experiment with a global climate model including a simple ocean. *J. Geophys. Res.*, **92**, 13315–13343.

Wilson, C. A., and J. F. B. Mitchell, 1987b: Simulated climate and CO_2 induced climate change over western Europe. *Clim. Change*, **10**, 11–42.

Wiscombe, W. J., and S. G. Warren, 1980: A model for the spectral albedo of snow. I: Pure snow. *J. Atmos. Sci.*, **37**, 2712–2733.

WMO, 1986: Dynamical processes, in Atmospheric ozone, assessment of our understanding of the processes controlling its present distribution and change. Report No. 16, WMO Global Ozone Research and Monitoring Project, Washington D.C., 241–347.

Wolfe, G. V., T. S. Bates and R. J. Charlson, 1991: Climatic and environmental implications of biogas exchange at the sea surface: modeling DMS and the marine biologic sulfur cycle. In *Ocean Margin Processes in Global Change*, R.F.C. Mantoura, J.-M. Martin, and R. Wollast (Eds.), John Wiley and Sons, 383–400.

Worthington, L. V., and W. R. Wright, 1970: *North Atlantic Ocean Atlas, II*. The Woods Hole Oceanogr. Inst. Atlas Ser. 2. Woods Hole, Ma.

Wright, D. G., and T. F. Stocker, 1991: A zonally averaged ocean model for the thermohaline circulation. Part I: Model development and flow dynamics. *J. Phys. Oceanogr.*, **21**, 1713–1724.

Wunsch, C., 1989: Tracer inverse problems. In *Oceanic Circulation Models: Combining Data and Dynamics*, D. L. T. Anderson and J. Willebrand (Eds.), Kluwer, pp. 1–79.

Wunsch, C., and B. Grant, 1982: Towards the general circulation of the North Atlantic Ocean. *Prog. Oceanogr.*, **11**, 1–59.

Wust, G., 1935: *The stratosphere of the Atlantic Ocean. Scientific Results of the German Atlantic Expedition of the Research Vessel "Meteor" 1925–27*. Vol. 6, 112 pp. English translation by W. J. Emery, Amering Publishing Co., Pvt. Ltd., 1987.

Wyrtki, K., 1975: El Niño – The dynamic response of the equatorial Pacific Ocean to atmospheric forcing. *J. Phys. Oceanogr.*, **5**, 572–584.

Wyrtki, K., 1979: The response of sea surface topography to the 1976 El Niño. *J. Phys. Oceanogr.*, **9**, 1223–1231.

References

Wyrtki, K., 1985: Water displacements in the Pacific and the genesis of El Niño cycles. *J. Geophys. Res.*, **90**, 11710–11725.

Xu, J.-S., and H. von Storch, 1990: Predicting the state of the Southern Oscillation using principal oscillation pattern analysis. *J. Climate*, **3**, 1316–1329.

Xu, K.-M., and S. K. Krueger, 1991: Evaluation of cloudiness parameterizations using a cumulus ensemble model. *Mon. Wea. Rev.*, **119**, 342–367.

Xue, Y-K, P. J. Sellers, J. L. Kinter and J. Shukla, 1991: A simplified Biosphere Model for global climate studies. *J. Climate*, **4**, 345–364.

Yamada, T., 1982: A numerical model study of turbulent airflow in and above a forest canopy. *J. Meteor. Soc. Japan*, **60**, 439–454.

Yamamoto, R., T. Iwashima and M. Hoshiai, 1975: Change of the surface air temperature averaged over the Northern Hemisphere and large volcanic eruptions during the years 1951–1972. *J. Meteor. Soc. Japan*, **53**, 482–485.

Yanai, M., S. Esbensen and J.-H. Chu, 1973: Determination of bulk properties of tropical cloud clusters from large-scale heat and moisture budgets. *J. Atmos. Sci.*, **30**, 611–627.

Yeh, T.-C., R. T. Wetherald and S. Manabe, 1983: A model study of the short-term climatic and hydrologic effects of sudden snow-cover removal. *Mon. Wea. Rev.*, **111**, 1013–1024.

Yeh, T.-C., R. T. Wetherald and S. Manabe, 1984: The effect of soil moisture on the short-term climate and hydrology change – A numerical experiment. *Mon. Wea. Rev.*, **112**, 474–490.

Yoshida, N., 1988: ^{15}N-depleted N_2O as a product of nitrification. *Nature*, **335**, 528–529.

Yoshida, N., H. Morimoto, M. Hirano, I. Koike, S. Matsuo, E. Wada, T. Saino and A. Hattori, 1989: Nitrification rates and ^{15}N abundances of N_2O and NO_3^- in the western North Pacific. *Nature*, **342**, 895–897.

Young, W. R., 1987: Baroclinic theories of the wind-driven circulation. In *General Circulation of the Ocean*, H. D. I. Abarbanel and W. R. Young (Eds.), Springer-Verlag, 291 pp.

Zebiak, S. E., 1986: Atmospheric convergence feedback in a simple model for El Niño. *Mon. Wea. Rev.*, **114**, 1263–1271.

Zebiak, S. E., 1989: Ocean heat content variability and El Niño cycles. *J. Phys. Oceanogr.*, **19**, 475–486.

Zebiak, S. E., and M. A. Cane, 1987: A model El Niño–Southern Oscillation. *Mon. Wea. Rev.*, **115**, 2262–2278.

Zebiak, S. E., and M. A. Cane, 1991: Natural climate variability in a coupled model. In *Workshop on Greenhouse-Gas-Induced Climate Change: A Critical Appraisal of Simulations and Observations*. M. E. Schlesinger (Ed.), Elsevier. 457–470.

Zhao, Y., M. Cane and S. Zebiak, 1991: Comparison of El Niño as recorded in the Quelccaya ice-cap and a coupled dynamical model. *J. Climate*, to be submitted.

Zienkiewicz, O. C., 1977: *The Finite Element Method*. McGraw-Hill.

Zubov, N. N., 1943: *Arctic Ice*. Translation, National Service (AD426972), USA, 491 pp.

Zwally, H. J., J. C. Comiso, C. L. Parkinson, W. J. Campbell, F. D. Carsey and P. Gloersen, 1983: Antarctic sea ice, 1973–1976: Satellite passive-microwave observations. NASA SP-459, Washington D.C. 206 pp.

Index

The principle entries for many topics in this index have been indicated by **bold-face** type.

Index

Index